D0138255

MATRIX STRUCTURAL ANALYSIS

MATRIX STRUCTURAL ANALYSIS

Lewis P. Felton
Richard B. Nelson

Civil and Environmental Engineering Department
School of Engineering and Applied Science
University of California, Los Angeles

John Wiley & Sons, Inc.

New York Chichester Brisbane Toronto Singapore Weinheim

ACQUISITIONS EDITOR Cliff Robichaud
ASSOCIATE EDITOR Sharon Smith
MARKETING MANAGER Jay Kirsch
SENIOR PRODUCTION EDITOR Tony VenGraitis
DESIGNER Kevin Murphy
MANUFACTURING MANAGER Mark Cirillo
ILLUSTRATION COORDINATOR Jaime Perea
PRODUCTION SERVICE Ruttle, Shaw & Wetherill, Inc.

COVER PHOTO: © Rafael Macia/Photo Researchers, Inc.

This book was set in Times Roman by *Ruttle, Shaw & Wetherill,* and printed and bound by Hamilton Printing. The cover was printed by Lehigh Press

Recognizing the importance of preserving what has been written, it is a policy of John Wiley & Sons, Inc. to have books of enduring value published in the United States printed on acid-free paper, and we exert our best efforts to that end.

The paper in this book was manufactured by a mill whose forest management programs include sustained yield harvesting of its timberlands. Sustained yield harvesting principles ensure that the number of trees cut each year does not exceed the amount of new growth.

Library of Congress Cataloging in Publication Data:

Felton, Lewis P.

Matrix structural analysis / Lewis P. Felton, Richard B. Nelson.

p. cm.

Includes bibliographical references.

ISBN 0–471–12324–2 (cloth : alk. paper)

1. Structural analysis (Engineering)—Matrix methods. I. Nelson, Richard B. II. Title.

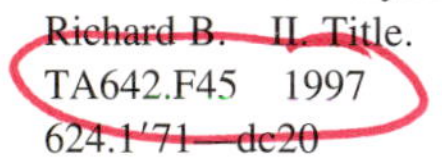

TA642.F45 1997
624.1′71—dc20 96–19454
CIP

Printed in the United States of America

10 9 8 7 6 5 4 3 2 1

To our families:
Tania Leovin, Debbie and Leah Felton, Eva Jabber, David and Maddie Uetrecht
Annabel, Kevin and Jeff Nelson

Preface

The field of structural analysis, like other branches of engineering, has experienced profound changes in the past three decades due to the development of powerful, affordable computers and computational software. In fact, the subject of structural analysis was the birthplace of the *finite element method,* the numerical solution technique that has revolutionized a number of other technical fields including solid and fluid mechanics, electrical and magnetic field theory, thermodynamics and even graphical visualization.

Today, by taking full advantage of the finite element method, structural engineers can investigate a wide range of complex problems with a level of precision and accuracy that could not have been imagined previously. Civil engineers are able to investigate the response of large structures, such as high-rise buildings and long-span bridges, to a variety of loads ranging from severe winds to earthquake shaking. In the mechanical and aerospace fields, engineers are able to analyze the structural behavior of vechicles and machinery under complex thermal, mechanical, and aerodynamic loads. In short, the finite element method has become an indispensable tool for structural analysis.

Yet, for all the power of the finite element method and computer-based analysis, the computational results are meaningful only if knowledgeable engineers can confirm their credibility. For this reason it is essential that engineers not only be able to understand the basic concepts that underlie the finite element method but, more to the point, be able to use these concepts to assess the quality of computer-based numerical analysis.

This book presents a rigorous, concise development of the concepts of modern matrix structural analysis, with particular emphasis on techniques and methods that are the basis for the finite element method. All relevant concepts are presented in this book in the context of two-dimensional (planar) structures composed of bar (truss) and beam (frame) elements, together with simple discrete axial, shear and moment resisting spring elements. The concepts presented require only some basic knowledge of matrix alegebra and fundamentals of the strength of materials.

The technical presentation is unified under the framework of the theorem of virtual work, which is used to develop, in matrix form, both the displacement method and the force method of structural analysis. Energy concepts are presented to establish classical theorems, including the theorems of stationary and minimum total potential energy. Most importantly, the energy theorems are used to develop a rational basis for evaluating the effects of approximations in structural analysis and, in particular, the types of approximations used in the finite element method itself.

The material is presented in a traditional manner, with basic structural definitions first, followed by the analysis of elementary statically determinate structures (Chapters 1–3). Next comes the theorem of virtual work and its various applications for these structures (Chapter 4). Formal matrix methods are introduced early in the analysis of indeterminate structures using the force method (Chapter 5); the basic conceptual layout and organization of data are emphasized in this treatment. The analysis is then recast in terms of the matrix displacement method (Chapter 8), accentuating the strong similarities between the force and displacement methods. The matrix displacement method is considered first in its simplest form and then (Chapter 9) in a form suitable for the direct stiffness method. Also, a useful discussion of structural symmetry is included (Chapter 6), as is a presentation of the use of influence lines as an application of virtual work in the form of the Müller-Breslau principle (Chapter 7). Finally, an Appendix is available to those who wish to pursue the traditional paper and pencil solution technique known as moment distribution. At this point students have a complete problem-solving capability for general two-dimensional truss/frame structures.

Since the basic analytic tools presented in the first nine chapters all employ the theorem of virtual work, students are well prepared for an introduction to the energy theorems and for working with the theorems in the context of approximate (truss/frame) analysis (Chapters 10 and 11). Upon completion of the text, students should have an excellent understanding of the fundamental concepts and numerical characteristics of the finite element method.

Throughout the text, we have attempted to provide meaningful examples that illustrate the underlying theories and procedures. The examples are presented in a step-by-step fashion as an aid to promoting understanding. Many examples are revisited in various sections of the text to illustrate and contrast solutions by different procedures. Most of the examples are presented in algebraic form because we believe that working with dimensionally homegeneous equations, rather than numbers, reinforces the concepts.

Today's students are fortunate to have a wide and ever-expanding range of relatively inexpensive and reliable PC software to assist them with computations. Any form of spreadsheet (e.g., Lotus 1-2-3, Excel) or computational math package (e.g., Mathcad, Matlab, Mathematica) can be used to solve the matrix equations associated with basic structural analysis problems of the type contained in this text. Manipulating the matrices at this level should give students a better appreciation of the capabilities of commercial structural analysis software (e.g., SAP 90, NASTRAN, ANSYS) when they encounter it in specialized computer analysis courses or in engineering practice.

This book is the result of many years of collaboration and experience in teaching the subject, but we are also grateful to a number of people who contributed to it in various ways. We particularly want to thank Professors Barry Goodno of Georgia Tech, Joseph Saliba of the University of Dayton, William Spillers of New Jersey Institute of Technology, John Lepore of the University of Pennsylvania, Michael Constantinou of SUNY at Buffalo, and Moshe Fuchs of Tel Aviv University for their comments and suggestions which helped us improve the manuscript; Mr. Kerop Janoyan, who was our TA for several offerings of the courses, corrected many typos and worked many problems; Mr. Cliff Robichaud, Acquisitions Editor of the College Division at John Wiley & Sons, saw enough merit in the proposed manuscript to offer us a contract; Ms. Catherine Beckham and, subsequently, Ms. Sharon Smith, Associate Editors at Wiley, encouraged us throughout the writing, revision, and publication of the final manuscript; Mr. Jamie Perea, who coordinated the preparation of the illustrations which were skillfully prepared by Mr. Boris Starosta; and, last but not least, Peg Markow, Senior Project Manager at Ruttle, Shaw & Wetherill, guided us through the galleys and page proofs with remarkable patience, expertise, and humor. We hope the end result will prove useful to others.

Finally, we would like to acknowledge the deep debt we owe to our own instructors and to the colleagues who taught us much of what we know of the subject. For the first author, Anthony Armenakas, Tung Au and F. R. Shanley; for the second author, Hans H. Bleich, Bruno Boley, Frank L. DiMaggio, Alfred M. Freudenthal, Raymond D. Mindlin, Mario G. Salvadori, and William R. Spillers of Columbia University. These were faculty who taught it right the first time, who enriched our understanding of the subject and set us on the path that led to this book. Last, but also far from least, we are both indebted to Moshe F. Rubinstein, one of our own faculty in the Civil and Environmental Engineering Department of UCLA, a master of the subject of structural analysis and a teacher of the highest distinction.

L. P. Felton
R. B. Nelson
UCLA
March, 1996

List of Symbols

A	Cross-sectional area
$A(x)$	Displacement pattern
$\boldsymbol{A}$	Flexibility matrix
$\boldsymbol{a}$	Matrix of element flexibilities, diagonal or block-diagonal
$a(x)$	Influence function
α	Coefficient of linear thermal expansion
$\boldsymbol{b}$	Equilibrium matrix
$B(x)$	Curvature pattern
$\boldsymbol{B}$	Vector of constants
β	Compatibility matrix
β_s	Compatibility matrix associated with $\boldsymbol{u}_s$
C, c	cos θ. cosh θ
Δ	Element deformation quantity
$\boldsymbol{\Delta}$	Vector of element deformations
δ	Displacement
$\boldsymbol{\delta}$	Vector of element end displacements
$\boldsymbol{\delta}_g$	Vector of element end dsiplacements in global coordinates
Δ_{IJ}, δ_{JI}	End deformations for beam element
$\boldsymbol{\Delta}_o$	Vector of unrestrained element deformations
ΔT	Temperature change

E	Modulus of elasticity (Young's modulus)
ϵ	Strain
$\boldsymbol{e}$	Unit vector
F	Discrete force magnitude
$\boldsymbol{F}$	Load vector
$\mathbf{F_q}$	Generalized force vector
$F_k^{(q)}$	Generalized force
$\mathbf{F}^*$	Effective load vector
$\mathbf{F^o}$	Force vector associated with interior loads
I	Moment of inertia
$\boldsymbol{i}, \boldsymbol{j}, \boldsymbol{k}$	Unit vectors in x, y, z-directions, respectively
K	Dimensionless parameter, I/Al^2
$\boldsymbol{K}$	Matrix of element stiffnesses, diagonal or block-diagonal
$\boldsymbol{k}$	Stiffness matrix
k	Spring stiffness; elastic foundation modulus
$\boldsymbol{K_g}$	Matrix of element stiffnesses in global coordinates
l	Length
λ	Length paramater
M	Discrete moment magnitude
$\boldsymbol{M}$	Moment vector
$M(x)$	Internal bending moment in beam
$M_I(x)$	Pseudo-bending moment associated with initial effects
M_{IJ}, M_{JI}	End moments on beam
$\widehat{M}_{\mathrm{IJ}}, \widehat{M}_{\mathrm{JI}}$	Fixed-end moments
$M_o(x)$	Bending moment in unrestrained beam
$M_T(x)$	Pseudo-bending moment associated with temperature effects
$\hat{M}(x)$	$M_o(x) + M_T(x) + M_I(x)$
$\phi(x)$	Slope [same as $v'(x)$ or $dv(x)/dx$]
ψ_{IJ}	Sidesway angle
P	Discrete axial force magnitude
$\boldsymbol{P}$	Vector of element internal forces
$P(x)$	Internal axial force in bar
$p(x)$	One-dimensional continuous axial force (force per unit length)
$\boldsymbol{P_g}$	Vector of element end forces in global coordinates
$P_I(x)$	Axial pseudo-force associated with initial effects
P_{IJ}, P_{JI}	Axial end forces on bar
$\widehat{P}_{\mathrm{IJ}}, \widehat{P}_{\mathrm{JI}}$	Fixed-end axial forces
$\boldsymbol{P_o}$	Vector of element forces due to applied loads (force method) *or* Vector of fixed-end forces (displacement method)

$P_T(x)$	Axial pseudo-force associated with temperature effects
$\widehat{P}(x)$	$P_o(x) + P_T(x) + P_I(x)$
$P_o(x)$	Axial force in unrestrained bar
Q	Response quantity
$q(x)$	One-dimensional continuous transverse force (force per unit length)
q_o	Uniform continous load
$\boldsymbol{r}$	Position vector
R	Reaction force (at support)
$\boldsymbol{\sigma}$	Stress
θ_{IJ}, θ_{JI}	End rotations for beam element
S, s	sin θ, sinh θ
$\boldsymbol{T}$	Transformation matrix
U	Strain energy
U_d	Deformation energy
U_e	External potential energy
U_o	Strain energy per unit volume (strain energy density)
U'	Strain energy per unit length of bar/beam
u	System displacement
$\bar{u}$	Approximate system displacement
$u(x)$	Axial displacement (x-direction)
$u'(x)$	Axial strain (same as ϵ)
$\boldsymbol{u}$	Vector of system displacement
$\boldsymbol{u}_o$	Vector of initial displacements
$\boldsymbol{u}_s$	Vector of known support settlements
$(...)_v$	Virtual response quantity, e.g., F_v, $v_v(x)$, etc.
V	Total potential energy
$V(x)$	Internal transverse shear in beam
$v(x)$	Transverse displacement (y-direction)
$v'(x)$	Slope [same as $\phi(x)$ or $dv(x)/dx$]
$v''(x)$	Curvature
$\bar{v}(x)$	Approximate tansverse displacement
V_{IJ}, V_{JI}	Transverse end forces on beam
$\widehat{V}_{IJ}$, $\widehat{V}_{JI}$	Fixed-end transverse forces
W	Work
X	Redundant force
$\boldsymbol{X}$	Vector of redundant forces

Contents

CHAPTER 1
BASIC CONCEPTS 1

1.0. Introduction 1

1.1. Types of Structural Elements 3
1.1.1. Typical One-Dimensional Elements 4
1.1.2. Typical Two-Dimensional Elements 7
1.1.3. Three-Dimensional Elements 9

1.2. Overview of the Concepts of Structural Analysis 9
1.2.1. The Theorem of Virtual Work 10

1.3. Loads 11

1.4. Material Behavior 14

1.5 Review of Equilibrium Equations 16
1.5.1. Concentrated Forces 16
1.5.2. Equilibrium of Forces 17
1.5.3. Moment of Force 18
1.5.4. Equilibrium of Moments 20
1.5.5. Free-Body Diagrams 22
1.5.6. Statically Equivalent Systems of Forces 22
1.5.7. Treatment of Distributed Forces 23

1.6. Analytic Models 27
1.6.1 Connections 28
1.6.2 Supports 34

1.7. Determinacy and Indeterminacy 54
1.8. Stability and Instability 63
1.9. Preview 66

CHAPTER 2
STATICALLY DETERMINATE TRUSSES 72

2.0. Introduction 72
2.1. Equilibrium of Axially Loaded Bars 72
2.2. Internal Forces in Statically Determinate Trusses 76
2.3. Statical Stability/Instability and Determinacy/Indeterminacy 95
2.4. Deformations of Axially Loaded Bars 97
2.5. Direct Calculation of Truss Displacements 104

CHAPTER 3
STATICALLY DETERMINATE FRAMES 119

3.0. Introduction 119
3.1. Equilibrium of Beams 119
3.2. Internal Forces in Statically Determinate Frames 121
3.2.1. Shear and Bending Moment Diagrams 127
3.2.2. Alternate Representations of Bending Moment 137
3.3. Deformations of Beams 139
3.4. Direct Calculation of Beam Displacements 144
3.4.1. Solution of Differential Equations 145
3.4.2. Moment-Area Method 156

CHAPTER 4
VIRTUAL WORK 169

4.0. Introduction 169
4.1. Virtual Work for Truss Joints 170
4.2. Virtual Work for Axially Loaded Bars 172
4.3. Virtual Work for Trusses 175
4.4. Virtual Work for Beams 192
4.5. Virtual Work for Frames 205
4.6. Summary 216

CHAPTER 5
INTRODUCTION TO ANALYSIS OF INDETERMINATE STRUCTURES: FORCE METHOD 229

5.0. Introduction 229
5.1. Redundant Forces 229
5.2. Trusses 232
5.2.1. Basic Formulation 234
5.2.2. Matrix Formulation 248
5.2.3. Matrix Formulation for Structures with General Element Flexibility 261
5.3. Frames 263
5.3.1. Basic Formulation 265
5.3.2. Flexibility Matrix for Beams 277
5.3.3. Matrix Formulation 284
5.3.4. Combined Bending and Axial Behavior 295
5.3.5. General Axial Behavior 300
5.4. Miscellaneous Considerations 307
5.4.1. Support Settlement 307
5.4.2. Structures Composed of Several Types of Elements 311
5.4.3. Initial Deformations 318
5.4.4. Effect of Moment-Bearing Joints on Truss Behavior 321
5.5. Computing Deflections 327

CHAPTER 6
SYMMETRY 348

6.0. Introduction 348
6.1. Behavioral Symmetry for Beams 348
6.1.1. Betti's Theorem for Linearly Elastic Bernoulli-Euler Beams 349
6.1.2. Maxwell's Reciprocal Relationship 351
6.2. Physical Symmetry 352
6.2.1. Structures Symmetric With Respect to a Line 353
6.2.2. Structures Symmetric With Respect to a Point 372

CHAPTER 7
INFLUENCE LINES 382

7.0. Introduction 382
7.1 Moving Loads and Influence Lines 383
7.2 Virtual Work and Influence Lines 388
7.2.1. Influence Lines for Statically Determinate Structures 388
7.2.2. Influence Lines for Indetermine Structures: The Müller-Breslau Principle 390
7.2.3. Influence Lines for Trusses 400

CHAPTER 8
INTRODUCTION TO THE DISPLACEMENT METHOD 408

8.0. Introduction 408
8.1. System Displacements 409
8.2. Analysis of Trusses 418
8.3. Analysis of Frames 434
8.3.1. Stiffness Matrix for Beams 434
8.3.2. Matrix Formulation 436
8.3.3. Equilibrium by Virtual Work 443
8.3.4. Fixed-End Moments From Betti's Theorem 468
8.4. Miscellaneous Considerations 475
8.4.1. Support Settlement 475
8.4.2. Structures Composed of Several Types of Elements 478

CHAPTER 9
DISPLACEMENT METHOD: ADVANCED FORMULATION 499

9.0. Introduction 499
9.1. Force-Displacement Stiffness Matrix for Beams 500
9.2. Equilibrium Equations 505
9.3. Force-Displacement Stiffness Matrix for Axially Loaded Bars 508
9.4. Combined Bending and Axial Behavior 509
9.5. Element Stiffness Matrix in Global Coordinates 522
9.6. Direct Stiffness Method 531
9.6.1. General Support Motions 538
9.6.2. Direct Stiffness Method for Trusses 540
9.7. Miscellaneous Considerations 546
9.7.1. Support Settlement 546
9.7.2. Structures Comosed of Several Types of Elements 550
9.7.3. Effect of Joint Dimensions 554
9.7.4. Releases 560

CHAPTER 10
ENERGY CONCEPTS AND APPROXIMATIONS 572

10.0 Introduction 572
10.1 Basic Energy Forms 573
10.1.1. Strain Energy 574
10.1.2. Symmetry of Stiffness and Flexibility Matrices 579
10.1.3. Positive Definiteness of Strain Energy 581
10.1.4. Convexity of Strain Energy 583

10.1.5. Partial Derivative of Strain Energy; Castigliano's Theorem, Part I 586
10.1.6. Deformation Energy; Castigliano's Theorem, Part II 587
10.1.7. External Potential Energy 588
10.1.8. Total Potential Energy; Stationary and Minimum Values 588

10.2. Effects of Approximations 590
10.2.1. Constrained Displacements 590
10.2.2. Incompatible Displacements 595
10.2.3. Approximations in the Force Method 596

10.3. Generalized Displacements and Forces 600
10.3.1. Approximate Displacement Patterns 604

10.4. Applications to Structures Composed of One-Dimensional Elements 606
10.4.1. Strain Energy of Bars and Beams 607
10.4.2. Examples of Rayleigh-Ritz Method 611

CHAPTER 11
INTRODUCTION TO FINITE ELEMENT ANALYSIS 632

11.0. Introduction 632

11.1. Basic Reqirements 632

11.2. Stiffness Matrix and Force Vector for Assumed Displacements 635
11.2.1. Development Using Energy 635
11.2.2. Development Using Virtual Work 638
11.2.3. Approximate Solutions as Best Fits 640

11.3. Examples of Finite Element Analysis 643

APPENDIX A
FIXED-END MOMENTS AND FORCES 673

APPENDIX B
MOMENT DISTRIBUTION METHOD FOR FRAME ANALYSIS 681

INDEX 697

Chapter 1

Basic Concepts

1.0 INTRODUCTION

People have been designing, building, and using structures for at least as long as history has been recorded. Indeed, the most rudimentary of early human's tools that have been discovered at archaeological sites, such as cutting stones, scrapers, and pottery, were themselves structures. Their design required some concept of intended usage, which in turn led to material selection, trial designs and, later, to design modification and refinement. In short, the same process we use today to produce our most sophisticated structures was carried out.

Many of the ancient artifacts are impressive even by modern standards, for example, the great pyramids of Egypt, the temples of Greece, the Roman aqueduct systems, and the great wall of China. Such structures could not have been built without some knowledge of strength of materials and the transmission of forces. No doubt, most of the knowledge of structural behavior developed in early times was strictly empirical in nature, and many of the successful designs that we see today are probably the outgrowth of others that failed.

Systematic studies of subjects leading to modern structural mechanics were first initiated in the Renaissance, most notably by Leonardo da Vinci. However, it was not until the seventeenth century, with the publication in 1638 of Galileo's famous book, *Two New Sciences,* that the modern era of structural mechanics and analysis can be said to have begun. Galileo was the first person to attempt a rational analytical study of the stresses in beams. Although his effort was not successful because of his limited knowledge of statics, he was able to draw many conclusions about the strength of structural components and to establish scaling relationships between size and strength.

It remained for Isaac Newton to set down in his *Mathematica Principia,* published in 1660, a set of concepts, now known as Newton's laws, which serve as the analytical basis for modern structural mechanics. With the advent of Newton's laws, the field of structural analysis progressed relatively steadily because of the contributions of many of the most famous scientists of the eighteenth and nineteenth centuries, among them such notables as Hooke, Bernoulli, Castigliano, Euler, Lagrange, Coulomb, Navier, Cauchy, Poisson, Lamé, Saint-Venant, Maxwell, Stokes, Kirchhoff, Kelvin, Rayleigh, Mohr, and Prandtl.[1]

By the end of the nineteenth century nearly all of the basic concepts needed to analyze most structural elements were known. However, a complete analysis capability for large structural assemblies, such as major buildings, remained a very formidable task. It was not until 1914 and 1915, when Bendixen and Maney developed the so-called slope-deflection equations, and later when Hardy Cross developed a simple iterative numerical technique based on these equations (*moment distribution,* 1932), that fundamental relations defining element behavior could be extended to large structural assemblies in a practical way. Still, in most cases, the computations involving large structures could not be dealt with effectively until the advent of high-speed digital computers.

As with other fields of engineering, structural analysis has advanced at an extraordinary rate since the introduction of high-speed computing. The development of the *Finite Element Method* of structural analysis in the early 1960s, together with tremendous increases in computing power and convenience, now makes it possible to understand structural behavior with levels of accuracy that could not have been imagined before the computer age. Yet, it is important to realize that these powerful engineering tools are all based on basic concepts that date back to the time of Galileo and Newton. A complete understanding of all underlying concepts is essential to the effective use of the modern computational methods of structural analysis.

Before beginning a detailed presentation of the various concepts that make up modern structural analysis, it is helpful to state what is meant by the term *structure*, define a number of the more common types of elements that make up a structure, and give a brief description of the types of structural behavior[2] we shall consider in this text.

For our purposes a *structure* may be defined as an assembly of one or more components or elements capable of supporting forces or other types of applied loads. The list of structures created by humankind is enormous, including such every-day items as kitchen utensils and ordinary hand tools, tables, chairs, and shelves; large stationary structures such as masonry walls, buildings, bridges, and dams; vehicles, such as ships, automobiles, airplanes, and spacecraft; all types of machines and, in fact, anything that must sustain loadings. In addition to structures created by humans, there is the study of the human body itself, the field of *biostructural mechanics,* which is a relatively new research area of major importance.

---

[1] For an exceptionally interesting treatment of the subject, the reader is referred to Timoshenko, S. P., *History of the Strength of Materials,* McGraw-Hill, New York, 1953. In fact, any of the many texts by Timoshenko is of real value. Timoshenko was one of the twentieth century's foremost authors in the field of structural mechanics.

[2] For an excellent introduction to structural forms and behavior of particular interest to civil engineers, see the book by M. Salvadori and R. Heller, *Structure in Architecture* (3rd Ed.), Prentice-Hall, Englewood Cliffs, New Jersey, 1986.

Even seemingly ordinary structures may, in reality, be complicated built-up assemblies of different types of elements and connections. For example, a simple table is made up of vertical columns (the table legs), plate elements (the table top), and beam elements (lateral supports and stiffeners), and also several different types of fasteners. If the table is properly designed, all these elements work together as a unit transmitting applied loads to the floor without damaging any of the elements that make up the table. It is also important to recognize that the floor might play an important role in determining the behavior of the table, especially if the floor were quite flexible. A building with its foundation on soil is an example of a situation in which this type of interaction might be significant.

In fact, defining how a structure interacts with its surroundings is a difficult and often very subjective part of structural analysis. The difficulty often extends to defining the types of possible loadings and their intensities. A designer must have a reasonable level of knowledge of the intended use of a structure, and of its expected environment, to produce a safe, efficient, and economical design.

The determination of how a given structure will perform under applied loads is the subject of structural analysis. Some sort of structural analysis is required at nearly every major decision point in the life of a structure, starting from concept development through preliminary design, detailed design, manufacture, and even to decisions regarding repair and major modifications after the structure is in service.

In earlier times, structural analysis was at best qualitative and relied almost entirely on experimental methods or on past experience. As a result, design methodology evolved very slowly, based on trial and error, good record keeping, and an ability to learn from past mistakes. Today, engineers are able to make quantitative assessments of designs quickly and efficiently, especially using computers to simulate structural behavior. Still, however, engineers rely on past mistakes and failures to improve designs. A notable example is learning from failures in earthquakes.

Modern structural analysis is a subject that is primarily mathematical in nature, and is based on the development and analysis of mathematical representations, or *analytical models,* of structures to simulate actual behavior. These numerical simulations may be used to predict structural response with sufficient accuracy to permit an assessment of structural performance.

1.1 TYPES OF STRUCTURAL ELEMENTS

In general terms, the various types of *elements* that make up a structure may be classified into major categories according to their geometric shapes and, within each category, classified according to the manner in which they transmit applied loads.

One-dimensional or *line elements* are slender rod-like elements, which may be represented by a line (straight or curved) termed the *reference axis* of the element. Cross sections of the element taken perpendicular to the reference axis of the element have small overall dimensions relative to the length of the axis itself. Many well-known structural components, including wires, bars, beams, cables, columns, arches, and ribs, may be classified and analyzed as one-dimensional or line elements.

As we shall see, the internal physics needed to describe the behavior of these elements can be greatly simplified, the primary reason being that internal forces and displacements depend on only one spatial variable, i.e., the position along the axis of the element. A less obvious reason is the fact that it is often possible to represent the stress-strain behavior of such elements using very simple one-dimensional stress-strain relationships.

Two-dimensional elements are objects that may be represented by flat or curved reference surfaces with a small thickness in the direction perpendicular to the reference surface. Here, the term ''two-dimensional'' is used in a very general context. Thus, flat plates, thin cylinders and spheres, and even more general shell structures, i.e., bodies with a ''thinness'' in one direction, are all termed two-dimensional elements. The mathematical theories needed to adequately describe the behavior of these bodies are much more complicated than those for simple one-dimensional elements, because the internal physics depends on two variables describing the position on the reference surface. Also, the more complicated geometry of these bodies produces more complicated physical behavior requiring more general concepts of stress and strain than for the one-dimensional elements. Consequently, the difficulty in obtaining accurate structural analyses of even the simplest two-dimensional structural elements (e.g., flat plates) increases by at least an order of magnitude compared to that for one-dimensional elements. As the geometric complexity of the reference surface increases, the difficulty in obtaining solutions increases even more rapidly. For such elements, numerical solution techniques provide the only practical means for obtaining useful engineering results.

Three-dimensional elements have no geometric features that make it possible to simplify the analysis to a one- or two-dimensional level. In other words, these elements must be analyzed as solid bodies using three-dimensional theory. Until recently, analyses of such objects were avoided if at all possible. Today, with the development of fast computers, finite element methods, and powerful graphical display algorithms, it is possible to prepare and analyze very complex computer models of structures involving three-dimensional elements without major conceptual difficulties. The analysis of a structure made of general three-dimensional (*3-D*) elements requires a much greater computer effort than for a comparable structure made of two-dimensional (*2-D*) or one-dimensional (*1-D*) elements. Therefore, analysts generally try to adopt some type of simplifying approximations to avoid the cost of a 3-D computer analysis.

The preceding classifications of elements can be further refined based on distinctions in basic physical behavior of different elements within each category. Some assumptions and approximations are almost always involved in the description of behavior. Various elements are examined in the next sections.

1.1.1 Typical One-Dimensional Elements

The subclassification of one-dimensional elements is based on the types of internal forces the elements are assumed to carry, especially with regard to the direction of the forces relative to the reference axis of the element, although to a lesser extent the sign of the loading may also influence these definitions. For example, bars and cables are assumed to be capable of transmitting axial forces only, i.e., forces directed along the reference axis. These elements are assumed to be incapable of carrying any other types of internal forces, such as lateral forces or bending moments. In order for the internal forces to be axial, the externally applied forces must be directed along the reference axis. In addition, we shall soon see that the force must pass through a particular point in the cross section known as the modulus-weighted centroid (Figure 1.1*a*).

Compressive forces in bars can result in entirely different behavior than tensile forces.

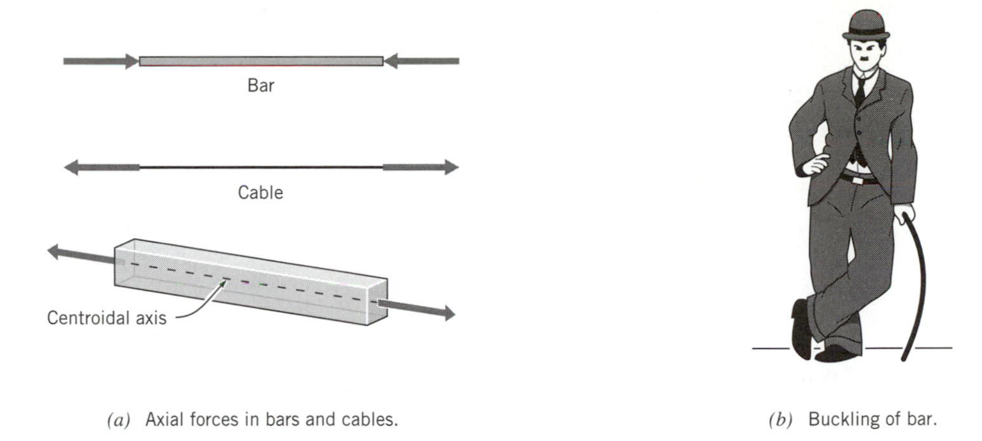

Figure 1.1 Axial force transmission. (Salvadori/Heller, *Structure in Architecture, 3/E.* © 1986; reprinted by permission of Prentice-Hall, Englewood Cliffs, NJ.)

Our ordinary experience indicates that when a slender bar is subjected to axial compression, *buckling* can occur (Figure 1.1*b*). Buckling involves transverse displacement or bending of the bar, usually beginning at load levels well below those that would cause material failure. However, once the buckling process begins, the element rapidly loses its load-carrying capability, and gross material failure and collapse can occur. Thus, buckling is of great concern to structural engineers. The problem of buckling is most acute in slender 1-D or thin 2-D structures subjected to compressive internal forces.

The behavior of a cable differs from that of a bar only in the inability of the cable to carry compressive forces. Stated another way, a cable buckles under arbitrarily small compressive loads. As a direct consequence of this restriction on its behavior, a cable can transmit arbitrarily directed concentrated external forces only if its shape is allowed to change radically and assume a truss-like geometry (Figure 1.2). A cable can also transmit distributed transverse external loads but, again, the shape of the cable is set by the precise form of the loading. For example, a constant vertical force per unit horizontal length is carried by a cable that assumes the shape of a parabola; a cable hanging under its own constant weight per unit length takes the shape of a catenary.

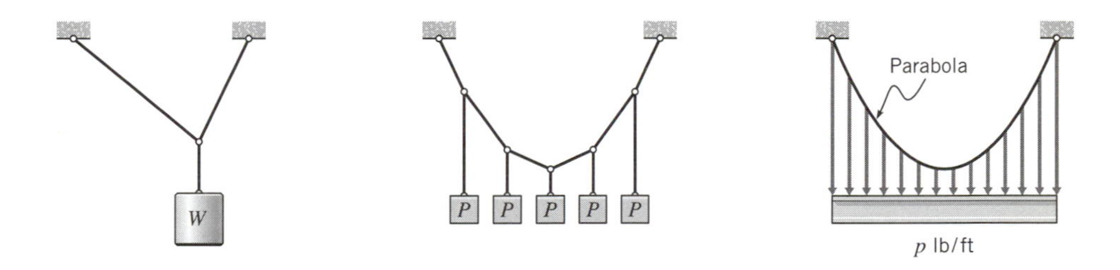

Figure 1.2 Cables. (Salvadori/Heller, *Structure in Architecture, 3/E,* © 1986; reprinted by permission of Prentice-Hall, Englewood Cliffs, NJ.)

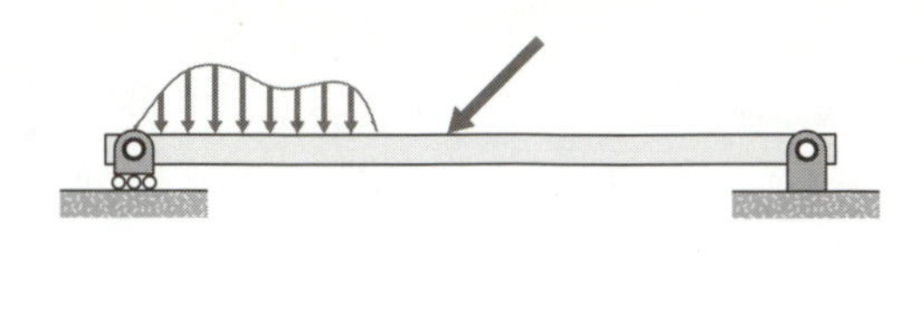

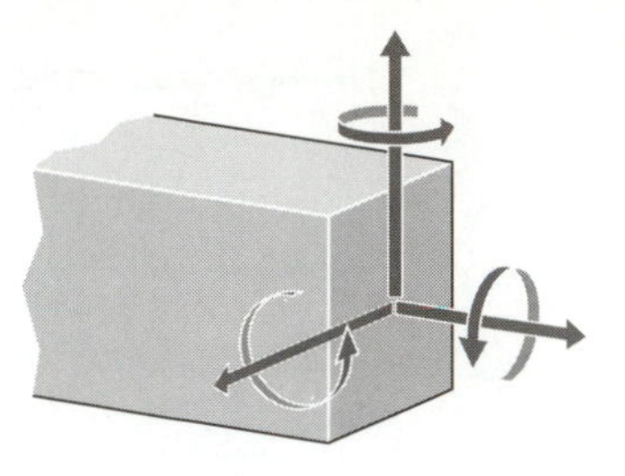

Figure 1.3 Beam.

Beams are one-dimensional components that are capable of withstanding not only axial forces (tensile or compressive), but a general internal force state[3] involving transverse shear forces together with bending and twisting moments (Figure 1.3). Beams are usually assumed to be straight and slender. The engineering analysis of beams requires a number of assumptions regarding the nature of the deformations due to applied loads, which will be examined in greater detail in Chapter 3.

One-dimensional elements may be assembled to form a variety of sophisticated structures. For example, if a structure is created by assembling elements in such a way that the predominant physical behavior of each element is axial force transmission, the resulting structure is referred to as a *truss*. The bars making up the structure are assumed to be joined at their ends by connections that behave like *pins* (Figure 1.4*a*), and external forces are assumed to be applied only at the joints. Truss behavior will be examined in greater detail in Chapter 2.

If at least some of the elements in a structure are required to carry loads in such a way that lateral internal forces and and/or internal moments exist, i.e., if the elements are beams, then the structure is referred to as a *frame* (Figure 1.4*b*). Connections between elements and to external supports may have a wide variety of characteristics, and therefore the overall behavior of frames is more complex than trusses.

Many truss and frame structures are three-dimensional, as in Figure 1.4*c*. However, the general difficulty of visualizing the geometry of a 3-D structure is such that, if possible, engineers try to design and analyze such structures as assemblies of 2-D trusses or frames. Then, each planar subassembly can be easily visualized and analyzed. This approach is widely used in practice, because most structures are rectilinear in form.

Arches are essentially beams with curved reference axes (Figure 1.5).[4] The curvature of an arch is usually designed so the structure develops internal axial forces and carries rather limited internal bending moments compared to a straight beam. The advantage of this is that structures that transmit external loads via axial internal forces can be more efficient than those that transmit external loads by internal moment action, such

[3] The word *force* is frequently used in a general sense to encompass all physical load quantities including *moments*.

[4] The distinction between an arch and a frame is not entirely clear. Arches can be assembled from many short, straight segments that give the overall curved shape of an arch. Therefore, an arch may be viewed as a special subclass of frame structures.

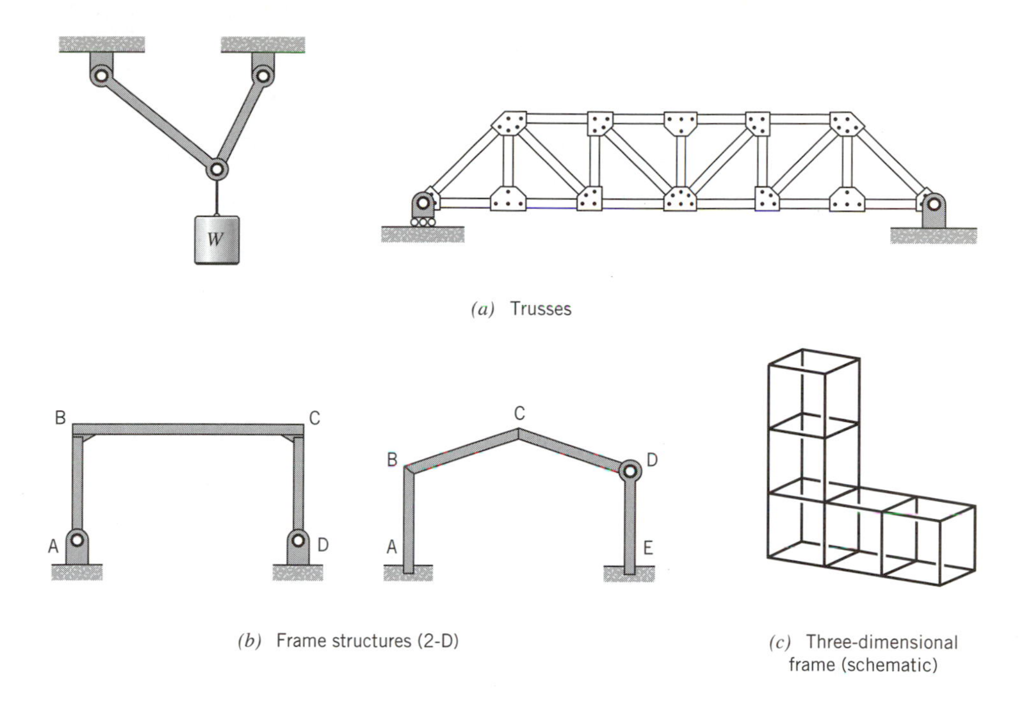

(*a*) Trusses

(*b*) Frame structures (2-D)

(*c*) Three-dimensional frame (schematic)

Figure 1.4 Trusses and frames.

as would be the case with a typical frame. The disadvantage is that fabrication may be more complex and costly.

1.1.2 Typical Two-Dimensional Elements

Membranes are the two-dimensional equivalents of cables and bars (Figure 1.6*a* and 1.6*b*). They are assumed to be so thin that they may be considered to be two-dimensional surfaces. Since membranes do, in fact, have thickness, the physical information needed to describe their behavior (e.g., internal stress resultants) is defined on the midsurface of the element. Internal stress resultants are assumed to act only in the plane tangential to the midsurface and are distributed over the midsurface, having dimensions of force per unit length of midsurface (Figure 1.6*c*). These stress resultants represent the effects

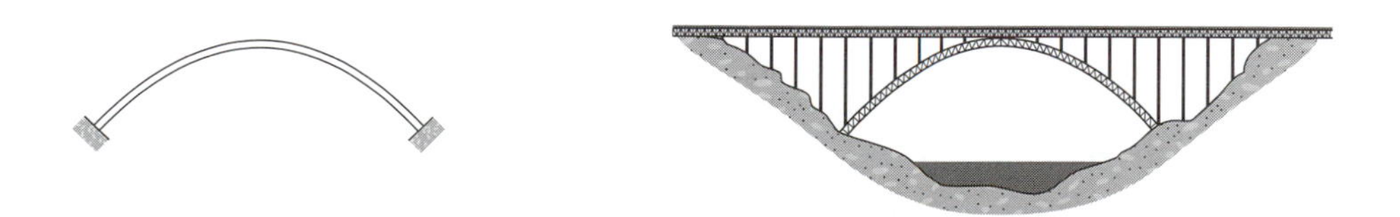

Figure 1.5 Arches. (Salvadori/Heller, *Structure in Architecture, 3/E,* © 1986; reprinted by permission of Prentice-Hall, Englewood Cliffs, NJ.)

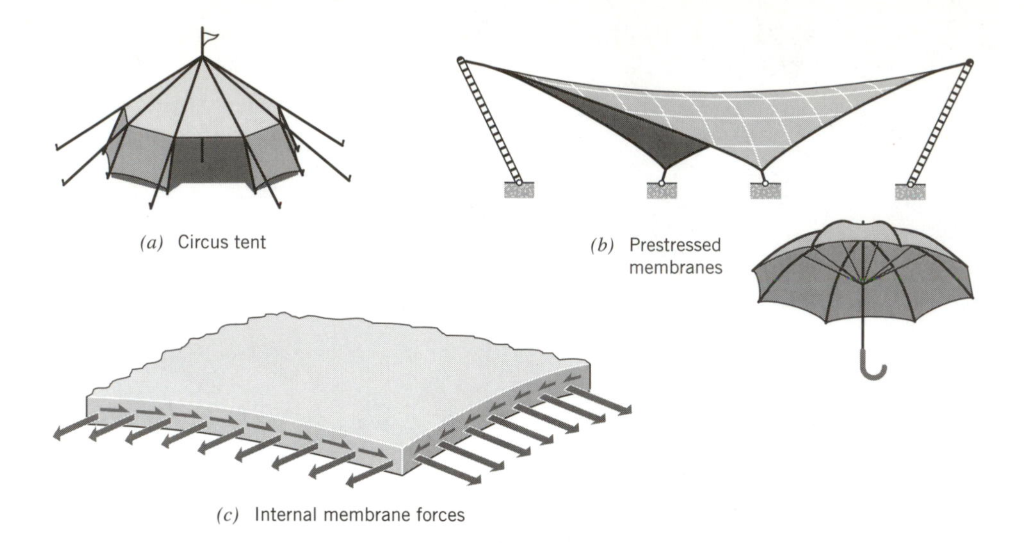

Figure 1.6 Membrane structures. (*a* and *b* from Salvadori/Heller, *Structure in Architecture, 3/E,* © 1986; reprinted by permission of Prentice-Hall, Englewood Cliffs, NJ.)

of internal tension, compression, and/or shear stresses acting throughout the membrane thickness in directions parallel to the midsurface of the membrane.

Membranes are capable of transmitting transverse pressures, although some, like cables, are incapable of developing internal compressive forces, e.g., balloons and fabric structures. Structures of this type are usually termed *tension* structures. Other 2-D membranes can develop internal compressive stresses. Such structures are most similar to 1-D arches in their tendency to support compressive loads by means of their curved shape. This is the case with many pressure vessels, for example, a thin sphere submerged in water. As with all thin structures in a state of compression, buckling is a possible failure mode for thin two-dimensional structures that develop compressive internal stress resultants.

Many two-dimensional structures have internal force states that are primarily, but not entirely, membrane-like. For example, the cylindrical portion of a soda can, except in the vicinity of the ends, can be expected to behave in a membrane-like manner and develop membrane forces in the axial and circumferential directions as a result of the internal pressure. Near the ends, some bending effects are unavoidable. In fact, it can be shown that any discontinuity or other abrupt change in a structure will disturb nominal membrane-like internal force distributions and lead to at least some localized bending effects.

Thin plates and *shells* (Figure 1.7) are the equivalents in two- and three-dimensional space of beam/frame systems, although perhaps the best description of typical shell structures may be as two-dimensional arch systems. In addition to in-surface (membrane) stress resultants, plates and shells can develop internal distributed transverse shear stress resultants and bending and twisting moment resultants, i.e., a complete multidirectional stress state.

The complexity of 2-D elements becomes apparent even if attention is restricted to flat plates. This ''simple'' structural configuration can be rectangular, circular, skew, arbitrary, folded, or stiffened, i.e., can assume a tremendous array of shapes. At the

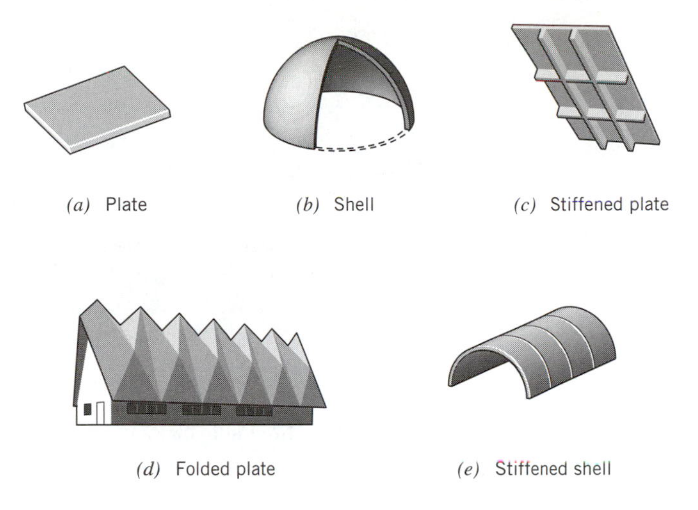

Figure 1.7 Plates and shells. (*c, d,* and *e* from Salvadori/Heller, *Structure in Architecture, 3/E,* © 1986; reprinted by permission of Prentice-Hall, Englewood Cliffs, NJ.)

same time, a large variety of different types of support conditions can be defined along edges.

Shells can be defined as surfaces of translation or revolution, as ruled surfaces, or may have more complex and arbitrary shapes. As with arches, the advantage of shells lies in their capability of transmitting external transverse loads largely by membrane action, and their disadvantage is in their complicated geometry, especially as it relates to design, fabrication, and utilization. The analysis of realistic plate or shell structures has only become practical since the advent of modern matrix computer solution methods, particularly the finite element method.

1.1.3 Three-Dimensional Elements

Structures made up of components that do not fit into a category for which simplifying assumptions are possible must be analyzed as general three-dimensional bodies. Such structures are usually analyzed using numerical computer-based methods.

One such approach, *finite element analysis,* has emerged as one of the most powerful tools for analyzing all types of structures regardless of whether or not the elements are one-, two-, or three-dimensional in form. In the finite element method, the structure is represented as an assembly of finite-sized subregions or elements, i.e., finite elements (Figure 1.8). The behavior of the elements is approximated, i.e., simplified greatly. As a result, the analysis of an assembly of such elements becomes quite similar to the analysis of frame/truss assemblies except that the matrix descriptions of the individual finite elements are much more complex than for bar or beam elements.

1.2 OVERVIEW OF THE CONCEPTS OF STRUCTURAL ANALYSIS

The analysis of structures involves a consideration of three different types of information, namely

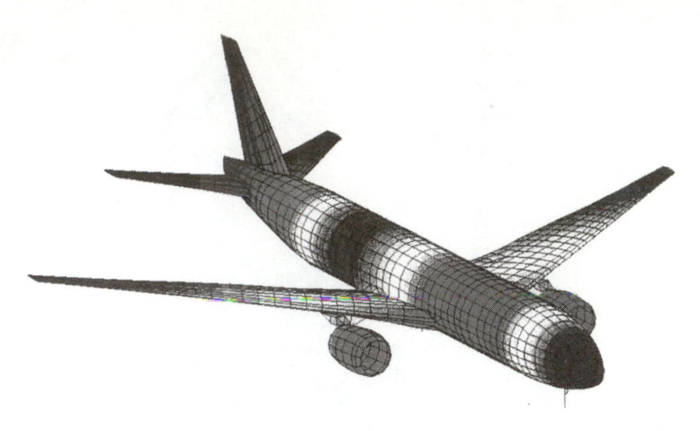

Figure 1.8 Finite elements.

1. *The state of equilibrium* between applied loads and the internal forces (stresses). Equilibrium is based on the application of Newton's laws and on the use of *free-body diagrams*[5] to analyze forces or stresses in the structure.
2. *The geometry of deformation (kinematics),* i.e., the relationship between displacements and the deformations (strains) that are produced. This information requires a careful analysis of the small strains that develop in a body under small but otherwise arbitrary motion. In principle, this analysis is an exercise in geometry, although in practice some apparently simple tasks prove to be quite challenging.
3. *The stress-strain relations,* also termed *constitutive relations,* are mathematical simulations of empirical force-deflection data converted to stress-strain form. Relationships between stress and strain may also depend on time (rate effects), temperature, moisture or other environmental effects, and on prior loading history. The particular form of these relations depends on the nature of the problem being analyzed. Mathematical models of stress-strain behavior often are the least accurate area of structural analysis.

In order to present these concepts concisely, and yet in a way that does not infer prior knowledge on the part of the reader, we will present the information in a rather traditional non-matrix format and then, later, present the same information in a matrix form. We shall see that the use of matrix algebra simplifies concept development considerably and allows us to ''see the forest rather than the trees.'' One very surprising and important result will emerge: for purposes of structural analysis the information in the equilibrium statement (1), and in the kinematics statement (2), are strongly related. This remarkable result will be proved with the aid of a key theorem, namely the *Theorem of Virtual Work.*

1.2.1 The Theorem of Virtual Work

The mathematical foundation on which nearly all of modern matrix methods of structural analysis is based, including finite element analysis, is the *Theorem of Virtual*

---

[5] The concept of free-body diagrams is extremely important in structural analysis, and will be discussed in greater detail shortly.

Work. The theorem is rather deceptive; it is nothing more than a combination of the equilibrium relations (1) and kinematical relations (2). Virtual work may be used to obtain information about either the physical states of equilibrium or deformation. However, the interpretation of either type of information from a *work* context is a powerful tool for understanding and solving problems of equilibrium or kinematics. In particular, the virtual work theorem provides the essential base for demonstrating the strong relationship between equilibrium and deformation.

Virtual work also serves as a foundation for deriving nearly all of the energy theorems of structural analysis. As we shall see later, these theorems lead to important approximation and bounding techniques in structural analysis.

Finally, and most importantly, the theorem provides an essential base for developing and understanding the finite element method of structural analysis, the matrix computer solution technique that has revolutionized not only structural analysis but all of modern engineering.

This text introduces the concepts of modern structural analysis that form the foundation for the finite element method. For the sake of simplicity and ease of understanding, these concepts are developed for the analysis of truss and frame structures that are primarily two-dimensional in form and made up of one-dimensional bar and beam elements. This approach will develop a solid foundation for the more advanced concepts needed to understand the finite element method. To accomplish this, our emphasis will be on developing a strong understanding of basic virtual work concepts, and especially on the development of these concepts using an approach based on matrix algebra. Specific forms of the Theorem of Virtual Work will be formulated in Chapter 4 following the introduction of required background material.

1.3 LOADS

Before analyzing any structure it is helpful to first define the external forces or other factors that may be expected to produce internal forces/stresses and/or deformations/strains in the structural elements. We shall place all factors that contribute to structural response into a category called "loads."

Some loads may be attributed to mechanical origins, such as forces associated with the weight of the structure itself, or those associated with external objects such as the contents of a building, snow on a roof, or weight of vehicles crossing a bridge. Forces may also arise from fluid-structure interaction, e.g., steady aerodynamic loads such as wind loads on buildings or pressure distributions on aircraft, or effects of ocean currents or waves on off-shore structures.

Some forces may be regarded as static, i.e., relatively constant over time. The weight of a building is static; the weight of people in the building is also assumed to be static even though the movement of people is clearly dynamic. In certain situations forces are dynamic, or varying with time, such as, for example, forces generated during impact of two automobiles or the air pressure due to shock waves from an explosion.

Other forces, such as those due to moving vehicles or fluid streams may be regarded as either static or dynamic depending on their characteristics and also possibly on the characteristics of the structure. A frequently used example involves the structural response of a bridge to a troop of soldiers. It may make a considerable difference if the soldiers march across the bridge in step or out of step. While the latter is most likely a static situation, the former would constitute a cyclic dynamic load if the pace of the marching soldiers should happen to excite a natural frequency of the structure. Similarly, deformations of flexible structures may interact with unsteady aerodynamic forces and produce oscillations (flutter). Wind on ordinary telephone lines may either produce a benign static response, or under special circumstances, lead to chaotic dynamic behavior. The famous Tacoma Narrows Bridge failure (Figure 1.9*a*), provides ample evidence that such effects are not confined to aircraft or to power lines, but to structures in general, and that dynamic response can lead to catastrophic consequences.

Other loads are due entirely to inertial effects, for example, earthquake loads on buildings (Figure 1.9*b*). Time-varying vertical and horizontal components of ground motion during earthquakes lead to oscillations of structures, which could range from being virtually unnoticeable to very destructive.

Even if dynamic oscillations do not immediately produce conditions of overstress leading to outright failure, the application of repeated cycles of loading can degrade the integrity of the structural materials. This general problem falls under the category of material fatigue and failure and, in recent years, has led to a subject known as *fracture mechanics.* Repetitive loads can conceivably lead to fractures even if the peak load levels are considerably below those that would cause failure under static conditions or those due to dynamic resonance. The analysis of fatigue and fracture strength is a subject of major importance in a variety of modern engineering structures, such as aircraft and power plants. The fatigue failure in the Aloha Airlines aircraft shown in Figure 1.10 provides ample support to the severity and importance of the problem.

Still other loads result from factors such as support settlement, component mismatch or residual stress, and thermal effects. Thermal effects are important in such structures as high-performance aircraft, combustion engines, nuclear containment vessels, piping systems, and boilers. They can also be significant in less dramatic but equally important situations, such as for buildings and bridges subjected only to changes in weather and climate (Figure 1.11).

Many of the basic loads on buildings and bridges (Figure 1.12), are specified in design codes. These loads are the variable static forces due for example to people, contents, vehicles, snow, and wind. Design codes also specify the manner in which the effects of earthquakes are to be dealt with, usually as a set of additional external forces

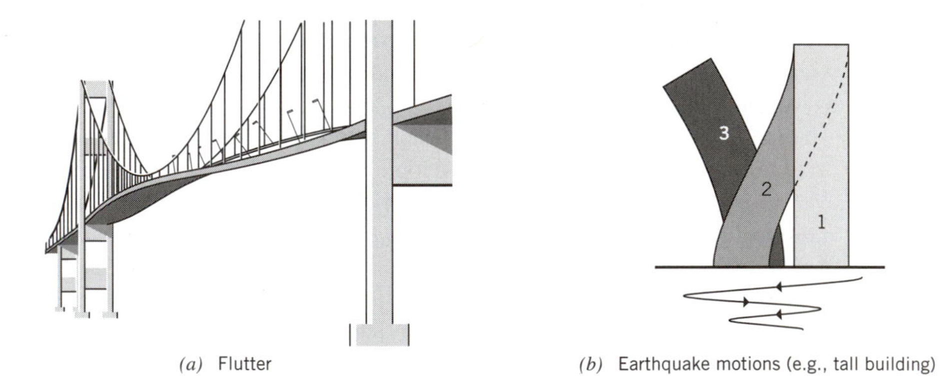

(a) Flutter

(b) Earthquake motions (e.g., tall building)

Figure 1.9 Dynamic leads. (Salvadori/Heller, *Structure in Architecture, 3/E,* © 1986; reprinted by permission of Prentice-Hall, Englewood Cliffs, NJ.)

Figure 1.10 Effects of fatigue: Aloha Airlines 737, Flight 243, April 28, 1988. The top half of an entire fuselage section tore loose on a flight from Hilo to Honolulu, Hawaii. One of the six crew members was swept overboard; the five other crew members and the 89 passengers survived. (Top photo © 1988, Robert N. Nichols, Black Star, New York, NY; bottom photo © 1988, Matthew Thayer.)

applied at various points in the structure. The loads specified by codes are most frequently static or pseudostatic regardless of their origins. Design codes also simplify load definition by assuming that they are deterministic in nature, i.e., are basically unvarying, whereas actual loads are more likely to be probabilistic in nature.

This book is concerned with the analysis of structures subjected to given loads and does not deal with the problem of defining such loads. Consequently, in what follows,

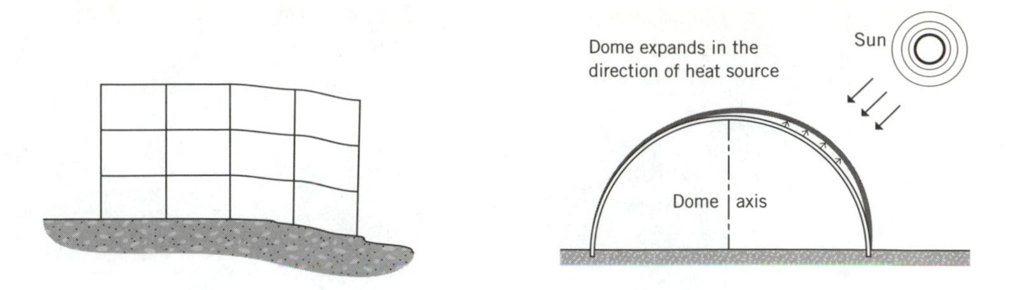

Figure 1.11 Support settlement and thermal effects. (Salvadori/Heller, *Structure in Architecture, 3/E,* © 1986; reprinted by permission of Prentice-Hall, Englewood Cliffs, NJ.)

we will assume that the loads have been specified to whatever extent is required for purposes of conducting a structural analysis. Extensive discussions of loads can be found in many reference books oriented toward design.

1.4 MATERIAL BEHAVIOR

Structures can be fabricated from a wide variety of materials. While every material has many different properties, those of primary interest to structural engineers are generally

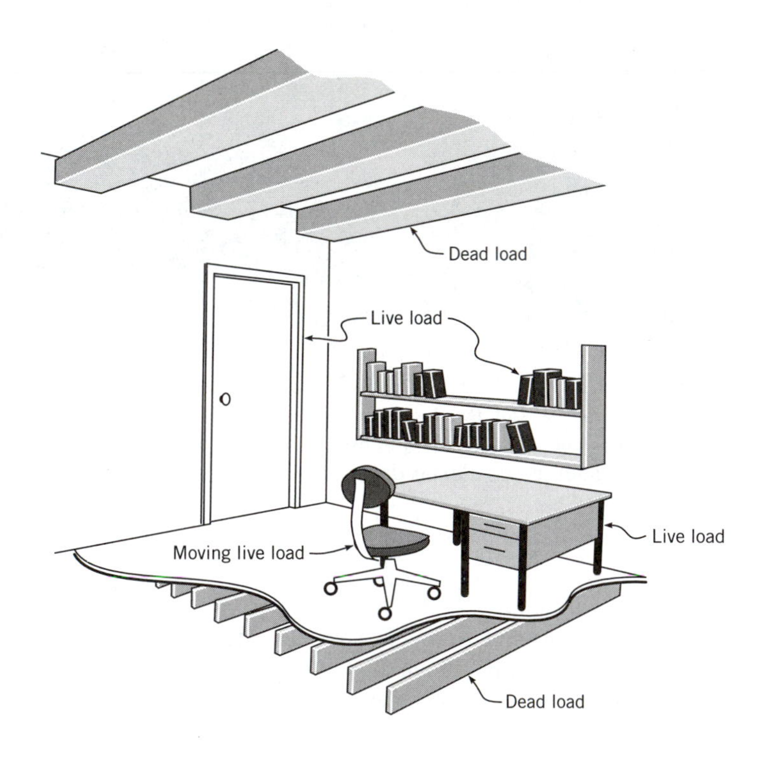

Figure 1.12 Loads. (Salvadori/Heller, *Structure in Architecture, 3/E,* © 1986; reprinted by permission of Prentice-Hall, Englewood Cliffs, NJ.)

related to stiffness, strength, and fatigue/fracture resistance, as well as their dependence on temperature and/or time.

The behavior of most materials is a function of the intensity of the loading. For example, most common structural materials exhibit elastic behavior at relatively low stresses, followed by inelastic behavior and eventual fracture as the stress level increases. *Elastic* behavior implies that no permanent deformations remain after loads are removed. *Inelastic* behavior implies the opposite, i.e., that there will be permanent deformations that remain after removal of stress.[6] Most structures are designed so that material behavior will be within the elastic range under nominal service conditions.

Under unusual or unlikely loading conditions, permanent deformation or damage is sometimes acceptable as long as catastrophic failure is prevented. For example, buildings are usually designed to survive a wide range of earthquake loadings without any permanent damage. It may not be practical or feasible, however, to design buildings to survive severe earthquakes without some damage. With careful design it may be possible to design a structure so that the damage is confined to areas that are easiest to repair. Under extreme loadings, the building might be so badly damaged that it becomes unusable. Such damage would be acceptable provided that such damage did not risk the safety of human occupants.

In this book we will deal only with normal situations and will assume elastic material behavior. Furthermore, we will assume a particular type of elastic material behavior, namely *linearly elastic* behavior. We will not deal with any form of nonlinear material behavior although the virtual work approach provides a strong foundation for investigating these more complex problems.

In one-dimensional elements of the type that will be considered in this book, the internal stresses are essentially also one-dimensional or *uniaxial*, dominated by a single stress. The assumption of linear elastic behavior means that we assume the existence of a *linear* material constitutive relationship between this stress and the corresponding strain. Also, we shall account for the effects of both thermal strains, due to small temperature changes, and initial strains, due to such effects as lack of fit or preexisting damage in the material. Even with this very simple characterization of material behavior, we will be able to investigate a wide variety of practical problems in structural engineering.

The assumption of linear elastic material behavior would appear to mean that deformations of structural elements will be linearly proportional to loads. This is not always the case, however. For example, a major exception to proportional behavior occurs when elements are subjected to loads approaching critical buckling values (see Figure 1.1*b*). In such cases elastic deformations can increase at a much greater rate than loads.

The assumption that deformations are proportional to loads is widely used in structural analysis. The assumption leads to a very important simplification. If structural behavior is truly linear, i.e., if deflections are linear functions of forces, the concept known as *superposition* can be utilized. Superposition means that if a linear elastic structure is subjected to a number of distinct but simultaneously applied loads the overall response can be determined by summing the response of the structure to the

---

[6] If inelastic deformation is much greater than elastic deformation then behavior is said to be *ductile*.

loads applied one at a time. As we shall see, the superposition property will be used extensively in theoretical developments and in problem-solving in structural analysis.

1.5 REVIEW OF EQUILIBRIUM EQUATIONS

The requirement that a structure be in a state of equilibrium under given loading is one of the fundamental concepts of structural analysis. The basic equations of equilibrium involve the interaction of forces and moments applied to the structure and the internal stresses that are needed to maintain equilibrium.

1.5.1 Concentrated Forces

Concentrated forces are *vector* quantities, i.e., they have both *magnitude* and *direction*. Such forces are most conveniently defined as the product of *scalar* and a *unit vector* that specifies the direction.[7] For example, if **e** denotes a unit vector with some specified direction, and F denotes the scalar value, then a force vector may be written

$$\mathbf{F} = F\mathbf{e} \tag{1.1}$$

The sense of **F** is defined by the *sign* of F, i.e., *positive* denotes a force in the direction of positive **e** and *negative* denotes a force in the direction of negative **e.**

Directions are most conveniently defined by a rectangular Cartesian coordinate system, e.g., an x, y, z coordinate system in three-dimensional space. Unit vectors in the positive x, y, and z directions are commonly denoted by the symbols **i**, **j**, and **k**, respectively. Thus, a force of magnitude F_x in the x direction is given by

$$\mathbf{F}_x = F_x\mathbf{i} \tag{1.2a}$$

Similarly, forces of magnitude F_y and F_z in the y and z directions are written

$$\mathbf{F}_y = F_y\mathbf{j} \tag{1.2b}$$

$$\mathbf{F}_z = F_z\mathbf{k} \tag{1.2c}$$

It is important to recognize that F_x, F_y, and F_z may be either positive or negative.

Forces add vectorially. The sum of vectors $\mathbf{F}_x$, $\mathbf{F}_y$, and $\mathbf{F}_z$, for example, is a *vector*, which in three dimensions provides an alternate representation of a general force such as **F** in Equation 1.1,

$$\mathbf{F} = \mathbf{F}_x + \mathbf{F}_y + \mathbf{F}_z \tag{1.3a}$$

or, using Equations 1.2*a*–1.2*c*, and Equation 1.3*a*

$$\mathbf{F} = F_x\mathbf{i} + F_y\mathbf{j} + F_z\mathbf{k} \tag{1.3b}$$

[7] The reader is assumed to have at least a brief introduction to the concept of vectors and some knowledge of vector algebra, including the scalar (dot) product and the vector (cross) product.

The quantities F_x, F_y, and F_z in Eqs. 1.2*a*–1.2*c* are referred to as the *components* of vector **F** (Fig. 1.13). The signs of F_x, F_y, and F_z define the senses of the individual components. The *magnitude* of **F** is related to the magnitudes of its components by

$$|\mathbf{F}| = F = \sqrt{(F_x)^2 + (F_y)^2 + (F_z)^2} \tag{1.3c}$$

i.e., magnitude is, by definition, a positive quantity.

1.5.2 Equilibrium of Forces

Now, consider a physical body that is acted upon by several concurrent forces, $\mathbf{F}_1$, $\mathbf{F}_2, \ldots, \mathbf{F}_n$, where

$$\begin{aligned} \mathbf{F}_1 &= F_{1_x}\mathbf{i} + F_{1_y}\mathbf{j} + F_{1_x}\mathbf{k} \\ \mathbf{F}_2 &= F_{2_x}\mathbf{i} + F_{2_y}\mathbf{j} + F_{2_z}\mathbf{k} \\ &\;\;\vdots \\ \mathbf{F}_n &= F_{n_x}\mathbf{i} + F_{n_y}\mathbf{j} + F_{n_z}\mathbf{k} \end{aligned} \tag{1.4}$$

The body is in a state of equilibrium whenever the sum of all the forces acting on it is zero, i.e.,

$$\sum_{i=1}^{n} \mathbf{F}_i = \mathbf{F}_1 + \mathbf{F}_2 + \cdots + \mathbf{F}_n = \mathbf{0} \tag{1.5}$$

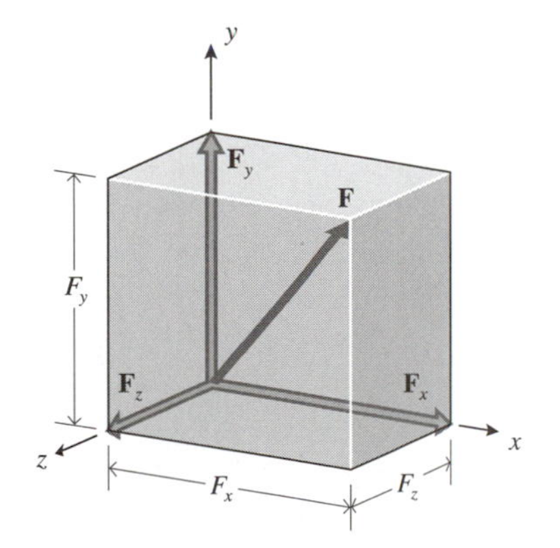

Figure 1.13 Components of force **F** (*positive* in directions shown).

If Eqs. 1.4 are substituted into Equation 1.5, the result is

$$(F_{1_x} + F_{2_x} + \cdots + F_{n_x})\mathbf{i} + (F_{1_y} + F_{2_y} + \cdots + F_{n_y})\mathbf{j} + (F_{1_z} + F_{2_z} + \cdots + F_{n_z})\mathbf{k} = \mathbf{0} \qquad \textbf{(1.6}\textbf{\textit{a}}\textbf{)}$$

or

$$\left(\sum_{i=1}^{n} F_{i_x}\right)\mathbf{i} + \left(\sum_{i=1}^{n} F_{i_y}\right)\mathbf{j} + \left(\sum_{i=1}^{n} F_{i_z}\right)\mathbf{k} = \mathbf{0} = 0\mathbf{i} + 0\mathbf{j} + 0\mathbf{k} \qquad \textbf{(1.6}\textbf{\textit{b}}\textbf{)}$$

Since **i**, **j**, and **k** are linearly independent, it follows that Equation 1.6*b* can be satisfied only if

$$\sum_{i=1}^{n} F_{i_x} = 0$$
$$\sum_{i=1}^{n} F_{i_y} = 0 \qquad \textbf{(1.7)}$$
$$\sum_{i=1}^{n} F_{i_z} = 0$$

i.e., the body will be in equilibrium only if the sum of the force components in each of the coordinate directions is zero. The three parts of Eq. 1.7 are referred to as the *equations of force equilibrium* in component form. In two-dimensional situations, one of these equations will be identically zero, and there will be only *two* nontrivial equations of force equilibrium.

1.5.3 Moment of Force

A study of the equilibrium of bodies subjected to forces that are not concurrent, and/or that are subjected to applied moments, requires an understanding of the concept

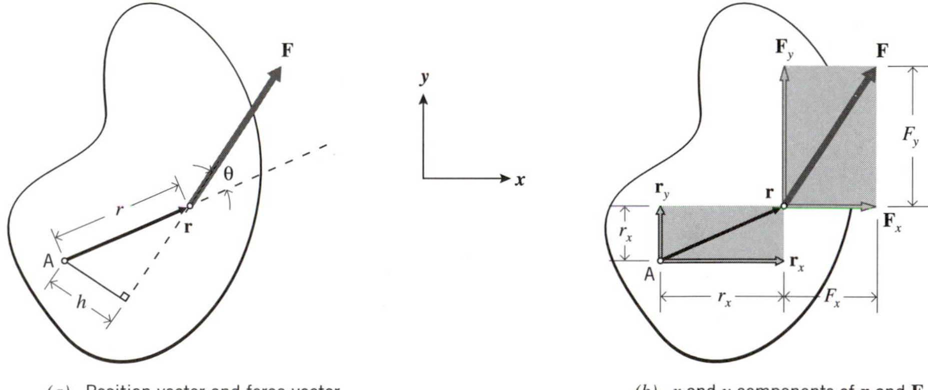

(*a*) Position vector and force vector

(*b*) *x* and *y* components of **r** and **F**

Figure 1.14 Moment of force **F** about point *A*.

of a *moment.* First, consider what is meant by the *moment of a force about a point.* This moment quantifies the tendency of the force to cause the body to rotate about some axis (or set of axes) through the point.

The moment of a force, such as force **F** in Figure 1.14*a*, about a point such as A in Figure 1.14*a*, may be defined as the *vector* cross-product of a position vector and the force vector **F**, i.e.,

$$\mathbf{M}_{F/\mathrm{A}} = \mathbf{r} \times \mathbf{F} \tag{1.8}$$

where **r**, the position vector, is a vector from point A to any point on the line of action of **F**. Vector $\mathbf{M}_{F/\mathrm{A}}$ is normal to the plane containing **r** and **F**, and is so directed that vectors **r**, **F**, and $\mathbf{M}_{F/\mathrm{A}}$, respectively, form a *right-handed* system.[8] The magnitude of the vector $\mathbf{M}_{F/\mathrm{A}}$ in Equation 1.8 is then defined by

$$M_{F/\mathrm{A}} = rF \sin \theta \tag{1.9}$$

where r is the magnitude of **r**, F is the magnitude of **F**, and $0 < \theta < 180°$ is the angle between **r** and **F**. Angle θ is defined as positive in the direction shown in Figure 1.14*a*.

Equation 1.9 provides a useful physical interpretation of the magnitude of $\mathbf{M}_{F/\mathrm{A}}$. The perpendicular distance from point A to the line of action of **F**, h in Figure 1.14*a*, is related to **r** and θ by

$$h = r \sin \theta \tag{1.10}$$

Consequently, Equation 1.9 is also equivalent to

$$M_{F/\mathrm{A}} = hF \tag{1.11}$$

i.e., the magnitude of the moment of a force **F** about point A equals the product of the magnitude of **F** and the perpendicular distance from point A to the line of action of **F**.

Vector **r** may also be expressed in terms of Cartesian components,

$$\mathbf{r} = r_x\mathbf{i} + r_y\mathbf{j} + r_z\mathbf{k} \tag{1.12}$$

as was **F** in Equation 1.3*b*. Moment **M** can then be expressed in terms of components of **r** and **F** by inserting Equations 1.12 and 1.3*b* into Equation 1.8 and using the identities relating cross products of unit vectors,

$$\mathbf{M}_{F/\mathrm{A}} = (r_yF_z - r_zF_y)\mathbf{i} + (r_zF_x - r_xF_z)\mathbf{j} + (r_xF_y - r_yF_x)\mathbf{k} \tag{1.13a}$$

or

$$\mathbf{M}_{F/\mathrm{A}} = M_x\mathbf{i} + M_y\mathbf{j} + M_z\mathbf{k} \tag{1.13b}$$

---

[8] The interpretation of *right-handedness* can be confusing; it is the direction of $\mathbf{M}_{F/A}$ defined by the thumb of the right hand when it is placed a) at the intersection of **r** and **F**, b) with the fingers pointing in the direction of increasing **r**, and c) curling the fingers in the direction of increasing **F** such that the angle between positive **r** and **F** is between 0 and 180°.

where M_x, M_y, and M_z are the three components of the moment of **F** about point A, i.e.,

$$M_x = r_y F_z - r_z F_y$$
$$M_y = r_z F_x - r_x F_z \qquad (1.14)$$
$$M_z = r_x F_y - r_y F_x$$

As with forces, the signs of the components of moments determine their senses; in particular, a positive sign implies that the component has the same direction as the positive coordinate axis, and a negative sign implies the opposite.

The components have a simple and useful physical interpretation. To see this consider the M_z component, and visualize the components of **r** and **F**, which appear when viewed looking down the z axis of the x, y, z coordinate system, namely the x and y components of each vector, as in Figure 1.14*b*. In this view, $r_x F_y$ and $-r_y F_x$ give the tendency of the body to rotate about the z axis in a counterclockwise manner (i.e., right-hand rule positive) due to the application of force components F_y and F_x, respectively. Thus, M_z is a measure of the tendency of **F** to rotate the body about the z axis. The other two axial components have analogous interpretations.

The physical interpretation of moment of force is particularly useful in two-dimensional situations; rather than using vector algebra to obtain moments, it is usually preferable to deal directly with the components of moments as defined in the last of Equations 1.14 or in Equation 1.11.

1.5.4 Equilibrium of Moments

Now, consider a body in a state of equilibrium under the action of several forces, $\mathbf{F}_1$, $\mathbf{F}_2, \ldots, \mathbf{F}_n$, such as those listed in Equations 1.4, applied at positions on the body defined by position vectors $\mathbf{r}_1, \mathbf{r}_2, \ldots, \mathbf{r}_n$, respectively. Consequently, each of the forces acting on the body in Figure 1.15*a* will generally have a moment about a point such as A. For the sake of convenience we will henceforth denote the moment of force $\mathbf{F}_i$ about point A by $\mathbf{M}_{i/A}$, or simply by $\mathbf{M}_i$ as long as it is clear about which point we are taking moments.

Furthermore, the body may also be subjected to *applied* moments, such as $\mathbf{M}_j$ in Figure 1.15*b*. Applied moments, which are commonly encountered in structural analysis, are convenient mathematical abstractions used to represent the net effects that certain types of force systems have on bodies. In particular, the components of an applied moment vector are measures of tendencies to cause rotations about axes parallel to the coordinate axes, exactly as discussed in the preceding section. Since the components have this inherent physical interpretation, it follows that the exact point of application of a pure moment is not significant insofar as equilibrium of the body is concerned.[9]

A body subjected to forces and applied moments will be in equilibrium whenever 1)

[9] The point of application of an applied moment is significant when we analyze internal forces in the body, cf. Chapter 3.

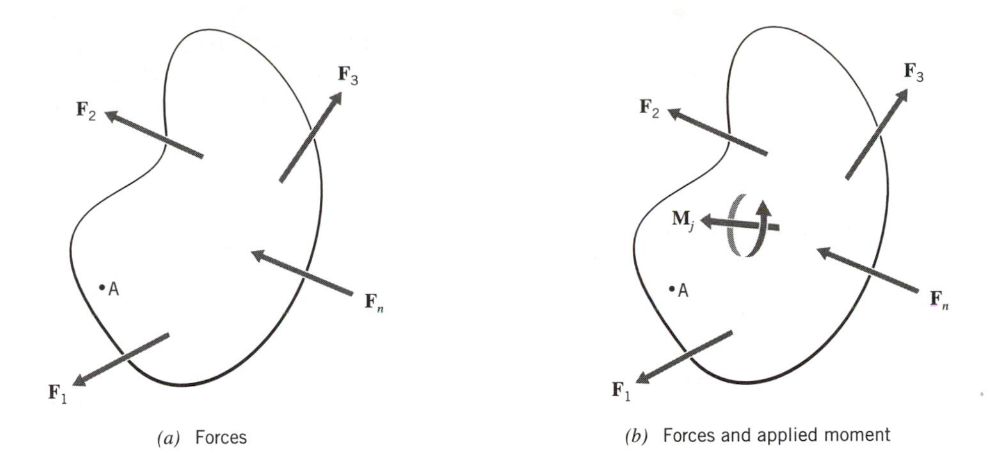

Figure 1.15 Body subjected to forces and moments.

the sum of all the forces is zero, and 2) the sum of all the moments of the forces about a point such as A, plus the sum of all the applied moments, is zero, i.e.,

$$\sum_{i=1}^{n} \mathbf{F}_i = \mathbf{0} \tag{1.15a}$$

and

$$\sum_{i=1}^{n} (\mathbf{r}_i \times \mathbf{F}_i) + \sum_{j=1}^{m} \mathbf{M}_j = \mathbf{0} \tag{1.15b}$$

where $\mathbf{r}_i \times \mathbf{F}_i$ is the moment of force i about point A, n is the total number of forces, and m is the total number of applied moments. Equation 1.15b is more conveniently written as

$$\sum_{k=1}^{N} \mathbf{M}_k = \mathbf{0} \tag{1.15c}$$

where $N = n + m$. Note that Equation 1.15a is identical to Equations 1.5 or 1.6 or 1.7. Equation 1.15c is equivalent to

$$(M_{1_x} + M_{2_x} + \cdots + M_{N_x})\mathbf{i} + (M_{1_y} + M_{2_y} + \cdots + M_{N_y})\mathbf{j} + (M_{1_z} + M_{2_z} + \cdots + M_{N_z})\mathbf{k} = \mathbf{0} \tag{1.16a}$$

or,

$$\sum_{k=1}^{N} M_{k_x} = 0$$

$$\sum_{k=1}^{N} M_{k_y} = 0 \tag{1.16b}$$

$$\sum_{k=1}^{N} M_{k_x} = 0$$

Thus, the body will be in equilibrium only if the sum of the force components and the sum of the moment components in each of the coordinate directions is zero.[10] In the remainder of this book, equilibrium will be assessed using the component forms of equilibrium given by Equations 1.7 and 1.16*b*.

1.5.5 Free-Body Diagrams

A sketch of a representation of a physical body, as in Figures 1.15*a* and 1.15*b*, which shows *all* of the forces acting on the body, is referred to as a *free-body diagram.* A *free-body* may consist of an entire structure, a single element, or any portion of a structure or element. The word *free* is particularly significant because it implies that the body under consideration is assumed to be isolated from all of its surroundings including contiguous portions of the structure and/or supports. In regions where the body is assumed to have been ''detached'' from contiguous portions of the structure or supports, the free-body diagram must include any and all forces which are exerted on the body by those contiguous portions or supports. (Such forces actually occur in equal and opposite groups on the contiguous surfaces, a manifestation of Newton's third law.) In other words, in all regions where the body is assumed to be ''cut'' in order to isolate it, forces that are originally internal to the actual body appear as external on the free-body. These forces must be included in the equilibrium equations! In short, the free-body diagram must show **all forces** that are applied to the free-body, regardless of their origins. **The proper representation of free-body diagrams is an absolutely essential prerequisite for equilibrium considerations in structural analysis.** The nature of free-body diagrams should become clearer as we progress to examples that illustrate the concept of equilibrium.

1.5.6 Statically Equivalent Systems of Forces

In the preceding sections we have seen, in essence, that the overall sum of the individual forces acting on a body that is in equilibrium must add up to zero. In other words, the net effect of all the forces is equivalent to a net force of magnitude zero and a net moment of magnitude zero. In many analysis situations it is necessary or convenient to deal with only a subset of the applied forces, and to replace this subset of forces with a single force and/or a single moment that will have the same effect in the equilibrium equations that hold for the body. The single force and moment with this property are said to be *statically equivalent* to the original subset of forces, and are referred to as the *resultant force* and *resultant moment* or, simply, as the statical *resultants* of the subset of forces.

Equivalence is achieved when the resultant force has the same magnitude and direction as the sum of the forces in the subset, and the resultant moment has the same magnitude and direction as the sum of the moments of the forces in the subset. If the magnitude of the resultant force is not zero, then equivalence can be achieved by locating the line of action of the resultant force so that its moment about some point is

---

[10] It follows that there are six *independent* equations of equilibrium in three dimensions and three independent equations in two dimensions. It is sometimes convenient to use alternate sets of independent equations to those given here, cf. Problem 1.P1.

the same as the sum of the moments of the forces in the subset.[11] This concept will be demonstrated shortly.

1.5.7 Treatment of Distributed Forces

In typical problems, structures are subjected to force and moment loads at many different locations on the body. Such *distributed* systems of forces and/or moments are often tedious and time consuming to deal with. Often, therefore, approximations are made to simplify the description of the loads. For example, the furniture and people in an office building in reality apply a very complicated and changing pattern of concentrated vertical forces on the floors of the building. Rather than face the task of trying to describe all the details of these forces, we will assume a uniformly applied vertical load having units of force per unit area. Thus, distributed forces can also encompass continuous force systems. On a 1-D beam/bar element the units of such a loading should be *force per unit length*; for a 2-D element the units would be *force per unit area*; and for the 3-D body the units are *force per unit volume* or *force per unit surface area.*

The equilibrium equations are very simple to apply for structures subjected to continuous forces. These continuous forces are simply multiplied by the appropriate differential length, area, or volume, and the resulting differential forces or moments are summed. The summation of an infinite number of differential-sized quantities is accomplished by the process of integration.

To see how continuous loads are treated in an equilibrium study, consider the one-dimensional continuous transverse force loading on the beam shown in Figure 1.16*a*. In the figure, $q(x)$ denotes the value of the force per unit length applied to the beam at station x.

Assume that the beam is in equilibrium under the action of the continuous force and concentrated forces at both ends of the beam, as shown in Figure 1.16*a*. These concentrated forces are vector quantities but, as noted previously, we will deal directly with the components of the equilibrium equations; it will therefore be convenient to represent the concentrated forces only by their magnitudes and directions without specifically invoking formal vector notation. Thus, in Figure 1.16*a*, the concentrated forces are designated by F_1 and F_2, and the directions are shown by vector arrows.

The beam will also be in equilibrium under the action of F_1, F_2, and R (Figure 1.16*b*), provided that R is statically equivalent to the continuous force; this means that R must have the same magnitude and direction as the resultant of the continuous force, and the same moment of force about a point such as A in Figure 1.16. (We have anticipated the result somewhat by showing R to be directed vertically in Figure 1.16*b*.) Note that the resultants of distributed force systems (including both concentrated forces and continuous forces) are *vector* quantities since they are obtained by a process of summation (or integration).

For this illustration, because the applied forces and forces per unit length are acting only in the y direction, we need to explicitly deal only with the equilibrium equation

[11] Note that two different force systems cannot be completely equivalent since any rearrangement of forces will move regions of high stress and low stress, and will greatly influence elastic behavior. The use of statically equivalent loads is primarily an aid in satisfying the equations of equilibrium.

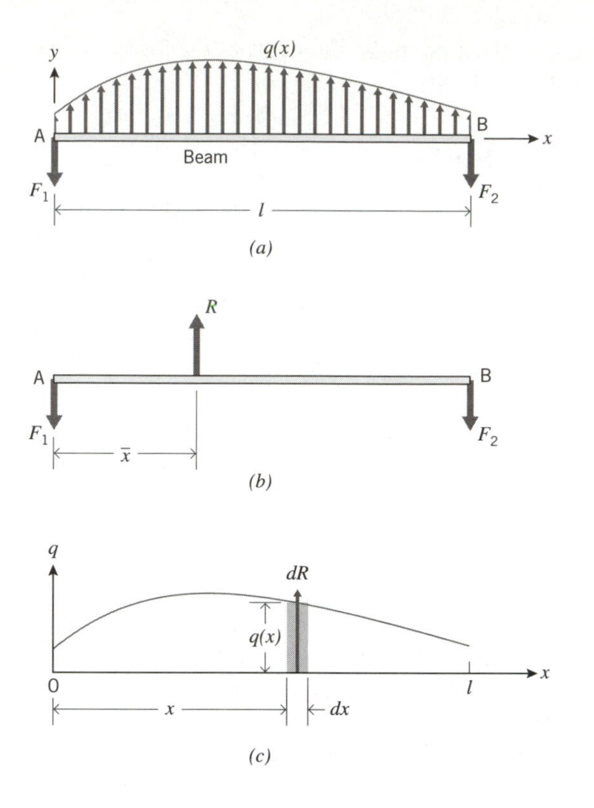

Figure 1.16 Resultant of distributed force.

of forces in the y direction, and as we shall see, only with the equilibrium of moments about the z axis.

The process of treating the continuous loading is straightforward. At station x the incremental force is

$$dR = q(x)dx \tag{1.17}$$

where dx is an incremental length in the x direction (Figure 1.16*c*). The sum of these incremental forces is obtained by integration, i.e.,

$$R = \int_0^l q(x)dx \tag{1.18}$$

Equation 1.18 defines the equivalent force R. A positive R is a force acting in the y direction because positive increments are acting in this direction. In this case, R is equal to the area under the graph of the function $q(x)$ (Figure 1.16*c*) between the limits $x = 0$ and $x = l$.

The resultant moment of a typical incremental force about point A (Figure 1.16*c*) is given by

$$dM = xdR = x[q(x)dx] \tag{1.19}$$

The sum of all the incremental moments about point A is therefore

$$M = \int_0^l xq(x)dx \tag{1.20}$$

where M is the equivalent moment of force about A of the continuous force $q(x)$; the direction of the moment is counterclockwise, which is defined as the positive direction by Equation 1.20.

If R is to be statically equivalent to the distributed force *and* have the same moment of force M, it must be placed to give the same moment about point A as the distributed force, i.e.,

$$\bar{x}R = \int_0^l xq(x)dx \tag{1.21}$$

where $\bar{x}$ is defined in Figure 1.16*b*. If Equation 1.18 is substituted for R, the result for the value of $\bar{x}$ is

$$\bar{x} = \frac{\int_0^l xq(x)dx}{\int_0^l q(x)dx} \tag{1.22}$$

This is identical to an expression that defines the x coordinate of the centroid of the area under the graph of the function $q(x)$ between the limits 0 and l.

Once the continuous force is replaced by the correct resultant, R, located at the correct position $\bar{x}$, then the values of F_1 and F_2 can be found by writing the equations of equilibrium for the free body in Figure 1.16*b*.

EXAMPLE 1

Consider the beam shown in Figure 1.17*a* subjected to a linearly varying continuous force defined by $q(x) = (q_0/l)x$ (Figure 1.17*b*). The quantity q_0 is the magnitude of the continuous force/length at $x = l$. Suppose we wish to replace the continuous force by its resultant R.

From Equation 1.18 we obtain

$$R = \int_0^l \frac{q_0}{l} x dx = \frac{q_0 l}{2} \tag{1.23}$$

and from Equation 1.22, the location of R is

$$\bar{x} = \frac{\int_0^l x\left(\frac{q_0}{l}x\right)dx}{R} = \frac{q_0 l^2/3}{q_0 l/2} = \frac{2}{3}l \tag{1.24}$$

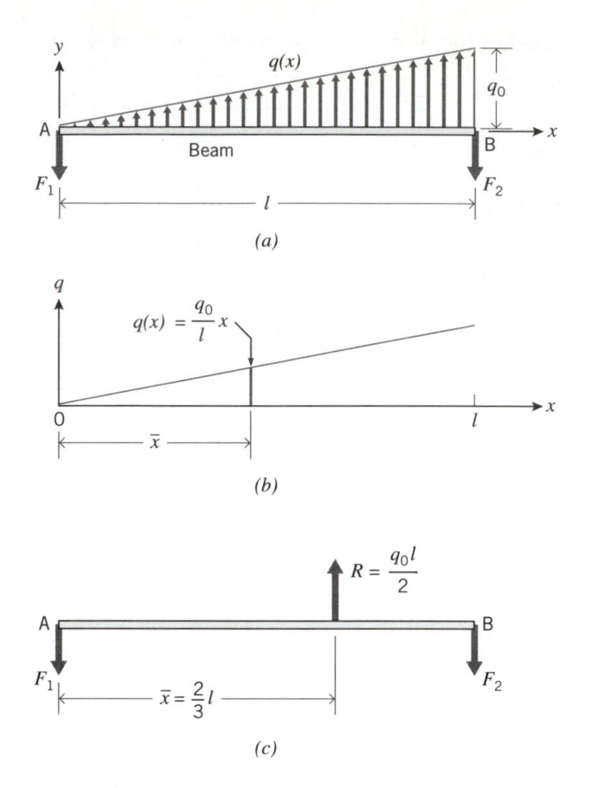

Figure 1.17 Example 1.

Again, we see that Equation 1.23 is identical to the area under the graph of $q(x)$ (Figure 1.17*b*), and that Equation 1.24 gives the x coordinate of the centroid of the area under the graph. Figure 1.17*c* shows the resultant of the distributed force.

EXAMPLE 2

Find the magnitudes of F_1 and F_2 required to maintain the equilibrium of the beam in Figure 1.17*a* (Example 1).

We apply the appropriate equations of equilibrium to the free body in Figure 1.17*c*. Because this is a two-dimensional problem with forces applied in the x-y plane, we need deal only with the first two of Equations 1.7 and the last of Equations 1.16*b*.

The first of Equations 1.7 is identically equal to zero because there are no applied forces acting in the x direction. From the second of Equations 1.7 we obtain

$$\sum F_y = 0 = R - F_1 - F_2 = \frac{q_0 l}{2} - F_1 - F_2 \tag{1.25}$$

and from the last of Equations 1.16b, taking moments about point A gives

$$\sum M_i = 0 = \bar{x}R - lF_2 = \left(\frac{2}{3}l\right)\left(\frac{q_0 l}{2}\right) - lF_2 \qquad \textbf{(1.26)}$$

or

$$F_2 = \frac{q_0 l}{3} \qquad \textbf{(1.27)}$$

Substituting F_2 from Equation 1.27 into Equation 1.25 gives

$$F_1 = \frac{q_0 l}{6} \qquad \textbf{(1.28)}$$

The directions shown in the free body diagram determined the signs of F_1 and F_2 in the equilibrium equations. The fact that F_1 and F_2 are both positive in Equations 1.27 and 1.28 implies that both were assumed to be acting in their respective positive directions in the free-body diagram (Figure 1.17a). If either direction had been opposite, e.g., if F_1 had been assumed to be acting upward rather than downward, then the result would have been negative rather than positive. This simply means that the direction assumed as positive was opposite to the direction the actual force is acting; the correct result is obtained regardless of which direction is arbitrarily chosen at the beginning of the analysis.

1.6 ANALYTIC MODELS

Unless we perform physical experiments, we never analyze "real" structures; rather, we analyze *analytic models* of structures. An analytic model is an idealized mathematical representation of the actual structure based on appropriate simplifications and assumptions regarding the nature of the structure and its modes of load transmission.

We have already implicitly introduced a number of assumptions in our set of examples that were necessary to carry out the equilibrium analyses, i.e., we made several basic assumptions regarding the treatment of forces and equilibrium. For example, a *concentrated force* is a force that is assumed to be applied at a *point*. This is an idealization that cannot exist in reality because any real force must actually be distributed over a finite area rather than a dimensionless point. However, it is both reasonable and appropriate to idealize a real force as concentrated if we can expect the differences between the effects of the real and idealized forces on the behavior of the body to be negligible.

Similarly, when writing equations of equilibrium for bodies, we generally ignore the effects of *deformations* on the magnitudes and directions of forces, i.e., we write the equations of equilibrium for the undeformed body and use the resulting force distributions to calculate the deformations. This is a reasonable approximation if the effects of the deformations on the magnitudes and directions of the forces are negligible, as is the case in most conventional structures.

In other words, under appropriate circumstances and for appropriate reasons, we may be able to obtain important information about the response of a structure by making some reasonable idealizations in the treatment of forces and in the definition of the analytic model of the structure itself. The ability to formulate appropriate idealizations, or to recognize the existence and consequences of idealizations, is a major requirement of engineering practice. Such idealizations require considerable experience and, in fact, make structural analysis an art as well as a science.

1.6.1 Connections

In addition to the idealizations previously discussed, another basic aspect of structural analysis that requires fundamental assumptions and idealizations is the representation of structural behavior at or near element connections. When we speak of an *assembly* of elements, we imply connections between the elements (''joints'') and/or connections of the elements to supports. An explicit definition of the behavior of the elements at the connections must be made to perform a structural analysis. This is a difficult problem in itself, because the behavior of a typical connection depends to a large extent on the level of force it carries, and to a lesser extent on the technique used to assemble the structure. However, the structural analyst has no choice but to set down some ''reasonable'' definition regarding the behavior of the structural connections and supports.

Consider, for example, a two-dimensional structure composed of beam elements such as in Figure 1.18. In this figure, corners at B and C represent *rigid* connections between contiguous elements, whereas corner D represents a *pin* or *hinge* connection. The distinctions between rigid and pin connections in the drawings are fairly subtle, but are significant because they represent important physical differences, which will be discussed shortly. Alternate representations are possible, including basic line drawings such as in Figure 1.18, which shows the same structure as in Figure 1.4*b*. None of the drawings purports to be a detailed depiction of an actual structure; rather, each drawing represents the essential analytic features of the structure, namely slender beam elements of specified lengths connected together at joints with particular characteristics, and connected to the ground at points A and E (''supports'').

A real connection is typically a very complicated contrivance requiring additional structural components such as plates, rivets, bolts, and/or welds, *and* involving discontinuities in the basic structural elements because of cut-outs or holes. These additional components and discontinuities induce stress distributions that are very complex, and that differ significantly from the nominal stress distributions in the basic structural elements. These complex stress distributions are usually confined to the immediate

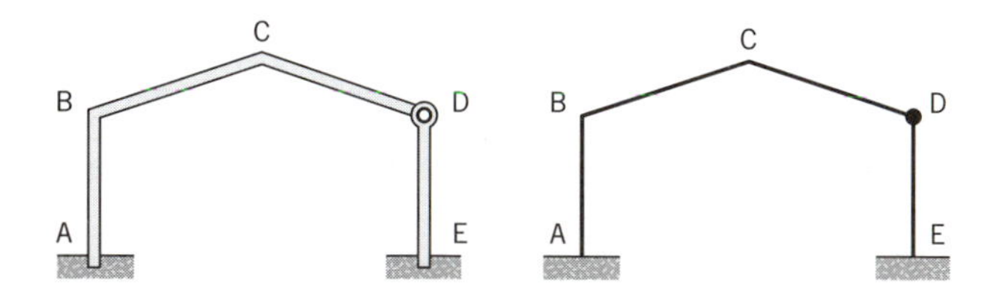

Figure 1.18 Line drawings of structure in Fig. 1.4*b*.

vicinity of the connection and do not extend to portions of the structure outside this local region.[12] As a consequence, the disruptions associated with connections are usually ignored for the purpose of determining nominal force distributions in the elements of a structure. This *does not* mean that these effects can be ignored in the detailed design or analysis of the connection. In fact, the nominal forces must be known *before* the connection can be designed or analyzed. Thus, if the analysis shows forces at connections that are significantly different from those on which the assumed connection behavior is based, the analyst must revise the connection definition and repeat the calculation. Alternatively, if there is great uncertainty in the quality of a given connection, then again the analyst must perform a set of calculations given different assumptions about the connection to make sure that the behavior of the structure is safe. Although the nature of stress distributions at connections is a subject that is not considered in this book, bear in mind that most real structural failures occur at connections, cf. Figure 1.19.

The preceding discussion leads us to the conclusion that, for purposes of structural analysis, in order to analytically model a connection it is best not to deal with stresses or resultant forces *within* the connection but, rather, bound the behavior of the connection by making general assumptions regarding the manner in which nominal element forces are transmitted across the connection. To do this, we can consider free-body diagrams of appropriate portions of the structure in the vicinity of the connections. Figure 1.20*a*, for example, shows a free-body diagram of a portion of the structure to either side of rigid joint B in Figure 1.18. Specifically, the free-body is assumed to consist of short segments of beams AB and BC, the two elements connected at joint B. The beam segments are assumed to be so short that their length dimensions are negligible. This implies that all concentrated forces acting on the free-body are to be treated as if they are concurrent and act through point B. For the sake of simplicity, we will henceforth refer to the free-body diagram in Figure 1.20*a* as the free-body diagram of joint B.

The force quantities shown acting on the ends of the element segments in Figure 1.20*a* are the general set of resultant forces that act on the cross section of a beam in a two-dimensional structure, namely an axial force (P), a transverse or shear force (V), and a bending moment (M).[13] These forces are actually internal to elements AB and BC, but must be shown as external forces on the free-body diagram of joint B. These forces are equal in magnitude and opposite in direction[14] to the forces on the contiguous ends of elements AB and BC, as shown in the partial free-body diagrams in Figure 1.20*b*.

Often, external forces, such as F and M_0 in Figure 1.21*a*, are assumed to be applied to structures at joints. In such cases, the free-body diagram of the joint must include these applied forces, as in Figure 1.21*b*. Equations of equilibrium applied to the free-body of the joint provide relations between the various force quantities. As an illustra-

---

[12] In slender beam elements, for example, stress distributions will be close to nominal at distances from discontinuities that are usually about equal to the maximum cross-sectional dimension of the beam. This qualitative concept is known as *St. Venant's principle.*

[13] The origin of these forces will be reviewed in Section 3.3.

[14] Newton's third law.

THE KANSAS CITY STAR.

SUSPENSION ROD TEARS FROM BOX BEAMS

CHANGE IN HYATT SKY WALK SUPPORTS

'ritical design change linked to collapse Hyatt's sky walks

Figure 1.19 The collapse of two suspended walkways in the Kansas City Hyatt Regency Hotel in 1981 killed over 100 people and was the worst accident due to a structural failure in the history of the United States. There were many theories offered for the cause of the collapse, including the presence of too many people dancing on the walkways, recalling the collapse of early suspension bridges under the feet of marching soldiers. The real cause of the failure was quickly traced to a single change in the design of a support detail, apparently made to faciliate the erection of the skywalks. *The Kansas City Star,* which had hired an engineer as a consultant on the story, revealed the true cause within days of the accident in this Pulitzer Prize-winning story. Later tests at the National Bureau of Standards confirmed the cause to be in the detail. (Reprinted from *To Engineer is Human,* Petroski, H., © 1982, 1983, 1984, 1985 by Henry Petroski, St. Martin's Press, New York, NY.)

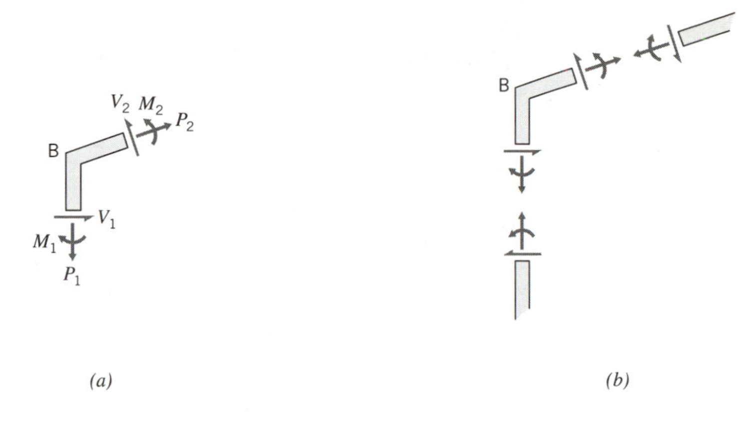

Figure 1.20 Forces near rigid joint in 2-D frame structure.

tive example, consider joint B of the rectangular frame structure shown in Figure 1.22*a*; the free-body diagram of joint B is shown in Figure 1.22*b*. Letting x and y denote horizontal and vertical directions, respectively, the equations of equilibrium are:

$$\sum F_x = 0 = F + V_1 + P_2$$

$$\sum F_y = 0 - P_1 + V_2 \qquad \textbf{(1.29)}$$

$$\sum M_i = 0 = M_2 - M_1 - M_0$$

Note that in writing the equation for moment equilibrium, forces F, P_1, P_2, V_1, and V_2 have all been assumed to pass through point B, and hence produce no moments about B; M_0, M_1, and M_2 are therefore the only quantities appearing in this equation.

In addition to the values of F and M_0, which are usually specified, Equations 1.29 contain six other force quantities. The equilibrium equations give us three independent relations that must be satisfied by these forces and moments. For example, the second of Equations 1.29 reveals, in this particular case, that the magnitudes of P_1 and V_2 must be equal. The other two equations provide similar information relating the remaining quantities. This type of information will be used extensively in subsequent work.

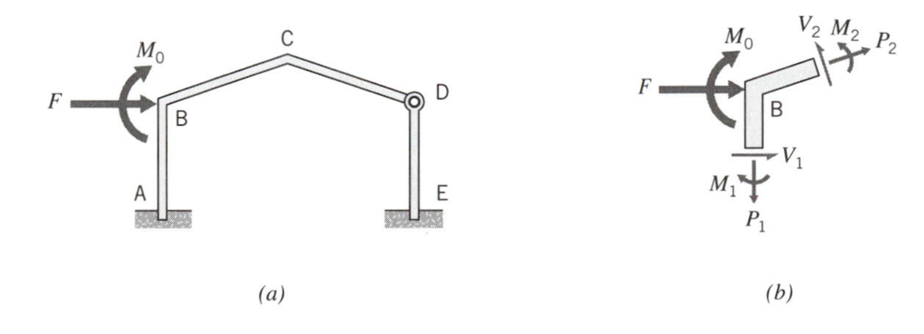

Figure 1.21 Frame structure with external forces applied to joint B; free-body diagram of joint B.

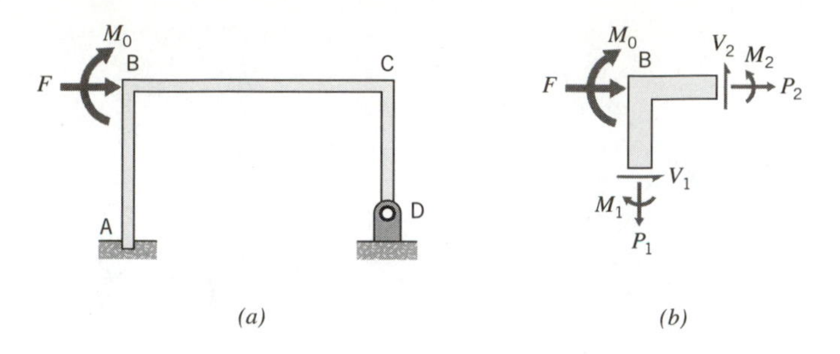

Figure 1.22 Rectangular frame; free-body diagram of joint B.

In all of the foregoing examples, joint B has been characterized as *rigid.* So far, we have discussed the *static* implications of this particular characterization, i.e., we have assumed that general sets of force quantities, including internal moments, act on the ends of the beam elements near the rigid joint, and we have related these force quantities via equations of equilibrium such as Equations 1.29. Of special importance is the fact that we assume there can be internal moments developed at the ends of the beam elements even when there are no external moments applied to the joint. In this sense rigid joint B is analogous to a sharp bend or "kink" in a beam; in general, we can expect to have internal moments in the beam near the "kink."

There is a second major implication of joint characterization involving *kinematics,* i.e., motion, or *deformations,* of the structure at and near the joint. In general, when a structure deforms the joints move. The movement of any joint can involve displacements of the joint itself, and rotations of the members that are connected to the joint. Each independent component of possible motion at a joint is referred to as a kinematic *degree of freedom.* The kinematic assumptions associated with a rigid joint, for example, are that the joint can displace and rotate, but that the angles between longitudinal axes of contiguous members at the joint remain unchanged when the structure deforms. This means that all members that are connected to a rigid joint must undergo the same rotation. Figure 1.23 shows a possible configuration of a rigid joint such as B before and after deformation. Here, u_1 and u_2 denote the horizontal and vertical components of displacement of point B, respectively, and ϕ denotes the angle of rotation of the

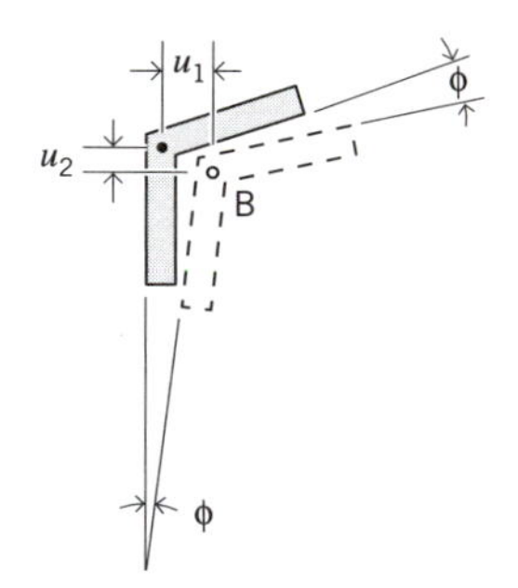

Figure 1.23 Displacement and rotation of rigid joint B.

joint. Thus, rigid joint B has three kinematic degrees of freedom, u_1, u_2, and ϕ in this case. As long as this joint is defined as rigid, the number of degrees of freedom would be unchanged even if additional beam elements were connected to the joint; there would, however, be additional force quantities associated with each extra beam element.

In Figure 1.23 both the displacement and the rotation are greatly exaggerated; deformations of real structures are usually very small in comparison with the dimensions of the structure, and are frequently not visible when superimposed on a scale drawing of the structure. In such circumstances, equations of equilibrium that are based on the geometry of the undeformed free-body of the joint are entirely appropriate.

The characteristics of all other types of connections are also defined by static and kinematic considerations. Consider a second basic type of connection, a *pinned*, or *hinged*, joint such as joint D in the frame structure of Figure 1.18 or Figure 1.21. Specifically, when two or more beam elements are connected by a frictionless pin, the static assumption is that all force quantities except internal moments can be developed in the beams in the immediate vicinity of the joint.[15] Thus, near the joint, the beam elements connected to a pinned joint each have one less possible internal force quantity than the beam elements connected to a rigid joint. A free-body diagram of pinned joint D is shown in Figure 1.24*a*.

The kinematic assumptions are that the joint can displace, and that the members connected by a pin can rotate relative to each other in the vicinity of the pin. Thus, a pinned joint has more kinematic degrees of freedom than a rigid joint has. A before-and-after picture of typical deformations is shown in Figure 1.24*b*. This figure shows the typical components of displacement of point D and the typical rotations of the beam segments that are connected to joint D. Note that the ends of the beams are assumed to rotate by different amounts. Consequently, there are four degrees of freedom: two associated with the displacements, u_1 and u_2, and one associated with the rotation of the end of each of the two beam elements, ϕ_1 and ϕ_2. If additional members are connected to the joint, there would be an additional rotational degree of freedom associated with each. As with rigid joints, displacements and rotations are usually small.

[15] *External moments* can be applied to beams at specific points *near* pinned joints; this will not change the basic static characteristics of the joint.

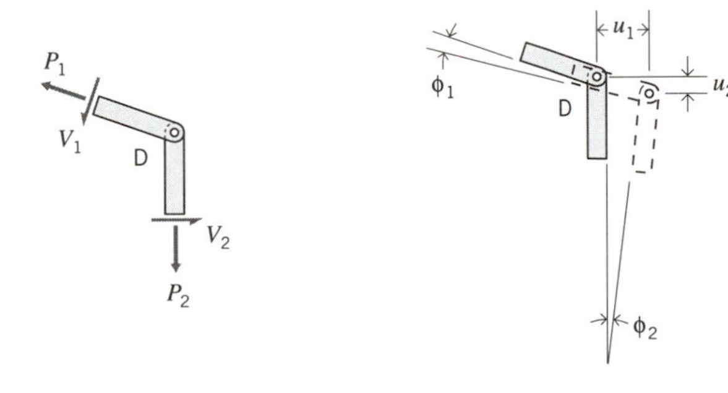

(*a*) Free-body diagram (*b*) Displacements

Figure 1.24 Pinned joint D.

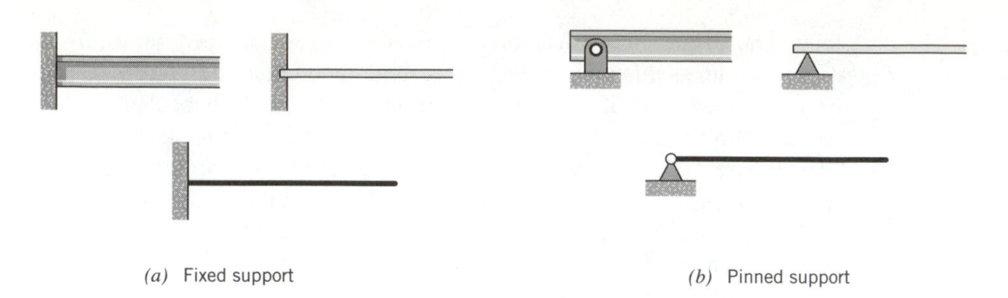

(a) Fixed support

(b) Pinned support

Figure 1.25 Basic support representation.

The static and kinematic assumptions associated with rigid and pinned joints are idealizations in the sense that no real joint will be perfectly rigid, and no real pin will be perfectly frictionless; however, the actual distinctions in behavior between an idealized joint and its real counterpart will often be inconsequential. In any case, real joints can be expected to exhibit behavior between these extremes; i.e., the rigid and pinned idealizations are actually limiting cases of real joint behavior.

A connection between a structural element and some ''external'' medium such as the ground is, as noted previously, referred to as a *support*. The characteristics of the supports with which we will deal are analogous to those of the joints considered previously, and are detailed in the following section.

1.6.2 Supports

As with the joints considered previously, the most basic types of supports are *rigid*, or *fixed*, and *pinned* supports, and their variations. Figure 1.25 illustrates typical representations of the basic forms of these supports for two-dimensional frame structures. Figure 1.25*a* shows some common, and equivalent, representations of fixed supports, and Figure 1.25*b* shows some common, and equivalent, representations of pinned supports. Similar representations have been used previously at points A and E of Figures 1.4*b*, 1.18, and 1.21*a*, and points A and D of Figure 1.22*a*. The static and kinematic implications depend only on the type of support being represented and not on the particular form of the representation.

The static implications of a fixed support are virtually identical to those of the rigid joint discussed previously. Specifically, we assume that a beam element can develop all possible internal force quantities near a fixed support (Figure 1.26*a*). Assuming that

(a) Internal forces near fixed support

(b) Free-body diagram of beam segment

Figure 1.26 Forces at a fixed support.

the length of the beam segment shown in Figure 1.26*a* is small, equilibrium requirements imply that the support must be applying, to the beam, a set of forces that are equal and opposite to the force quantities shown in the figure. We can justify this statement by considering the free-body diagram in Figure 1.26*b*, which shows the same small portion of the beam as Figure 1.26*a*; this free-body will be in equilibrium only if force quantities equivalent to those denoted by R_{A_x}, R_{A_y}, and M_A are applied to the beam by the support, and then only if these force quantities are related to the force quantities P, V and M by equilibrium equations, i.e.,

$$\begin{aligned} \sum F_x &= 0 = R_{A_x} + P \\ \sum F_y &= 0 = R_{A_y} + V \\ \sum M_i &= 0 = M_A - M \end{aligned} \tag{1.30}$$

As before, we have assumed that all forces in Figure 1.26*b* are concurrent.

The actual forces applied to the beam by a real fixed support are quite complex in nature, consisting of forces distributed nonuniformly over the portion of the beam that is imbedded in the support; however, the equilibrium equations require that these forces have resultants that are equivalent to the quantities R_{A_x}, R_{A_y}, and M_A, and it is these resultants with which we will deal in all analytic formulations.

The quantities R_{A_x}, R_{A_y}, and M_A, i.e., the resultant forces exerted by the support on the structure, are usually termed the support *reactions*. These force quantities are usually evaluated in the course of any structural analysis. In most cases, these reactions are described or determined first, and the internal beam forces are then related to the reactions through equilibrium equations such as Equations 1.30. In any event, equations such as Equations 1.30 show the equivalence of the reactions and the internal forces near the support.

The kinematic assumptions associated with a fixed support reflect exactly what the name implies: unless otherwise specified or modified, a point of fixed support for a beam element undergoes no displacement, and the beam axis undergoes no rotation near the support. Thus, a basic fixed support has no kinematic degrees of freedom.

The static characteristics of a pinned support are analogous to those of a pinned joint: we assume that all force quantities except internal moment can be developed in the beam near the support. In other words, near the support, the beam is assumed capable of developing only internal axial and shear forces.

Consider, for example, the pinned support at point A of the frame structure shown in Figure 1.27. Figure 1.28*a* shows a short segment of beam AB near support A. The force quantities that act on the end of the beam segment in this case are the axial force,

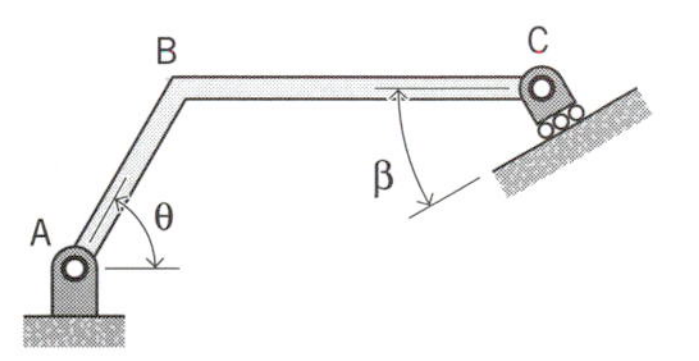

Figure 1.27 Frame structure with pinned and roller supports.

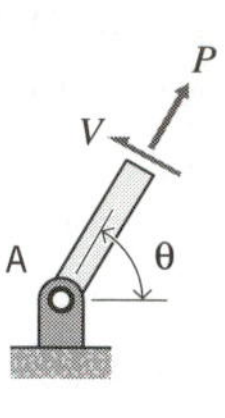

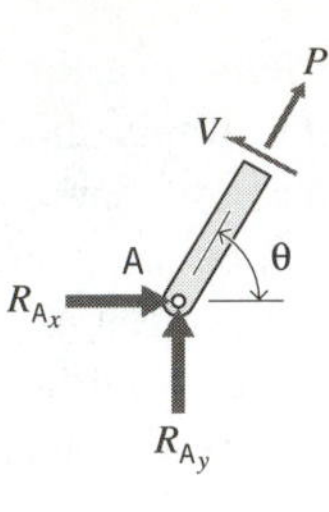

(*a*) Internal forces near pinned support (*b*) Free-body diagram

Figure 1.28 Forces and reactions near pinned support.

P, and the shear force, V. The corresponding free-body diagram, which includes the resultant support reactions, is shown in Figure 1.28*b*. As in Figure 1.26*b*, the reactions are shown in terms of equivalent horizontal and vertical components of force.

The two equations of equilibrium that relate the force quantities in Figure 1.28*b* can be written in a manner similar to the first two of Equations 1.30, but must now be expressed in terms of components of P and V,

$$\sum F_x = 0 = R_{A_x} + P\cos\theta - V\sin\theta$$
$$\sum F_y = 0 = R_{A_y} + P\sin\theta + V\cos\theta \qquad \textbf{(1.31)}$$

The pinned support provides one kinematic degree of freedom, namely a capacity of the beam axis to rotate near the support; we assume that the support does not otherwise displace. Figure 1.29 illustrates the possible rotation; again, angle ϕ is assumed to be small.

Both fixed and pinned supports can be modified by the addition of *rollers*. A roller is equivalent to a set of frictionless wheels confined to tracks that are parallel to the supporting surface. A support with rollers is assumed capable of small movement along the supporting surface, and is assumed to be incapable of exerting any reaction force that would tend to prevent such displacement. Thus, a roller will reduce the number of possible reaction forces by one, and will increase the kinematic degrees of freedom at a support by one.

Figure 1.30 illustrates various representations of the most common type of roller support, namely the *pinned support with roller*, such as is shown at point C of the structure in Figure 1.27. This particularly common support is often referred to simply as a *roller support*, or just a *roller*. Regardless of terminology, the figures represent a

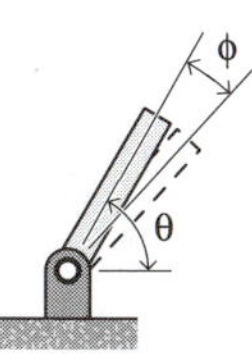

Figure 1.29 Rotation at pinned support.

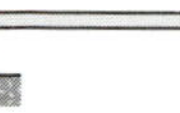

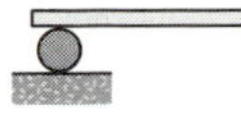

Figure 1.30 Roller support.

pinned support that can displace in the direction of the supporting surface, and that *cannot* exert any force on the structure parallel to this surface. In other words, a roller support can only exert a reaction force on a structure in the direction perpendicular to the supporting surface.[16]

Consider, again, the structure in Figure 1.27; a free-body diagram of a segment of beam BC near the roller support at point C is shown in Figure 1.31*a*. Note that the reaction force, R, is shown with a line of action normal to the support surface but, as with a pinned support, it is assumed that R can act in either of the directions shown. In other words, we assume that the roller is constrained against lifting off the surface and can, therefore, exert a reaction force "up" or "down" as required by the loads that are applied to the structure.[17]

The kinematic degrees of freedom for the roller support at C are shown in Figure 1.31*b*. Note that there are *two* degrees of freedom, u and ϕ, but that u may be expressed in terms of its horizontal and vertical components, u_1 and u_2, which are related to u by the angle β.

A representation of a fixed support with rollers[18] is shown in Figure 1.32*a*. As with the pinned support, the addition of a roller to a fixed support reduces the number of possible reaction forces by one, and increases the kinematic degrees of freedom at the

[16] In this respect, a roller support is similar to a frictionless surface, i.e., a body that is in point contact with a frictionless surface will be subjected to a reaction force at the contact point that is perpendicular to the surface.

[17] In this respect, a roller support differs from a frictionless surface since a frictionless surface can exert only an "upward" reaction force on the body.

[18] This type of support is not likely to be encountered in a physical structure; it does, however, provide an appropriate analytic representation of conditions at points that lie on axes of symmetry in some structures (see Chapter 6).

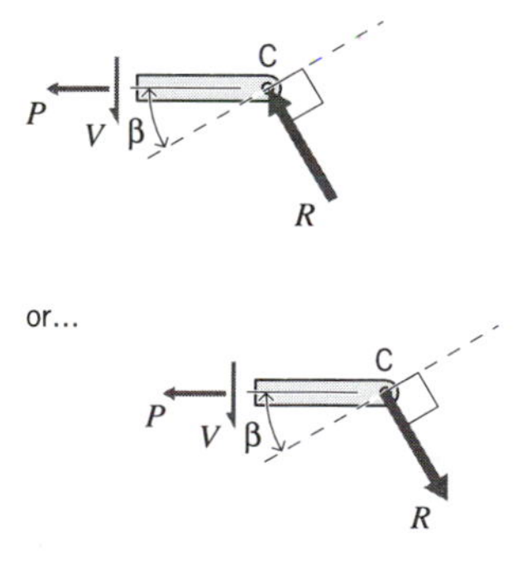

(*a*) Free-body diagram

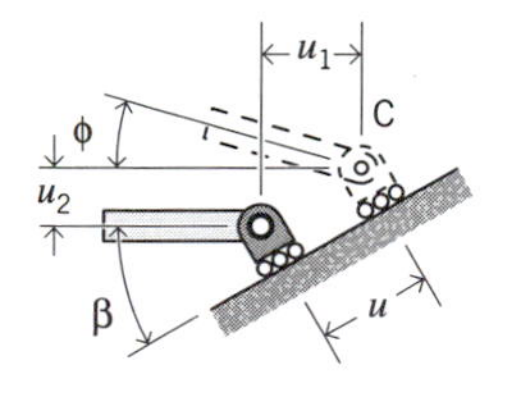

(*b*) Degrees of freedom

Figure 1.31 Reactions and displacements at point C of structure.

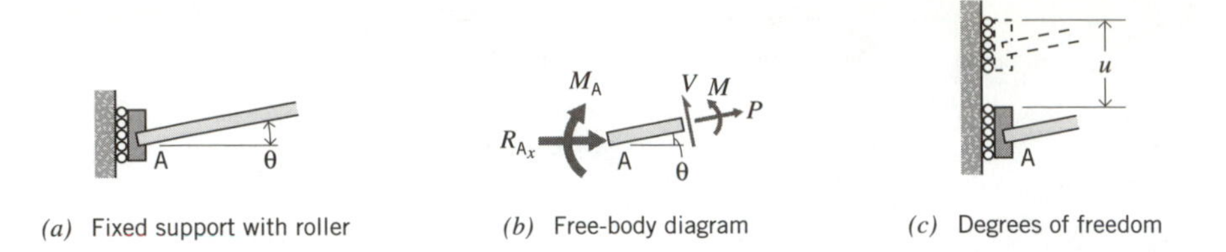

Figure 1.32 Reactions and displacement at fixed support with roller.

support by one. Again, the roller support *cannot* exert a reaction force on the beam in the direction parallel to the support surface. For the portion of a structure shown in Figure 1.32*a*, it follows that the reaction forces at the support consist of a moment M_A and a force perpendicular to the support surface, R_{A_x} (horizontal in this particular case), as shown in Figure 1.32*b*. In general, the internal forces in the beam near this support consist of the three usual components, *P*, *V*, and *M*. These forces are related to R_{A_x} and M_A by the equations of equilibrium,

$$\sum F_x = 0 = R_{A_x} + P\cos\theta - V\sin\theta$$

$$\sum F_y = 0 = P\sin\theta + V\cos\theta \qquad \textbf{(1.32)}$$

$$\sum M_i = 0 = M - M_A$$

As with a basic roller support, R_{A_x} may act in either direction normal to the support surface.

The fixed support with roller has *one* kinematic degree of freedom, namely the displacement in the direction of the roller; there is no rotation of the beam axis near the support. Figure 1.32*c* illustrates this degree of freedom.

Figure 1.33*a* shows a fixed support with two sets of rollers[19] on orthogonal surfaces. The reaction forces and degrees of freedom for this type of support are shown in Figures 1.33*b* and 1.33*c*, respectively. In this case, there is only one reaction force, M_A. It follows from the equilibrium equations (Equations 1.32 with $R_{A_x} = 0$) that *P* and *V*

[19] *ibid.*

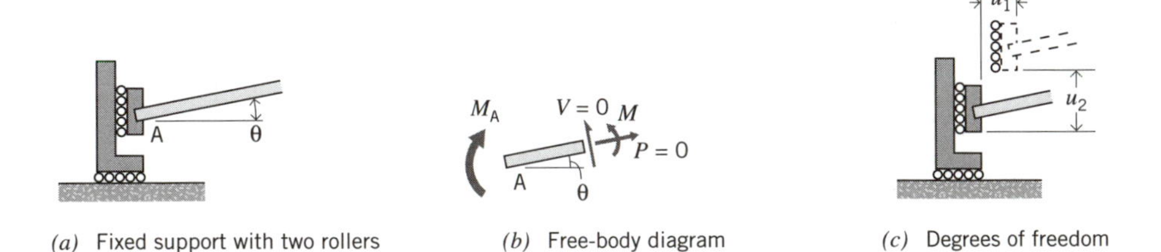

Figure 1.33 Reactions and displacements at fixed support with two rollers.

must equal zero in the beam near the support. There are two kinematic degrees of freedom, u_1 and u_2; it is still assumed that there is no rotation of the beam axis near the support.

Finally, we note that there are three-dimensional equivalents, analogues, and combinations of these two-dimensional supports, but none will be discussed here. Practice in modelling and analyzing two-dimensional structures should help facilitate any transition to three-dimensional cases. The remainder of this section will be devoted to the analysis of reactions for several two-dimensional structures.

EXAMPLE 3

Find reactions for the beam in Figure 1.34*a*. This beam has a pin support at the left end and a roller support at the right end.[20] As in Example 1 of Section 1.5.7, the quantity q_0 is the magnitude of the continuous force (force/length) at the right end of the beam.

A free-body diagram of the entire beam is shown in Figure 1.34*b*. The free-body diagram includes the specified continuous load and the reaction forces, which have unknown magnitudes at this stage. The pin support can exert two components of reaction force, denoted by R_{A_x} and R_{A_y}, and the roller can exert only one component of reaction force, denoted by R_B. These three unknown forces can be evaluated from the equations of equilibrium. Note that the vertical components of reaction forces have been shown as acting downward; this choice of directions is arbitrary on the free-body, but the equations of equilibrium must be consistent with these assumed directions.

The continuous force is replaced by its statically equivalent resultant, as in Example 1, and the equations of equilibrium may then be written for the forces shown in Figure 1.34*c*. From Equation 1.7*a* we obtain

$$\sum F_x = 0 = R_{A_x} \tag{1.33}$$

i.e., R_{A_x} is identically equal to zero in this case because there are no other components of horizontal force acting on the body. The remaining two equations of equilibrium can be written exactly as in Equations 1.25 and 1.26 because this example is identical to Example 2 in Section 1.5.7; note, however, that it is almost always desirable to write the equations in the simplest possible way. Thus, Equation 1.26 should be written and solved *before* Equation 1.25 because it contains only one unknown reaction, R_B. Also, bear in mind that, although the solution is unique, the exact form of the three independent equilibrium equations is not unique; we could, for example, replace Equation 1.26 by an equation of moment equilibrium about point B since this equation would also contain only one unknown reaction, R_{A_y}.[21]

The final free-body diagram is shown in Figure 1.34*d*. Note that the continuous load is again shown acting on the beam. It is important to remember that the resultant of

[20] A beam with one pin and one roller support is referred to as *simply supported.*

[21] See also Problem 1.P1.

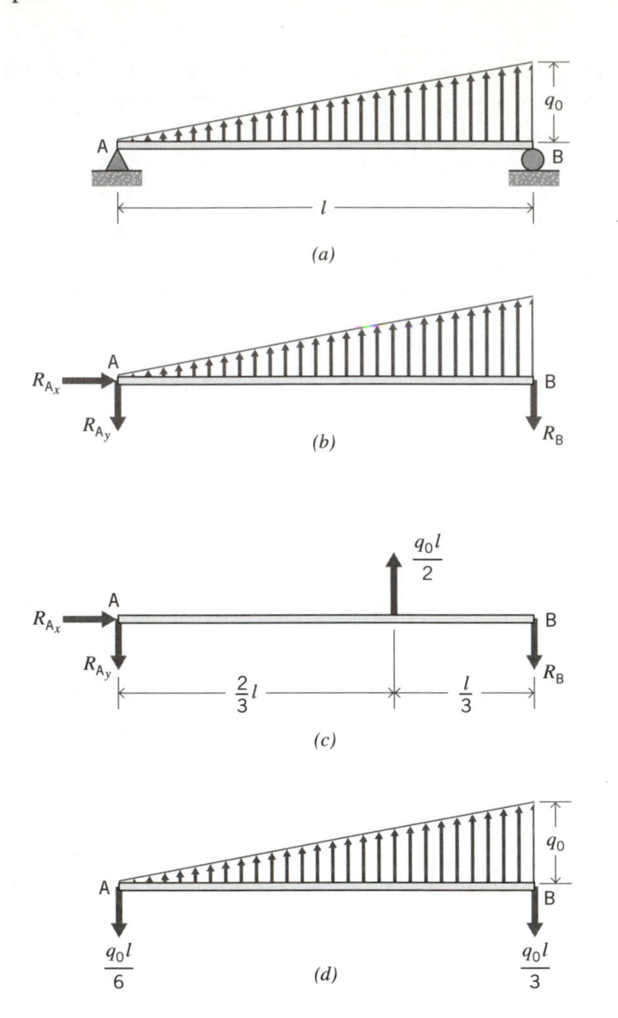

Figure 1.34 Example 3.

the distributed load, which was shown in Figure 1.34*c*, is used only for the purpose of writing the equilibrium equations. The continuous force, not its resultant, is the actual load on the structure.

EXAMPLE 4

Find the reactions for the simply supported beam in Figure 1.35*a*. This beam has an external moment, M_0, applied at a distance a from the left end. In this two-dimensional case, the applied moment is assumed to have an axial component only in the z direction; this component is given a common ''scalar'' representation in Figure 1.35*a*, i.e., it is

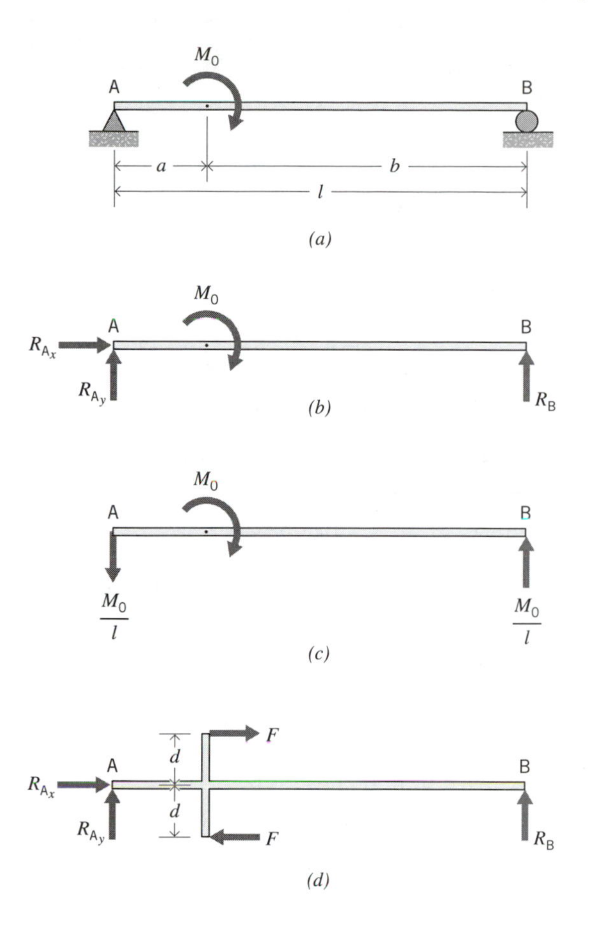

Figure 1.35 Example 4.

represented by a ‘‘circular arrow,’’ which denotes the tendency to cause rotation about an axis in the z direction.

A free-body diagram of the beam is shown in Figure 1.35*b*. As in the preceding example, summing forces in the x direction gives $R_{A_x} = 0$. Summing moments[22] about point A gives

$$\sum M = 0 = R_B l - M_0 \tag{1.34a}$$

or

$$R_B = \frac{M_0}{l} \tag{1.34b}$$

[22] Because we are dealing exclusively with the equations of equilibrium in component form, rather than with the *vector* equations, we will no longer specify what is implied, namely that we are ‘‘. . . summing *components* of moments. . . .’’ The same applies with forces, i.e., we sum *components*.

and summing forces in the y direction then gives

$$\sum F_y = 0 = R_{A_y} + \frac{M_0}{l} \qquad \textbf{(1.35b)}$$

or

$$R_{A_y} = -\frac{M_0}{l} \qquad \textbf{(1.35b)}$$

i.e., R_{A_y} is *opposite* to the direction assumed in the free-body of Figure 1.35*b*. The computed reactions are shown in Figure 1.35*c*. In this case the reactions are independent of dimension a.[23]

Figure 1.35*d* shows a free-body diagram with an alternate representation of the applied moment M_0; here, the moment is represented by a *couple* of magnitude $2Fd$. Summing forces in the x direction gives

$$\sum F_x = 0 = R_{A_x} + F - F \qquad \textbf{(1.36)}$$

i.e., we still get $R_{A_x} = 0$. Summing moments about point A gives

$$\sum M = 0 = R_B l - Fd - Fd \qquad \textbf{(1.37)}$$

which is identical to Equation 1.34*a*.

EXAMPLE 5

Find the reactions for the frame structure in Figure 1.36*a*. This structure is identical to the one in Figure 1.27 except that angles are specified, and dimensions and magnitudes of applied loads are included as algebraic quantities.[24] The choice of $l_2 = 2l_1$ is arbitrary and made for the sake of convenience only.

Figure 1.36*b* shows a free-body diagram of the entire structure. The pin support at A produces two unknown components of reaction denoted by R_{A_x} and R_{A_y}; the roller support at C produces a reaction of unknown magnitude R_C with the line of action shown.

Before writing the equilibrium equations, we replace the continuous force by its resultant, and resolve R_C into horizontal and vertical components, as shown in Figure

[23] Remember, however, that internal forces are a function of dimension a.

[24] At this stage of development we will continue to work examples in terms of such algebraic quantities rather than numerical values. This has the advantages of providing clarity in equations, and results that are *dimensionally homogeneous* (a dimensionally homogeneous equation is one whose form does not depend on the system of units of measurement employed); the disadvantage is that we may end up with some complicated algebraic expressions. In subsequent examples, we will employ U.S. customary units and/or SI units when specifying numerical values for dimensional quantities.

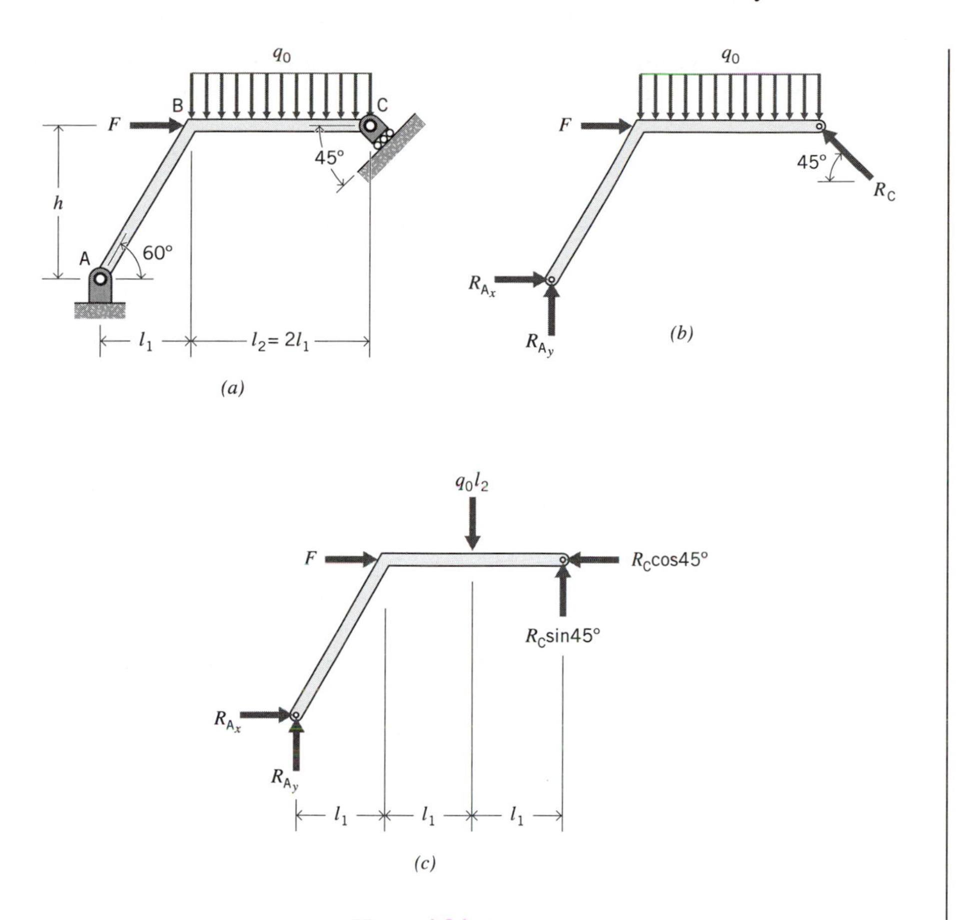

Figure 1.36 Example 5.

1.36*c*. The components of R_C are assumed to act in directions that are consistent with the assumed direction of R_C.

The equilibrium equations can be written in any order; however, an appropriate choice of order can often reduce the overall computational effort. In this case, for example, starting with moment equilibrium about point A gives us an equation with R_C as the only unknown quantity, i.e.,

$$\sum M = 0 = -Fh - (2q_0 l_1)(2l_1) + (R_C \cos 45°)h + (R_C \sin 45°)(3l_1) \quad \textbf{(1.38a)}$$

Substituting $h = l_1 \tan 60°$ and solving for R_C, leads to

$$R_C = \frac{\sqrt{6}F + 4\sqrt{2}q_0 l_1}{3 + \sqrt{3}} \quad \textbf{(1.38b)}$$

Note that R_C is a linear combination of F and q_0, which demonstrates the applicability of the principle of superposition (Section 1.4) in the calculation of reactions.

With R_C determined, we can now write either of the force equilibrium equations to evaluate another unknown magnitude of reaction; for example,

$$\sum F_x = 0 = F - R_C \cos 45° + R_{A_x} \tag{1.39a}$$

Substituting Equation 1.38*b* into Equation 1.39*a* leads to

$$R_{A_x} = \frac{4q_0 l_1 - 3F}{3 + \sqrt{3}} \tag{1.39b}$$

Similarly,

$$\sum F_y = 0 = R_{A_y} + R_C \sin 45° - q_0 l_2 \tag{1.40a}$$

or

$$R_{A_y} = \frac{-\sqrt{3}F + 2(1 + \sqrt{3})q_0 l_1}{3 + \sqrt{3}} \tag{1.40b}$$

Equation 1.39*b* indicates that the direction of R_{A_x} depends on the relative values of F and $q_0 l_1$. If $4q_0 l_1 > 3F$ then R_{A_x} is positive, and hence directed to the right as assumed in Figure 1.36*b* and the equilibrium equations; if $4q_0 l_1 < 3F$ then R_{A_x} is negative, and hence directed oppositely to what was assumed, i.e., to the left in the figure. The final free-body diagram of the entire structure is the same as in Figure 1.36*b*.

EXAMPLE 6

Find the reactions for the cantilever beam shown in Figure 1.37*a*. The beam is subjected to a linearly varying continuous load similar to that in Figure 1.34 (Example 3).

The beam is free at end A and fixed at end B; therefore, there are no reactions at A, and the possible reactions at B consist of two components of force with magnitudes denoted by R_{B_x} and R_{B_y}, and a moment with magnitude denoted by M_B. These reactions are shown in Figure 1.37*b*.

The continuous load is replaced by its resultant, Figure 1.37*c*, and the equations of equilibrium are written for the free-body shown in this figure, i.e.,

$$\sum F_x = 0 = R_{B_x} \tag{1.41}$$

$$\sum F_y = 0 = R_{B_y} - \frac{q_0 l}{2} \tag{1.42a}$$

or

$$R_{B_y} = \frac{q_0 l}{2} \tag{1.42b}$$

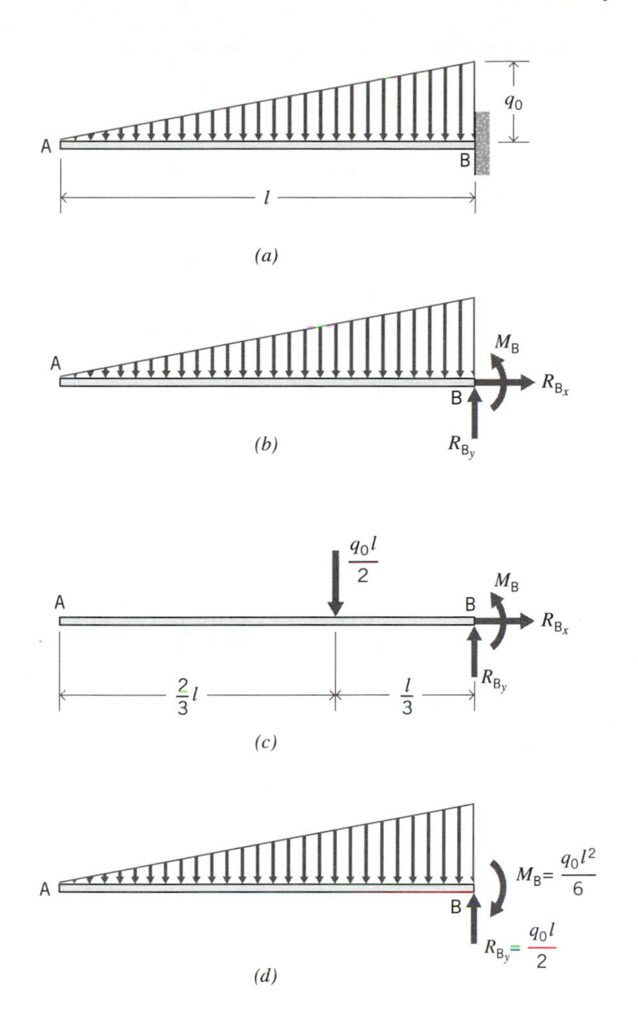

Figure 1.37 Example 6.

Summing moments about point B gives

$$\sum M = 0 = M_B + \left(\frac{q_0 l}{2}\right)\left(\frac{l}{3}\right) \quad \textbf{(1.43a)}$$

or

$$M_B = -\frac{q_0 l^2}{6} \quad \textbf{(1.43b)}$$

The reactions are shown in Figure 1.37*d*. Note that Figure 1.37*d* shows the proper direction of M_B.

EXAMPLE 7

The beam shown in Figure 1.38*a* is pinned at end A and supported by a fixed support with two rollers at end B. Find the reactions.

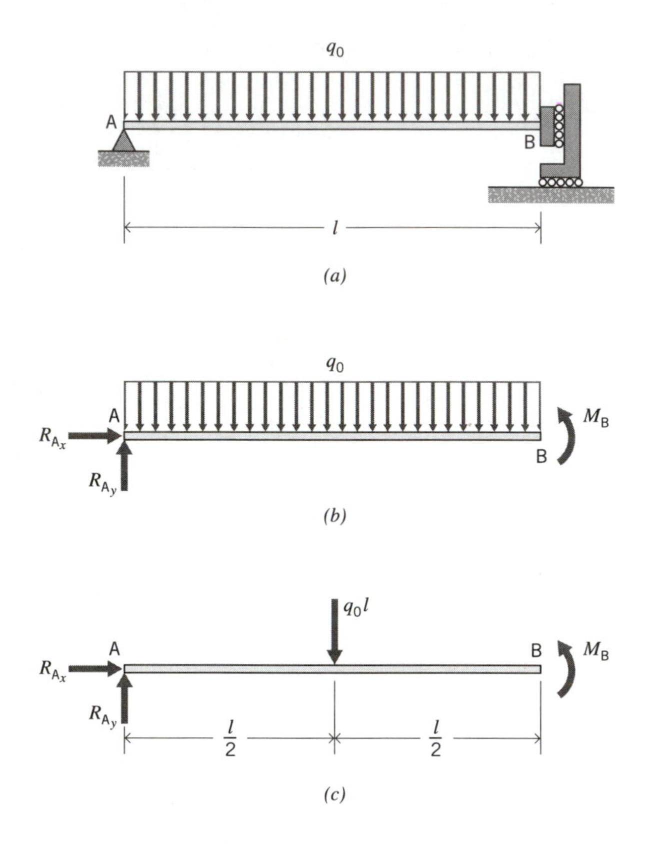

Figure 1.38 Example 7.

The possible reactions are shown in Figure 1.38*b*. The possible reactions at A consist of two components of force denoted by R_{A_x} and R_{A_y}. At B the possible reaction is a moment denoted by M_B.

Referring to Figure 1.38*c*, which shows the resultant of the continuous load, we write the equilibrium equations for sum of forces in the *x* direction, sum of forces in the *y* direction, and sum of moments about point A.

$$\sum F_x = 0 = R_{A_x} \tag{1.44}$$

$$\sum F_y = 0 = R_{A_y} - q_0 l \tag{1.45a}$$

or

$$R_{A_y} = q_0 l \tag{1.45b}$$

and

$$\sum M = 0 = M_B - (q_0 l)\left(\frac{l}{2}\right) \tag{1.46a}$$

or

$$M_B = \frac{q_0 l^2}{2} \tag{1.46b}$$

Because M_B is positive, it acts in the direction assumed in Figure 1.38*b*.

EXAMPLE 8

The frame structure in Figure 1.39*a* has beam EB rigidly connected to beam AC at B; beam CD is connected to beam AC by a hinge at C. There is a pin support at A, and roller supports at both D and E.

A free-body diagram of the entire structure is shown in Figure 1.39*b*. The pin support at A produces two components of reaction, and the rollers at D and E each produce one component of reaction. In other words, there are four unknown reactions acting on the free-body in Figure 1.39*b*. We can not solve for these four unknown quantities from three independent equations of equilibrium. In this case, however, because there is a hinge in the structure we *will* be able to evaluate the reactions by considering equilibrium of different free-bodies.[25]

In particular, we know that the moment in the beams in the immediate vicinity of a hinge is zero. As we have seen, this means that there are only *two* internal forces in the beams near a hinge, i.e., a shear force and an axial force. We can use this information to solve for the reactions in this example. We proceed by sketching a free-body diagram of some appropriate portion of the structure to one side or the other of the hinge.

For example, Figure 1.39*c* shows the portion of the entire structure that lies to the right of the hinge. We should be very specific here and observe that ''point C'' in Figure 1.39*c* is actually a point in the beam very close to the hinge; it can be a point either to the left of the hinge or to the right of the hinge. The figures will generally indicate the exact location of points such as C. Thus, Figure 1.39*c* actually shows point C just to the right of the hinge, while Figure 1.39*d* shows point C just to the left of the hinge. In this case, the unknown internal forces, i.e., shear V_C and axial force P_C, are the same in either location. We know this from a consideration of a free-body diagram of a small portion of the beam near the hinge (Figure 1.39*e*).[26] The length of this

[25] This implies that if there were no hinge in the structure we would not be able to evaluate the reactions by using only the equilibrium equations. In other words, the structure without the hinge would be statically indeterminate. This concept is discussed in more detail in the next sections.

[26] In other words, a free-body diagram of joint C, similar to the joint in Figure 1.24*a*.

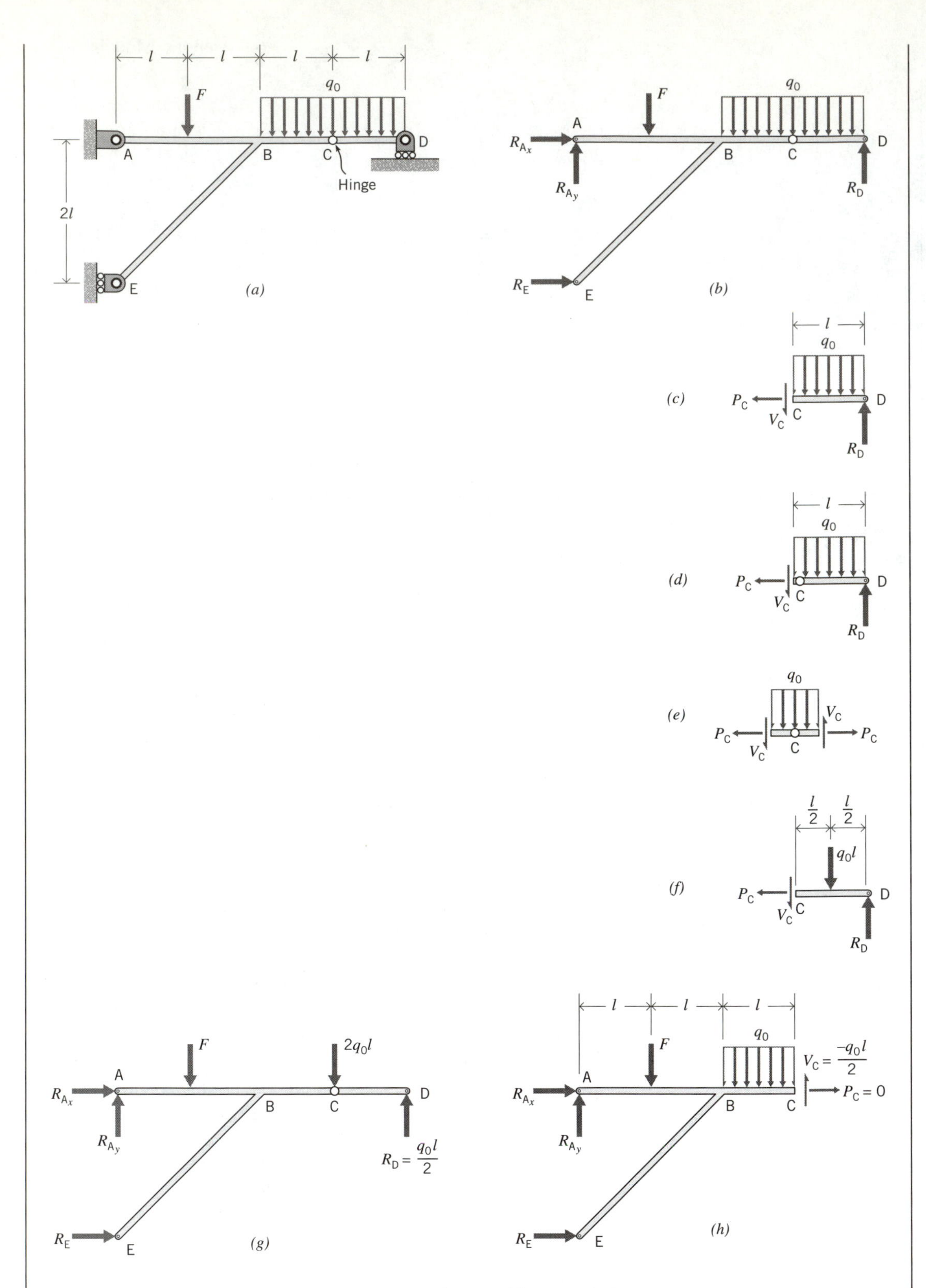

Figure 1.39 Example 8.

portion of the beam is very small and, consequently, the resultant of the continuous force acting on this free-body is negligible; therefore, equilibrium requires equal magnitudes and opposite directions for forces on both ends of the free-body in Figure 1.39*e*. These forces are, in turn, related to the forces on the ends of contiguous elements of adjacent free-bodies; e.g., forces on the *right* end of the beam in Figure 1.39*e* must be equal and opposite to forces on the *left* end of the beam in Figure 1.39*c*, whereas forces on the *left* end of the free-body in Figure 1.39*e* are identical to the forces on the left end of the free-body in Figure 1.39*d*.

We may now write the equilibrium equations for either free-body (Figure 1.39*c* or Figure 1.39*d*). As usual, however, we first replace the continuous force by its resultant, e.g., Figure 1.39*f*, which shows the same free-body as Figure 1.39*c*.[27] Writing the equation for sum of moments about point C equals zero gives

$$0 = R_D l - (q_0 l)\left(\frac{l}{2}\right) \quad \textbf{(1.47a)}$$

or

$$R_D = \frac{q_0 l}{2} \quad \textbf{(1.47b)}$$

If we are only interested in the reactions at points A, D, and E we don't need to evaluate P_C and/or V_C; however, if we wish to evaluate them, we can do so from the remaining equations of equilibrium. For example,

$$\sum F_x = 0 = -P_C \quad \textbf{(1.48)}$$

$$\sum F_y = 0 = R_D - q_0 l - V_C \quad \textbf{(1.49a)}$$

Substituting Equation 1.47*b* into Equation 1.49*a* gives

$$V_C = -\frac{q_0 l}{2} \quad \textbf{(1.49b)}$$

i.e., V_C is opposite to the direction shown in Figure 1.39*f*.

With R_D given by Equation 1.47*b*, we can return to the free-body diagram of the entire structure shown in Figure 1.39*b* and use the equations of equilibrium to evaluate R_{A_x}, R_{A_y}, and R_E. Again, we proceed by replacing the continuous load by its resultant, Figure 1.39*g*. The equation for sum of moments about point E is

$$\sum M = 0 = R_D 4l - (q_0 2l)(3l) - Fl - R_{A_x} 2l \quad \textbf{(1.50a)}$$

Substituting Equation 1.47*b* gives

$$R_{A_x} = -2q_0 l - \frac{F}{2} \quad \textbf{(1.50b)}$$

[27] It is not necessary to sketch a separate free-body diagram showing such resultant forces as long as the equilibrium equations are written correctly. We will dispense with these distinct free-body diagrams in most examples in subsequent chapters.

Also,

$$\sum F_x = 0 = R_{A_x} + R_E \qquad \textbf{(1.51a)}$$

or

$$R_E = -R_{A_x} = 2q_0 l + \frac{F}{2} \qquad \textbf{(1.51b)}$$

and

$$\sum F_y = 0 = R_{A_y} - F - q_0 2l + R_D \qquad \textbf{(1.52a)}$$

so

$$R_{A_y} = F + \frac{3q_0 l}{2} \qquad \textbf{(1.52b)}$$

Thus, all the reactions in Figure 1.39*b* have been determined. Also, all the forces in Figure 1.39*c* were determined. Between these two figures there was a total of *six* unknown forces: the four reactions, R_{A_x}, R_{A_y}, R_D, R_E, and the two "internal forces," P_C, V_C. We used *six* equations of equilibrium to solve for these unknown forces: three independent equations for the free-body in Figure 1.39*b* and three independent equations for the free-body in Figure 1.39*c*.

The procedure that we followed in the solution described here is not unique. We did have to consider the free-body in Figure 1.39*c* (or Figure 1.39*d*) because that is the sole free-body on which only three unknown forces act. (Remember that we had to consider the internal forces near joint C because that was a point at which we knew from the outset that the internal moment was zero.) Even for this free-body, however, we could have written an alternate set of three independent equilibrium equations. Furthermore, once we had evaluated P_C and V_C we could have used the free-body in Figure 1.39*h*, instead of that in Figure 1.39*b*, to solve for the remaining unknown reactions, R_{A_x}, R_{A_y}, and R_E. The results would have been exactly the same as described here even though the equations would be different. In other words, even though the solution is unique the path to it is not. This fact will become even more salient when we introduce alternate formulations of the analysis procedures in subsequent chapters. In any event, we could use the results of an analysis of Figure 1.39*h* as an independent check of our prior calculations; any inconsistencies would indicate an error in some computations. *The performance of such checks is a highly recommended procedure.*

EXAMPLE 9

The uniform beam in Figure 1.40*a* is supported by three rollers with the orientations shown, and is subjected to a continuous force per unit length of magnitude q_0 which acts vertically. This continuous force may be viewed as similar to a load associated with the mass of the beam.

The three possible reactions are shown in Figure 1.40*b*. The resultant of the continuous force is a vertical force of magnitude $q_0 l$ as shown in the free-body of Figure 1.40*c*. We write the equilibrium equations for sum of moments about A equals zero,

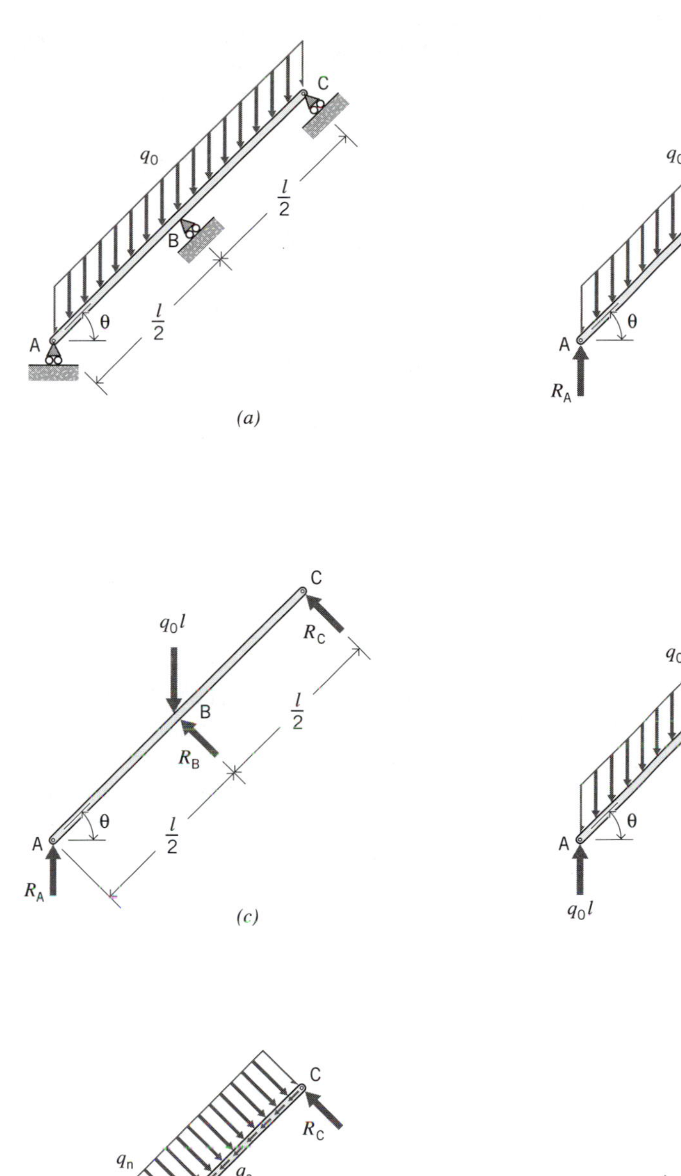

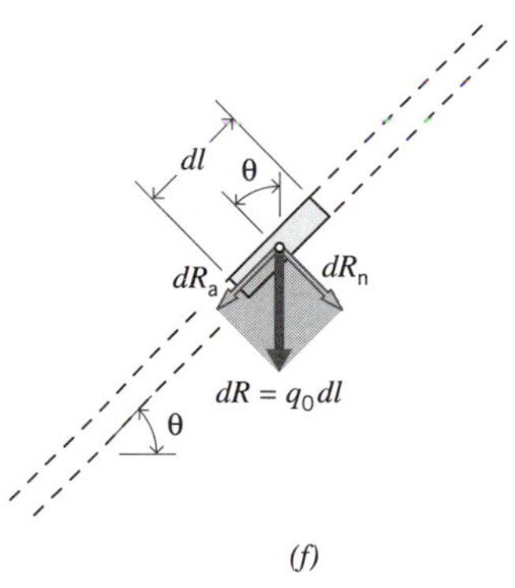

Figure 1.40 Example 9.

sum of forces in the x direction equals zero, and sum of forces in the y direction equals zero, as follows:

$$\sum M = 0 = \frac{R_B l}{2} - q_0 l\left(\frac{l}{2}\cos\theta\right) + R_c l \tag{1.53a}$$

$$\sum F_x = 0 = -R_B \sin\theta - R_c \sin\theta \tag{1.53b}$$

$$\sum F_y = 0 = R_A + R_B \cos\theta + R_C \cos\theta - q_0 l \tag{1.53c}$$

Solving Equations 1.53*a* and 1.53*b* simultaneously leads to

$$R_B = -q_0 l \cos\theta \tag{1.54a}$$

$$R_C = q_0 l \cos\theta \tag{1.54b}$$

Substituting these values into Equation 1.53*c* gives

$$R_A = q_0 l \tag{1.54c}$$

These results are shown in Figure 1.40*d*.

In this example, it was convenient to write equations of equilibrium that all contained two or more unknown reactions. This illustrates the frequent necessity of solving simultaneous equations in the course of evaluating reactions. We should note, however, that in this particular example we could have avoided this by writing moment equilibrium equations about points of intersection of lines of action of reactions.

In subsequent developments it will be necessary to deal with components of continuous forces of the type in this example, specifically with normal and axial components of the continuous force (Figure 1.40*e*). We can determine these components by considering the components of an increment of the resultant, i.e., a force acting over an incremental length dl, as shown in Figure 1.40*f*. This increment of force has magnitude $q_0 dl$ and acts in the vertical direction. It may be resolved into components normal and tangential to the beam. The component of this incremental force that acts in a direction normal to the beam axis is

$$dR_n = q_0 dl \cos\theta \tag{1.55a}$$

and the component parallel to the beam axis is

$$dR_a = q_0 dl \sin\theta \tag{1.55b}$$

Dividing each of these forces by the length dl gives the components of the continuous force, i.e.,

$$q_n = q_0 \cos\theta \tag{1.56a}$$

$$q_a = q_0 \sin\theta \tag{1.56b}$$

Solving the equilibrium equations for the free-body in Figure 1.40*e* will lead to the same results as those obtained from the free-body of Figure 1.40*b*.

EXAMPLE 10

The load on the beam in Example 9 is changed to a uniform force per unit of horizontal length, as shown in Figure 1.41*a*. This type of load is representative of the effects of snow on structures.

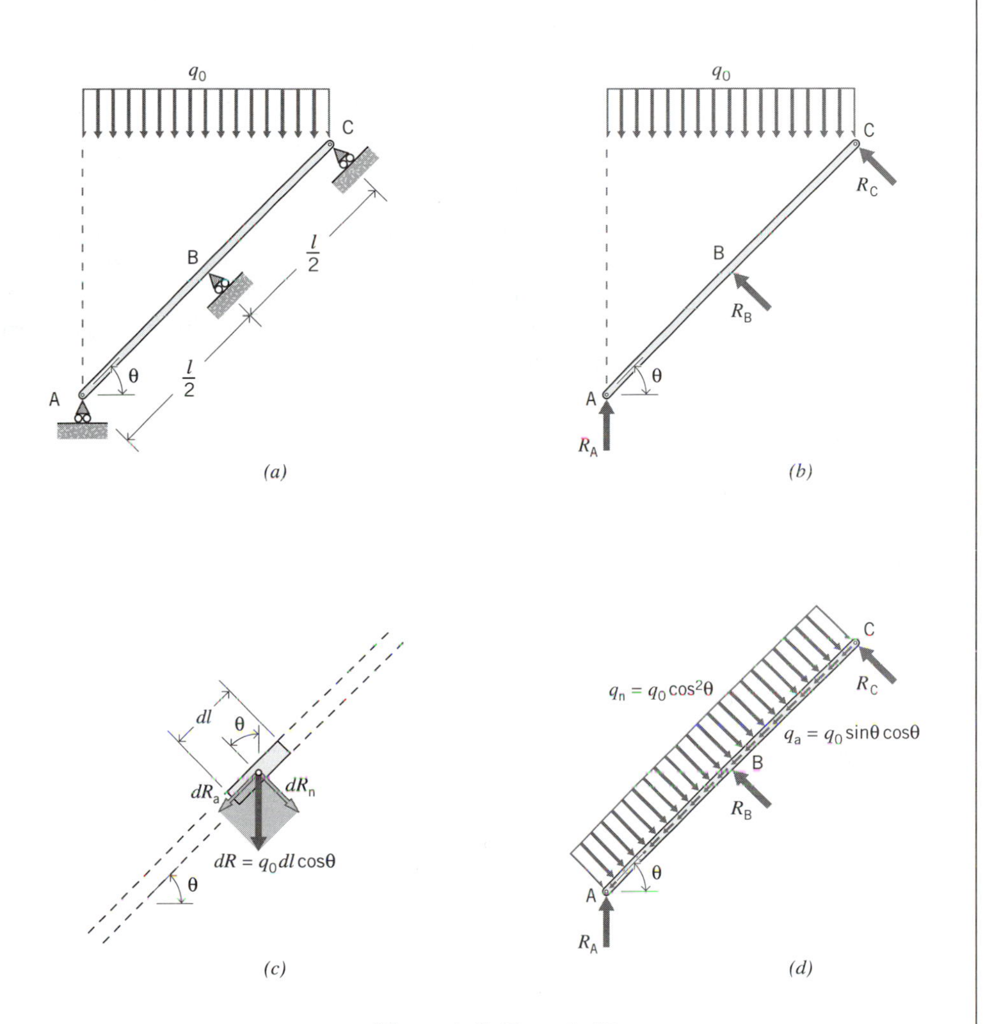

Figure 1.41 Example 10.

The equilibrium equations for the free-body in Figure 1.41*b* may be written like those for the beam in Figure 1.40*b*, i.e.,

$$\sum M = 0 = \frac{R_B l}{2} - [q_0(l\cos\theta)]\left(\frac{l}{2}\cos\theta\right) + R_C l \tag{1.57a}$$

$$\sum F_x = 0 = -R_B \sin\theta - R_C \sin\theta \tag{1.57b}$$

$$\sum F_y = 0 = R_A + R_B \cos\theta + R_C \cos\theta - q_0(l \cos\theta) \qquad \textbf{(1.57c)}$$

The only distinction between Equations 1.57*a*–*c* and Equations 1.53*a*–*c* is that the resultant of the continuous force is a vertical force of magnitude $q_0(l \cos\theta)$ rather than $q_0 l$.

Solving Equations 1.57*a* through 1.57*c* gives

$$\begin{aligned} R_A &= q_0 l \cos\theta \\ R_B &= -q_0 l \cos^2\theta \\ R_C &= q_0 l \cos^2\theta \end{aligned} \qquad \textbf{(1.58)}$$

As was pointed out in the preceding example, it is frequently necessary to deal with normal and axial components of a continuous force of this type. Referring to Figure 1.41*c*, we can again resolve an incremental force into components. The increment of force here has magnitude $q_0 dl \cos\theta$. The component of this incremental force that acts in a direction normal to the beam axis is

$$dR_n = q_0 dl \cos^2\theta \qquad \textbf{(1.59a)}$$

and the component parallel to the beam axis is

$$dR_a = q_0 dl \sin\theta \cos\theta \qquad \textbf{(1.59b)}$$

Dividing each of these forces by the length dl gives the components of the continuous force, i.e.,

$$\begin{aligned} q_n &= q_0 \cos^2\theta \\ q_a &= q_0 \sin\theta \cos\theta \end{aligned} \qquad \textbf{(1.60)}$$

as shown in Figure 1.41*d*.

1.7 DETERMINACY AND INDETERMINACY

The examples in the preceding section illustrate structures for which we were able to determine all *reactions* from equations of static equilibrium. In other words, the equations of equilibrium alone define a unique set of external forces acting on each of the structures we analyzed. In cases such as these, the reactions are said to be *statically determinate*, or simply *determinate.*

As was pointed out in Example 8, however, it is possible to have more reactions than we are able to evaluate from a set of independent equilibrium equations. For instance, the frame structure in Figure 1.42*a*, which is the same as that in Example 8 but without the hinge at point C, has the same four reactions as in Figure 1.39*b* (also shown in Figure 1.42*b*). In this case, we do not have any prior knowledge of internal forces anywhere in the structure (other than near the supports), i.e., we do not know

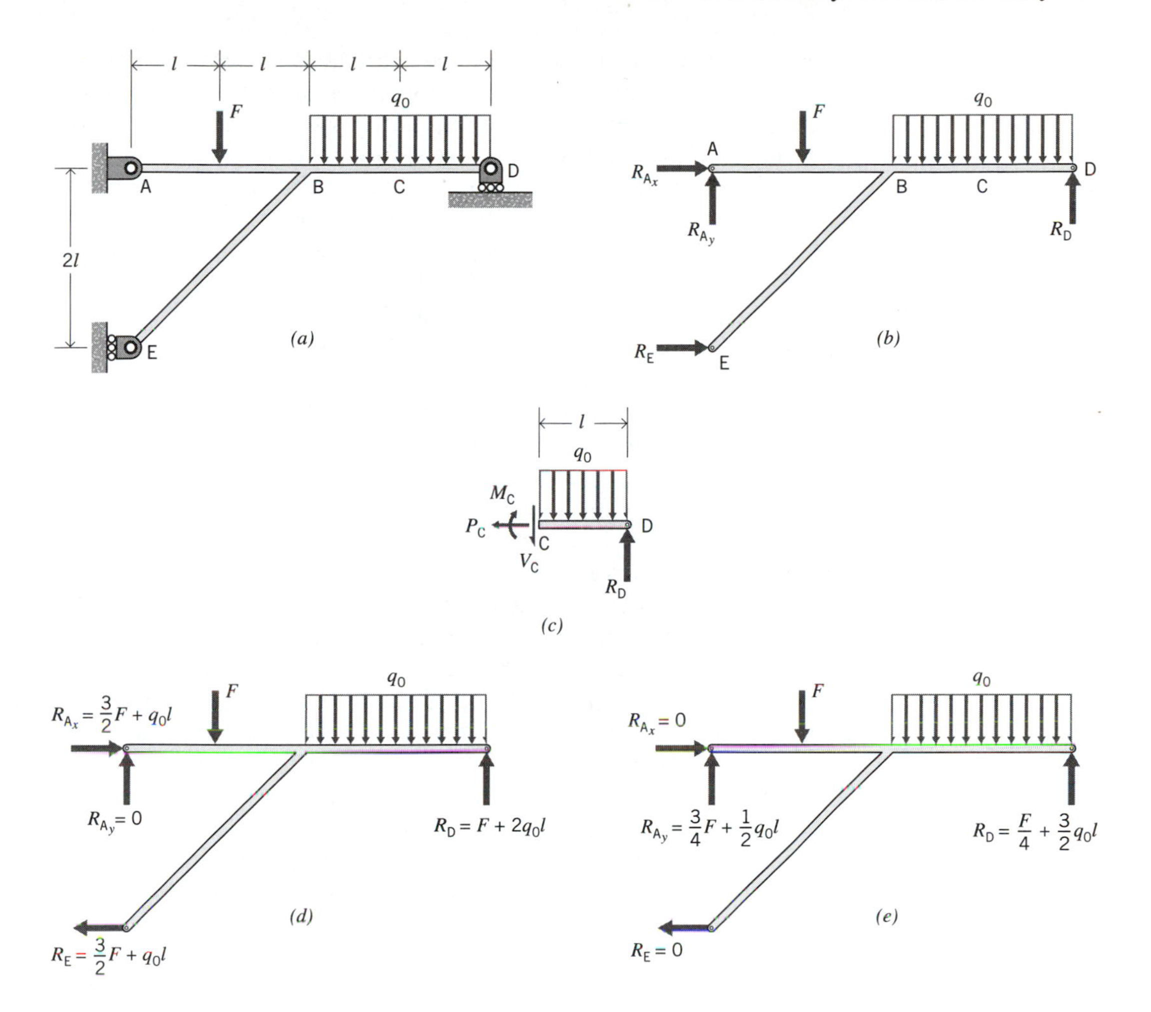

Figure 1.42 Indeterminate structure; reactions in (*d*) and (*e*) satisfy equilibrium but are *not* correct solutions.

anything about the value of the internal moment at point C or about any other internal force. Any free-body such as that in Figure 1.42*c*, which shows the portion of the entire structure to the right of point C, also has four (or more[28]) unknown forces, three (or more) of which are ''internal'' forces rather than reactions (P_C, V_C, M_C). Therefore, we do not gain anything by writing a set of equations of equilibrium for any other free-body besides the one in Figure 1.42*b*; the total number of unknown force magnitudes will still exceed the total number of independent equilibrium equations by one. This structure is said to be *statically indeterminate*, or simply *indeterminate*.

This does not mean that we will be unable to evaluate the reactions. In fact, most of the subject of structural analysis deals with indeterminate structures. To analyze inde-

[28] For example, a free-body of the portion of the structure to the left of point C will contain six unknown forces.

terminate structures we will still have to impose equilibrium as we have been doing; equilibrium is a necessary condition that must be satisfied in any structural analysis. For indeterminate structures, equilibrium alone does not provide sufficient information from which to obtain a complete solution. The reason for this is that for indeterminate structures, there is more than one force distribution[29] that will satisfy the equilibrium equations; only *one* of these force distributions is correct. For example, the sets of reactions shown in Figures 1.42*d* and 1.42*e* both satisfy the equilibrium equations but neither is correct.[30] As we will see subsequently, the correct force distribution is the one for which a condition of compatibility of structural deformations is also satisfied.

Compatibility is the term used to denote the idea that deformations of structures must make physical/analytical sense, i.e., deformed structural components must fit together in a manner that is consistent with the ways in which they are joined and supported.

The concept of compatibility has been implicitly introduced in the prior discussion of connections and supports, Sections 1.6.1 and 1.6.2. There, we defined the basic analytic representations of displacements of various connections and supports. For an indeterminate structure, compatibility equations are somewhat more involved than this, but it is the coupling of equilibrium and compatibility equations that leads to the unique set of forces that is consistent with a unique set of appropriate displacements.

At this stage of development, we are not yet concerned with the equations of compatibility; that will come later. Here, we wish only to introduce the concepts of *indeterminacy* and *degree of indeterminacy*. By degree of indeterminacy we mean the number of unknown forces in excess of the number of independent equations of equilibrium. Thus, the structure in Figure 1.42 is 1-degree indeterminate; this implies that we will eventually need one compatibility equation, in addition to three independent equilibrium equations, to solve for the four unknown quantities in Figure 1.42*b*. The number of compatibility equations required for analysis of any indeterminate structure is the same as its degree of indeterminacy. It is, therefore, essential to be able to properly determine the degree of indeterminacy of a structure.

There is no limit to the number of degrees of indeterminacy in a structure. It is not unusual for a real structure to be several hundred, or even several thousand, degrees indeterminate. In this book, we will deal with only relatively few degrees of indeterminacy because the analysis concepts are the same regardless of the actual number. For example, Figure 1.43 illustrates some variations of the supports for the structure in Figure 1.42 that increase the degree of indeterminacy. In Figure 1.43*a*, the support at D has been changed from a roller to a *fixed support with roller*, thereby adding a component of reaction. As may be seen from the free-body in Figure 1.43*b*, there are now five components of reaction forces (R_{A_x}, R_{A_y}, R_E, R_D, M_D); since we still have only three independent equations of equilibrium this structure is 2-degrees indeterminate. Similarly, changing the roller at E to a pin (Figure 1.43*c*) adds another component of reaction (Figure 1.43*d*) making the structure 3-degrees indeterminate.

So far, we have discussed structures for which reactions were indeterminate.[31] Fur-

---

[29] This concept will be useful in connection with the application of the virtual force form of the theorem of virtual work, Sections 4.3 and 4.4.

[30] These solutions were obtained by arbitrarily specifying one of the unknown forces, and then applying the equilibrium equations.

[31] Such structures are sometimes referred to as *externally* indeterminate because the indeterminacy is viewed as associated with forces that are external to a free-body of the entire structure. This type of distinction is unimportant.

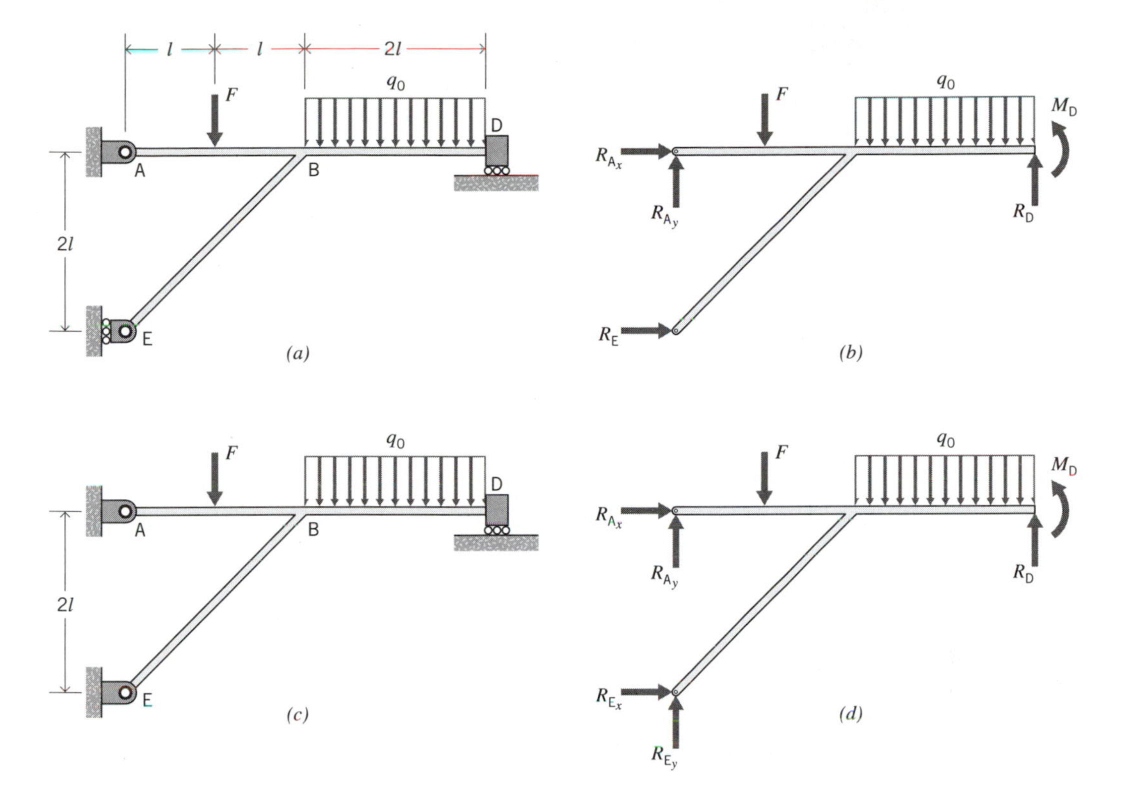

Figure 1.43 Various indeterminate structures.

thermore, for the particular indeterminate structures in the preceding examples (Figures 1.42 and 1.43), once the reactions are known the internal forces (i.e., axial force, shear force, and bending moment) can be determined everywhere in the structure.[32] This can be seen from a free-body such as the one in Figure 1.42*c*: once the reaction at D is known we can then solve for the three unknown internal forces at C by writing the equilibrium equations for this free-body. We can perform a similar analysis to find the internal forces at any other cross section.[33]

It is also possible to have situations in which reactions are determinate but internal forces are indeterminate.[34] Such a case is shown in Figure 1.44*a*. Here, the frame structure[35] is pinned at A and supported by a roller at B. Consequently, there are two components of reaction at A and one component of reaction at B, three in all. Any loads applied to the structure will be specified, even though none are shown in Figure 1.44*a*. Regardless of the loads, then, there will be only three reactions, and these can be evaluated from the equations of equilibrium applied to a free-body of the entire structure.

[32] Such structures are sometimes referred to as internally determinate.

[33] This will be discussed more fully in Chapter 3.

[34] It is worth emphasizing the point that our study involves the determination of both external *and* internal forces, and the ''type'' of indeterminacy really doesn't matter because the analysis procedures are the same for structures that are externally and/or internally indeterminate.

[35] A structure of this type is commonly referred to as a two-story, single-bay frame.

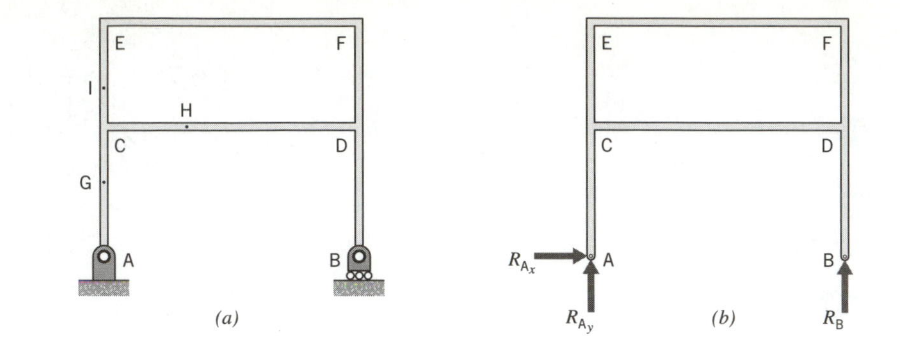

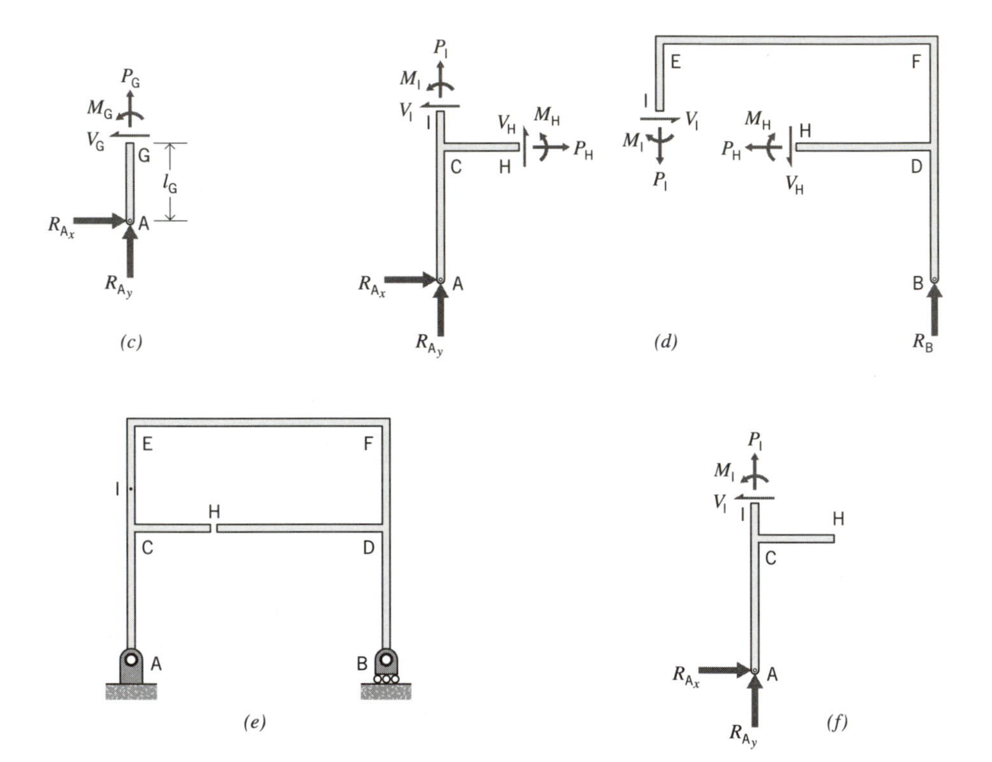

Figure 1.44 Internally indeterminate structure.

Suppose, now, that the reactions have been evaluated for some given set of loads, so that R_{A_x}, R_{A_y}, and R_B in Figure 1.44*b* are known. Despite knowing the reactions, it is *not* possible to determine the internal forces at every point in this structure using only equilibrium, and the structure is therefore indeterminate. We can determine internal forces at all points in components AC and BD, but cannot do the same anywhere in components CD, CE, EF, or DF.

To illustrate these statements, consider first a point such as G, which represents some typical cross section in component AC. A free-body of AG is shown in Figure 1.44*c*. (Although no specific loads are shown applied between points A and G, remember that arbitrary loads are implied in this discussion.) Since R_{A_x} and R_{A_y} are known, the three internal force components at G, denoted by P_G, V_G, and M_G, can be evaluated from equilibrium equations. (This is similar to the analysis of the free-body in Figure 1.42*c*.) We can do the same for internal forces in BD.

Now, consider a point such as H, which is shown in component CD, but which is representative of cross sections anywhere in components CD, CE, EF, or DF. To evaluate internal forces at H, we must consider a free-body of some portion of the entire structure in which the cross section at H is exposed, as we did in the previous case (Figure 1.44*c*). In this case, however, there is no free-body that will have only three unknown force components. To expose the cross section at H, it is also necessary to expose another cross section somewhere else, and there will be unknown forces at both of these cross sections.

For example, Figure 1.44*d* shows free-bodies of a left and a right portion of the structure, both of which expose the cross section at H. Since a free-body must represent a distinct part of the structure, it is necessary to include a cross section at another location such as I. Point I is representative of points anywhere in components CD, CE, EF, or DF. In general, there will be three unknown forces at each of sections H and I. Since any other free-body will be similar, we conclude that this structure is 3-degrees indeterminate, i.e., there are three more unknown forces than independent equations of equilibrium.[36]

A variation of this type of analysis is useful for assessing the degree of indeterminacy of complex frame structures.[37] Specifically, we can ask the question, "How many unknown forces must be removed from the structure in order to make it statically determinate[38] (or, more precisely, statically determinate and stable[39])?" The answer to this question corresponds to the degree of indeterminacy.

This approach is fairly direct when dealing with external indeterminacy. Consider the structure in Figure 1.43*c–d*; removing three reactions, such as M_D, R_D, and R_{E_y}, leaves a statically determinate (and stable) structure (see Example 11, which follows), and we therefore conclude that the original structure is 3-degrees indeterminate. Note that removing more than three reactions, or removing an inappropriate combination of three reactions, will result in a structure that is unstable, i.e., a structure for which we are unable, in general, to satisfy equilibrium. For example, removing R_{A_y}, R_{E_y}, and R_D will make it impossible to satisfy equilibrium of vertical forces. It is actually quite important to be able to identify an appropriate set of reactions to remove (or release)[40] because one of the basic approaches to the analysis of indeterminate structures, the so-called *force method*[41] of analysis, requires this; the analysis actually begins with equilibrium equations for a statically determinate (and stable) configuration of the structure.

---

[36] Note that the degree of indeterminacy is independent of loads; it is a characteristic of the structure.

[37] Trusses will be considered in the next chapter.

[38] It is also possible to develop a somewhat more formal approach based on counting the number of components, number of rigid connections, number of reactions, and so on. We believe that for frame structures such an approach tends to obscure important physical aspects of the problem, and is basically nonessential. On the other hand, a formal approach is convenient for use in conjunction with trusses, Chapter 2.

[39] We will discuss the concept of *stability* briefly in this section and in greater detail in the next section.

[40] The reactions that are "released" are referred to as *redundant* forces.

[41] The other basic method of analysis is known as the *displacement method.*

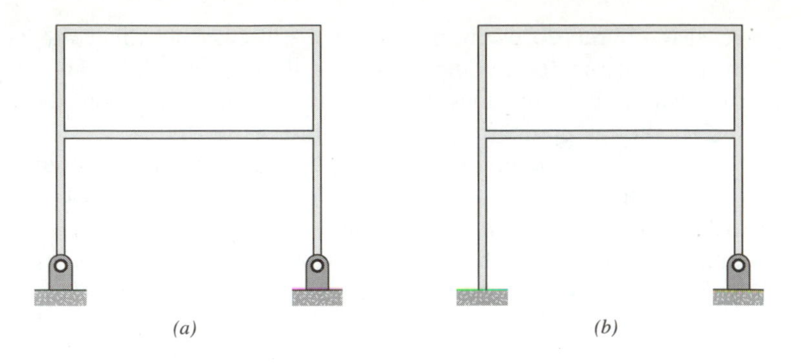

Figure 1.45 Indeterminate 2-story, single-bay frames: (*a*) 4-degrees indeterminate and (*b*) 5-degrees indeterminate.

When we deal with internal indeterminacy, the procedure becomes a bit more abstract. Consider the structure in Figure 1.44*a*, and imagine component CD "cut" at cross section H, as in Figure 1.44*e*. If the structure were actually cut at this point there could be no internal force components (P_H, V_H, M_H) in the immediate vicinity of H. A free-body exposing any other cross section such as I (Figure 1.44*f*), would then contain only three unknown forces (at I) which could be evaluated from equilibrium equations. In other words, removing three internal forces makes this structure determinate, so we conclude that the original structure must be 3-degrees indeterminate. As in the previous case, we could also modify the supports and increase the number of reactions, thereby increasing the degree of indeterminacy. For example, Figures 1.45*a* and 1.45*b* show structures that are 4-degrees indeterminate and 5-degrees indeterminate, respectively.

EXAMPLE 11

Use the preceding approach to find the degree of indeterminacy of the structure in Figure 1.43*c*; assume no prior knowledge of the structure.

We ask ourselves how many unknown forces must be removed from the structure to make it statically determinate (and stable). Removing three forces can make this structure determinate, and there is a variety of combinations of three forces that we can remove to accomplish this. For example, in Figure 1.46*a* reaction forces R_D, M_D, and R_{Ey} (see Figure 1.43*d*) have been removed; in Figure 1.46*b* reaction forces R_{Ay}, M_D, and R_{Ey} have been removed; in Figure 1.46*c* reaction forces R_{Ay}, R_D, and R_{Ex} have been removed; in Figure 1.46*d* reaction forces M_D and R_{Ey}, and internal force M_C (see Figure 1.39*c*) have been removed. Each of the structures in Figure 1.46*a*–*d* is determinate[42] (and stable), and each has been obtained by removing three components of force from the structure in Figure 1.43*c*. We conclude that the structure in Figure 1.43*c* is 3-degrees indeterminate.

[42] The structure in Figure 1.46*d* is identical to the structure in Example 8.

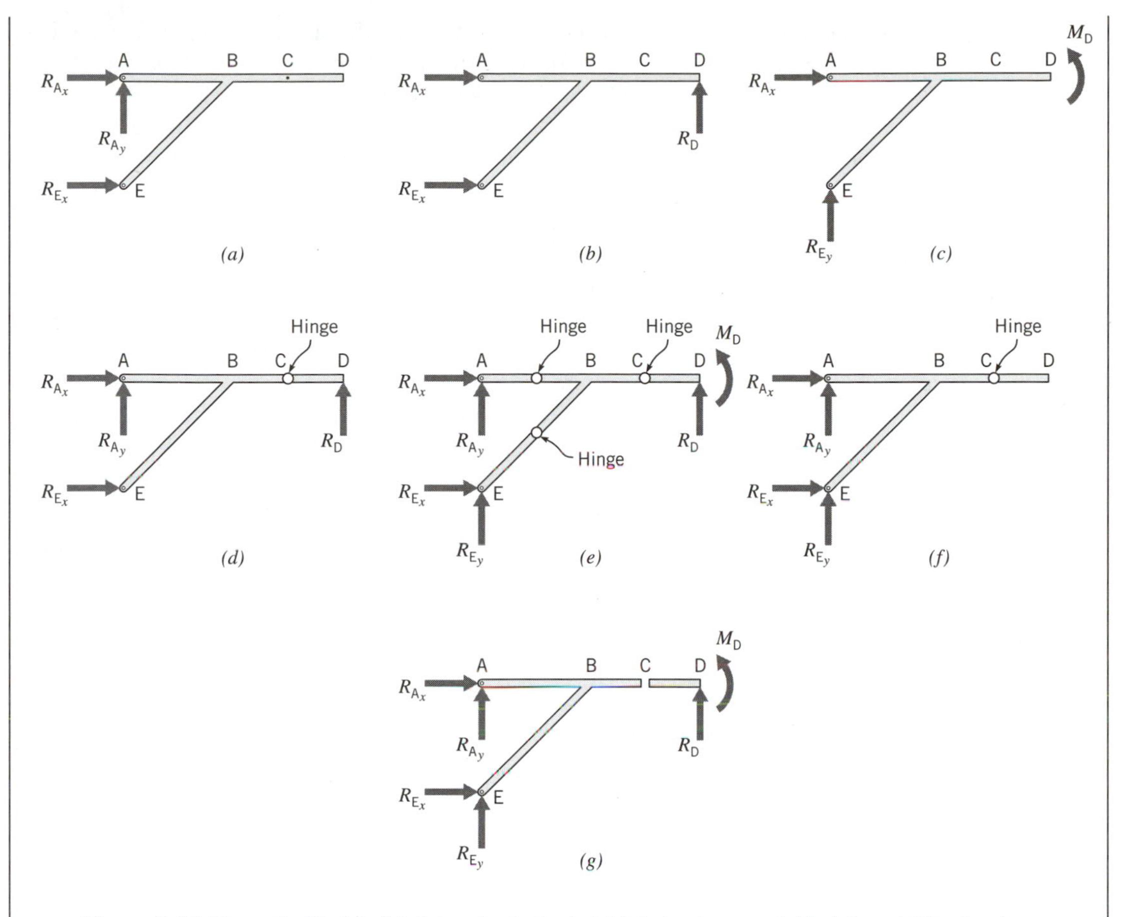

Figure 1.46 Example 11. (*a*)–(*e*) determinate (and stable) structures and (*f*)–(*g*) unstable structures.

Note that in the last case (Figure 1.46*d*) we removed an internal force, M_C, from the structure. We represented this by the addition of a hinge at point C in the figure. Alternately, we could have removed internal force V_C, or P_C, and accomplished the same thing; this would have been a bit more difficult to represent pictorially, but it is conceptually analogous to removing M_C.

Furthermore, since point C is actually representative of *any* internal point in the structure, the implication is that there are an infinite number of choices of internal force that we could remove. In fact, we could analyze the degree of indeterminacy by considering only the number of internal forces that must be removed to obtain a determinate structure. Figure 1.46*e*, for instance, shows the structure with all six reactions intact, but with three internal forces removed. This structure is statically determinate.

Although we have a choice of an infinite number of internal forces that we can consider, only three forces can be chosen. This does not mean that we are entirely free to select the three forces to remove in a *totally arbitrary* manner. As noted previously, removing R_{A_y}, R_{E_y}, and R_D will make it impossible to satisfy equilibrium of vertical forces; furthermore, Figures 1.46*f* and 1.46*g* show additional cases for which three forces have been removed, but where the resulting structures are still indeterminate. In Figure 1.46*f*, the reactions at D (M_D and R_D) and the internal moment at C (M_C) have been removed; nevertheless, internal forces in components AB and EB cannot be determined from equilibrium equations. Similarly, in Figure 1.46*g*, all internal forces at C (P_C, V_C, M_C) have been removed (imagine the structure ''cut'' at C), but we still cannot determine internal forces in components AB and EB. We know that this structure is 3-degrees indeterminate; so why is it that we have removed three forces and have still not obtained a determinate structure? The answer is that the structures in Figures 1.46*f* and 1.46*g* are unstable, i.e., as was stated above, we cannot, in general, satisfy equilibrium for these cases; the sets of forces which we removed were inappropriate. Specifically, component CD in Figure 1.46*f* cannot satisfy the moment equilibrium equation under an arbitrary set of applied loads; similarly, component CD in Figure 1.46*g* cannot satisfy the horizontal force equilibrium equation. Thus, in order to assess the degree of indeterminacy by removing forces, we must be careful to only remove a set of forces sufficient to produce a determinate and stable structure.

If the frame structure under consideration has any connections between components that are not rigid, i.e., have releases such as hinges, it is usually best to analyze the degree of indeterminacy by first assuming all internal connections to be rigid; the special conditions associated with releases at connections will then reduce the overall degree of indeterminacy. For example, the frame structure in Figure 1.47 is similar to the structure in Figure 1.43*c*, except for the hinge at point C in Figure 1.47. Since the structure in Figure 1.43*c* is 3-degrees indeterminate, it follows that the structure in Figure 1.47 must be only 2-degrees indeterminate, i.e., our knowledge of the fact that the internal moment near the hinge is zero is a piece of information that will enable us to write an additional independent equilibrium equation. Thus, we can subtract one degree of indeterminacy from the structure in Figure 1.43*c* to obtain the degree of indeterminacy of the structure in Figure 1.47. In general, we can subtract one degree of indeterminacy for each internal force removed by a release at a connection.

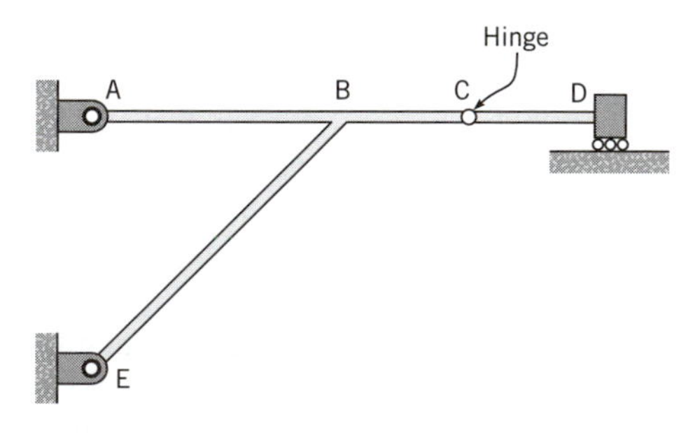

Figure 1.47 2-degree indeterminate structure.

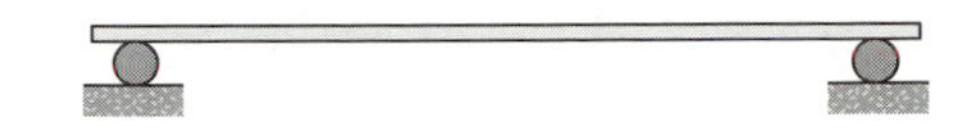

Figure 1.48 Beam with 2 roller supports (unstable).

1.8 STABILITY AND INSTABILITY

In the preceding section, an unstable structure was defined as one for which equations of equilibrium could not be satisfied for arbitrary loads.[43] This is not necessarily always detrimental; witness the structure in Figure 1.48, which is supported by two rollers. This structure is capable of equilibrating any set of vertical forces, but will not be in equilibrium if any horizontal force is applied to it; however, if the figure represents an analytic model of something like a skateboard, these supports would be entirely appropriate. On the other hand, the figure better not represent an analytic model of a bridge.

The skateboard example illustrates the fact that not all structures are fixed in space. Automobiles, aircraft, and spacecraft are other examples of structures that, like the skateboard, have rigid-body degrees of freedom, i.e., can undergo displacements that are independent of any deformations. Rigid-body motion of these structures implies acceleration under the application of arbitrary forces, and analysis may, therefore, require consideration of inertia effects. Even though there is nothing inherently unacceptable about the idea of a structure with rigid-body degrees of freedom, we will nevertheless classify such a structure as unstable in accordance with our elementary definition, i.e., because it cannot satisfy equations of static equilibrium in general.[44]

By extension of this concept, we will also classify as unstable any structure that is not capable of satisfying static equilibrium equations in its initial (undisplaced and/or undeformed) configuration. Thus, the structure in Figure 1.49*a* is classified as unstable because it can equilibrate vertical forces only in a deformed/displaced configuration (Figure 1.49*b*). Here, we can intuitively accept the idea that, even if beam AB was

[43] This definition is adequate for structures that are composed of components that can be considered to be rigid, but is not adequate for structures composed of components that exhibit stiffness-dependent behavior, e.g., the axially loaded bar in Figure 1.1*b*. Although the latter is not considered in this book, structural analysts must always be aware of the possibility of buckling due to compressive stresses. The type of stability we are considering, namely that associated with satisfaction of equations of static equilibrium of ''rigid'' bodies, is usually referred to as *statical stability*.

[44] The main reason for classifying a structure of this type as statically unstable is that we wish to avoid this form of instability when we analyze indeterminate structures using the force method, i.e., we begin by removing redundant forces in order to obtain a statically determinate and stable structure.

Figure 1.49 Unstable structure.

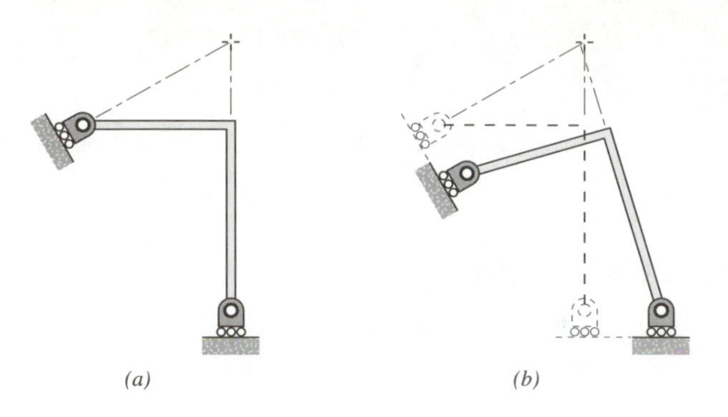

Figure 1.50 Unstable frame.

considered to be rigid, under the action of vertical loads it would rotate and spring BC would elongate in order to reach an equilibrium configuration. Note that the displacement may or may not be large, depending on the stiffness of the spring; regardless, the structure is classified as unstable because equilibrium equations that are written for the undeformed configuration have no solution.[45]

The beam in Figure 1.48 is unstable because it has only two reaction forces, an insufficient number to keep it in equilibrium under arbitrary loads. Thus, this particular beam has the rigid-body degree of freedom discussed previously, i.e., it can move in the horizontal direction. A similar statement can be made for the structure in Figure 1.50*a*; under arbitrary loads this structure also has a rigid-body degree of freedom, i.e., it can rotate about the point of intersection of the two reactions, as shown in Figure 1.50*b*. A minimum of three reaction forces is necessary to provide stability of a two-dimensional structure, but even this is not always sufficient; for example, the beams in Figures 1.51*a* and 1.51*b* have three and four reaction forces, respectively, but both are unstable because they can still not equilibrate horizontal loads.

Similarly, the structures in Figures 1.52*a* and 1.52*b* each have three nonparallel reaction forces, but both are unstable because all reactions are concurrent, i.e., have lines of action that intersect at a common point. As with the structure in Figure 1.50, under arbitrary loads these structures can rotate about the point of intersection of the reactions, point A in the figures; in other words, the equation of moment equilibrium cannot be satisfied. Furthermore, we can generalize the preceding cases, and have structures with more than three nonparallel reaction forces that are unstable, provided only that all reactions are concurrent; imagine the structures in Figures 1.52*a* and 1.52*b*

[45] This is an example of stable nonlinear behavior, i.e., behavior in which displacements and internal forces are not proportional to applied loads; thus, we have another situation for which our definition of statical instability is shown to be less than rigorous.

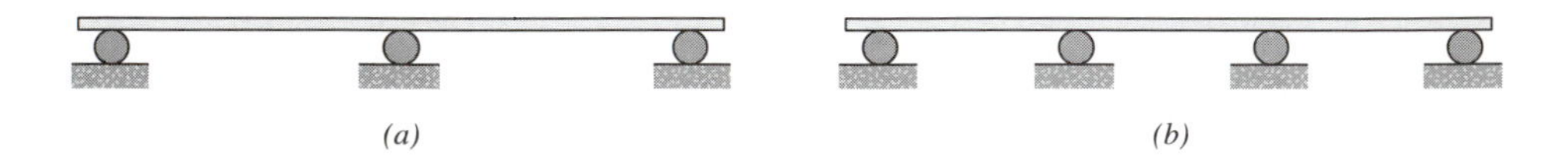

Figure 1.51 Unstable beams.

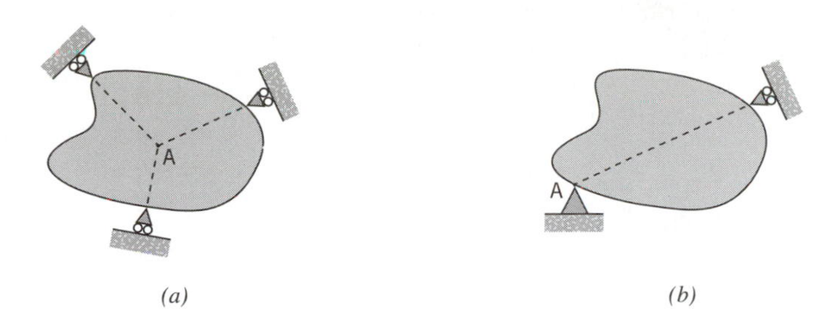

Figure 1.52 Concurrent reactions.

with additional roller supports that provide reaction forces with lines of action directed through point A.

In most cases, it should be reasonably apparent when a structure has a set of reactions that are improper, i.e., that do not provide a stable structure. In general, there is no point in attempting to write the equilibrium equations for a structure we recognize as unstable because the equilibrium equations have no solution. This would become apparent soon enough if we did attempt a solution.

Structures with a proper set of external supports can still be unstable if the internal connections are inadequate. This idea has already been introduced in the previous section (see Figures 1.46*f* and 1.46*g*). Additional examples are shown in Figures 1.53*a* and 1.53*b*; here, both are mechanisms rather than rigid frames even though both have four nonconcurrent reactions. The instability of these ''structures'' can often be revealed by analyzing their degrees of indeterminacy, as discussed previously. For example, if these structures did not have internal pin connections, each would be 1-degree indeterminate; if we remove two internal forces (moments) by introducing hinges at the corners it follows that the structures must be unstable.

Unfortunately, the preceding type of approach doesn't always work, as may be seen from the structure in Figure 1.54. If there were no internal hinges in this frame it would be 6-degrees indeterminate. Adding the four internal hinges shown produces an unstable structure with a stable portion that is still 3-degrees indeterminate rather than a stable structure that is only 2-degrees indeterminate. The reason for this is that the hinges are arranged in such a manner that the upper story is unstable, i.e., it is similar to the structure in Figure 1.53*a*. In other words, the structure is classified as unstable even if

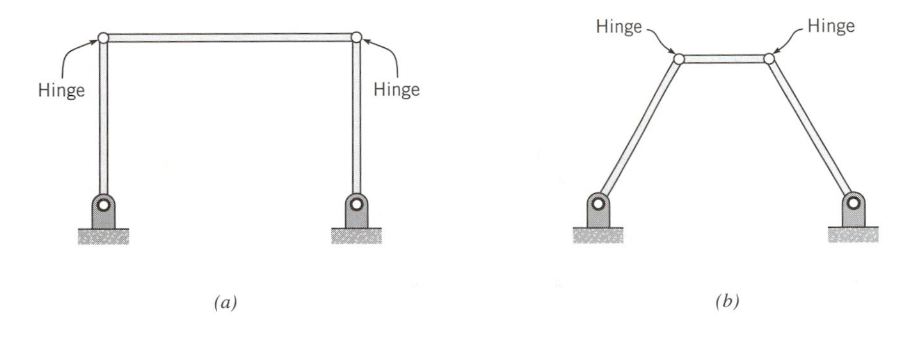

Figure 1.53 Mechanisms.

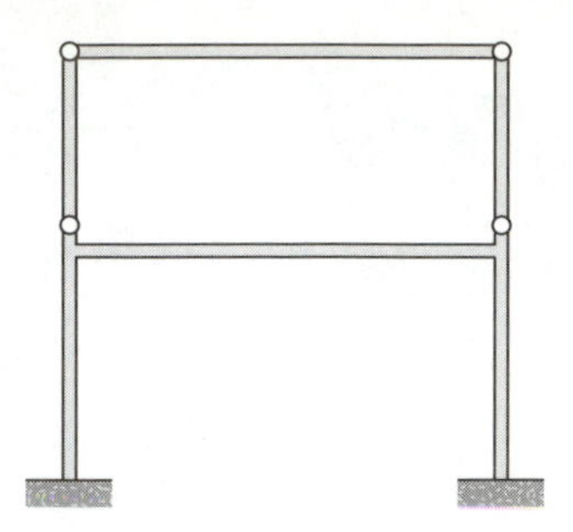

Figure 1.54 Unstable frame.

only a local portion, or substructure, is unstable. As with much of the field of structural analysis, it is necessary to proceed cautiously in the categorization of behavior, and trust that practice and experience will provide the insight required for proficiency.

1.9 PREVIEW

This chapter has been devoted primarily to a general orientation to the field of structural analysis, a review of equilibrium, and an introduction to analytic models and to some of the fundamental concepts of structural behavior, notably concepts related to indeterminacy and geometric stability.

In the following chapters we will present expanded examinations of truss and frame analysis, including the use of virtual work for calculation of equilibrium and compatibility relations. As we have noted, equilibrium and compatibility are two of the basic ingredients of the analysis procedure for any indeterminate structure.

There are two distinct approaches to the analysis of indeterminate structures, usually called the *force method* and the *displacement method*, which will be discussed subsequently. The force method will be introduced first because it is pedagogically convenient to do so, but both approaches are closely related; in fact, one may be regarded as the inverse of the other. More specifically, the force method follows the type of procedure discussed in this chapter, i.e., write equilibrium equations for a statically determinate and stable structure that is obtained by identifying a set of "redundant" forces, and then impose compatibility. The displacement method involves starting with the equivalent of the compatibility relations and *then* imposing equilibrium. Regardless of which approach is used, a complete understanding of equilibrium and compatibility is absolutely essential to a command of the subject.

At the outset, we will be able to demonstrate most important concepts without the necessity of solving many simultaneous equations; there will, therefore, be no immediate advantage to the use of matrices. As we progress through the topics and the problems become more involved computationally, we will find it convenient and conceptually helpful to introduce matrix formulations into our procedures.

PROBLEMS

1.P1. Consider a general set of forces applied in the xy plane. From Equations 1.7 and 1.16*b* we recognize that these forces will be in equilibrium provided

$\Sigma F_x = \Sigma F_y = \Sigma M_z = 0$. Show that each of the following alternative sets of independent equations also implies equilibrium of the forces.

(a) $\Sigma F_r = \Sigma F_s = \Sigma M_z = 0$, where r and s are any two nonparallel directions in the plane.

(b) $\Sigma M_A = \Sigma M_B = \Sigma F_{A/B} = 0$, where ΣM_A is the sum of the components of moments about a z axis through some point A, ΣM_B is the sum of the components of moments about a z axis through some point B, and $\Sigma F_{A/B}$ is the sum of the components of force in a direction parallel to the line joining A and B.

(c) $\Sigma M_A = \Sigma M_B = \Sigma M_C = 0$, where the summations are about z axes through any three noncollinear points A, B, and C.

1.P2–P3. Write the equations of equilibrium for the rigid joint shown.

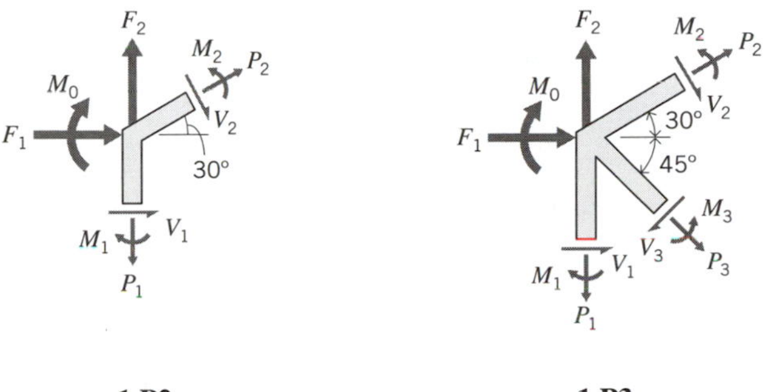

1.P2

1.P3

1.P4–P9. Find the reactions for the structure shown.

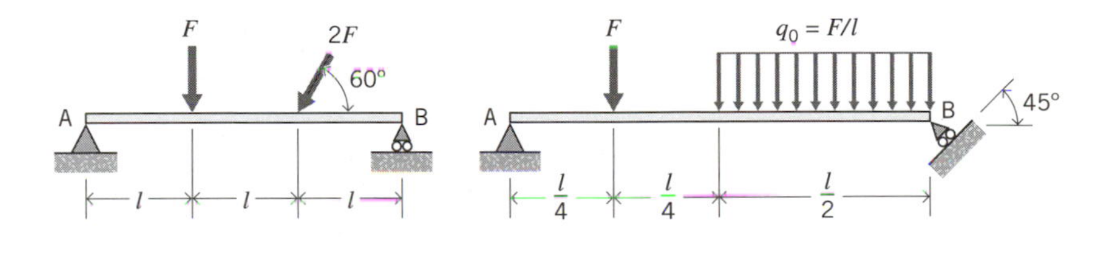

1.P4 **1.P5**

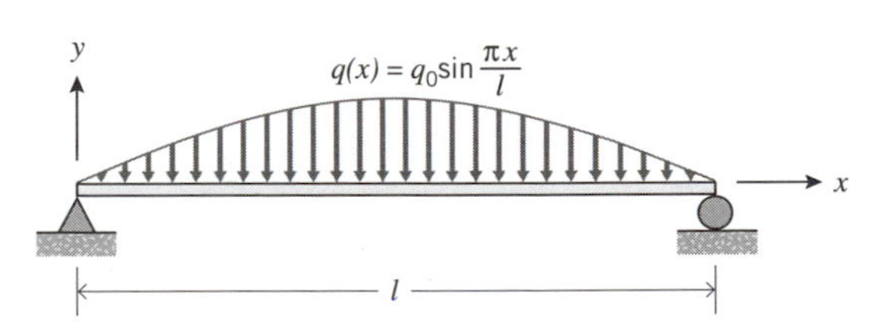

1.P6

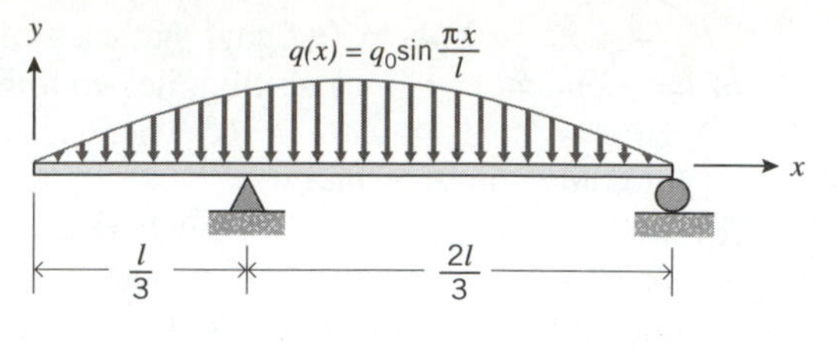

1.P7

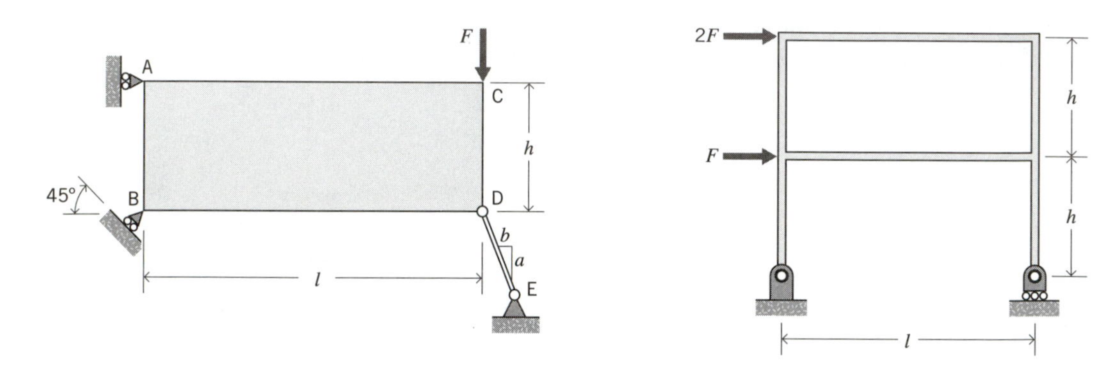

1.P8

1.P9

1.P10. Find the reaction force vector and moment vector at point A for the highway sign structure shown. Assume that the weight of the sign is 3000 lb, the weight of the vertical post is 2000 lb, and the horizontal wind pressure on the sign is 2 lb/ft^2.

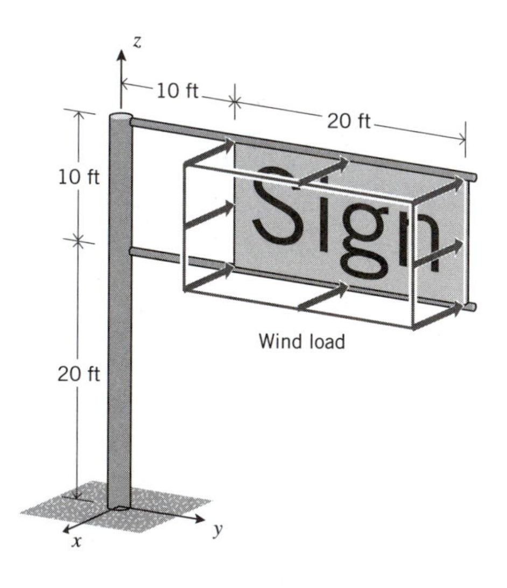

1.P10

1.P11. Discuss the differences between forces acting on the ends of the beam in Figure 1.19 for the two different support systems shown. In both cases, the upper rod supports the beam/floor shown; assume the lower rod also supports a similar beam/floor structure.

1.P12. Write and solve the equilibrium equations for the free-body in Figure 1.40*e*. Are the results the same as in Equations 1.54?

1.P13. Change the load on the beam in Figure 1.41*a* to a *uniform force per unit of vertical length* and find the transverse and axial components of the load q_0. Find the reactions.

1.P14–P23. Find the reactions for the structure shown.

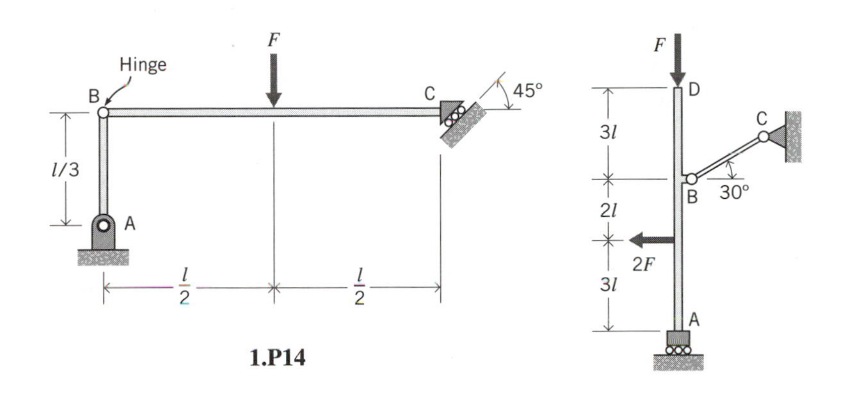

1.P14

1.P15

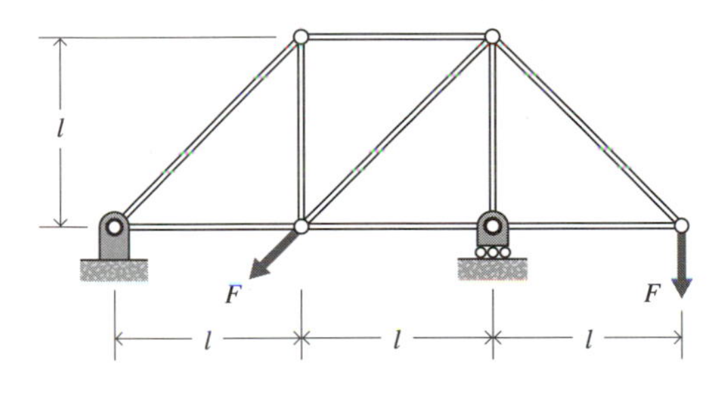

1.P16

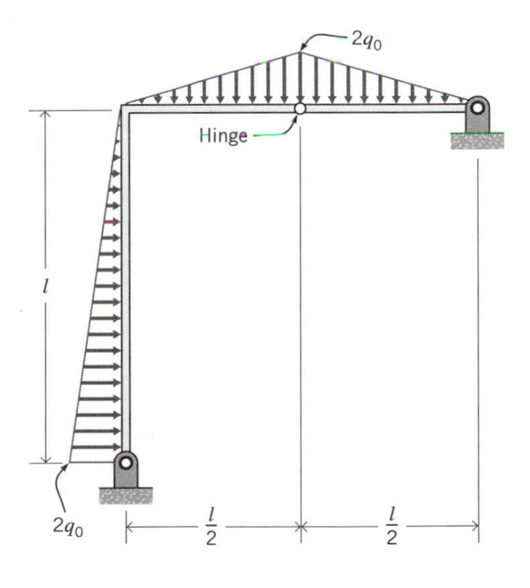

1.P17

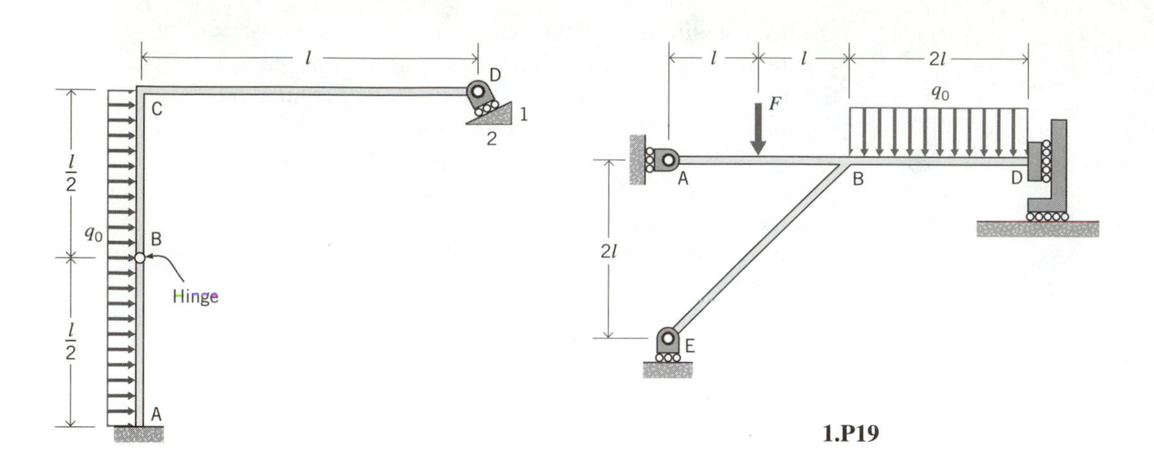

1.P19

1.P18

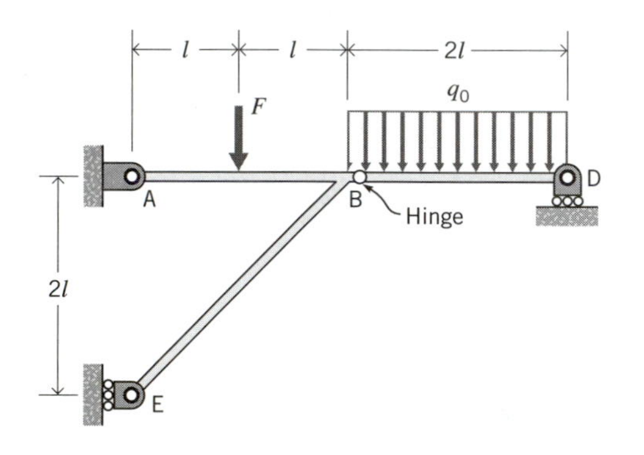

1.P20

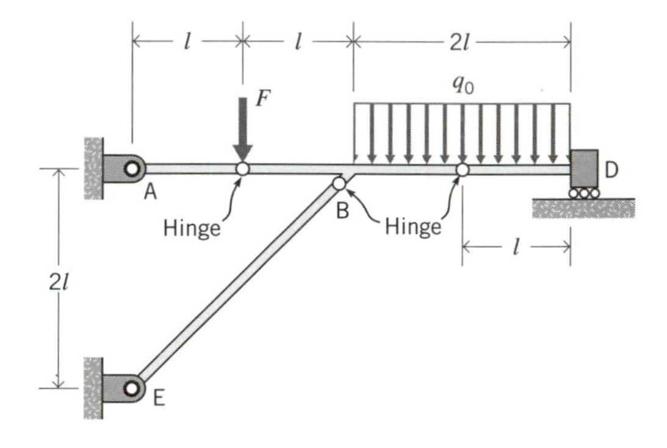

1.P21

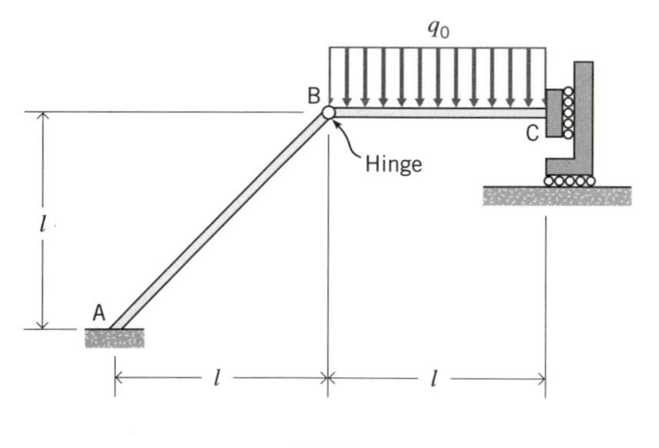

1.P22

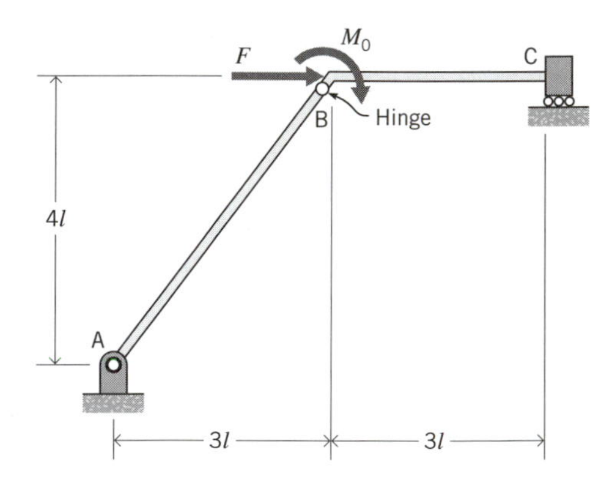

1.P23

1.P24–P30. Specify whether the following structures are stable or unstable, determinate or indeterminate. If stable and indeterminate, specify the degree of indeterminacy.

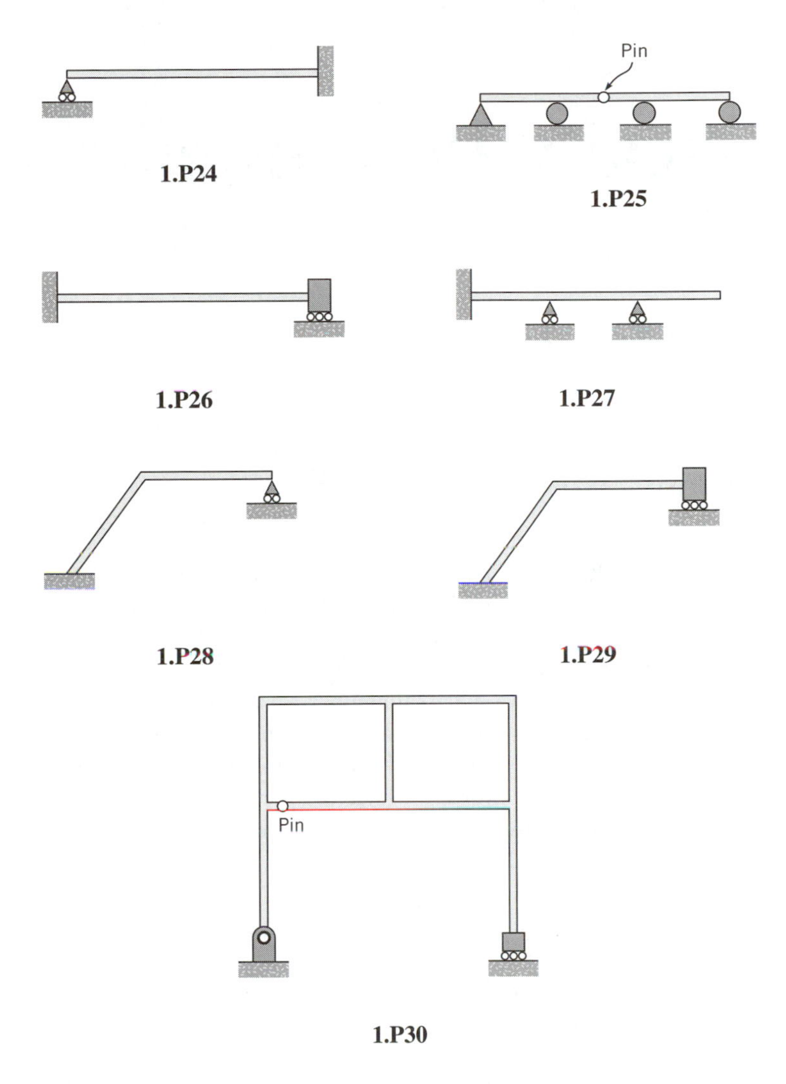

1.P24

1.P25

1.P26

1.P27

1.P28

1.P29

1.P30

1.P31. Is the structure in 1.P8 stable for all values of a and b? If not, specify the values of a and b for which the structure is unstable.

Chapter 2

Statically Determinate Trusses

2.0 INTRODUCTION

As was pointed out in Section 1.1.1, trusses are assemblies of one-dimensional elements such that the predominant physical behavior of each element is axial force transmission. Truss-type behavior is obtained when the elements (bars) are connected at each end by pins and when the external loads are applied only at the joints. Under such conditions, the internal axial forces in each truss bar will be constant.

In this chapter we will examine basic aspects of truss behavior with an emphasis on the computation of internal forces in statically determinate trusses. Before beginning a discussion of truss analysis, however, we will first establish a general equilibrium relationship for bar forces, which will be useful in later developments, and we will then review the analytic justification for the assumption of constant axial forces in truss bars.

2.1 EQUILIBRIUM OF AXIALLY LOADED BARS

We begin by considering equilibrium of a general slender,[1] straight axially loaded bar. By straight we shall mean that the bar axis that passes through the centroids of all cross-sectional areas is straight. Cross-sectional areas need not be uniform. Assume that the bar is in equilibrium under the action of concentrated axial loads at its ends and a nonuniform intermediate axial load, as shown in Figure 2.1. The end loads are P_A and P_B, respectively, and the intermediate load is taken to be a continuous loading, i.e., *a force per unit length*,[2] denoted by $p(x)$. As shown in Figure 2.1, these applied

[1] A *slender* element is usually defined as one whose length is at least ten times greater than a characteristic cross-sectional dimension.

[2] Intermediate axial loads are included here for the sake of generality even though they are not usually considered in connection with truss bars. Their inclusion here is useful because the continuous axial load is analogous to a continuous transverse load on a beam, and we will follow the same type of developments for beam elements in the next chapter. Furthermore, beam elements can be subjected to both transverse loads and axial loads of this type.

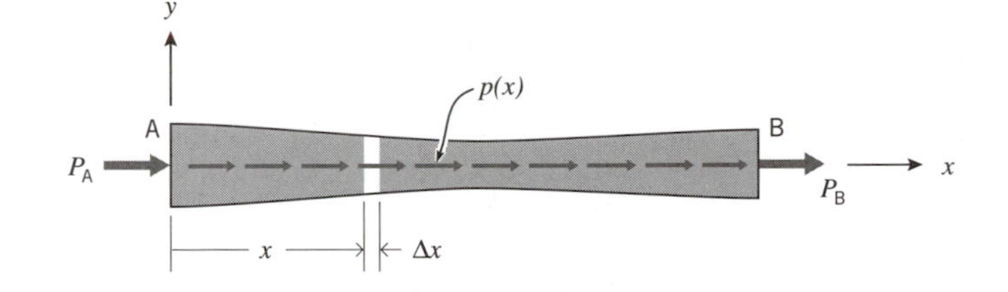

Figure 2.1 Axially loaded bar.

loads are assumed to be positive when acting in the positive x direction. This assumption is arbitrary.[3]

As we shall show from equilibrium considerations, an internal axial force resultant, denoted by $P(x)$,[4] will exist throughout the bar, i.e., at every station x along the bar there will be a resultant axial force. As shown in Figure 2.2, we shall define this force to be positive if it is tensile. Note the distinction between applied (external) loads and resultant (internal) forces in the sign conventions; in particular, we define as a positive *resultant* force one that acts in the positive x direction on a surface whose outward normal is the positive x direction. It follows that the equal and opposite resultant force on the contiguous surface (i.e., the surface whose outward normal is the negative x direction) is also defined as positive. Also note that we are treating the end loads as *external*; thus, even though equilibrium requires that P_A be equal and opposite to $P(x)$ at $x = 0$, we still choose to call P_A positive when it acts in the direction of increasing x, as shown.

Now, consider a free-body diagram of a small portion of the bar in Figure 2.1, specifically a typical portion of length Δx identified in Figure 2.1 and shown in detail in Figure 2.3. The forces acting on this *element*[5] are the resultant $P(x)$ on its left cross section, located x units from the left end of the bar, the distributed axial force per unit length $p(x)$, and a resultant force on its right cross section, $P(x + \Delta x)$. Recognizing that any changes in the axial force that develop over the change in position from x to $x + \Delta x$ must be very small, we shall express $P(x + \Delta x)$ as being equal to $P(x) +$

[3] The establishment of sign conventions is an integral part of the analysis process. Sign conventions are not universal.

[4] The relationship between this resultant axial force and the internal stresses will be reviewed in Section 2.4.

[5] Here, the word *element* is used to denote a small (sometimes infinitesimal) portion of a larger body. It should also be noted that the forces in Figure 2.3 have all been defined as positive when they act in the directions shown.

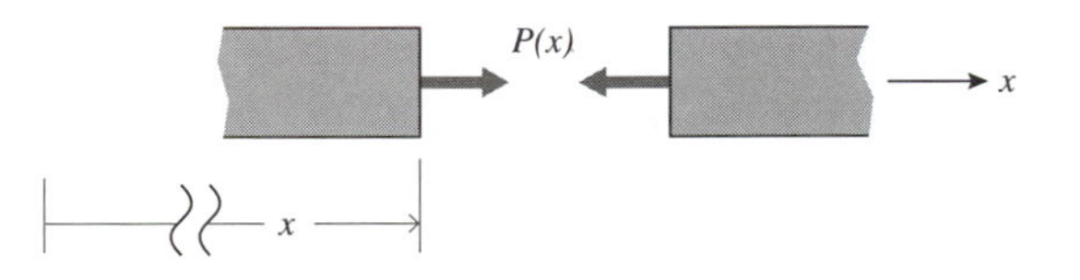

Figure 2.2 Positive resultant axial force at a cross section.

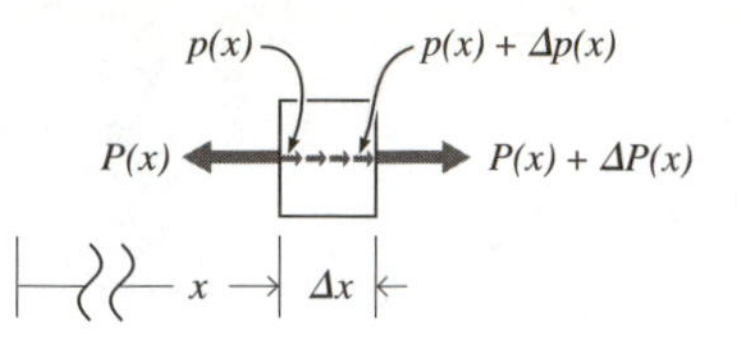

Figure 2.3 Free-body diagram of element of bar of length Δx.

$\Delta P(x)$.[6] The distributed force may also vary over the length of the element, from $p(x)$ at the left cross section to $p(x) + \Delta p(x)$ at the right cross section; because the length of the element is assumed to be small we can also assume that the variation of $p(x)$ is approximately linear between cross sections.

Now, because the free-body in Figure 2.3 is in a state of equilibrium we must have $\Sigma F_x = 0$, i.e.,

$$[P(x) + \Delta P(x)] - P(x) + \left[p(x)\Delta x + \frac{1}{2}\Delta p(x)\Delta x\right] = 0 \qquad \textbf{(2.1}\boldsymbol{a}\textbf{)}$$

The term in the last set of square brackets is the resultant of the distributed load, which, as stated above, we have assumed varies linearly over Δx. Canceling the positive and negative $P(x)$ terms, and dividing by Δx gives

$$\frac{\Delta P(x)}{\Delta x} + p(x) + \frac{1}{2}\Delta p(x) = 0 \qquad \textbf{(2.1}\boldsymbol{b}\textbf{)}$$

In the limit as Δx approaches zero, Δp approaches zero and $\Delta P/\Delta x$ approaches the derivative dP/dx; therefore, we find that[7]

$$\frac{dP(x)}{dx} + p(x) = 0 \qquad \textbf{(2.1}\boldsymbol{c}\textbf{)}$$

or, in alternate mathematical notation using *primes* to denote differentiation with respect to x,

$$P'(x) + p(x) = 0 \qquad \textbf{(2.1}\boldsymbol{d}\textbf{)}$$

Equation 2.1c or 2.1d is the differential equation of equilibrium of a slender, axially loaded bar. This equation provides the relationship between applied distributed axial

---

[6] This expression for the variation over length Δx may also be justified mathematically by expanding $P(x)$ about x using a *Taylor series*; if Δx is small, we need consider only the linear terms of the expansion, i.e., $P(x + \Delta x) \approx P(x) + (dP/dx)\Delta x$. The last term is equivalent to an incremental force, $\Delta P(x)$.

[7] This result can also be obtained by writing the equilibrium equation using the expression for $P(x + \Delta x)$ given in the preceding footnote.

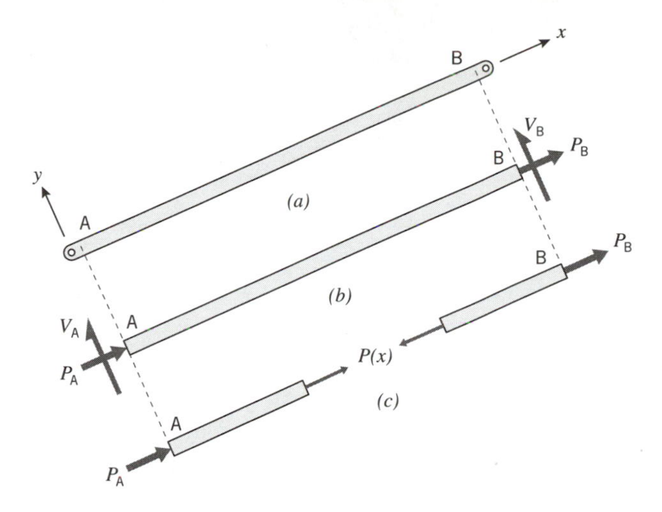

Figure 2.4 (*a*) Truss element; (*b*) free-body diagram; (*c*) resultant force.

forces (per unit length) and internal resultant forces that satisfies the condition of equilibrium. For example, in the special case when $p(x) = 0$ (i.e., a truss bar), the solution of the differential equation is $P(x) = Constant$. This differential equation has no further applications in this chapter, but its relationship to analogous equations for beams should be noted for future reference. Equations of this type are the essential mathematical building blocks in the derivation and application of virtual work, the most powerful theorem in modern structural analysis.

The fact that truss bars are considered to be axially loaded is, as implied previously, a consequence of the characteristics associated with the analytic model of this class of structural components. To illustrate, consider a typical truss bar, which by definition is assumed to have pinned ends. Such a bar is shown schematically in Figure 2.4*a*;[8] a free-body diagram of the portion of the bar between the ends is shown in Figure 2.4*b*. Near each end of this bar, in the vicinity of the pin, there can be, at most, two components of force, as in Figure 2.4*b*.[9] There are no intermediate applied loads, either axial or transverse, again by definition. If the free-body in Figure 2.4*b* is to be in equilibrium, the transverse end forces must be zero, and the axial end forces must be equal and opposite. This may be seen, for example, by summing moments about A in Figure 2.4*b*; equilibrium requires that $V_B = 0$. Equilibrium of forces in the y direction then requires that $V_A = 0$, and equilibrium of forces in the x direction requires that $P_B = -P_A$. Furthermore, from Figure 2.4*c*, we see that $P(x) = -P_A = P_B$, i.e., the internal resultant force at each cross section has the same magnitude as the applied loads acting on the ends of the bar; thus, for the truss bar, there is really no need to draw a distinction between the internal force and the external loads.

[8] Most modern trusses actually have riveted or bolted joints, but the predominant mode of force transmission in the bars is usually axial nevertheless. Also, truss bars usually have uniform cross sections.

[9] See Section 1.5.

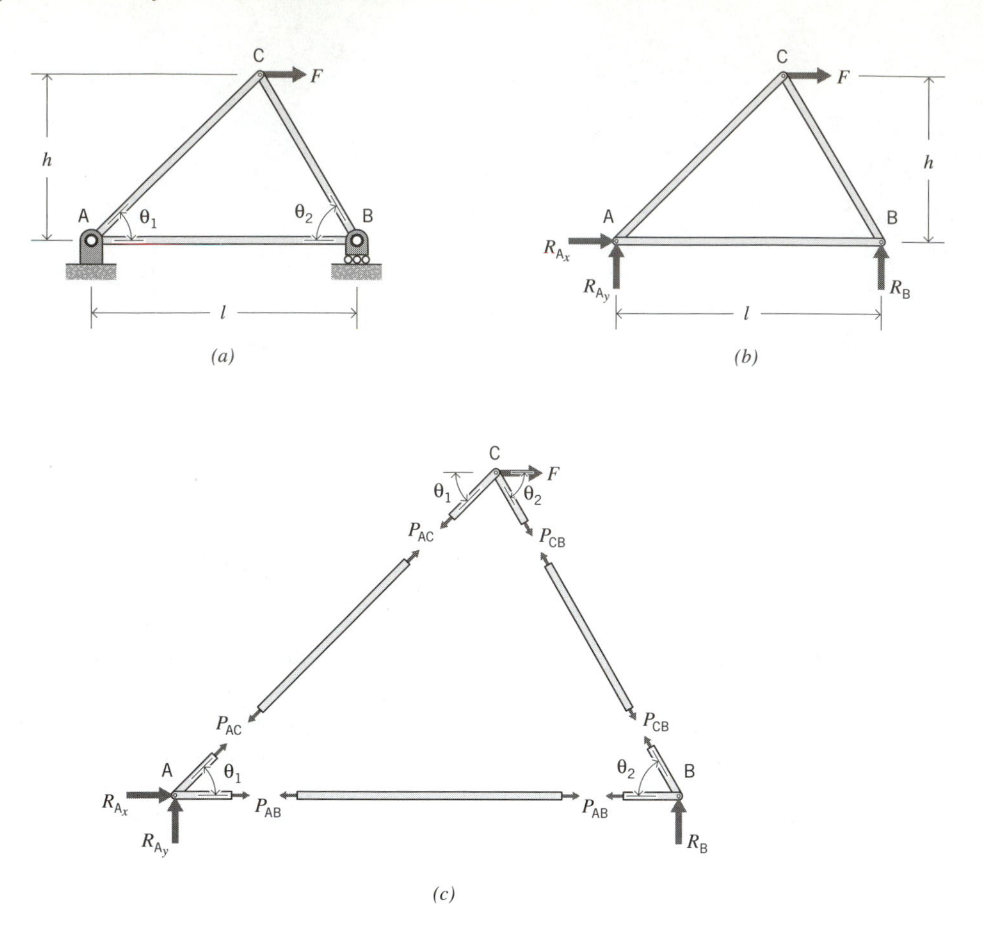

Figure 2.5 Three-bar truss; $0 < \theta_1, \theta_2 < 90°$.

2.2 INTERNAL FORCES IN STATICALLY DETERMINATE TRUSSES

With respect to a truss assembly, any number of bars may be connected at a given pinned joint, but the axes of all bars are assumed to be concentric at the joint.[10] There is an infinite variety of possible truss configurations. A few special configurations of historic interest are identified by specific names, and sometimes trusses are classified into categories based on certain topological characteristics.[11] For the sake of conciseness, we shall concentrate only on the basic concepts of analysis. We will also not concern ourselves here with any discussion of how external loads might be applied to joints.

Figure 2.5*a* shows a typical example of one of the most basic trusses, a simple

[10] Otherwise, some additional lateral loads must be applied to each bar to maintain equilibrium. These loads would, in turn, induce some bending stresses, which are not taken into account in truss theory.

[11] For example, Warren or Howe truss; simple, compound, and complex trusses.

triangular arrangement of bars. This particular truss is assumed to have a pin support at joint A, a roller support at joint B, and a specified horizontal force F applied at joint C; we will also assume that angles θ_1 and θ_2 are acute, i.e., $0 < \theta_1, \theta_2 < 90°$. The reactions, shown in Figure 2.5*b*, are statically determinate, and the structure is statically stable. We wish to find the internal forces in the bars of this truss. We will denote these three forces by P_{AB}, P_{AC}, and P_{CB}, i.e., we identify the constant force in each bar by the symbol P_{IJ} with subscripts I and J corresponding to the joints at its end points; the order of the subscripts is unimportant.

Figure 2.5*c* shows free-body diagrams of various portions of the truss, specifically the portions of the truss in the vicinity of each joint, and the bars between these joints. Each of these free-body diagrams shows all forces that act on the free-body. Note that forces on contiguous surfaces are equal and opposite, and that all internal bar forces have been arbitrarily assumed to be tensile as shown in the figure.

Each free-body in Figure 2.5*c* must be in equilibrium. Bars AB, AC, and BC are already in equilibrium because we have implicitly utilized the equilibrium equations to obtain the equal and opposite axial end forces that are shown acting on these bars. The free-bodies of the joints A, B, and C, on the other hand, are each acted on by a set of concurrent forces, some of which have unknown magnitudes. As with every component or subsection of the structure, each of the joints[12] must satisfy the equations of equilibrium. Since the forces acting on each joint are concurrent, equilibrium of moments is automatically satisfied. However, for the planar truss joint, two nontrivial equations, $\Sigma F_x = 0$ and $\Sigma F_y = 0$, must be satisfied, i.e., we must satisfy two independent equilibrium equations for each joint. Since there are three joints in this truss, we have a total of six independent equilibrium equations. There are also six unknown forces that exist on or within the truss, namely, R_{A_x}, R_{A_y}, R_B, P_{AB}, P_{AC}, and P_{CB}. We expect (but at this time without any complete assurance) that we can solve these equations for all six unknown forces.

Fortunately, it is nearly always possible to find the unknowns without having to solve a 6 by 6 set of simultaneous linear equations, which would usually involve matrix solution procedures. The simplest strategy for finding unknown forces is known as the *method of joints*. The method consists of solving the joint equilibrium equations one joint at a time, i.e., in a sequential manner. The success of this solution procedure depends on selecting and analyzing specific joints that contain no more than two unknown forces; the two independent equilibrium equations at the joint can then be used to determine the unknown forces. The procedure carries with it the advantage that no more than two simultaneous equations must be solved. This has the effect of circumventing the inconvenience of dealing with large systems of simultaneous equations and usually requires relatively little computational effort. On the other hand, when the problems become more complex, for example, when the geometric layout of the truss is complicated, then the method of joints will not provide significant reduction in solution effort as compared with matrix solution techniques.

To illustrate the method of joints, we return to the truss in Figure 2.5; initially, this truss has only one joint with as few as two unknown forces, namely joint C with

[12] From this point on, reference to a free-body of a joint actually implies a free-body of a portion of the truss in the immediate vicinity of that joint.

unknown forces P_{AC} and P_{CB}. Summing forces that act in the x direction on the free-body of this joint gives

$$\sum F_x = 0 = F + P_{CB} \cos \theta_2 - P_{AC} \cos \theta_1 \tag{2.2a}$$

Similarly, summing forces in the y direction gives

$$\sum F_y = 0 = -P_{CB} \sin \theta_2 - P_{AC} \sin \theta_1 \tag{2.2b}$$

These two equations can be solved simultaneously; the result is

$$P_{AC} = \frac{F \sin \theta_2}{\sin \theta_1 \cos \theta_2 + \cos \theta_1 \sin \theta_2} = \frac{F \sin \theta_2}{\sin(\theta_1 + \theta_2)} \tag{2.3a}$$

$$P_{CB} = -\frac{F \sin \theta_1}{\sin \theta_1 \cos \theta_2 + \cos \theta_1 \sin \theta_2} = -\frac{F \sin \theta_1}{\sin(\theta_1 + \theta_2)} \tag{2.3b}$$

The negative sign in Equation 2.3*b* shows that P_{CB} is actually a compressive force.

The magnitudes and senses of P_{AC} and P_{CB} are now known on the free-bodies of joints A and C, respectively, as seen in Figure 2.5*c*. Joint A still has three other unknown forces P_{AB}, R_{A_x}, and R_{A_y}, whereas joint B has only two, namely R_B and P_{AB}. Therefore, we can now proceed to analyze joint B.

Since we now know that P_{CB} is compressive, we have two alternative ways of representing P_{CB} in the free-body diagram of joint B; these choices are illustrated in Figure 2.6. Figure 2.6*a* shows P_{CB} in the same direction as was assumed in Figure 2.5*c* (tension), and leaves its value negative as in Equation 2.3*b*; Figure 2.6*b* shows the direction of P_{CB} reversed (compression), and changes its value to positive. Thus, both free-bodies identify P_{CB} as a compressive force and either is appropriate; however, if we use Figure 2.6*b*, we should be consistent and also change the direction of P_{CB} in all other free-body diagrams. Perhaps the best strategy to follow with regard to sign convention is to assume one sense of loading, say tension, as positive and to simply accept minus signs (and compression forces) without continually trying to update sign conventions so that everything turns out positive!

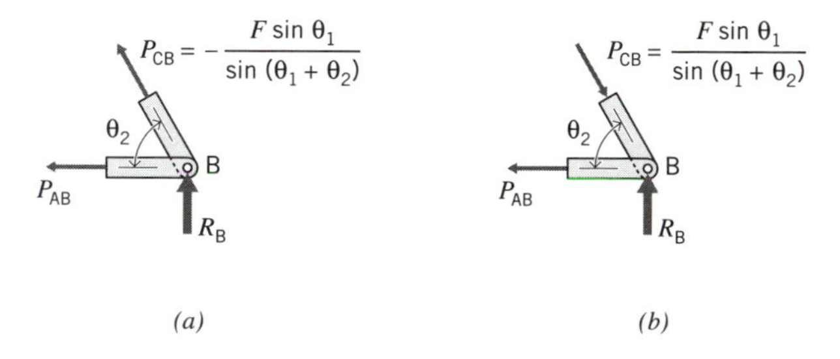

Figure 2.6 Joint B of three-bar truss.

Proceeding with the equilibrium equations for the free-body of joint B in Figure 2.6*a*, we get

$$\sum F_x = 0 = -P_{AB} - P_{CB} \cos \theta_2 \tag{2.4a}$$

$$\sum F_y = 0 = R_B + P_{CB} \sin \theta_2 \tag{2.4b}$$

Note that, in this case, each of the equations contains only one unknown force, so each equation can be solved directly; e.g., solving Equation 2.4*b* for R_B gives

$$R_B = -P_{CB} \sin \theta_2 \tag{2.5a}$$

and substituting Equation 2.3*b* for P_{CB} leads to

$$R_B = -\left[\frac{-F \sin \theta_1}{\sin(\theta_1 + \theta_2)}\right] \sin \theta_2 = \frac{F \sin \theta_1 \sin \theta_2}{\sin(\theta_1 + \theta_2)} \tag{2.5b}$$

Similarly, from Equation 2.4*a* we get

$$P_{AB} = \frac{F \sin \theta_1 \cos \theta_2}{\sin(\theta_1 + \theta_2)} \tag{2.5c}$$

Since the values of both R_B and P_{AB} are positive, the directions assumed in Figures 2.5 and 2.6 are consistent with positive answers. We will get exactly the same results for R_B and P_{AB} from the equilibrium equations applied to the free-body in Figure 2.6*b*.

With P_{AB} determined in Equation 2.5*c*, we can now evaluate the last two unknown forces, R_{A_x} and R_{A_y}, by writing the equilibrium equations for joint A, Figure 2.5*c*. Summing forces in the x and y directions gives

$$\sum F_x = 0 = R_{A_x} + P_{AB} + P_{AC} \cos \theta_1 \tag{2.6a}$$

$$\sum F_y = 0 = R_{A_y} + P_{AC} \sin \theta_1 \tag{2.6b}$$

Solving each equation provides

$$R_{A_x} = -F \tag{2.7a}$$

$$R_{A_y} = -\frac{F \sin \theta_1 \sin \theta_2}{\sin(\theta_1 + \theta_2)} \tag{2.7b}$$

Note that R_{A_x} is equal and opposite to the applied load F, and that R_{A_y} is equal and opposite to R_B; this is a consequence of the fact that the entire truss (Figure 2.5*b*), must also satisfy all equations of equilibrium. In fact, we could have evaluated R_{A_x}, R_{A_y}, and R_B at the outset. For example, from the free-body in Figure 2.5*b*, summing moments about A gives

$$\sum M_A = 0 = -Fh + R_B l \tag{2.8a}$$

where l and h are the dimensions identified in the figure. From the geometry of the truss, we obtain

$$h = \frac{l \tan \theta_1 \tan \theta_2}{\tan \theta_1 + \tan\theta_2} = \frac{l \sin \theta_1 \sin \theta_2}{\sin(\theta_1 + \theta_2)} \tag{2.8b}$$

It follows from Equations 2.8*a* and 2.8*b* that

$$R_B = \frac{F \sin \theta_1 \sin \theta_2}{\sin(\theta_1 + \theta_2)} \tag{2.8c}$$

in other words, R_B is exactly the same as in Equation 2.5*b*. Similarly, summing forces in the x direction gives $R_{A_x} = -F$, as in Equation 2.7*a*, and summing forces in the y direction gives $R_{A_y} = -R_B$, as in Equation 2.7*b*. If we had evaluated these reactions at the outset, there would have been only three remaining unknown forces, P_{AB}, P_{AC}, and P_{CB}. These forces could then have been evaluated from three independent equilibrium equations written for any two of the joints. For instance, we could have written the two equilibrium equations at joint A to obtain P_{AB} and P_{AC}, and the equation for $\Sigma F_y = 0$ at joint B to obtain P_{CB}.

From this discussion it is apparent that there are several alternate paths to the same solution; in fact, it is a very good idea to use some of these alternate computations as a means of checking results whenever possible.

EXAMPLE 1

The 2-bar truss in Figure 2.7*a* has pin supports at both A and B. This truss is otherwise similar to the one in Figure 2.5 except for the fact that there is no member connecting points A and B.[13]

There are two internal forces, P_{AC} and P_{CB}, and four components of reaction, R_{A_x}, R_{A_y}, R_{B_x}, and R_{B_y} (Figure 2.7*b*). Since there are three joints, we expect to be able to write six independent equilibrium equations from which to evaluate all unknown forces. In fact, the internal forces are identical to those found for the truss in Figure 2.5 because the equilibrium equations for joint C are exactly the same as those in Equations 2.2, i.e., P_{AC} and P_{CB} are given by Equations 2.3*a* and 2.3*b*, respectively. The remaining four independent equilibrium equations will reveal nothing more than the fact that R_{A_x} and R_{A_y} are equal and opposite to the horizontal and vertical components of P_{AC}, and that R_{B_x} and R_{B_y} are equal and opposite to the horizontal and vertical components of P_{CB}. For example, from the free-body of joint A in Figure 2.7*c* we get

$$\sum F_x = 0 = R_{A_x} + P_{AC} \cos \theta_1 \quad \textit{or} \quad R_{A_x} = -P_{AC} \cos \theta_1 \tag{2.9a}$$

$$\sum F_y = 0 = R_{A_y} + P_{AC} \sin \theta_1 \quad \textit{or} \quad R_{A_y} = -P_{AC} \sin \theta_1 \tag{2.9b}$$

[13] A bar connecting two immovable pin supports cannot contribute to the transmission of applied loads.

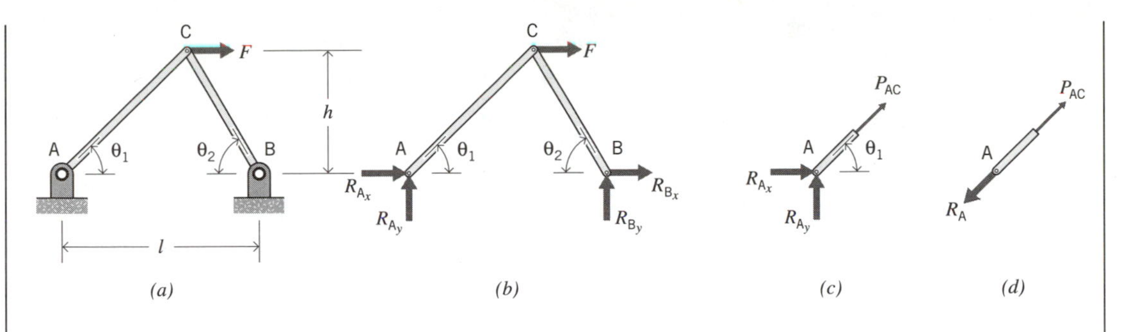

Figure 2.7 Example 1.

The implication of Equations 2.9*a* and *b* is that the vector resultant of R_{A_x} and R_{A_y} must be equal and opposite to P_{AC}, as expected for an axially loaded bar and as shown in Figure 2.7*d*. The magnitude of this resultant, denoted by R_A in Figure 2.7*d*, is given by[14]

$$R_A = \sqrt{(R_{A_x})^2 + (R_{A_y})^2} = \sqrt{(-P_{AC}\cos\theta_1)^2 + (-P_{AC}\sin\theta_1)^2} = P_{AC} \quad \textbf{(2.10)}$$

A similar result holds at joint B.

Note that R_{A_y} and R_{B_y} are exactly the same as for the truss in Figure 2.5; this is consistent with the equilibrium requirements for the free-body of the entire truss (Figure 2.7*b*). Here again, summing moments about A leads to an expression for R_{B_y} that is identical to the one for R_B in Equation 2.8*c*, and summing forces in the y direction shows that R_{A_y} must be equal and opposite to R_{B_y}, as in Equation 2.7*b*. In this case, however, R_{B_x} is not zero, and R_{A_x} is not equal and opposite to F; rather, $R_{A_x} + R_{B_x} = -F$ (see Equation 2.9).

Although the internal forces are the same for the bars of this truss and the one in Figure 2.5, the differences in supports will lead to different displacements;[15] this will be discussed briefly in Section 2.5, and in more detail in subsequent chapters.

EXAMPLE 2

The truss in Figure 2.8*a* has a pin support at A and a roller support at B; find the internal bar forces.

There are three reaction forces (Figure 2.8*b*), seven bar forces, and five joints; therefore, we can write 10 equilibrium equations from which we can evaluate the 10 unknown forces. In this case, however, there is no joint that initially has fewer than

[14] See Equation 1.3*c*.

[15] When we speak of displacements of a truss, we are referring to the x and y components of displacement of the joints, as in Figure 1.24. If needed, rotations of the bars can be easily calculated from the joint displacements.

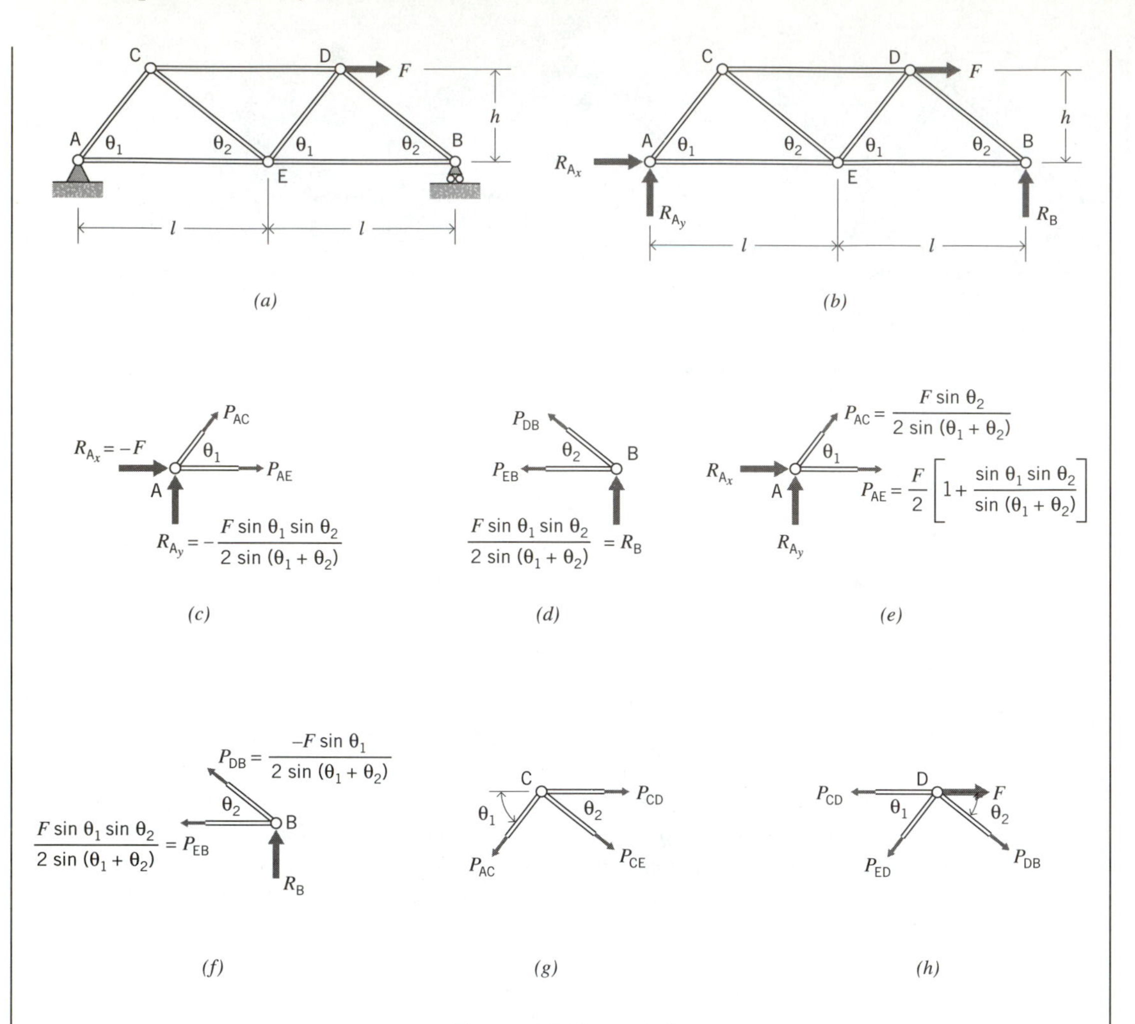

Figure 2.8 Example 2.

three unknown forces; consequently, we are forced to determine the reaction forces before we consider equilibrium of individual joints.

The geometry of this truss is closely related to that of the two trusses considered previously; in particular, dimension h is still given by Equation 2.8*b*, so summing moments about A for the free-body in Figure 2.8*b* now gives

$$\sum M_A = 0 = -Fh + R_B(2l) \tag{2.11}$$

By comparing this equation to Equation 2.8*a*, we see that R_B must be one-half of the value computed previously in Equation 2.5*b*. It also follows from the other two equilibrium equations that $R_{A_y} = -R_B$ and $R_{A_x} = -F$, as before.

With the reactions now known, we can proceed to analyze either joint A or joint B because the free-body of each has only two unknown forces, namely P_{AC} and P_{AE} at joint A and P_{DB} and P_{EB} at joint B, as shown in Figures 2.8*c* and 2.8*d*. The free-bodies of these two joints are similar to those in Figure 2.5*c*, except that the bar forces have different designations because of the additional joints in the truss. Also, as noted above, the vertical reactions are one-half their previous values.

Consider the free-body of joint B (Figure 2.8*d*); since R_B is one-half of its previous value, it follows that P_{EB} must be one-half of the value computed for P_{AB} in Equation 2.5*c* and P_{DB} must be one-half of the value computed for P_{CB} in Equation 2.3*b*.[16] The correct values of P_{EB} and P_{DB} are shown in Figure 2.8*f*.

Similarly, at joint A (Figure 2.8*c*), we see that P_{AC} is proportional to R_{A_y} but not proportional to R_{A_x}; because R_{A_y} is one-half of its previous value, it follows that P_{AC} will be one-half of the value computed in Equation 2.3*a*. On the other hand, P_{AE} is proportional to both R_{A_x} and P_{AC} (which is proportional to R_{A_y}); because R_{A_x} is still equal and opposite to F then P_{AE} will not be one-half of the value previously computed for P_{AB}.

To illustrate the last point, we will write the equilibrium equation for sum of forces in the x direction at joint A, i.e.,

$$\sum F_x = 0 = P_{AE} + P_{AC}\cos\theta_1 - F \tag{2.12a}$$

from which, after substituting the expression for P_{AC}, we obtain

$$P_{AE} = F\left[1 - \frac{\cos\theta_1 \sin\theta_2}{2\sin(\theta_1 + \theta_2)}\right] = \frac{F}{2}\left[1 + \frac{\sin\theta_1\cos\theta_2}{\sin(\theta_1 + \theta_2)}\right] \tag{2.12b}$$

The values of P_{AC} and P_{AE} are shown in Figure 2.8*e*.

In the preceding computations we used seven equilibrium equations, three to determine reactions and two each for the analysis of joints A and B. We can write some appropriate combination of three more equilibrium equations at joints C, D, or E to determine the remaining forces. For example, Figure 2.8*g* shows a free-body diagram of joint C; the equilibrium equations for this joint are

$$\sum F_x = 0 = P_{CD} + P_{CE}\cos\theta_2 - P_{AC}\cos\theta_1 \tag{2.13a}$$

$$\sum F_y = 0 = -P_{AC}\sin\theta_1 - P_{CE}\sin\theta_2 \tag{2.13b}$$

Substituting the previously computed value of P_{AC} leads to

$$P_{CE} = -\frac{F\sin\theta_1}{2\sin(\theta_1 + \theta_2)} \tag{2.14a}$$

$$P_{CD} = \frac{F}{2} \tag{2.14b}$$

Finally, we can write one equilibrium equation at either joint D or joint E to evaluate the only remaining unknown force, P_{ED}. For example, consider joint D (Figure 2.8*h*)

[16] In other words, internal forces are proportional to applied loads, a consequence of the fact that the equations of equilibrium are linear.

Summing forces in the y direction provides

$$\sum F_y = 0 = -P_{DB} \sin \theta_2 - P_{ED} \sin \theta_1 \qquad \textbf{(2.15a)}$$

and substituting the appropriate value of P_{DB} leads to

$$P_{ED} = \frac{F \sin \theta_2}{2 \sin (\theta_1 + \theta_2)} \qquad \textbf{(2.15b)}$$

We would obtain the same result from the equation for the sum of forces in the x direction on joint D. As noted previously, we can use the extra equilibrium equations as a means of checking our computations; the equations should be identically equal to zero when we substitute the computed values of forces.

In the preceding discussion, we have tacitly assumed that we were interested in finding the force in every bar of the truss and, consequently, we did the equivalent of analyzing every joint of the truss. Sometimes we may be interested in only a few isolated bar forces; in many such cases it may not be necessary to go through a complete analysis. Rather, we may be able to restrict our analysis to the particular forces of interest by considering free-bodies of larger portions of the structure than the single joints studied previously. The approach based on this concept is known as the *method of sections.*[17] The key to this procedure is the selection of a particular free-body on which the unknown bar forces of interest, together with some known forces, will appear as external to the free-body; this implies that we imagine the truss to be "cut" through the bars of interest. At the same time, the free-body cuts must not also reveal undesired (unknown) bar forces that will prevent the calculation of the forces of interest.

Consider, for illustrative purposes, the same truss as in Example 2 (Figure 2.8), and suppose we want to find only the forces in bars CD and ED; in this case it is possible to find P_{CD} and P_{ED} without first evaluating any of the other bar forces. As in Example 1, however, we must again determine the reaction forces before doing anything else. We will use the results obtained in Example 2 for these values, as shown in Figure 2.9*a*.

Figures 2.9*b* and 2.9*c* show typical free-bodies that we consider in order to evaluate P_{CD} and P_{ED}. In this particular illustration, the free-bodies consist of the portions of the structure to the left or right, respectively, of a cut[18] through bars CD, DE, and EB. Note that it is necessary to cut three bars to obtain an appropriate free-body, even though we are not necessarily interested in bar EB; i.e., a free-body must represent a complete portion of the original truss. As with the free-bodies of single joints, the free-bodies in Figures 2.9*b* and 2.9*c* include the forces that act on the ends of all the cut bars.

The free-body in Figure 2.9*c* contains three unknown forces, P_{CD}, P_{ED}, and P_{EB}, as

[17] The preceding method of joints may be regarded as a special case of the method of sections.

[18] In this respect, the procedure is analogous to the one discussed in Section 1.6, *cf* Figure 1.44.

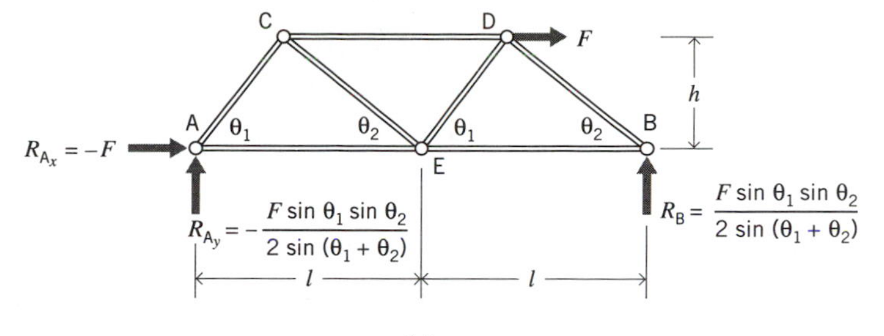

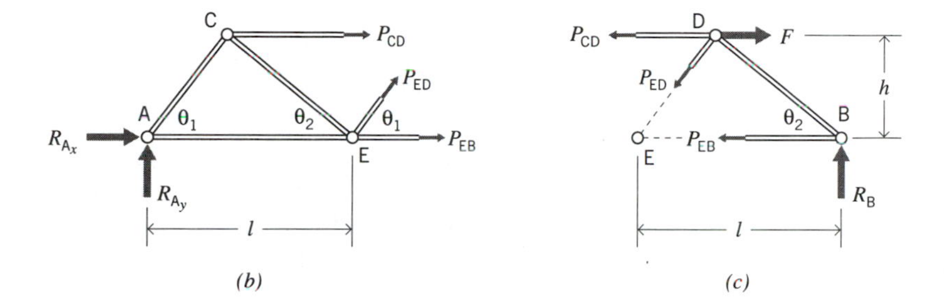

Figure 2.9 Analysis of selected bar forces.

well as two known forces, F and R_B. Since these forces are neither all concurrent nor parallel, we can write three equilibrium equations from which we can solve for P_{CD}, P_{ED}, and P_{EB}. Fortunately, it is possible in this case to avoid solving simultaneous equations provided we apply the equilibrium equations according to a careful plan. Specifically, we may begin by writing the equilibrium equation for sum of moments about the intersection of the lines of action of P_{ED} and P_{EB}, point E in Figure 2.9*c*; this eliminates two of the unknown forces from the equation, i.e.,

$$\sum M_E = 0 = R_B l + P_{CD} h - F h \tag{2.16a}$$

Substituting the known value of R_B, and the expression for h in Equation 2.8*b*, and solving for P_{CD}, leads to

$$P_{CD} = \frac{F}{2} \tag{2.16b}$$

which is exactly the same result obtained in Equation 2.14*b*.

The simplest way to obtain P_{ED} is to write the equilibrium equation for sum of forces in the y direction for the free-body in Figure 2.9*c*, i.e.,

$$\sum F_y = 0 = R_B - P_{ED} \sin \theta_1 \tag{2.17a}$$

which leads to

$$P_{ED} = \frac{F \sin \theta_2}{2 \sin (\theta_1 + \theta_2)} \tag{2.17b}$$

as in Equation 2.15*b*.

The same results can be obtained from the free-body in Figure 2.9*b*; in fact, the equilibrium equations can be written in the same way. Specifically, summing moments about point E in Figure 2.9*b* gives

$$\sum M_E = 0 = -R_{A_y} l - P_{CD} h \tag{2.18}$$

which again provides the previous result for P_{CD}.

Finally, it should be noted that we could have chosen a different free-body from either of the ones in Figures 2.9*b* and 2.9*c* from which to evaluate P_{CD}. For example, we could have isolated a portion of the truss to either side of a cut through bars CD, CE, and AE. The choice of free-bodies is frequently a matter of convenience rather than of necessity. Of course proficiency in choosing the ''best'' free body, or sequence of free-body diagrams, requires practice and some effort even for experienced analysts. However, the pay off is an ability to essentially go directly into a structure (using the correct free-body diagram) and solve for the force of interest. For this reason the method of sections is preferred over the method of joints.

EXAMPLE 3

Find the forces in bars DB and FH of the truss in Figure 2.10*a*; this truss, although unusual, is statically determinate and geometrically stable. The first computation will be fairly routine, the second somewhat less so. It is not necessary to evaluate the reactions in either case.

To find P_{DB} directly, we can imagine a free-body cut through members DB, BE, and EC. The free-body of the portion of the truss above this cut is shown in Figure 2.10*b*; the unknown forces on this free-body are P_{DB}, P_{BE}, and P_{EC}. (We avoid analyzing the free-body of the portion of the structure below this cut because, including the three reactions, there would be a total of six unknown forces on that free-body.)

Consider summing moments about the intersection of the lines of action of P_{BE} and P_{EC}, point E in Figure 2.10*b*. This is similar to the procedure followed for the calculation of P_{CD} in the previous illustration, Figure 2.9*c*. There is one distinction, however; the perpendicular distance from point E to the line of action of P_{DB} is not obvious, so the magnitude of the moment of P_{DB} about E is not as apparent as it was previously. In this case, it is probably more straightforward to deal with the moments of the components of P_{DB}.[19] We will denote the horizontal and vertical components of P_{DB} by P_{DB_x} and P_{DB_y}, respectively. From the geometry of the truss we find the components to be

[19] See Equation 1.14*c*.

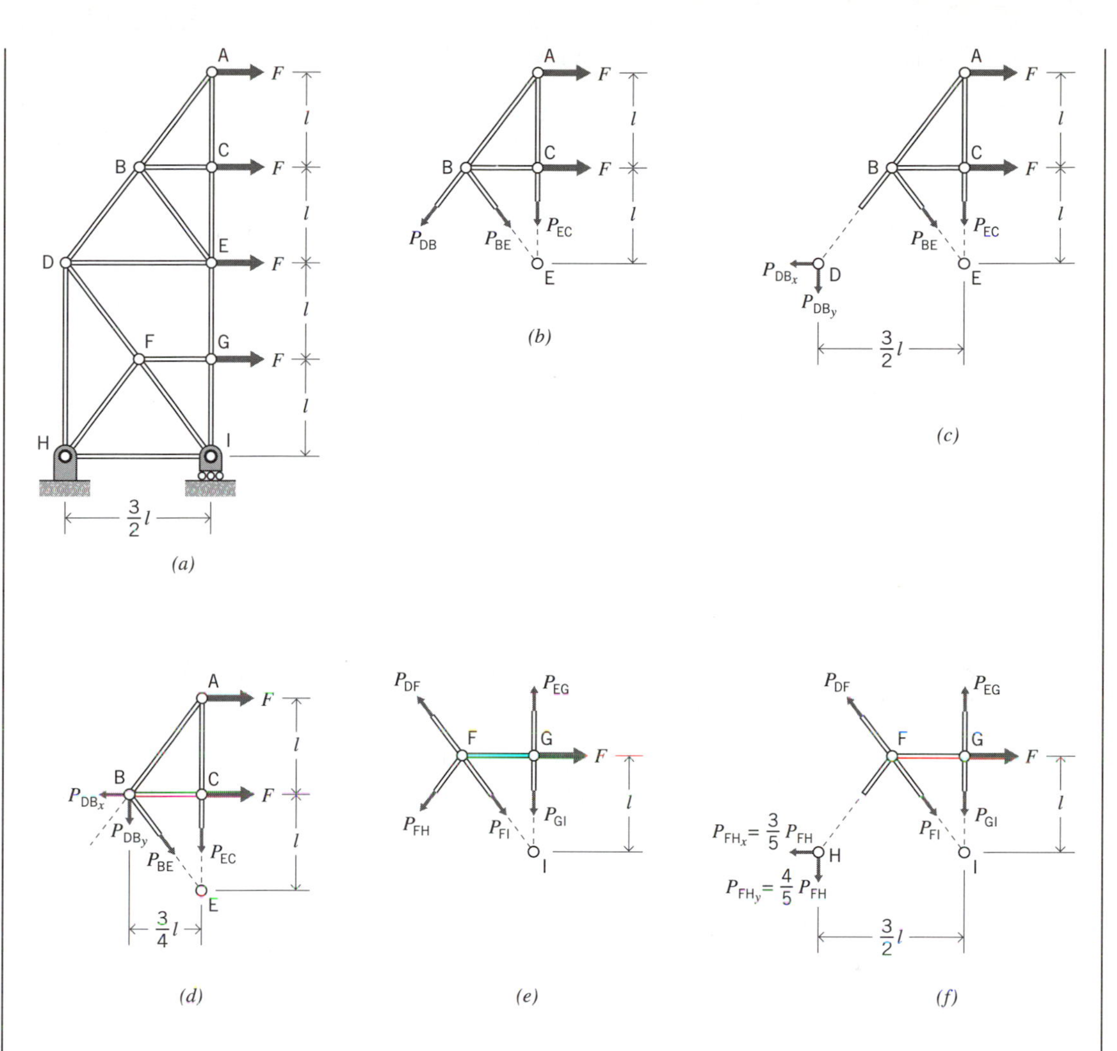

Figure 2.10 Example 3.

$$P_{DB_x} = \frac{3}{5} P_{DB} \tag{2.19a}$$

$$P_{DB_y} = \frac{4}{5} P_{DB} \tag{2.19b}$$

Now, P_{DB} can be replaced by these two components at any point along its line of action. Figures 2.10*c* and 2.10*d* show two equivalent representations of the free-body acted on by the components of P_{DB}. Summing moments about E in Figure 2.10*c* gives

$$\sum M_E = 0 = P_{DB_y}\left(\frac{3}{2}l\right) - Fl - F(2l) \tag{2.20a}$$

while summing moments about E in Figure 2.10*d* gives

$$\sum M_{\mathrm{E}} = 0 = P_{\mathrm{DB}_y}\left(\frac{3}{4}l\right) + P_{\mathrm{DB}_x}(l) - Fl - F(2l) \tag{2.20b}$$

Both equations give the same result for P_{DB}, namely

$$P_{\mathrm{DB}} = \frac{5}{2}F \tag{2.20c}$$

Free-body cuts may sometimes require considerably more subtlety than any discussed thus far. As an indication, consider the free-body shown in Figure 2.10*e,* which we will use to determine P_{FH}; it is obtained by imagining a cut through bars EG, DF, FH, FI, and GI. Admittedly, this is a very unusual choice of free-body; it is useful in this case because of the rather special geometry involved. Even though there are five unknown forces on this free-body, it is possible to evaluate P_{FH} because of the fact that the other four forces are concurrent. Forces P_{EG}, P_{DF}, P_{FI}, and P_{GI} all have lines of action that pass through point I; consequently, the equilibrium equation for sum of moments about point I contains only one unknown force, P_{FH}. Proceeding as described before, we resolve P_{FH} into horizontal and vertical components as shown in Figure 2.10*f*. The equilibrium equation is then

$$\sum M_{\mathrm{I}} = 0 = P_{\mathrm{FH}_y}\left(\frac{3}{2}l\right) - Fl = \left(\frac{4}{5}P_{\mathrm{FH}}\right)\left(\frac{3}{2}l\right) - Fl \tag{2.21a}$$

from which we obtain

$$P_{\mathrm{FH}} = \frac{5}{6}F \tag{2.21b}$$

EXAMPLE 4

Find the forces in bars GJ and GH of the truss in Figure 2.11*a*. Again some careful planning of the proper free-body diagram is required for the first bar force, and a different strategy is necessary for the second because P_{GH} cannot be determined as directly as in previous examples.

To find P_{GJ} we use the free-body in Figure 2.11*b*, obtained by passing a cut through bars GJ, GH, HI, and IK. There are four unknown forces on this free-body but, as in the previous example, three are concurrent. The equilibrium equation for sum of moments about point I is

$$\sum M_{\mathrm{I}} = 0 = P_{\mathrm{GJ}}\left(\frac{3}{2}l\right) - 2Fl - F(2l) \tag{2.22a}$$

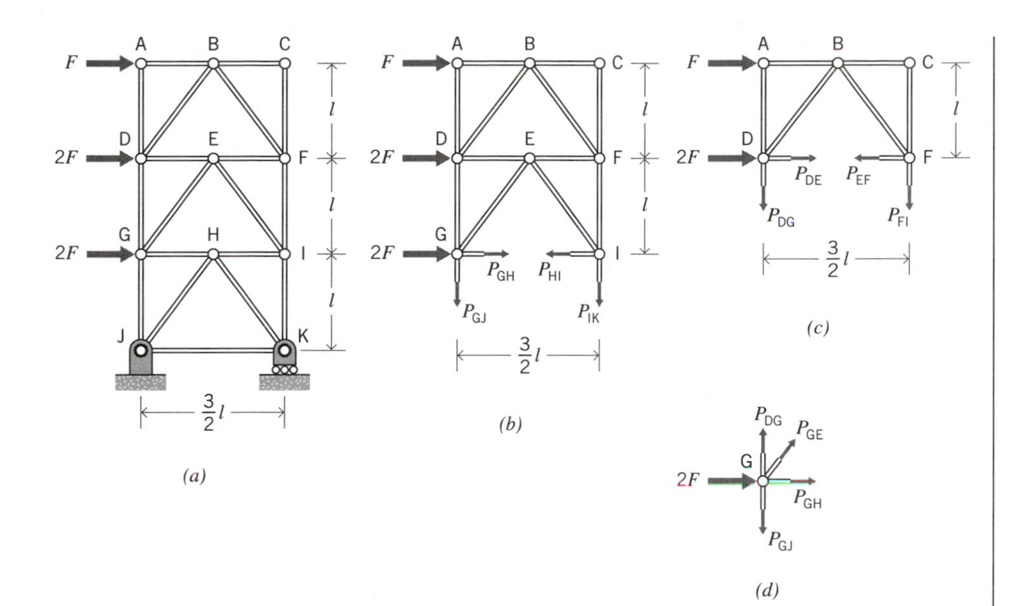

Figure 2.11 Example 4.

from which we get

$$P_{GJ} = \frac{8}{3}F \tag{2.22b}$$

There are several approaches that can be followed to determine P_{GH}, only one of which will be presented here. Since it was a fairly direct matter to calculate P_{GJ}, we can calculate P_{DG} in an analogous manner using the free-body in Figure 2.11*c*; then we will be able to determine P_{GH} from a free-body of joint G (Figure 2.11*d*). From the equilibrium equation for sum of moments about point F (Figure 2.11*c*) we get

$$\sum M_F = 0 = P_{DG}\left(\frac{3}{2}l\right) - Fl \tag{2.23a}$$

or

$$P_{DG} = \frac{2}{3}F \tag{2.23b}$$

The equilibrium equations for joint G (Figure 2.11*d*) are

$$\sum F_x = 0 = P_{GH} + 2F + P_{GE}\left(\frac{3}{5}\right) \tag{2.24a}$$

$$\sum F_y = 0 = P_{DG} - P_{GJ} + P_{GE}\left(\frac{4}{5}\right) \tag{2.24b}$$

from which, after substituting the expressions for P_{GJ} and P_{DG}, we find,

$$P_{GE} = \frac{5}{2}F \tag{2.24c}$$

$$P_{GH} = -\frac{7}{2}F \tag{2.24d}$$

EXAMPLE 5

The truss in Figure 2.12*a* has roller supports at A and B, and a pin support at D; this truss presents some special problems, which are worthy of examination.

There are four reactions (Figure 2.12*b*) and eight bar forces for a total of 12 unknowns. By writing two equilibrium equations at each of the six joints we will be able to determine all the unknowns. However, it is not possible to evaluate the four reactions using the equilibrium equations for the entire truss. Consequently, there is no joint that initially has fewer than three unknown forces. So, if we hope to avoid the necessity of solving 12 simultaneous equations, we will have to approach the problem using a slightly different strategy than we have used previously.

First, let's review what we do know about the reactions. The best we can do initially is determine R_{D_x} from equilibrium of forces in the x direction, i.e., $R_{D_x} = -F$; we cannot explicitly determine the three vertical reactions from the remaining two independent equilibrium equations. Knowing R_{D_x} doesn't help much because, including the vertical reaction, there are still three unknown forces at joint D. Similarly, as noted above, there are at least three unknown forces at every other joint. Apparently, then, attempting to analyze individual joints will be fruitless.

Consider, instead, the free-bodies in Figures 2.12*c* and 2.12*d*. The free-body in Figure 2.12*c* is obtained by cutting bars CB, CG, and ED; the free-body in Figure 2.12*d* is obtained by cutting bars CB, CG, and DG. Note that F is shown as an external load on the free-body in Figure 2.12*c*, but *not* on the free-body in Figure 2.12*d*; as long as we clearly delineate the free-bodies, there should be no ambiguity regarding such external loads.

Each of these free-bodies has four unknown forces, namely the three bar forces and a vertical reaction force. We cannot completely evaluate either set of forces from the three equilibrium equations applicable to each free-body. However, because two bar forces, P_{CB} and P_{CG}, are common to *both* free-bodies, it is possible, for this special geometry, to evaluate those two forces. We accomplish this by writing one equilibrium equation for *each* free-body, with the only unknown forces in each equation being P_{CB} and P_{CG}; because the two equations are independent, they can be solved simultaneously for the two unknowns.

We avoid determining the two unknown forces in which we are not interested by writing the equilibrium equation for sum of moments about the intersection of these two forces, point O in Figure 2.12*c* and point Q in Figure 2.12*d*. We will deal with the x and y components of P_{CB} and P_{CG}, which are given by

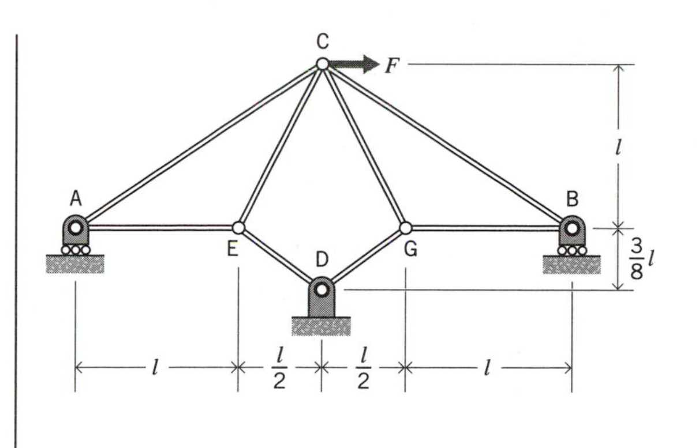

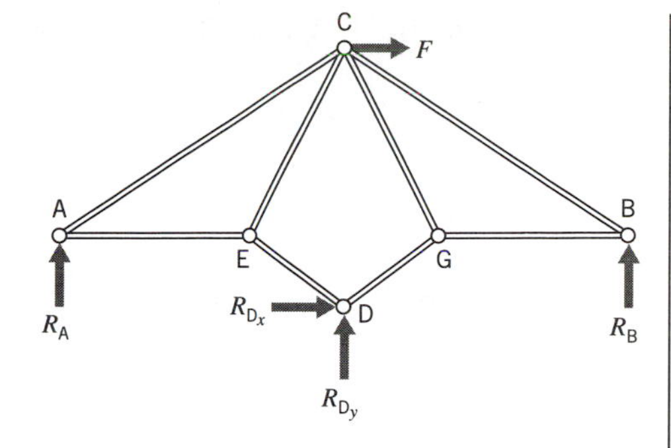

(a) (b)

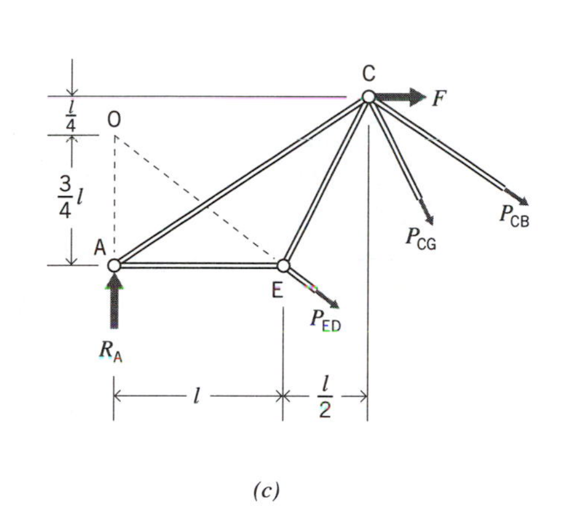

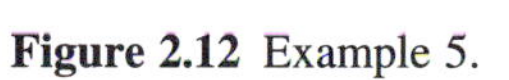

(c) (d)

Figure 2.12 Example 5.

$$P_{CB_x} = \frac{3}{\sqrt{13}} P_{CB} \tag{2.25a}$$

$$P_{CB_y} = \frac{2}{\sqrt{13}} P_{CB} \tag{2.25b}$$

$$P_{CG_x} = \frac{1}{\sqrt{5}} P_{CG} \tag{2.25c}$$

$$P_{CG_y} = \frac{2}{\sqrt{5}} P_{CG} \tag{2.25d}$$

From Figure 2.12*c* we get

$$\sum M_O = 0 = -F\frac{l}{4} - (P_{CB_x} + P_{CG_x})\frac{l}{4} - (P_{CB_y} + P_{CG_y})\frac{3l}{2} \tag{2.26a}$$

and from Figure 2.12*d*,

$$\sum M_Q = 0 = (P_{CB_x} + P_{CG_x})\frac{l}{4} - (P_{CB_y} + P_{CG_y})\frac{3l}{2} \qquad \textbf{(2.26b)}$$

In both equations, the components of the bar forces have been assumed to act at point C. Solving simultaneously leads to

$$P_{CB} = -\frac{11\sqrt{13}}{48}F \qquad \textbf{(2.27a)}$$

$$P_{CG} = \frac{3\sqrt{5}}{16}F \qquad \textbf{(2.27b)}$$

With P_{CB} and P_{CG} determined, it is a relatively straightforward matter to evaluate the remaining unknown forces by using either the method of joints or the method of sections.

EXAMPLE 6

The truss in Figure 2.13*a* has a slightly different geometry from the one in Figure 2.12*a*. (Note the vertical dimension $l/2$ which replaces $3l/8$.) The loads shown are representative of any arbitrary loads.

Figures 2.13*b* and 2.13*c* show the same free-bodies as Figures 2.12*c* and 2.12*d*, respectively; however, points O and Q now lie along the horizontal line through point C.

Summing moments about point O in Figure 2.13*b* now gives

$$\sum M_O = 0 = -F_2\left(\frac{3l}{2}\right) - (P_{CB_y} + P_{CG_y})\frac{3l}{2} \qquad \textbf{(2.28a)}$$

or

$$(P_{CB_y} + P_{CG_y}) = -F_2 \qquad \textbf{(2.28b)}$$

and summing moments about point Q in Figure 2.13*c* now gives

$$\sum M_Q = 0 = -(P_{CB_y} + P_{CG_y}) \qquad \textbf{(2.28c)}$$

These results do not appear to be consistent! We cannot have two different values for a quantity such as $P_{CB_y} + P_{CG_y}$. Because the two equations cannot both be true, the

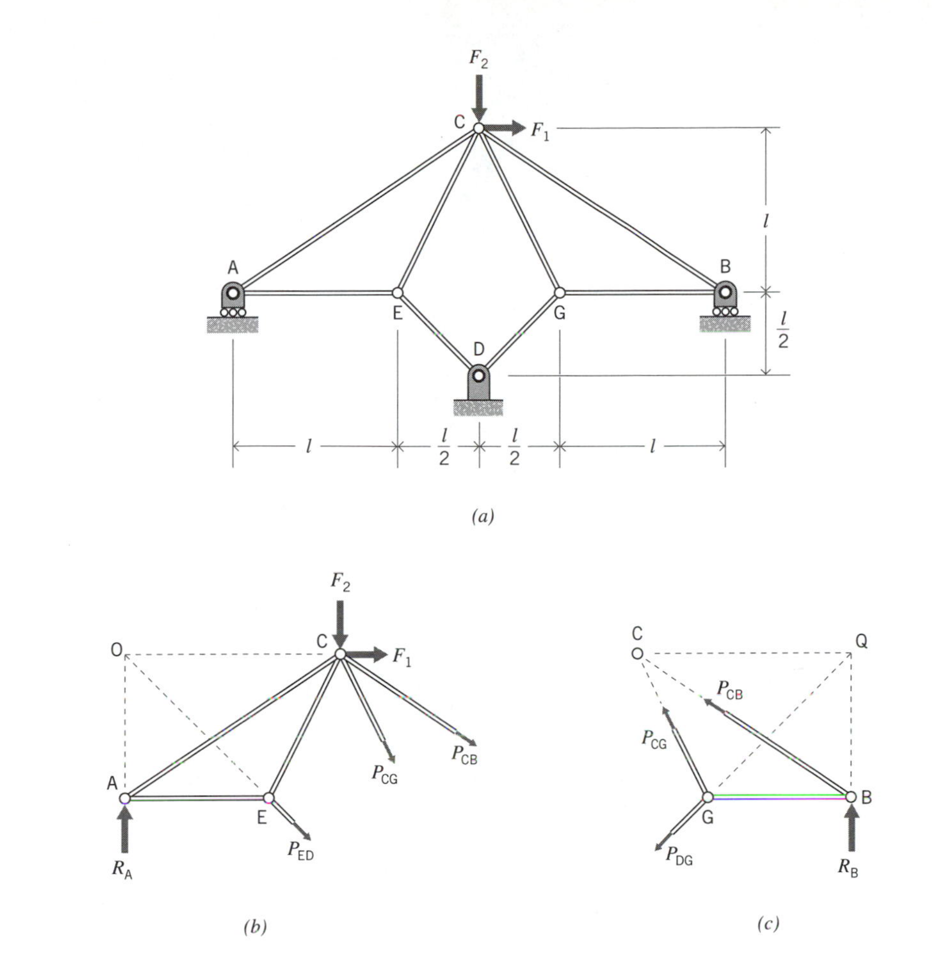

Figure 2.13 Example 6.

implication is that this structure must be statically unstable.[20] This is not obvious without some analysis (or, alternatively, experience with problems of this type); the instability is due to the particular geometry of the truss.

EXAMPLE 7

Consider the truss in Figure 2.14*a*; bars CB and AD are not connected at their midpoints. This truss has three reactions and six bars, for a total of nine unknowns. There are four joints, so we only have eight equilibrium equations. Therefore, we conclude that the

[20] This concept has been noted in Section 1.7.

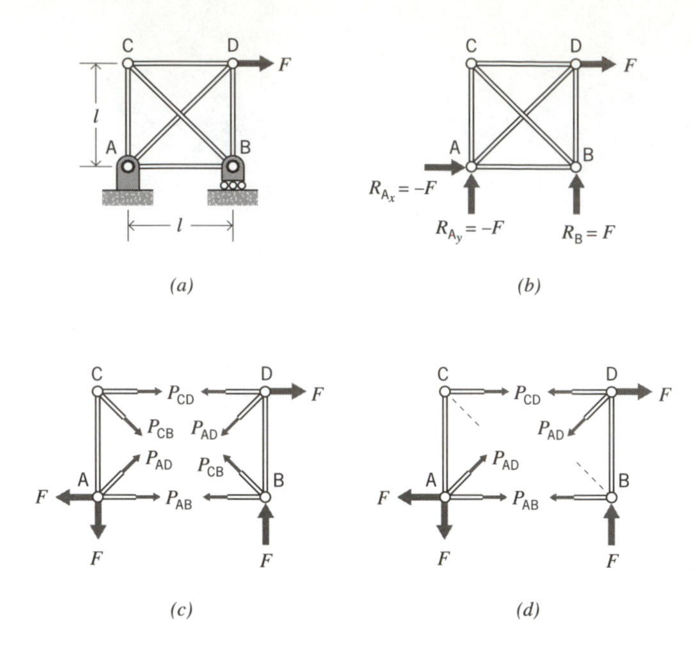

Figure 2.14 Example 7.

structure is 1-degree indeterminate. Even though we are able to evaluate the reactions, we cannot evaluate any of the bar forces from equilibrium considerations alone. We will need one additional equation to find the additional unknown. This "extra" unknown bar force, in fact, can be thought of as surplus or redundant. In other words, from the viewpoint of maintaining equilibrium of all the joints, the extra bar could be eliminated.

Even though the truss in Figure 2.14*a* is statically indeterminate, there is a special circumstance under which we can sometimes treat the structure as if it were statically determinate; that is, if we define bars AD and CB as "counters." *Counters* are pairs of diagonal bars that are assumed to buckle at very low values of compressive force and are regarded as incapable of transmitting compression, much like cables. Analysis requires only that we qualitatively identify which of the two counters would likely be subjected to compression under the specified loads, and effectively delete that bar from the truss.[21] We will illustrate the analysis of the truss in Figure 2.14*a* assuming that bars AD and CB are counters.

Figure 2.14*b* shows the reactions for the truss, and Figure 2.14*c* shows free-bodies obtained by imagining cuts through bars AB, CD, AD, and CB. By qualitatively examining either of these free-bodies, we conclude that bar AD must be in tension and bar CB would probably be in compression. We reach this conclusion by considering equilibrium of forces in the y direction on the free-body; in other words, we can satisfy

[21] It is possible for both bars to be in tension under some load conditions, in which case the truss must be treated as statically indeterminate.

equilibrium if bar AD is subjected to tension and bar CB is deleted, as in Figure 2.14*d*. For the free-body in Figure 2.14*d* we see that equilibrium requires $P_{AD_y} = F$, or $P_{AD} = \sqrt{2}\,F$. Analysis of the remaining bar forces is straightforward.

2.3 STATICAL STABILITY/INSTABILITY AND DETERMINACY/INDETERMINACY

As we have seen in the last two examples of the preceding section, it is important to be able to distinguish statically indeterminate trusses from determinate trusses and unstable trusses from stable trusses. In this section we will amplify these concepts, and restate a prior conjecture as a necessary condition for statical stability and determinacy.

Most of the basic concepts pertaining to indeterminacy and instability that were introduced in Chapter 1 are still applicable here. For example, we know that we must have at least three reactions for statical stability, and that these reactions must be arranged properly. Similarly, we would generally expect that a truss having supports that produce more than three reactions would be statically indeterminate; this is true for the truss shown in Figure 2.15*a* but, as we have seen, is not true for the special trusses in Figures 2.12 or 2.13. Despite such anomalies, it is generally easier to assess the characteristics of trusses than it is to assess frames of the type examined in Chapter 1; in fact, as was noted in Section 1.6, it is frequently convenient to use somewhat formal criteria for assessing basic characteristics of trusses.

The criteria we seek are based on the previously stated relationship between the number of reactions, number of bars, and number of joints. As a prelude to the formal statement of such criteria, let

$$NR = \textit{Number of Reactions}$$

$$NB = \textit{Number of Bars}$$

$$NJ = \textit{Number of Joints}$$

Based on what we have observed about the trusses that have been analyzed thus far, it is possible to conclude that if a truss is statically determinate and stable we must have

$$NR + NB = 2 \cdot NJ \qquad \textbf{(2.29)}$$

In other words, in order for there to exist a unique solution to the equations of static equilibrium, it is necessary for the number of independent equilibrium equations ($2 \cdot NJ$) to equal the total number of unknown forces ($NR + NB$). Although this may seem evident, it is not a sufficient condition to guarantee statical determinacy and stability; it is only a necessary condition.

To illustrate the latter point, we need only recall once again the truss in Figure 2.13*a*, which we found to be unstable despite the fact that the criterion stated in Equation 2.29 was satisfied. This is not the only possible type of violation of Equation 2.29, however; among many conceivable illustrations that are much less subtle is the truss shown in Figure 2.15*b*. The left portion of this truss is identical to the indeterminate truss in Figure 2.14*a*; the right portion of this truss is a mechanism that is incapable of main-

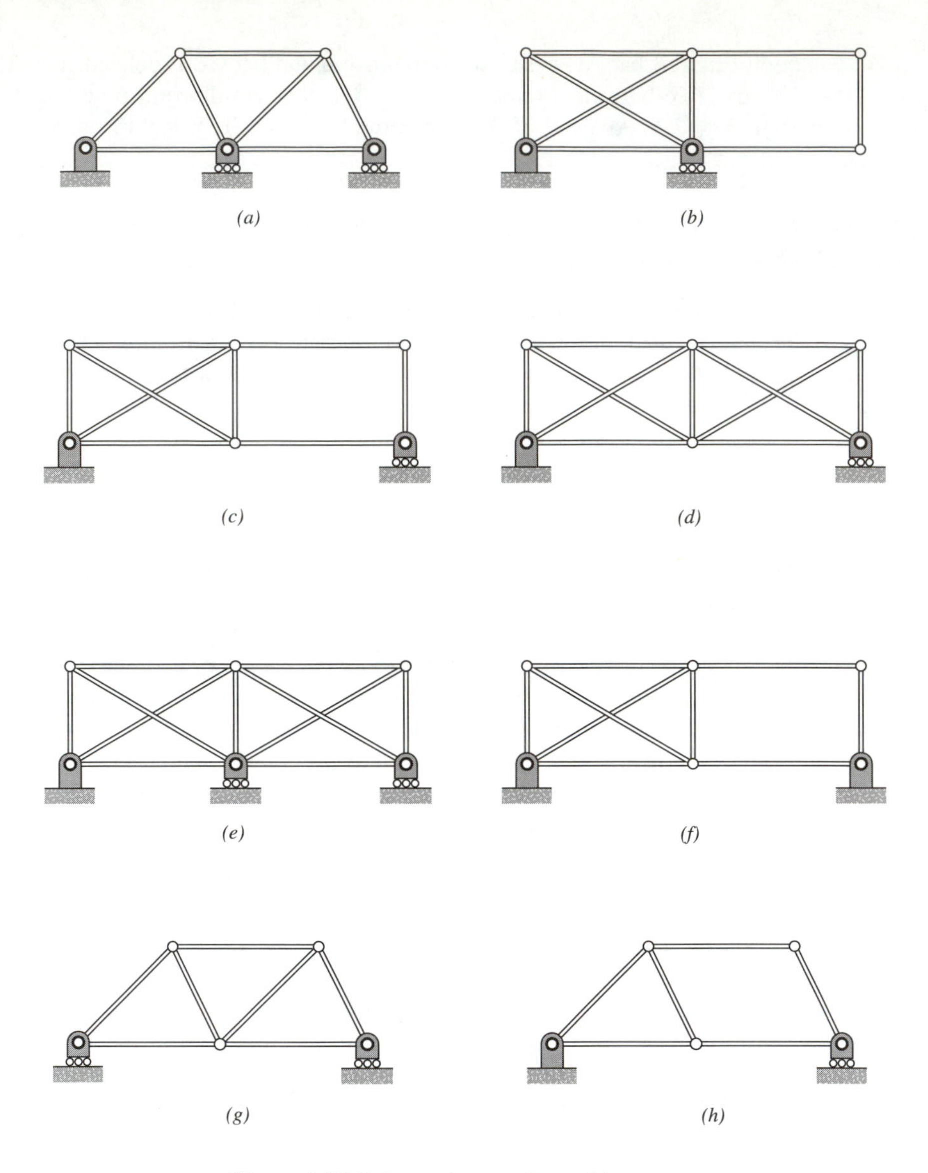

Figure 2.15 Indeterminate and unstable trusses.

taining its rectangular configuration under vertical loads. Consequently this structure is both indeterminate and unstable despite the fact that $NR = 3$, $NB = 9$, and $NJ = 6$, a set of numbers that satisfies Equation 2.29. Moving a support, as in Figure 2.15*c*, will not eliminate the instability;[22] in short, the criterion fails to reveal the true nature of this type of truss.

[22] See, for example, the free-body in Figure 2.9*c*; if the diagonal bar is removed, it becomes impossible to satisfy the equilibrium equation for sum of forces in the y direction.

As inferred previously, when there are more forces than equilibrium equations, it must follow that the truss is statically indeterminate, i.e., if

$$NR + NB > 2 \cdot NJ \tag{2.30}$$

then the truss is indeterminate. For example, the truss in Figure 2.14*a*, which has been examined previously, or the truss in Figure 2.15*a* for which $NR = 4$, $NB = 7$, and $NJ = 5$, both satisfy relation 2.30. Furthermore, the relation indicates the degree of indeterminacy as 1 in both these cases; i.e., for stable trusses, the integer by which $NR + NB$ exceeds $2 \cdot NJ$ is equal to the degree of indeterminacy. Thus, we are able to conclude that the truss in Figure 2.15*d* is 2-degrees indeterminate, and the one in Figure 2.15*e* is 3-degrees indeterminate. Although the trusses referred to here are stable, this still does not imply that all trusses that satisfy inequality 2.30 must be stable; as we have seen in Figures 2.15*b* and 2.15*c*, trusses can be both statically indeterminate and unstable. Figure 2.15*f* shows a truss with these characteristics; one reaction has been added to the right support of the truss in Figure 2.15*c*, but this does not change the fact that the truss is unstable even though it now satisfies the inequality above.

The converse of inequality 2.30 is

$$NR + NB < 2 \cdot NJ \tag{2.31}$$

and it follows that trusses that satisfy this inequality must be statically unstable because there are not enough forces to satisfy the equilibrium equations under general loading conditions. For example, the truss in Figure 2.15*g* has an insufficient number of reactions, $NR = 2$; since $NR + NB = 9$, and $NJ = 5$, the inequality reveals some type of instability. Similarly, for the truss in Figure 2.15*h*, $NR = 3$, $NB = 6$, and $NJ = 5$, revealing an instability, which in this case is due to an insufficient number of bars.[23] Although it may be largely irrelevant, a truss satisfying inequality 2.31 could still be statically indeterminate; an example is the truss in Figure 2.15*c* with the pin support replaced by a roller.

Despite their limitations, the foregoing criteria are actually quite useful for the assessment of truss characteristics. They are most useful when we are confident that we are dealing with stable trusses; then, we are able to easily identify determinate trusses or determine the degree of indeterminacy of indeterminate trusses. The latter is particularly important for subsequent developments.

2.4 DEFORMATIONS OF AXIALLY LOADED BARS

In the next section and in subsequent chapters we will undertake systematic examinations of truss displacements, which will require an understanding of the relationship between bar forces and deformations. For the sake of completeness, this basic relationship will be reviewed here. The development will include the effects of thermal and initial strains on deformation; an understanding of how to properly deal with these effects is becoming increasingly important in structural analysis. In addition to their

---

[23] *ibid.*

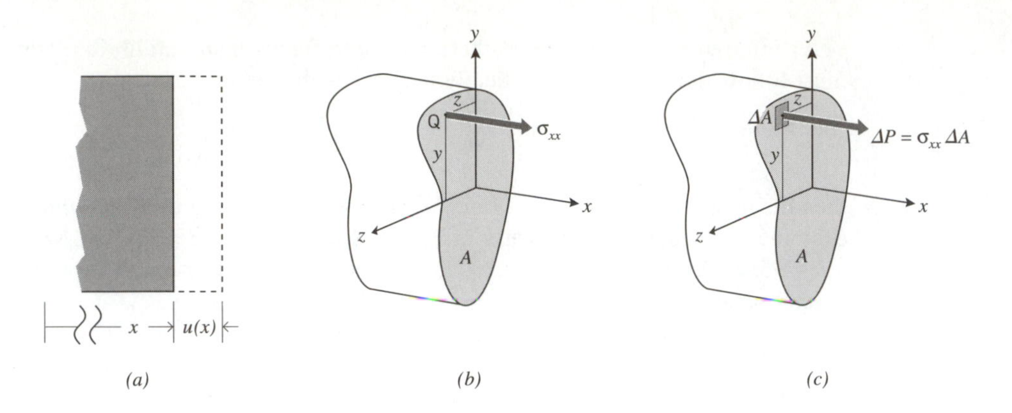

Figure 2.16 Displacement and stress in bar.

direct significance in truss analysis, the formulations will also provide a foundation for analogous developments, which will be essential in connection with the study of beams.

As is the case with the analysis of entire structures, the analysis of single bars requires some consideration of the three fundamental factors mentioned at the beginning of this book, i.e., 1) forces, 2) motion, and 3) the relation between forces and motion.

We will begin with the motion of a bar of the type shown in Figure 2.1, which may have variable cross section. We introduce a hypothesis regarding the general nature of bar deformations; specifically, we will assume that cross sections that are initially plane and parallel in the undeformed bar will remain plane and parallel after the bar is deformed.[24] Then the distance between any two cross sections will change when the bar deforms. We describe this relative movement of cross sections in terms of displacements of points on the cross sections measured with respect to the end of bar. Thus, for a typical cross section shown in Figure 2.16*a*, we will assume that as a result of deformation every point moves from its original coordinate, x, to a new coordinate $x + u(x)$, where $u(x)$ is the displacement in the direction of the x axis; we assume that points move only in the x direction, and that there are no components of displacement in the y or z directions. The displacement in the axial direction may be different at different cross sections, but the displacement of any point in the bar is a function only of its x coordinate.

Recall, now, that strains are defined in terms of partial derivatives of displacements;[25] it follows that the normal strain in the x direction, ϵ_{xx}, at any point in the bar is given by

$$\epsilon_{xx} = \frac{\partial u(x)}{\partial x} = \frac{du(x)}{dx} = u'(x) \tag{2.32}$$

---

[24] This is analogous to the Bernoulli-Euler hypothesis for beam deformations which will be reviewed in the next chapter.

[25] Specifically, if u, v, and w denote components of displacement in the x, y, and z directions, respectively, and if derivatives of displacements are small, then $\epsilon_{xx} = \partial u/\partial x$, $\epsilon_{yy} = \partial v/\partial y$, $\epsilon_{zz} = \partial w/\partial z$, $\epsilon_{xy} = (1/2)(\partial u/\partial y + \partial v/\partial x)$, $\epsilon_{xz} = (1/2)(\partial u/\partial z + \partial w/\partial x)$, and $\epsilon_{yz} = (1/2)(\partial v/\partial z + \partial w/\partial y)$. Here, terms of form ϵ_{ii} are normal, or axial, strains and terms of form ϵ_{ij}, $j \neq i$, are shear strains; related quantities, $\gamma_{ij} = 2\epsilon_{ij}$, are frequently referred to as engineering shear strains.

Since displacement is a function of x only, the partial derivative has been replaced by an ordinary derivative. Bear in mind that nothing has yet been said about the source of this strain; Equation 2.32 should be regarded as providing only a definition of total normal strain, ϵ_{xx}, at any point in the bar due to deformation. It is a consequence of our hypothesis that all components of strain other than ϵ_{xx} are equal to zero.[26]

The next factor to be considered is the nature of the resultant forces, specifically the relationship between internal stresses and resultant forces. The resultant of the internal stresses on any cross section must be exactly what has been assumed for the bar, namely the axial force illustrated in Figure 2.2 and nothing else.

Figure 2.16*b* shows a normal stress, σ_{xx}, which is assumed to act at a typical point Q on a cross section; the location of point Q is expressed in terms of y and z coordinates defined in the plane of the cross section, as shown in the figure. If we define an incremental area, ΔA, in the immediate vicinity of point Q, it follows that an increment of resultant force acting on this area is expressible as $\Delta P = \sigma_{xx}\Delta A$, as shown in Figure 2.16*c*. The resultant of all incremental forces acting on the cross section must equal $P(x)$ (Figure 2.2). In the limit as $\Delta A \rightarrow 0$, this resultant is given by

$$P(x) = \int_A \sigma_{xx} dA \qquad \textbf{(2.33a)}$$

where the integration is performed over the entire cross-sectional area A. There are also two other incremental force quantities of interest associated with ΔP, namely incremental moments about the y and z axes, i.e., $\Delta M_z = -y\Delta P$ and $\Delta M_y = z\Delta P$.[27] As noted, the resultant moments must equal zero, i.e.,

$$M_z(x) = -\int_A y\sigma_{xx} dA = 0 \qquad \textbf{(2.33b)}$$

$$M_y(x) = \int_A z\sigma_{xx} dA = 0 \qquad \textbf{(2.33c)}$$

As we will see shortly, a consequence of the formulation is that these two equations are identically equal to zero.

For a 1-dimensional element such as a bar, we require a relation between the single strain component, ϵ_{xx}, and the stress, σ_{xx}. In accordance with the concepts discussed in Section 1.3, we will assume linearly elastic material behavior. Here, however, we must recall that ϵ_{xx}, as defined in Equation 2.32, is the total axial strain, which may actually be a sum of several distinct constituents. In fact, we will assume that ϵ_{xx} is composed of three independent components, i.e.,

$$\epsilon_{xx} = \epsilon_E + \epsilon_T + \epsilon_I \qquad \textbf{(2.34a)}$$

where ϵ_E is *strain due to stress*, ϵ_T is *thermal strain*, and ϵ_I is *initial strain*.

---

[26] The hypothesis, therefore, gives incorrect values for ϵ_{yy} and ϵ_{zz}, because for an isotropic material under uniaxial stress, both must actually be equal to $-\nu\epsilon_{xx}$, where ν is Poisson's ratio.

[27] Here, moments are defined as positive using a right-hand rule with the coordinate axes shown in Figure 2.16*c*; consequently, a positive force ΔP acting in the 1st quadrant, as shown, causes a negative moment about the z axis and a positive moment about the y axis.

The relation between stress and the various components of strain is illustrated in Figure 2.17,[28] i.e., stress σ_{xx} is directly proportional to ϵ_E only. Specifically, $\sigma_{xx} = E\epsilon_E$, where E is Young's modulus. Also, thermal strain is assumed to be directly proportional to temperature change, ΔT, such that $\epsilon_T = \alpha\Delta T$; the constant α is the coefficient of linear thermal expansion, a property of a given material.[29] Initial strain may be regarded as any strain not related to stress or temperature; e.g., strain associated with an incorrect length of a bar. Thus, Equation 2.34*a* can be rewritten as

$$\epsilon_{xx} = \frac{\sigma_{xx}}{E} + \alpha\Delta T + \epsilon_I \tag{2.34b}$$

Solving for σ_{xx} leads to

$$\sigma_{xx} = E(\epsilon_{xx} - \alpha\Delta T - \epsilon_I) \tag{2.34c}$$

Equations 2.32 through 2.34 must now be combined to obtain the desired results. We begin by substituting Equation 2.32 into Equation 2.34*c* to get an expression for stress in terms of displacement,

$$\sigma_{xx} = E[u'(x) - \alpha\Delta T - \epsilon_I] \tag{2.35}$$

Next, substitute Equation 2.35 into Equation 2.33*a*,

$$\begin{aligned} P(x) &= \int_A E[u'(x) - \alpha\Delta T - \epsilon_I]dA \\ &= \int_A Eu'(x)dA - \int_A E\alpha\Delta T dA - \int_A E\epsilon_I dA \\ &= u'(x)\int_A E dA - \int_A E\alpha\Delta T dA - \int_A E\epsilon_I dA \end{aligned} \tag{2.36a}$$

Now, we will introduce another assumption about the material properties of the bar, namely that the cross section is homogeneous; consequently, we can assume that neither E nor α is a function of y or z, so both are constant with respect to the integration in Equation 2.36*a*. Equation 2.36*a* then becomes

$$\begin{aligned} P(x) &= Eu'(x)\int_A dA - E\alpha\int_A \Delta T dA - E\int_A \epsilon_I dA \\ &= EAu'(x) - E\alpha\int_A \Delta T dA - E\int_A \epsilon_I dA \end{aligned} \tag{2.36b}$$

---

[28] For elastic behavior, the order of occurrence of the strain components is irrelevant, e.g., Figure 2.17 could show ϵ_E first, followed by ϵ_I and ϵ_T. Also, stress-strain relationships are assumed to be identical for tension and comprehension.

[29] Use of the term "coefficient" is something of a misnomer since α is not a dimensionless quantity; the dimension of α is degrees^{-1}. It should also be remembered that the temperature range over which α is actually constant is limited and depends on the particular material.

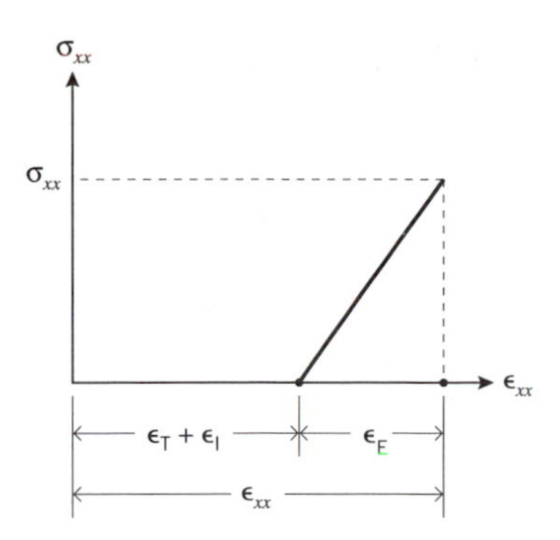

Figure 2.17 Stress-strain relation.

The two integrals in this expression cannot be evaluated until ΔT, ϵ_I, and the cross-sectional shape are specified; note that, in general, ΔT and ϵ_I may be functions of x, y, and z. To simplify the form of Equation 2.36*b*, we will establish the following definitions,

$$P_T(x) = E\alpha \int_A \Delta T dA$$
$$P_I(x) = E \int_A \epsilon_I dA \qquad \textbf{(2.36c)}$$

By these definitions, $P_T(x)$ and $P_I(x)$ denote resultant axial pseudoforces associated with thermal and initial strains, respectively; $P(x)$, on the other hand, is the actual resultant axial force associated with applied loads, i.e., the force that appears in the equilibrium equation. As a consequence of the integration process, P_T and P_I are functions of x only, as indicated.

Equation 2.36*b* can now be rewritten as

$$P(x) = EAu'(x) - P_T(x) - P_I(x) \qquad \textbf{(2.36d)}$$

or, solving for $u'(x)$,

$$u'(x) = \frac{P(x) + P_T(x) + P_I(x)}{EA} \qquad \textbf{(2.36e)}$$

Equation 2.36*e* shows the relation we have been seeking;[30] it is a differential equation relating strains and axial forces that can be integrated to obtain total deformation. We will perform the integration shortly, but first we must examine Equations 2.33*b* and 2.33*c* to complete the analysis.

[30] If we were interested in stress in addition to deformation, we would substitute Equation 2.36*e* into Equation 2.35 to obtain $\sigma_{xx} = [P(x) + P_T(x) + P_I(x)]/A - E\alpha\Delta T - E\epsilon_I$; when thermal and initial strains are zero, this reduces to the familiar expression $\sigma_{xx} = P(x)/A$.

Substituting Equation 2.35 into Equation 2.33*b* gives

$$\begin{aligned} M_z(x) = 0 &= -\int_A yE[u'(x) - \alpha \Delta T - \epsilon_I]dA \\ &= -Eu'(x)\int_A ydA + E\alpha \int_A y\Delta T dA + E\int_A y\epsilon_I dA \end{aligned} \tag{2.37}$$

The first integral above will be identically equal to zero provided we assume that the y and z axes are centroidal, which is equivalent to assuming that the x axis is a centroidal axis; this has been a tacit assumption all along. The second and third integrals above will reduce to the first integral if ΔT and ϵ_I are independent of y and z; for simplicity we will assume this to be the case for truss bars.[31] Thus, for truss bars of the type in Figure 2.1, Equation 2.33*b* will be identically equal to zero as long as ΔT and ϵ_I are functions of x only. A similar conclusion will be reached for Equation 2.33*c*.

For the special case of truss bars (Figure 2.4), we know that $P(x)$ is constant; it is also reasonable to assume that $P_T(x)$ and $P_I(x)$ are constant and, in fact, that A and E are also not functions of x. Under these circumstances, the integration of Equation 2.36*e* is straightforward, i.e.,

$$u'(x) = \frac{du}{dx} = \frac{P + P_T + P_I}{EA} \tag{2.38a}$$

or

$$\int_0^l du = \frac{P + P_T + P_I}{EA}\int_0^l dx \tag{2.38b}$$

So

$$u|_0^l = u(l) - u(0) = \Delta l = \frac{(P + P_T + P_I)l}{EA} \tag{2.38c}$$

Here, we identify $u(l) - u(0)$, the difference in displacements between the ends of the bar, as the change in length (deformation), Δl. Equation 2.38*c* may be rewritten as

$$\Delta l = \frac{P^* l}{EA} \tag{2.38d}$$

where

$$P^* = P + P_T + P_I \tag{2.38e}$$

If P_T and P_I are both equal to zero, $P^* = P$, and Equation 2.38*d* is the equation for the change in length of an elastic bar of uniform cross section due to a constant axial load.

---

[31] We will relax this assumption when we consider general beam/truss behavior.

EXAMPLE 8

A uniform bar of length l has a linear temperature variation along its length, i.e.,

$$T = T_o(1 + Cx) \tag{2.39a}$$

where T_o is ambient temperature, C is a constant (having dimension of length^{-1}), and coordinate x is measured from one end of the bar. Find the change in length.

We begin by finding the change in temperature as a function of x, i.e.,

$$\Delta T = T - T_o = T_o(1 + Cx) - T_o = T_oCx \tag{2.39b}$$

Then, from the first of Equations 2.36*c* we obtain

$$P_T(x) = E\alpha T_oCx \int_A dA = EA\alpha T_oCx \tag{2.40a}$$

and from Equation 2.36*e* we get the expected expression for $\epsilon_{xx}(x)$, i.e.,

$$u'(x) = \frac{P_T(x)}{EA} = \alpha T_oCx \tag{2.40b}$$

Integrating gives the desired result,

$$u|_0^l = \Delta l = \alpha T_oC \int_0^l xdx = \alpha T_oC \frac{x^2}{2}\Bigg|_0^l = \frac{\alpha T_oCl^2}{2} \tag{2.40c}$$

EXAMPLE 9

A uniform bar of length l, which is subjected to a uniform temperature increase ΔT, is prevented from expanding by being placed between immovable fixed supports at its ends. Find the force in the bar.

In this case, Equation 2.36*c* gives

$$P_T = EA\alpha\Delta T \tag{2.41}$$

Since ΔT is not a function of x, we use Equation 2.38*c*, i.e.,

$$\Delta l = 0 = \frac{(P + P_T)l}{EA} \tag{2.42a}$$

or

$$P = -P_T = -EA\alpha\Delta T \tag{2.42b}$$

Note that this bar is actually statically indeterminate, and that we could not have calculated the axial force without using knowledge of the deformation (zero in this case).

The solution given by Equation 2.42*b* is entirely consistent with an elementary analysis that can be performed using only the most basic concepts. Notably, we could first assume the bar is free to expand; the change in length would equal thermal strain times length, i.e., $\Delta l = \alpha\Delta T\, l$. Next, we impose an equal and opposite change in length by applying a compressive axial force, $P = -EA\Delta l/l = -EA\alpha\Delta T$ (Equation 2.38*d*).

The same type of analysis would be used for initial effects rather than thermal effects, except that initial effects would most likely be expressed directly in terms of a length change relative to a nominal value.

2.5 DIRECT CALCULATION OF TRUSS DISPLACEMENTS

Joints of trusses displace as a result of the bars of the truss changing length and, as we have seen, length changes are a consequence of forces or thermal or initial effects. Regardless of the source of the individual length changes, we continue to assume that all deformations are relatively small and do not significantly alter the basic geometry of the structure.

To gain some initial insight into the nature of truss displacements, we will demonstrate how these displacements may be determined directly from geometric considerations alone. The difficulties inherent in even the most basic of these computations should provide an appreciation of the need for the alternate approach, which virtual work will provide subsequently.

The geometric analysis of truss displacements generally begins at a pinned support; consider a single bar, AC, connected to such a support at its end A, as in Figure 2.18*a*. Assume that this bar undergoes a length change Δ_{AC},[32] which is small compared to l; Figure 2.18*a* shows the deformation greatly exaggerated for the sake of clarity. The unsupported end of the bar is now C′, as shown; however, in a truss, point C′ will not necessarily remain along the original line of bar AC. In general, the bar will rotate about point A; consequently, point C′ will end up somewhere on the arc, which point C′ describes when the bar rotates. Since we are assuming small deformations and displacements, it follows that the arc can be accurately represented by its tangent in the immediate vicinity of point C′; thus, the end of the bar can be assumed to lie on the line through point C′ perpendicular to the original orientation of AC. Figure 2.18*b* shows a typical orientation of the displaced bar. As with the deformation, the rotation as shown is greatly exaggerated. As we will see shortly, the actual location of point C′ in a truss depends on the deformations and displacements of all other bars that are

[32] Length changes are not restricted to elastic deformations. Note that Equation 2.38*d* applies only to elastic deformations.

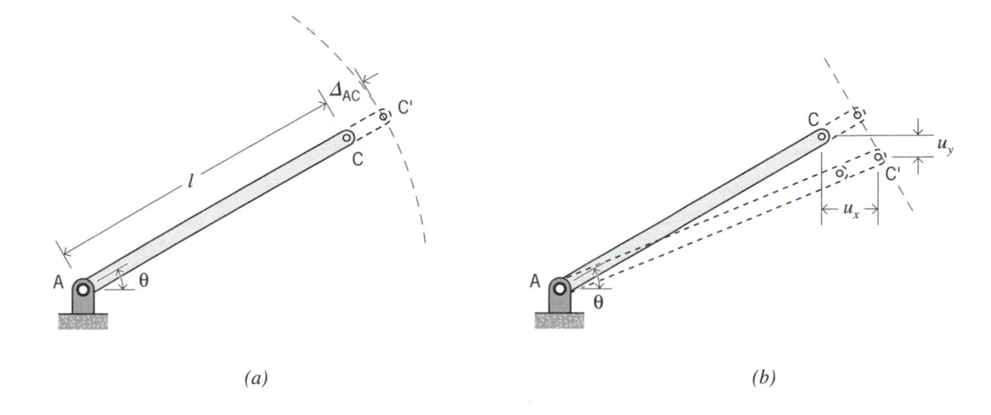

Figure 2.18 Displacement of bar.

connected to that end of bar AC. In any event, the x and y components of displacement, denoted in Figure 2.18b by u_x and u_y, respectively, are the quantities of interest at each joint.

Consider, now, the two-bar truss in Figure 2.7a, Example 1, for which there are pin supports at A and B. Assume that something (not necessarily load F) causes bars AC and CB to deform by amounts Δ_{AC} and Δ_{CB}, respectively. Figure 2.19a shows an exaggerated view of the displacement of joint C, which moves to C′ as a result of the bar deformations; as indicated in the figure, point C′ is the intersection of the perpendicular lines drawn through the end of each deformed bar. The actual relationship between the components of displacement of point C and the bar deformations requires a detailed analysis of the geometry of the quadrilateral that describes the displacement.

Figure 2.19b is an enlarged view of the quadrilateral in question; edge Cb corresponds to Δ_{CB}, the deformation of bar CB. As shown, this deformation is in the direction of bar CB. Similarly, edge Ca corresponds to Δ_{AC}, the deformation of bar AC. Edge bC′ is perpendicular to the end of deformed bar CB and, as noted before, represents the tangent direction in which the end of the bar moves. Edge aC′ is the direction in which

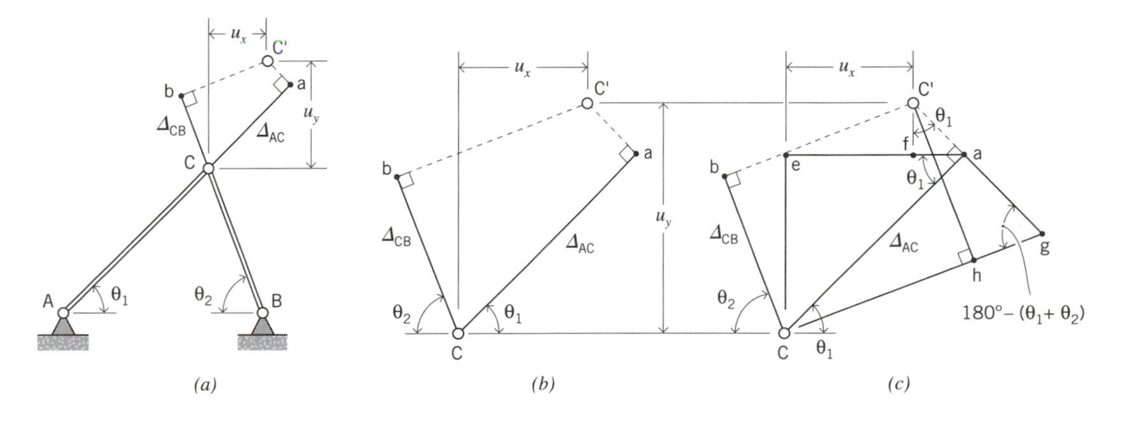

Figure 2.19 Truss displacements.

the end of deformed bar AC moves; point C′, the intersection of the two tangents, is thus the location of the displaced joint. We seek to describe the components of displacement, u_x and u_y, in terms of bar deformations, Δ_{AC} and Δ_{CB}, and truss geometry.

To accomplish this, we use the construction in Figure 2.19*c*; here, several ancillary lines have been added to quadrilateral CbC′a. Specifically, Ch and hC′ have been added parallel to bC′ and Cb, respectively, forming rectangle CbC′h in which hC′ is equal to Δ_{CB}; also Ch and C′a have been extended to meet at point g, and horizontal and vertical lines efa, Ce, and fC′ have been added, as shown. The magnitudes of several important angles are indicated. From this figure, we see that $u_y = \overline{Ce} + \overline{fC'}$ and $u_x = \overline{ae} - \overline{af}$, where the overbars denote lengths of line segments.

From triangle Cea, we obtain

$$\overline{Ce} = \Delta_{AC} \sin \theta_1 \tag{2.43a}$$

$$\overline{ae} = \Delta_{AC} \cos \theta_1 \tag{2.43b}$$

Similarly, from triangle ghC′ we get

$$\overline{gC'} = \frac{\Delta_{CB}}{\sin [180° - (\theta_1 + \theta_2)]} = \frac{\Delta_{CB}}{\sin (\theta_1 + \theta_2)} \tag{2.43c}$$

and, from triangle Cag,

$$\overline{ga} = \frac{\Delta_{AC}}{\tan [180° - (\theta_1 + \theta_2)]} = -\frac{\Delta_{AC}}{\tan (\theta_1 + \theta_2)} = -\frac{\Delta_{AC} \cos (\theta_1 + \theta_2)}{\sin (\theta_1 + \theta_2)} \tag{2.43d}$$

It follows that

$$\overline{aC'} = \overline{gC'} - \overline{ga} = \frac{\Delta_{CB} + \Delta_{AC} \cos (\theta_1 + \theta_2)}{\sin (\theta_1 + \theta_2)} \tag{2.43e}$$

Then, from triangle afC′ we get

$$\overline{fC'} = \overline{aC'} \cos \theta_1 = \frac{\Delta_{CB} \cos \theta_1 + \Delta_{AC} \cos \theta_1 \cos (\theta_1 + \theta_2)}{\sin (\theta_1 + \theta_2)} \tag{2.43f}$$

$$\overline{af} = \overline{aC'} \sin \theta_1 = \frac{\Delta_{CB} \sin \theta_1 + \Delta_{AC} \sin \theta_1 \cos (\theta_1 + \theta_2)}{\sin (\theta_1 + \theta_2)} \tag{2.43g}$$

From these equations we obtain the desired results, i.e.,

$$\begin{aligned} u_y = \overline{Ce} + \overline{fC'} &= \Delta_{AC} \sin \theta_1 + \frac{\Delta_{CB} \cos \theta_1 + \Delta_{AC} \cos \theta_1 \cos (\theta_1 + \theta_2)}{\sin (\theta_1 + \theta_2)} \\ &= \frac{\Delta_{AC} \cos \theta_2 + \Delta_{CB} \cos \theta_1}{\sin (\theta_1 + \theta_2)} \end{aligned} \tag{2.44a}$$

$$u_x = \overline{ae} - \overline{af} = \Delta_{AC} \cos\theta_1 - \frac{\Delta_{CB} \sin\theta_1 + \Delta_{AC} \sin\theta_1 \cos(\theta_1 + \theta_2)}{\sin(\theta_1 + \theta_2)}$$
$$= \frac{\Delta_{AC} \sin\theta_2 - \Delta_{CB} \sin\theta_1}{\sin(\theta_1 + \theta_2)} \tag{2.44b}$$

A considerable amount of geometric manipulation and computational effort was required for the derivation of u_x and u_y, even though this truss has only a single joint that undergoes displacement. The direct approach employed here becomes quite cumbersome even for minor increases in the complexity of the truss geometry. For example, the three-bar truss in Figure 2.5*a* has two joints that displace, C and B; the analysis of displacements of joint B is straightforward, but that of joint C is significantly more involved than in the previous case.

The major complication in any geometric analysis of truss displacements is probably in the definition of the displacement pattern. Figure 2.20*a* shows an overview of the displacements of the truss in Figure 2.5*a*. The roller support at joint B has one kinematic degree of freedom, i.e., displacement in the direction of the horizontal surface; since joint A is pinned, this displacement must be equal to the deformation of bar AB, Δ_{AB}. The deformations of bars AC and CB are also shown in the original directions of the bars; the final position of joint C will once again be at the intersection of the tangents to the ends of these bars, point C″ in Figure 2.20*a*.

Figure 2.20*b* is a detailed sketch of the geometry from which the displacement components u_x and u_y can be calculated; here, Cdb′C″a is an enlarged view of the motion depicted at joint C in Figure 2.20*a*. This displacement is shown superimposed on the quadrilateral from Figure 2.19*b*. This enables us to determine the location of point C″ relative to point C′ from the previous case. Specifically, for the truss in Figure 2.20*a*, horizontal displacement of joint C equals the horizontal displacement of point C′ plus increment Δu_x shown, and vertical displacement equals the vertical displacement of point C′ minus increment Δu_y shown. To calculate Δu_x and Δu_y, we need only consider triangle rC′C″, which is enlarged in Figure 2.20*c*.

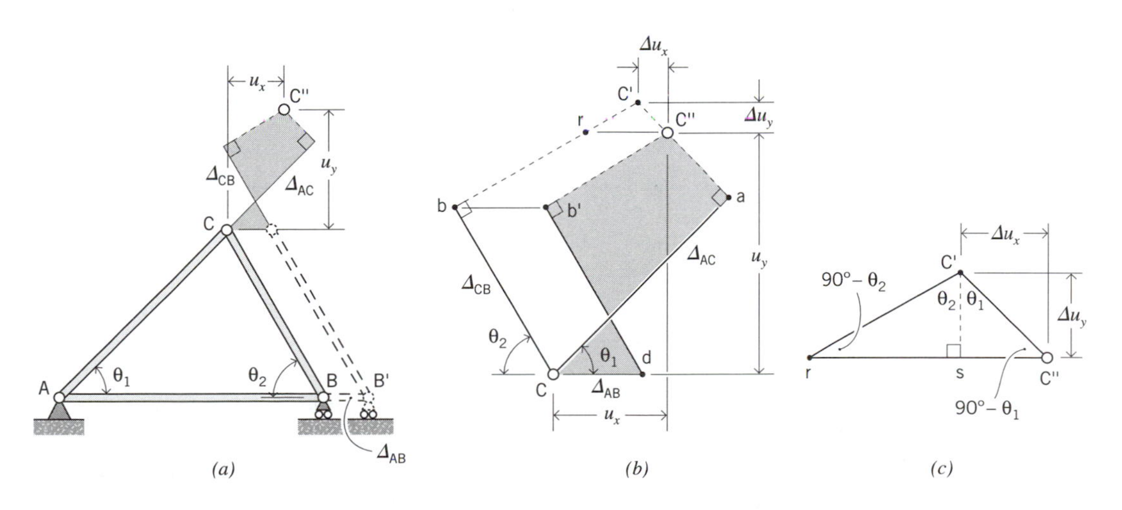

Figure 2.20 Three-bar truss displacements.

Since db′ in Figure 2.20*b* is parallel to Cb, and b′C″ is parallel to bC′, it follows that rC″ is equal to Cd, which is the same as Δ_{AB}. Therefore, from Figure 2.20*c* we obtain

$$\overline{rs} = \Delta u_y \tan\theta_2 = \Delta u_y \frac{\sin\theta_2}{\cos\theta_2} \tag{2.45a}$$

$$\overline{sC''} = \Delta u_y \tan\theta_1 = \Delta u_y \frac{\sin\theta_1}{\cos\theta_1} \tag{2.45b}$$

so,

$$\Delta_{AB} = \overline{rs} + \overline{sC''} = \Delta u_y\left(\frac{\sin\theta_2}{\cos\theta_2} + \frac{\sin\theta_1}{\cos\theta_1}\right) = \Delta u_y \frac{\sin(\theta_1 + \theta_2)}{\cos\theta_1 \cos\theta_2} \tag{2.45c}$$

or

$$\Delta u_y = \frac{\Delta_{AB}\cos\theta_1\cos\theta_2}{\sin(\theta_1 + \theta_2)} \tag{2.45d}$$

Also,

$$\Delta u_x = \Delta u_y \tan\theta_1 = \Delta u_y \frac{\sin\theta_1}{\cos\theta_1} = \frac{\Delta_{AB}\sin\theta_1\cos\theta_2}{\sin(\theta_1 + \theta_2)} \tag{2.45e}$$

As noted before, the final result for u_x is obtained by adding Δu_x to the right side of Equation 2.44*b*, i.e.,

$$u_x = \frac{\Delta_{AC}\sin\theta_2 - \Delta_{CB}\sin\theta_1 + \Delta_{AB}\sin\theta_1\cos\theta_2}{\sin(\theta_1 + \theta_2)} \tag{2.46a}$$

and for u_y, by subtracting Δu_y from the right side of Equation 2.44*a*, i.e.,

$$u_y = \frac{\Delta_{AC}\cos\theta_2 + \Delta_{CB}\cos\theta_1 - \Delta_{AB}\cos\theta_1\cos\theta_2}{\sin(\theta_1 + \theta_2)} \tag{2.46b}$$

EXAMPLE 10

The truss in Figure 2.5*a* is composed of bars of identical material and cross-sectional area, *A*. Find the displacement of joint C due to load *F* shown. Assume elastic behavior.

This is the same truss for which we have just derived relations for displacements due to arbitrary bar deformations, Equations 2.46*a* and 2.46*b*. For this case, with load *F* applied to the truss, the length changes are due to the internal bar forces; these internal forces were previously determined in Equations 2.3*a*, 2.3*b*, and 2.5*c*. For elastic behavior, the length changes are found from Equation 2.38*d*. We will express results in terms of modulus of elasticity, *E*, and dimension *l* defined in the figure.

From the geometry of the truss and Equation 2.8*b*, we obtain

$$l_{AB} = l$$

$$l_{AC} = \frac{h}{\sin \theta_1} = \frac{l \sin \theta_2}{\sin (\theta_1 + \theta_2)} \qquad \textbf{(2.47}\textbf{\textit{a}}\textbf{)}$$

$$l_{CB} = \frac{h}{\sin \theta_2} = \frac{l \sin \theta_1}{\sin (\theta_1 + \theta_2)}$$

Substituting these relations and Equations 2.3*a*, 2.3*b*, and 2.5*c* into Equation 2.38*d* leads to

$$\Delta_{AB} = \frac{Fl}{EA}\left[\frac{\sin \theta_1 \cos \theta_2}{\sin (\theta_1 + \theta_2)}\right]$$

$$\Delta_{AC} = \frac{Fl}{EA}\left[\frac{\sin^2 \theta_2}{\sin^2 (\theta_1 + \theta_2)}\right] \qquad \textbf{(2.47}\textbf{\textit{b}}\textbf{)}$$

$$\Delta_{CB} = -\frac{Fl}{EA}\left[\frac{\sin^2 \theta_1}{\sin^2 (\theta_1 + \theta_2)}\right]$$

Substituting Equations 2.47*b* into Equations 2.46 will provide the desired solution.

EXAMPLE 11

Assume that the truss in Figure 2.5*a* has no external loads applied to it, but that the length of member AB is shorter than its nominal value l by a small amount, Δ_o. Find the displacement of joint C from its nominal position.

This example involves an initial deformation due to unspecified causes. The initial deformation is stated directly in terms of a length change, so it is not necessary to use Equation 2.38. In this case, $\Delta_{AC} = \Delta_{CB} = 0$, and $\Delta_{AB} = -\Delta_o$ where the minus sign denotes a shortening of the bar rather than an elongation. Substituting these values into Equations 2.46*a* and 2.46*b* gives the desired results. Note that u_x is a negative quantity, as might be expected from a contraction of bar AB. However, u_y is not zero despite the fact that there is no deformation of bars AC and CB; in fact, u_y is a positive quantity. In other words, joint C displaces to the left and upwards as a result of the deformation of bar AB.

As noted previously, the total amount of computation required to obtain analytic expressions for displacements, such as those in Equations 2.46*a* and 2.46*b*, is quite considerable. The geometric aspects of the analysis are not only more formidable than the equilibrium and deformation analyses that are used to determine length changes of individual bars, but are also totally dissociated from these analyses. The character and laboriousness of the geometric analysis raises the prospect that direct calculation of displacements will be infeasible for all but the simplest trusses.[33]

Fortunately, there are alternative approaches to the analysis of truss displacements, most notably virtual work, which will be introduced in Chapter 4; as will be seen, virtual work requires only equilibrium and deformation analyses, thus completely obviating any need for the particular type of geometric analysis undertaken here. Despite its disadvantages, however, results obtained from this direct approach provide a useful basis for comparison with subsequent analyses. Furthermore, the inverse of this direct approach is utilized in the displacement method (Chapter 8); there, element deformations will be related to joint motions by a similar but generally simpler type of geometric analysis.

PROBLEMS

2.P1–P8. Find all the bar forces for the structure shown.

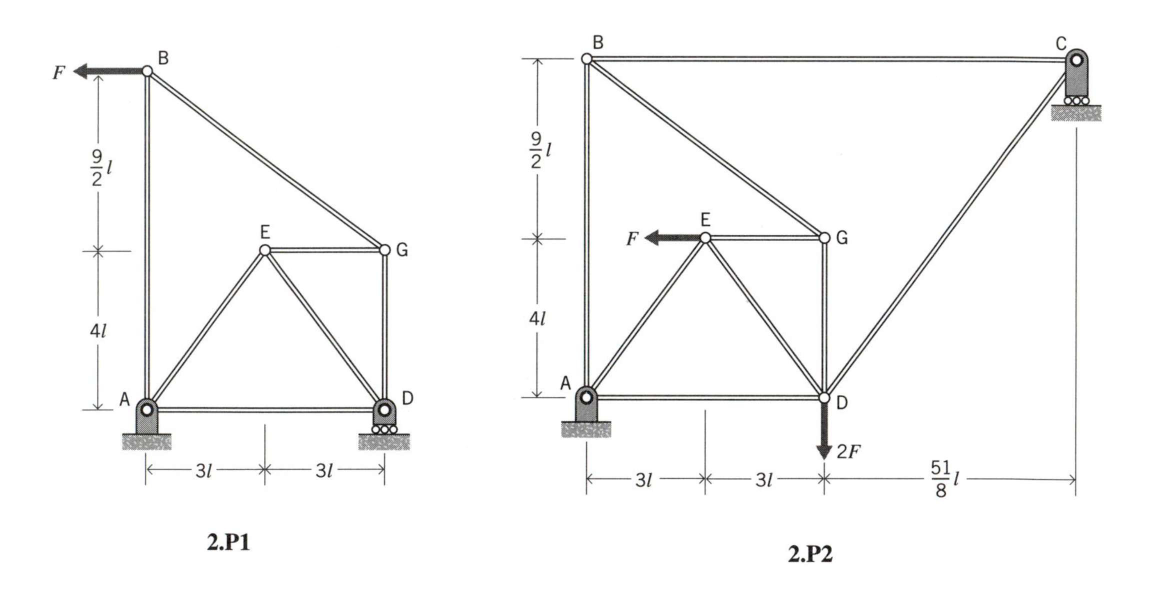

2.P1

2.P2

[33] Before the advent of modern analytic techniques, this very complex problem of finding joint displacements from bar elongations was approached using a graphical construction technique due to Williot (1877) and Mohr (1887). Construction of Williot-Mohr diagrams was one of the most difficult portions of courses on elementary structural analysis for students and instructors alike!

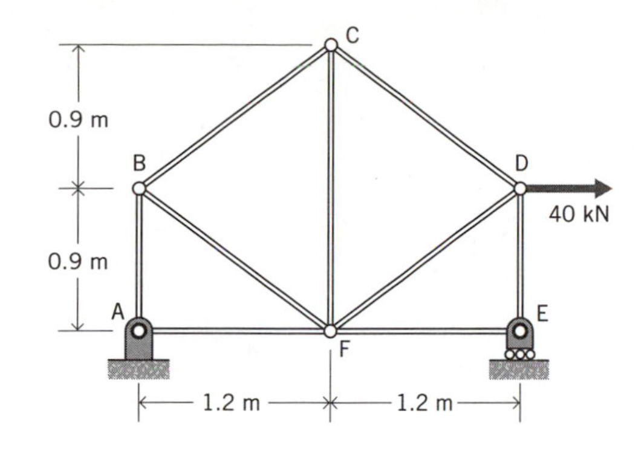
C
0.9 m
B
D
40 kN
0.9 m
A
E
F
1.2 m
1.2 m

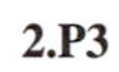
2.P3

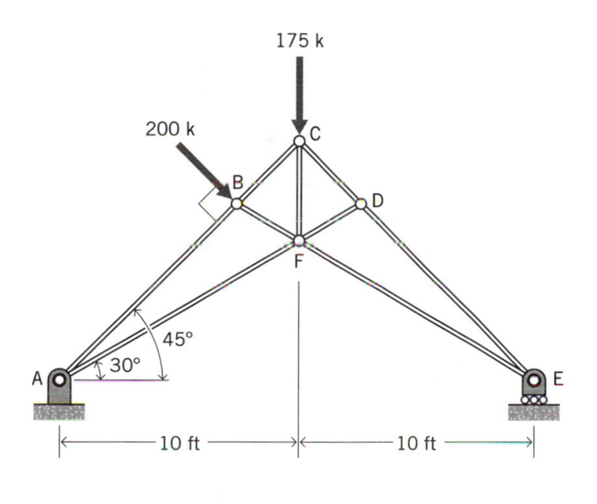
175 k
200 k
C
B
D
F
45°
30°
A
E
10 ft
10 ft

2.P4

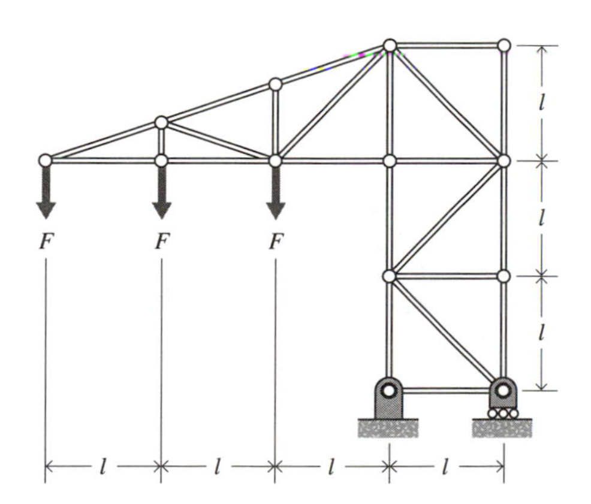
F
F
F
l
l
l
l
l
l
l

2.P5

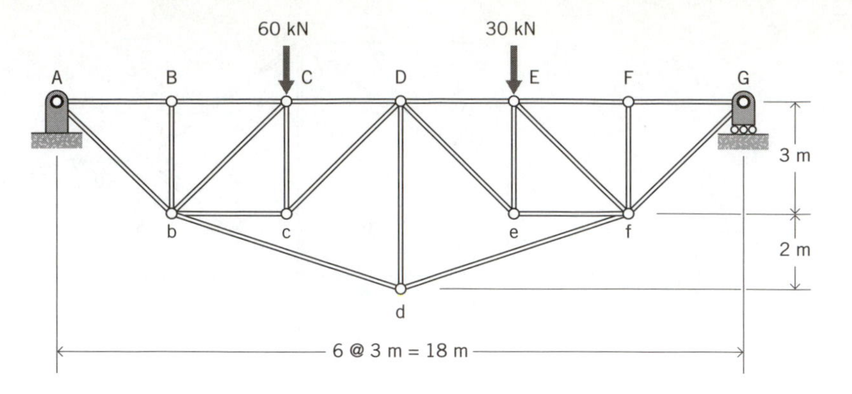
60 kN
30 kN
A
B
C
D
E
F
G
3 m
b
c
e
f
2 m
d
6 @ 3 m = 18 m

2.P6

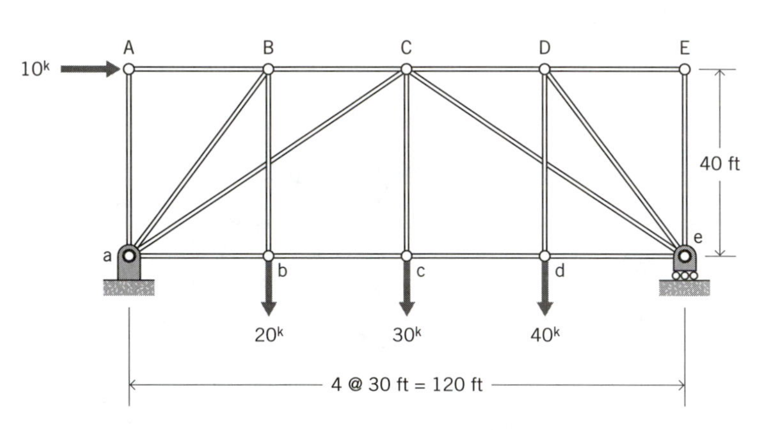
10k
A
B
C
D
E
40 ft
a
b
c
d
e
20k
30k
40k
4 @ 30 ft = 120 ft

2.P7

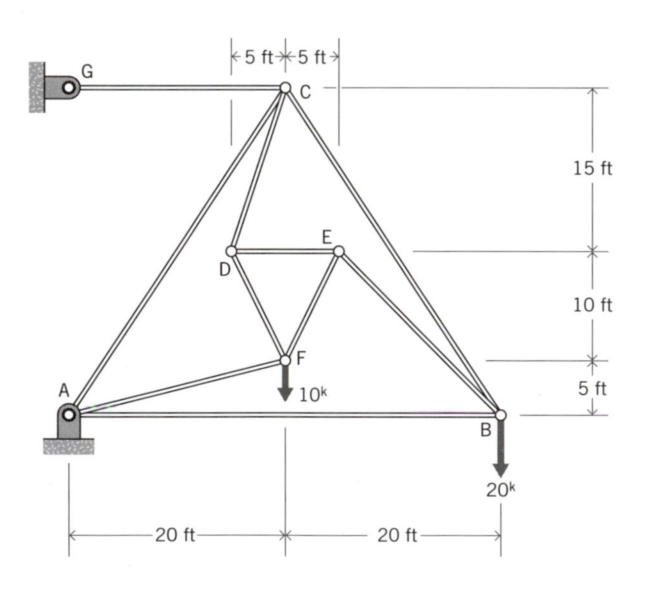
G
5 ft
5 ft
C
15 ft
E
D
10 ft
F
10k
5 ft
A
B
20k
20 ft
20 ft

2.P8

2.P9–P13. Find all the bar forces for the truss.

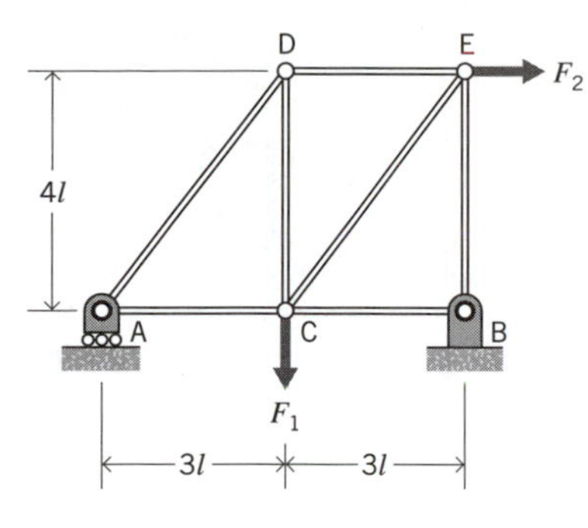

2.P9

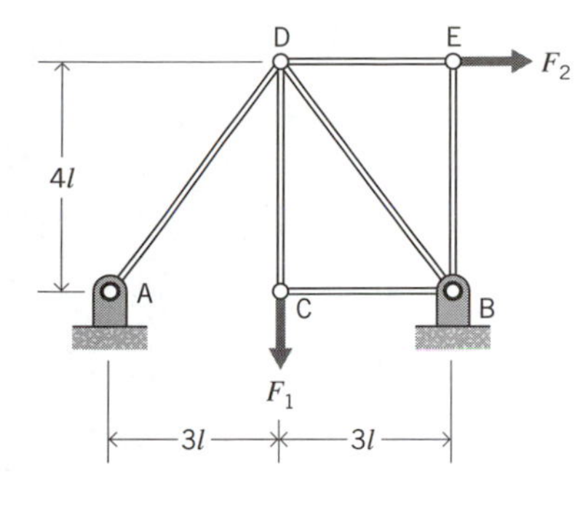

2.P10

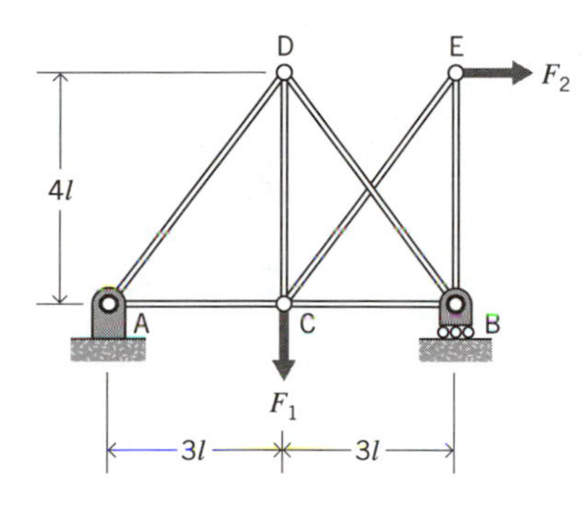

2.P11

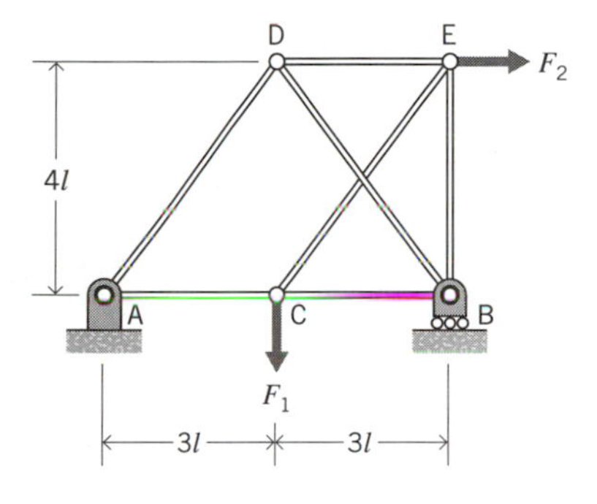

2.P12

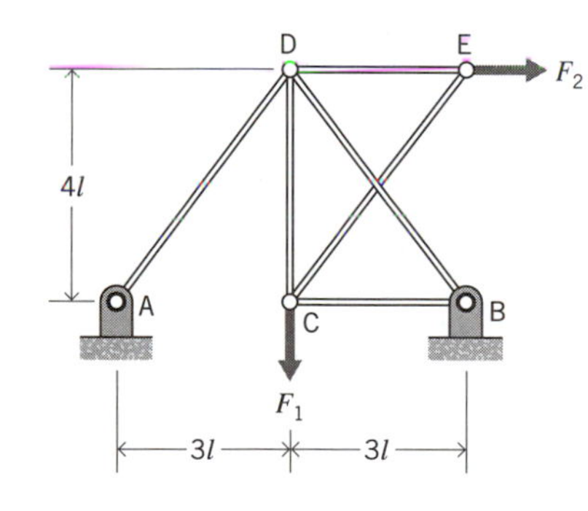

2.P13

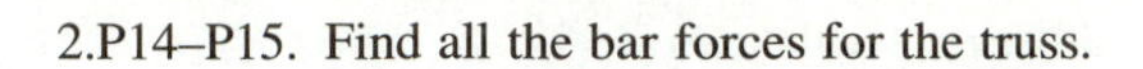

2.P14–P15. Find all the bar forces for the truss.

2.P14

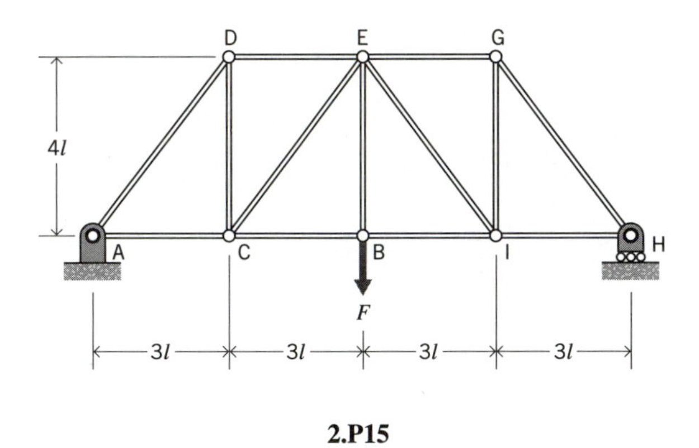

2.P15

2.P16–P19. Evaluate the forces in the lettered bars of the truss.

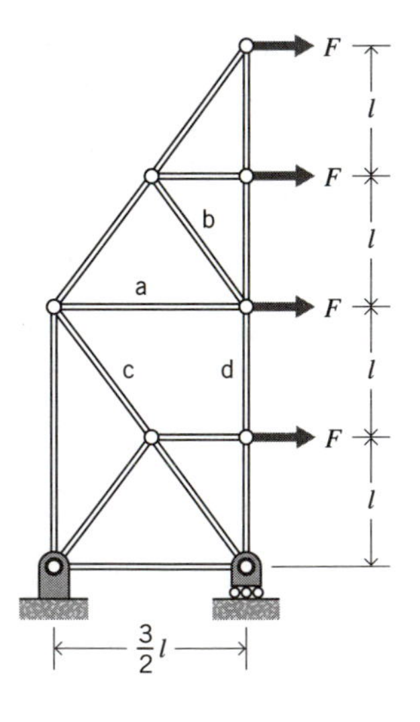

2.P16

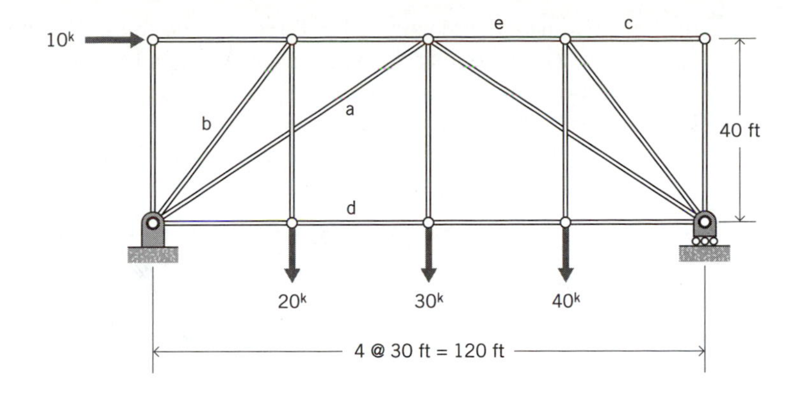

2.P17

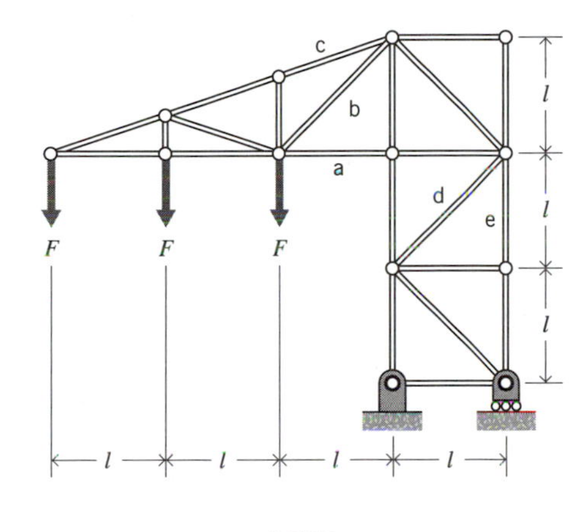

2.P18

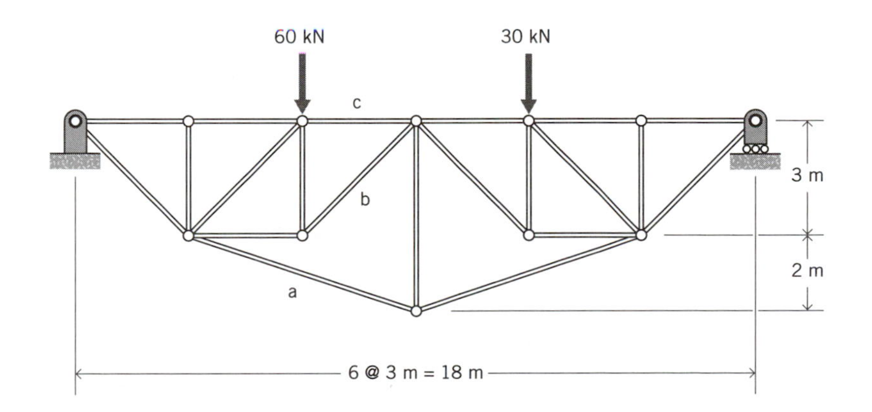

2.P19

2.P20. Evaluate the remaining bar forces for the truss of Example 5 (Figure 2.12). Verify that the reactions at A, B, and D satisfy equations of equilibrium for the entire truss.

2.P21. (a) Evaluate all the bar forces.
(b) Find the bar forces if the dimension h is changed to 0.55 m.

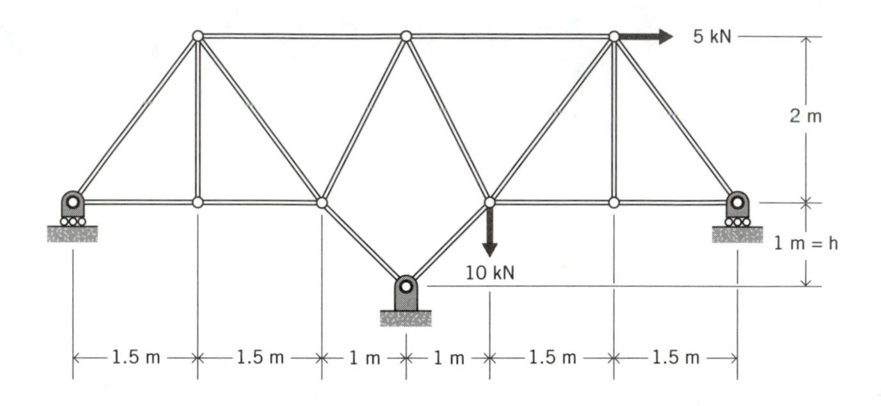

2.P22. (a) Is the structure statically determinate? Explain your answer.
(b) Is the structure statically stable? Explain your answer.
(c) Find the forces in bars FG, GK, and GJ.

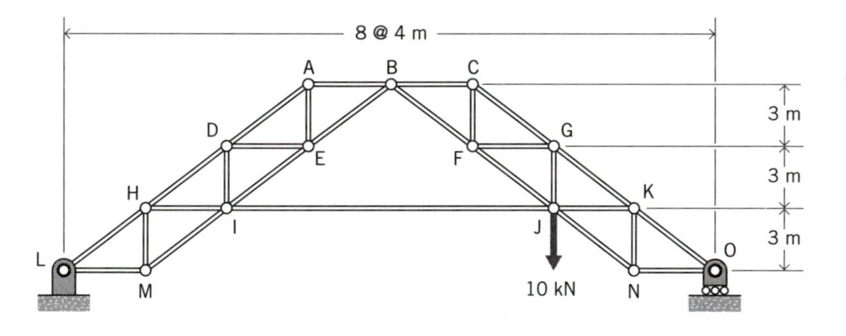

2.P23. Are the trusses stable? Explain your answer.

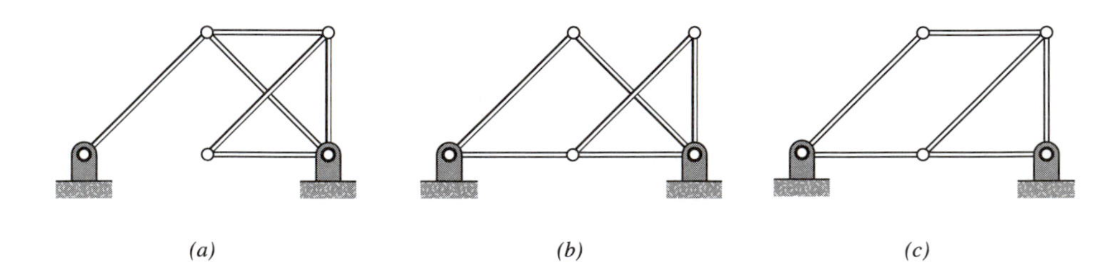

2.P24. Find the degree of indeterminacy for the structures shown.

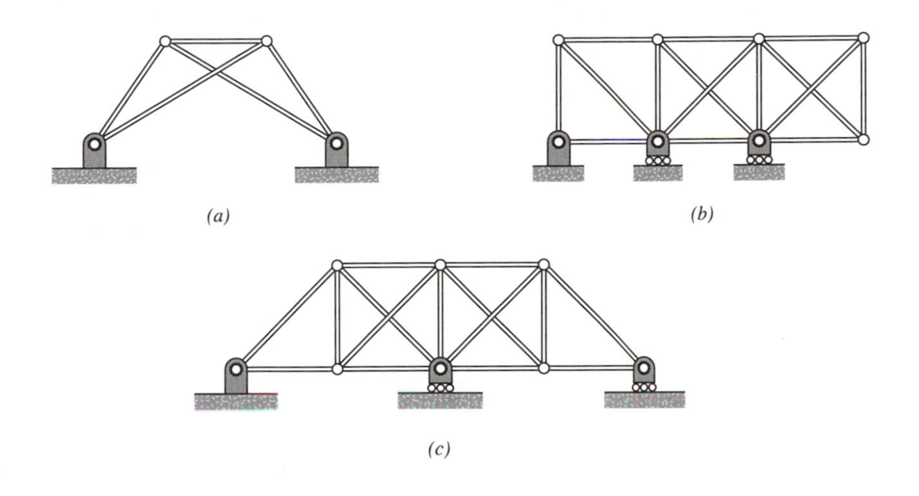

2.P25. Find the degree of indeterminacy for the structures shown.

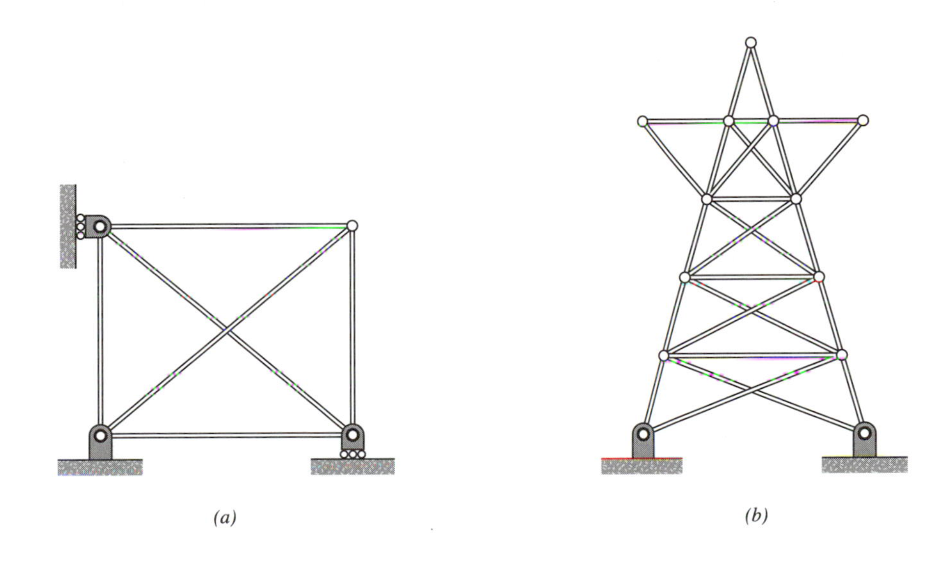

2.P26. The uniform bar shown is subjected to a continuous axial load that varies linearly from 0 at the left end to p_o at the right end. If the bar is restrained only at the left end, find the elongation.

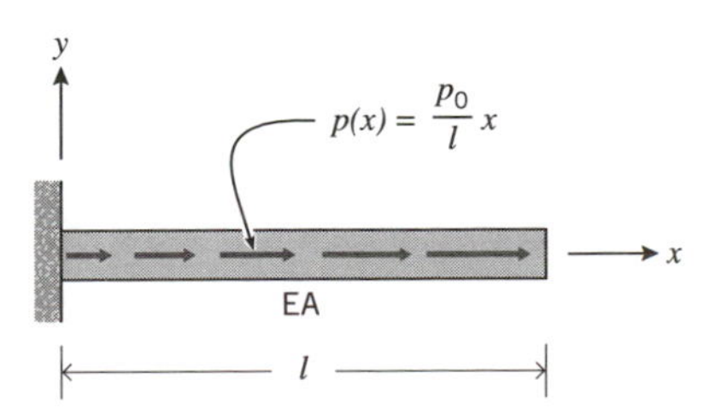

2.P27. The structure shown is composed of three uniform bars (AD, AE, and BE) having the same cross-sectional properties EA, and a rigid beam ABC. The beam may be assumed to be weightless.

(a) Find the forces in the bars.

(b) Find the displacement of point C.

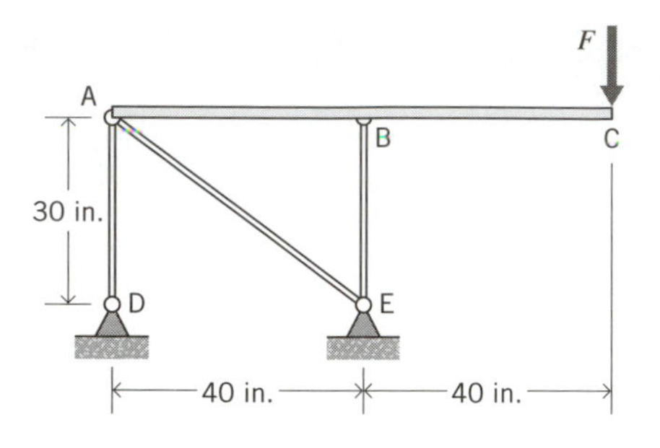

2.P28. (a) For the truss in Figure a, it is known that bar 1 elongates by an amount Δ_1 and bar 2 elongates by an amount Δ_2. Find the horizontal and vertical components of displacement, u_x and u_y, of joint A in terms of Δ_1 and Δ_2.

(b) For the truss in Figure b, it is known that the horizontal and vertical components of displacement of joint A are u_x and u_y, respectively. Find the corresponding length changes of bars 1, 2, and 3 in terms of u_x and u_y. (Hint: consider each component of joint displacement separately.)

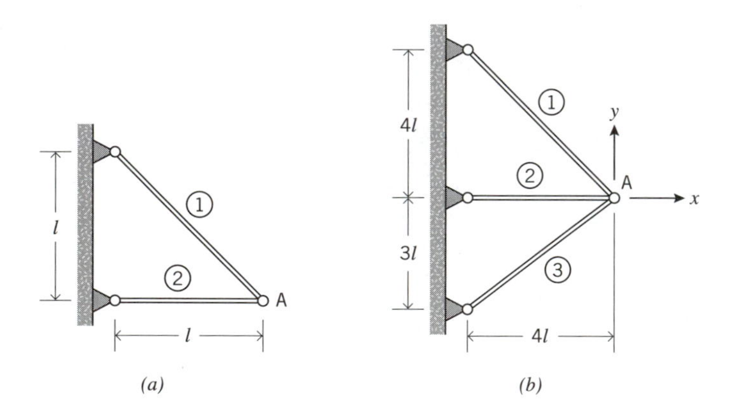

Chapter 3

Statically Determinate Frames

3.0 INTRODUCTION

We now turn our attention to a study of internal forces and deformations for statically determinate frame structures. These structures are the same type of 2-dimensional frames we discussed qualitatively in portions of Chapter 1. As was done for bars in the preceding chapter, we will first review the differential equations of equilibrium for beams; this will lead naturally to a review of procedures for evaluating and displaying shears and bending moments. We will then consider beam deformations, specifically the differential equation for transverse displacement as a function of internal forces, the equivalent of Equation 2.36*e*. Finally, this equation will be used for the direct computation of beam and frame displacements.

3.1 EQUILIBRIUM OF BEAMS

Consider a general slender beam with straight centroidal axis, x, similar to the bar in Figure 2.1, which is in equilibrium under the action of transverse loads, including moments. We assume end loads consisting of moments about the z axis and transverse forces in the y direction, and an intermediate nonuniform continuous load that acts in the y direction.[1] As shown in Figure 3.1, the end moments are denoted by M_A and M_B, and the end forces by V_A and V_B; the continuous load is denoted by $q(x)$. Figure 3.1 defines the positive directions for these quantities. Note, in particular, that V_A, V_B, and $q(x)$ are defined as positive when they act in the positive y direction, and that clockwise moments, M_A and M_B, are defined as positive.

Equilibrium requires that there be internal resultant forces at each cross section of the beam. These resultant forces consist of a bending moment (about the z axis), $M(x)$, and a transverse shear (in the y direction), $V(x)$. Figure 3.2 shows what we are defining as the positive senses of these internal force quantities.[2] The sign convention established for the internal forces is independent of the sign convention for the end forces.

---

[1] Intermediate concentrated loads will be routinely incorporated in ensuing computations, but are not included here because they are not material to immediate developments.

[2] As with the sign convention for internal forces in bars, we define as a positive resultant force one that acts in a positive coordinate direction on a surface whose outward normal is the positive x direction; thus, on the surface whose outward normal is the positive x direction, $V(x)$ is positive because it acts in the positive y direction, and $M(x)$ is positive because its vector acts in the positive z direction according to the right-hand rule.

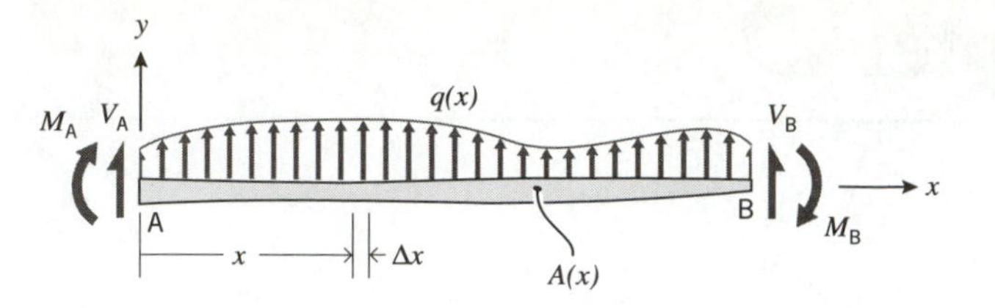

Figure 3.1 Transversely loaded beam.

Consider a free-body diagram of a portion of the beam of small length Δx (Figures 3.1 and 3.3). The forces acting on this element are the resultants $M(x)$ and $V(x)$ on the left cross section, $M(x) + \Delta M(x)$ and $V(x) + \Delta V(x)$ on the right cross section, and the continuous load, which is assumed to vary linearly from $q(x)$ to $q(x) + \Delta q(x)$, as shown.

Because no forces act in the x direction on the free-body, it is only necessary to consider equilibrium of forces in the y direction and moments about a z axis. From the former we get

$$\sum F_y = 0 = [V(x) + \Delta V(x)] - V(x) + \left[q(x)\Delta x + \frac{1}{2}\Delta q(x)\Delta x\right] \qquad \textbf{(3.1a)}$$

In the limit as $\Delta x \rightarrow 0$ this reduces to

$$\frac{dV(x)}{dx} + q(x) = 0 \qquad \textbf{(3.1b)}$$

or

$$V'(x) = -q(x) \qquad \textbf{(3.1c)}$$

Similarly, summing moments about a z axis in the plane of the right cross-section gives[3]

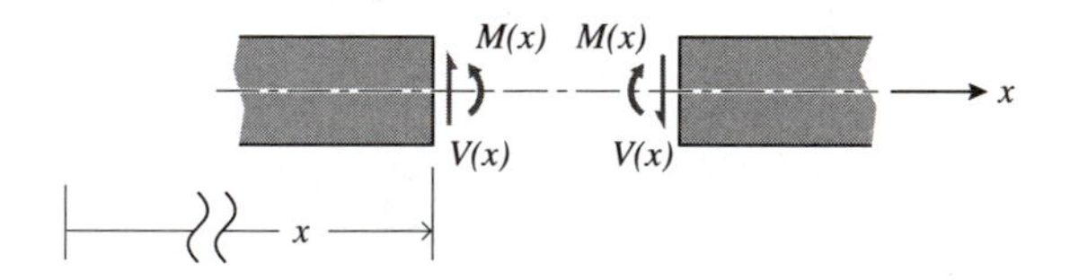

Figure 3.2 Positive resultant shear and bending moment at a cross section.

[3] Note that the sign convention for positive and negative internal moments is separate and distinct from the sign convention for positive and negative moments in an equilibrium equation; thus, although both bending moments acting on this free-body are defined as positive, in the equilibrium equation the two moments must have opposite signs because they have opposite directions.

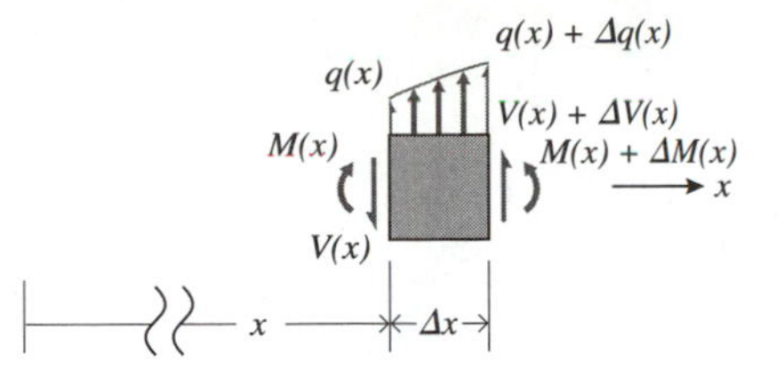

Figure 3.3 Free-body of beam element.

$$\sum M = 0 = [M(x) + \Delta M(x)] - M(x) + V(x)\Delta x - [q(x)\Delta x]\left(\frac{1}{2}\Delta x\right) - \left[\frac{1}{2}\Delta q(x)\Delta x\right]\left(\frac{1}{3}\Delta x\right) \quad \textbf{(3.2a)}$$

which, in the limit, reduces to

$$\frac{dM(x)}{dx} + V(x) = 0 \quad \textbf{(3.2b)}$$

or

$$M'(x) = -V(x) \quad \textbf{(3.2c)}$$

Equations 3.1*b* or 3.1*c* and 3.2*b* or 3.2*c* are the differential equations of equilibrium of a general transversely loaded beam. The solutions of these equations for specific beams with specific boundary conditions are equations for V and M as functions of x.[4] In subsequent developments, it will be essential for us to have such equations but, as we will see, it is usually more convenient to find $V(x)$ and $M(x)$ directly from equilibrium of free-bodies than it is to solve the differential equations. Regardless, the differential equations are vital for the development of virtual work in Chapter 4.

3.2 INTERNAL FORCES IN STATICALLY DETERMINATE FRAMES

As was stated in Chapter 1, the internal forces of interest in the components of frame structures are the transverse shear, V, bending moment, M, and axial force, P. The differential equation for axial forces, Equation 2.1*c*, is developed from the bending equations, Equations 3.1*b* and 3.2*b*. Thus, the effects of axial forces may be superposed on the effects of transverse forces whenever necessary; therefore, for simplicity, we will begin with a review of basic procedures for the computation of shears and bending moments in some elementary beams in which there are no axial forces.

Consider the uniformly loaded, simply supported beam in Figure 3.4*a*. Figure 3.4*b* is a free-body diagram of the entire beam showing reactions. As noted previously, we

[4] It should be recognized that beams are fundamentally different from the truss bars considered in the preceding chapter in that $V(x)$ and $M(x)$ are not usually constant.

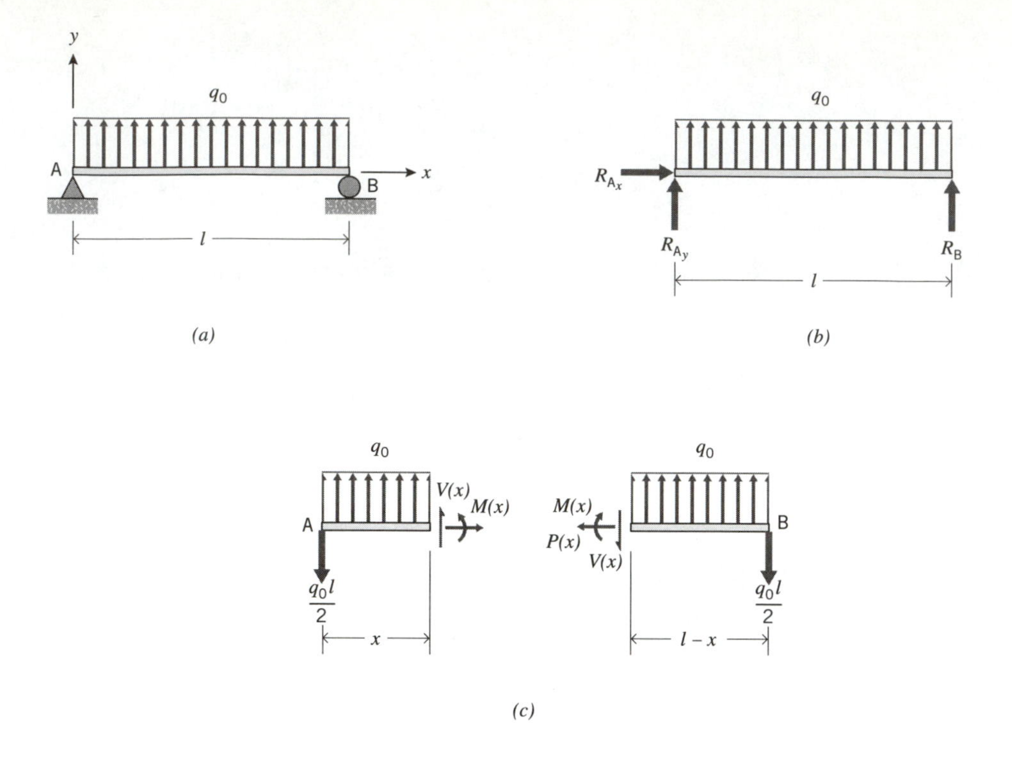

Figure 3.4 Uniformly loaded simply supported beam.

may compute $V(x)$ and $M(x)$ in either of two ways: 1) directly from equilibrium considerations or 2) by solving the differential equations, Equations 3.1*b* and 3.2*b*. We will demonstrate both.

To compute $V(x)$ and $M(x)$ directly, we first determine the reactions from equilibrium of the free body in Figure 3.4*b*, i.e., $R_{A_x} = 0$, $R_{A_y} = -q_o l/2$, $R_B = -q_o l/2$. Next, we consider equilibrium of the portion of the beam either to the left or to the right of some arbitrary cross section; assume that the cross section is a distance x from the left end of the beam, as in Figure 3.4*c*. Note that the nature of the free-bodies in Figure 3.4*c* is as shown for any value of x between 0 and l.

Since either of the free-bodies in Figure 3.4*c* is appropriate for illustrative purposes, let us arbitrarily begin with the equilibrium equations for the free-body of the left portion of the beam. From $\Sigma F_x = 0$, we obtain $P(x) = 0$, as expected. Summing forces in the y direction gives

$$\sum F_y = 0 = -\frac{q_o l}{2} + q_o x + V(x) \tag{3.3a}$$

from which we obtain

$$V(x) = \frac{q_o l}{2} - q_o x = q_o\left(\frac{l}{2} - x\right) \tag{3.3b}$$

Summing moments about a z axis in the plane of the cross section gives

$$\sum M = 0 = \frac{q_o l}{2}x - q_o x\left(\frac{x}{2}\right) + M(x) \quad \textbf{(3.4a)}$$

or

$$M(x) = \frac{q_o x}{2}(x - l) \quad \textbf{(3.4b)}$$

Equations 3.3*b* and 3.4*b* are the desired expressions for shear and bending moment, respectively, and are valid for all values of x, $0 < x < l$. Identical expressions are obtained from the equilibrium equations for the other free-body in Figure 3.4*c*, i.e.,

$$\sum F_y = 0 = -\frac{q_o l}{2} + q_o(l - x) - V(x) \quad \textbf{(3.5a)}$$

$$\sum M = 0 = -\frac{q_o l}{2}(l - x) + q_o(l - x)\left(\frac{l - x}{2}\right) - M(x) \quad \textbf{(3.5b)}$$

Equation 3.5*a* is identical to Equation 3.3*a*, and Equation 3.5*b* is identical to Equation 3.4*a*. Note that Equation 3.5*b* is again an equation for sum of moments about a z axis in the plane of the cross section.

The differential equations of equilibrium, Equations 3.1*c* and 3.2*c* will also lead to the same results. In this case, we are given $q(x) = q_o$, a constant, for $0 < x < l$; from Equation 3.1*c* we have

$$\frac{dV(x)}{dx} = -q_o, \qquad 0 < x < l \quad \textbf{(3.6a)}$$

Integrating leads to

$$V(x) = -q_o x + C_1 \qquad 0 < x < l \quad \textbf{(3.6b)}$$

where C_1 is a constant of integration. This type of constant is evaluated using boundary conditions, i.e., known values of shear (or bending moment) at specific points in the beam. For example, in this case, if we had previously evaluated the reactions, as in Figure 3.4*b*, we would know from the free-body in Figure 3.4*c* that as $x \rightarrow 0$ equilibrium requires that $V(x) \rightarrow q_o l/2$. In other words, $V(0^+)$[5] is equal and opposite to R_{A_y} and, from Equation 3.6*b*, we would conclude that $C_1 = V(0^+) = q_o l/2$. Here, we specifically refer to the shear at a value of x slightly greater than zero because $x = 0$ is the point at which the reaction, R_{A_y}, is assumed to act, and shear is undefined at such a point; in fact, shear and bending moment are only defined immediately to one side or the other of points of application of concentrated loads.

---

[5] We will use the notation 0^+ to denote a value of a coordinate slightly greater than zero; similarly, the notation l^- denotes a value slightly less than l.

Usually, however, we would not bother to compute the reactions because C_1 is easily evaluated from boundary conditions for bending moments in a case such as this one. To illustrate, assume that C_1 is unknown, and substitute Equation 3.6*b* into Equation 3.2*c*,

$$\frac{dM(x)}{dx} = q_o x - C_1 \quad \textbf{(3.6c)}$$

Integrating leads to

$$M(x) = \frac{q_o x^2}{2} - C_1 x + C_2 \quad \textbf{(3.6d)}$$

Here, we can see from inspection of the free-body in Figure 3.4*c* that as $x \rightarrow 0$, $M(x) \rightarrow 0$, i.e., $M(0^+) = 0$. (This is also verified by Equation 3.4*b*.) Similarly, we can see from the free-body of the right portion of the beam in Figure 3.4*c* that as $x \rightarrow l$, $M(x) \rightarrow 0$, i.e., $M(l^-) = 0$. We can evaluate C_1 and C_2 from these two conditions; specifically, from Equation 3.6*d* we get

$$M(0^+) = 0 = C_2 \quad \textbf{(3.6e)}$$

and

$$M(l^-) = 0 = \frac{q_o l^2}{2} - C_1 l$$

or

$$C_1 = \frac{q_o l}{2} \quad \textbf{(3.6f)}$$

as expected.

Both approaches were relatively simple for the preceding beam. Unfortunately, the computations become more involved whenever there are discontinuities in $V(x)$ and/or $M(x)$ resulting from discontinuities in $q(x)$, or from concentrated loads. Solution of the differential equations, in particular, becomes an arduous task even when the number of discontinuities is small. Consider, for example, the simply supported beam in Figure 3.5*a*, which is subjected to the single concentrated load F applied at $x = a$, as shown. Reactions are shown in Figure 3.5*b*.

Direct computation of $V(x)$ and $M(x)$ requires that we consider two different free-bodies. Two pairs of appropriate free-bodies are shown in Figures 3.5*c* and 3.5*d*; the free-bodies in Figure 3.5*c* expose a cross section for which $0 < x < a$, while those in Figure 3.5*d* expose a cross section for which $a < x < l$. From equilibrium equations for either of the free-bodies in Figure 3.5*c* we obtain

$$V(x) = -\frac{Fb}{l}, \quad 0 < x < a \quad \textbf{(3.7a)}$$

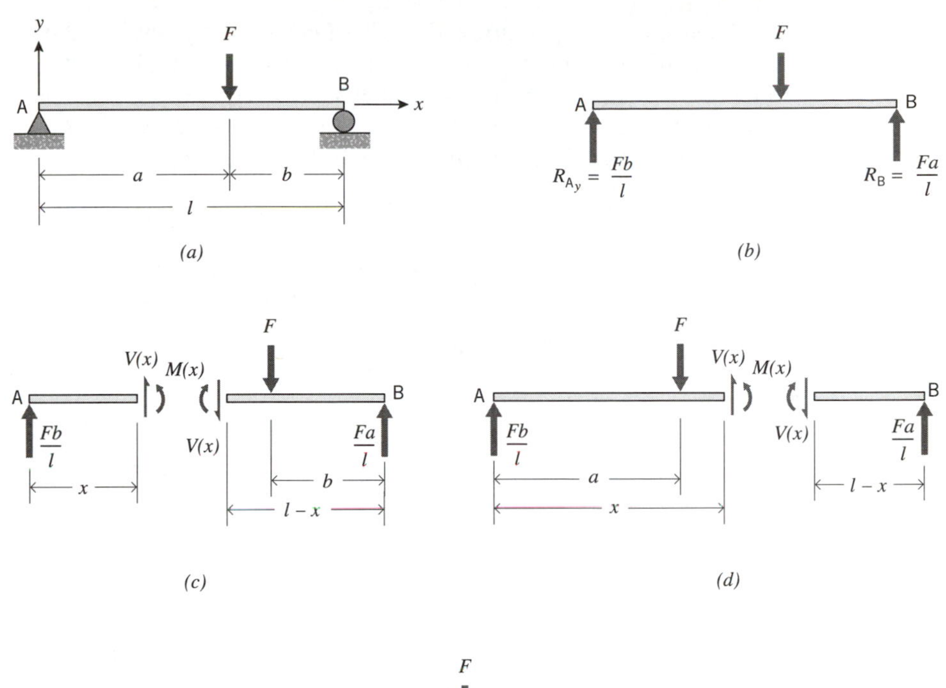

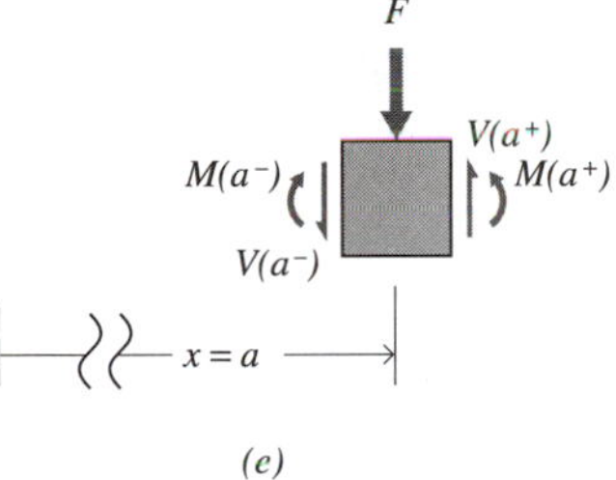

Figure 3.5 Simply supported beam with concentrated load.

$$M(x) = \frac{Fb}{l}x, \quad 0 < x < a$$

and from equilibrium equations for either of the free-bodies in Figure 3.5d we obtain

$$V(x) = \frac{Fa}{l}, \quad a < x < l$$

$$M(x) = \frac{Fa}{l}(l - x), \quad a < x < l \qquad \textbf{(3.7}b\textbf{)}$$

These two sets of equations for $V(x)$ and $M(x)$ yield the desired results. Note that the equations for $V(x)$ are discontinuous at $x = a$, i.e., Equation 3.7a gives $V(a^-) =$

$-Fb/l$ and Equation 3.7b gives $V(a^+) = Fa/l$; similarly, the equations for $M(x)$ are continuous at $x = a$, but have discontinuous derivatives at this point.

Proper integration of the differential equations requires *a priori* recognition of the existence of the discontinuities at $x = a$. This implies that we must establish distinct sets of mathematical functions for $V(x)$ and $M(x)$ in each of the regions $0 < x < a$ and $a < x < l$,[6] i.e., from Equation 3.1c

$$\frac{dV(x)}{dx} = 0, \quad 0 < x < a$$
$$\frac{dV(x)}{dx} = 0, \quad a < x < l \tag{3.8a}$$

Integrating gives

$$V(x) = C_1, \quad 0 < x < a$$
$$V(x) = C_1', \quad a < x < l \tag{3.8b}$$

Similarly, substituting Equations 3.8b into Equation 3.2c and integrating (leaving evaluation of constants aside temporarily) leads to

$$M(x) = -C_1x + C_2, \quad 0 < x < a$$
$$M(x) = -C_1'x + C_2', \quad a < x < l \tag{3.8c}$$

Equations 3.8b and 3.8c contain four constants of integration, C_1, C_2, C_1', and C_2'; therefore, we require four boundary conditions for their evaluation. As in the previous case, we still have the two conditions $M(0^+) = 0$ and $M(l^-) = 0$; substituting the first of these conditions into the first of Equations 3.8c provides

$$M(0^+) = 0 = C_2 \tag{3.8d}$$

and substituting the second condition into the second of Equations 3.8c gives

$$M(l^-) = 0 = -C_1'l + C_2' \tag{3.8e}$$

Two more boundary conditions can be obtained from $V(0^+)$ and $V(l^-)$ if we know R_{A_y} and R_B or, preferably, from an analysis of the relationships between $V(a^-)$ and $V(a^+)$ and $M(a^-)$ and $M(a^+)$; the latter are completely general and can be used independently of any knowledge of R_{A_y} and R_B. To illustrate these general boundary

---

[6] Alternately, it is possible to make use of singularity functions in order to circumvent the necessity of writing separate sets of expressions; however, since solution by integration of differential equations will receive only limited treatment herein, there is no compelling reason to discuss this approach.

conditions we make use of a free-body of a small length of the beam in the vicinity of $x = a$, as in Figure 3.5*e*; from the equilibrium equation for sum of forces in the y direction we obtain

$$\sum F_y = 0 = -V(a^-) - F + V(a^+) \tag{3.8f}$$

and from the equation of moment equilibrium,

$$\sum M = 0 = -M(a^-) + M(a^+) \tag{3.8g}$$

Substituting Equations 3.8*b* into Equation 3.8*f* gives

$$-C_1 - F + C_1' = 0 \tag{3.8h}$$

and substituting Equations 3.8*c* into Equation 3.8*g* gives

$$C_1 a - C_2 - C_1' a + C_2' = 0 \tag{3.8i}$$

Solving Equations 3.8*d*, 3.8*e*, 3.8*h*, and 3.8*i* simultaneously provides

$$C_1 = -\frac{Fb}{l}, \quad C_2 = 0, \quad C_1' = \frac{Fa}{l}, \quad C_2' = Fa$$

which, when substituted into Equations 3.8*b* and 3.8*c*, yields results identical to those previously obtained in Equations 3.7*a* and 3.7*b*. The previous results were realized with much less computational effort than was required here. Since this is fairly representative of such computations, we will not pursue the integration approach any further in connection with the determination of shears and bending moments; we will, however, find the differential equations useful as an aid in sketching the shear and bending moment functions, the next topic to be reviewed. We will also return to integration of differential equations when we examine beam and frame deflections.

3.2.1 Shear and Bending Moment Diagrams

Shear and bending moment diagrams are plots of the functions $V(x)$ and $M(x)$; they can be drawn directly from the mathematical expressions, or their shapes can be sketched based on a qualitative assessment of beam equilibrium. These diagrams are useful primarily as a visual aid in identifying extreme values of shear and bending moment. Bending moment diagrams are also an important aid for estimating deformed shapes of beams and frames; the moment-area method, which will be introduced later in this chapter, uses the diagrams in connection with the computation of specific deformation quantities.

It is customary, and useful, to show shear and bending moment diagrams in relation to the particular beam under consideration. For example, Figure 3.6 shows shear and bending moment diagrams for the uniformly loaded beam in Figure 3.4*a*, and Figure 3.7 shows the diagrams for the beam in Figure 3.5*a*; each set of diagrams is drawn directly under a sketch of the beam. In particular, the shear diagram in Figure 3.6*b*

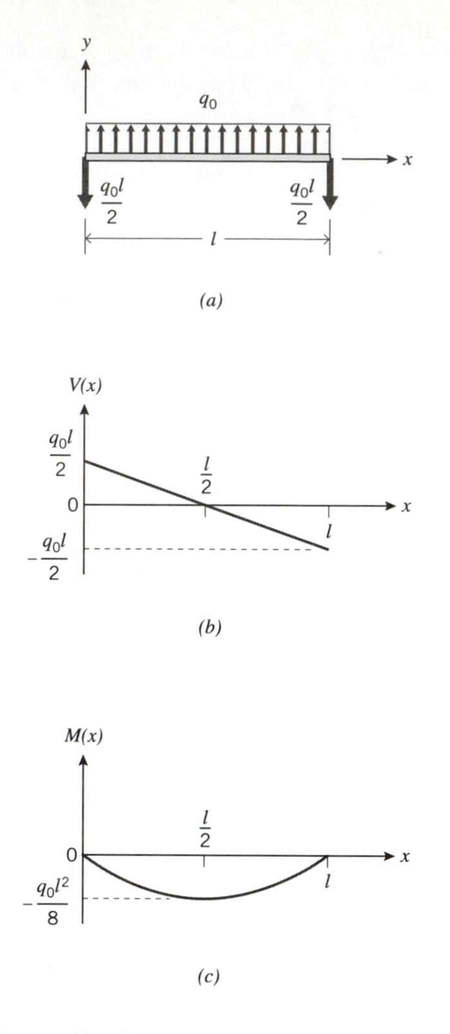

Figure 3.6 Shear and bending moment diagrams for beam in Figure 3.4.

corresponds to Equation 3.3*b*, and the moment diagram in Figure 3.6*c* corresponds to Equation 3.4*b*; similarly, the shear and moment diagrams in Figures 3.7*b* and 3.7*c*, respectively, are obtained from the two pairs of expressions in Equations 3.7*a* and 3.7*b*.

While explicit mathematical functions $V(x)$ and $M(x)$ are required for many subsequent applications, it is not always necessary to have such functions; for some applications it is sufficient to know the general shape of the moment diagram and to assess values only at extreme points. In such circumstances, the differential equations provide a useful geometric interpretation. Specifically, note that the quantities on the left of Equations 3.1*c* and 3.2*c* are slopes of the shear and moment diagrams, respectively. Thus, Equation 3.1*c* indicates that slope of the shear function at any coordinate x is equal to the negative of the continuous load at that coordinate; similarly, Equation 3.2*c*

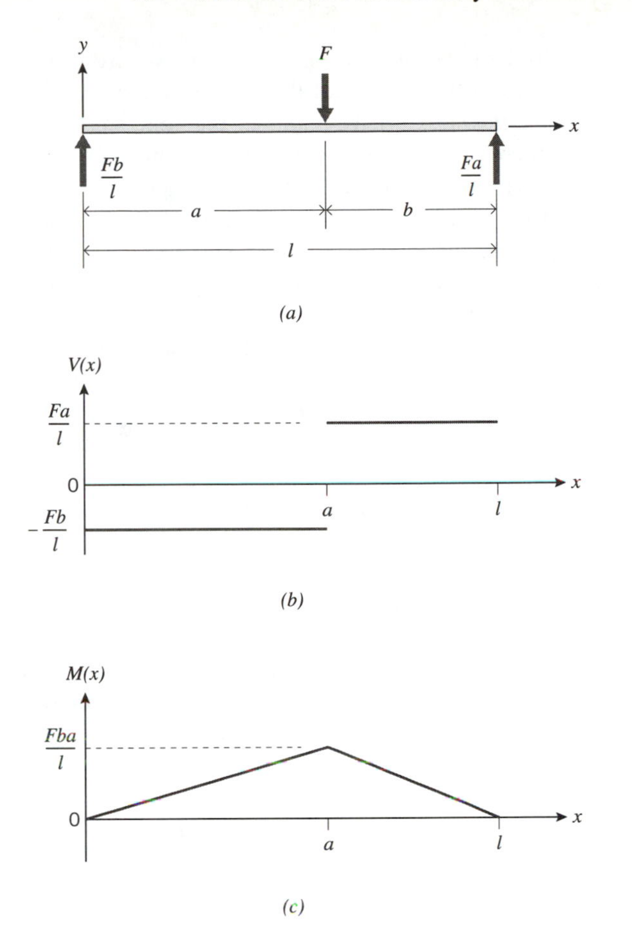

Figure 3.7 Shear and bending moment diagrams for beam in Figure 3.5.

indicates that slope of the moment function at any coordinate x is equal to the negative of the shear at that coordinate.

These relations can be used to independently verify the basic correctness of the diagrams in Figures 3.6 and 3.7. For example, for the beam in Figure 3.6, we see that $q(x) = q_0$, a positive constant for $0 < x < l$; it follows that the slope of the shear diagram must be a negative constant for $0 < x < l$. Furthermore, for this load we should expect R_{A_y} to be directed downward, so we can also expect $V(0^+)$ to be positive even if we don't know its magnitude; consequently, the shear function must have the general shape shown in Figure 3.6*b*. Finally, because R_B must equal R_{A_y} (from symmetry), we conclude that $V(l^-)$ must be equal and opposite to $V(0^+)$.

Similarly, to sketch the general shape of the moment diagram in Figure 3.6*c*, we begin with our prior observation that $M(0^+) = 0$ and $M(l^-) = 0$, which defines the end points. Then, from the shear diagram, we see that $V(0^+)$ has a relatively large positive value, which implies that the slope of the moment diagram at $x = 0$ must have a relatively large negative value; as x increases from 0 to $l/2$, $V(x)$ gets less positive, which implies that the slope of the moment diagram gradually becomes less negative. At $x = l/2$ the slope of the moment diagram must be zero. For $l/2 < x < l$, $V(x)$ assumes increasingly negative values as x increases, and the slope of the moment diagram must therefore assume increasingly positive values. This is all consistent with the plot shown in Figure 3.6*c*. Note also, that since $M(x)$ is the integral of a linear function, we should recognize that $M(x)$ must be a quadratic function.

The shear and moment diagrams in Figures 3.7*b* and 3.7*c* are also consistent with this type of interpretation of Equations 3.1*c* and 3.2*c*. In particular, since $q(x) = 0$ it follows that the slope of the shear diagram must be zero everywhere, which is consistent with constant magnitudes of shear; similarly, the slope of the moment diagram must be a positive constant for $0 < x < a$ since $V(x)$ is a negative constant in this region, and must be a negative constant for $a < x < l$ since $V(x)$ is a positive constant there. We should also discern (from a free-body like the one in Figure 3.5*e*) that there is no discontinuity in the moment at $x = a$.

Once the shapes of the diagrams are established, it is a relatively straightforward matter to determine the magnitudes of the extreme values[7] of shears and moments. These extreme values can always be determined from equilibrium equations for appropriate free-bodies. In some cases, it is also convenient to determine extreme values by computing changes in values between specific points; this is usually accomplished by the equivalent of integration of the differential equations between those points, i.e., from Equations 3.1*c* and 3.2*c* we obtain

$$V(x_2) - V(x_1) = -\int_{x_1}^{x_2} q(x)dx$$
$$M(x_2) - M(x_1) = -\int_{x_1}^{x_2} V(x)dx \qquad \textbf{(3.9)}$$

The integrals on the right-hand sides of these equations correspond to the areas under the respective diagrams between the limits x_1 and x_2.[8] Thus, we can conclude, for example, that the change in bending moments between $x = 0$ and $x = l/2$ in Figure 3.6*c* is equal to the negative of the area under the shear diagram between those limits, i.e., $M(l/2) - M(0^+) = -q_o l^2/8$, and since $M(0^+) = 0$ it follows that $M(l/2) = -q_o l^2/8$. The same result could also be obtained from equilibrium of a free-body of half the beam, i.e., $x = l/2$ in Figure 3.4*c*.

We may also conclude from the preceding discussion that, in general, extreme values of bending moment will occur either at coordinates where shear equals zero, or at coordinates where there is a discontinuity in shear; likewise, extreme values of shear will occur either at coordinates where continuous load equals zero, or at coordinates where there is a discontinuity in load.

---

[7] Extreme values include relative maxima or minima.

[8] It is implied in Equations 3.9 that there are no singularities in $q(x)$ or $V(x)$ between the limits x_1 and x_2, i.e., no concentrated transverse forces. A concentrated transverse force causes a discontinuity in shear.

EXAMPLE 1

Sketch the shear and bending moment diagrams for the cantilever beam with the linearly varying continuous load shown in Figure 3.8*a* (Example 6, Chapter 1). Reactions are shown in Figure 3.8*b*.

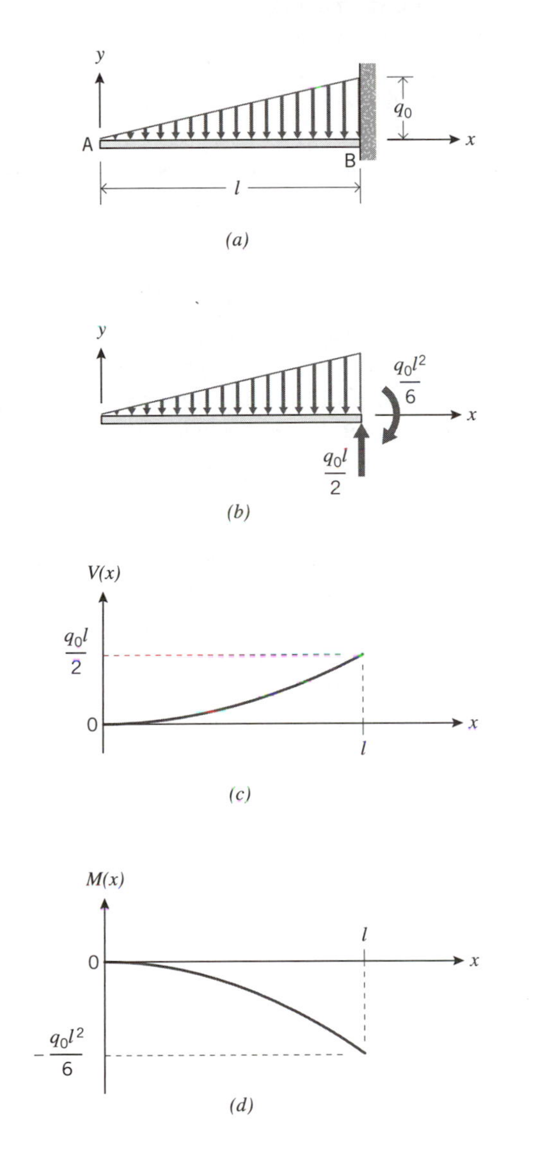

Figure 3.8 Example 1.

In this case, we observe that $V(0^+)$ and $M(0^+)$ are both equal to zero. Since $q(0) = 0$, it also follows that the slope of the shear diagram must equal zero at $x = 0$. The slope must become increasingly positive as x increases; consequently, the shear diagram must be a quadratic function (Figure 3.8*c*). From the shear diagram, we deduce that the slope of the moment diagram must become increasingly negative as x increases; consequently, the moment diagram must be a cubic function (Figure 3.8*d*).

Note that we have not derived explicit expressions for $V(x)$ or $M(x)$; rather, we have sketched the shapes of each function, identified the nature of the functions, and shown the extreme values which, in this case, occur near the end $x = l$. At $x = l^-$, these extreme values must have magnitudes equal to the reactions at B. As stated above, this information is often sufficient for many applications.

EXAMPLE 2

Sketch the shear and bending moment diagrams for the beam in Figure 3.9*a* (Example 4, Chapter 1). Reactions are shown in Figure 3.9*b*.

In this case, we start with the observations that $V(0^+) = V(l^-) = M_o/l$, and $M(0^+) = M(l^-) = 0$. Since $q(x) = 0$ it follows that the slope of the shear diagram must equal *zero*; consequently, we conclude that the shear has the constant value M_o/l for all x. Note that there is no discontinuity in shear at $x = a$. This may be verified from a free-body of a portion of the beam in the vicinity of $x = a$ (Figure 3.9*c*); here, the applied external moment at $x = a$ does not produce a change in internal shear between $x = a^-$ and $x = a^+$. The shear diagram is shown in Figure 3.9*d*.

On the other hand, the applied moment M_o will produce a discontinuity in bending moment. Specifically, equilibrium requires that $M(a^+) - M(a^-) - M_o = 0$, or $M(a^+) = M(a^-) + M_o$. Furthermore, since $V(x)$ is a positive constant, it follows that the slope of the moment diagram must be a negative constant for all values of x. The result is shown in Figure 3.9*e*. Note that the two segments of the moment diagram are parallel lines.

As before, the extreme values are determined from appropriate free-bodies, or by

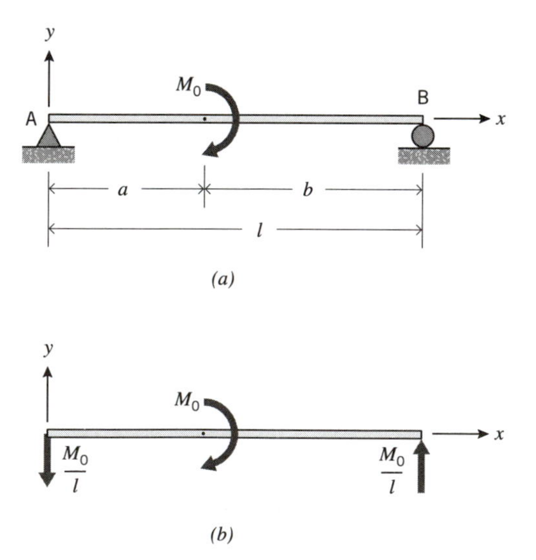

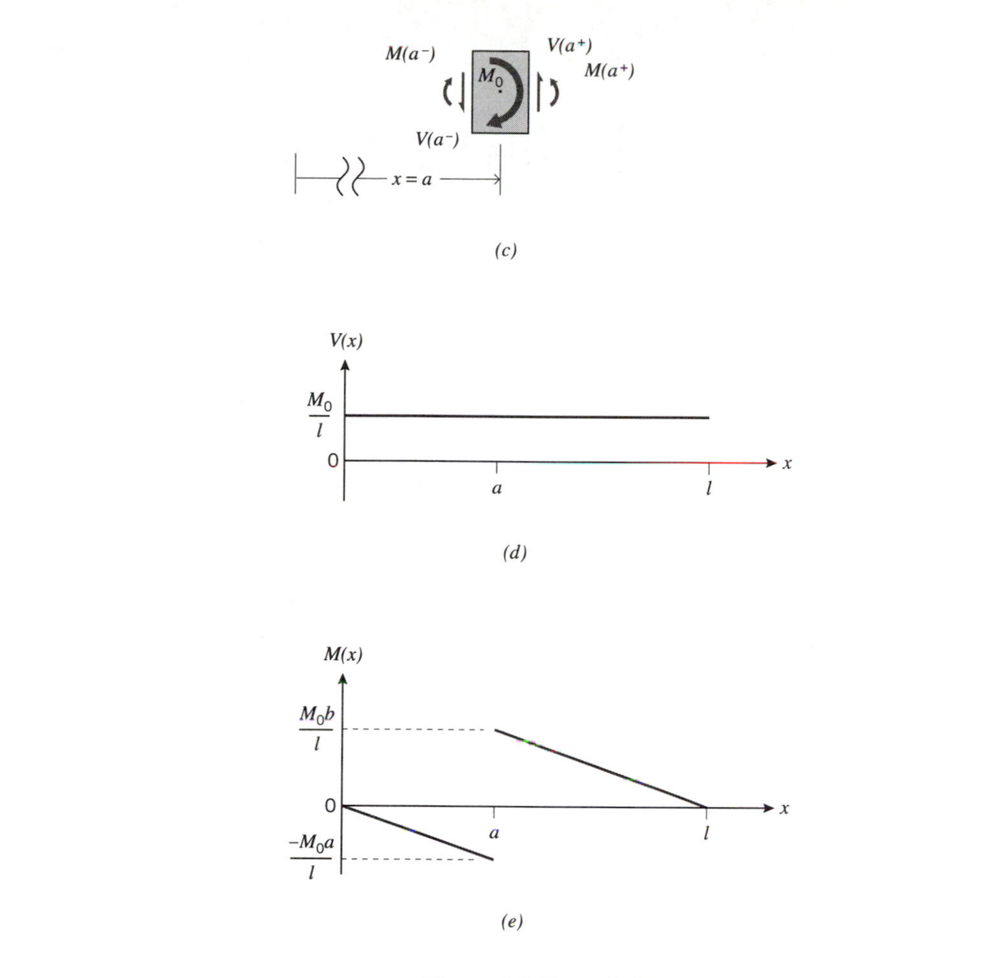

Figure 3.9 Example 2.

computing changes between significant points. Using the latter approach, for instance, we get $M(a^-) - M(0^+) = M(a^-) = -M_o a/l$, where $M_o a/l$ is the *area* under the shear diagram between the limits $x = 0$ and $x = a^-$; similarly, $M(l^-) - M(a^+) = -M(a^+) = -M_o b/l$, or $M(a^+) = M_o b/l$. This is consistent with the previous result from Figure 3.9*c*, i.e., $M(a^+) = M(a^-) + M_o = -M_o a/l + M_o = M_o(-a/l + 1) = M_o b/l$.

EXAMPLE 3

The uniformly loaded beam structure in Figure 3.10 is fixed at A, has a roller support at C and has an internal hinge at point B. Find an expression for bending moment in the structure as a function of coordinate x, and sketch the shear and moment diagrams.

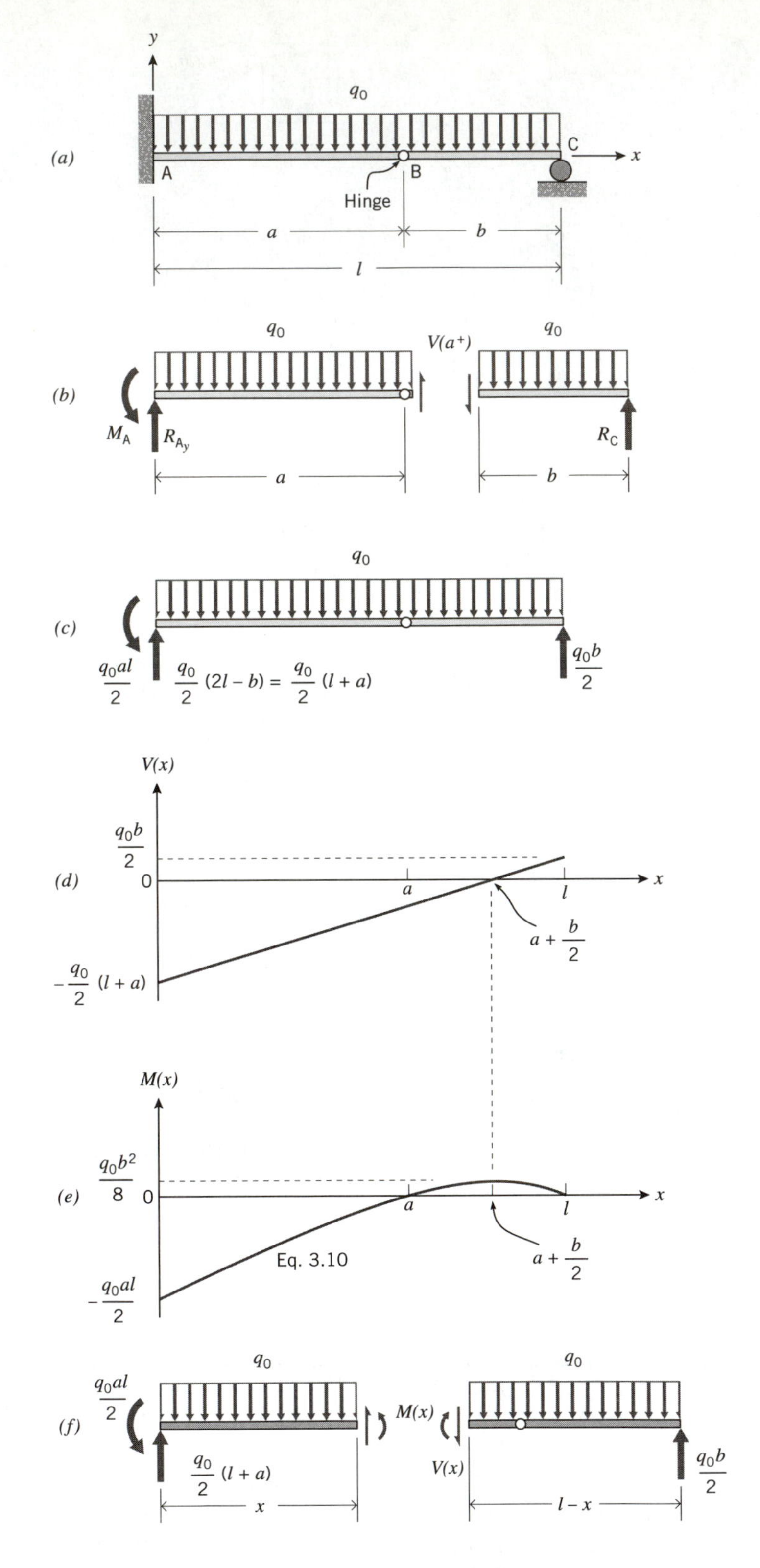

Figure 3.10 Example 3.

This structure bears some resemblance to the frame structure in Example 8 of Chapter 1. Because it is known that the internal bending moment is zero in the vicinity of the hinge at point B, it follows that the structure is statically determinate. Reactions are determined from the free-bodies in Figure 3.10*b*; in this case, the figures show a cross section just to the right of point B, but a cross section just to the left of point B would produce the same results. Reactions obtained from equilibrium equations are shown in Figure 3.10*c*.

Now, because $q(x)$ is a negative constant, we must have a shear diagram with constant positive slope, as in Figure 3.10*d*. Note that $V(0^+)$ is related to the known reaction at end *A* and, from equilibrium of the free-bodies in Figure 3.10*b*, that $V(a^-) = V(a^+) = -q_o b/2$. It follows that $V(x) = 0$ at $x = a + b/2$, as shown in Figure 3.10*d*, and that this must be a point at which $M(x)$ has an extreme value.

Since the shear diagram is a linear function of x, the moment diagram must be a quadratic function. In this case, we also observe that $M(0^+) = -q_o al/2$, and we know that $M(a^-) = M(a^+) = M(l^-) = 0$; therefore, the moment diagram must have the shape shown in Figure 3.10*e*. The desired analytic expression for $M(x)$ may be obtained from equilibrium of the free-body in Figure 3.10*f*, i.e.,

$$M(x) = -\frac{q_o al}{2} + \frac{q_o(l + a)}{2}x - \frac{q_o}{2}x^2 \tag{3.10}$$

It should also be noted that $M(x)$ is a continuous function of x despite the presence of the hinge at point B. The maximum magnitude of bending moment occurs either at point A or at the point where $x = a + b/2$, depending on the relative magnitudes of lengths a and b. The bending moment at $x = a + b/2$ can be determined either from Equation 3.10 or from equilibrium of the free-body in Figure 3.10*f* or by application of Equation 3.9; the result is $M(a + b/2) = q_o b^2/8$.

EXAMPLE 4

Sketch the shear and bending moment diagrams for beams AB and BC of the frame structure in Figure 3.11*a*. There is a pin support at A, a rigid connection at B, and a fixed support with two rollers at C; the latter is identical to the type of support examined in Example 7 of Chapter 1. The only applied load is a horizontal force *F* at B.

The three components of reaction are shown in Figure 3.11*b*, and are evaluated from three independent equations of equilibrium written for this free-body. We will omit the details of such computation and note the results, which are $R_{A_y} = 0$, $R_{A_x} = -F$ and $M_C = Fl$.

Figure 3.11*c* shows free-body diagrams of each beam and of joint B; each free-body must satisfy the equations of equilibrium. Consider, for example, the free-body of beam BC; here, end B corresponds to a cut just to the right of joint B. The free-body shows the previously determined reaction at C, namely M_C, and the complete set of possible forces at cut end B; these possible forces are an axial force, denoted P_{BC}, a shear force, denoted V_{BC}, and a moment, denoted M_{BC}. From the equilibrium equations, we conclude

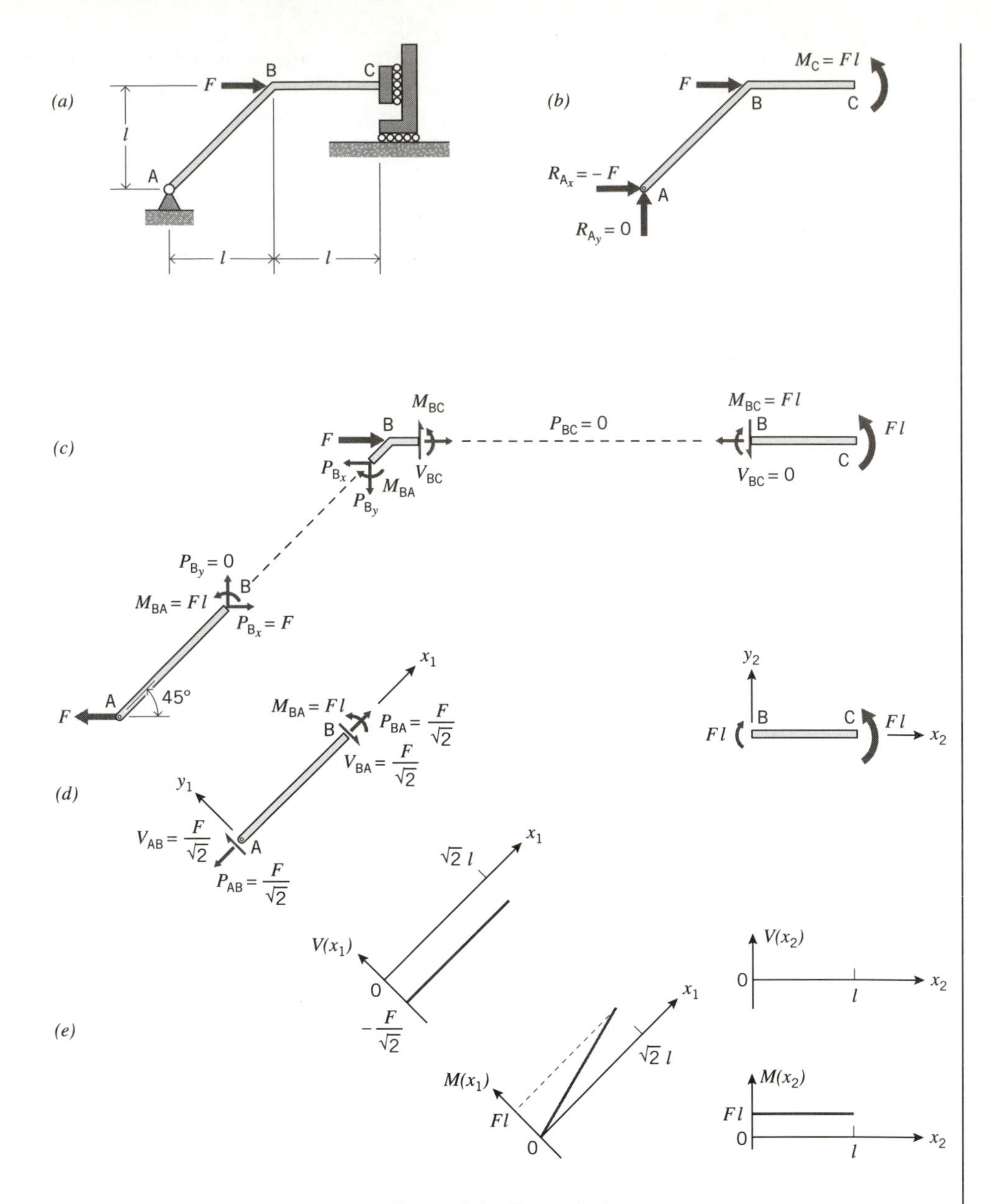

Figure 3.11 Example 4.

that $P_{BC} = V_{BC} = 0$, and $M_{BC} = M_C = Fl$. Equal and opposite forces must act on the contiguous end of joint B.

We may proceed from the analysis of beam BC to an analysis of either joint B or beam AB. We will arbitrarily choose beam AB. Here, end B corresponds to a cut just below joint B. We represent the possible forces at end B by horizontal and vertical components rather than by axial and shear components; this is done for the sake of convenience only. The horizontal force at end B is denoted by P_{B_x} and the vertical force is denoted by P_{B_y}; there is also a possible moment which will be denoted by

M_{BA}. From the equilibrium equations we will conclude that $P_{B_y} = 0$, $P_{B_x} = F$, and $M_{BA} = Fl$, as shown in Figure 3.11*c*. Once again, equal and opposite forces will act on the opposite side of the cut at joint B. Joint B is, in fact, in equilibrium under the forces shown, thereby verifying the correctness of the preceding analyses.

The horizontal force F acting on the ends of beam AB must now be resolved into axial and transverse components. The result is shown in Figure 3.11*d*. This figure also shows a set of axial and transverse coordinate axes, which will serve as reference axes for plotting shear and bending moment diagrams. Note that the axial force, which has the constant value $F/\sqrt{2}$ in this case, could, in general, also be plotted as a function of coordinate x_1; however, at this time we have no particular need for such an axial force diagram.

Shear and bending moment diagrams are sketched in Figure 3.11*e*. Here, a second set of coordinate axes has been established for beam BC. Thus, we may conclude that, for beam AB, $M(x_1) = Fx_1/\sqrt{2}$, $0 < x_1 < \sqrt{2}l$, and for beam BC, $M(x_2) = Fl$, a constant for $0 < x_2 < l$.

3.2.2 Alternate Representations of Bending Moment

The procedures for the determination and representation of bending moments presented in the previous sections are generally sufficient for all applications. However, there are two areas where some minor modifications of procedures or representations can frequently lead to a significant reduction in computational effort.

The first such area is the selection of coordinate axes; in particular, a careful selection of coordinate axes can frequently simplify the analytic expression for bending moment. For example, consider the beam in Figure 3.10*a* for which the origin of coordinates was taken at the left end. Choosing the origin of coordinates at the right end, as in Figure 3.12*a*, leads to the following expression for bending moment,

$$M(\eta) = \frac{q_o b}{2}\eta - \frac{q_o}{2}\eta^2 \tag{3.11}$$

Note that, to be consistent with the definitions of positive resultant force quantities in Figure 3.2, the choice of axes in Figure 3.12*a* now implies that positive resultants on a cross section a distance η from the right end of the beam are defined as in Figure 3.12*c*; thus, in a plot of $V(\eta)$, the direction of the shears would be opposite to that in Figure 3.10*d*, but $M(\eta)$ would appear identical to Figure 3.10*e*. Most importantly, however, Equation 3.11 represents exactly the same function as Equation 3.10 but contains one less term, and is otherwise generally a somewhat simpler algebraic expression.

The second area of potential improvement in computational efficiency has to do with representations of moment diagrams. This is especially significant in connection with the moment-area method for computation of displacements, which will be introduced shortly. The alternate representation at issue is equivalent to plotting each term of a polynomial expression for bending moment separately.[9] For example, the moment

[9] This type of representation is frequently referred to as the construction of bending moment diagrams by parts.

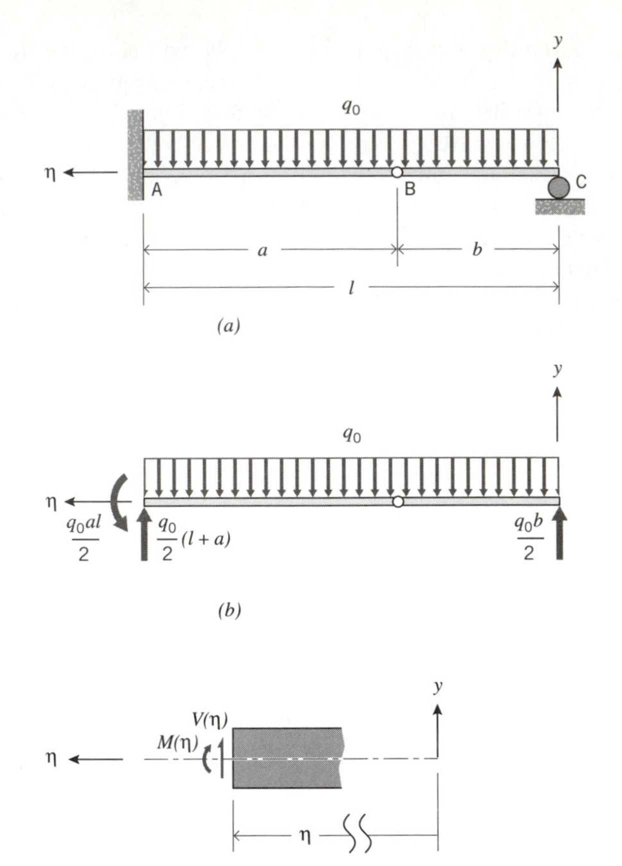

Figure 3.12 Alternate coordinated axes.

diagram in Figure 3.10*e*, which is a plot or sketch of Equation 3.10, may be viewed as the sum of the three diagrams in Figures 3.13*b*, 3.13*c*, and 3.13*d*. Figure 3.13*b* is a sketch of the first term in Equation 3.10, i.e., a sketch of $M(x)$ due to M_A; Figure 3.13*c* is a sketch of the second term, i.e., $M(x)$ due to R_{A_y}; Figure 3.13*d* is a sketch of the third term, i.e., $M(x)$ due to the continuous load. Note that it is not necessary to have the analytic expression for $M(x)$, Equation 3.10, to sketch these constituent bending moment diagrams; the bending moment due to each force component may simply be sketched independently of the others. It is only necessary to bear in mind that the actual moment diagram is obtained by superposition of the individual diagrams. The significance of the individual diagrams is that their shapes are relatively simple, i.e., the area in Figure 3.13*b* is rectangular, that in Figure 3.13*c* is triangular; the area under the parabola in Figure 3.13*d* is also fairly rudimentary. Each of these diagrams is inherently more basic than the composite diagram in Figure 3.10*e*. It should also be noted that moment diagrams can be drawn like this using an alternate η-y coordinate system, e.g., each term of Equation 3.11 could be sketched separately.

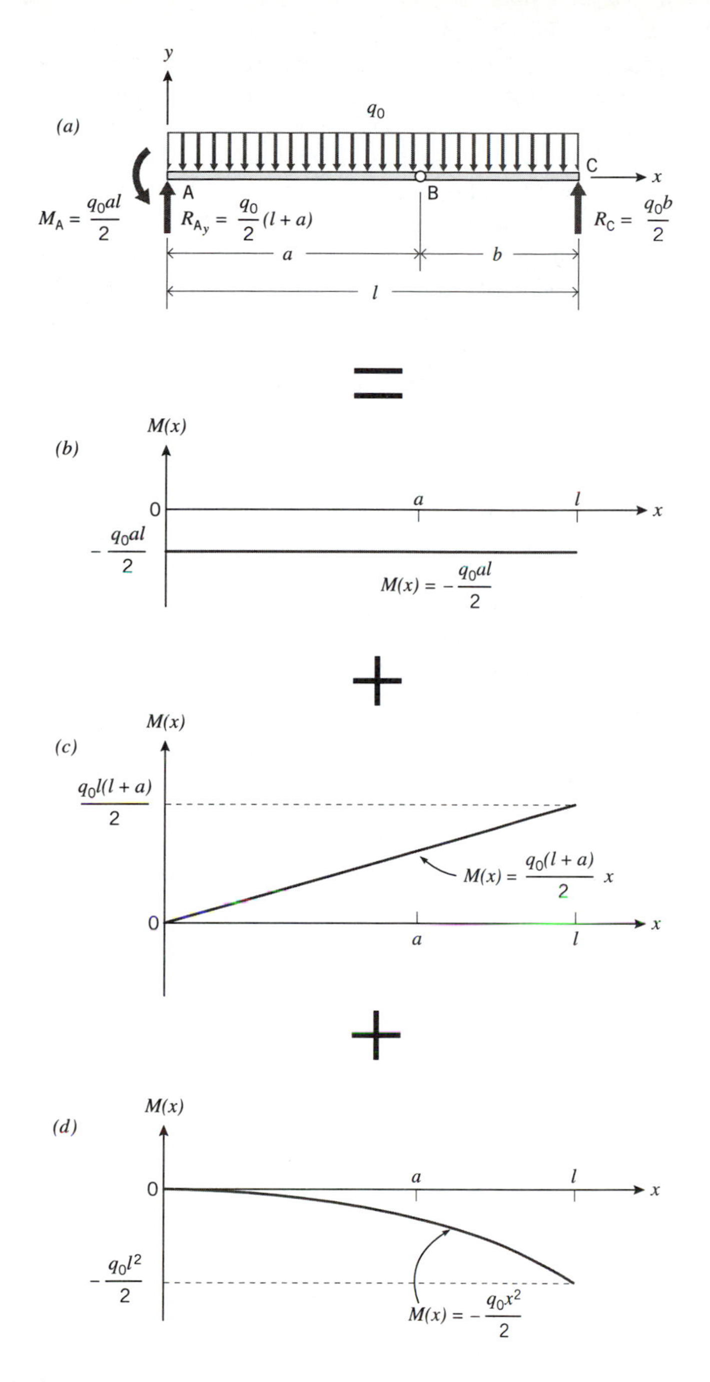

Figure 3.13 Alternate representation of moment diagram (moment diagram by parts).

3.3 DEFORMATIONS OF BEAMS

The analysis of deformations of beams is analogous to that of bars in Section 2.4. We begin with the motion of a beam of the type in Figure 3.1, and introduce a suitable hypothesis regarding the general nature of the deformations. Specifically, we will hypothesize that cross sections that are initially plane and perpendicular to the x axis in an undeformed beam will remain plane and will be perpendicular to the deformed x-axis.[10] We will investigate the displacement of a typical point on a cross section, such as point Q in Figures 3.14*a* and 3.14*b*. The location of point Q is defined by the coordinates shown in Figure 3.14*b*. We wish to relate the displacement of point Q to that of point O, which is the point on the x axis in the cross section. Because we are restricting our analysis to two dimensions, we assume that the beam displaces only in the xy plane. Thus, we will assume that point O displaces from its original coordinates, $(x, 0, 0)$, to new coordinates, $[x + u(x), v(x), 0]$, i.e., $u(x)$ and $v(x)$ are components of displacement of point O in the x and y directions, respectively; these components are functions of x only.

Figure 3.14*a* shows a view of the cross section before and after deformation; specifically, point O is assumed to displace to O′, and point Q to Q′. Since transverse displacement of point O is a function of x, i.e., $v(x)$, it follows that the x axis will, in general, undergo a small rotation, $\phi(x) \approx dv(x)/dx$, at the cross section. Consequently, according to our hypothesis, the cross-section remains perpendicular to the rotated x axis, and therefore makes an angle $\phi(x)$ with the vertical; this is equivalent to an inference that the angle $\phi(x)$ represents a rotation of the cross section about the z axis. Therefore, point Q is actually representative of all points on the cross section that are located a distance y above the z axis.

Referring to Figure 3.14*a*, we may now determine the components of displacement of point Q. The quantity $\Delta u(x,y)$ in Figure 3.14*a* is the difference in displacements in the x direction of points Q and O. Since $\phi(x)$ is small, it follows that $\Delta u(x,y) \approx y\phi(x) = ydv(x)/dx$; it also follows that the vertical distance of point Q′ above the displaced xz plane is still approximately equal to y. Therefore, we conclude that the x component of displacement of point Q is the same as that of point O minus $\Delta u(x,y)$, and the y component of displacement is approximately the same as that of point O; we also assume that the z component of displacement of point Q is zero. In analytic terms, the x, y, and z components of displacement of point Q, denoted by u_Q, v_Q and w_Q, respectively, are given by

$$\begin{aligned} u_Q &= u(x) - yv'(x) \\ v_Q &= v(x) \\ w_Q &= 0 \end{aligned} \tag{3.12}$$

From the definitions of strains as partial derivatives of displacements, it follows that the components of strain at point Q are

$$\begin{aligned} \epsilon_{xx} &= \frac{\partial u_Q}{\partial x} = u'(x) - yv''(x) \\ \epsilon_{yy} &= \frac{\partial v_Q}{\partial y} = 0 \end{aligned} \tag{3.13}$$

[10] This is usually referred to as the Bernoulli-Euler hypothesis, and sometimes as the Navier hypothesis.

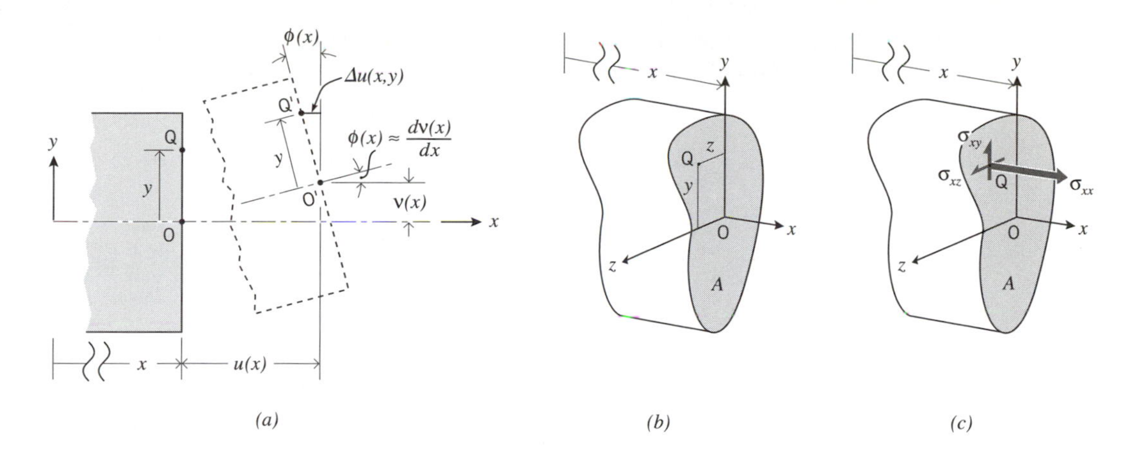

Figure 3.14 Displacements and stresses in beam cross section.

$$\epsilon_{zz} = \frac{\partial w_Q}{\partial z} = 0$$

$$\epsilon_{xy} = \frac{1}{2}\left[\frac{\partial u_Q}{\partial y} + \frac{\partial v_Q}{\partial x}\right] = \frac{1}{2}[-v'(x) + v'(x)] = 0$$

$$\epsilon_{xz} = \frac{1}{2}\left[\frac{\partial u_Q}{\partial z} + \frac{\partial w_Q}{\partial x}\right] = 0$$

$$\epsilon_{yz} = \frac{1}{2}\left[\frac{\partial v_Q}{\partial z} + \frac{\partial w_Q}{\partial y}\right] = 0$$

Note that, as a consequence of the original hypothesis regarding deformations, all strains except ϵ_{xx} are zero. As in the case of bars, ϵ_{xx} is total normal strain at any point in the cross section of a beam due to deformation.

The possible components of stress in the plane of the cross section at point Q are shown in Figure 3.14*c*, and consist of the normal stress, σ_{xx}, and transverse shear stresses, σ_{xy} and σ_{xz}. In general, the resultants of these stresses must be statically equivalent to the shear and bending moment in Figure 3.2, plus the axial force in Figure 2.2. The resultants of σ_{xx} have been defined previously in Equation 2.33.

The only component of stress that can be related to the strains in Equation 3.13 is σ_{xx};[11] consequently, the relationship between stress and strain will once again be taken as that given in Equation 2.34*c*. Substituting the expression for ϵ_{xx} in Equation 3.13 into Equation 2.34*c* now gives

$$\sigma_{xx} = E[u'(x) - yv''(x) - \alpha\Delta T - \epsilon_I] \qquad \textbf{(3.14)}$$

[11] Shear stresses are computed from alternate equilibrium considerations; see any text on strength of materials.

where v″ (x) is termed the beam curvature. We assume that the cross section is homogeneous; substitution of Equation 3.14 into Equation 2.33*a* leads to

$$
\begin{aligned}
P(x) &= Eu'(x)\int_A dA - Ev''(x)\int_A ydA - E\alpha\int_A \Delta TdA - E\int_A \epsilon_I dA \\
&= EAu'(x) - E\alpha\int_A \Delta \mathrm{T}d\mathrm{A} - \mathrm{E}\int_A \epsilon_I dA \qquad \textbf{(3.15}\boldsymbol{a}\textbf{)} \\
&= EAu'(x) - P_T(x) - P_I(x)
\end{aligned}
$$

where $P_T(x)$ and $P_I(x)$ have been defined in Equations 2.36*c*. The second integral above has been deleted because we will stipulate that the coordinate axes are centroidal, in which case the integral is identically equal to zero. Equation 3.15*a* therefore reduces to Equation 2.36*b*.

Substituting Equation 3.14 into the integral in Equation 2.33*b* leads to

$$
\begin{aligned}
M_z(x) &= -\int_A yE[u'(x) - yv''(x) - \alpha\Delta T - \epsilon_I]dA \\
&= -Eu'(x)\int_A ydA + Ev''(x)\int_A y^2dA + E\alpha\int_A y\Delta TdA + E\int_A y\epsilon_I dA \\
&= EI_{zz}v''(x) - M_{z_T}(x) - M_{z_I}(x) \qquad \textbf{(3.15}\boldsymbol{b}\textbf{)}
\end{aligned}
$$

where

$$
I_{zz} = \int_A y^2 dA
$$

$$
M_{z_T}(x) = -E\alpha\int_A y\Delta TdA
$$

$$
M_{z_I}(x) = -E\int_A y\epsilon_I dA
$$

The quantity I_{zz} is the second moment of the area about the z axis, i.e., moment of inertia of the cross section about the z axis; it should be noted that both I_{zz} and A may be functions of x if the beam has variable cross section. The quantities $M_{z_T}(x)$ and $M_{z_I}(x)$ are pseudomoments[12] about the z axis associated with thermal and initial strains, respectively. Equation 3.15*b*, unlike Equation 2.36*b*, is not identically equal to zero because $M_z(x)$ is generally not equal to zero. Equation 3.15*b* is the major new relation in the present development; it is a differential equation relating curvature and moment, which can be integrated to obtain transverse displacements of the beam axis.

[12] Resultants of stress distributions associated with thermal and initial effects are zero; consequently, as with axial pseudoforces, these quantities do not appear in free-body diagrams.

Before discussing the solution of Equation 3.15*b*, we will complete the present development by substituting Equation 3.14 into Equation 2.33*c*; the result is

$$M_y(x) = 0 = \int_A zE[u'(x) - yv''(x) - \alpha\Delta T - \epsilon_I]dA$$

$$= Eu'(x)\int_A zdA - Ev''(x)\int_A yzdA - E\alpha\int_A z\Delta TdA - E\int_A z\epsilon_I dA$$

$$= -EI_{yz}v''(x) - M_{y_T}(x) - M_{y_I}(x) \qquad \textbf{(3.15c)}$$

Under conditions where

$$I_{yz} = \int_A yzdA = 0$$

$$M_{y_T}(x) = E\alpha\int_A z\Delta TdA = 0$$

$$M_{y_I}(x) = E\int_A z\epsilon_I dA = 0$$

then Equation 3.15*c* is zero. The expression for I_{yz} will equal zero provided that the y and z axes in Figure 3.14*b* are principal axes of the cross section. This is the case for most commonly used cross sections such as I beams, rectangular tubes, and other shapes when one of the axes of symmetry is aligned with the z axis. We will henceforth assume $I_{yz} = 0$ and also that the expressions for M_{y_T} and M_{y_I} are equal to zero, as indicated.

Our analysis of beam deformations has therefore led to the two uncoupled differential equations, Equations 3.15*a* and 3.15*b*. These two equations may be rewritten as

$$u'(x) = \frac{P^*(x)}{EA} \qquad \textbf{(3.16a)}$$

$$v''(x) = \frac{M_z^*(x)}{EI_{zz}} \qquad \textbf{(3.16b)}$$

where

$$P^*(x) = P(x) + P_T(x) + P_I(x)$$

$$M_z^*(x) = M_z(x) + M_{z_T}(x) + M_{z_I}(x) \qquad \textbf{(3.16c)}$$

Equation 3.16*b* is the moment-curvature relation for the beam. We refer to the quantities $P^*(x)$ and $M^*(x)$ as *effective* axial force and bending moment, respectively, although only $P(x)$ and $M(x)$ are the true axial force and bending moment, i.e., the quantities appearing in free–body diagrams. Since the equations are uncoupled, Equation 3.16*b* may be solved independently of Equation 3.16*a*; it is Equation 3.16*b* in which we are primarily interested at this stage of developments. Furthermore, because we will henceforth be dealing only with moments and moments of inertia about the z axis, we will drop the subscript z in all subsequent equations. The resulting notation will be consistent with that used previously.

Equation 3.16*b* may be solved by integrating twice; however, it is usually convenient to replace this second-order equation by two first-order equations because each first-order equation has a pertinent physical interpretation. One of the two first-order equations has already been introduced in Figure 3.14*a*, namely the relation between slope and displacement, $\phi(x) \approx v'(x)$. It follows from this relation that $\phi'(x) = v''(x)$. Substituting the latter into Equation 3.16*b* gives the second equation, which is a relation between slope and curvature, i.e.,

$$v''(x) = \phi'(x) = \frac{M^*(x)}{EI} \qquad \textbf{(3.17a)}$$

Integration of Equation 3.17*a* gives $\phi(x)$; this may then be substituted into the former equation,

$$v'(x) = \phi(x) \qquad \textbf{(3.17b)}$$

which may be solved for $v(x)$. Equations 3.17*a* and 3.17*b* are analogous to Equations 3.1*c* and 3.2*c*. The four equations, Equations 3.16*a*, 3.16*b*, 3.17*a*, and 3.17*b*, may be combined into a single fourth-order differential equation relating displacement and load, i.e.,

$$[EIv''(x)]'' = q(x) + M''_T(x) + M''_I(x) \qquad \textbf{(3.17c)}$$

which, for E and I constant, reduces to

$$v''''(x) = \frac{q(x)}{EI} + \frac{M''_T(x)}{EI} + \frac{M''_I(x)}{EI} \qquad \textbf{(3.17d)}$$

Finally, substituting Equations 3.16*a* and 3.16*b* into Equation 3.14 gives an expression for normal stress, i.e.,

$$\sigma_{xx} = \frac{P^*(x)}{A} - \frac{M^*(x)y}{I} - E\alpha\Delta T - E\epsilon_I \qquad \textbf{(3.17e)}$$

3.4 DIRECT CALCULATION OF BEAM DISPLACEMENTS

In the remainder of this chapter, we will deal with two direct procedures for the computation of beam and frame displacements. The first is the solution of the differential equations, Equations 3.17*a* and 3.17*b*; this leads to expressions for $\phi(x)$ and $v(x)$, i.e., functions that define slope and displacement at all points of a beam axis. This procedure involves the same type of computational effort as does finding $V(x)$ and $M(x)$

by solving Equations 3.1*c* and 3.2*c*. The second procedure, the moment-area method, involves relationships between geometric/algebraic properties of moment diagrams and specific displacements, i.e., displacement quantities at specific points on the beam axis. To the degree that the moment-area method involves geometry, it can be regarded as analogous to the procedure used in Chapter 2 for analysis of truss displacements. The moment-area method is, however, more generally useful than the geometric truss analysis procedure, not only as a procedure for finding displacements but also as an aid in interpreting results of subsequent developments.

3.4.1 Solution of Differential Equations

As noted above, the procedure for solution of Equations 3.17*a* and 3.17*b* is similar to that used in conjunction with Equations 3.1*c* and 3.2*c*. The major distinctions are that we start with a knowledge of $M^*(x)/EI$ rather than $q(x)$, and constants of integration are evaluated from kinematic boundary conditions, i.e., from specified values of slope and/or displacement at specific points.

Consider the uniformly loaded beam in Figures 3.4*a*, 3.6*a* and 3.15*a*, for which $M(x)$ has been determined previously and is given in Equation 3.4*b*. If we assume that E and I are constant, and that there are no thermal or initial effects, then the right side of Equation 3.17*a* is identical to Equation 3.4*b* divided by the constant EI; it is often helpful to sketch this relationship as in Figure 3.15*b*. Given this relationship, we now have from Equation 3.17*a*,

$$\phi'(x) = \frac{q_o x}{2EI}(x - l) \qquad \textbf{(3.18a)}$$

Integrating gives

$$\phi(x) = \frac{q_o x^3}{6EI} - \frac{q_o l x^2}{4EI} + C_1 = v'(x) \qquad \textbf{(3.18b)}$$

Integrating Equation 3.18*b* leads to

$$v(x) = \frac{q_o x^4}{24EI} - \frac{q_o l x^3}{12EI} + C_1 x + C_2 \qquad \textbf{(3.18c)}$$

Here, we know that $v(0) = v(l) = 0$, two conditions from which we determine that $C_2 = 0$ and $C_1 = q_o l^3/24EI$. Substituting these values into Equations 3.18*b* and 3.18*c* gives the final results,

$$\phi(x) = \frac{q_o x^3}{6EI} - \frac{q_o l x^2}{4EI} + \frac{q_o l^3}{24EI} \qquad \textbf{(3.19a)}$$

$$v(x) = \frac{q_o x^4}{24EI} - \frac{q_o l x^3}{12EI} + \frac{q_o l^3 x}{24EI} \qquad \textbf{(3.19b)}$$

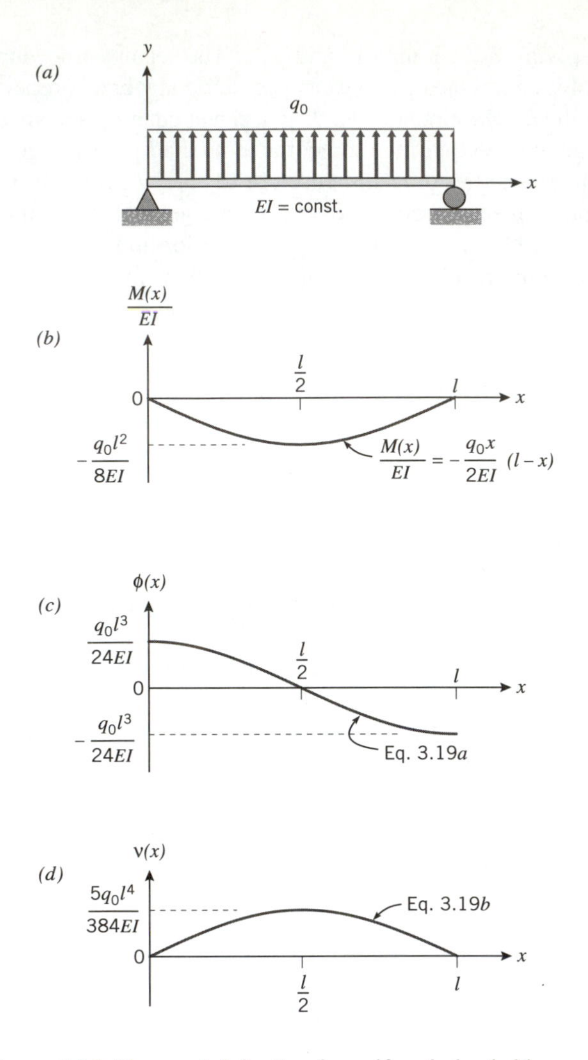

Figure 3.15 Slope and deflection for uniformly loaded beam.

We observe from Equation 3.18*b* that $\phi(0) = C_1 = q_o l^3/24EI$, which is a dimensionless quantity, as an angle in radians must be. We may also obtain $v(l/2) = 5q_o l^4/384EI$, a well-known result for maximum deflection of a uniform and uniformly loaded simply supported beam.

Plots of Equations 3.19*a* and 3.19*b* provide slope and deflection diagrams (Figures 3.15*c* and 3.15*d*, respectively). As with shear and moment diagrams, Equations 3.17*a* and 3.17*b* may be used as aids in sketching the slope and deflection diagrams: Equation 3.17*a* indicates that slope of the function $\phi(x)$ at any coordinate x_o is equal to $M^*(x_o)/EI$, and Equation 3.17*b* indicates that slope of the function $v(x)$ at any coordinate x_o is equal to $\phi(x_o)$. Note that the slope of the function $\phi(x)$ in Figure 3.15*c* is zero at $x = 0^+$ and $x = l^-$, points at which $M(x)/EI = 0$. Also, $\phi(l/2) = 0$, which implies that $v(l/2)$ must be an extreme point of the function $v(x)$.

EXAMPLE 5

The uniform cantilever beam in Figure 3.16*a* is heated along its top surface. Assume that, as a result, the temperature varies linearly from ambient (T_o) on the bottom surface to twice ambient ($2T_o$) on the top surface. Find the transverse deflection of the beam.

The change in temperature from ambient is obtained as a function of coordinate y in Figure 3.16*a*, and is given by

$$\Delta T = \frac{T_o}{2H}(2y + H) \qquad \textbf{(3.20a)}$$

where H is the depth of the rectangular cross section. Then, from Equation 3.15*b* we obtain thermal moment as

$$\begin{aligned} M_T &= -E\alpha \int_A y\Delta T dA = -\frac{E\alpha T_o}{2H}\int_A y(2y + H)dA \\ &= -\frac{E\alpha T_o}{2H}\left[2\int_A y^2 dA + H\int_A y dA\right] \qquad \textbf{(3.20b)} \\ &= -\frac{EI\alpha T_o}{H} \end{aligned}$$

Note that M_T is constant in this case. Since there are no internal bending moments due to forces, effective moment equals thermal moment, i.e., from Equation 3.16*c* we have

$$M^*(x) = M_T = -\frac{EI\alpha T_o}{H} \qquad \textbf{(3.20c)}$$

Dividing by the constant EI gives the term on the right side of Equation 3.17*a*; this is illustrated in Figure 3.16*b*.

Thus, Equation 3.17*a* is

$$\phi'(x) = -\frac{\alpha T_o}{H} \qquad \textbf{(3.20d)}$$

Integrating leads to

$$\phi(x) = -\frac{\alpha T_o}{H}x + C_1 \qquad \textbf{(3.20e)}$$

For this cantilever beam we must have $\phi(0) = 0$, from which we conclude $C_1 = 0$. Substituting Equation 3.20*e* into Equation 3.17*b* and integrating then provides

$$v(x) = -\frac{\alpha T_o}{2H}x^2 + C_2 \qquad \textbf{(3.20f)}$$

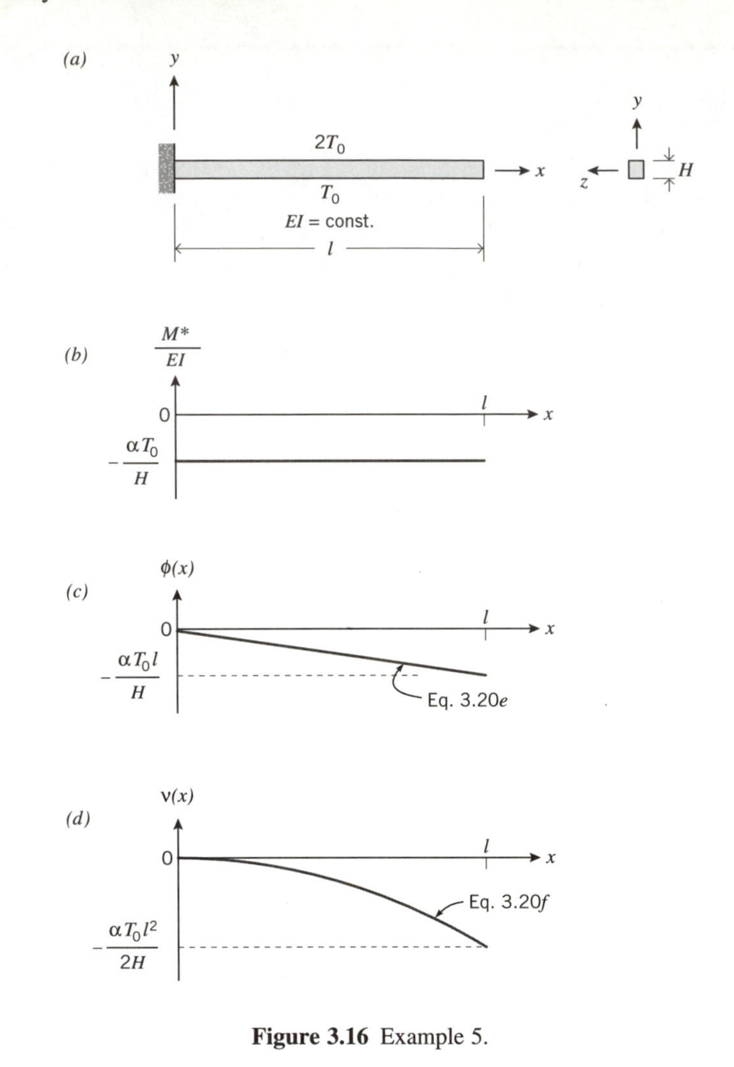

Figure 3.16 Example 5.

Since $v(0) = 0$, it follows that $C_2 = 0$. Equations 3.20*e* and 3.20*f* are shown in Figures 3.16*c* and 3.16*d*, respectively. Maximum transverse deflection occurs at $x = l$, i.e., $v(l) = \alpha T_o l^2/2H$. There is also an axial deflection that can be obtained from Equation 3.16*a* in a similar manner.

EXAMPLE 6

Consider a heated cantilever beam similar to that in the preceding example, but assume the beam is made of two segments with different depths, namely $2H$ for $0 < x < l/2$ and H for $l/2 < x < l$, as shown in Figure 3.17*a*. Then M^*/EI is a discontinuous function of x as illustrated in Figure 3.17*b*.

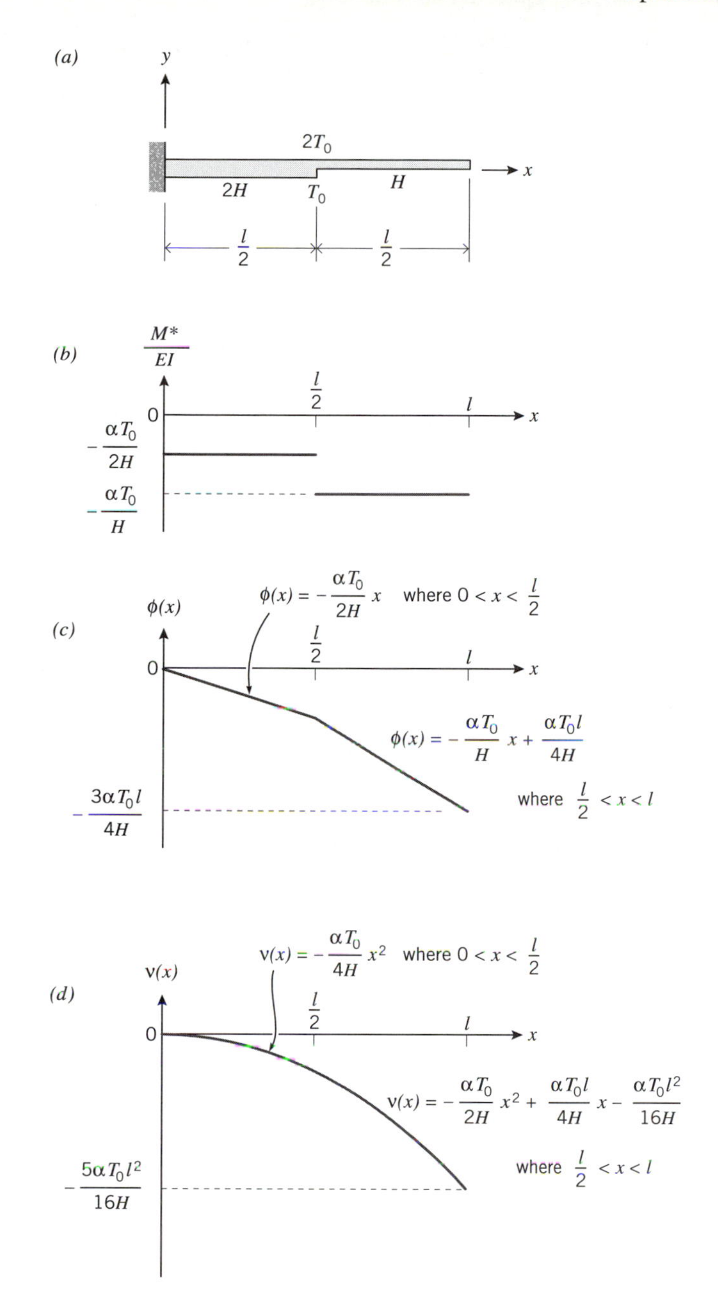

Figure 3.17 Example 6.

Therefore, we write Equation 3.17*a* as follows.

$$\phi'(x) = -\frac{\alpha T_o}{2H}, 0 < x < \frac{l}{2} \qquad \textbf{(3.21a)}$$

$$\phi'(x) = -\frac{\alpha T_o}{H}, \frac{l}{2} < x < l$$

Integrating gives

$$\phi(x) = -\frac{\alpha T_o}{2H}x + C_1, 0 < x < \frac{l}{2}$$
$$\phi(x) = -\frac{\alpha T_o}{H}x + C_1', \frac{l}{2} < x < l \tag{3.21b}$$

Because $\phi(0) = 0$, we have $C_1 = 0$. In order to evaluate C_1' we must impose the condition that $\phi(l^-/2) = \phi(l^+/2)$, i.e., the slope of the beam must be continuous at the midpoint, even though H is discontinuous there. This leads to

$$-\frac{\alpha T_o l}{4H} = -\frac{\alpha T_o l}{2H} + C_1'$$
$$C_1' = \frac{\alpha T_o l}{4H} \tag{3.21c}$$

Substituting Equations 3.21*b* into Equation 3.17*b* and integrating then gives

$$v(x) = -\frac{\alpha T_o}{4H}x^2 + C_2, 0 < x < \frac{l}{2}$$
$$v(x) = -\frac{\alpha T_o}{2H}x^2 + \frac{\alpha T_o l}{4H}x + C_2', \frac{l}{2} < x < l \tag{3.21d}$$

The condition $v(0) = 0$ gives $C_2 = 0$ and, as above, setting $v(l^-/2) = v(l^+/2)$ gives

$$-\frac{\alpha T_o l^2}{16H} = -\frac{\alpha T_o l^2}{8H} + \frac{\alpha T_o l^2}{8H} + C_2'$$
$$C_2' = -\frac{\alpha T_o l^2}{16H} \tag{3.21e}$$

The final results are obtained by substituting the values of C_1' and C_2' into the equations for $\phi(x)$ and $v(x)$, i.e.,

$$\phi(x) = -\frac{\alpha T_o}{2H}x, 0 < x < \frac{l}{2}$$
$$\phi(x) = -\frac{\alpha T_o}{H}\left(x - \frac{l}{4}\right), \frac{l}{2} < x < l \tag{3.21f}$$

$$v(x) = -\frac{\alpha T_o}{4H}x^2, 0 < x < \frac{l}{2} \tag{3.21g}$$

$$v(x) = -\frac{\alpha T_o}{2H}\left(x^2 - \frac{l}{2}x + \frac{l^2}{8}\right), \frac{l}{2} < x < l$$

Lastly, the maximum values of slope and displacement are given by

$$\phi(l) = -\frac{3\alpha T_o l}{4H}$$

$$v(l) = -\frac{5\alpha T_o l^2}{16H} \tag{3.21h}$$

EXAMPLE 7

Find $\phi(x)$ and $v(x)$ for the beam structure in Figure 3.10 (Example 3). Assume that EI is constant. The structure and moment diagram are also shown in Figures 3.18*a* and 3.18*b*.

$M(x)$ has been determined in Equation 3.10, and is a continuous function of x for $0 < x < l$; however, the presence of the hinge at $x = a$ requires an *a priori* recognition of the existence of a discontinuity in slope at $x = a$, i.e., $\phi(a^-) \neq \phi(a^+)$. Consequently, it is necessary to write two sets of differential equations, one for $0 < x < a$ and one for $a < x < l$, i.e.,

$$\phi'(x) = -\frac{q_o}{2EI}x^2 + \frac{q_o(l + a)}{2EI}x - \frac{q_o al}{2EI}, \quad 0 < x < a$$

$$\phi'(x) = -\frac{q_o}{2EI}x^2 + \frac{q_o(l + a)}{2EI}x - \frac{q_o al}{2EI}, \quad a < x < l \tag{3.22a}$$

Integrating gives

$$\phi(x) = -\frac{q_o}{6EI}x^3 + \frac{q_o(l + a)}{4EI}x^2 - \frac{q_o al}{2EI}x + C_1, \quad 0 < x < a$$

$$\phi(x) = -\frac{q_o}{6EI}x^3 + \frac{q_o(l + a)}{4EI}x^2 - \frac{q_o al}{2EI}x + C_1', \quad a < x < l \tag{3.22b}$$

From $\phi(0) = 0$ we get $C_1 = 0$. Substituting Equations 3.22*b* into Equation 3.17*b* and integrating leads to

$$v(x) = -\frac{q_o}{24EI}x^4 + \frac{q_o(l + a)}{12EI}x^3 - \frac{q_o al}{4EI}x^2 + C_2, \quad 0 < x < a \tag{3.22c}$$

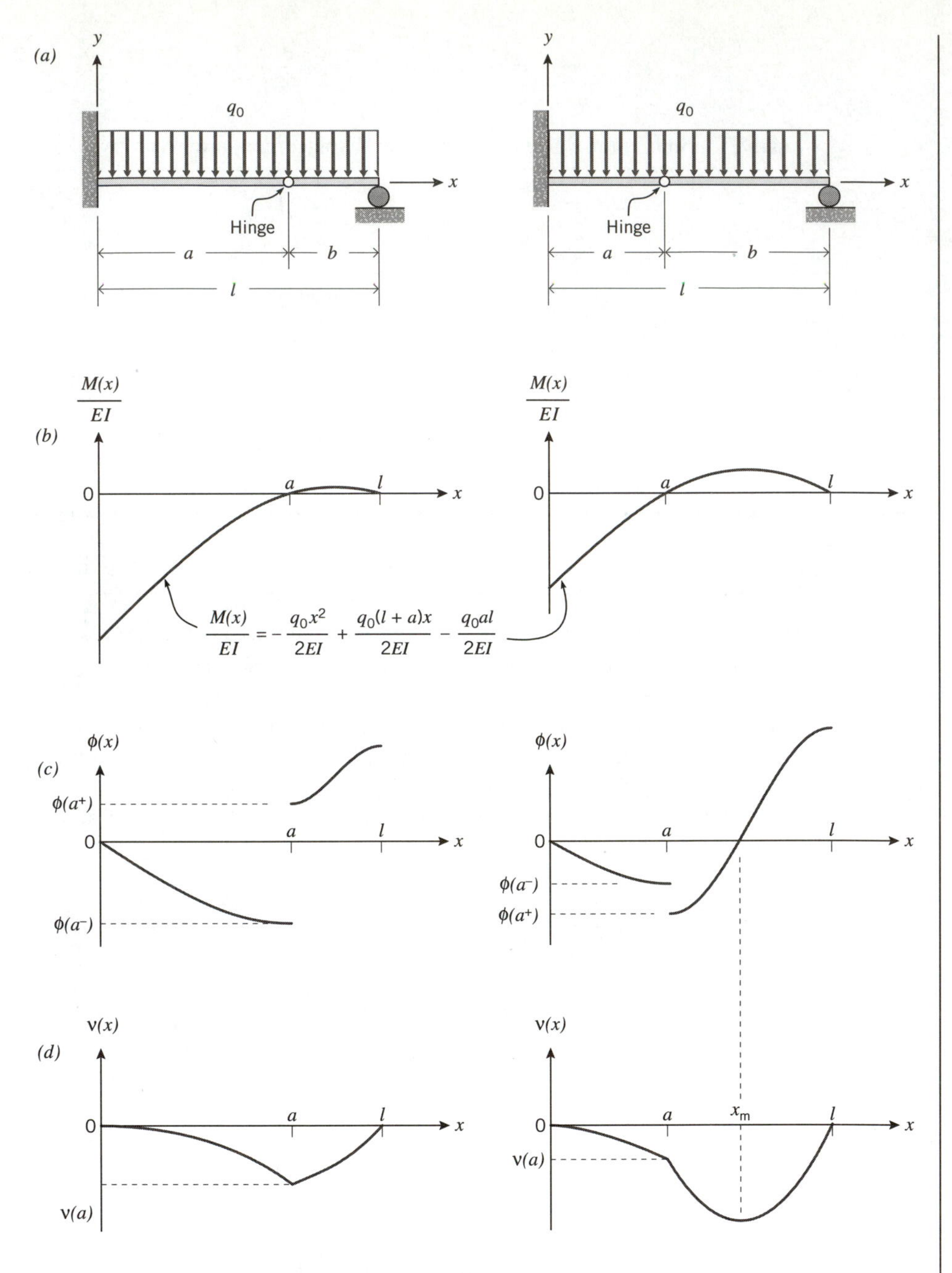

Figure 3.18 Example 7.

$$v(x) = -\frac{q_o}{24EI}x^4 + \frac{q_o(l + a)}{12EI}x^3 - \frac{q_o al}{4EI}x^2 + C_1'x + C_2', \quad a < x < l$$

From $v(0) = 0$ we get $C_2 = 0$. The remaining two constants, C_1' and C_2', are evaluated from the conditions $v(l) = 0$ and $v(a^-) = v(a^+)$. The results are

$$C_1' = \frac{q_o al^3}{6bEI} - \frac{q_o l^4}{24bEI}$$
$$C_2' = -\frac{q_o a^2 l^3}{6bEI} + \frac{q_o al^4}{24bEI} \tag{3.22d}$$

For $0 < x < a$, the final expressions for $\phi(x)$ and $v(x)$ are given by the first of Equations 3.22*b* and 3.22*c* (with $C_1 = C_2 = 0$); for $a < x < l$ the corresponding expressions are

$$\phi(x) = -\frac{q_o}{6EI}x^3 + \frac{q_o(l + a)}{4EI}x^2 - \frac{q_o al}{2EI}x + \frac{q_o al^3}{6bEI} - \frac{q_o l^4}{24bEI} \tag{3.22e}$$

$$v(x) = -\frac{q_o}{24EI}x^4 + \frac{q_o(l + a)}{12EI}x^3 - \frac{q_o al}{4EI}x^2 + \left(\frac{q_o al^3}{6bEI} - \frac{q_o l^4}{24bEI}\right)(x - a) \tag{3.22f}$$

These two sets of expressions for $\phi(x)$ and $v(x)$ are sketched in Figures 3.18*c* and 3.18*d*. Two different sets of slope and displacement diagrams are illustrated here because the exact shapes of the functions in the region $a < x < l$ depend on the ratio a/b.

The two sets of diagrams can be considered as limiting cases of possible behavior. The first set of diagrams represents a situation in which maximum displacement occurs at the hinge; this is consistent with a change in sign of slopes from negative at $x = a^-$ to positive at $x = a^+$. The second set of diagrams represents all situations in which $\phi(a^+)$ is negative; here $\phi(a^+)$ can be either less than or greater than $\phi(a^-)$. In this case, maximum displacement occurs at the coordinate designated as x_m in Figure 3.18*d*, i.e., at the point where $\phi(x) = 0$, $a < x < l$.

The preceding discussion implies that there is a location of the hinge for which $\phi(a^-) = \phi(a^+)$, i.e., a location of the hinge for which slope is continuous at $x = a$. It can be shown from Equations 3.22*b* that this will be the case for $a = b/3 = l/4$. A hinge at $a = l/4$ would actually be superfluous since this would correspond to an inflection point in a continuous beam. In other words, a continuous beam having no hinge, which would be one-degree indeterminate, will have an inflection point or point of zero moment at $a = l/4$. Thus, the behavior of a continuous (indeterminate) beam is identical to that of the structure with the hinge at $a = l/4$. We usually do not know the locations of inflection points before analysis, but it is sometimes desirable to conduct approximate analyses by estimating the locations of inflection points, thereby effectively reducing the degree of indeterminacy of the structure.

Finally, we note that the analysis of this particular structure would be basically similar even if beams AB and BC were not collinear; however, because of the different geometry, relations matching displacement conditions in the vicinity of an internal joint, such as at B, will generally require modification. This will be demonstrated in the next example.

EXAMPLE 8

Determine and sketch the deflected shape of the frame structure in Figures 3.11*a* (Example 4) and 3.19*a*. This structure has a rigid connection at B. Assume that *EI* is constant and equal for beams AB and BC.

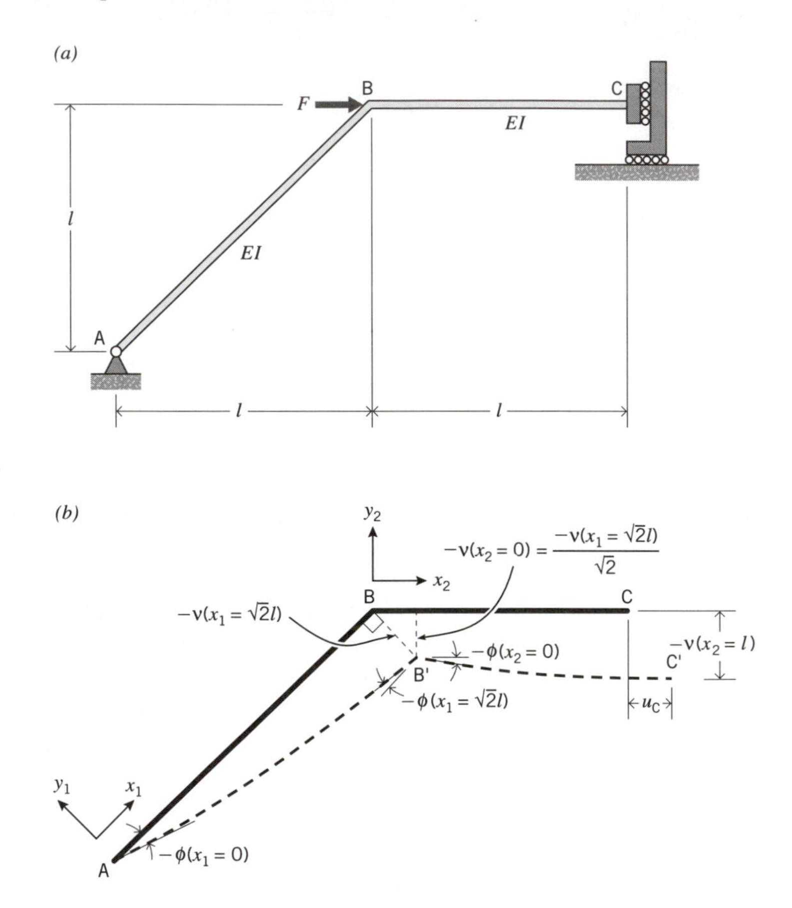

Figure 3.19 Example 8.

We will use the two sets of coordinate axes shown in Figure 3.11*e*. Based on the previous solution for internal moments in Example 4, we now have

$$\phi'(x_1) = \frac{F}{\sqrt{2}EI}x_1,\ 0 < x_1 < \sqrt{2}l$$

$$\phi'(x_2) = \frac{Fl}{EI},\ 0 < x_2 < l \tag{3.23a}$$

Integrating gives

$$\phi(x_1) = \frac{F}{2\sqrt{2}EI}x_1^2 + C_1, 0 < x_1 < \sqrt{2}l$$

$$\phi(x_2) = \frac{Fl}{EI}x_2 + C_1', 0 < x_2 < l \tag{3.23b}$$

We must have $\phi = 0$ at $x_2 = l$, i.e., at end C of beam BC; this gives $C_1' = -Fl^2/EI$. We do not know the magnitude of the slope at any other point in this structure; however, we do know that the rotation of end B of beam AB must be the same as the rotation of end B of beam BC. In other words, ϕ at $x_1 = \sqrt{2}l$ must equal ϕ at $x_2 = 0$; from Equations 3.23*b* this becomes

$$\frac{Fl^2}{\sqrt{2}EI} + C_1 = -\frac{Fl^2}{EI} \tag{3.23c}$$

from which we obtain

$$C_1 = -\frac{(1 + \sqrt{2})}{\sqrt{2}}\frac{Fl^2}{EI} \tag{3.23d}$$

Substituting Equations 3.23*b* into Equation 3.17*b* and integrating again gives

$$v(x_1) = \frac{F}{6\sqrt{2}EI}x_1^3 - \frac{(1 + \sqrt{2})Fl^2}{\sqrt{2}EI}x_1 + C_2$$

$$v(x_2) = \frac{Fl}{2EI}x_2^2 - \frac{Fl^2}{EI}x_2 + C_2' \tag{3.23e}$$

For $x_1 = 0$ we have $v(0) = 0$, which leads to $C_2 = 0$. The remaining constant, C_2', is evaluated by matching displacements at end B of beams AB and BC. In this case, it is not proper to set v at $x_1 = \sqrt{2}l$ equal to v at $x_2 = 0$; the reason is that the transverse displacement of end B of beam BC equals the transverse displacement of end B of beam AB divided by $\sqrt{2}$. This geometric relationship is illustrated in Figure 3.19*b*; here, we are neglecting the effects of axial change in length of bar AB due to the axial force $F/\sqrt{2}$. This is equivalent to assuming that the beams are axially rigid; for slender beams, axial deformations are usually negligible in comparison with transverse deformations. As usual, we also assume small displacements, which are shown greatly exaggerated in Figure 3.19*b*.

Some further discussion of the nature of the displacements discussed above may be worthwhile here. Since beam AB is taken to be axially rigid, joint B is assumed to displace to B′ along a line from B perpendicular to the axis of AB, as shown in Figure 3.19*b*; this is what we mean by transverse displacement of end B of beam AB, i.e., displacement in the direction of the y_1 axis. It follows that end B of beam BC must also displace to B′; therefore, we see that end B of beam BC undergoes displacement in both the transverse and the axial directions. In this particular case, the transverse displacement of end B of beam BC is equal to the vertical component of the displacement of end B of beam AB; from geometry, we see that this vertical component of displacement equals the transverse displacement divided by $\sqrt{2}$, as stated previously.

For beam AB, we obtain the transverse displacement at end B from the first of Equations 3.23*e* as

$$v(\sqrt{2}l) = -\frac{(2 + 3\sqrt{2})Fl^3}{3EI} \tag{3.23f}$$

Then, for end B of beam BC, the second of Equations 3.23*e* and the geometric relation between displacements at joint B leads to

$$v(0) = -\frac{(2 + 3\sqrt{2})Fl^3}{3\sqrt{2}EI} = C_2' \tag{3.23g}$$

which completes the solution.

The axial component of displacement of end B of beam BC equals the transverse component of displacement in this case; furthermore, since beam BC is regarded as axially rigid, the axial displacement of end C, denoted by u_C in Figure 3.19*b*, is the same as that of end B. This and other displacement quantities of interest may be determined from the equations above.

3.4.2 Moment-Area Method

The moment-area method involves two distinct relations between displacement quantities and curvature, M^*/EI. The first is obtained directly from Equation 3.17*a* by integrating between limits, i.e.,

$$\phi(x_2) - \phi(x_1) = \int_{x_1}^{x_2} \frac{M^*(x)}{EI} dx \tag{3.24a}$$

This equation[13] is analogous to Equation 3.9. In essence, the equation indicates that the difference in slopes between two points x_1 and x_2 on a beam axis is equal to the area under the M^*/EI diagram between the points.

The second moment-area relation requires some derivation and physical interpretation. Consider a deformed beam axis between the points x_1 and x_2, as shown in Figure 3.20*a*; here, for convenience, all displacements and slopes are shown as positive relative to the *x*-*y* coordinate system. The figure includes tangents to the deformed axis at x_1

[13] This equation is sometimes referred to as the first moment-area theorem. As with Equation 3.9, it is implied that there are no concentrated curvatures between x_1 and x_2, such as due to hinges or kinks, that would cause discontinuities in slope.

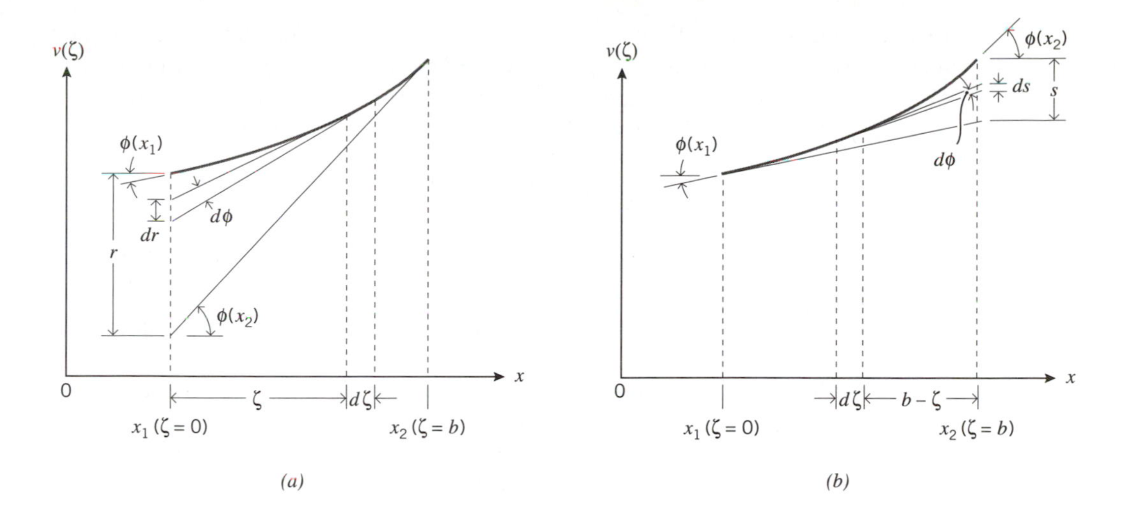

Figure 3.20 Second moment-area relation.

and x_2; the former has slope $\phi(x_1)$ and the latter has slope $\phi(x_2)$. We are interested in the vertical distance between these tangent lines at x_1, which is denoted by r in Figure 3.20*a*.

We define a local coordinate $\zeta = x - x_1$ ranging in value from zero to $b = x_2 - x_1$. To derive an expression for r we consider two adjacent points located at distances ζ and $\zeta + d\zeta$, respectively, from x_1, as shown. The difference in slopes between these two points is $d\phi$, where, for consistency, ϕ is now assumed to be expressed as a function of ζ. Since all slopes are small, it follows that dr, shown in Figure 3.20*a*, is approximately equal to $\zeta d\phi$. Furthermore, from Equation 3.17*a*, we have $d\phi = [M^*(\zeta)/EI]d\zeta$, so

$$dr = \zeta \frac{M^*(\zeta)}{EI} d\zeta$$

and

$$r = \int_0^b \zeta \frac{M^*(\zeta)}{EI} d\zeta \tag{3.24b}$$

The integral in Equation 3.24*b* is equivalent to the first moment, about x_1, of the area of the M^*/EI diagram between x_1 and x_2.[14] It must be noted that positive increments of slope, $d\phi$, between x_1 and x_2 produce increments dr that lie below the tangent to the beam axis at x_1; therefore, a positive value of r is a distance in the negative y direction from the beam axis at x_1.

The same type of argument as used previously will lead to the conclusion that the

[14] This equation is sometimes referred to as the second moment-area theorem. Note that the coordinate translation does not change the shape of the M^*/EI diagram.

vertical distance s at x_2 between tangent lines (Figure 3.20*b*) is given by an expression similar to Equation 3.24*b*, i.e., s is given by the first moment, about x_2, of the area of the M^*/EI diagram between x_1 and x_2. Furthermore, as before, a positive value of s is a distance in the negative y direction from the beam axis at x_2. The result is

$$s = \int_0^b (b - \zeta)\frac{M^*(\zeta)}{EI}\,d\zeta \qquad \textbf{(3.24c)}$$

Equations having the forms of 3.24*b* and 3.24*c* will be encountered frequently in subsequent developments. It is important to understand that these equations define the quantities r and s shown in Figure 3.20.

Actual applications of the moment-area method are almost always restricted to situations where formal integration of Equations 3.24*a* through 3.24*c* can be avoided. Thus, the method is useful primarily where the results of the integrations can be determined directly from geometric properties of the relevant M^*/EI diagrams. The properties of interest are areas and the locations of centroids of areas; consequently, the moment-area method is most useful when M^*/EI diagrams can be represented in terms of rectangles, triangles and other simple power functions.

The pertinent geometric properties of power functions are available from references or may be derived. To illustrate, consider the class of power functions of form $f(\zeta) = k\zeta^n$, where k is a constant, over the interval $0 \le \zeta \le b$, as shown in Figure 3.21. The area, A, shown shaded in Figure 3.21, is given by the integral of $f(\zeta)$; a coordinate defining the location of the centroid of the area, $\bar{\zeta}$, is given by the integral of $\zeta f(\zeta)$ divided by the integral of $f(\zeta)$, i.e.,

$$A = k\int_0^b \zeta^n\,d\zeta = \frac{kb^{n+1}}{n+1} = \frac{b(kb^n)}{n+1} = \frac{bh}{n+1} \qquad \textbf{(3.25a)}$$

$$\bar{\zeta} = \frac{k\int_0^b \zeta^{n+1}\,d\zeta}{A} = \frac{kb^{n+2}/(n+2)}{bh/(n+1)} = \left(\frac{n+1}{n+2}\right)b \qquad \textbf{(3.25b)}$$

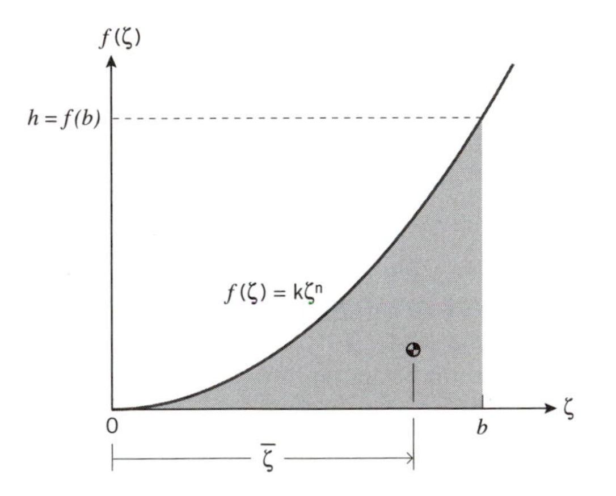

Figure 3.21 Power function.

Note that $h = f(b) = kb^n$ in Equations 3.25. These equations are analogous to Equations 1.18 and 1.22 for finding the resultant and a centroidal coordinate of a continuous load function. The significance of Equations 3.25 is similar to that of Equations 1.18 and 1.22; specifically, in this case, if M^*/EI is given by a power function of form $k\zeta^n$, then the integral in Equation 3.24*a* is given by Equation 3.25*a* and the integral in Equation 3.24*b* is given by the product of Equations 3.25*a* and 3.25*b*, i.e.,

$$r = \frac{hb^2}{n + 2} \qquad \textbf{(3.25c)}$$

EXAMPLE 9

Find the slope and transverse displacement at the free end of the heated beam from Example 6. The beam is shown again in Figure 3.22*a*; Figure 3.22*b* is a sketch of the anticipated deflected shape. Figure 3.22*c* shows the M^*/EI diagram, which was obtained previously.

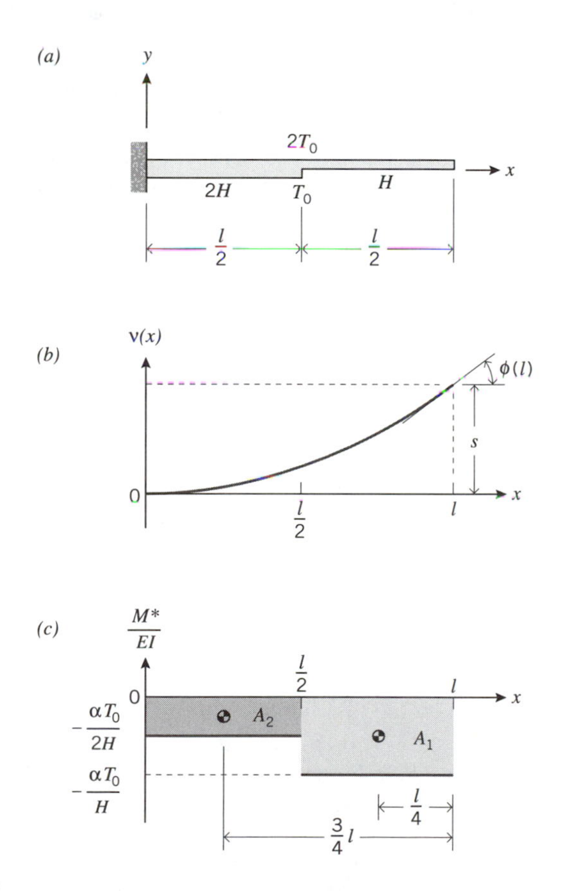

Figure 3.22 Example 9.

The slope at the free end is obtained from Equation 3.24*a* with $x_2 = l$, $x_1 = 0$; we know that $\phi(0) = 0$, so it follows that $\phi(l)$ is given by the area under the M^*/EI diagram between 0 and l in Figure 3.22*c*, i.e.,

$$\phi(l) = A_1 + A_2 = \left(-\frac{\alpha T_o}{H}\right)\frac{l}{2} + \left(-\frac{\alpha T_o}{2H}\right)\frac{l}{2} = -\frac{3\alpha T_o l}{4H} \qquad \textbf{(3.26a)}$$

which agrees with Equation 3.21*h*.

Since $\phi(0) = 0$, the tangent to the left end of the beam ($x = 0$) coincides with the x axis; therefore, the distance s in Figure 3.22*b*, which is the vertical distance between tangents, corresponds to the transverse displacement at the end in this case. The quantity s has the significance discussed in connection with Figure 3.20*b*; however, since the geometric shapes of the M^*/EI diagram are known, we can proceed with the appropriate computations without performing detailed integrations. We simply find s by taking the first moment of area of the M^*/EI diagram about the right end. The pertinent coordinates of the centroids of the areas are shown in Figure 3.22*c*, and the resulting expression for s is

$$s = \frac{l}{4}A_1 + \frac{3l}{4}A_2 = \frac{l}{4}\left(-\frac{\alpha T_o l}{2H}\right) + \frac{3l}{4}\left(-\frac{\alpha T_o l}{4H}\right) = -\frac{5\alpha T_o l^2}{16H} \qquad \textbf{(3.26b)}$$

which again agrees with Equation 3.21*h*. In this case, the negative sign indicates that the vertical distance s from the deformed beam axis to the tangent extended from ($x = 0$) is measured in the positive y direction . In other words, in this case, the negative value of s coincidentally indicates that the deformed beam axis lies below the coordinate axis.

EXAMPLE 10

For the structure in Example 7, find $v(a)$ and $\phi(a^+)$. The structure is shown in Figure 3.23*a* and, as in Figure 3.13, its M/EI diagram is represented by the sum of the diagrams in Figures 3.23*b*, 3.23*c*, and 3.23*d*. Each of the shapes here is defined by a simple mathematical function.

Figure 3.23*e* shows the deformed beam axis between $x = 0$ and $x = a$.[15] As in the previous example, the vertical distance between tangents to the end points, s_1, coincides with $v(a)$; s_1 is given by the first moment, about $x = a$, of the areas of the constituent M/EI diagrams between $x = 0$ and $x = a$. The constituent areas in question, denoted A_1, A_2, and A_3, are shown in Figures 3.23*b*, 3.23*c*, and 3.23*d*. Area A_3 is obtained from Equation 3.25*a* with $n = 2$. Equation 3.25*b* gives the result that the centroid of A_3 is

[15] In drawing the deformed shapes of structures, it is important to show the beams with positive curvatures and with distances r or s consistent with the basic definitions in Figures 3.20*a* and 3.20*b*.

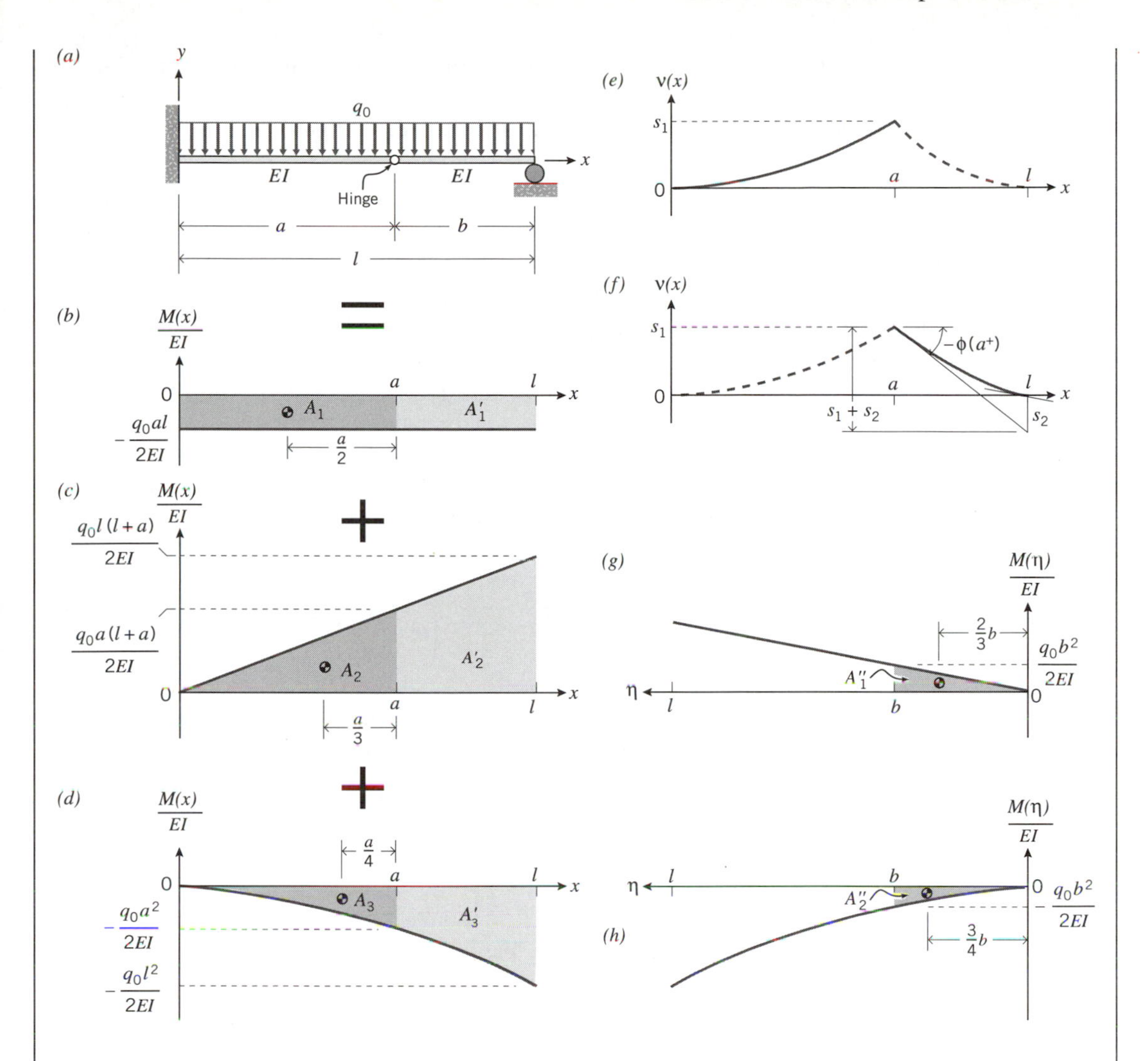

Figure 3.23 Example 10.

located at a distance $3a/4$ from the left end, which corresponds to $a/4$ from the right end, as shown.

$$s_1 = \frac{a}{2}A_1 + \frac{a}{3}A_2 + \frac{a}{4}A_3$$

$$= \frac{a}{2}\left[\left(-\frac{q_o al}{2EI}\right)(a)\right] + \frac{a}{3}\left\{\left(\frac{1}{2}\right)\left[\frac{q_o a(l + a)}{2EI}\right](a)\right\} + \frac{a}{4}\left[\left(\frac{1}{3}\right)\left(-\frac{q_o a^2}{2EI}\right)(a)\right] \quad \textbf{(3.27a)}$$

$$= -\frac{q_o a^3}{6EI}\left(l - \frac{a}{4}\right)$$

This is in agreement with the expression for $v(a)$ that can be obtained from Equations 3.22*c* and 3.22*f*.

The slope $\phi(a^+)$ is found from the quantities illustrated in Figure 3.23*f*. Specifically,

$$\phi(a^+) = -\frac{s_1 + s_2}{b} \qquad \textbf{(3.27b)}$$

Here, s_2 is the vertical distance, at $x = l$, from the deformed beam axis to the tangent extended from $x = a^+$. Note that, in this instance, s_2 is not directly equivalent to a transverse displacement of the beam axis; it is only a quantity having some geometric significance in connection with the computation of $\phi(a^+)$. Consequently, Equation 3.27*b* is not a general expression for slope, but is, rather, specific to this particular situation.

The quantity s_2 is given by the first moment, about $x = l$, of the areas of the *M/EI* diagrams between $x = a$ and $x = l$. The areas in question are denoted by A_1', A_2', and A_3' in Figures 3.23*b*, 3.23*c*, and 3.23*d*. We have

$$\begin{aligned} s_2 &= +\left[\frac{l}{2}(A_1 + A_1') - A_1\left(b + \frac{a}{2}\right)\right] \\ &= +\left[\frac{l}{3}(A_2 + A_2') - A_2\left(b + \frac{a}{3}\right)\right] \\ &= +\left[\frac{l}{4}(A_3 + A_3') - A_3\left(b + \frac{a}{4}\right)\right] \end{aligned} \qquad \textbf{(3.27c)}$$

In Equation 3.27*c* we have expressed the first moments of the desired areas as the differences in first moments of the total and partial areas; this has been done to make it possible to use Equations 3.25. Expanding Equation 3.27*c* now gives

$$s_2 = \frac{l}{2}(A_1 + A_1') + \frac{l}{3}(A_2 + A_2') + \frac{l}{4}(A_3 + A_3') - b(A_1 + A_2 + A_3) - s_1 \qquad \textbf{(3.27d)}$$

Unfortunately, the algebra is still very involved; rather than substitute expressions for areas into Equation 3.27*d*, we will find it much easier to obtain s_2 by using the expression for $M(\eta)$ in Equation 3.11. The two components of this expression, divided by *EI*, are shown in Figures 3.23*g* and 3.23*h*. The quantity s_2 in Figure 3.23*f* is now given by

$$\begin{aligned} s_2 &= \frac{2}{3}bA_1'' + \frac{3}{4}bA_2'' \\ &= \frac{2b}{3}\left[\left(\frac{b}{2}\right)\left(\frac{q_o b^2}{2EI}\right)\right] + \frac{3b}{4}\left[\left(\frac{b}{3}\right)\left(-\frac{q_o b^2}{2EI}\right)\right] \\ &= \frac{q_o b^4}{24EI} \end{aligned} \qquad \textbf{(3.27e)}$$

Finally, we obtain $\phi(a^+)$ by substituting Equations 3.27*a* and 3.27*e* into Equation 3.27*b*; the result is

$$\phi(a^+) = \frac{q_o}{24bEI}(4a^3l - a^4 - b^4) \qquad \textbf{(3.27f)}$$

As the preceding examples illustrate, the moment-area method offers a fairly direct and computationally flexible alternative to the solution of differential equations. The method can become arduous when we seek algebraic rather than numeric solutions, but that is true of all analysis procedures. The moment-area method does require careful attention to signs of computed quantities and, in this respect, is much more difficult to apply effectively to large structural problems than virtual work, our next topic. It should be reemphasized, however, that an understanding of the physical significance of integrals having the forms of Equations 3.24*b* and 3.24*c* will be important in connection with subsequent theoretical developments.

PROBLEMS

3.P1–P14. Derive the analytic expressions for internal shear and bending moment. Sketch the shear and bending moment diagrams.

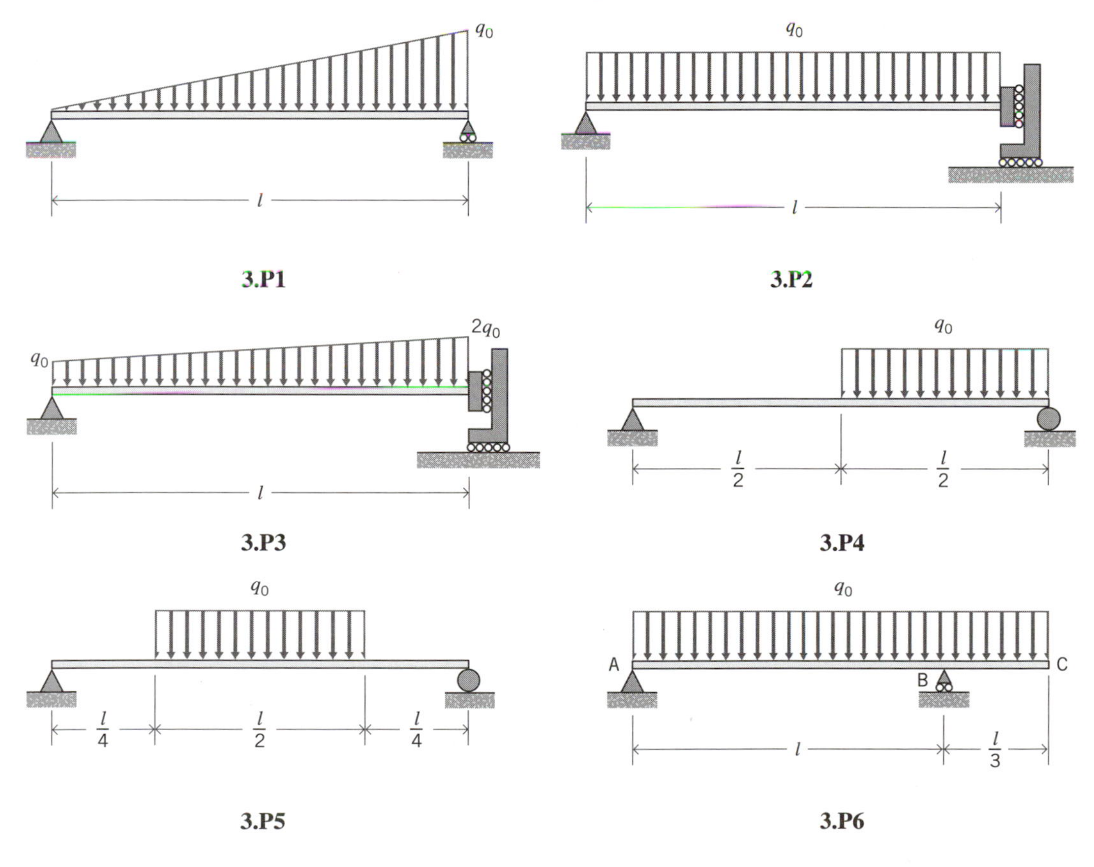

3.P1

3.P2

3.P3

3.P4

3.P5

3.P6

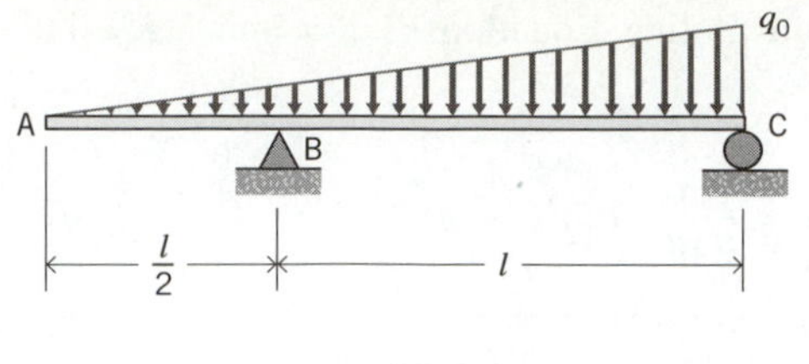

3.P7

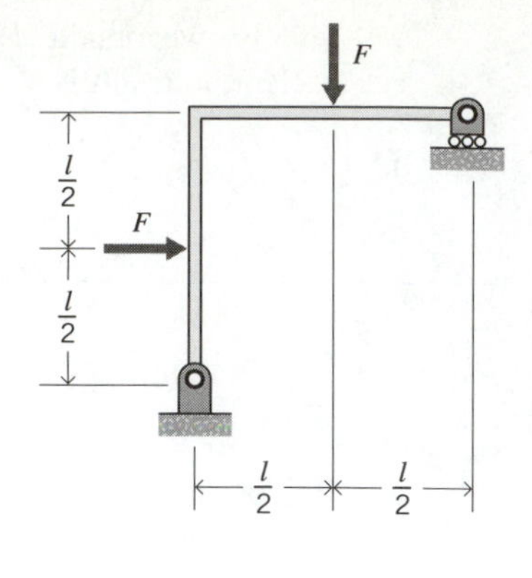

3.P8

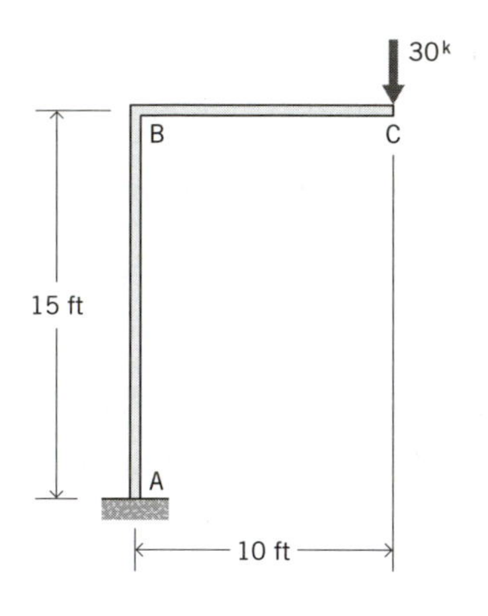

3.P9

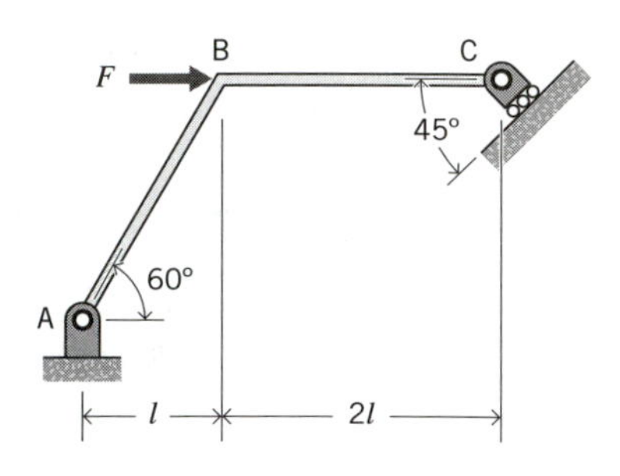

3.P10

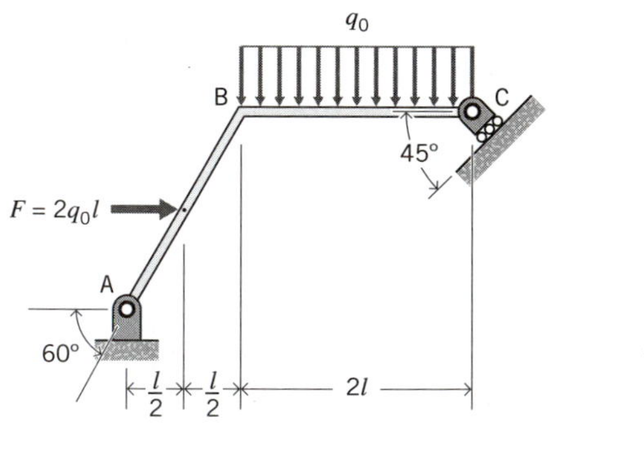

3.P11

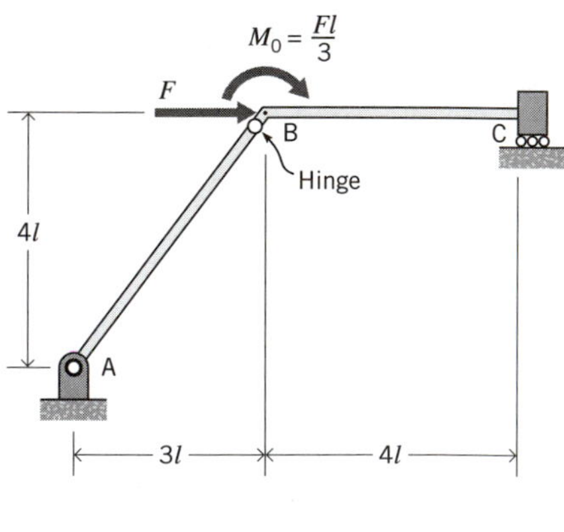

3.P12

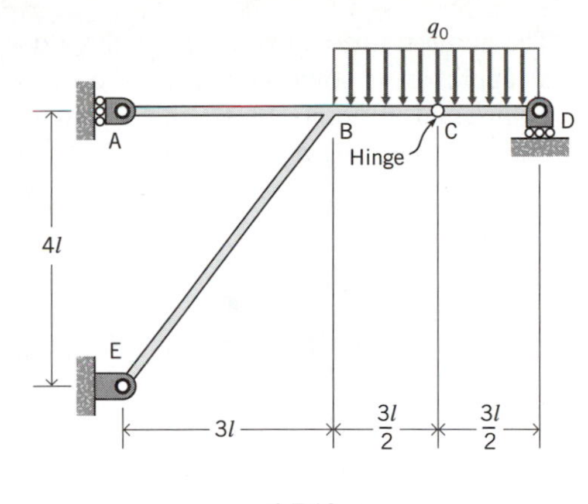

3.P13

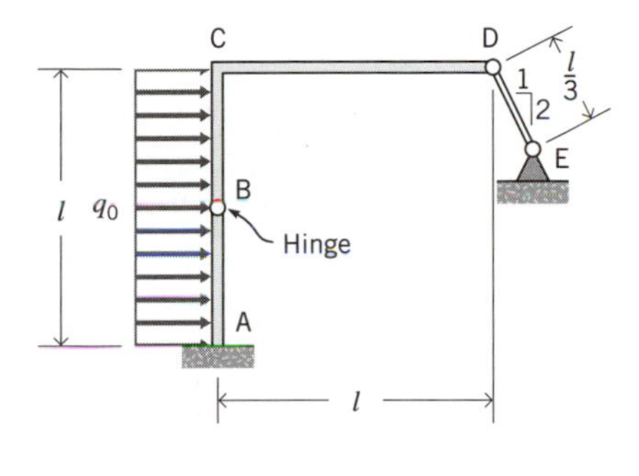

3.P14

3.P15. Evaluate the forces in the truss bars and in beam/truss element BCD. Sketch the shear and bending moment diagrams for element BCD.

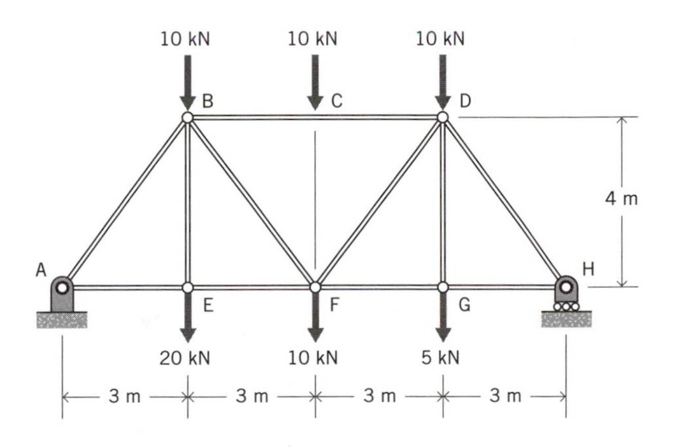

3.P15

3.P16. The portal frame shown below is three degrees indeterminate. To obtain an approximate solution, hinges (shown as dotted circles) are assumed to be inserted at points B, D, and F. Sketch the approximate shear, axial force, and bending moment diagrams for the frame.

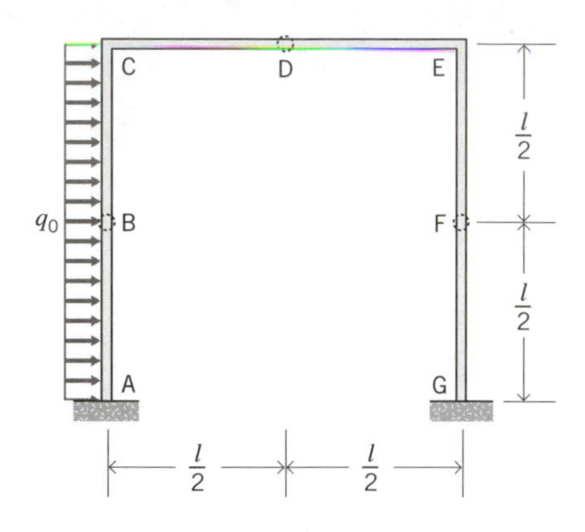

3.P17–P22 Find the analytic expressions for slope and transverse displacement for the beam in Problem 3.P1–P6. Sketch the slope and displacement diagrams. Find the maximum transverse displacement. Assume that EI is constant.

3.P23. (a) Find $\phi(x)$ and $v(x)$ for the beam in Figure 3.5a. Assume that EI is constant.

(b) For the same beam, find $\phi(x_1)$, $\phi(x_2)$, $v(x_1)$, $v(x_2)$, where the coordinates x_1 and x_2 are shown in the figure below. Compare to the solution in (a).

(c) Find the maximum displacement.

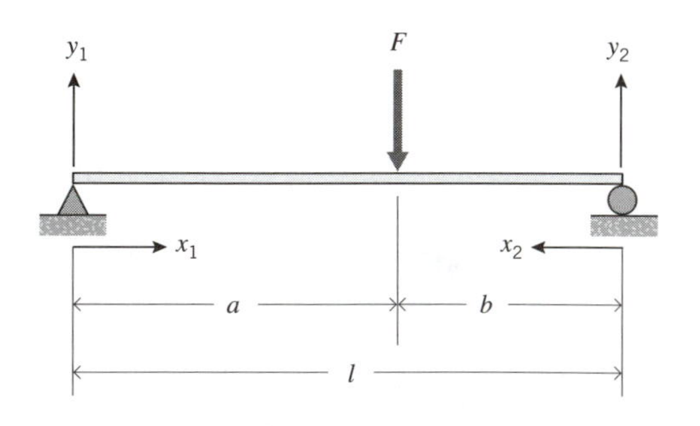

3.P24. Consider the solution given in Example 7.
(a) Solve for the value of a for which $\phi(a^+) = \phi(a^-)$;
(b) Discuss the physical significance of the solution that is obtained from Equation 3.22c as $a \to l$.

3.P25–P27. Find the end slopes for the beam in Problem 3.P1, 3.P4, 3.P5. Assume that EI is constant.

3.P28. The cantilever structure below consists of two identical beams connected by a hinge with a torsional spring joining the contiguous ends of the beams. The torsional spring rotates under the application of a moment, i.e., $\phi = M/k$ where k is the linear spring constant. (This is analogous to an axial spring, which elongates under the application of an axial force such that $\Delta = P/k_a$.) Find the deflection at the free end. Discuss the physical significance of the solution that is obtained as $k \to \infty$.

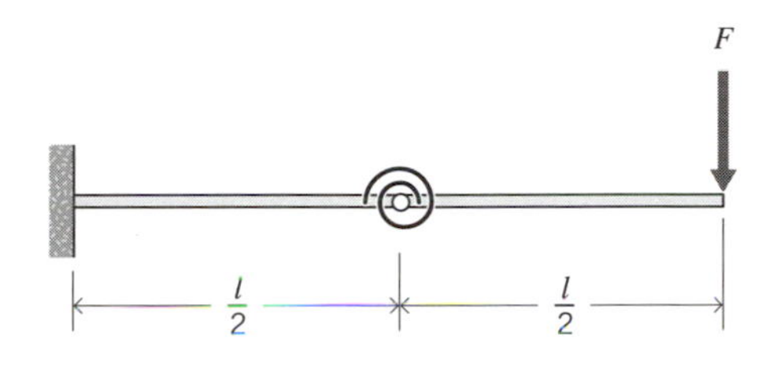

3.P29–P34. Sketch the deflected shape of the structure in Problem 3.P8–P13. Assume that EI is constant and identical for all beams. Show the magnitudes of key slopes and displacements.

3.P35. (a) Find the transverse displacement at the midpoint.
(b) Find the maximum transverse displacement.

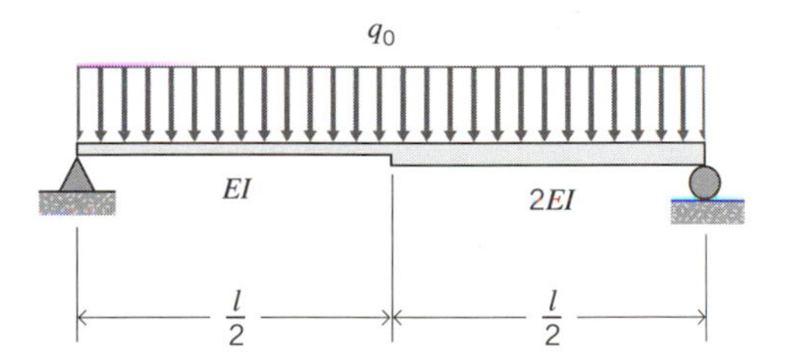

3.P36. Use the moment-area method to calculate the vertical displacement at C for the structure in Problem 3.P9. Assume that EI is constant and identical for both beams.

3.P37. Use the moment-area method to find the transverse displacement at $x = a$ for the beam in Figure 3.9a. Assume that EI is constant.

3.P38–P39. Use the moment-area method to find the transverse displacement at the midpoint, and the end slopes, of the beam in Problem 3.P4–P5. Assume that EI is constant.

3.P40. The beam shown below is one-degree indeterminate; find $v(x)$. Hint: consider the reaction at A as a force of unknown magnitude and use the kinematic boundary conditions to evaluate the constants of integration *and* the unknown reaction.

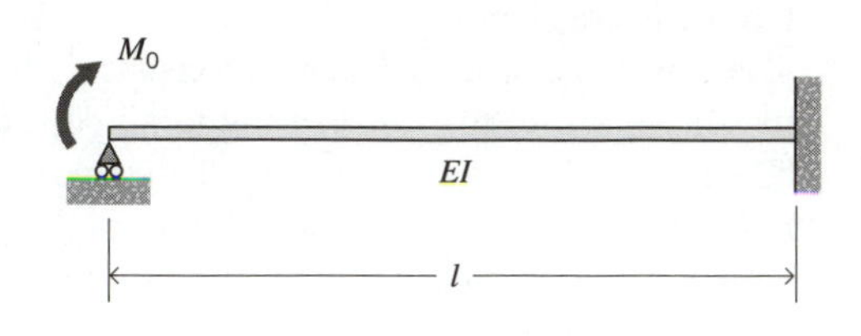

Chapter 4

Virtual Work

4.0 INTRODUCTION

Virtual work is the single most powerful and versatile analytic concept used in modern structural analysis. Among other things, it is a key link in the formulations of the two basic procedures in which we are interested, namely the force method and the displacement method. It may also be shown to provide a basis for most of the energy theorems employed in structural analysis.

Virtual work is essentially a combination of two of the three fundamental concepts that are basic to any structural analysis, i.e., information about equilibrium of forces and information about motion (displacement) of the body. Virtual work does not utilize the relationship between forces and displacements. In fact, analytic expressions of virtual work do not even require the existence of a relationship between forces and displacements. The surprising implication of this last statement is that the equilibrium and displacement information contained in a virtual work theorem can, therefore, be totally independent of each other, i.e., totally unrelated in the sense that neither is a consequence of the other. This ''unrelatedness'' leads to two distinct forms of virtual work, i.e., *virtual forces* and *virtual displacements.* In the former, displacements are real and forces are ''arbitrary'' or virtual; in the latter, forces are real and displacements are ''arbitrary'' or virtual. In either case, the virtual state is chosen in such a way as to provide some desired information about the real state. As might be anticipated, these two forms of virtual work ultimately relate to the force method and the displacement method, respectively.

There are several possible ways of deriving and expressing virtual work. In the approach taken here we will develop virtual work theorems that are specific to structures composed of truss bars and beams. As in previous developments, we will begin with a discussion of virtual work for trusses, and then use analogous procedures to establish the corresponding relationships for frames.

Specifically, we will begin by describing the concept of virtual work[1] for bodies that

---

[1] A rudimentary form of the principle of virtual work was first formulated by the mathematician John Bernoulli in the early 1700s.

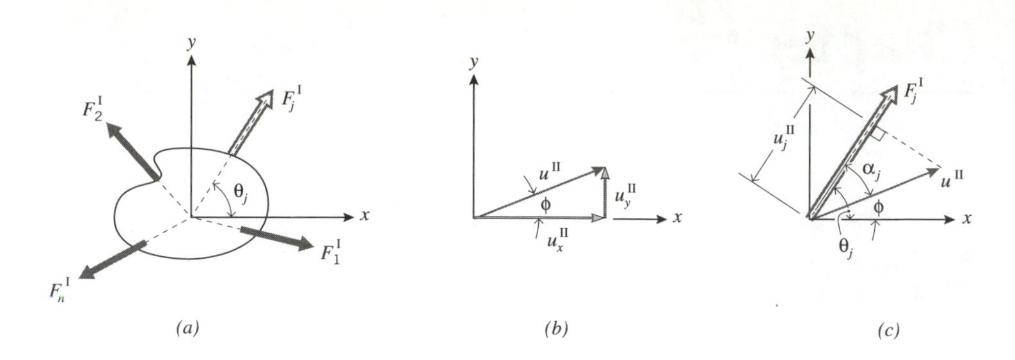

Figure 4.1 Force and displacement states, concurrent forces.

are acted upon by a set of concurrent forces in the same manner as truss joints. We will then derive a virtual work expression for axially loaded bars, and the virtual work expressions for joints and bars will be combined into a virtual work theorem for trusses.

4.1 VIRTUAL WORK FOR TRUSS JOINTS

In accordance with the preceding introductory comments, our developments begin with sets of unrelated forces and displacements. Here, and in all subsequent developments, we will emphasize this ''unrelatedness'' by using distinct superscripts to distinguish quantities associated with a ''force state'' from quantities associated with a ''displacement state''; specifically, we will attach the superscript I to all force quantities, and the superscript II to all displacement quantities.

Now, assume we have a body in two-dimensional space that is acted upon by a set of concurrent forces with magnitudes $F_1^I, F_2^I, \ldots, F_j^I, \ldots, F_n^I$ (Figure 4.1*a*); the direction of each force F_j is defined by an angle θ_j measured from the positive x axis, as shown. Further, and most importantly, we assume that the forces are in equilibrium, a condition that we express by the familiar equations

$$\sum_{j=1}^{n} F_{jx}^{I} = 0 \tag{4.1a}$$

$$\sum_{j=1}^{n} F_{jy}^{I} = 0 \tag{4.1b}$$

where F_{jx}^I and F_{jy}^I denote the x and y components of F_j^I, respectively.

Next, assume that the point at which the forces intersect undergoes a displacement u^{II}, with components u_x^{II} and u_y^{II} in the x and y directions, respectively (Figure 4.1*b*). Multiplying Equation 4.1*a* by u_x^{II} and Equation 4.1*b* by u_y^{II}, and adding, gives

$$\left(\sum_{j=1}^{n} F_{jx}^{I}\right)u_x^{II} + \left(\sum_{j=1}^{n} F_{jy}^{I}\right)u_y^{II} = 0 \tag{4.2}$$

Equation 4.2 is typical of the initial relationships from which we will derive the desired specific analytic expressions that we call theorems of virtual work; however, in their initial forms, equations like Equation 4.2 are of no particular use, and do not reveal anything we don't already know. It is only after some manipulation that such equations become interesting and useful. One significant feature of equations like Equation 4.2, which is worth emphasizing at this point, is that the equations contain terms that are products of forces and displacements, i.e., terms that have dimensions of mechanical work. However, because the forces and displacements are unrelated, this work is generally not real; consequently, we refer to it as virtual work.

To proceed in this particular case, consider the interpretation of terms in Equation 4.2; for a typical force F_j^{I}, we have

$$F_{jx}^{\mathrm{I}} = F_j^{\mathrm{I}} \cos \theta_j \qquad F_{jy}^{\mathrm{I}} = F_j^{\mathrm{I}} \sin \theta_j \tag{4.3a}$$

Also, the components of u^{II} are given by

$$u_x^{\mathrm{II}} = u^{\mathrm{II}} \cos \phi \qquad u_y^{\mathrm{II}} = u^{\mathrm{II}} \sin \phi \tag{4.3b}$$

After substituting Equations 4.3*a* and 4.3*b*, Equation 4.2 can be written as

$$\sum_{j=1}^{n} (F_j^{\mathrm{I}} \cos \theta_j)(u^{\mathrm{II}} \cos \phi) + \sum_{j=1}^{n} (F_j^{\mathrm{I}} \sin \theta_j)(u^{\mathrm{II}} \sin \phi) = 0 \tag{4.3c}$$

or

$$\sum_{j=1}^{n} (F_j^{\mathrm{I}} u^{\mathrm{II}})(\cos \theta_j \cos \phi + \sin \theta_j \sin \phi) = 0 \tag{4.3d}$$

which is equivalent to

$$\sum_{j=1}^{n} F_j^{\mathrm{I}} u^{\mathrm{II}} \cos (\theta_j - \phi) = 0 \tag{4.3e}$$

Angle $\theta_j - \phi$ is the angle between force $\mathrm{F}_j^{\mathrm{I}}$ and the direction of u^{II}, which is denoted by α_j in Figure 4.1*c*. So, Equation 4.3*e* becomes

$$\sum_{j=1}^{n} F_j^{\mathrm{I}} u^{\mathrm{II}} \cos \alpha_j = 0 \tag{4.3f}$$

But $u^{\mathrm{II}} \cos \alpha_j$ is *the component of* u^{II} *in the direction of* F_j^{I}; we will denote this component by u_j^{II}. Thus, the final form of Equation 4.2 is

$$\boxed{\sum_{j=1}^{n} F_j^{\mathrm{I}} u_j^{\mathrm{II}} = 0} \tag{4.3g}$$

This is the theorem of virtual work for a body subjected to concurrent forces. This specific form of the theorem is still not particularly useful by itself; this is such an elementary type of body that the most we can hope to obtain from the theorem is what we started with, i.e., the equivalent of Equation 4.2. Equation 4.3*g* is, however, an important ingredient in the development of the theorem of virtual work for structures composed of assemblies of deformable components, such as trusses. We will see shortly that the theorem becomes much more powerful when it is developed for such a structure.

4.2 VIRTUAL WORK FOR AXIALLY LOADED BARS

For the sake of generality, we will once again begin with a bar of the type shown in Figure 2.1. Regardless of whether or not the force state is "real," we will assume that this state is consistent with the types of forces discussed for bars in Section 2.1, and that these forces are *in equilibrium.* This implies that the forces satisfy Equation 2.1*d*, which we now rewrite as

$$P'^{\mathrm{I}}(x) + p^{\mathrm{I}}(x) = 0 \tag{4.4}$$

We also assume the existence of an unrelated axial displacement field $u^{\mathrm{II}}(x)$, Figure 2.16*a*. Again, regardless of whether or not the displacement state is "real" we will assume that this state is consistent with the type of displacements discussed for bars in Section 2.4; the axial strain, which is the only significant component of strain, is therefore related to displacement by Equation 2.32.

Following a formulation analogous to that used in the preceding section, we multiply Equation 4.4 by $u^{\mathrm{II}}(x)$ and, because the terms in Equation 4.4 have dimensions of force/length, we multiply by dx to obtain "force." Furthermore, since forces and displacements are functions of x, we integrate over x to obtain an expression of virtual work for the entire bar. The initial relationship in this case is thus

$$\int_0^l [P'^{\mathrm{I}}(x) + p^{\mathrm{I}}(x)]u^{\mathrm{II}}(x)dx = 0 \tag{4.5a}$$

Equation 4.5*a* must hold true for any finite $u^{\mathrm{II}}(x)$ because the terms in the bracket are always zero due to equilibrium. Expanding Equation 4.5*a* gives

$$\int_0^l P'^{\mathrm{I}}(x)u^{\mathrm{II}}(x)dx + \int_0^l p^{\mathrm{I}}(x)u^{\mathrm{II}}(x)dx = 0 \tag{4.5b}$$

The first integrand above can be rewritten as the difference of two terms, i.e.,[2]

$$\int_0^l \{[P^{\mathrm{I}}(x)u^{\mathrm{II}}(x)]' - P^{\mathrm{I}}(x)u'^{\mathrm{II}}(x)\}dx + \int_0^l p^{\mathrm{I}}(x)u^{\mathrm{II}}(x)dx = 0 \tag{4.5c}$$

Expanding the first integral and replacing $u'^{\mathrm{II}}(x)$ in the second integral by strain (Equation 2.32) leads to

$$\int_0^l [P^{\mathrm{I}}(x)u^{\mathrm{II}}(x)]'dx - \int_0^l P^{\mathrm{I}}(x)\epsilon^{\mathrm{II}}(x)dx + \int_0^l p^{\mathrm{I}}(x)u^{\mathrm{II}}(x)dx = 0 \tag{4.5d}$$

---

[2] This procedure is equivalent to integration by parts. Equation 4.5*c* is valid provided $u^{\mathrm{II}}(x)$ can be differentiated throughout the interval $0 \leq x \leq l$. This is a slight restriction on $u^{\mathrm{II}}(x)$.

The first integrand is now a quantity that is differentiated with respect to x; the integral of this term is therefore the quantity itself. Since this is a definite integral we obtain

$$P^{\mathrm{I}}(x)u^{\mathrm{II}}(x)\big|_0^l + \int_0^l p^{\mathrm{I}}(x)u^{\mathrm{II}}(x)dx = \int_0^l P^{\mathrm{I}}(x)\epsilon^{\mathrm{II}}(x)dx \qquad \textbf{(4.5e)}$$

or

$$P^{\mathrm{I}}(l)u^{\mathrm{II}}(l) - P^{\mathrm{I}}(0)u^{\mathrm{II}}(0) + \int_0^l p^{\mathrm{I}}(x)u^{\mathrm{II}}(x)dx = \int_0^l P^{\mathrm{I}}(x)\epsilon^{\mathrm{II}}(x)dx \qquad \textbf{(4.5f)}$$

The force quantities $P^{\mathrm{I}}(l)$ and $P^{\mathrm{I}}(0)$ are resultant internal forces at the ends of the bar; we choose to replace these internal forces by the external end loads, P_{B} and P_{A}. Recall that equilibrium requires that $P^{\mathrm{I}}(l) = P^{\mathrm{I}}_{\mathrm{B}}$ and $P^{\mathrm{I}}(0) = -P^{\mathrm{I}}_{\mathrm{A}}$; the equation now becomes

$$P^{\mathrm{I}}_{\mathrm{B}}u^{\mathrm{II}}(l) + P^{\mathrm{I}}_{\mathrm{A}}u^{\mathrm{II}}(0) + \int_0^l p^{\mathrm{I}}(x)u^{\mathrm{II}}(x)dx = \int_0^l P^{\mathrm{I}}(x)\epsilon^{\mathrm{II}}(x)dx \qquad \textbf{(4.5g)}$$

or, replacing $u^{\mathrm{II}}(l)$ and $u^{\mathrm{II}}(0)$ by $u^{\mathrm{II}}_{\mathrm{B}}$ and $u^{\mathrm{II}}_{\mathrm{A}}$, respectively

$$\boxed{P^{\mathrm{I}}_{\mathrm{B}}u^{\mathrm{II}}_{\mathrm{B}} + P^{\mathrm{I}}_{\mathrm{A}}u^{\mathrm{II}}_{\mathrm{A}} + \int_0^l p^{\mathrm{I}}(x)u^{\mathrm{II}}(x)dx = \int_0^l P^{\mathrm{I}}(x)\epsilon^{\mathrm{II}}(x)dx} \qquad \textbf{(4.5h)}$$

which is the desired form of the theorem of virtual work for a bar. In this form, we observe that all forces on the left side of Equation 4.5*h* are external applied loads; consequently, we refer to the terms on the left side of Equation 4.5*h* collectively as external virtual work. The force on the right side of Equation 4.5*h* is the internal (resultant) force; consequently, we refer to the term on the right side of Equation 4.5*h* as internal virtual work. In the context of these definitions, Equation 4.5*h* states that external virtual work equals internal virtual work. We may occasionally state this symbolically as

$$W_{EV} = W_{IV} \qquad \textbf{(4.5i)}$$

In the special case of a truss bar for which $p^{\mathrm{I}}(x) = 0$, we have $P^{\mathrm{I}}(x) = P^{\mathrm{I}}$, a constant; consequently, Equation 4.5*h* reduces to

$$P^{\mathrm{I}}_{\mathrm{B}}u^{\mathrm{II}}_{\mathrm{B}} + P^{\mathrm{I}}_{\mathrm{A}}u^{\mathrm{II}}_{\mathrm{A}} = P^{\mathrm{I}}\int_0^l \epsilon^{\mathrm{II}}(x)dx \qquad \textbf{(4.5j)}$$

or, since the integral is the change in length of the bar, which we will denote by Δ^{II},

$$\boxed{P^{\mathrm{I}}_{\mathrm{B}}u^{\mathrm{II}}_{\mathrm{B}} + P^{\mathrm{I}}_{\mathrm{A}}u^{\mathrm{II}}_{\mathrm{A}} = P^{\mathrm{I}}\Delta^{\mathrm{II}}} \qquad \textbf{(4.5k)}$$

Once again, the specific form of the theorem of virtual work in Equation 4.5*k* is not particularly useful by itself. A truss bar is still a very elementary component whose

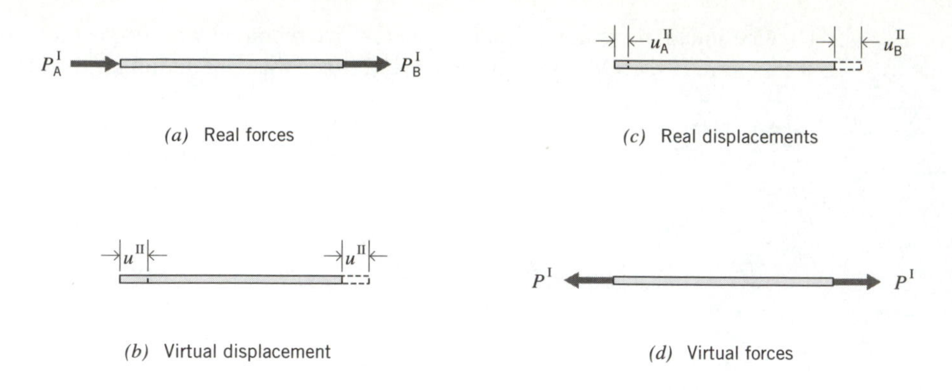

Figure 4.2 ''Trivial'' application of virtual work for bars: (*a*) and (*b*) = virtual displacements, (*c*) and (*d*) = virtual forces.

behavior is described by simple relations, and the most we can hope to obtain from the theorem is the equivalent of what we started with. However, to illustrate the direction in which our discussion of virtual work is heading, we will demonstrate some elementary applications of Equation 4.5*k*.

Suppose we have a bar subjected only to real end loads (Figure 4.2*a*) and suppose we wish to use Equation 4.5*k* to determine the relation between P_A^I and P_B^I rather than use equations of equilibrium directly. We may accomplish this by using virtual displacements; specifically, we may choose as the virtual displacement a rigid-body translation of the bar. For example, imagine that the virtual displacement consists of moving the bar a distance u^{II} to the right of its original position (Figure 4.2*b*). This implies that every point on the bar undergoes the same displacement; consequently, there is no change in length of the bar, i.e., $\Delta^{II} = 0$. From Equation 4.5*k* we obtain

$$P_B^I u^{II} + P_A^I u^{II} = 0 \tag{4.6a}$$

Here, note that both terms are positive because forces in Figure 4.2*a* act in the same directions as displacements in Figure 4.2*b*. Dividing Equation 4.6*a* by u^{II} leaves $P_B^I + P_A^I = 0$, exactly as would be obtained from an equation of equilibrium. In fact, this should be expected because virtual displacements must provide information about the force state, and the force state has been assumed to be in equilibrium from the outset.

Next, consider the bar in Figure 4.2*c*, which is assumed to undergo real displacements such that the displacement at end A is u_A^{II} and the displacement at end B is u_B^{II}. We wish to ascertain the type of information that Equation 4.5*k* will provide about these displacements. We choose the virtual force state shown in Figure 4.2*d*, which consists of forces in equilibrium, i.e., the end forces and internal resultant force all have magnitude P^I. From Equation 4.5*k* we now obtain

$$P^I u_B^{II} - P^I u_A^{II} = P^I \Delta^{II} \tag{4.6b}$$

Here, note that the sign of the second term is negative because the external virtual force P^I at end A and u_A^{II} are opposite in direction. Dividing by P^I now leaves the familiar result $u_B^{II} - u_A^{II} = \Delta^{II}$. In other words, in this case, the virtual force form of the theorem of virtual work has provided information about the relation between end displacements and bar deformation.

4.3 VIRTUAL WORK FOR TRUSSES

We now have sufficient information on which to base the derivation of the theorem of virtual work for trusses; in particular, we will use Equation 4.3*g* to account for truss joints, and Equation 4.5*k* to account for truss bars.

Let the truss in Figure 4.3*a* be representative of this class of structures. Assume that we have a force state *in equilibrium*, and an unrelated truss-type displacement state. In the force state, equilibrium is satisfied at every joint and in every bar (Figure 4.3*b*). The displacement state is subject to the same ''small strain'' limitation discussed for bars in Equation 2.32; in addition, the displacements are assumed to be ''compatible'' in the sense that bar deformations are related to joint displacements (Figure 4.3*c*).

At a typical joint, such as A in Figure 4.3*b*, there may be loads acting external to the truss, which we will now denote by F_{Aj}^{I}, and forces that are internal to the truss bars, denoted by P_{Ak}^{I}. For the sake of clarity and to emphasize the state of equilibrium, we have chosen to show the truss bars with equal and opposite end forces rather than with the end forces shown in Figure 2.1. Therefore, $P_{Ak}^{I} = P_{Bk}^{I} \equiv P_k^{I}$ in Figure 4.36. Regardless of which representation is used, a force such as P_{Ak}^{I} occurs in equal and opposite pairs on the contiguous ends of the bar at the joint.

A general displacement state is illustrated in Figure 4.3*c*. Note that this figure shows all joints displacing in an arbitrary manner. Such arbitrary displacements are permissible if the displacement state is not real, i.e., is virtual; if the displacement state is real, displacements at supports must be consistent with the kinematic constraints imposed by the supports, e.g., joint C in Figure 4.3*c* could not displace and joint B could displace only in the direction tangential to the support surface. Now consider, in particular, the displacement of joint A, which is denoted by u_A^{II} in Figure 4.3*c*. Associated with u_A^{II} are components of displacement in the directions of each load F_{Aj}^{I} and each force P_k^{I}; we will denote these components by u_{Aj}^{II} and u_{Ak}^{II}, respectively (Figure 4.3*d*).

The virtual work expression for joint A (Equation 4.3*g*) can now be modified to distinguish between applied loads $F_A^I j$ and bar forces P_k^I; we will write the modified equation in the form

$$\sum_{j=1}^{n_A} F_{Aj}^{I} u_{Aj}^{II} + \sum_{k=1}^{m_A} P_k^{I} u_{Ak}^{II} = 0 \tag{4.7a}$$

where n_A denotes the number of applied loads at joint A, and m_A denotes the number of bar forces at joint A.

The virtual work expression for a bar such as AB (Equation 4.5*k*) must also be modified to be consistent with the directions and notations of the end forces shown in Figure 4.3*b*; the result is

$$P_k^{I} u_{Bk}^{II} - P_k^{I} u_{Ak}^{II} = P_k^{I} \Delta_k^{II} \tag{4.7b}$$

i.e., we have replaced P_B^I by P_k^I and P_A^I by $-P_k^I$.

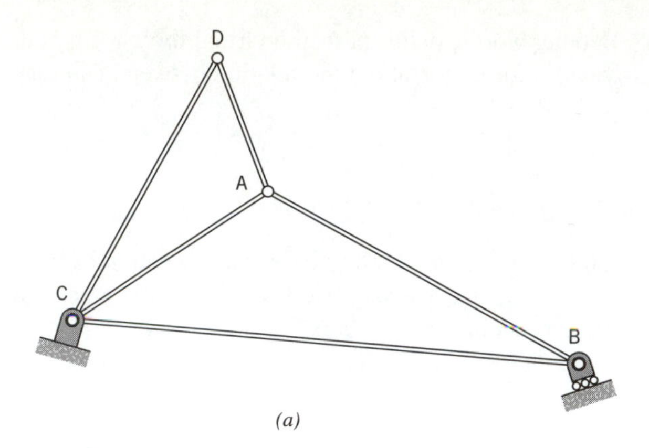

(a)

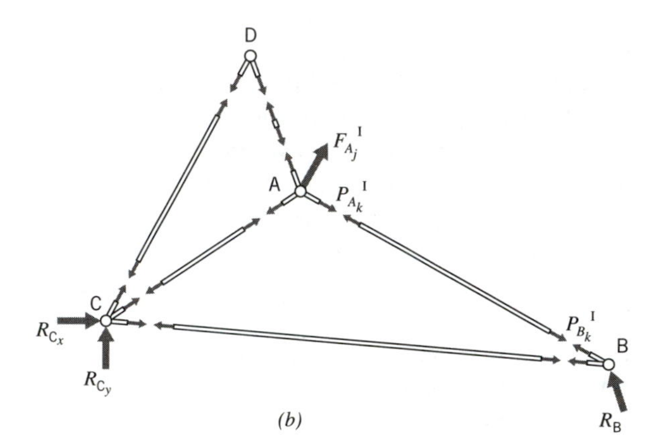

(b)

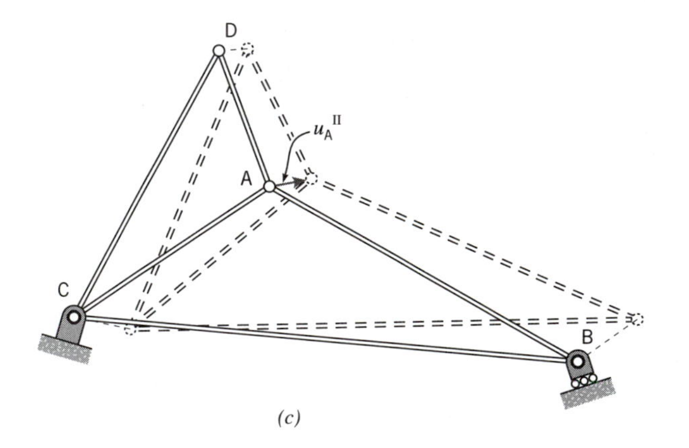

(c)

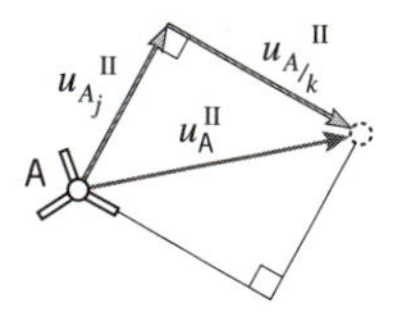

(d)

Figure 4.3 Virtual work for trusses.

To obtain an expression of virtual work for the entire truss, we must add together the virtual work expressions for all joints and all bars. Thus, we add equations of form 4.7*a* and 4.7*b* for each joint and each bar. Note, however, that both Equations 4.7*a* and 4.7*b* contain a term of form $P_k^{\mathrm{I}} u_{\mathrm{A}k}^{\mathrm{II}}$, but that the terms are of opposite sign; the opposite sign is a result of oppositely directed forces but a single direction of displacement. Consequently, when these equations are added, these terms cancel each other. Similarly, terms of form $P_k^{\mathrm{I}} u_{\mathrm{B}k}^{\mathrm{II}}$ cancel each other when the expression for virtual work at joint B is added. It follows that the virtual work for the entire truss will contain only terms involving the applied loads and the internal virtual work for the bars; we may write this as

$$\boxed{\sum_{j=1}^{NL} F_j^{\mathrm{I}} u_j^{\mathrm{II}} = \sum_{k=1}^{NB} P_k^{\mathrm{I}} \Delta_k^{\mathrm{II}}} \tag{4.8}$$

where *NL* now denotes the total number of applied (joint) loads (including reactions) on the entire truss u_j^{II} is the component of joint displacement in the direction of load F_j^{I}, *NB* is the total number of bars in the truss, and Δ_k^{II} is the change in length of bar *k*. Equation 4.8 is the theorem of virtual work for trusses. Once again, we identify the terms on the left side of the equation as external virtual work, and those on the right side of the equation as internal virtual work.

Equation 4.8 is a very important and useful relation with applications to both statically determinate and indeterminate trusses. Before illustrating some of these applications, it is worth reiterating the major points raised in the introduction to this chapter. First, and foremost, the force state and the displacement state are unrelated; generally, we consider one of these states to be "real," in which case the other is "virtual." When the force state is virtual, we refer to Equation 4.8 as the theorem of virtual forces; when the displacement state is virtual we refer to Equation 4.8 as the theorem of virtual displacements. In either application, the virtual state is chosen in such a way as to provide some desired information about the real state.

We will conclude by noting that when the displacement state is real (virtual forces), the theorem, Equation 4.8, requires the real length change of each truss bar, Δ_k^{II}. These length changes are a consequence of the factors discussed in Chapter 2, i.e., thermal effects, initial effects, or real forces (Equation 2.38*d*); in the latter case, the real forces are separate and distinct from the virtual forces.

EXAMPLE 1 APPLICATION OF VIRTUAL DISPLACEMENTS.

(a) Use Equation 4.8 to find R_{B} for the truss in Figure 2.8*b* and Figure 2.9*a*.
(b) Use Equation 4.8 to find the force in bar CD of this truss.

(a) The first application is elementary because the result has already been easily obtained from equilibrium equations in Chapter 2; however, it provides a useful introduction to the use of Equation 4.8. Here, we are asked to find the value of a real force, R_B (henceforth denoted R_B^I); consequently, we must utilize virtual displacements.

For the sake of convenience, the free-body diagram of the truss subjected to the real loads is repeated in Figure 4.4*a*. An appropriate virtual displacement is shown in Figure 4.4*b*. This virtual displacement state consists of an arbitrary small rigid-body rotation of the truss about point A, which we define by the angle θ^{II} shown.[3] We have chosen this particular virtual displacement because, as we will demonstrate, the only unknown force that will then appear in Equation 4.8 is R_B^I. Furthermore, since we have chosen a rigid-body displacement, it follows that there are no length changes Δ_k^{II}.

Consider the application of Equation 4.8 given the force state in Figure 4.4*a* and the displacement state in Figure 4.4*b*. Since $\Delta_k^{II} = 0$, the internal virtual work, i.e., the term on the right side of Equation 4.8, is zero; this is the case even though the real bar forces in Figure 4.4*a*, P_k^I, are not zero. We are left with the external virtual work terms, i.e., those on the left side of Equation 4.8. These terms involve the real external (applied) loads, which in this case consist of F^I applied at joint D and the reactions at joints A and B; each of these real loads is multiplied by a corresponding component of virtual joint displacement in the same direction as the load. In particular, R_B^I is multiplied by the vertical component of displacement of joint B, denoted by u_{By}^{II} in Figure 4.4*b*, and F^I is multiplied by the horizontal component of displacement of joint D, denoted by u_{Dx}^{II} in Figure 4.4*b*. Note that there are no components of displacement of joint A in Figure 4.4*b* and, consequently, there is no virtual work associated with the reactions at A (Figure 4.4*a*). Also note that there is a vertical component of displacement of joint D, but this does not contribute to the external virtual work unless there is a vertical component of real load at joint D.[4]

The pertinent displacements in Figure 4.4*b* may be related to the angle θ^{II} and the dimensions of the truss, i.e., $u_{By}^{II} \approx 2l\theta^{II}$ for θ^{II} small; similarly, $u_{Dx}^{II} \approx h\theta^{II}$. Before writing Equation 4.8, note that u_{Dx}^{II} in Figure 4.4*b* is a displacement of joint D to the left of its original position, whereas load F^I in Figure 4.4*a* is directed to the right; therefore, the product of these two quantities is a negative virtual work. On the other hand, R_B^I and u_{By}^{II} are in the same direction, and their product is a positive virtual work. Thus, Equation 4.8 reduces to

$$R_B^I(2\,l\theta^{II}) - F^I(h\theta^{II}) = 0 \qquad \textbf{(4.9a)}$$

On cancelling θ^{II}, this is seen to be identical to Equation 2.11, i.e., to the equilibrium equation for sum of moments about point A for the free-body in Figure 4.4*a*. This is actually quite remarkable considering that nothing even remotely resembling a moment-equilibrium equation went into the derivation of Equation 4.8. However, we see that the theorem of virtual displacements does provide an exact equilibrium equation.

(b) Real internal forces, such as P_{CD}^I, may be determined from Equation 4.8 in a related manner; it is, however, necessary to include an internal virtual work term involving P_{CD}^I in the equation. Since we have determined R_B^I above, we may evaluate P_{CD}^I by

[3] The rotation can be chosen either counterclockwise, as shown, or clockwise; the results will be identical.

[4] The horizontal component of virtual displacement at joint B is negligible for small θ; so, even if there were a horizontal component of real load at joint B, there would be no corresponding virtual work.

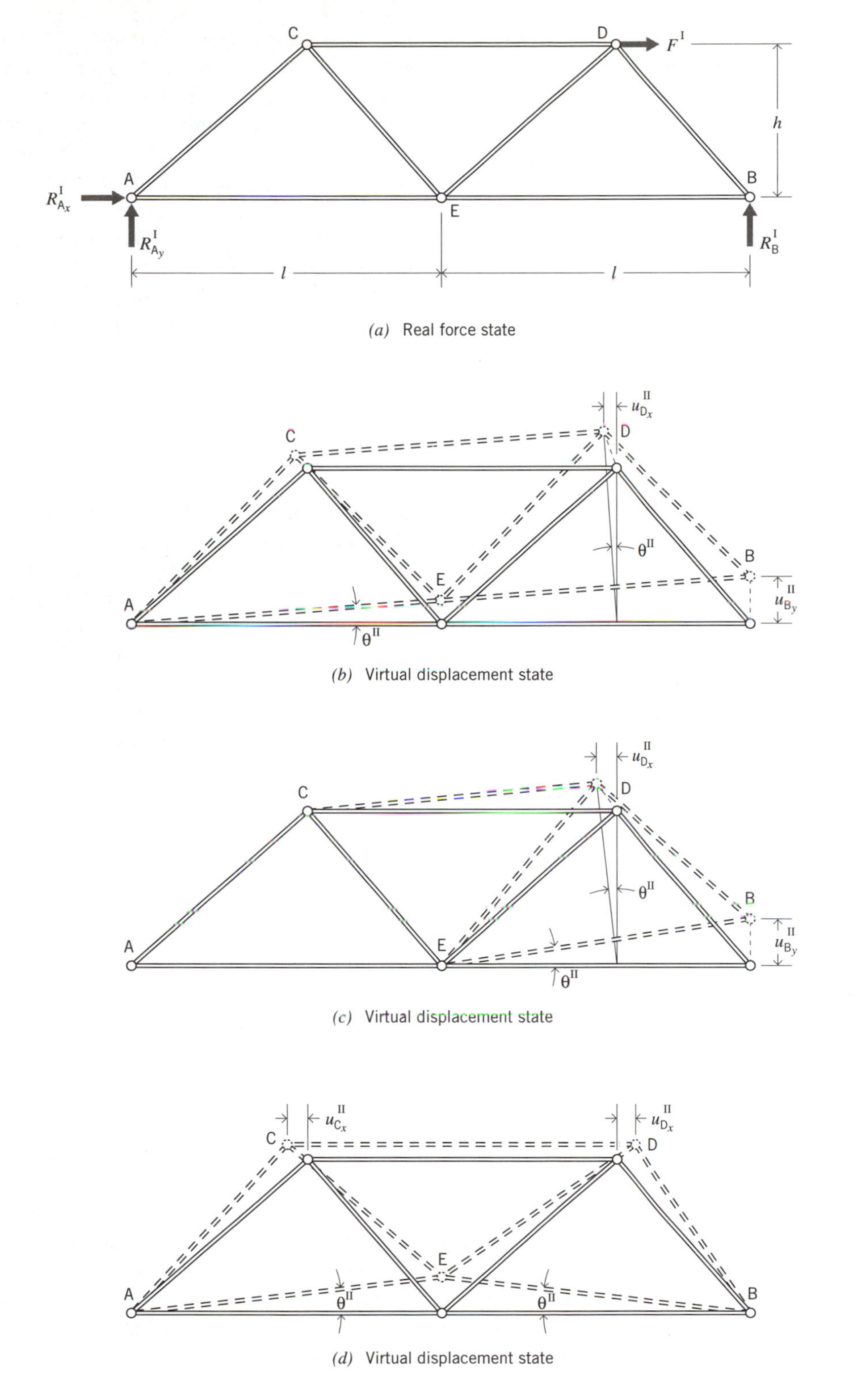

Figure 4.4 Example 1.

choosing a virtual displacement state of the type illustrated in Figure 4.4*c*. This virtual displacement state consists of a rigid-body rotation of triangle BDE, about E, through the arbitrary small angle θ^{II} shown; triangle ACE is assumed to undergo no displacement. As a result of the displacement specified in Figure 4.4*c*, joint B displaces upward an amount $u^{II}_{By} \approx l\theta^{II}$, and joint D displaces to the left by an amount $u^{II}_{Dx} \approx h\theta^{II}$. Since joint C does not displace, it follows that the change in length of bar CD, Δ^{II}_{CD}, is given by $\Delta^{II}_{CD} \approx -u^{II}_{Dx} \approx -h\theta^{II}$, i.e., bar CD gets shorter by an amount equal to the horizontal component of displacement of joint D. This is the only bar that changes length in Figure 4.4*c*; therefore, the only internal virtual work will involve the product of Δ^{II}_{CD} and P^{I}_{CD}. If we assume the unknown force, P^{I}_{CD}, is positive (tensile), this will be opposite in sense to Δ^{II}_{CD}, and the resulting internal virtual work will be negative.

Equation 4.8 can now be written as

$$R^{I}_{B}(l\theta^{II}) - F^{I}(h\theta^{II}) = -P^{I}_{CD}(h\theta^{II}) \tag{4.9b}$$

This is identical to Equation 2.16*a*, and is again an equation of moment equilibrium, in this case for the portion of the truss shown in Figure 2.9*c*.

The preceding computation of P^{I}_{CD} assumed a prior knowledge of R^{I}_{B}, resulting in an equation that has been demonstrated to be equivalent to that obtained in Chapter 2 from an application of the method of sections. It is a remarkable feature of the theorem of virtual work that it is versatile enough to actually enable us to directly compute P^{I}_{CD} without a prior knowledge of any reactions, something that is not possible utilizing the procedures in Chapter 2. To demonstrate, consider the virtual displacement state in Figure 4.4*d*, which is obtained from rigid-body rotations of triangles ACE and BDE, as shown. This state does not involve any displacements of joints A or B; consequently, there is no external virtual work involving reactions. Furthermore, there are no changes in lengths of bars other than bar CD, so there is still no internal virtual work other than the product of P^{I}_{CD} and Δ^{II}_{CD}.

For this virtual displacement state, joint D has a horizontal component of virtual displacement, to the right, of magnitude $h\theta^{II}$, and joint C has a horizontal component of virtual displacement, to the left, also of magnitude $h\theta^{II}$. Therefore, there is a virtual elongation of bar CD of magnitude $2h\theta^{II}$. In this case, if we assume P^{I}_{CD} is tension, the internal virtual work will be positive. We also note that the external virtual work is positive because F^{I} is in the same direction as u^{II}_{Dx}. Equation 4.8 now gives

$$F^{I}(h\theta^{II}) = P^{I}_{CD}(2h\theta^{II}) \tag{4.9c}$$

which is identical to the result in Equation 2.14*b* and Equation 2.16*b*! Equation 4.9*c* is an equilibrium equation that has no direct counterpart in Chapter 2, i.e., we cannot identify a single free-body from which we can obtain this equation directly. In other words, this particular form of equilibrium equation is a unique consequence of virtual work. Bear in mind, however, that there can be only $2NJ$ independent equilibrium equations (NJ = number of joints), and virtual work cannot provide any additional independent equilibrium equations beyond that number. We still cannot analyze indeterminate trusses without considering compatibility relations.

EXAMPLE 2 APPLICATION OF VIRTUAL DISPLACEMENTS.

The truss in Figure 4.5*a* has a slightly different geometry than the truss in the preceding example. Find the force in bar CD if the real applied loads are as shown in Figure 4.5*b*. In this case there is a horizontal load, F_1^{I}, at joint D and a vertical load, F_2^{I}, at joint E.

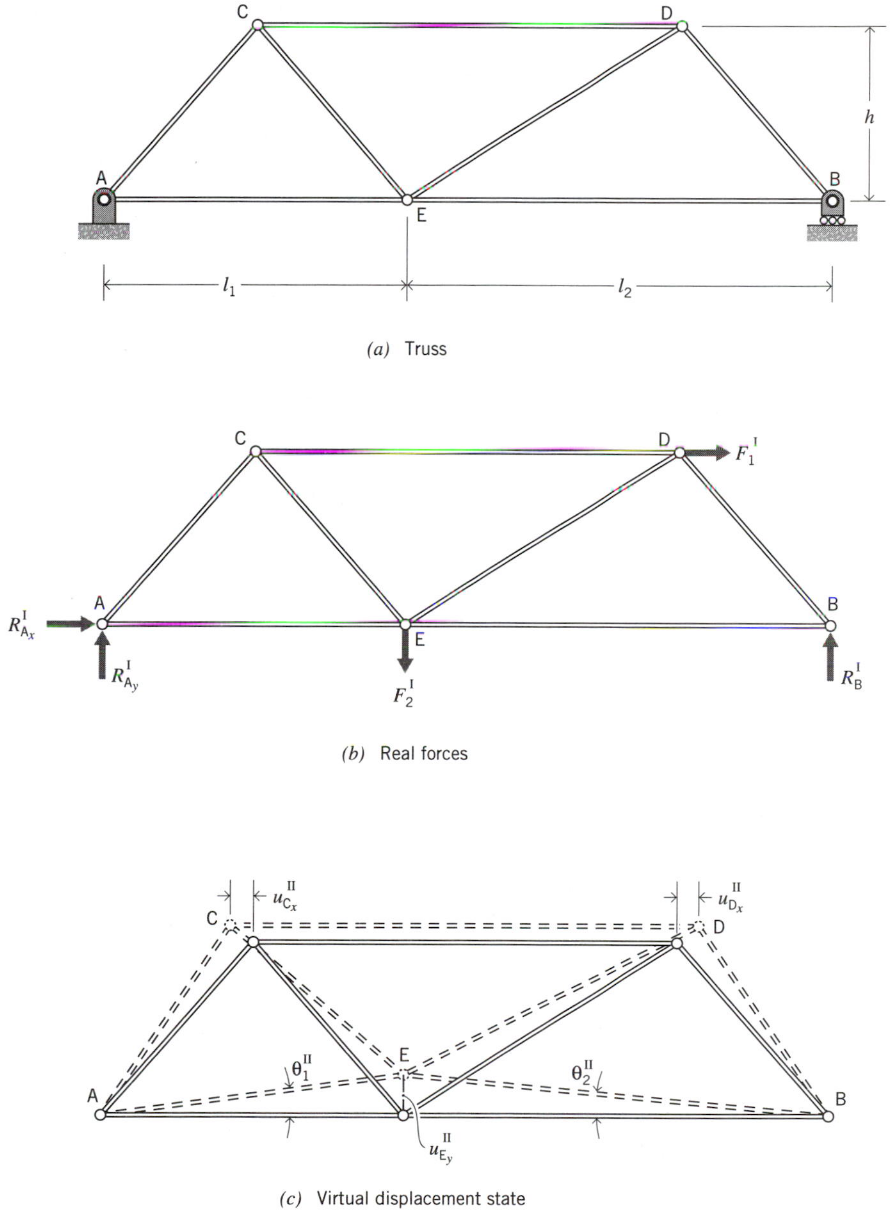

(a) Truss

(b) Real forces

(c) Virtual displacement state

Figure 4.5 Example 2.

The virtual displacement state is shown in Figure 4.5*c*; this state is similar to the one in Figure 4.4*d* except that the small angles θ_1^{II} and θ_2^{II} are not equal. We denote the vertical displacement at joint E by u_{Ey}^{II}, and observe that $u_{Ey}^{II} = l_1\theta_1^{II} = l_2\theta_2^{II}$. It follows that $\theta_2^{II} = (l_1/l_2)\theta_1^{II}$. As in the previous case, this virtual displacement involves horizontal and vertical components of displacement of joints C and D; we are interested in the horizontal components of displacement because they relate to the change in length of bar CD and, also, because the product of the horizontal component of displacement at joint D and F_1^I is part of the external virtual work.

Joint C moves to the left an amount $u_{Cx}^{II} = h\theta_1^{II}$, and joint D moves to the right an amount $u_{Dx}^{II} = h\theta_2^{II}$. This implies an elongation of bar CD, i.e., $\Delta_{CD}^{II} = h(\theta_1^{II} + \theta_2^{II})$; internal virtual work is therefore $P_{CD}^{I}\Delta_{CD}^{II} = P_{CD}^{I}h(\theta_1^{II} + \theta_2^{II})$. In this case, the total external virtual work is $F_1^I u_{Dx}^{II} - F_2^I u_{Ey}^{II}$, so Equation 4.8 becomes

$$F_1^I\left(h\frac{l_1}{l_2}\theta_1^{II}\right) - F_2^I(l_1\theta_1^{II}) = P_{CD}^{I}\left[h\left(1 + \frac{l_1}{l_2}\right)\theta_1^{II}\right] \tag{4.10}$$

which may be solved for P_{CD}^{I}. Equation 4.10 reduces to Equation 4.9*c* when $l_2 = l_1$ and $F_2^I = 0$.

EXAMPLE 3 APPLICATION OF VIRTUAL FORCES.

Use Equation 4.8 to find the horizontal and vertical components of displacement of joint C of the truss in Figure 2.19*a*; the truss is also shown in Figure 4.6*a*. As in Chapter 2, we assume that the change in length of bar AC is Δ_{AC}, now denoted Δ_{AC}^{II}, and the change in length of bar CB is Δ_{CB}, now denoted Δ_{CB}^{II}. Note that Figure 4.6*a* shows all the real displacements of this truss, namely the aforementioned bar elongations, and the joint displacements, u_{Cx}^{II} and u_{Cy}^{II}; there are no displacements of joints A or B. As in Chapter 2, we are not yet specifying what causes the bar elongations; it is sufficient to note that these bar elongations could be a consequence of either thermal or initial effects, or real applied forces.

Because we are dealing with real displacements, we must use virtual forces. We use two distinct virtual force states, one to evaluate u_{Cx}^{II} and one to evaluate u_{Cy}^{II}.

We will begin with the evaluation of u_{Cx}^{II}; Figure 4.6*b* shows the truss subjected to the required external virtual force, namely a horizontal load F^I applied at joint C.[5] We have chosen this force to be in the same direction as that assumed for the displacement component we seek to evaluate, u_{Cx}^{II}, because this gives rise to an external virtual work $F^I u_{Cx}^{II}$. Furthermore, this is the only external virtual work because the only other external virtual forces on the truss in Figure 4.6*b* are the reactions at A and B and, regardless of their magnitudes, there are no real displacements at A or B (Figure 4.6*a*).

Now, regardless of whether it is virtual or real, state I always consists of forces in

[5] This external virtual force is sometimes assigned a magnitude of 1 for ease of computation of internal forces; we will continue to use the general designation F^I for the sake of clarity.

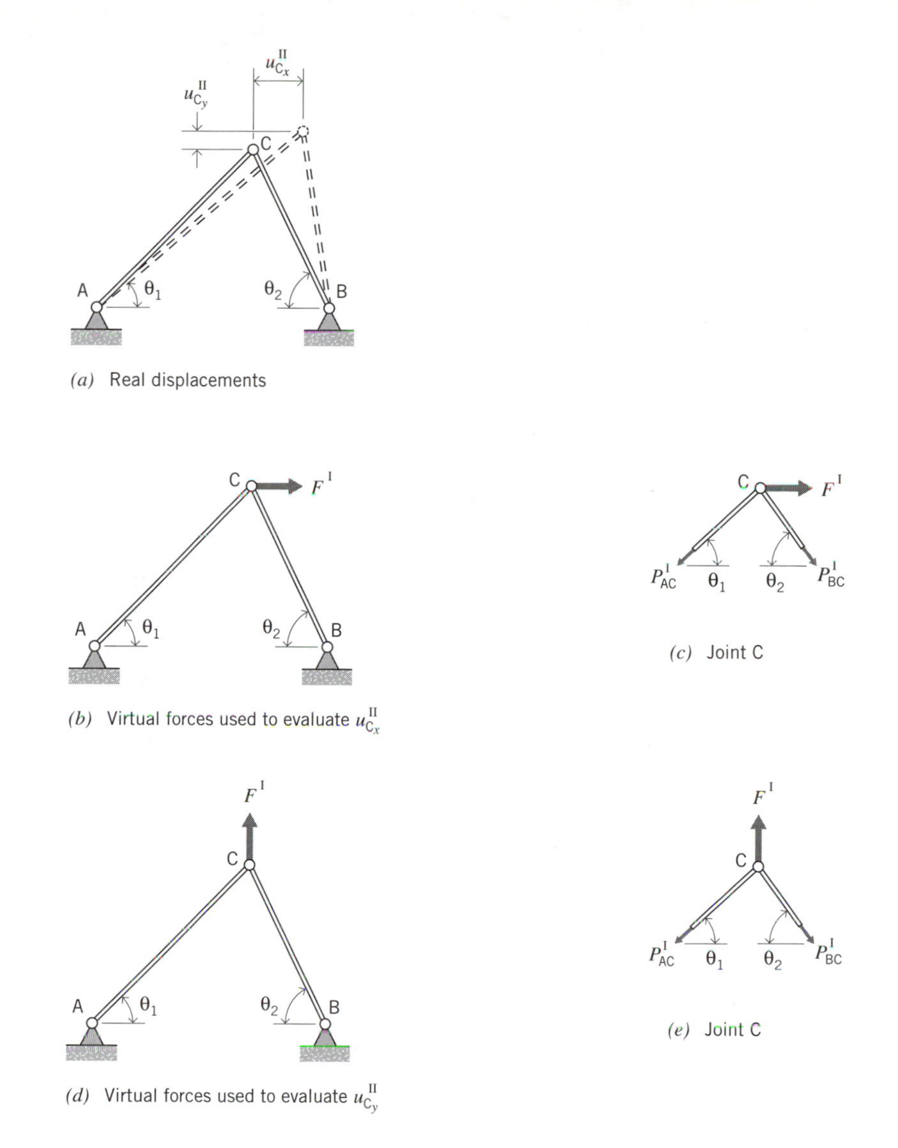

Figure 4.6 Example 3.

equilibrium; therefore, the virtual forces in Figure 4.6*b* must satisfy all equations of equilibrium. This means that joint C (Figure 4.6*c*) must be in equilibrium. For this particular case, bar forces P_{AC}^{I} and P_{BC}^{I} have previously been evaluated in Chapter 2 (see Equations 2.3*a* and 2.3*b*, or Equation 2.10), i.e.,

$$P_{AC}^{I} = \frac{F^{I} \sin \theta_2}{\sin(\theta_1 + \theta_2)} \tag{4.11a}$$

Therefore, internal virtual work is $P^{I}_{AC}\Delta^{II}_{AC} + P^{I}_{BC}\Delta^{II}_{BC}$, and Equation 4.8 becomes

$$F^{I}u^{II}_{Cx} = P^{I}_{AC}\Delta^{II}_{AC} + P^{I}_{BC}\Delta^{II}_{BC} \tag{4.12a}$$

Substituting Equations 4.11*a* and 4.11*b* gives

$$F^{I}u^{II}_{Cx} = \left[\frac{F^{I}\sin\theta_2}{\sin(\theta_1+\theta_2)}\right]\Delta^{II}_{AC} + \left[-\frac{F^{I}\sin\theta_1}{\sin(\theta_1+\theta_2)}\right]\Delta^{II}_{BC} \tag{4.12b}$$

and cancelling F^{I} leaves

$$u^{II}_{Cx} = \frac{\Delta^{II}_{AC}\sin\theta_2 - \Delta^{II}_{BC}\sin\theta_1}{\sin(\theta_1+\theta_2)} \tag{4.12c}$$

which is identical to Equation 2.44*b*. As was indicated in Chapter 2, the forgoing was a much easier means of computing u^{II}_{Cx} than the geometric procedure employed in Section 2.5.

To evaluate the vertical component of displacement, u^{II}_{Cy} in Figure 4.6*a*, we use a different virtual force state; specifically, we choose a virtual load at joint C in the same direction as u^{II}_{Cy} (Figure 4.6*d*). We will use the same designation for the virtual load as before, namely F^{I}, with the understanding that this is not the same virtual force state as in Figures 4.6*b* and 4.6*c*.

In this case, equilibrium of joint C will lead to different values of P^{I}_{AC} and P^{I}_{BC} (Figure 4.6*e*), i.e.,

$$\begin{aligned}\sum F_x &= 0 = P^{I}_{BC}\cos\theta_2 - P^{I}_{AC}\cos\theta_1\\ \sum F_y &= 0 = -P^{I}_{BC}\sin\theta_2 - P^{I}_{AC}\sin\theta_1 + F^{I}\end{aligned} \tag{4.13a}$$

from which we obtain

$$\begin{aligned}P^{I}_{AC} &= \frac{F^{I}\cos\theta_2}{\sin(\theta_1+\theta_2)}\\ P^{I}_{BC} &= \frac{F^{I}\cos\theta_1}{\sin(\theta_1+\theta_2)}\end{aligned} \tag{4.13b}$$

Equation 4.8 now gives

$$\begin{aligned}F^{I}u^{II}_{Cy} &= P^{I}_{AC}\Delta^{II}_{AC} + P^{I}_{BC}\Delta^{II}_{BC}\\ &= \left[\frac{F^{I}\cos\theta_2}{\sin(\theta_1+\theta_2)}\right]\Delta^{II}_{AC} + \left[\frac{F^{I}\cos\theta_1}{\sin(\theta_1+\theta_2)}\right]\Delta^{II}_{BC}\end{aligned} \tag{4.13c}$$

or

$$u^{II}_{Cy} = \frac{\Delta^{II}_{AC}\cos\theta_2 + \Delta^{II}_{BC}\cos\theta_1}{\sin(\theta_1+\theta_2)} \tag{4.13d}$$

which is identical to Equation 2.44*a*.

EXAMPLE 4 APPLICATION OF VIRTUAL FORCES.

Find the vertical component of displacement of joint C of the truss in Example 10, Chapter 2; the truss is shown in Figures 2.5*a*, 2.20*a*, and 4.7*a*. The bars of the truss may be assumed to be composed of identical materials and to have identical cross-sectional areas, as in Example 10, Chapter 2.

In this case, we are asked to find real displacements that are due to applied forces. The applied forces in Figures 2.5*a* and 4.7*a* cause internal forces in the bars and produce length changes; these length changes have been determined previously (Equations 2.47*b*). Note that we have labeled Figure 4.7*a* as "real displacements," and we have shown the real forces that produce the displacements; it is not necessary to attempt to show an accurate sketch of the deformed truss as long as we know what we are doing. The forces in Figure 4.7*a* must not be confused with the virtual force state (Figure 4.7*b*), which is similar to that in Figure 4.6*d*.

We reiterate that the virtual force state consists of forces in equilibrium; consequently, each joint of the truss in Figure 4.7*b* must be in equilibrium. The equilibrium equations for joint C are identical to Equations 4.13*a*; therefore, internal forces P_{AC}^{I} and P_{BC}^{I} are the same as in Equations 4.13*b*. We may obtain P_{AB}^{I} from equilibrium of joint B (Figure 4.7*c*), i.e.,

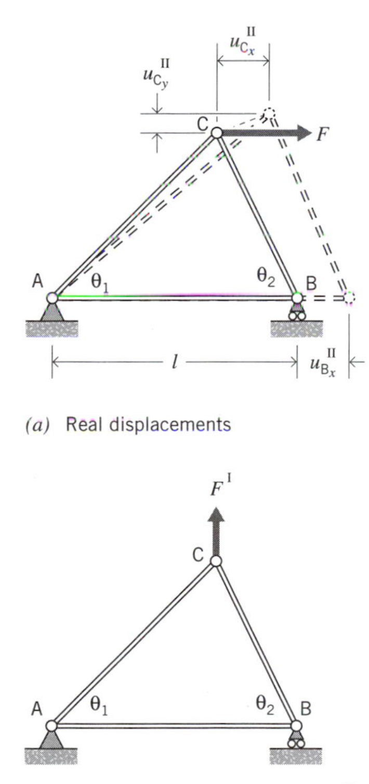

(a) Real displacements

(b) Virtual forces used to evaluate $u_{C_y}^{II}$

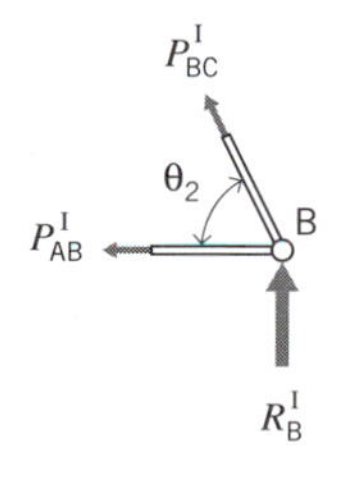

(c) Joint B

Figure 4.7 Example 4.

$$\sum F_x = 0 = -P^{\mathrm{I}}_{\mathrm{AB}} - P^{\mathrm{I}}_{\mathrm{BC}} \cos \theta_2 \qquad \textbf{(4.14a)}$$

Substituting the second of Equations 4.13*b* gives

$$P^{\mathrm{I}}_{\mathrm{AB}} = -P^{\mathrm{I}}_{\mathrm{BC}} \cos \theta_2 = -\frac{F^{\mathrm{I}} \cos \theta_1 \cos \theta_2}{\sin (\theta_1 + \theta_2)} \qquad \textbf{(4.14b)}$$

Note that there is also a vertical reaction at joint B, $R^{\mathrm{I}}_{\mathrm{B}}$ (Figure 4.7*c*); however, there is no real vertical displacement at joint B in Figure 4.7*a*. Consequently, there is no need to evaluate $R^{\mathrm{I}}_{\mathrm{B}}$ because there is no external virtual work associated with $R^{\mathrm{I}}_{\mathrm{B}}$ even though joint B displaces horizontally.

The theorem of virtual work, Equation 4.8, now becomes

$$F^{\mathrm{I}} u^{\mathrm{II}}_{\mathrm{C}y} = P^{\mathrm{I}}_{\mathrm{AC}} \Delta^{\mathrm{II}}_{\mathrm{AC}} + P^{\mathrm{I}}_{\mathrm{BC}} \Delta^{\mathrm{II}}_{\mathrm{BC}} + P^{\mathrm{I}}_{\mathrm{AB}} \Delta^{\mathrm{II}}_{\mathrm{AB}}$$

$$= \left[\frac{F^{\mathrm{I}} \cos \theta_2}{\sin (\theta_1 + \theta_2)}\right] \Delta^{\mathrm{II}}_{\mathrm{AC}} + \left[\frac{F^{\mathrm{I}} \cos \theta_1}{\sin (\theta_1 + \theta_2)}\right] \Delta^{\mathrm{II}}_{\mathrm{BC}} + \left[-\frac{F^{\mathrm{I}} \cos \theta_1 \cos \theta_2}{\sin (\theta_1 + \theta_2)}\right] \Delta^{\mathrm{II}}_{\mathrm{AB}} \qquad \textbf{(4.14c)}$$

or

$$u^{\mathrm{II}}_{\mathrm{C}y} = \frac{\Delta^{\mathrm{II}}_{\mathrm{AC}} \cos \theta_2 + \Delta^{\mathrm{II}}_{\mathrm{BC}} \cos \theta_1 - \Delta^{\mathrm{II}}_{\mathrm{AB}} \cos \theta_1 \cos \theta_2}{\sin (\theta_1 + \theta_2)} \qquad \textbf{(4.14d)}$$

which, once again, is identical to a previous equation, Equation 2.46*b*. As in Example 10, Chapter 2, we obtain $u^{\mathrm{II}}_{\mathrm{C}y}$ by substituting the real length changes given by Equations 2.47*b*.

EXAMPLE 5 APPLICATION OF VIRTUAL FORCES.

The truss in Figure 4.8*a* is composed of bars of identical linearly elastic materials and cross-sectional areas. Bars AD and BC are not connected at their midpoints, i.e., the only joints are at A, B, C, and D. (This truss is similar to the one in Figure 2.14*a*, except that here the supports at A and B are both pinned.) The truss is loaded at joints C and D as shown. These loads give rise to internal forces and consequent length changes, resulting in displacements of joints C and D, the components of which are also indicated schematically in Figure 4.8*a*. For illustrative purposes, we will determine $u^{\mathrm{II}}_{\mathrm{D}y}$.

To find the length changes we need to know the real bar forces; however, this truss is 1-degree redundant ($NJ = 4$, $NR = 4$, $NB = 5$; see Equation 2.20) and we are not yet able to evaluate the real internal forces in a redundant truss. To enable us to proceed, we will provide the solution, i.e., the internal forces in the bars of the truss in Figure 4.8*a* have the following values:[6]

[6] See Chapter 5.

$$P_{CD} = \left(\frac{2 + 4\sqrt{2}}{3 + 4\sqrt{2}}\right)F$$

$$P_{AC} = -\left(\frac{1}{3 + 4\sqrt{2}}\right)F$$

$$P_{BD} = -\left(\frac{1}{3 + 4\sqrt{2}}\right)F \qquad \textbf{(4.15}\textbf{\textit{a}}\textbf{)}$$

$$P_{BC} = \left(\frac{\sqrt{2}}{3 + 4\sqrt{2}}\right)F$$

$$P_{AD} = \left(\frac{\sqrt{2}}{3 + 4\sqrt{2}}\right)F$$

The changes in lengths of the bars are obtained from Equation 2.38*d*, i.e., $\Delta_k^{II} = P_k l_k / EA$, where E is modulus of elasticity and A is cross-sectional area. Note that $l_{CD} = l_{AC} = l_{BD} = l$, and $l_{BC} = l_{AD} = \sqrt{2}l$, so in this case we get

$$\Delta_{CD}^{II} = \left(\frac{2 + 4\sqrt{2}}{3 + 4\sqrt{2}}\right)\frac{Fl}{EA}$$

$$\Delta_{AC}^{II} = -\left(\frac{1}{3 + 4\sqrt{2}}\right)\frac{Fl}{EA}$$

$$\Delta_{BD}^{II} = -\left(\frac{1}{3 + 4\sqrt{2}}\right)\frac{Fl}{EA} \qquad \textbf{(4.15}\textbf{\textit{b}}\textbf{)}$$

$$\Delta_{BC}^{II} = \left(\frac{2}{3 + 4\sqrt{2}}\right)\frac{Fl}{EA}$$

$$\Delta_{AD}^{II} = \left(\frac{2}{3 + 4\sqrt{2}}\right)\frac{Fl}{EA}$$

To evaluate u_{Dy}^{II}, we must choose an appropriate virtual force state. As in the preceding examples, we take the external virtual force as a vertical force at joint D, (Figure 4.8*b*). Now, we must find internal virtual forces that satisfy equilibrium.

The feature of this example that makes it different from the preceding cases is that this truss is statically indeterminate; consequently, because there is no requirement that the virtual force state satisfy compatibility, *more than one set of forces can be found to satisfy all the equations of equilibrium.*

Two sets of virtual forces that satisfy equilibrium are shown in Figures 4.8*b* and 4.8*c*. Note that F^I is the same in both cases, but that the internal virtual forces are totally different. As in Figures 1.42*d* and 1.42*e*, these force states have been obtained

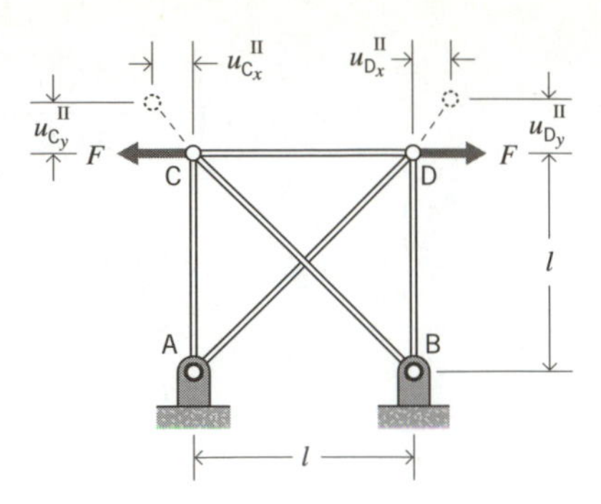

(a) Real displacements

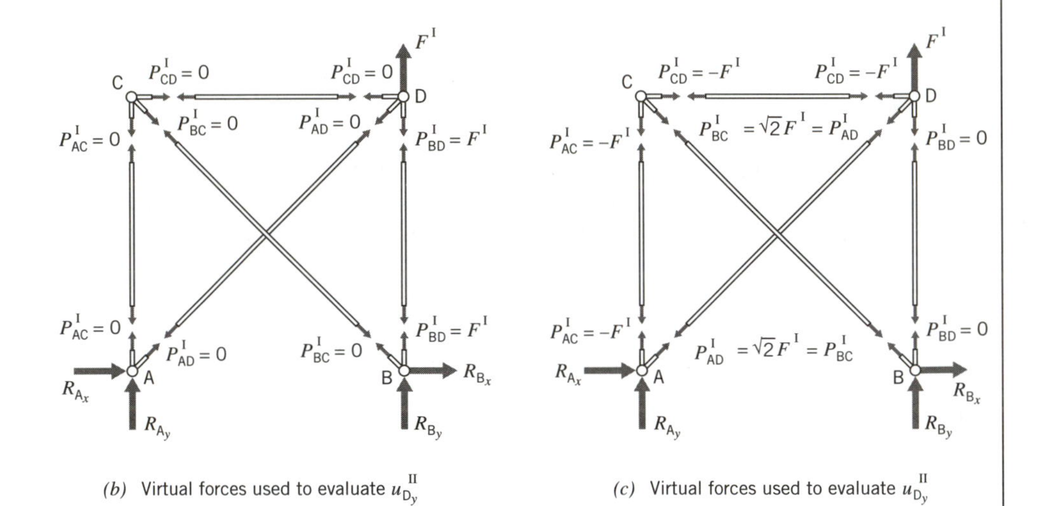

(b) Virtual forces used to evaluate $u_{D_y}^{II}$

(c) Virtual forces used to evaluate $u_{D_y}^{II}$

Figure 4.8 Example 5.

by arbitrarily setting a redundant force equal to zero.[7] In particular, in Figure 4.8*b*, P_{AD}^{I} has been set equal to zero; it follows from the equilibrium equations for joint D that $P_{CD}^{I} = 0$ and $P_{BD}^{I} = F^{I}$. Equilibrium equations at joint C then give $P_{AC}^{I} = P_{BC}^{I} = 0$. In Figure 4.8*c*, P_{BD}^{I} has been set equal to zero, and it follows from the equilibrium equations for joint D that $P_{AD}^{I} = \sqrt{2}F^{I}$ and $P_{CD}^{I} = -F^{I}$. Equilibrium equations at joint C then give $P_{AC}^{I} = -F^{I}$ and $P_{BC}^{I} = \sqrt{2}F^{I}$. Once again, we do not need to know the reactions at A and B but, if desired, they can be determined from equilibrium equations for joints A and B.

Both of these virtual force states (Figures 4.8*b* and 4.8*c*), can be used to determine u_{Dy}^{II}. Furthermore, if we do not get the same answer from both it must mean that there is an error in Equations 4.15*b*, i.e., that the real displacement state does not actually

[7] The selected redundant force can be assigned any arbitrary value (e.g., $0.63F^{I}$, $-1.92F^{I}$, etc.); we have chosen zero for the sake of convenience.

satisfy compatibility conditions. In other words, if we do not obtain a unique set of displacements then there is something wrong with the displacement state; this can only mean a lack of compatibility. To demonstrate this point we will compute u_{Dy}^{II} using both virtual force states.

Consider the virtual work associated with the forces in Figure 4.8*b* and the real displacements in Figure 4.8*a*. From Equation 4.8 we obtain

$$F^{I}u_{Dy}^{II} = P_{BD}^{I}\Delta_{BD}^{II} = F^{I}\Delta_{BD}^{II} \tag{4.15c}$$

Here, all internal virtual forces except P_{BD}^{I} are zero. Substituting Δ_{BD}^{II} from Equation 4.15*b* gives

$$u_{Dy}^{II} = -\left(\frac{1}{3 + 4\sqrt{2}}\right)\frac{Fl}{EA} \tag{4.15d}$$

Similarly, for the virtual force state in Figure 4.8*c*, Equation 4.8 gives[8]

$$\begin{aligned} F^{I}u_{Dy}^{II} &= P_{AD}^{I}\Delta_{AD}^{II} + P_{CD}^{I}\Delta_{CD}^{II} + P_{BC}^{I}\Delta_{BC}^{II} + P_{AC}^{I}\Delta_{AC}^{II} \\ &= (\sqrt{2}F^{I})\Delta_{AD}^{II} + (-F^{I})\Delta_{CD}^{II} + (\sqrt{2}F^{I})\Delta_{BC}^{II} + (-F^{I})\Delta_{AC}^{II} \end{aligned} \tag{4.15e}$$

Substituting the appropriate values of Δ_{k}^{II} from Equation 4.15*b* leads to

$$\begin{aligned} u_{Dy}^{II} = {} & \sqrt{2}\left[\left(\frac{2}{3 + 4\sqrt{2}}\right)\frac{Fl}{EA}\right] - \left[\left(\frac{2 + 4\sqrt{2}}{3 + 4\sqrt{2}}\right)\frac{Fl}{EA}\right] + \sqrt{2}\left[\left(\frac{2}{3 + 4\sqrt{2}}\right)\frac{Fl}{EA}\right] \\ & - \left[-\left(\frac{1}{3 + 4\sqrt{2}}\right)\frac{Fl}{EA}\right] = -\left(\frac{1}{3 + 4\sqrt{2}}\right)\frac{Fl}{EA} \end{aligned} \tag{4.15f}$$

which is identical to Equation 4.15*d*! We observe that if any one of the deformations in Equations 4.15*b* were different, the two answers could not have agreed. Thus, we have reason to believe that the deformations given in Equations 4.15*b* are, in fact, correct.

EXAMPLE 6 APPLICATION OF VIRTUAL FORCES.

We will end this section with a few additional applications involving the truss in Figure 4.8*a* to provide some insight into the role virtual work will play in the subsequent analysis of indeterminate structures.

[8] When there are large numbers of virtual forces and real bar deformations, it is more convenient to tabulate data than to write them in the form of an equation; however, at this stage of developments, we are not as concerned about the efficiency of the computations as we are about demonstrating the concepts.

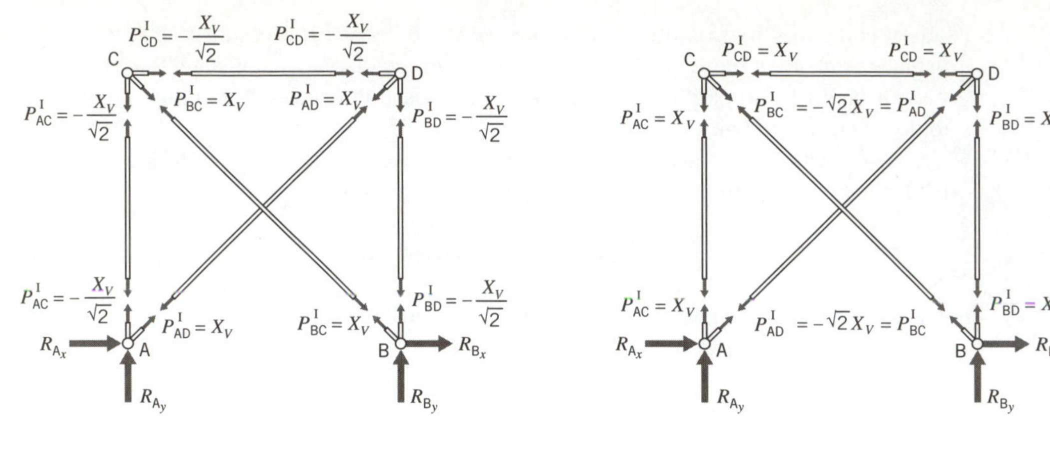

(a) Virtual forces used to find compatibility equation

(b) Virtual forces used to find compatibility equation

Figure 4.9 Example 6.

Recall that the preceding example (Figure 4.8*a*) involved a truss that is 1-degree redundant. As we will see in Chapter 5, the choice of redundant force in any indeterminate structure is not unique. In this particular truss, we can choose any one of the internal forces or reaction components as the redundant force; e g., in Figure 4.8*b* we referred to P^{I}_{AD} as a redundant force, while in Figure 4.8*c* we referred to P^{I}_{BD} as a redundant force.

Now, consider Figure 4.8*a* to again be representative of a real displacement state, and suppose that we choose a virtual force state for which there are no applied external virtual forces other than reactions at supports; assume that in this virtual force state a redundant force is initially assigned an arbitrary nonzero value. For example, in Figure 4.9*a*, P^{I}_{AD} has been assigned the value X_V; the remaining internal forces are determined from equilibrium of joints D and C, resulting in the values shown in Figure 4.9*a*.

For the real displacements in Figure 4.8*a* and the virtual forces in Figure 4.9*a*, the theorem of virtual work, Equation 4.8, gives

$$
\begin{aligned}
0 &= P^{I}_{AD}\Delta^{II}_{AD} + P^{I}_{BC}\Delta^{II}_{BC} + P^{I}_{AC}\Delta^{II}_{AC} + P^{I}_{BD}\Delta^{II}_{BD} + P^{I}_{CD}\Delta^{II}_{CD} \\
&= (X_V)\Delta^{II}_{AD} + (X_V)\Delta^{II}_{BC} + \left(-\frac{X_V}{\sqrt{2}}\right)\Delta^{II}_{AC} + \left(-\frac{X_V}{\sqrt{2}}\right)\Delta^{II}_{BD} + \left(-\frac{X_V}{\sqrt{2}}\right)\Delta^{II}_{CD}
\end{aligned}
\tag{4.16a}
$$

or

$$
0 = \Delta^{II}_{AD} + \Delta^{II}_{BC} - \frac{\Delta^{II}_{AC}}{\sqrt{2}} - \frac{\Delta^{II}_{BD}}{\sqrt{2}} - \frac{\Delta^{II}_{CD}}{\sqrt{2}} \tag{4.16b}
$$

Note that the external virtual work is zero in Equation 4.16*a* because there are no external virtual forces other than reactions in Figure 4.9*a*, and there are no real displacements at the supports in Figure 4.8*a*. Equation 4.16*b* is a relationship that the real bar deformations must satisfy for the deformed bars to remain connected at the displaced joints, i.e., it is the *compatibility equation* for the truss in Figure 4.8*a*. This is the additional relationship that, together with the $2NJ = 8$ equilibrium equations, enables us to solve for the $NB + NR = 9$ unknown real forces.

The compatibility equation (Equation 4.16*b*), is unique to the structure and does not depend on the choice of redundant force. This is illustrated by the alternate virtual force state in Figure 4.9*b*, in which $P^{\mathrm{I}}_{\mathrm{CD}}$ has been assumed to be the redundant force and has been assigned the value X_V. Equation 4.8 now gives

$$0 = P^{\mathrm{I}}_{\mathrm{AD}}\Delta^{\mathrm{II}}_{\mathrm{AD}} + P^{\mathrm{I}}_{\mathrm{BC}}\Delta^{\mathrm{II}}_{\mathrm{BC}} + P^{\mathrm{I}}_{\mathrm{AC}}\Delta^{\mathrm{II}}_{\mathrm{AC}} + P^{\mathrm{I}}_{\mathrm{BD}}\Delta^{\mathrm{II}}_{\mathrm{BD}} + P^{\mathrm{I}}_{\mathrm{CD}}\Delta^{\mathrm{II}}_{\mathrm{CD}} \qquad \mathbf{(4.16}\boldsymbol{c}\mathbf{)}$$

$$= (-\sqrt{2}X_V)\Delta^{\mathrm{II}}_{\mathrm{AD}} + (-\sqrt{2}X_V)\Delta^{\mathrm{II}}_{\mathrm{BC}} + X_V\Delta^{\mathrm{II}}_{\mathrm{AC}} + X_V\Delta^{\mathrm{II}}_{\mathrm{BD}} + X_V\Delta^{\mathrm{II}}_{\mathrm{CD}}$$

which is the same as Equation 4.16*b*. This utilization of virtual forces to obtain compatibility equations is a key step in the force method of analysis. Compatibility will be discussed more fully in Chapter 5.

EXAMPLE 7 APPLICATION OF VIRTUAL DISPLACEMENTS.

We will demonstrate the converse of the application in Example 6, namely a typical use of virtual displacements to obtain equilibrium equations of the type utilized in the displacement method of analysis. Assume that Figure 4.8*a* represents a real force state rather than a real displacement state. We will choose the virtual displacement state as one that involves only one of the kinematic degrees of freedom of the structure. The truss in Figure 4.8*a* has four kinematic degrees of freedom, represented by the four components of joint displacement shown in the figure. Therefore, we will choose a virtual displacement state with only a single component of displacement at one of the joints, for example the displacement state in Figure 4.10.

The only component of joint displacement in Figure 4.10 is $u^{\mathrm{II}}_{\mathrm{D}x}$, but this displacement pattern does involve elongations of bars AD and CD, i.e., $\Delta^{\mathrm{II}}_{\mathrm{AD}} = u^{\mathrm{II}}_{\mathrm{D}x}/\sqrt{2}$ and $\Delta^{\mathrm{II}}_{\mathrm{CD}} = u^{\mathrm{II}}_{\mathrm{D}x}$; no other bars change length in this virtual displacement state. Consequently, Equation 4.8 becomes

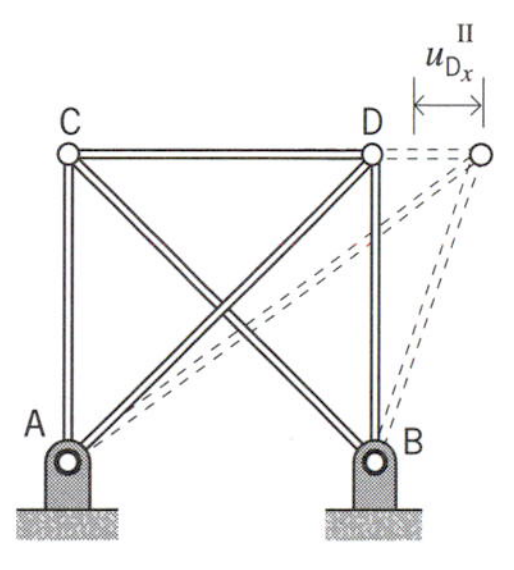

Figure 4.10 Example 7 (virtual displacements).

$$Fu_{Dx}^{II} = P_{AD}\Delta_{AD}^{II} + P_{CD}\Delta_{CD}^{II} \tag{4.16d}$$

$$= P_{AD}\frac{u_{Dx}^{II}}{\sqrt{2}} + P_{CD}u_{Dx}^{II}$$

or

$$F = \frac{P_{AD}}{\sqrt{2}} + P_{CD} \tag{4.16e}$$

which is the equilibrium equation, $\Sigma F_x = 0$, for joint D. The significance of this type of equation will be discussed more fully in connection with the displacement method.

4.4 VIRTUAL WORK FOR BEAMS

The development of the theorem of virtual work for frame structures follows a procedure that is analogous to the corresponding development for trusses, i.e., virtual work expressions for joints and beams are combined. The frame joint will be regarded as a rigid body with negligible dimensions subjected to ''forces'' that include moment quantities.

A typical rigid joint is illustrated in Figure 4.11; the end of each beam at the joint is, in general, subjected to an axial force, P_k, a transverse shear force, V_k, and a moment, M_k. The joint may also be subjected to applied external forces F_j and moments M_i. The forces P_k and V_k are still all regarded as concurrent; therefore, a virtual work expression for this joint will differ from that for the truss joint in Equation 4.7*a* primarily by the addition of terms involving the moments and a joint rotation.

We will begin with a general beam subjected to a transverse continuous load and end loads of the types considered originally in Figure 3.1. Internal resultant forces are identical to those defined in Figure 3.2. As in the preceding sections, the external loads in Figure 3.1 and the internal forces in Figure 3.2 are defined as positive when they have the directions shown relative to the *x-y* coordinates axes, and are assumed to be in equilibrium. We will again identify all forces in this equilibrium state by a superscript I.

We will also assume the existence of an unrelated displacement state, the kinematics of which is consistent with the Bernoulli-Euler hypothesis (Figure 3.14*a*). Furthermore, we will restrict our attention to deformations associated with curvature, and exclude

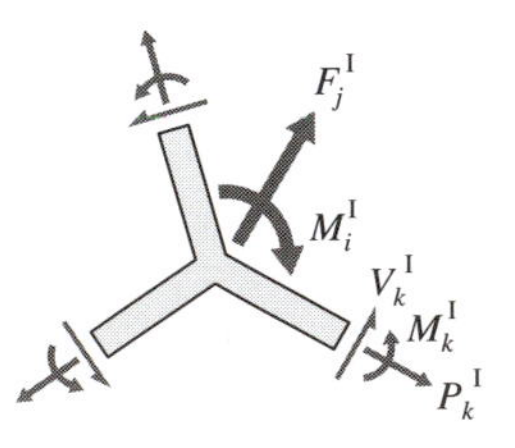

Figure 4.11 Forces on rigid joint.

any deformations associated with length changes;[9] therefore, we need consider only the transverse displacement of the beam axis, $v^{\mathrm{II}}(x)$. We will introduce a set of end displacements and end rotations, with terminology and positive directions as defined in Figure 4.12*a*. Note that, in comparison with Figure 3.14*a*, we have $v_{\mathrm{A}}^{\mathrm{II}} = v^{\mathrm{II}}(0)$, $v_{\mathrm{B}}^{\mathrm{II}} = v^{\mathrm{II}}(l)$, $\phi_{\mathrm{A}}^{\mathrm{II}} = -v'^{\mathrm{II}}(0)$ and $\phi_{\mathrm{B}}^{\mathrm{II}} = -v'^{\mathrm{II}}(l)$.

As in the case of bars, the derivation of the theorem of virtual work for beams utilizes equilibrium equations, Equations 3.1*b* and 3.2*b*, which we now rewrite as

$$\begin{aligned} V'^{\mathrm{I}}(x) + q^{\mathrm{I}}(x) &= 0 \\ M'^{\mathrm{I}}(x) + V^{\mathrm{I}}(x) &= 0 \end{aligned} \tag{4.17a}$$

It is probably easiest to proceed by combining Equations 4.17*a* into a single equation. This may be accomplished by differentiating the second equation once, and substituting $V'^{\mathrm{I}}(x)$ from the first equation; the result is

$$M''^{\mathrm{I}}(x) - q^{\mathrm{I}}(x) = 0 \tag{4.17b}$$

As was done in connection with bar elements (Equation 4.5*a*) we multiply Equation 4.17*b* by $v^{\mathrm{II}}(x)$ and dx, and integrate from 0 to l, i.e.,

$$\int_0^l [M''^{\mathrm{I}}(x) - q^{\mathrm{I}}(x)]v^{\mathrm{II}}(x)dx = 0 \tag{4.17c}$$

The procedure from this point on is entirely analogous to that followed in the derivation of Equation 4.5*h*; the first term above is replaced by the difference of two terms,

$$\int_0^l \{[M'^{\mathrm{I}}(x)v^{\mathrm{II}}(x)]' - M'^{\mathrm{I}}(x)v'^{\mathrm{II}}(x)\}dx - \int_0^l q^{\mathrm{I}}(x)v^{\mathrm{II}}(x)dx = 0 \tag{4.17d}$$

The first term above is a differential quantity. The second term may be replaced by the difference in two terms, so we now have

$$M'^{\mathrm{I}}(x)v^{\mathrm{II}}(x)|_0^l - \int_0^l \{[M^{\mathrm{I}}(x)v'^{\mathrm{II}}(x)]' - M^{\mathrm{I}}(x)v''^{\mathrm{II}}(x)\}dx - \int_0^l q^{\mathrm{I}}(x)v^{\mathrm{II}}(x)dx = 0 \tag{4.17e}$$

From the second of Equations 4.17*a*, we may replace $M'^{\mathrm{I}}(x)$ by $-V^{\mathrm{I}}(x)$, which leads to

$$V^{\mathrm{I}}(x)v^{\mathrm{II}}(x)|_0^l + M^{\mathrm{I}}(x)v'^{\mathrm{II}}(x)|_0^l + \int_0^l q^{\mathrm{I}}(x)v^{\mathrm{II}}(x)dx = \int_0^l M^{\mathrm{I}}(x)v''^{\mathrm{II}}(x)dx \tag{4.17f}$$

or

$$\begin{aligned} V^{\mathrm{I}}(l)v^{\mathrm{II}}(l) - V^{\mathrm{I}}(0)v^{\mathrm{II}}(0) + M^{\mathrm{I}}(l)v'^{\mathrm{II}}(l) - M^{\mathrm{I}}(0)v'^{\mathrm{II}}(0) + \\ \int_0^l q^{\mathrm{I}}(x)v^{\mathrm{II}}(x)dx = \int_0^l M^{\mathrm{I}}(x)v''^{\mathrm{II}}(x)dx \end{aligned} \tag{4.17g}$$

[9] As with forces, these effects can be included by superposition of axial bar-type behavior.

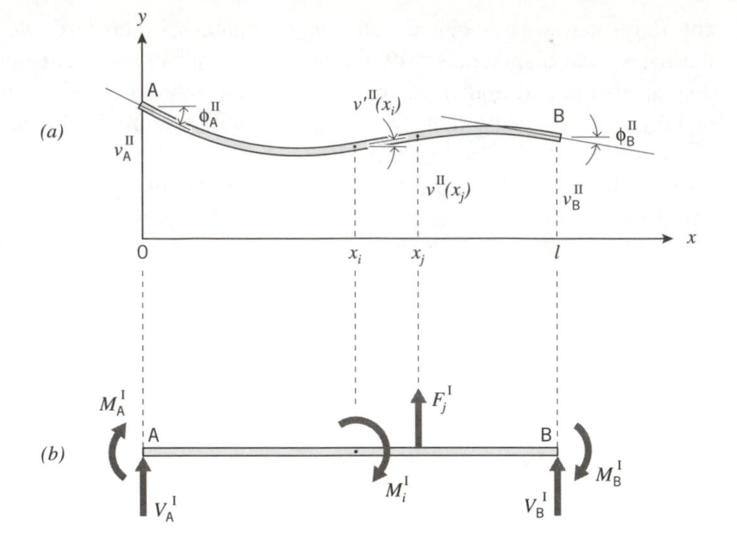

Figure 4.12 (*a*) Displacement state and (*b*) force state for beam.

Finally, substituting end loads from Figure 3.1 and end displacements from Figure 4.12*a* gives

$$V_B^I v_B^{II} + V_A^I v_A^{II} + M_B^I \phi_B^{II} + M_A^I \phi_A^{II} + \int_0^l q^I(x) v^{II}(x) dx = \int_0^l M^I(x) v''^{II}(x) dx \qquad \textbf{(4.17h)}$$

which is the basic theorem of virtual work for a beam, the analogue of Equation 4.5*h*. Once again, we may associate the terms on the left side of the equation with external virtual work, and the term on the right side of the equation with internal virtual work. Note that the first four external virtual work terms in Equations 4.17*h* or 4.17*g* have exactly the same form as virtual work terms involving V_k^I and M_k^I for the joint in Figure 4.11.

In its present form, the theorem reflects only the effects of continuous intermediate loads, whereas beams may also be subjected to concentrated intermediate loads, as in Figure 4.12*b*.[10] We recognize that concentrated intermediate loads such as F_j^I and M_i^I will, in general, contribute additional external virtual work terms to the basic theorem. The forms of the additional terms are identical to those for external virtual work associated with the concentrated end loads. For example, if $v^{II}(x_j)$ denotes the transverse displacement at $x = x_j$, the point of application of F_j^I, the associated external virtual work is $F_j^I v^{II}(x_j)$. Similarly if $v'^{II}(x_i)$ denotes the slope of the beam at $x = x_i$, the point of application of M_i^I, the associated external virtual work is $-M_i^I v'^{II}(x_i)$; the

[10] Concentrated loads F_j^I are assumed to act in the y direction.

negative sign is a consequence of the fact that we define clockwise applied moments as positive (Figure 4.12*b*), whereas beam slopes are defined as positive if the axis undergoes a counterclockwise rotation (Figure 3.14*a*). Consequently, we may append terms of the forgoing types to Equation 4.17*h* to obtain a more general expression of the theorem of virtual work for a beam, i.e.,

$$V_B^I v_B^{II} + V_A^I v_A^{II} + M_B^I \phi_B^{II} + M_A^I \phi_A^{II} + \sum_{j=1}^{n} F_j^I v^{II}(x_j) - \sum_{i=1}^{m} M_i^I v'^{II}(x_i) + \int_0^l q^I(x) v^{II}(x) dx = \int_0^l M^I(x) v''^{II}(x) dx \tag{4.18}$$

where n is the number of intermediate applied concentrated loads, and m is the number of intermediate applied moments. Equation 4.18 is what we will henceforth refer to as the theorem of virtual work for a beam. As usual, when the displacement state is real and the force state virtual, we refer to Equation 4.18 as the theorem of virtual forces; when the force state is real and the displacement state virtual, we refer to Equation 4.18 as the theorem of virtual displacements.

The theorem is actually less complicated than it may appear at first. The external virtual work terms on the left side of Equation 4.18, with the exception of the integral, are all basically similar. There is a large number of these terms only because we have chosen not to consolidate them into any more general type of representation; the form of Equation 4.18 should allow us to clearly delineate the contributions of all quantities in subsequent analyses.

EXAMPLE 8 APPLICATION OF VIRTUAL DISPLACEMENTS.

We will begin our examination of applications with some basic illustrations of virtual displacements.

Consider the simply supported beam in Figure 4.13*a*; the beam is subjected to a uniform load, q_o, a concentrated load, F, applied at $x = a$, and a moment, M_o, applied at $x = l/2$. A free-body diagram of the beam is shown in Figure 4.13*b*. We will use virtual displacements to find the reaction at B. We choose as the virtual displacement the small rigid-body rotation in Figure 4.13*c*, which is similar to what was done with the truss in Example 1, Figure 4.4*b*. Here, θ_v replaces θ^{II}.

In this case, there is a transverse virtual displacement given by $v^{II}(x) = v_v(x) = \theta_v x$, and a virtual slope $v'_v(x) = \theta^{II}$, but the virtual curvature, $v''_v(x)$, is zero; consequently, there is no internal virtual work. Equation 4.18 therefore becomes

$$R_B v_{BV} + F v_V(a) - M_o v'_V(l/2) - \int_0^l q_o v_V(x) dx = 0 \tag{4.19a}$$

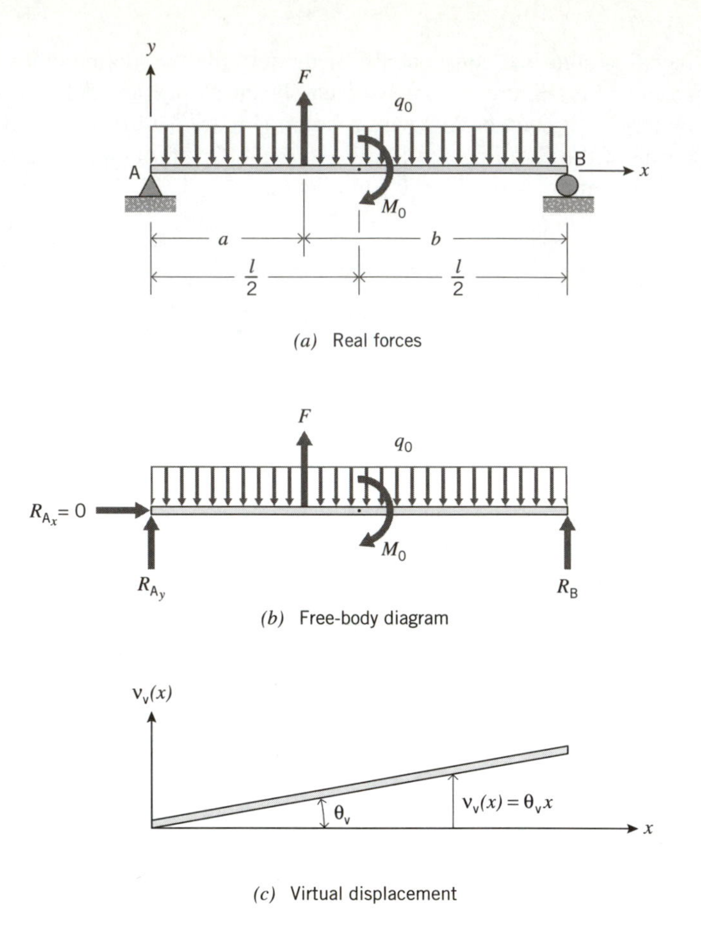

Figure 4.13 Example 8.

Note the negative sign preceding the integral, which is a consequence of the fact that q_o is negative according to our sign convention. Substituting virtual displacements in terms of angle θ_V, i.e., $v^{II}_{BV} = \theta_V l$, $v_V(a) = \theta_V a$, $v'_V(l/2) = \theta_V$, leads to[11]

$$R_B(\theta_V l) + F(\theta_V a) - M_o(\theta_V) - q_o \int_0^l (\theta_V x)dx = 0 \tag{4.19b}$$

from which we obtain

$$R_B = \frac{M_o}{l} - \frac{Fa}{l} + \frac{q_o l}{2} \tag{4.19c}$$

[11] The superscripts have been dropped because Equation 4.19 should be clear without them.

which is the equilibrium equation for sum of moments about A. Equation 4.19c again illustrates the principle of superposition, i.e., R_B is given by the sum of the effect of each applied load.

EXAMPLE 9 APPLICATION OF VIRTUAL DISPLACEMENTS.

Find the internal moment at $x = l/2$ for the uniformly loaded beam in Figure 4.14a. To accomplish this, we will use the virtual displacement state shown in Figure 4.14b. This virtual displacement state consists of small rigid-body rotations of the two halves of the beam, thereby implying the existence of a "concentrated curvature" at $x = l/2$. (A concentrated curvature is conceptually equivalent to a hinge.) The transverse displacement of the beam is defined in terms of angle θ_v, i.e., $v_v(x) = \theta_v x$, $0 < x < l/2$; the displacement is symmetric about the midpoint. Note that there are no transverse displacements at the supports, so there can be no virtual work associated with reactions.

In this case, the only external virtual work is that associated with the applied load, q_o; therefore, the external virtual work (the terms on the left side of Equation 4.18) reduces to

$$W_{EV} = \int_0^l q(x)v_v(x)dx = 2q_o \int_0^{l/2} (\theta_v x)dx = \frac{q_o\theta_v l^2}{4} \qquad \textbf{(4.20a)}$$

We compute the internal virtual work as follows: the virtual curvature of the beam is zero everywhere except in the immediate vicinity of the midpoint (Figure 4.14b); the real internal moment in the immediate vicinity of the midpoint is assumed to be a constant, which we will denote by M_C. Consequently, the integral on the right side of Equation 4.18, the internal virtual work, must have magnitude equal to the product of M_C and the change in slope at the "hinge," i.e.,

$$W_{IV} = \int_0^l M(x)v_v''(x)dx = -M_C(2\theta_v) \qquad \textbf{(4.20b)}$$

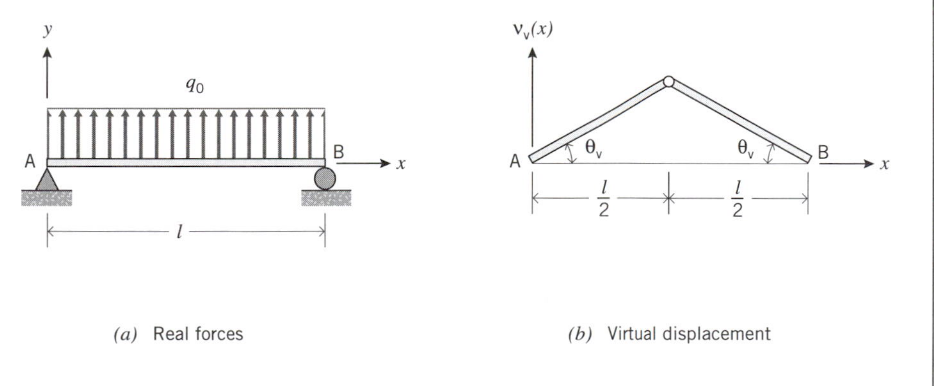

(*a*) Real forces (*b*) Virtual displacement

Figure 4.14 Example 9.

The negative sign is a consequence of the fact that a positive internal moment has the direction defined in Figure 3.2, which is "opposite" to the direction of the curvature in Figure 4.14*b*.

Equating internal and external virtual work leads to

$$W_{IV} = W_{EV}$$

$$-M_C(2\theta_v) = \frac{q_o \theta_v l^2}{4} \tag{4.20c}$$

or

$$M_C = -\frac{q_o l^2}{8} \tag{4.20d}$$

which is the expected result (see Figure 3.6). Again, note that we were able to compute the internal moment without knowledge of the reactions, something that could not be accomplished from a consideration of equilibrium of a single free-body.

EXAMPLE 10 APPLICATION OF VIRTUAL DISPLACEMENTS.

Consider the simply supported beam in Figure 4.15*a*, which is subjected to a uniform load q_o. Find the vertical reaction at A, R_{A_y} (Figure 4.15*b*). Note that the internal (resultant) moment in the beam (Figure 4.15*c*) is a function of R_{A_y}, i.e.,

$$M(x) = R_{A_y} x + \frac{q_o x^2}{2} \tag{4.21a}$$

Although a procedure for computing R_{A_y} has already been illustrated in Example 8, we will demonstrate an alternate analysis involving a virtual displacement state having nonzero curvature (Figure 4.15*d*). This displacement is taken to be

$$v_v(x) = a \sin \frac{\pi x}{l} \tag{4.21b}$$

where a is an arbitrary constant (which, in this case, is the magnitude of the displacement at $x = l/2$). Note that, although this equation satisfies the kinematic boundary conditions, this is definitely not an expression for the actual displacement of the beam in Figure 4.15*a*. (The real displacement is given by Equation 3.19*b*.) Once again, the virtual displacement is totally unrelated to the real force state.

Differentiating Equation 4.21*b* twice gives the virtual curvature

$$v''_v(x) = -\frac{a\pi^2}{l^2} \sin \frac{\pi x}{l} \tag{4.21c}$$

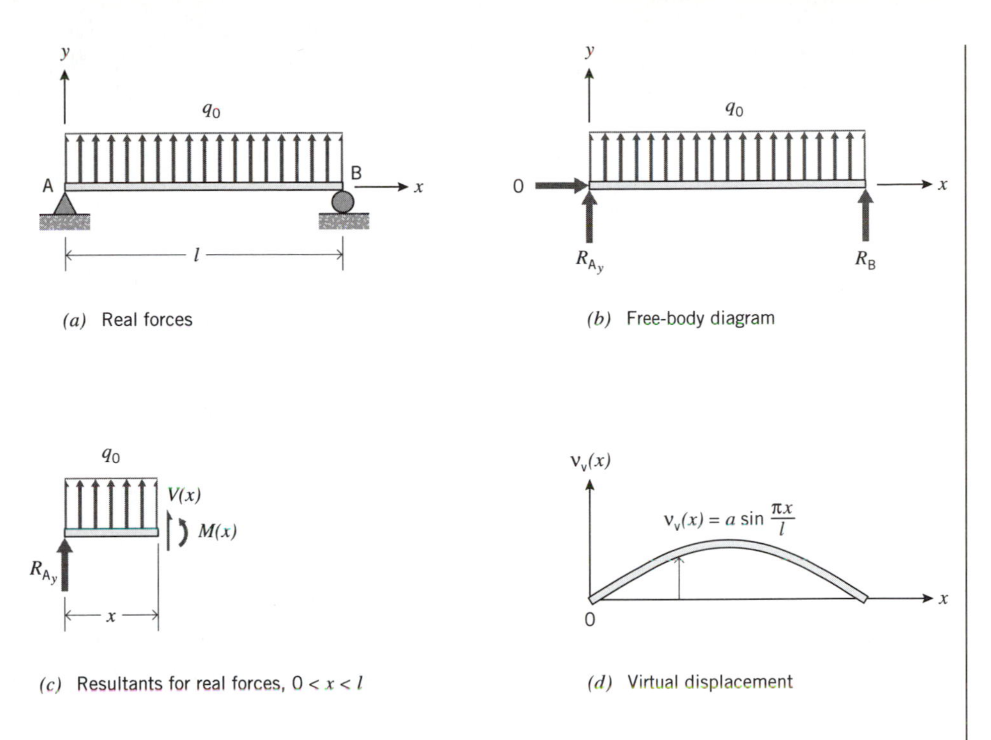

Figure 4.15 Example 10.

and we now have all the information required to evaluate external and internal virtual work. The former is given by

$$
\begin{aligned}
W_{EV} &= \int_0^l q_o v_v(x)dx = \frac{q_o al}{\pi}\int_0^l \sin\frac{\pi x}{l}\frac{\pi}{l}dx \\
&= -\frac{q_o al}{\pi}\cos\frac{\pi x}{l}\bigg|_0^l \\
&= \frac{2q_o al}{\pi}
\end{aligned}
\tag{4.21d}
$$

and the latter by

$$
W_{IV} = \int_0^l M(x)v''_v(x)dx = \int_0^l \left(R_{A_y}x + \frac{q_o}{2}x^2\right)\left(-\frac{a\pi^2}{l^2}\sin\frac{\pi x}{l}\right)dx \tag{4.21e}
$$

By expanding Equation 4.21e, and integrating, we obtain

$$
W_{IV} = -R_{A_y}a\int_0^l \left(\frac{\pi x}{l}\right)\left(\sin\frac{\pi x}{l}\right)\frac{\pi}{l}dx - \frac{q_o al}{2\pi}\int_0^l \left(\frac{\pi x}{l}\right)^2\left(\sin\frac{\pi x}{l}\right)\frac{\pi}{l}dx
$$

$$= -R_{A_y}a\left[\sin\frac{\pi x}{l} - \left(\frac{\pi x}{l}\right)\cos\frac{\pi x}{l}\right]_0^l - \frac{q_o al}{2\pi}\left[2\left(\frac{\pi x}{l}\right)\sin\frac{\pi x}{l} - \left(\frac{\pi^2 x^2}{l^2} - 2\right)\cos\frac{\pi x}{l}\right]_0^l$$

$$= -R_{A_y}a\pi - \frac{q_o al}{2\pi}(\pi^2 - 4) \tag{4.21f}$$

Equating external and internal virtual work, Equation 4.21*d* and Equation 4.21*f*, respectively, gives

$$\frac{2q_o al}{\pi} = -R_{A_y}a\pi - \frac{q_o al\pi}{2} + \frac{2q_o al}{\pi} \tag{4.21g}$$

or

$$R_{A_y} = -\frac{q_o l}{2} \tag{4.21h}$$

as expected.

Although the forgoing procedure is considerably more involved than the one used in Example 8, it should help clarify the interpretation of the internal virtual work term in Equation 4.18.

EXAMPLE 11 APPLICATION OF VIRTUAL FORCES.

We will begin our illustration of the use of virtual forces with the uniformly loaded, simply supported beam considered in the preceding two examples and shown again in Figure 4.16*a*.[12] Specifically, we will use virtual forces to (a) compute the midpoint deflection, $v(l/2)$ and (b) compute the slope at end A, ϕ_A. The displacements, shown greatly exaggerated in Figure 4.16*a*, have been previously computed in Chapter 3 in conjunction with Equations 3.19*a* and 3.19*b* (see Figure 3.15). As in Chapter 3, we assume linear elastic behavior, and a beam with constant *EI*.

Because we are using virtual forces the displacement state (state II) is real. Consequently, the expression for internal virtual work on the right side of Equation 4.18 involves the real curvature, $v''(x)$; this real curvature is related to the real internal resultant moment $M(x)$ by Equation 3.16*b*, i.e., $v''(x) = M(x)/EI$.[13] The real internal moment, $M(x)$, is obtained from an analysis of equilibrium of the free-body in Figure 4.16*b*; the result, given in Equation 3.4*b*, leads to

$$\frac{M(x)}{EI} = \frac{q_o}{2EI}x^2 - \frac{q_o l}{2EI}x \tag{4.22a}$$

[12] In subsequent examples, we will generally not show the real loads and their consequent displacements in the same figure, as we have done for the sake of clarity in Figure 4.16*a*.

[13] This relationship is analogous to Equation 2.38*d* for bars.

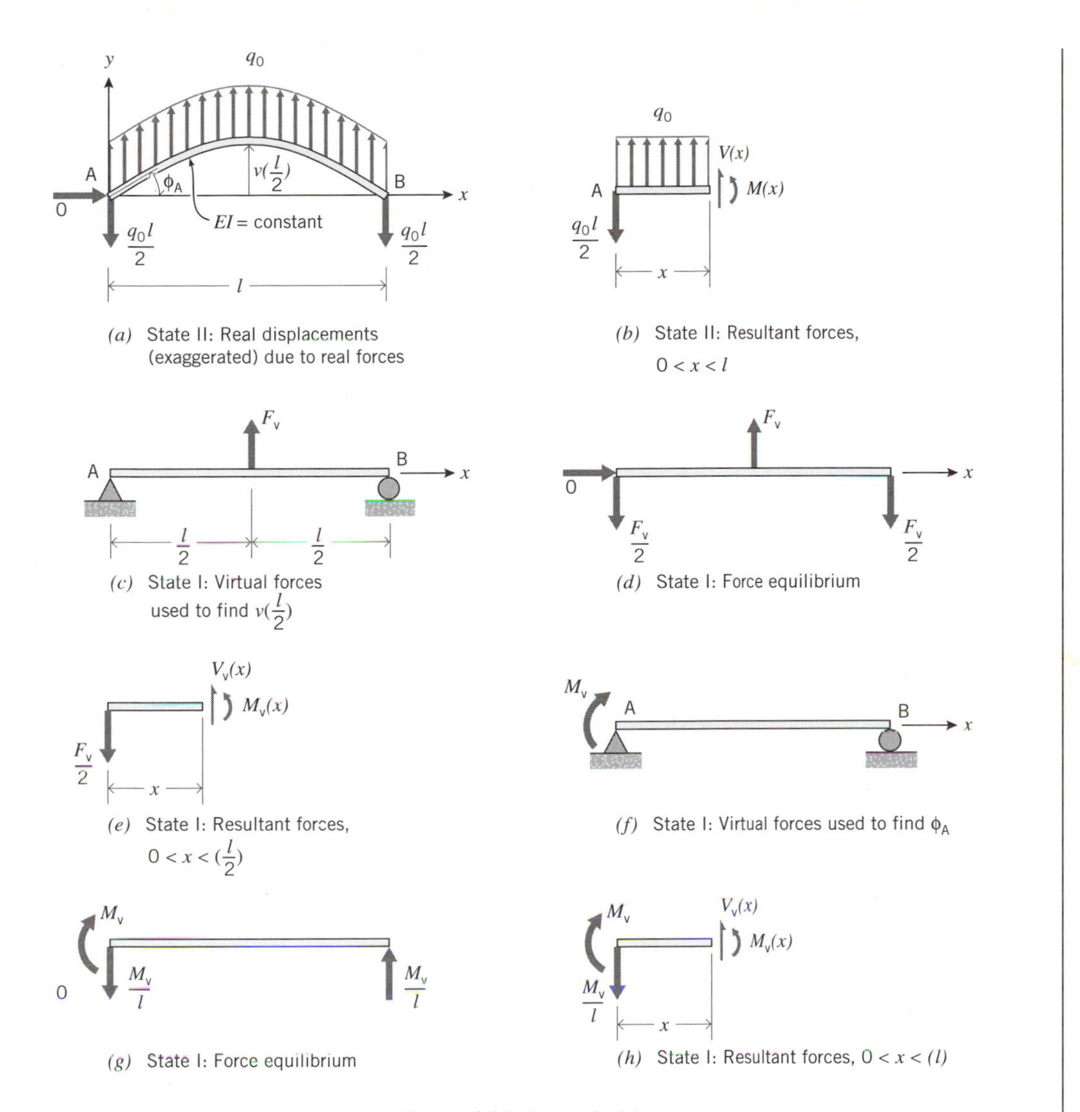

Figure 4.16 Example 11.

(a) To compute the midpoint displacement, we choose virtual forces (state I) as shown in Figure 4.16*c*, i.e., a concentrated virtual force F_V applied at the midpoint;[14] since this force state must be in equilibrium, we must have the reactions shown in Figure 4.16*d*. For this particular choice of virtual forces, the external virtual work in Equation 4.18 reduces to the product of F_V and $v(l/2)$; note that there are no real displacements at the end supports (A and B) and, consequently, there cannot be any external virtual work associated with the reactions in Figure 4.16*d*.

The internal virtual work is a function of the resultant internal virtual moment, Figure 4.16*e*; this resultant virtual moment is given by

[14] Here, for the sake of clarity, we use F_V rather than F^I to denote the virtual force.

$$M_V(x) = -\frac{F_V}{2}x \tag{4.22b}$$

Equation 4.22*b* is valid only for $0 < x < l/2$; we will utilize symmetry, thereby eliminating the need for an expression for $M_V(x)$ in the range $l/2 < x < l$.

Substituting Equations 4.22*a* and 4.22*b* into Equation 4.18 leads to

$$F_V v(l/2) = 2\int_0^{l/2}\left(-\frac{F_V}{2}x\right)\left(\frac{q_o}{2EI}x^2 - \frac{q_o l}{2EI}x\right)dx$$

$$= -\frac{F_V q_o}{2EI}\int_0^{l/2}(x^3 - lx^2)dx \tag{4.22c}$$

$$= \frac{F_V 5 q_o l^4}{384EI}$$

or

$$v(l/2) = \frac{5q_o l^4}{384EI} \tag{4.22d}$$

which is the desired result.

(b) To compute the slope at end A, we choose virtual forces (state I) as shown in Figure 4.16*f*, namely, a virtual moment M_V applied at end A; since this force state must be in equilibrium, we must have the reactions shown in Figure 4.16*g*. For this particular choice of virtual forces, the magnitude of the external virtual work in Equation 4.18 reduces to the product of M_V and ϕ_A; the sign of this external virtual work term is negative because M_V and ϕ_A have opposite assumed directions in their respective figures.

The internal virtual work is a function of the resultant internal virtual moment (Figure 4.16*h*); this resultant virtual moment is given by

$$M_V(x) = M_V - \frac{M_V}{l}x \tag{4.22e}$$

Since the real curvature is still given by Equation 4.22*a*, substituting this and Equation 4.22*e* into Equation 4.18 leads to

$$-M_V\phi_A = \int_0^{l}\left(M_V - \frac{M_V}{l}x\right)\left(\frac{q_o}{2EI}x^2 - \frac{q_o l}{2EI}x\right)dx$$

$$= \frac{M_V q_o}{2EI}\int_0^{l}\left(1 - \frac{x}{l}\right)(x^2 - lx)dx \tag{4.22f}$$

$$= -\frac{M_V q_o l^3}{24EI}$$

or

$$\phi_A = \frac{q_o l^3}{24EI} \tag{4.22g}$$

as expected. The positive sign here (and in Equation 4.22*d*) means that the displacement is in the direction originally assumed in Figure 4.16*a*. Similar procedures can be used for finding slope and/or displacement at any point on the beam axis. The use of virtual forces to compute specific values of slopes or displacements is generally quite a bit less involved than direct solution by integration of differential equations. In fact, as we will see in subsequent examples, the theorem of virtual forces is basically identical to the moment-area method, but is considerably more systematic because it does not depend on the type of graphical constructions that are required by the moment-area method.

EXAMPLE 12 APPLICATION OF VIRTUAL FORCES.

Compute the transverse displacement (state II) at the free end of the uniform cantilever beam in Example 5, Chapter 3 (Figure 3.16). The beam, shown in Figure 4.17*a*, is subjected to a temperature variation through its depth as defined in the previous example. For the given temperature variation, the real curvature (Equation 3.16*b*) is found by dividing Equation 3.20*c* by *EI*, i.e.,

$$v''(x) = \frac{M_T(x)}{EI} = -\frac{\alpha T_o}{H} \tag{4.23a}$$

We choose a virtual force state (state I) as in Figure 4.17*b*, for which the corresponding reactions required for equilibrium are shown in Figure 4.17*c*. For this virtual force state, the internal resultant moment (Figure 4.17*d*), is given by

$$M_V(x) = F_V l - F_V x \tag{4.23b}$$

The external virtual work is given by the product of F_V and $v(l)$, so Equation 4.18 can be written as

$$\begin{aligned} F_V v(l) &= \int_0^l M_V(x) v''(x) dx \\ &= F_V \int_0^l (l - x) \left[\frac{M_T(x)}{EI} \right] dx \end{aligned} \tag{4.23c}$$

If we divide by F_V, Equation 4.23*c* may be seen to have a form identical to Equation 3.24*c*, the so-called second moment-area theorem. This relationship between the virtual force theorem and the moment-area theorem is particularly clear when we are dealing with displacements of cantilever beams. The relationship exists in all cases, but some algebraic manipulation is required to demonstrate this fact in most other cases.

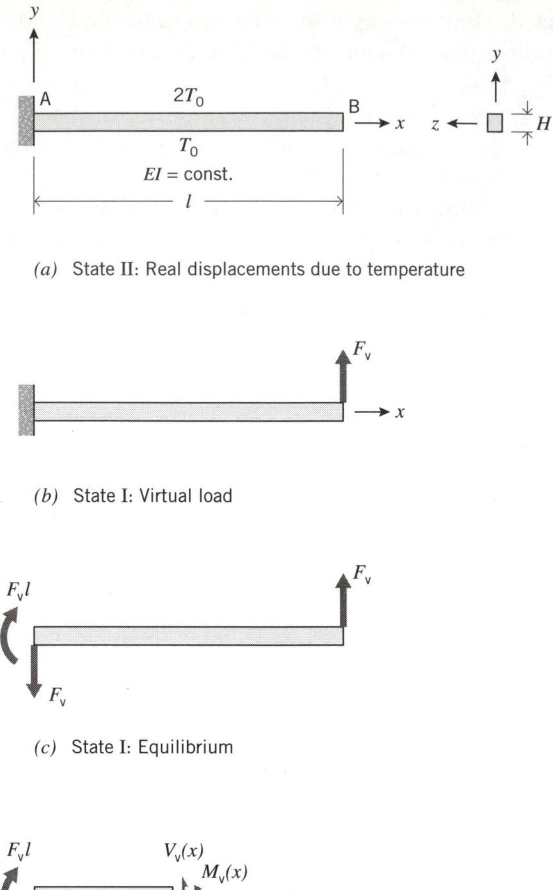

(*a*) State II: Real displacements due to temperature

(*b*) State I: Virtual load

(*c*) State I: Equilibrium

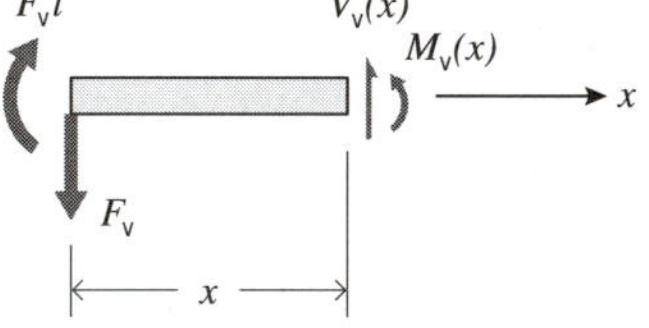

(*d*) Stat I: Internal resultant forces

Figure 4.17 Example 12.

Substituting the expression for M_T in Equation 4.23*a* leads to

$$\begin{aligned} v(l) &= \int_0^l (l - x)\left(-\frac{\alpha T_o}{H}\right)dx \\ &= -\frac{\alpha T_o}{H}\int_0^l (l - x)dx \qquad \textbf{(4.23}\boldsymbol{d}\textbf{)} \\ &= -\frac{\alpha T_o l^2}{2H} \end{aligned}$$

as was obtained in Chapter 3.

We will now consider more complex frame structures for which virtual work provides significant gains in problem solving power and ease of use.

4.5 VIRTUAL WORK FOR FRAMES

The theorem of virtual work for frames is developed in a manner analogous to the theorem of virtual work for trusses, i.e., we combine theorems for joints and beams. In the case of frames, we continue to ignore, for the time being, virtual work associated with axial deformations. This is equivalent to assuming (temporarily) that beams are axially rigid. As noted previously, the axial effects can be included by superposition.

Since the conceptual development of the theorem for trusses has been discussed at length, we will omit the analogous details here. It should be sufficient to reiterate that all virtual work terms associated with internal end forces must cancel; in other words, the first four terms in Equation 4.18 will be equal and opposite to corresponding terms for contiguous joints. Virtual work terms associated with external loads on joints are given by terms identical to the fifth and sixth terms of Equation 4.18. Furthermore, if we regard reaction forces on the frame as concentrated forces of the types encompassed by the fifth and sixth terms of Equation 4.18, we can simplify the overall form of the theorem. The desired form of the theorem is then obtained by summing external virtual work associated with all concentrated forces, summing external virtual work associated with continuous loads over all beams, and summing internal virtual work over all beams. The resulting theorem of virtual work for frames is

$$\sum_{j=1} F_j^{\mathrm{I}} v_j^{\mathrm{II}} - \sum_{i=1} M_i^{\mathrm{I}} v_i'^{\mathrm{II}} + \sum_{k=1}^{NB} \int_0^{l_k} q_k^{\mathrm{I}}(x_k) v_k^{\mathrm{II}}(x_k) dx_k = \sum_{k=1}^{NB} \int_0^{l_k} M_k^{\mathrm{I}}(x_k) v_k''^{\mathrm{II}}(x_k) dx_k \qquad \textbf{(4.24)}$$

where v_j^{II} is the displacement at the point of application, and in the direction, of concentrated force F_j, $v_i'^{\mathrm{II}}$ is the slope at the point of application of pure moment M_i^{I}, NB is the total number of beams in the frame, and subscript k denotes quantities associated with the kth beam. As with all prior forms of the theorem, we refer to terms on the left side as external virtual work and terms on the right side as internal virtual work.

EXAMPLE 13 APPLICATION OF VIRTUAL FORCES.

Find $v'(a^+)$ for the structure shown in Figure 3.23*a* and repeated in Figure 4.18*a*; here, $v'(a^+)$ denotes the slope at $x = a^+$. The real displacement state in which we are interested (state II) is fully described in Figures 3.18*c* and 3.18*d*; these real displacements are a consequence of the applied loads shown in Figure 4.18*a*. The slope $v'(a^+)$, previously denoted by $\phi(a^+)$, has been determined in Example 10, Chapter 3, by means of the moment-area method. To illustrate the application of Equation 4.24 for the calculation of $v'(a^+)$ we will now regard AB and BC as two distinct beams.

As in Example 11(*b*), to determine $v'(a^+)$ we apply the virtual moment M_V shown

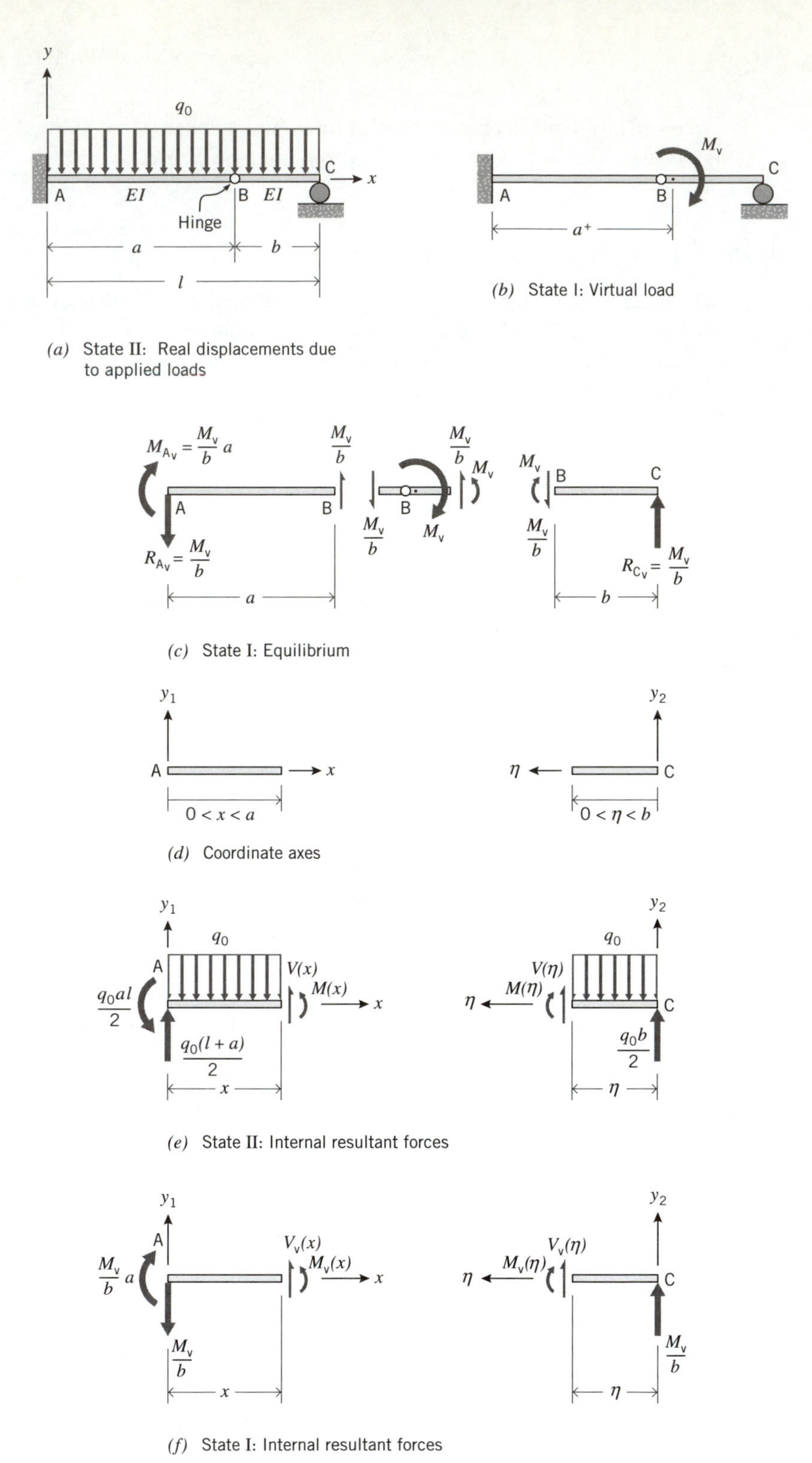

Figure 4.18 Example 13.

in Figure 4.18*b* just to the right of the hinge at B, i.e., at $x = a^+$. We determine reactions by considering equilibrium of the free-bodies in Figure 4.18*c*. Recall that there can be no internal moment in the beam just to the left of the hinge at B; consequently, we conclude that the internal virtual moment just to the right of the point of application of M_V must be equal and opposite to M_V. It follows that $R_{C_V} = M_V/b$; hence, $R_{A_V} = M_V/b$ (in the direction shown) and $M_{A_V} = M_V a/b$.

To emphasize the distinctness of the two beams, we will establish separate coordinate systems for each, as in Figure 4.18*d*. Specifically, we will determine the real curvature and resultant internal virtual moment in beam AB as a function of coordinate x, and the real curvature and resultant internal virtual moment in beam BC as a function of coordinate η.

Figure 4.18*e* shows free-body diagrams of portions of the beams subjected to the real loads, based on data compiled in Example 3, Chapter 3. For beam AB, the real curvature is obtained as a function of x, i.e.,

$$v''(x) = \frac{M(x)}{EI} = -\frac{q_o}{2EI}x^2 + \frac{q_o(l + a)}{2EI}x - \frac{q_o al}{2EI}, \quad 0 < x < a \qquad \textbf{(4.25a)}$$

and, for beam BC, the real curvature is obtained as a function of η,

$$v''(\eta) = \frac{M(\eta)}{EI} = \frac{q_o b}{2EI}\eta - \frac{q_o}{2EI}\eta^2, \quad 0 < \eta < b \qquad \textbf{(4.25b)}$$

Figure 4.18*f* shows free-body diagrams of portions of the beams subjected to the virtual forces. The resultant internal virtual moments in the two beams are,

$$M_V(x) = \frac{M_V a}{b} - \frac{M_V}{b}x, \quad 0 < x < a$$

$$\textbf{(4.25c)}$$

$$M_V(\eta) = \frac{M_V}{b}\eta, \quad 0 < \eta < b$$

We have now compiled all the information needed for application of Equation 4.24. Since there are no real transverse displacements at A or C of the structure in Figure 4.18*a*, it follows that there can be no external virtual work associated with the virtual forces at A or C in Figure 4.18*b*; consequently, the only external virtual work is that associated with the product of M_V (Figure 4.18*b*) and $v'(a^+)$ (Figures 4.18*a* or 3.18*c*). We have assumed positive M_V in Figure 4.18*b*, so if we also assume positive $v'(a^+)$, as in Figure 3.14*a*, then the external virtual work term is negative in accordance with the derivation of Equation 4.24. Internal virtual work is the sum of the integral on the right side of Equation 4.24 for each of the two beams, AB and BC.

To examine the relationship between the theorem of virtual work and the moment-area method, we will write the initial form of Equation 4.24 as

$$-M_V v'(a^+) = \int_0^a \left(\frac{M_V a}{b} - \frac{M_V}{b}x\right)\left[\frac{M(x)}{EI}\right]dx + \int_0^b \left(\frac{M_V}{b}\eta\right)\left[\frac{M(\eta)}{EI}\right]d\eta \qquad \textbf{(4.25d)}$$

Dividing by M_V and rearranging slightly gives

$$v'(a^+) = -\frac{1}{b}\left\{\int_0^a (a - x)\left[\frac{M(x)}{EI}\right]dx + \int_0^b \eta\left[\frac{M(\eta)}{EI}\right]d\eta\right\} \quad \textbf{(4.25e)}$$

The first integral above is the first moment of the area of the $M(x)/EI$ diagram, $0 < x < a$, about $x = a$; the second integral is the first moment of the $M(\eta)/EI$ diagram, $0 < \eta < b$, about $\eta = 0$. Thus, since beam AB is a cantilever beam, the first integral has magnitude equal to $v(a)$, but the sign of this displacement will turn out to be negative, i.e., downward. The second integral is equal to the quantity s_2 in Figure 3.23*f*. In other words, Equation 4.25*e* is exactly the same as Equation 3.27*b*! This can be confirmed by completing the evaluation of the integrals, i.e., by substituting Equations 3.25*a* and 3.25*b* into Equation 4.25*e*. We will not bother with the algebra here; it is sufficient to state that the result will be

$$v'(a^+) = \frac{q_o}{24bEI}(4a^3l - a^4 - b^4) \quad \textbf{(4.25f)}$$

A positive value of $v'(a^+)$ implies a positive slope in accordance with the established sign convention (Figure 3.14*a*). It should be emphasized that this result is obtained without any of the geometric interpretations that were required in Chapter 3. Thus, the theorem of virtual forces provides a direct, straightforward procedure that automatically reveals the directions of frame slopes/displacements within the context of a rigorous sign convention.

EXAMPLE 14 APPLICATION OF VIRTUAL FORCES.

Let's find the transverse displacement at end *C* of the frame in Figure 4.19*a*. This frame has been examined in Examples 4 and 8 of Chapter 3. We will use information generated in these previous examples, especially information in Figures 3.11*e* and 3.19*b*. Figure 3.19*b* shows a sketch of the deflected shape of the frame. The particular displacement in which we are interested has been denoted $v(x_2 = l)$ in Figure 3.19*b*; here, we will denote this displacement by v_C.

The displacement state (state II) in Figure 3.19*b* is real, and is a consequence of real force *F* (and reactions) in Figure 4.19*a*; consequently, we must determine the real curvatures of beams AB and BC. These curvatures can be obtained from the moment diagrams in Figure 3.11*e*. As in the previous example, we will use a separate coordinate system for each beam, in this case the x_1-y_1 and x_2-y_2 axes shown. For beam AB we obtain

$$v''(x_1) = \frac{M(x_1)}{EI} = \frac{F}{\sqrt{2}EI}x_1, \quad 0 < x_1 < \sqrt{2}l \quad \textbf{(4.26a)}$$

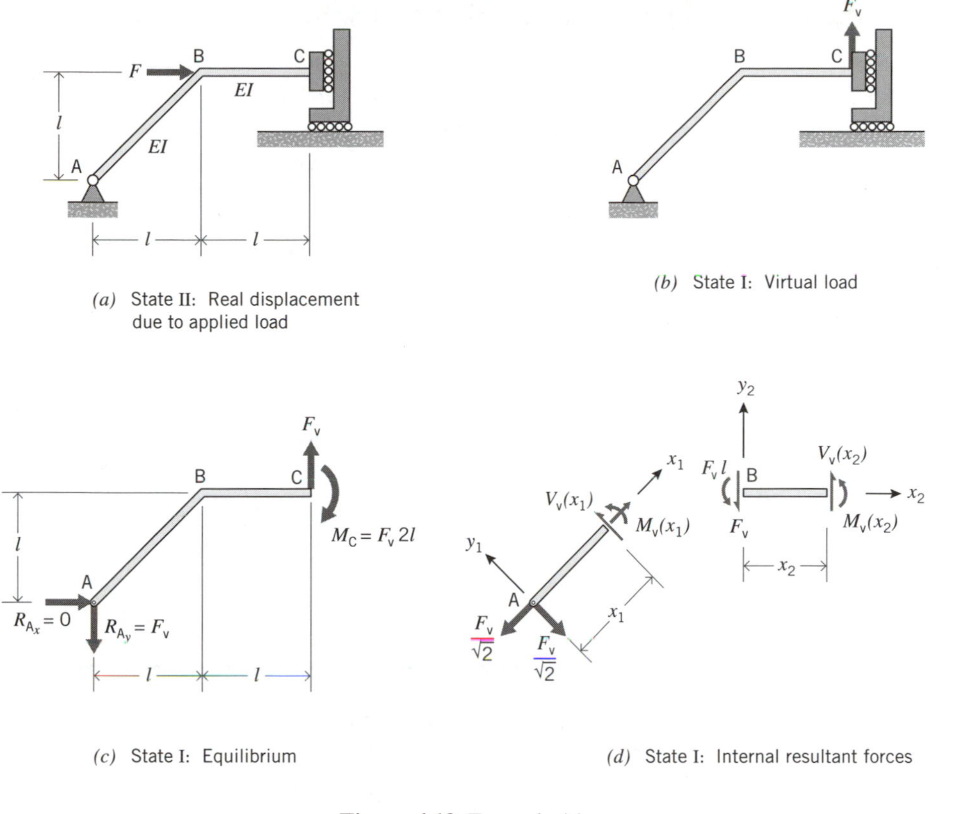

Figure 4.19 Example 14.

and for beam BC we obtain

$$v''(x_2) = \frac{M(x_2)}{EI} = \frac{Fl}{EI}, \quad 0 < x_2 < l \qquad \textbf{(4.26b)}$$

We choose a virtual force F_V applied at point C, as shown in Figure 4.19*b*; the corresponding reactions, which have been determined in a manner similar to that employed in Example 4 of Chapter 3, are shown in Figure 4.19*c*. Note that we have chosen F_V in the same direction as positive v_C.

The external virtual forces in Figure 4.19*c*, together with the internal virtual moments, Figure 4.19*d*, comprise state I. The internal virtual moments for beams AB and BC, respectively, are given by

$$\begin{aligned} M_V(x_1) &= -\frac{F_V}{\sqrt{2}}x_1, \quad 0 < x_1 < \sqrt{2}l \\ M_V(x_2) &= -F_V x_2 - F_V l, \quad 0 < x_2 < l \end{aligned} \qquad \textbf{(4.26c)}$$

The theorem of virtual work, Equation 4.24, now becomes

$$F_V v_C = \int_0^{\sqrt{2}l} \left(-\frac{F_V}{\sqrt{2}}x_1\right)\left(\frac{F}{\sqrt{2}EI}x_1\right)dx_1 + \int_0^l (-F_V x_2 - F_V l)\left(\frac{Fl}{EI}\right)dx_2 \quad \textbf{(4.26d)}$$

which, after manipulation, yields

$$\begin{aligned} v_C &= -\frac{F}{2EI}\int_0^{\sqrt{2}l} x_1^2 dx_1 - \frac{Fl}{EI}\int_0^l (x_2 + l)dx_2 \\ &= -\left(\frac{9 + 2\sqrt{2}}{6}\right)\frac{Fl^3}{EI} \end{aligned} \quad \textbf{(4.26e)}$$

This is identical to the result obtained from Equations 3.23*e* and 3.23*g*.

EXAMPLE 15 APPLICATION OF VIRTUAL FORCES.

We will now demonstrate an application of the theorem of virtual forces to the computation of displacements of an indeterminate frame. Specifically, we will compute the slope at end A, v'_A, of the 1-degree indeterminate beam in Figure 4.20*a*. This structure is similar to the one in Example 13, but without the intermediate hinge. This application of virtual forces to a frame is analogous to the application to a truss in Example 5.

The structure in Figure 4.20*a* is pinned at end A and fixed at end B; consequently, there is one component of reaction at A (vertical force) and three components of reaction at B (vertical force, horizontal force, and moment). We will assume that the correct reactions have been determined (Figure 4.20*b*).[15] Note that the vertical reaction at end A is $3q_o l/8$, not $q_o l/2$ as would be the case if the beam were simply supported. The forces in Figure 4.20*b* produce the real curvature given by

$$\begin{aligned} v''(x) &= \frac{M(x)}{EI} = \frac{R_A}{EI}x - \frac{q_o}{2EI}x^2 \\ &= \frac{3q_o l}{8EI}x - \frac{q_o}{2EI}x^2 \end{aligned} \quad \textbf{(4.27a)}$$

Because we are interested in computing the real slope v'_A, we apply a virtual load consisting of a pure moment, M_V, at end A, as shown in Figure 4.20*c*. The external virtual work in Equation 4.24 will then consist only of the negative product $-M_V v'_A$.

[15] See Example 16, which follows.

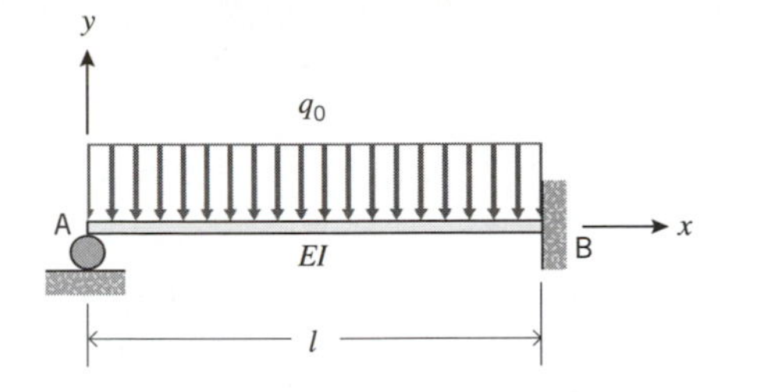

(a) State II: Real displacements due to applied loads

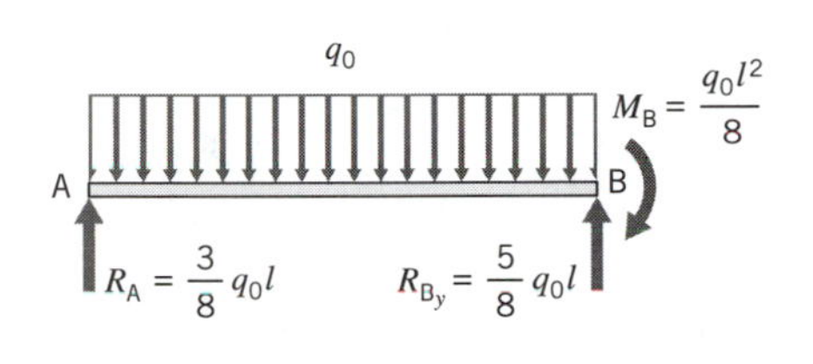

(b) State II: Reactions

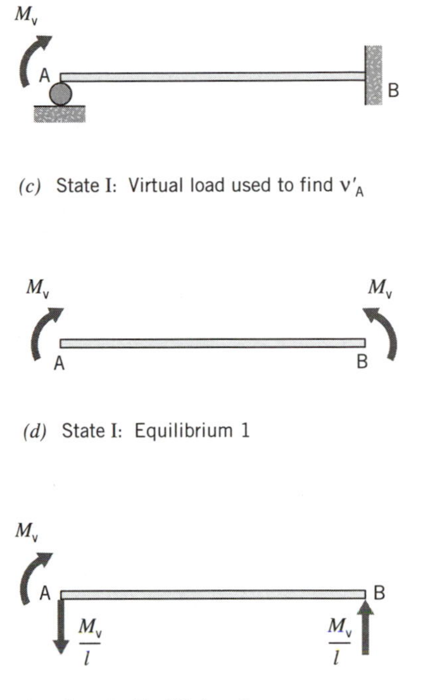

(c) State I: Virtual load used to find v'_A

(d) State I: Equilibrium 1

(e) State I: Equilibrium 2

Figure 4.20 Example 15.

Now, because the structure is indeterminate, there is more than one set of reactions that can equilibrate M_V. Two such sets of reactions that satisfy equilibrium are shown in Figures 4.20*d* and 4.20*e*. For the virtual forces in Figure 4.20*d*, the internal resultant moment is given by

$$M_V(x) = M_V \tag{4.27b}$$

while for the virtual forces in Figure 4.20*e*,

$$M_V(x) = M_V - \frac{M_V}{l}x \tag{4.27c}$$

Substituting Equations 4.27*a* and 4.27*b* into Equation 4.24 gives

$$\begin{aligned} -M_V v'_A &= \int_0^l M_V\left(\frac{3q_o l}{8EI}x - \frac{q_o}{2EI}x^2\right)dx \\ &= \frac{M_V q_o}{2EI}\int_0^l \left(\frac{3l}{4}x - x^2\right)dx \end{aligned} \tag{4.27d}$$

from which we obtain

$$v'_A = -\frac{q_o l^3}{48EI} \tag{4.27e}$$

Similarly, substituting Equations 4.27*a* and 4.27*c* into Equation 4.24 gives

$$\begin{aligned} -M_V v'_A &= \int_0^l \left(M_V - \frac{M_V}{l}x\right)\left(\frac{3q_o l}{8EI}x - \frac{q_o}{2EI}x^2\right)dx \\ &= \frac{M_V q_o}{2EI}\int_0^l \left(1 - \frac{x}{l}\right)\left(\frac{3l}{4}x - x^2\right)dx \\ &= \frac{M_V q_o}{2EI}\int_0^l \left(\frac{3l}{4}x - x^2\right)dx - \frac{M_V q_o}{2EI}\int_0^l \left(\frac{x}{l}\right)\left(\frac{3l}{4}x - x^2\right)dx \end{aligned} \tag{4.27f}$$

The first integral in Equation 4.27*f* is identical to the integral in Equation 4.27*d*; the second integral is zero. Thus, Equation 4.27*f* gives the same result as Equation 4.27*e*! As in Example 5, the fact that two independent virtual force states lead to the same value of displacement gives us confidence that the real curvature (Equation 4.27*a*) has been correctly determined; this, in turn, implies that the reactions in Figure 4.20*b* are correct.

EXAMPLE 16 APPLICATION OF VIRTUAL FORCES.

The theorem of virtual forces plays a role in the analysis of indeterminate frames that is analogous to its role in the analysis of indeterminate trusses, i.e., it is used for the determination of compatibility equations (as previously demonstrated in Example 6).

Here, we will illustrate the determination of a compatibility equation for the indeterminate structure in Figure 4.20*a*. Since the structure is 1-degree indeterminate, we must select one appropriate force quantity as the ''redundant force''; in this case, we can select as the redundant force any of the unknown reactions or any unknown internal resultant force quantity (at any cross section).

Suppose, for example, that the reaction at A has been designated as the redundant force. We wish to ascertain what we get from the theorem of virtual work, Equation 4.24, when we apply to the structure a virtual force ''like'' the reaction at A; we will now denote this virtual force by X_V (Figure 4.21*a*). It follows that there exists a unique set of reactions at B which equilibrate X_V, as shown in Figure 4.21*b*. Also, the internal resultant moment in this virtual force state, Figure 4.21*c*, is given by

$$M_V(x) = X_V x \tag{4.28a}$$

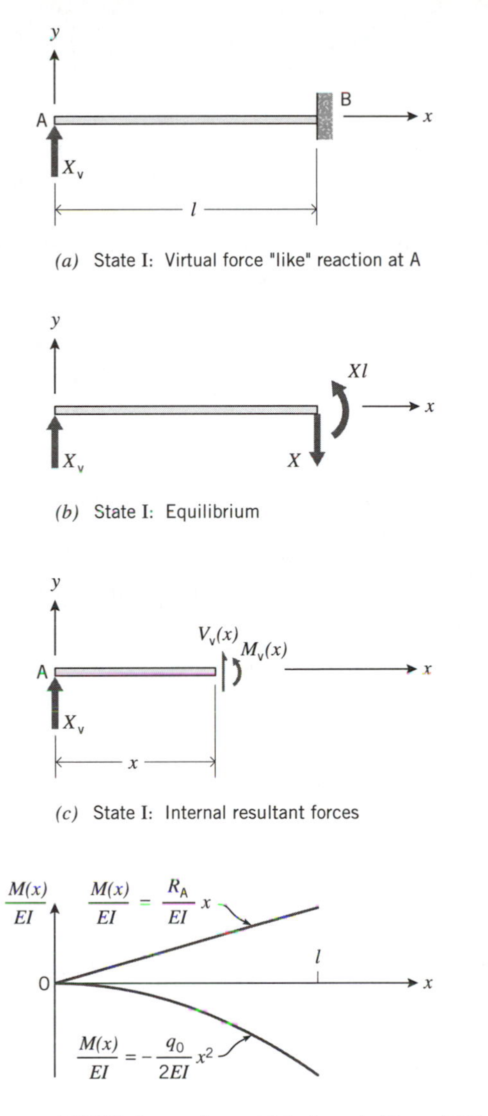

(a) State I: Virtual force "like" reaction at A

(b) State I: Equilibrium

(c) State I: Internal resultant forces

(d) M/EI diagram (by parts) for beam in Figure 4.20*b*

Figure 4.21 Example 16.

Note that the real displacement state must be consistent with the kinematic constraints (supports) in Figure 4.20*a*, i.e., there is no transverse displacement at A or B, and no rotation at B; therefore, the external virtual work in Equation 4.24 must be zero. The real curvature is still given by Equation 4.27*a*, so we will now write Equation 4.24 as

$$0 = \int_0^l (X_V x)\left(\frac{R_A}{EI}x - \frac{q_o}{2EI}x^2\right)dx \qquad \textbf{(4.28}\textbf{\textit{b}}\textbf{)}$$

$$= \int_0^l x\left(\frac{R_A}{EI}x\right)dx + \int_0^l x\left(-\frac{q_o}{2EI}x^2\right)dx$$

If we construct the *M/EI* diagram for the structure in Figure 4.20*b* by parts, as in Figure 4.21*d*, we can identify the first integral above as the first moment, about $x = 0$, of the area of that part of the *M/EI* diagram that is associated with R_A, and the second integral as the first moment, about $x = 0$, of that part of the *M/EI* diagram that is associated with q_o. In other words, we identify, from the moment-area method, the first integral as the transverse deflection of the free end of a cantilever beam due to R_A, and the second integral as the transverse deflection of the free end of a cantilever beam due to q_o. Thus, Equation 4.28*b* states the condition that

$$0 = v_{A(X)} + v_{Aqo} \qquad \textbf{(4.28c)}$$

i.e., that the net deflection at A, which is obtained by the superposition of deflections due to the redundant force and the other loads, acting independently on the beam, must be zero. This is the form of the compatibility equation that results from this particular choice of redundant force. Note that the solution of Equation 4.28*b* gives $R_A = 3q_o l/8$, as stated in the previous example. Thus, Equation 4.28*b* actually provides the value of the redundant force. (We are able to solve Equation 4.28*b* for R_A because the equilibrium equations have been implicitly incorporated via Equation 4.27*a*.) This is typical of procedures we will follow for the analysis of indeterminate structures in the next chapter.

EXAMPLE 17 APPLICATION OF VIRTUAL DISPLACEMENTS.

For the final example of this chapter, we will take a brief look at the use of virtual displacements to obtain equilibrium equations of the type utilized in the displacement method of analysis of frame structures. Again, the procedure differs somewhat from that illustrated for trusses, Example 7, because the relevant force quantities are continuous functions of position coordinates rather than constants.

In the displacement method of analysis of frames, we will use equilibrium equations written in terms of beam end forces, such as M_A and M_B in Figures 3.1 or 4.12*b*. For simple indeterminate structures, such as the beam in Figure 4.20*a*, this type of equilibrium equation will have the trivial form $M_A = 0$, i.e., the equation of moment equilibrium for joint A. For a frame structure composed of several beams the equilibrium equations can be more involved. To distinguish between end forces in various beams, we will find it convenient to modify our notation to clearly identify the specific beam with which each end force is associated. In particular, we will add a second subscript to each end force, e.g., M_A will be replaced by M_{AB}, and M_B will be replaced by M_{BA}. The first subscript continues to identify a particular end of the beam, while the two subscripts together identify the beam itself.

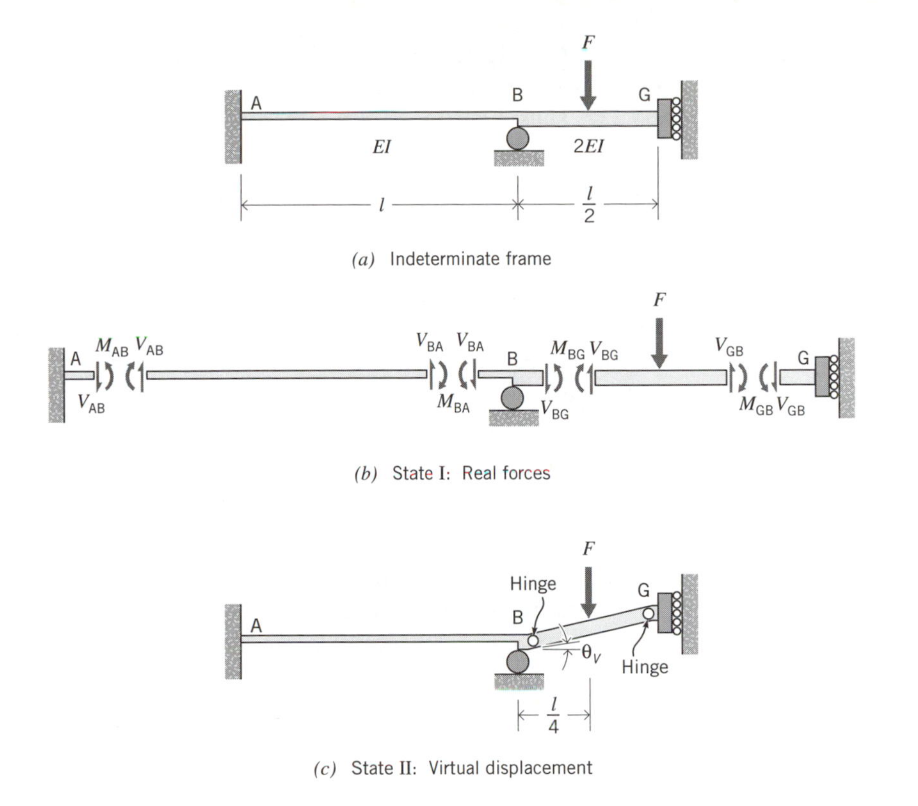

Figure 4.22 Example 17.

Consider, for example, the continuous beam shown in Figure 4.22*a*; the structure has a fixed support at A, a roller support at B, and a fixed support with a roller at G. The load consists of a concentrated force F applied at the midpoint of span BG. We will analyze this structure as though it is composed of two beams, AB and BG, rigidly connected at B; at joint B we can now distinguish between the two moments in the contiguous beams, M_{BA} and M_{BG} (Figure 4.22*b*). As we will see subsequently in our discussion of the displacement method, we require two equilibrium equations for analysis of this structure (one for each kinematic degree of freedom at the joints). One of these equations is obvious, i.e., $M_{BA} + M_{BG} = 0$ (moment equilibrium for joint B). The second equation is less obvious, and it is this equation that we will obtain by use of virtual displacements.

The virtual displacement state we will employ here is conceptually similar to the one used in Example 9, i.e., it will involve ''concentrated curvatures'' of the type originally introduced in Figure 4.14*b*. In this particular case, the virtual displacement state will be taken as in Figure 4.22*c*: we will assume two concentrated curvatures in beam BG, one near joint B, and one near joint G; otherwise, the curvature of beam BG is zero everywhere.

For this example, then, the only external virtual work is that associated with applied force F, Figure 4.22*a*, and the corresponding transverse virtual displacement, $\theta_V l/4$, Figure 4.22*c*. The only internal virtual work is that associated with the real internal moments M_{BG} and M_{GB}, Figure 4.22*b*, and the virtual changes in slopes at the points where M_{BG} and M_{GB} act (the points of concentrated curvature); the latter are both equal to θ_V, Figure 4.22*c*. Thus, Equation 4.24 becomes

$$-\frac{Fl}{4}\theta_V = M_{BG}\theta_V + M_{GB}\theta_V \tag{4.29a}$$

The external virtual work is negative because F and the virtual displacement are opposite in direction, whereas the internal virtual work terms are both positive because the internal moments are in the same directions as the slope changes.

Dividing by θ gives

$$-\frac{Fl}{4} = M_{BG} + M_{GB} \tag{4.29b}$$

which is the desired equilibrium equation. This equation could also be obtained from a consideration of equilibrium of a free-body of beam BG, Figure 4.22*b*.

4.6 SUMMARY

In this chapter, we have presented a development of the theorem of virtual work in special forms applicable to trusses and frames. We have illustrated a wide variety of applications of the two basic derivatives of the theorem, namely the theorem of virtual displacements and the theorem of virtual forces. A qualitative summary of these applications is given in Table 4.1.

Table 4.1. Summary of Virtual Work

Theorem	State I (forces in equilibrium)	State II (displacements)	Applications
Virtual forces	Virtual	Real	Find displacements; compatibility equations for force method of analysis
Virtual displacements	Real	Virtual	Find equilibrium equations; equilibrium equations for displacement method of analysis

As will be seen in greater detail in subsequent chapters, the theorem will play a very prominent role in the analysis of indeterminate structures, particularly when we examine the matrix formulations of the force method and the displacement method.

PROBLEMS

4.P1. In Example 1 the horizontal motion of joint D due to the virtual displacement (rigid-body rotation) in Figure 4.4*b* was said to be equal to $h\theta$ (for θ small). Prove this by considering a general case of a rigid body in the x-y plane, which undergoes a small rotation θ about a z axis through some point O as shown in the figure below. Let η define an arbitrary axis in the x-y plane through O, and let D be any point a distance h from axis η. Show that the magnitude of the component of displacement of point D in the direction parallel to η is $h\theta$. Hint: determine components of motion at D by considering the motion of line OD when the body rotates through a small angle θ.

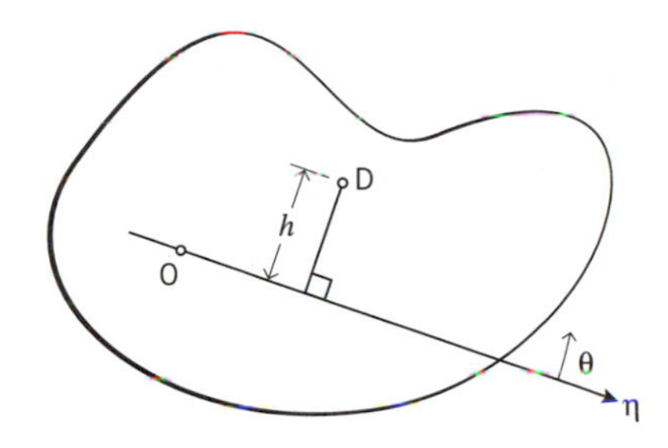

4.P2. Use a virtual displacement to find the internal force in bar EG. Hint: consider coupled motions of joints B and G that involve no length change of any bar except EG.

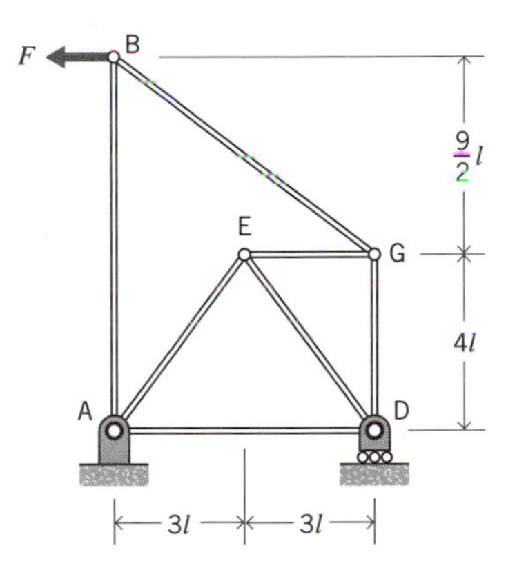

4.P3. It is known that bar 1 elongates by an amount Δ_1 and bar 2 elongates by an amount Δ_2. Find the horizontal and vertical components of displacement of joint A in terms of Δ_1 and Δ_2.

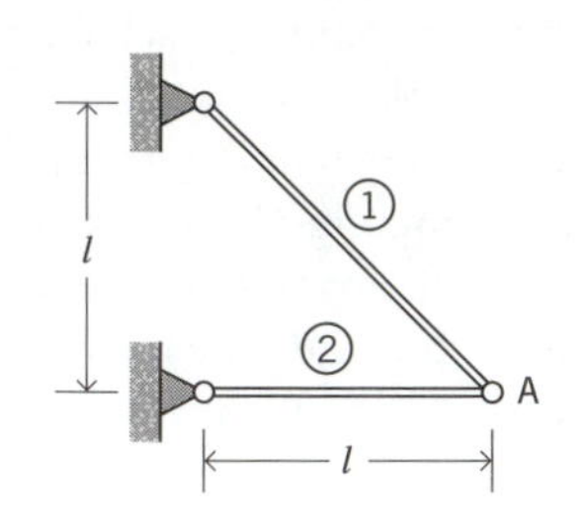

4.P4. For the truss in Problem 4.P2
(a) Find the horizontal displacement of joint B;
(b) Find the horizontal displacement of joint E;
(c) Find the vertical displacement of joint E.
Assume that EA is the same for all bars.

4.P5. (a) Find the horizontal displacement of joint E;
(b) Find the vertical displacement of joint E;
(c) Find the vertical displacement of joint D.
Assume that EA is the same for all bars.

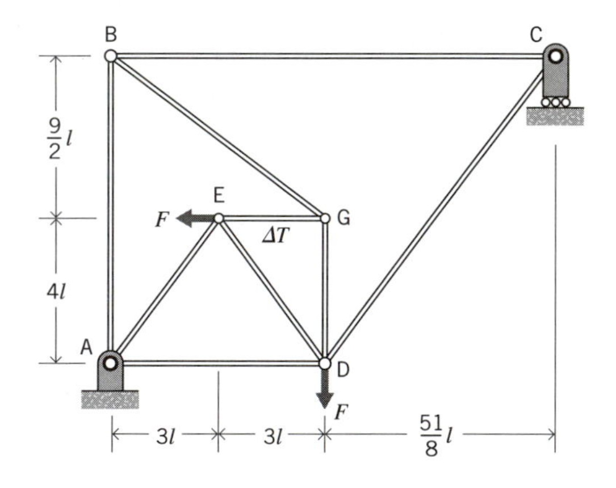

4.P6. (a) Find the horizontal displacement of joint D;
(b) Find the vertical displacement of joint C.
Assume that EA is the same for all bars.

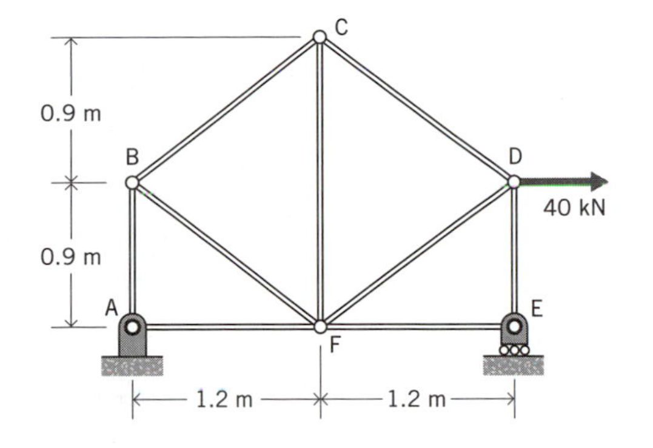

4.P7. (a) Find the vertical displacement of joint C;
(b) Find the horizontal displacement of joint C.
Assume that EA is the same for all bars.

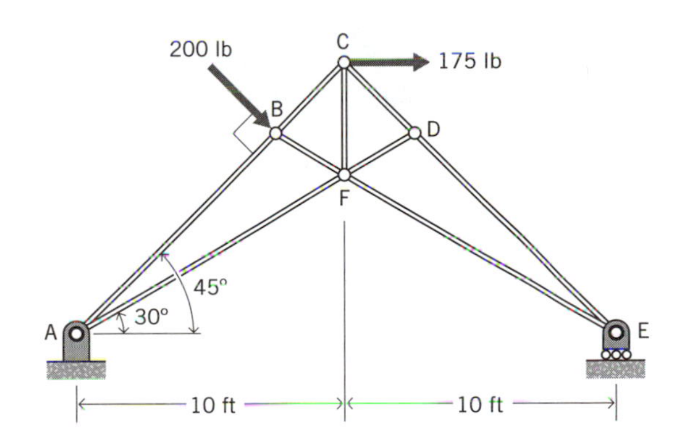

4.P8. (a) Find the horizontal displacement of joint G;
(b) Find the vertical displacement of joint D.
Assume that EA is the same for all bars.

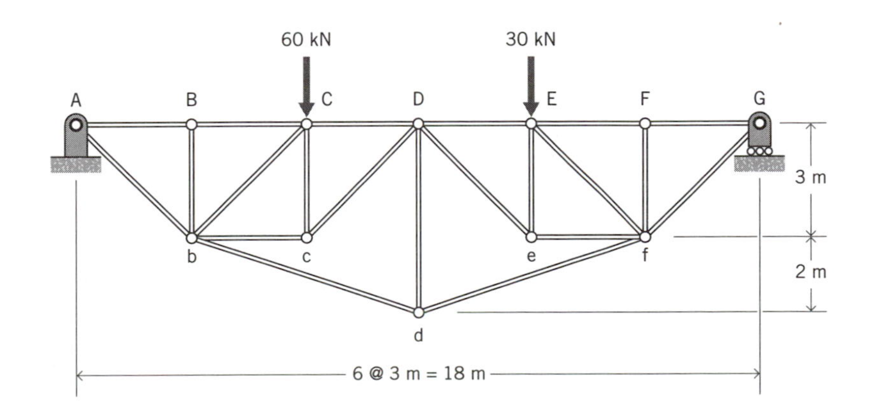

4.P9. (a) Find the horizontal displacement of joint G;
(b) Find the vertical displacement of joint G.
Assume that EA is the same for all bars.

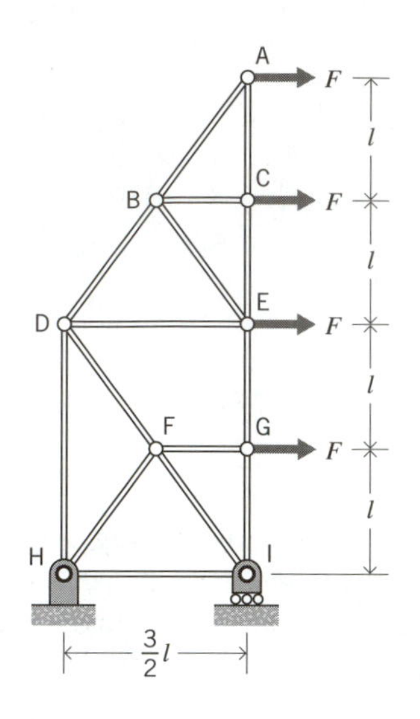

4.P10. The temperature of bars BC and CG is changed by an amount ΔT. Find the horizontal component of displacement of joint C. Assume that EA and α are the same for all bars.

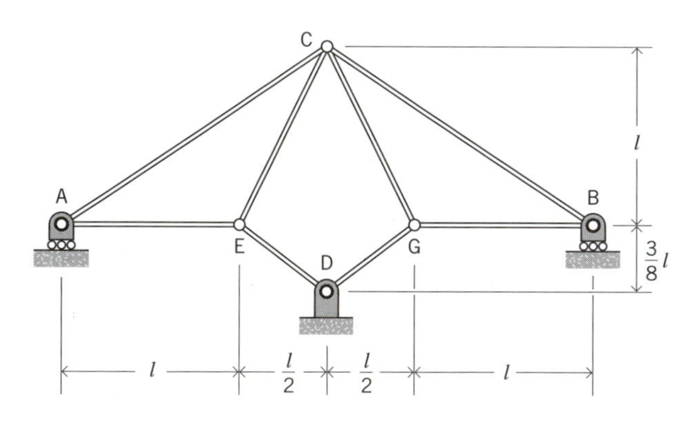

4.P11. The truss is subjected to an applied force F. In addition, bar IJ has both a temperature change ΔT and initial length change Δl_o.

(a) Find the vertical displacement of joint J;

(b) Find the horizontal displacement of joint O;

(c) What length change Δl_o leads to zero vertical displacement at J under the action of F and ΔT?

Assume that EA is the same for all bars.

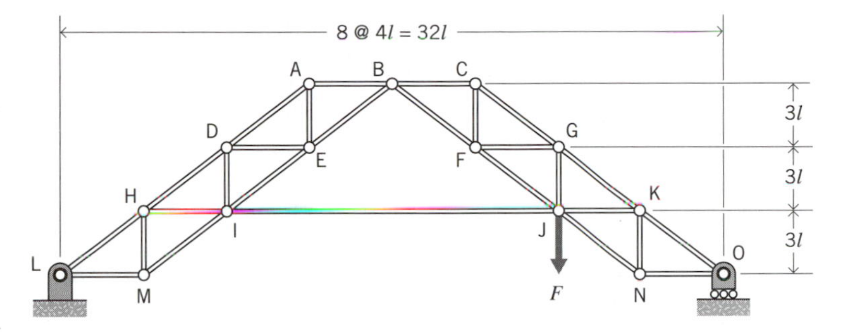

4.P12. The structure shown is composed of three uniform bars (AD, AE, and BE) having the same cross-sectional properties EA, and a rigid beam ABC. The beam may be assumed to be weightless. Find the vertical displacement at C.

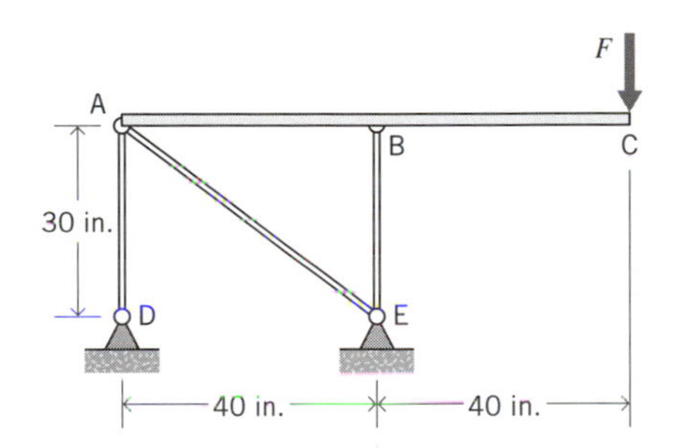

4.P13. The figure below shows internal forces in the bars of an indeterminate truss. All bars have the same EA. Find the vertical displacement of joint D using two different virtual force distributions.

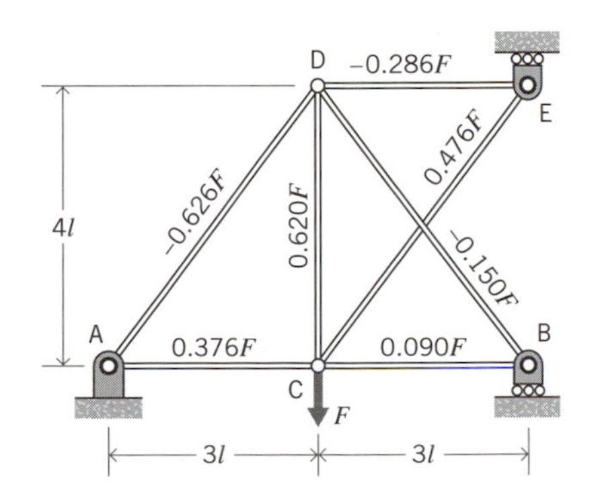

4.P14. The figure below shows internal forces in the bars of an indeterminate truss. All bars have the same EA. Find the vertical displacement of joint B using two different virtual force distributions.

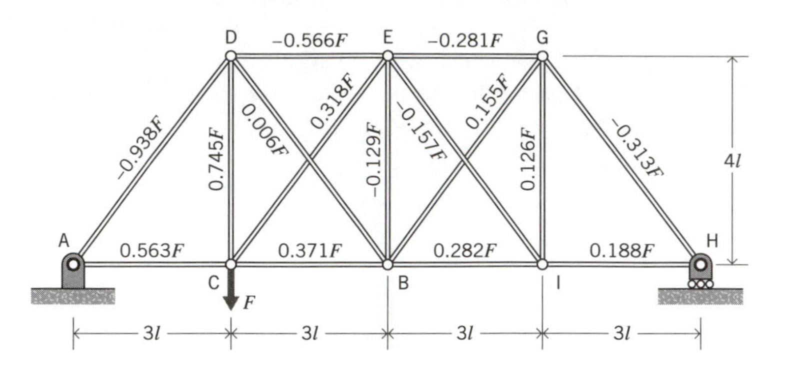

4.P15. Figure a shows real forces on a simply supported beam. Use the virtual displacement in Figure b to evaluate R_A. Assume that EI is constant.

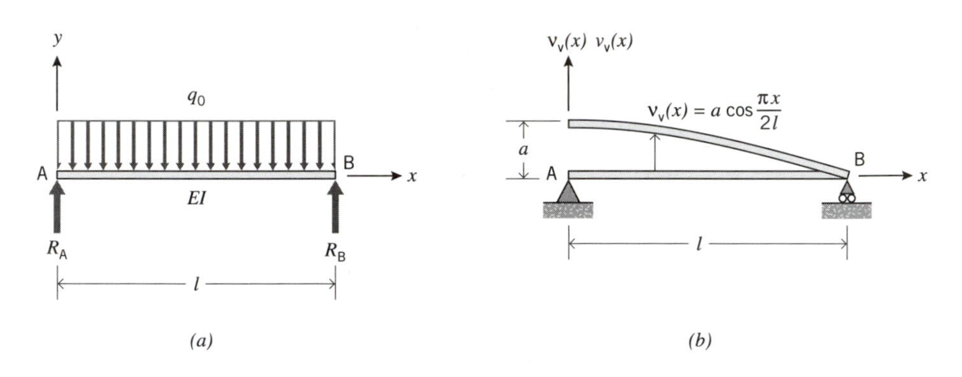

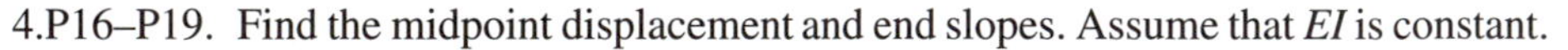

4.P16–P19. Find the midpoint displacement and end slopes. Assume that EI is constant.

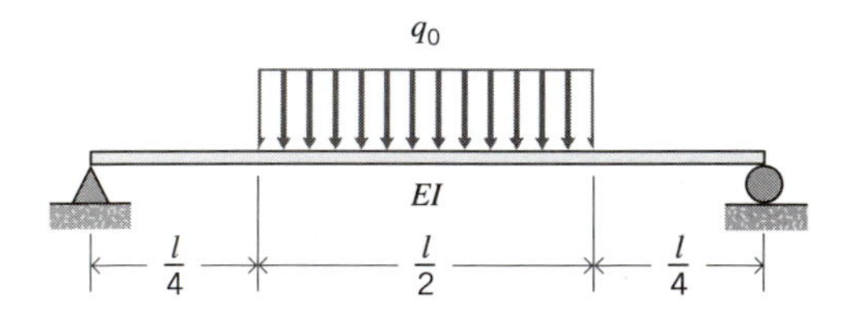

4.P16

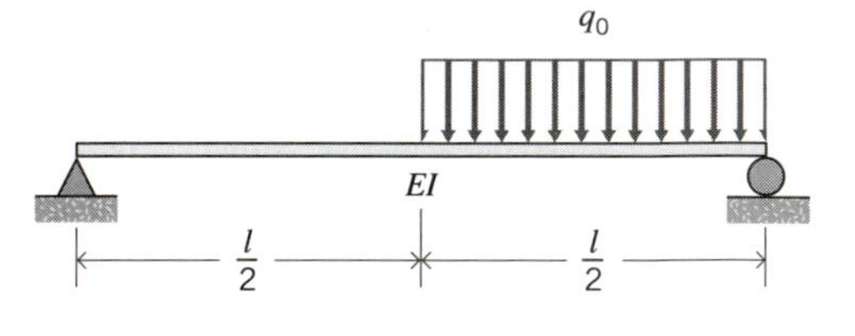

4.P17

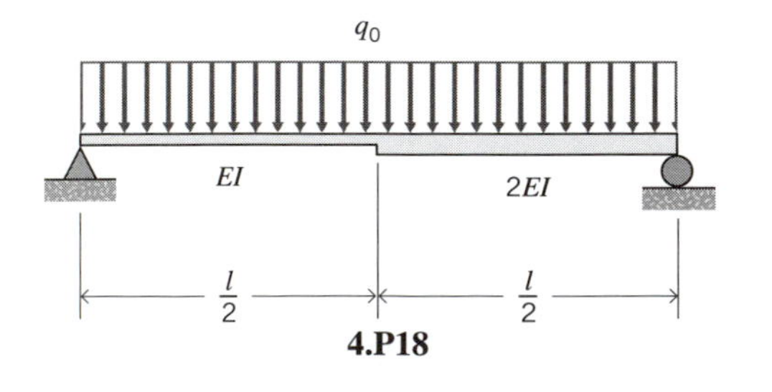

4.P18

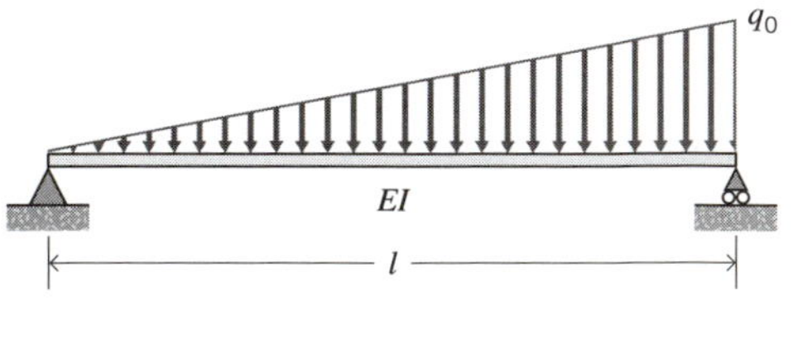

4.P19

4.P20–P21. Find the transverse displacement of B and rotation at A. Assume that EI is constant.

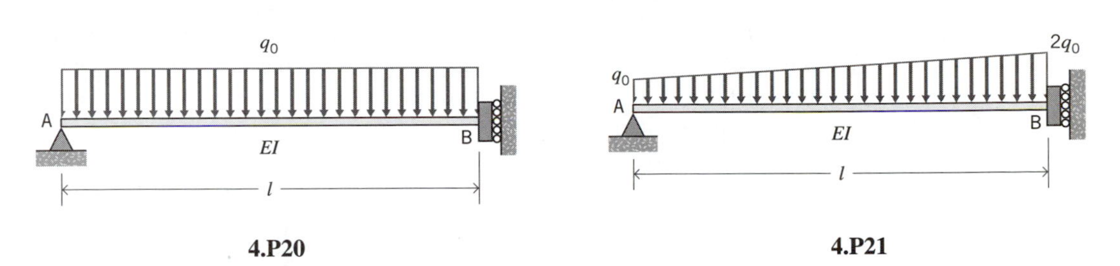

4.P20 **4.P21**

4.P22–P24. Find the transverse displacement at C. Assume that EI is constant.

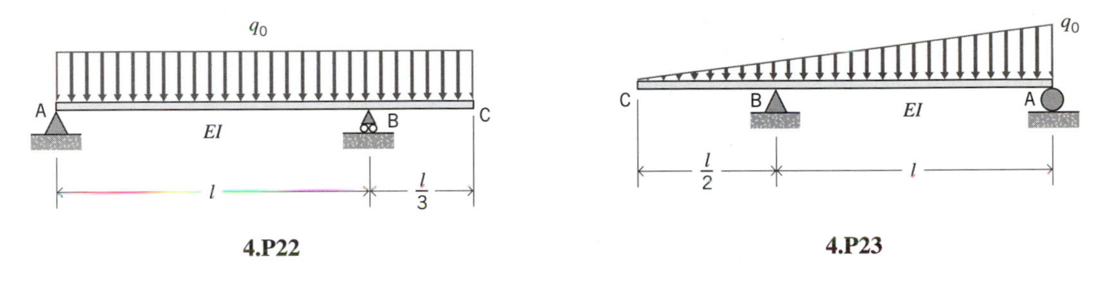

4.P22 **4.P23**

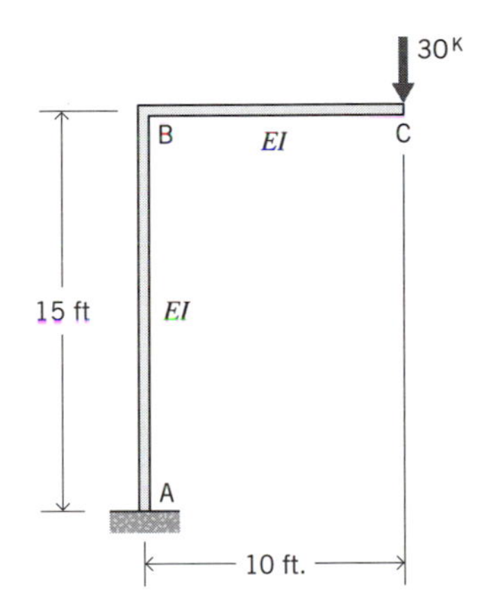

4.P24

4.P25. The cantilever structure below consists of two identical uniform beams connected by a hinge with a torsional spring joining the contiguous ends of the beams. The torsional spring undergoes an angular offset ϕ due to an applied moment M such that $\phi = M/k$ where k is the linear spring constant. (This is analogous to an axial spring that elongates under the application of an axial force such that $\Delta = P/k_s$.) Find the transverse displacement at the free end.

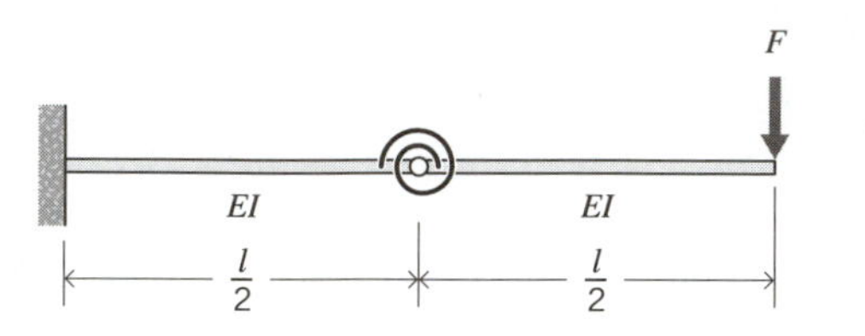

4.P26–P27. (a) Find the horizontal and vertical displacement at B;
(b) Find the rotation at A and C;
(c) Find the rotation at B.

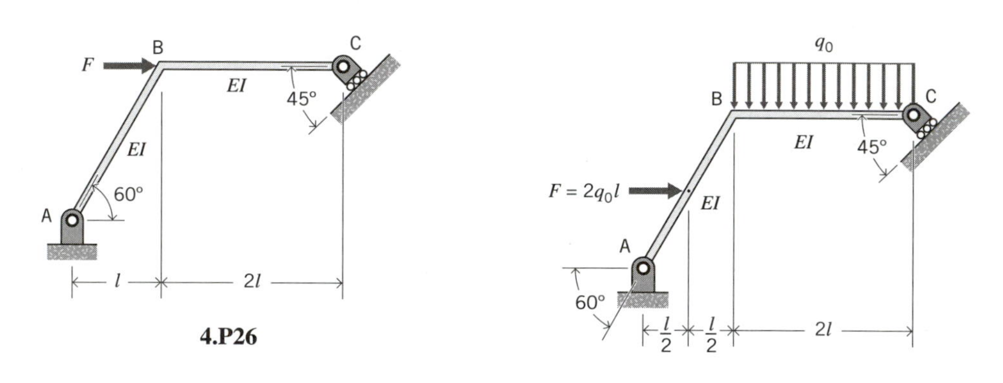

4.P26

4.P27

4.P28. (a) Find the horizontal and vertical displacement at B;
(b) Find the rotations on each side of the hinge.

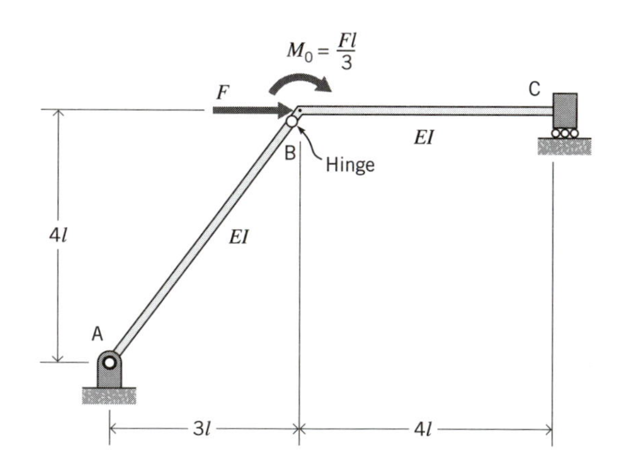

4.P29. Find the vertical displacement at the hinge.

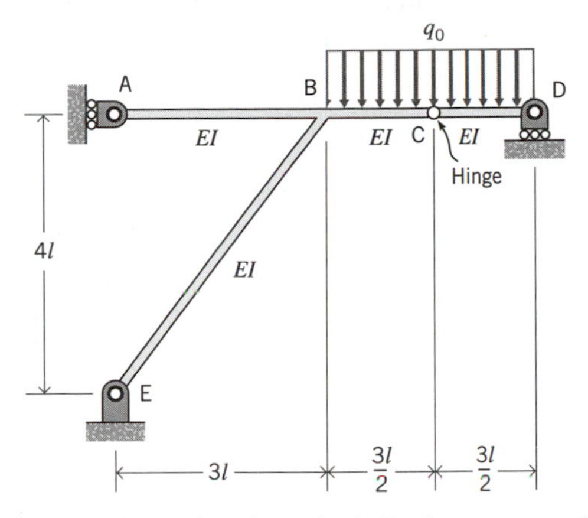

4.P30. (a) Find the horizontal and vertical displacement at C;
(b) Find the rotation at C.

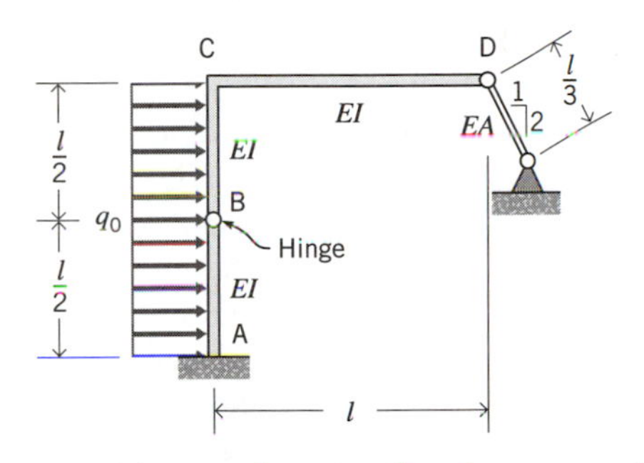

4.P31. The frame has a hinge and torsional spring connecting beams AB and BC at joint B. The relationship between moment and angle change at the spring is $M = k\phi$ where k is the spring constant.

(a) Find the horizontal component of displacement at joint C. Assume that $k = EI/2l$.
(b) It is found from part (a) that the horizontal displacement at C is twice the maximum allowable value, so it is proposed to use a different spring at B for which $k = 2EI/l$. Will this new spring lead to an acceptable displacement?

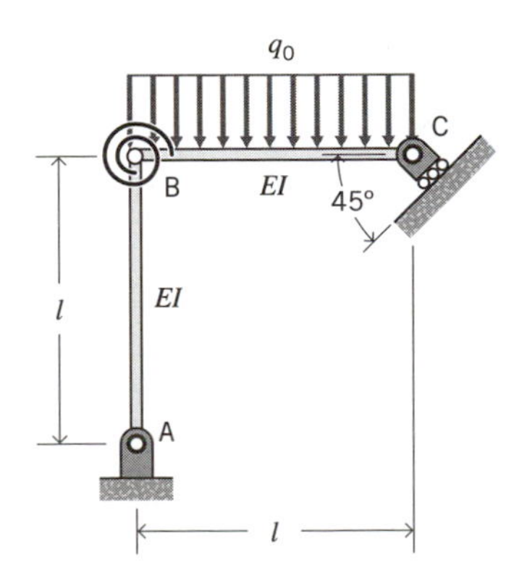

4.P32. Find the vertical deflection at point C in the truss-supported beam shown. Assume that all truss bars have the same EA, and that the bending stiffness of beam ABC is EI.

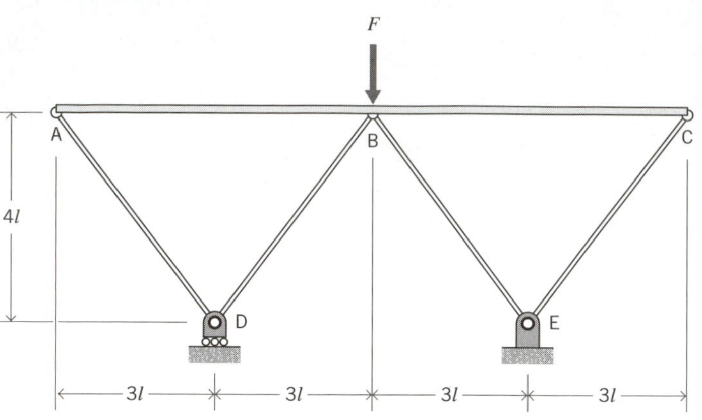

4.P33. For the structure in Problem 4.P32, find the virtual force pattern, including the necessary internal forces/moments, needed to compute the following displacements:

(a) vertical displacement at B;
(b) rotation of the beam at C;
(c) horizontal displacement of the roller at D.

4.P34. Figure a shows an indeterminate structure that is subjected to a force F. Figures b and c show the results of two different analyses by students. Use virtual work to determine which, if any, of the solutions is correct. Hint: are the kinematic conditions at the supports satisfied?

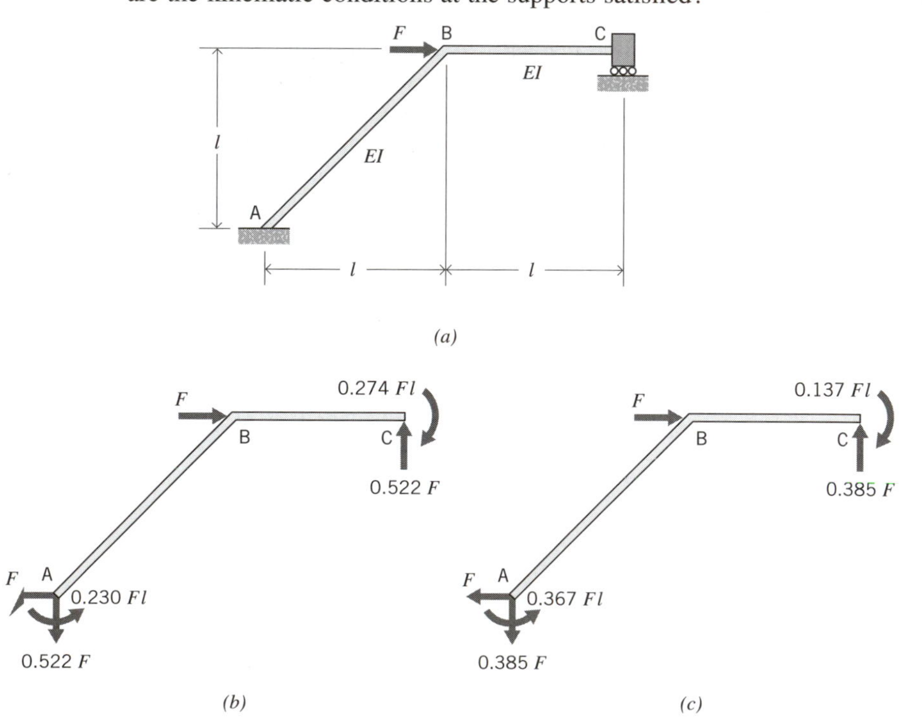

4.P35. The beam is one-degree indeterminate. A student claims that $R_A = -M_o/l$ (negative sign indicates downward). Use virtual work to demonstrate that this value is *not* correct. Hint: are the kinematic conditions at the supports satisfied?

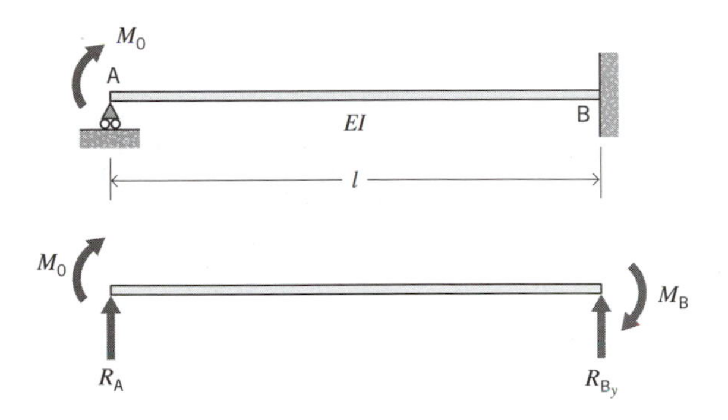

4.P36. The correct value of R_A in Problem 4.P35 is $-3M_o/2l$. Find the slope at A; use two different virtual force patterns.

4.P37. In addition to the force F, the structure in Figure 4.22a is subjected to a temperature change $\Delta T = T_1 y$ in beam AB only. The internal moments, identified in the figure below, may be shown to be given by

$$M_{AB} = \frac{3Fl}{64} + \frac{5EI\alpha T_1}{4}$$

$$M_{BA} = \frac{3Fl}{32} - \frac{EI\alpha T_1}{2}$$

$$M_{BG} = -\frac{3Fl}{32} + \frac{EI\alpha T_1}{2}$$

$$M_{GB} = -\frac{5Fl}{52} - \frac{EI\alpha T_1}{2}$$

where α is the coefficient of thermal expansion. Use virtual work to find the vertical displacement at G and the rotation at B.

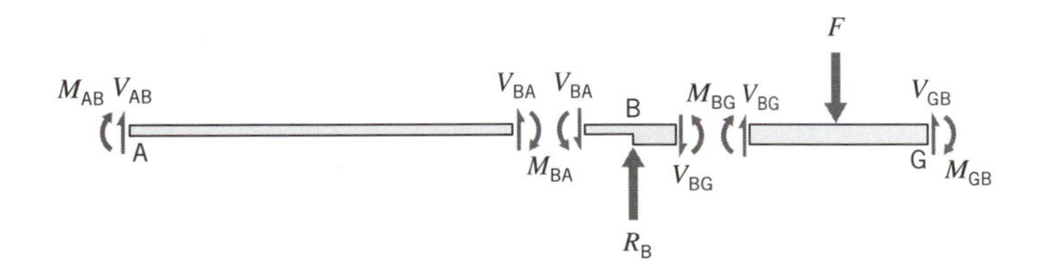

4.P38. The portal frame is subjected to two independent sets of loads, namely a uniform downward loading q_o on the horizontal span, and a concentrated horizontal force F applied at point D.

(a) Find the horizontal motion, u_d, of point D due to each load.

(b) Find the value of the horizontal force F which, together with the load q_o, causes $u_d = 0$.

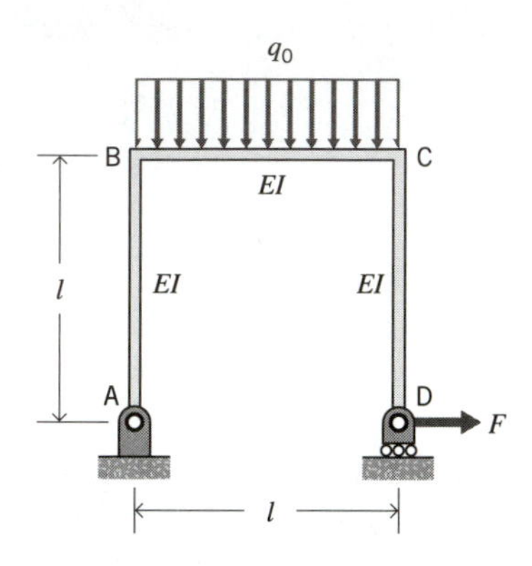

Chapter 5

Introduction to Analysis of Indeterminate Structures: Force Method

5.0 INTRODUCTION

When the concept of indeterminacy was introduced in Chapter 1[1] we observed that a structure is classified as statically indeterminate whenever the number of unknown forces is greater than the number of independent equations of equilibrium. To get the additional equations needed to completely determine the forces we will employ the concept of geometric compatibility. This approach to analyzing structures is termed the force method. In the force method we first write equilibrium equations in terms of applied loads and a set of ''redundant forces,'' then impose compatibility by using the theorem of virtual forces.

The force method is closely related to the statical analysis procedures presented in Chapters 2 and 3 for statically determinate structures. In fact, all of the basic concepts required for application of the force method have already been discussed in the preceding chapters; all we will have to do here is to assemble the concepts into a coherent procedure. As part of the development, the procedure will be expressed in matrix form, which will allow some generalization and simplification.

5.1 REDUNDANT FORCES

Before beginning a detailed presentation of the force method, it may be helpful to again review the term ''redundant force.'' As noted above, an indeterminate structure is one that, in essence, contains more forces than are required to satisfy equilibrium. This implies that we may temporarily delete these ''extra'' forces and still satisfy the equilibrium equations. Any force that can be temporarily removed in this manner is referred to as a *redundant force*. Recall from discussions of some examples in Chapter 4 that

[1] See Section 1.7.

the choice of forces to consider as redundants is not unique; any internal force or reaction that can be removed without causing statical instability may be regarded as a redundant force.

As an illustration, consider the truss in Figure 5.1*a,* which is one-degree indeterminate because $NR + NB - 2NJ = 1$ ($NR = 4$, $NB = 7$, $NJ = 5$).[2] The fact that this truss is one-degree indeterminate means that it is possible to delete *one* force and be left with a statically determinate and stable structure. It is essential, however, to choose this force appropriately, i.e., in such a way that its removal does, in fact, leave the structure statically stable. Figures 5.1*b*–5.1*e* show the only four appropriate choices of this force. In Figure 5.1*b* the horizontal component of reaction has been removed from the right pin support; this is indicated schematically by replacement of the pin support by a roller support. Then, $NR = 3$ and the truss is now both determinate and stable. Thus, we refer to the horizontal component of reaction as the *redundant force.* Alternately, the horizontal component of reaction at the left pin support may be regarded as the redundant force, as indicated in Figure 5.1*c*. Other choices are equally valid; Figures 5.1*d* and 5.1*e* show acceptable choices of internal forces that may be removed from the truss in Figure 5.1*a*. In each of these two cases the removal of a particular bar force, denoted by a dotted line, leaves a statically determinate and stable structure; thus, either of the indicated bar forces may also be regarded as the redundant force for this truss.

[2] See Section 2.3.

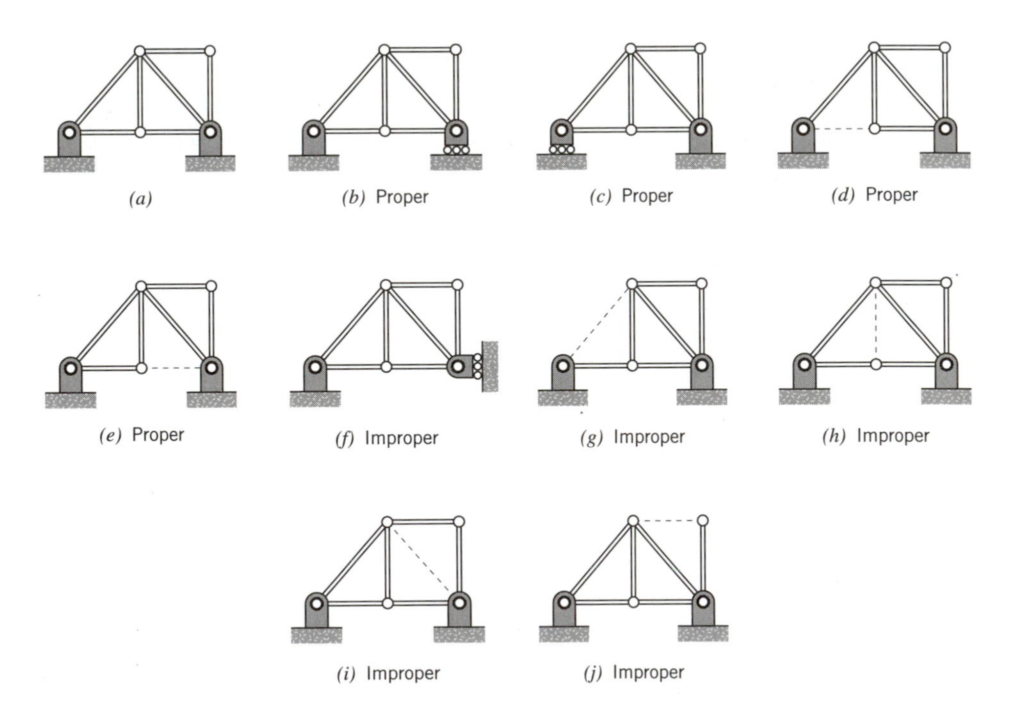

Figure 5.1 (*a*) Truss that is one-degree indeterminate; (*b*)–(*e*) acceptable choices of redundant force; (*f*)–(*i*) improper choices of redundant force.

There is no force, other than those noted above, that may be regarded as a redundant force, because the removal of any other component of reaction or internal force renders the truss in Figure 5.1*a* statically unstable. For example, the removal of a vertical component of reaction, as in Figure 5.1*f*, or any one of the internal bar forces shown in Figures 5.1*g*–5.1*j*, is incorrect for the reason stated, namely that the truss configurations shown are incapable of satisfying equilibrium for arbitrary loads and are therefore classified as statically unstable.[3]

In general, for an indeterminate structure the total number of redundant forces will be equal to the degree-of-indeterminacy. As we will see shortly, in the force method of analysis we begin by identifying a set of n redundant forces where n is the degree-of-indeterminacy of the structure. In other words, removal of the n forces in this set would leave the structure statically determinate and stable. To identify these forces we proceed as above, except that the forces are selected in a sequential manner. To demonstrate, consider the truss in Figure 5.2*a* which is two-degrees indeterminate ($NR = 4$, $NB = 8$, $NJ = 5$; there is no connection at the point where the two diagonal bars cross). Figures 5.2*b*–5.2*h* show some of the acceptable combinations of two redundant forces, and Figures 5.2*i* and 5.2*j* show only two of the numerous possible incorrect combinations. Note that although either horizontal component of reaction may be selected as a redundant, it is improper to select *both* because the result would again

[3] This discussion is based on the assumption that all joints are necessary for good reason, e.g., every joint may be subjected to applied loads.

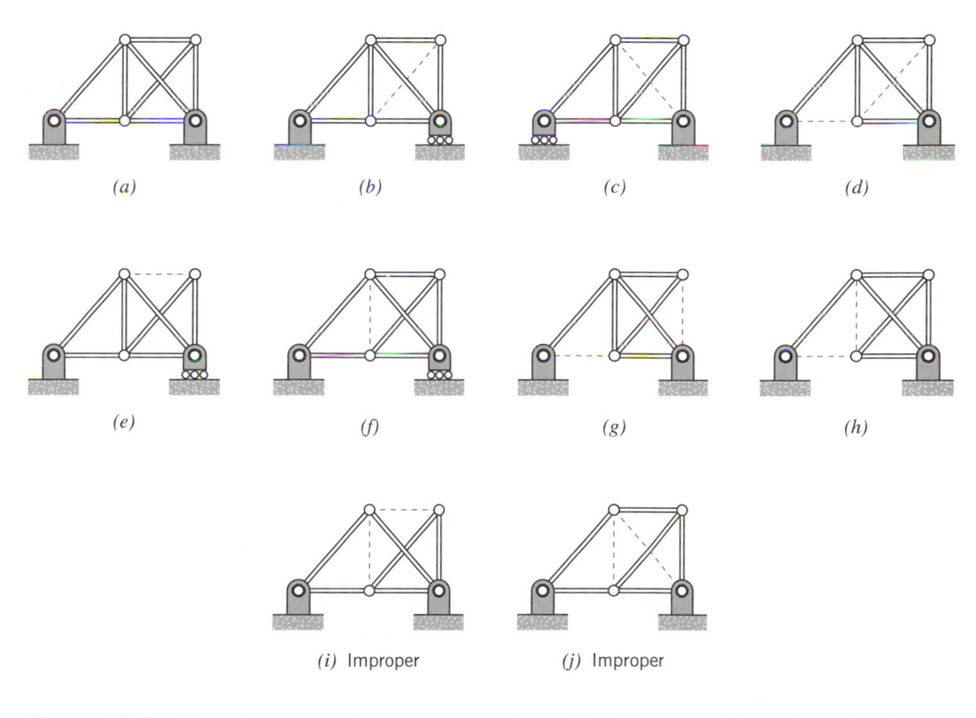

Figure 5.2 (*a*) Truss that is two-degrees indeterminate; (*b*)–(*h*) acceptable choices of redundant forces; (*i*)–(*j*) improper choices of redundant forces.

be a statically unstable structure. Similarly, either of the diagonal bar forces may be selected, but the simultaneous choice of both is improper because the result would be an unstable truss like the one in Figure 5.1*i*. The somewhat less obvious situation depicted in Figure 5.2*j* suffers from the same type of shortcoming as the truss in Figure 5.1*i* and illustrates the fact that the avoidance of incorrect choices of redundant forces is not always a clear-cut proposition.

The independent redundant forces in an indeterminate structure are not actually equal to zero; rather, their magnitudes are treated as ''unknown'' mathematical quantities in the statical (equilibrium) analysis for the remaining forces. This is essentially the second step in a five-step process that characterizes the force method of analysis for both trusses and frames. A complete overview of the process may be summarized as follows:

Step 1. Identify a set of n unknown redundant forces;
Step 2. Solve equilibrium equations;
Step 3. Write element deformation-force relations;
Step 4. Find compatibility equations; and
Step 5. Solve compatibility equations for redundant forces.

Step 1 has been described above in some detail. Step 2 involves finding *all* internal forces and reactions in terms of redundant forces and known applied loads using only the equations of statics. This leads to expressions for all internal forces and reactions in terms of the known forces and unknown redundant forces. Step 3, in turn, leads to a set of element deformations, which are also expressed in terms of both the known and unknown forces. The compatibility equations, Step 4, are relations that the element deformations must satisfy in order to ensure that the components of the structure fit together properly and are properly connected to supports. As we shall see, this concept will yield exactly the additional information needed to determine the unknown redundant forces in Step 5. We then back-substitute the results into Step 2 to find component forces and into Step 3 to get element deformations. Details of the procedure will become clear as we consider typical applications for trusses and frames.

5.2 TRUSSES

The starting point for the analysis of statically indeterminate trusses is the identification of a set of n redundant forces, some of which may be components of reactions and some of which may be internal bar forces. As was observed in Examples 5 and 6 in Chapter 4, redundant forces can be assigned arbitrary values; the equilibrium equations can be solved to obtain all other force quantities in terms of these arbitrary values. Because the equilibrium equations are linear, all force quantities will be linear combinations of the redundant forces and the specified loads. This linearity allows the use of *superposition* in analysis.

Consider Figure 5.3, which illustrates superposition of forces for the statical analysis of the truss in Figure 5.2*a*. (The force F shown in Figure 5.3 is intended to be representative of any arbitrary load.) Figure 5.3*a*, for example, shows superposition of forces based on the choice of redundants in Figure 5.2*b*; here, the quantity X_1 denotes the unknown magnitude of the horizontal component of the reaction at the right support and X_2 denotes the unknown magnitude of the internal force in the diagonal bar. In essence, we may regard the forces in the truss of Figure 5.2*a* as the sum of (1) forces

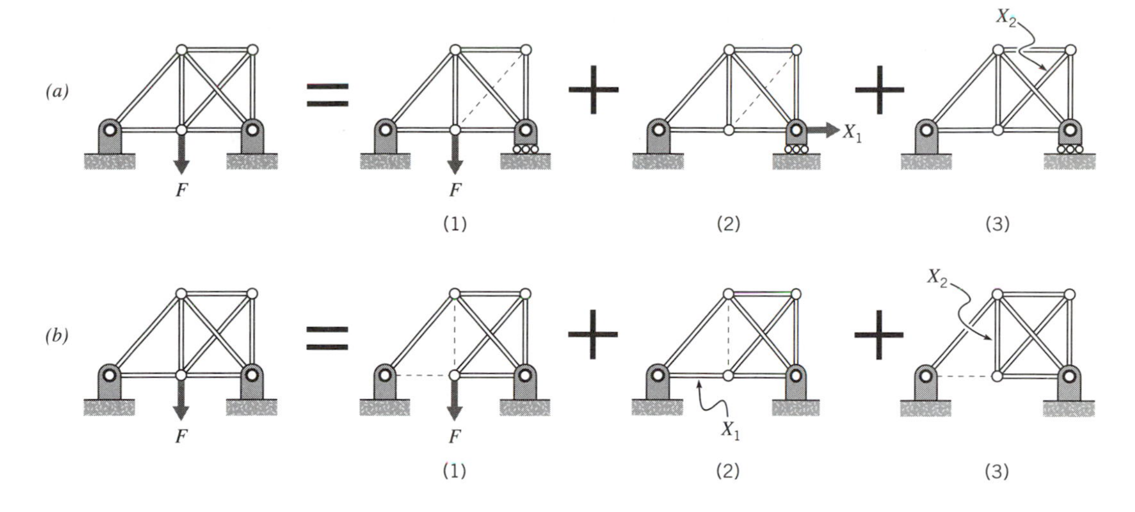

Figure 5.3 Superposition of forces.

due to the applied loads when the values of the redundants are set to zero, (2) forces due to the redundant X_1 acting alone, and (3) forces due to the redundant X_2 acting alone. Note, again, that the forces in each of these three cases are statically determinate and that the actual forces are a sum of the three different effects.

The equilibrium portion of the analysis of this truss may be conducted using superposition of forces based on any other suitable choice of redundants, such as those shown in Figures 5.2*c*–5.2*h*. For instance, Figure 5.3*b* illustrates the nature of the superposition for the choice of redundants in Figure 5.2*h*.

One additional point is worth discussing before we proceed to illustrate the details of the actual computational procedure. That point concerns the ways in which "temporarily deleted" redundant forces are represented. The representation of redundants that are components of reactions is a fairly straightforward proposition. As we have seen, the removal of the capacity of a support to transmit a component of force can be portrayed schematically by a physically meaningful deletion of the support, i.e., the change of a pin support to a roller support. Analogous "releases" are possible for bars. For example, it is possible to represent the deletion of the capacity of a truss bar to transmit an axial force by introducing the roller-type release shown in Figure 5.4*a*. (This is also frequently referred to as "cutting" the bar.) This type of device is a useful schematic aid for the subsequent interpretation of the significance of compatibility equations, but it is not convenient to show it on sketches of trusses. Consequently, we will continue to use dotted lines, or "cuts" of the type shown in Figure 5.4*b* as simplified representations of this hypothesized behavior. Note that when the internal redundant force X_2 is applied to the "cut" bar, the force will cause relative displacements of the "cut" ends of the bar, e.g., Figure 5.4*c*. Unless the value of the redundant is chosen correctly, relative displacements will cause either a gap or an overlap to develop, neither of which is possible in the physical bar. This concept, that the "cuts" are not actually opening or closing, will lead to the development of compatibility equations, one for each redundant force.

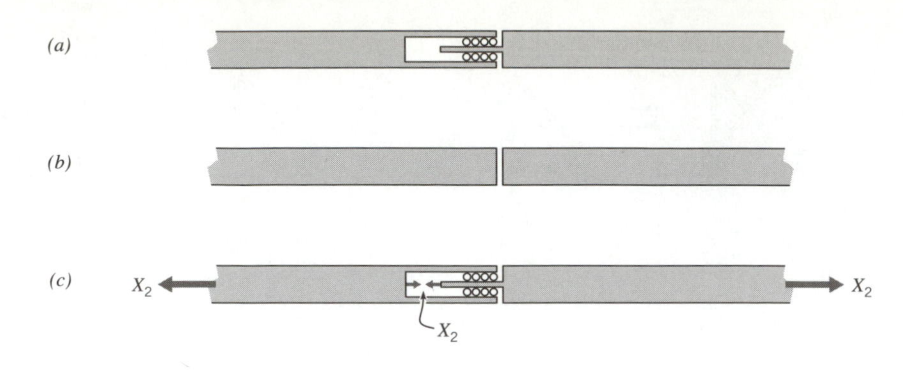

Figure 5.4 (*a*), (*b*) ''Cut'' bar; (*c*) internal redundant force.

5.2.1 Basic Formulation

To facilitate a discussion of compatibility equations, we will begin the detailed applications of the force method of truss analysis with a one-degree indeterminate truss of relatively simple geometry, Figure 5.5*a*. Assume the truss has loads applied symmetrically at nodes C and D, and that a uniform temperature change ΔT is applied to bar CD. Also, for simplicity, let all bars have the same cross-sectional area A and modulus of elasticity E. We will go through the five-step procedure defined in Sec. 5.2.

Step 1. We will choose the redundant force as the horizontal component of reaction at B and denote this force by X.

Step 2. As a result of the choice made in Step 1, the statical analysis must be conducted using the superposition of forces indicated in Figure 5.5*b*. Figure 5.5*c* shows the results of the equilibrium analysis of the truss in Figure 5.5*b*(1), and Figure 5.5*d* shows the results of the equilibrium analysis of the truss in Figure 5.5*b*(2). It should be emphasized that the forces shown in these figures are obtained from an equilibrium analysis using only the equations of statics.[4] Note that all bar forces in Figure 5.5*b*(1) except for P_{CD} are zero in this particular case. Note the important fact that the temperature change does not cause forces in any of the bars.[5] Thus, the bar forces for the truss in Figure 5.5*a* are given by

$$
\begin{aligned}
P_{AC} &= -X \\
P_{AD} &= \sqrt{2}X \\
P_{BC} &= \sqrt{2}X
\end{aligned}
\qquad (5.1)
$$

[4] Equilibrium analysis shows that reactions at A and B are zero for the truss in Figure 5.5*b*(1) or Figure 5.5*c*; similarly, the vertical components of reaction at A and B are zero for the truss in Figure 5.5*b*(2) or Figure 5.5*d*.

[5] This last point is sometimes a source of confusion. The truss in Figure 5.5*b*(1) is statically determinate and, consequently, bars can change length freely without inducing forces. The only way internal forces can be developed in a statically determinate structure is if mechanical forces are applied (so that they appear in the equations of equilibrium).

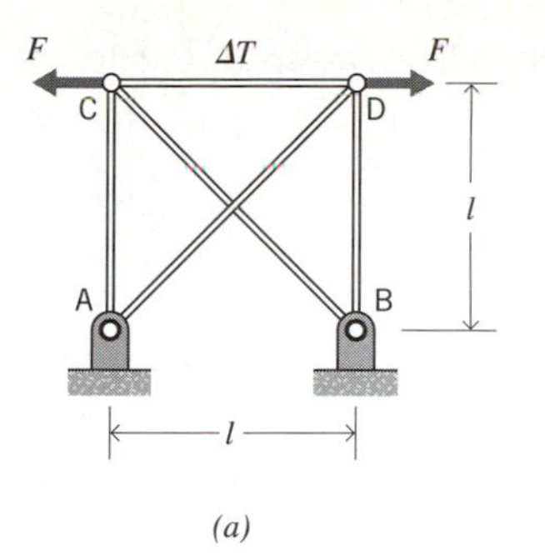

(a)

$=$ (1) $+$ (2)

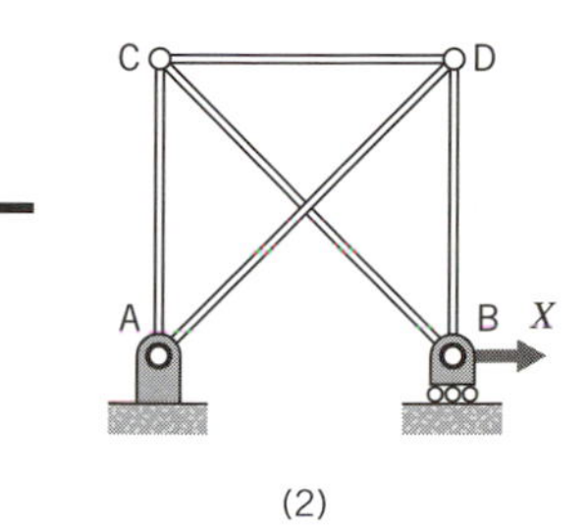

(1) (2)

(b)

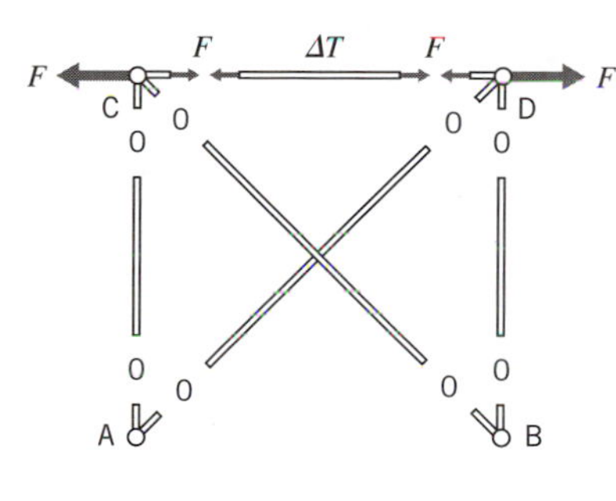

(c)

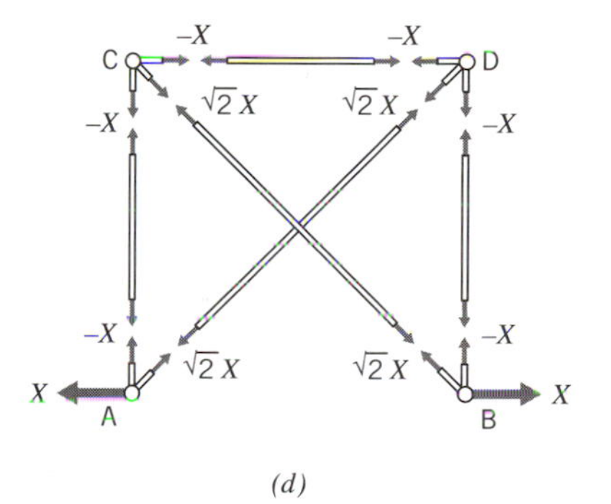

(d)

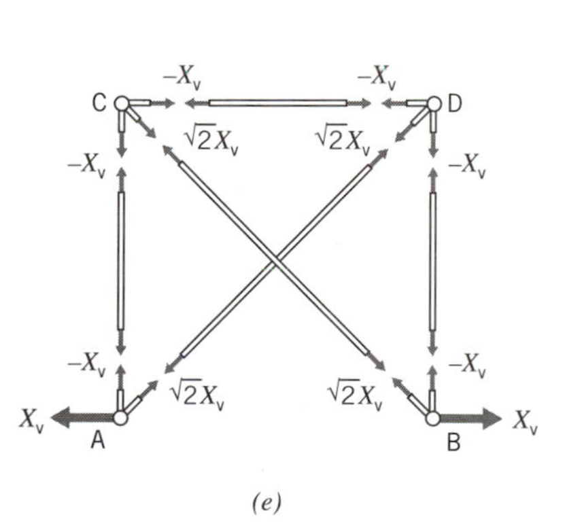

(e)

Figure 5.5 Statical analysis of one-dgree redundant truss.

$$P_{BD} = -X$$
$$P_{CD} = -X + F$$

Step 3. For linear elastic truss bars, deformations (length changes) are related to bar forces and temperature changes by Equation 2.38*d*. For bar CD there is a thermal load ΔT which will produce a net elongation $\alpha \Delta T l$. Bar deformations are therefore given by

$$\begin{aligned}
\Delta_{AC} &= -\frac{Xl}{AE} \\
\Delta_{AD} &= \frac{(\sqrt{2}X)(\sqrt{2}l)}{AE} \\
\Delta_{BC} &= \frac{(\sqrt{2}X)(\sqrt{2}l)}{AE} \\
\Delta_{BD} &= -\frac{Xl}{AE} \\
\Delta_{CD} &= \frac{(-X + F)l}{AE} + \alpha \Delta T l
\end{aligned} \tag{5.2}$$

Step 4. Because this truss is one-degree indeterminate we will need one compatibility equation in addition to the equilibrium equations in order to find the unknown X and complete the analysis. We will obtain this compatibility equation by an application of the theorem of virtual forces. For reasons that will be explained shortly, we will generate this equation by choosing a virtual force state that is identical to the force state associated with the redundant force X. Such a virtual force state is shown in Figure 5.5*e*. This virtual force state contains information that is identical to the information in Figure 5.5*d*; thus, no new computations are required. The real element deformations and joint displacements are those that result from the loads in Figure 5.5*a*. Therefore, we write the equation of virtual work, Equation 4.8, using the real bar deformations and joint displacements and the virtual forces in Figure 5.5*e*.

The external virtual work involves the products of the external virtual forces (including reactions) at the nodes of the truss in Figure 5.5*e* and the real displacements at the nodes in Figure 5.5*a*. There are no virtual forces at nodes C and D in Figure 5.5*e* and no displacements at nodes A and B in Figure 5.5*a*; consequently, the external virtual work is zero!

The internal virtual work involves products of virtual bar forces in Figure 5.5*e* and real bar deformations in Figure 5.5*a*. Equating external and internal virtual work leads to

$$\begin{aligned}
0 &= (-X_V)\Delta_{AC} + (\sqrt{2}X_V)\Delta_{AD} + (\sqrt{2}X_V)\Delta_{BC} + (-X_V)\Delta_{BD} + (-X_V)\Delta_{CD} \\
&= -\Delta_{AC} + \sqrt{2}\Delta_{AD} + \sqrt{2}\Delta_{BC} - \Delta_{BD} - \Delta_{CD}
\end{aligned} \tag{5.3}$$

Equation 5.3 is the compatibility equation for this truss.[6] It is the relationship that the real bar deformations must satisfy in order for the deformed bars to remain properly connected at all nodes. We will examine the geometric significance of this equation in more detail after we complete the five-step solution process.

Step 5. Substituting the expressions for the real bar deformations given by Equations 5.2 into Equation 5.3 leads to

$$0 = \frac{l}{AE}X + \frac{2\sqrt{2}l}{AE}X + \frac{2\sqrt{2}l}{AE}X + \frac{l}{AE}X - \left[\frac{(-X + F)l}{AE} + \alpha \Delta T l\right] \quad \textbf{(5.4)}$$

from which we obtain

$$X = \frac{F}{3 + 4\sqrt{2}} + \frac{AE\alpha\Delta T}{3 + 4\sqrt{2}} \quad \textbf{(5.5)}$$

Substituting this value of X into Equation 5.1 provides the bar forces,

$$\begin{aligned}
P_{AC} &= -\frac{F}{3 + 4\sqrt{2}} - \frac{AE\alpha\Delta T}{3 + 4\sqrt{2}} \\
P_{AD} &= \frac{\sqrt{2}F}{3 + 4\sqrt{2}} + \frac{\sqrt{2}AE\alpha\Delta T}{3 + 4\sqrt{2}} \\
P_{BC} &= \frac{\sqrt{2}F}{3 + 4\sqrt{2}} + \frac{\sqrt{2}AE\alpha\Delta T}{3 + 4\sqrt{2}} \\
P_{BD} &= -\frac{F}{3 + 4\sqrt{2}} - \frac{AE\alpha\Delta T}{3 + 4\sqrt{2}} \\
P_{CD} &= \left(\frac{2 + 4\sqrt{2}}{3 + 4\sqrt{2}}\right)F - \frac{AE\alpha\Delta T}{3 + 4\sqrt{2}}
\end{aligned} \quad \textbf{(5.6)}$$

which completes the analysis.[7]

The procedure followed above was quite direct except, perhaps, for Step 4 where the virtual work theorem was used to impose compatibility. We shall find that this method is the same for all subsequent truss analyses using the force method. Other applications differ primarily in the number of redundant forces and the number of simultaneous equations that must therefore be solved, but no new concepts will be used beyond those already presented.

Before proceeding to any other applications, it is helpful to examine the physical

---

[6] Also see Equation 4.16*b*.

[7] These values of bar forces (for the case $\Delta T = 0$) were used previously in Example 5, Chapter 4 (Equations 4.15*a*).

significance of the compatibility equation, Equation 5.3. Recall that at the outset we arbitrarily chose the redundant force in this truss as the horizontal component of reaction at B. This led to the statical analyses of the trusses shown in Figures 5.5*b*(1) and 5.5*b*(2), each of which is analyzed as if there were a roller support at B. This, in turn, implies that the loads on these trusses may cause a horizontal displacement at B. The actual truss (Figure 5.5*a*) has a pin support at B which cannot displace horizontally; as we will see, the compatibility equation is a mathematical statement of this criterion.

To illustrate this point we will begin with a direct geometric analysis[8] of the effects of bar deformations on the movement of a roller support at B (Figure 5.6) without taking advantage of the virtual work theorem. This straightforward approach will be shown to yield Equation 5.3. We assume that node B is free to move horizontally, and we will determine the effects of individual bar deformations on this movement. Figure 5.6*a*, for example, shows the effect of a deformation of bar AC, Δ_{AC}, on the displacement at B; here, all bars except for AC are assumed to remain undeformed. The dotted lines show an exaggerated picture of the deformed geometry that results from Δ_{AC}. Node C moves to C′ and, because bars CD and AD are undeformed, node D must remain in its original position; also, because bars BC and BD are undeformed, it follows that node B must move to B′ as shown. From geometry (the two lengths denoted by *a* in Figure 5.6*a* must be equal), we conclude that the magnitude of this horizontal displacement is equal to Δ_{AC}; the negative sign is added in Figure 5.6*a* because we are arbitrarily defining displacements to the right as positive.

The remaining figures are analogous; Figure 5.6*b* shows the effect of a deformation Δ_{AD}, Figure 5.6*c* shows the effect of a deformation Δ_{BC}, Figure 5.6*d* shows the effect of a deformation Δ_{BD}, and Figure 5.6*e* shows the effect of a deformation Δ_{CD}. The compatibility equation, Equation 5.3, states mathematically that the components of displacement at B that develop due to all the different element deformations in Figure 5.6 must add up to zero. In other words, the compatibility equation is a relation between the bar deformations that must be satisfied to assure that the truss displacements are consistent with the actual physical support conditions.

Alternate interpretations of the physical significance of the compatibility equation are also possible. For example, consider Equation 5.4, which is the form of the compatibility equation after the expressions for bar deformations (Equations 5.2) have been substituted. The terms in Equation 5.4 may be rearranged to give

$$\left[\frac{(3 + 4\sqrt{2})l}{AE}\right]X + \left(-\frac{Fl}{AE} - \alpha\Delta Tl\right) = 0 \qquad \textbf{(5.7)}$$

The first term in the above equation is equivalent to the horizontal displacement at B in Figure 5.5*b*(2) and the second term is equivalent to the horizontal displacement at B in Figure 5.5*b*(1). Each of these terms may be obtained by using virtual work based on the virtual forces in Figure 5.5*e* and the real displacements in Figures 5.5*b*(2) and 5.5*b*(1), respectively. Thus, Equation 5.7 indicates that the sum of the horizontal displacements at B in Figures 5.5*b*(2) and 5.5*b*(1) must add up to zero because of the

---

[8] See Section 2.5.

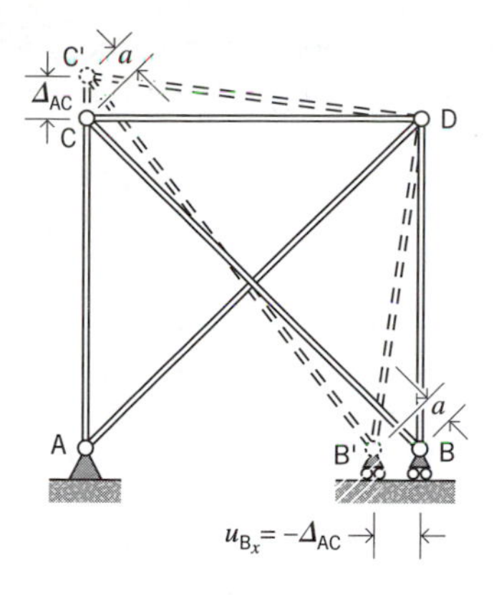

(a)

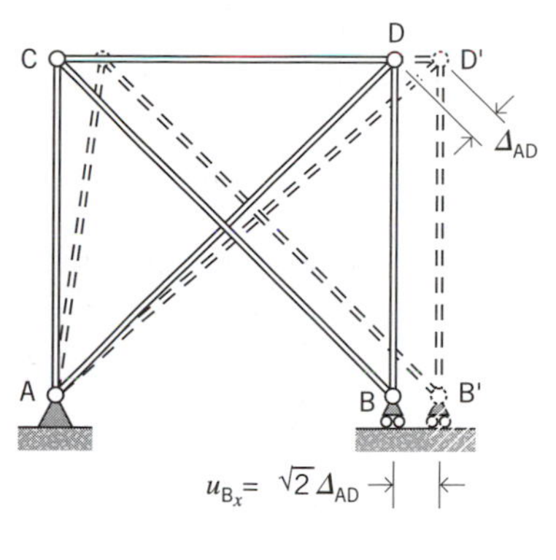

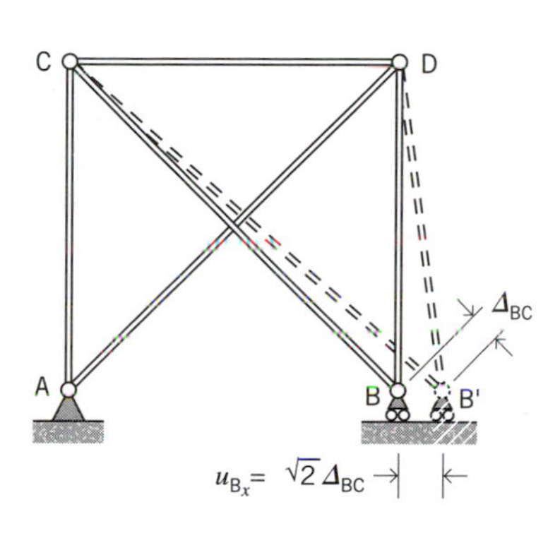

(c)

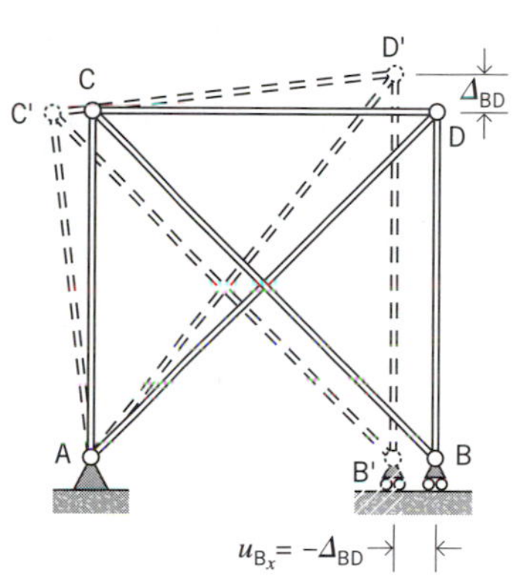

(d)

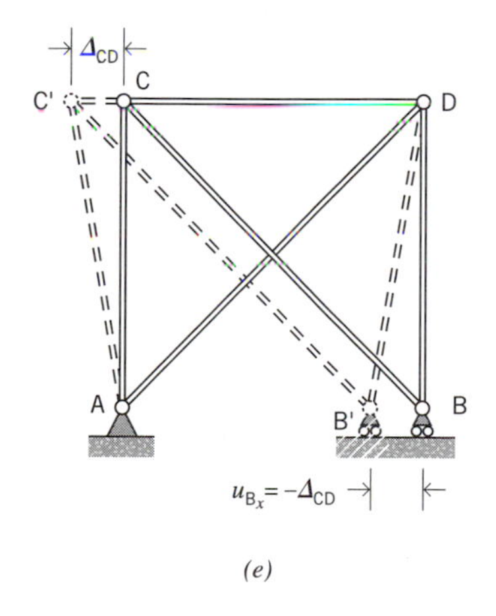

(e)

Figure 5.6 Truss compatibility.

immovable support at B.[9] The first term in brackets above is referred to as a flexibility coefficient, i.e., it is a displacement at a certain point in the structure (in this case a horizontal displacement at B) due to a unit value of X.

Regardless of which of the preceding interpretations is attached to the compatibility equation, the important fact is that the equation was generated directly by the procedure followed in Step 4 above. This procedure, in effect, "borrowed" a portion of the equilibrium information that had already been obtained in Step 2; this relationship between Step 2 and Step 4 is quite remarkable. In some way, the equilibrium study provided all the necessary information to establish the compatibility equation. This result will play a very significant role in the matrix formulation, which will be developed in Section 5.2.2.

As a second illustration of the procedure for the basic force method, which will also allow further interpretation of the preceding compatibility equation, we will reanalyze the truss in Figure 5.5*a* with an alternate choice of redundant force.

Step 1. This time we will choose the redundant force as the internal force in bar AD. Again, we denote the magnitude of this redundant force by X.

Step 2. The statical analysis is now conducted using the superposition of forces indicated in Figure 5.7*b*. Figure 5.7*b*(1) shows the truss with diagonal bar AD "cut," and Figure 5.7*b*(2) shows the truss with the internal force X applied to the bar (as in Figure 5.4*c*). The results of the equilibrium analyses are displayed in Figures 5.7*c* and 5.7*d*, respectively; note that in both figures the equilibrium analyses start at joint D because the internal force in bar AD is specified (either 0 or X). The bar forces are now found to be

$$\begin{aligned} P_{AC} &= -\frac{X}{\sqrt{2}} \\ P_{AD} &= X \\ P_{BC} &= X \\ P_{BD} &= -\frac{X}{\sqrt{2}} \\ P_{CD} &= -\frac{X}{\sqrt{2}} + F \end{aligned} \tag{5.8}$$

These equations look much different from Equations 5.1 because "X" has an entirely different meaning than in the previous example.

Step 3. Bar deformations are

$$\Delta_{AC} = -\frac{l}{\sqrt{2}AE}X \tag{5.9}$$

[9] This form of compatibility equation is frequently generated as part of a variation of the force method referred to as the method of consistent displacements.

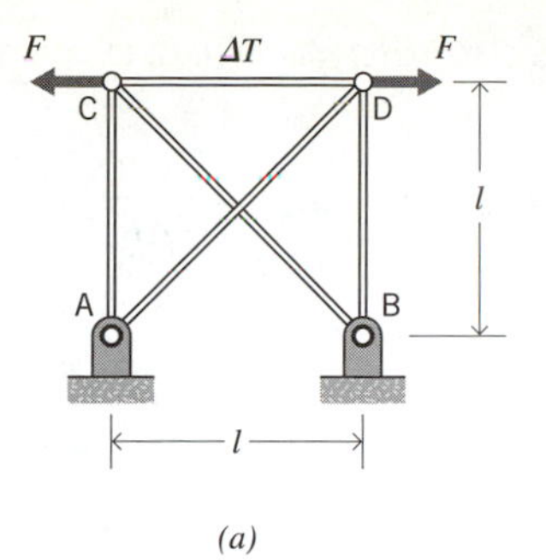

(a)

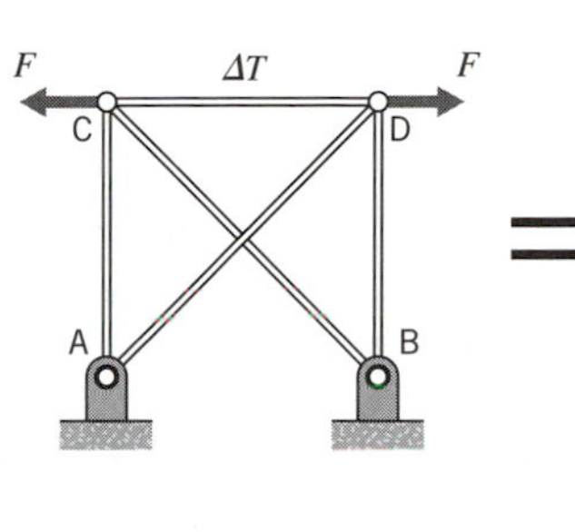

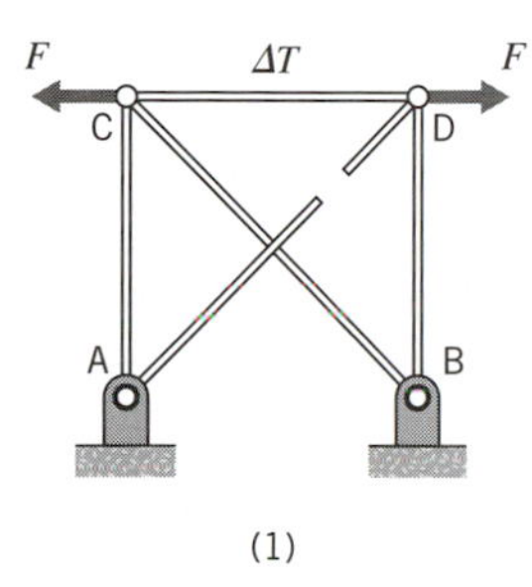

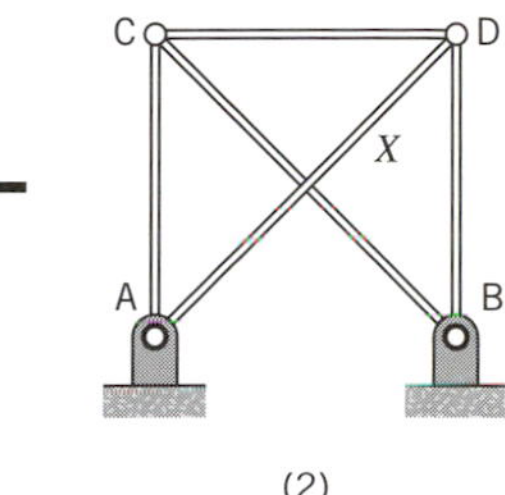

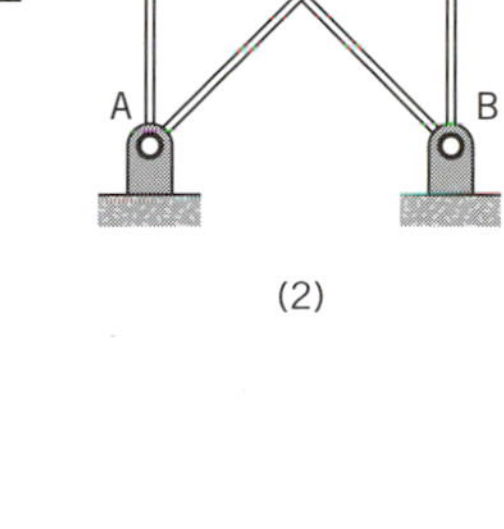

(1) (2)

(b)

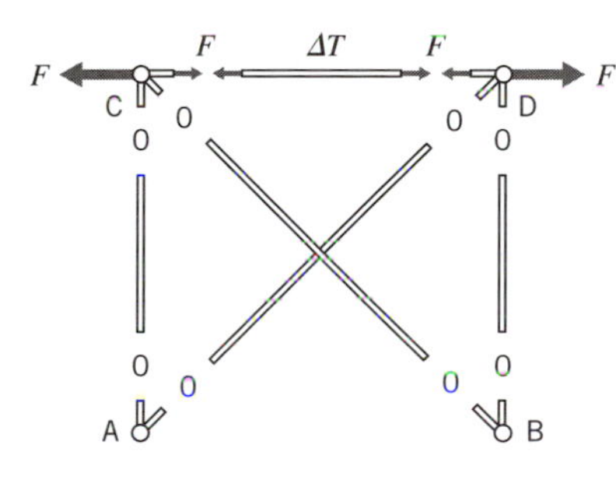

(c)

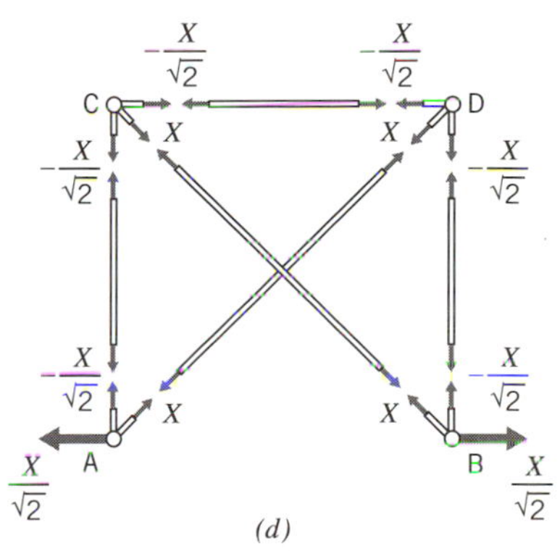

(d)

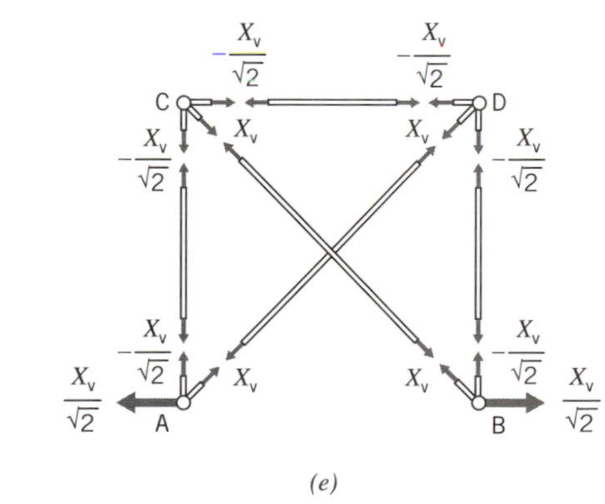

(e)

Figure 5.7 Statical analysis of one-degree redundant truss.

$$\Delta_{AD} = \frac{\sqrt{2}l}{AE}X$$

$$\Delta_{BC} = \frac{\sqrt{2}l}{AE}X$$

$$\Delta_{BD} = -\frac{l}{\sqrt{2}AE}X$$

$$\Delta_{CD} = \frac{l}{AE}\left(-\frac{X}{\sqrt{2}} + F\right) + \alpha\Delta Tl$$

Step 4. As in the previous case, we generate the compatibility equation by using a virtual force state that is identical to the force state associated with X, i.e., we use the virtual force state in Figure 5.7*e*, which is identical to the real force state in Figure 5.7*d*. Because the real displacements at A and B are zero, the external virtual work is again zero.

The internal virtual work still involves the products of the virtual bar forces in Figure 5.7*e* and the real bar deformations in Figure 5.7*a*, so the virtual work equation becomes

$$0 = -\frac{\Delta_{AC}}{\sqrt{2}} + \Delta_{AD} + \Delta_{BC} - \frac{\Delta_{BD}}{\sqrt{2}} - \frac{\Delta_{CD}}{\sqrt{2}} \tag{5.10}$$

This equation is the same as Equation 5.3 when multiplied by the scale factor $\sqrt{2}$. Evidently, the same relation between bar deformations still applies; we will reexamine the significance of this equation shortly.

Step 5. Substituting the expressions for the real bar deformations given by Equations 5.9 into Equation 5.10 leads to

$$0 = \left[\left(\frac{3 + 4\sqrt{2}}{2}\right)\frac{l}{AE}\right]X + \left(-\frac{Fl}{\sqrt{2}AE} - \frac{\alpha\Delta Tl}{\sqrt{2}}\right) \tag{5.11a}$$

or

$$X = \frac{\sqrt{2}F}{3 + 4\sqrt{2}} + \frac{\sqrt{2}AE\alpha\Delta T}{3 + 4\sqrt{2}} \tag{5.11b}$$

which is identical to the previous expression for P_{AD} in Equation 5.6. Substitution of this value of X into Equations 5.8 will provide the same forces as in Equations 5.6.

The physical significance of the compatibility equation, Equation 5.10, may be related to the relative displacement between "cut" ends of the bar whose force is chosen as the redundant, as was inferred in connection with Figure 5.4. In this case the redundant force was chosen in bar AD, and the associated interpretation of the geometry of the truss deformation is depicted in Figure 5.8. Figures 5.8*a*–5.8*d* show the effects of individual bar deformations on the relative displacements between "cut" ends of bar AD. For example, Figure 5.8*a* shows that a deformation Δ_{AC} causes a relative displace-

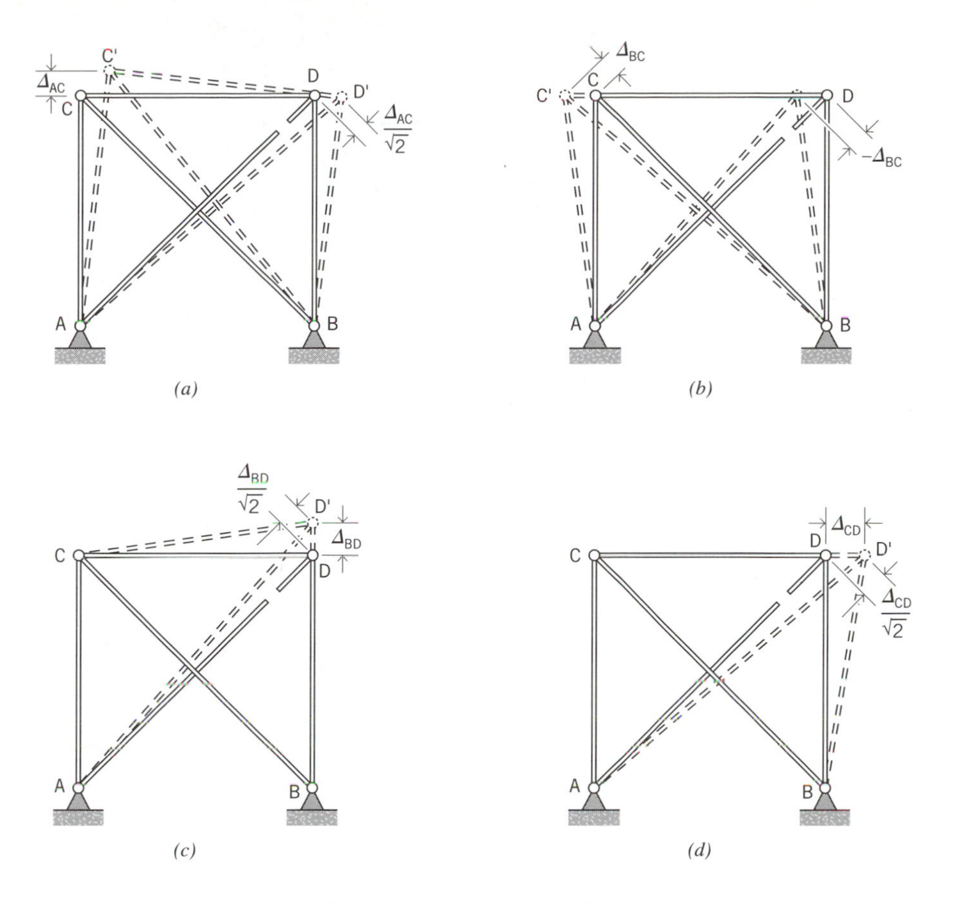

Figure 5.8 Truss compatibility.

ment $\Delta_{AC}/\sqrt{2}$. The sum of all such relative displacements must add up to the actual deformation of bar AD, Δ_{AD}.[10] This is exactly what is expressed by the compatibility equation!

The alternate form of the compatibility equation, Equation 5.11*a*, has an interpretation analogous to that discussed for Equation 5.7; in this case, however, the first term is the relative displacement between cut ends of bar AD due to X and the second term is the relative displacement due to loads. Thus, the compatibility equation may always be related to displacements associated with a particular choice of redundant forces. An important point must be made here, however: *The procedure used in Step 4 to obtain the compatibility equation is entirely mathematical and is not dependent on any physical interpretation of displacements.* This should become more apparent as we proceed to analyze trusses with more than one redundant force. We shall see that this will greatly simplify analysis of redundant structures.

[10] In Equation 5.10 all the bar deformations Δ_{ij} are shown on the same side of the equation. The above interpretation is only Equation 5.10 with Δ_{AD} written on one side and the other terms written on the other side.

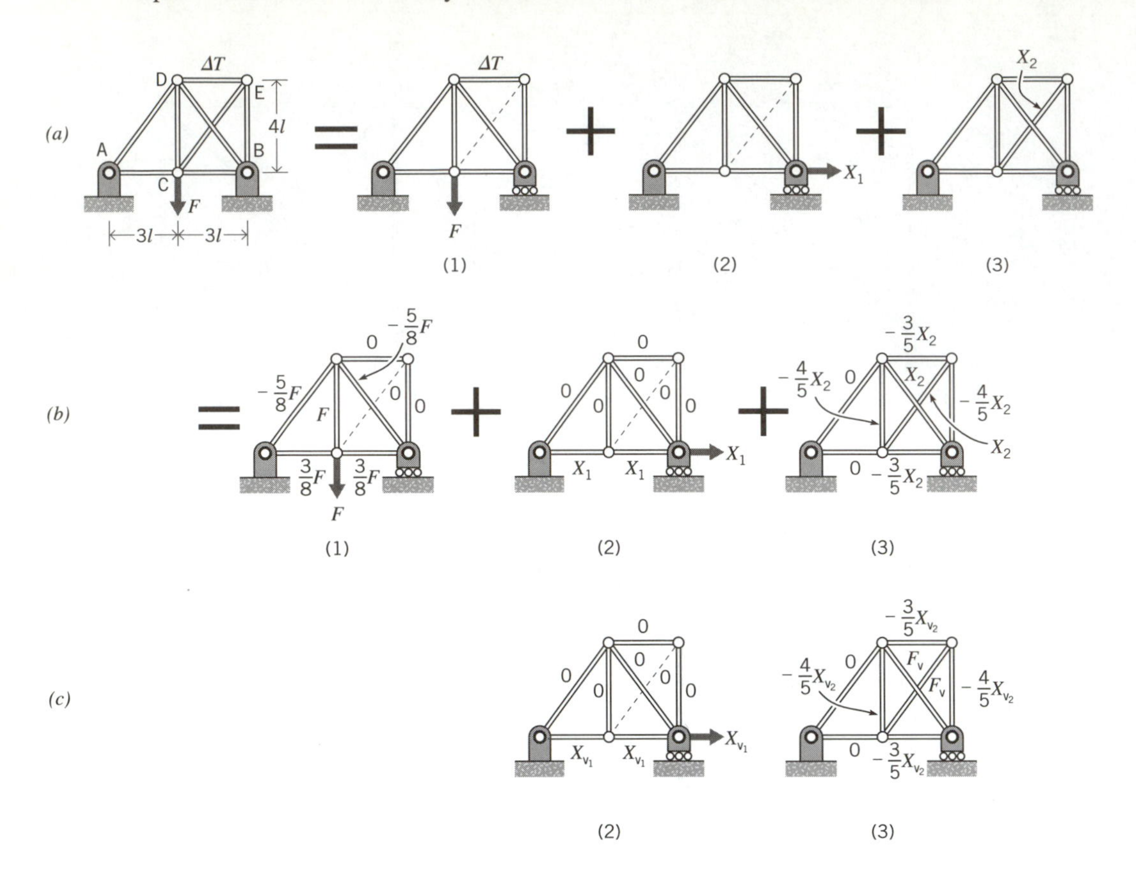

Figure 5.9 Statical analysis of two-degree redundant truss.

Let us now consider the two-degree redundant truss in Figure 5.3*a*, which we will analyze using the choice of redundant forces shown (Step 1). We will assume the geometry specified in Figure 5.9*a* and that all bars have the same value of *AE*. For illustrative purposes, we will also assume a uniform temperature change ΔT in bar DE.

Step 2. Figures 5.9*b*(1)–5.9*b*(3) show the results of the statical analyses of the corresponding trusses in Figure 5.9*a*. Here, for the sake of conciseness, we show only the bar forces rather than all the free-body diagrams. As with the previous truss, direct equilibrium analyses of joint E are possible because the forces in bar CE are specified as either 0 or X_2 in each of the three sketches; also, the reactions are all determinate and, consequently, the remaining bar forces can be obtained from equilibrium of joints A, B, and C. As noted previously, the temperature change in bar DE has no effect on the forces in the statically determinate truss in Figure 5.9*b*(1). Superposition of the forces in Figs. 5.9*b*(1)–5.9*b*(3) gives the individual bar forces as functions of the two redundant forces, X_1 and X_2, and the applied force *F*,

$$P_{AC} = X_1 + \frac{3}{8}F \tag{5.12}$$

$$P_{CB} = X_1 - \frac{3}{5}X_2 + \frac{3}{8}F$$

$$P_{AD} = -\frac{5}{8}F$$

$$P_{CD} = -\frac{4}{5}X_2 + F$$

$$P_{CE} = X_2$$

$$P_{DE} = -\frac{3}{5}X_2$$

$$P_{BE} = -\frac{4}{5}X_2$$

$$P_{BD} = X_2 - \frac{5}{8}F$$

Step 3. The bar deformations are

$$\begin{aligned}
\Delta_{AC} &= \frac{(3l)}{AE}P_{AC} \\
\Delta_{CB} &= \frac{(3l)}{AE}P_{CB} \\
\Delta_{AD} &= \frac{(5l)}{AE}P_{AD} \\
\Delta_{CD} &= \frac{(4l)}{AE}P_{CD} \\
\Delta_{CE} &= \frac{(5l)}{AE}P_{CE} \\
\Delta_{DE} &= \frac{(3l)}{AE}P_{DE} + 3l\alpha\Delta T \\
\Delta_{BE} &= \frac{(4l)}{AE}P_{BE} \\
\Delta_{BD} &= \frac{(5l)}{AE}P_{BD}
\end{aligned} \tag{5.13}$$

Note that the length change of bar DE due to temperature is included here. (To keep each step of the analysis as clear as possible, the values of bar forces, Equations 5.12, will not be substituted until the last step.)

Step 4. We require two compatibility equations to find X_1 and X_2. These are obtained from two virtual work equations, one based on virtual forces identical to those associated with X_1 in Figure 5.9*b*(2), and one based on virtual forces identical to those associated with X_2 in Figure 5.9*b*(3). It is not necessary to sketch these virtual force states because they embody the same information contained in Figures 5.9*b*(2) and 5.9*b*(3); however, to avoid any ambiguity, the virtual force patterns are shown in Figures 5.9*c*(2) and 5.9*c*(3). In Figure 5.9*c*(2) X_{V_1} is ''like'' X_1 and in Figure 5.9*c*(3) X_{V_2} is ''like'' X_2. For the case $X_{V_1} = 1$,[11] the virtual bar forces have the following values

$$
\begin{aligned}
P_{\mathrm{AC}_V} &= 1 \\
P_{\mathrm{CB}_V} &= 1 \\
P_{\mathrm{AD}_V} &= 0 \\
P_{\mathrm{CD}_V} &= 0 \\
P_{\mathrm{CE}_V} &= 0 \\
P_{\mathrm{DE}_V} &= 0 \\
P_{\mathrm{BE}_V} &= 0 \\
P_{\mathrm{BD}_V} &= 0
\end{aligned}
\tag{5.14a}
$$

and for $X_{V_2} = 1$ the virtual bar forces are

$$
\begin{aligned}
P_{\mathrm{AC}_V} &= 0 \\
P_{\mathrm{CB}_V} &= -\frac{3}{5} \\
P_{\mathrm{AD}_V} &= 0 \\
P_{\mathrm{CD}_V} &= -\frac{4}{5} \\
P_{\mathrm{CE}_V} &= 1 \\
P_{\mathrm{DE}_V} &= -\frac{3}{5}
\end{aligned}
\tag{5.14b}
$$

[11] Because X_V cancels out of the virtual work equation, it is convenient to let $X_V = 1$.

$$P_{\mathrm{BE}_V} = -\frac{4}{5}$$

$$P_{\mathrm{BD}_V} = 1$$

Once again, the external virtual work involves the external virtual forces (including reactions) in either of Figures 5.9*c* and the corresponding real displacements in the actual truss of Figure 5.9*a*; as in the preceding illustrations, in each case this external virtual work is equal to zero. There are circumstances under which external virtual work based on these virtual forces (i.e., virtual force states chosen "like" redundant force states) may not be zero, e.g., if there is support movement (settlement[12]) in the actual structure.

Thus, from the virtual work equation for each of the virtual force states in Figure 5.9*c* and the real truss deformations in Figure 5.9*a* we obtain

$$0 = \Delta_{\mathrm{AC}} + \Delta_{\mathrm{CB}} \tag{5.14c}$$

$$0 = -\frac{3}{5}\Delta_{\mathrm{CB}} - \frac{4}{5}\Delta_{\mathrm{CD}} + \Delta_{\mathrm{CE}} - \frac{3}{5}\Delta_{\mathrm{DE}} - \frac{4}{5}\Delta_{\mathrm{BE}} + \Delta_{\mathrm{BD}} \tag{5.14d}$$

which are the desired compatibility equations. The interpretation of Equation 5.14*c* is quite direct, namely that the sum of the deformations of bars AC and CB must be zero if joint B is not to displace horizontally. The second compatibility equation, Equation 5.14*d*, is most easily related to the relative displacements between "cut" ends of bar CE, but a geometric assessment of the type performed in Figure 5.8 would require a rather involved graphical construction that will not be presented here.

Step 5. The solution proceeds by substituting Equations 5.12 and 5.13 into Equations 5.14*c* and 5.14*d*; this leads to

$$0 = \left(\frac{6l}{AE}\right)X_1 + \left(-\frac{9l}{5AE}\right)X_2 + \left(\frac{9l}{4AE}F\right) \tag{5.15a}$$

$$0 = \left(-\frac{9l}{5AE}\right)X_1 + \left(\frac{432l}{25AE}\right)X_2 + \left(-\frac{7l}{AE}F - \frac{9}{5}l\alpha\Delta T\right) \tag{5.15b}$$

Solving for X_1 and X_2 yields

$$\begin{aligned} X_1 &= -0.262F + 0.032AE\alpha\Delta T \\ X_2 &= 0.378F + 0.108AE\alpha\Delta \mathrm{T} \end{aligned} \tag{5.16}$$

Substituting this result back into Equations 5.12 gives the bar forces,

$$P_{\mathrm{AC}} = 0.113F + 0.032AE\alpha\Delta T \tag{5.17}$$

---

[12] In this case the support settlement must be of known magnitude.

$$P_{CB} = -0.113F - 0.032AE\alpha\Delta T$$

$$P_{AD} = -0.625F$$

$$P_{CD} = 0.698F - 0.086AE\alpha\Delta T$$

$$P_{CE} = 0.378F + 0.108AE\alpha\Delta T$$

$$P_{DE} = -0.227F - 0.065AE\alpha\Delta T$$

$$P_{BE} = -0.302F - 0.086AE\alpha\Delta T$$

$$P_{BD} = -0.247F + 0.108AE\alpha\Delta T$$

The significance of the compatibility equations, Equations 5.15, is similar to those in the previous two illustrations. The first term in Equation 5.15*a* is the horizontal component of displacement at B in Figure 5.9*b*(2), the second term is the horizontal displacement at B in Figure 5.9*b*(3), and the third term is the horizontal displacement at B in Figure 5.9*b*(1). The terms in Equation 5.15*b* are analogous, except that they all involve the relative displacement of bar CE.

5.2.2 Matrix Formulation

The compatibility equations, Equations 5.15, are linear equations in the variables X_1 and X_2. Whenever we deal with sets of simultaneous linear equations, it is logical to think of the use of matrices[13] in the solution process. For instance, Equations 5.15*a* and 5.15*b* may be rewritten in matrix notation as

$$\begin{bmatrix} \dfrac{6l}{AE} & -\dfrac{9l}{5AE} \\ -\dfrac{9l}{5AE} & \dfrac{432l}{25AE} \end{bmatrix} \begin{Bmatrix} X_1 \\ X_2 \end{Bmatrix} = \begin{Bmatrix} -\dfrac{9Fl}{4AE} \\ \dfrac{7Fl}{AE} + \dfrac{9}{5}l\alpha\Delta T \end{Bmatrix} \quad \textbf{(5.18)}$$

Equations of this type are usually represented by the general linear form[14]

$$\mathbf{AX} = \mathbf{B} \quad \textbf{(5.19)}$$

where **X** is the vector[15] of unknown variables, **A** is the matrix of coefficients of the unknown variables, and **B** is the vector of constants.

The **A** matrix is referred to as a flexibility matrix and the individual terms in the matrix are ''flexibility coefficients'' with the significance noted previously. Note, in

[13] The reader is assumed to be familiar with basic concepts of matrix algebra.

[14] In the remainder of this book boldface type will be used to denote a general form of a matrix. Note that boldface symbols are used for both the 2 × 2 and 2 × 1 matrices in Equation 5.18.

[15] We refer to a matrix with a single column as a vector.

particular, that the **A** matrix in Equation 5.18 is square and symmetric. The term in the second column of the first row is the horizontal displacement at B due to a unit value of redundant X_2, and the term in the first column of the second row is the relative displacement between "cut" ends of bar CE due to a unit value of redundant X_1. That these particular flexibility coefficients are equal is not a coincidence; as will be proved in Chapter 6, the equality of such symmetrically positioned coefficients in the flexibility matrix is a general property of linear elastic structures.

The solution of equations of the type represented by Equation 5.19 is obtained by premultiplying both sides by the inverse of matrix **A**, i.e.,

$$\mathbf{X} = \mathbf{A}^{-1}\mathbf{B} \tag{5.20}$$

where $\mathbf{A}^{-1}$ denotes the inverse of **A**. In the particular case of Equation 5.18, this operation would yield results identical to Equations 5.16.

Matrix algebra is useful for much more than just solving simultaneous equations. The entire analysis procedure encompassed by Steps 2–5 may be formulated in matrix notation. This does not require any new concepts; the equations encountered at each step of the analysis are simply rewritten in matrix form. The result is a significant generalization and simplification of the entire analysis procedure. The overall concepts become quite clear, and the procedure becomes very easy to follow. We will illustrate this by presenting the data in matrix notation for the preceding two-degree redundant truss.

Step 2. Equations 5.12 may be written in the form

$$\begin{Bmatrix} P_{AC} \\ P_{CB} \\ P_{AD} \\ P_{CD} \\ P_{CE} \\ P_{DE} \\ P_{BE} \\ P_{BD} \end{Bmatrix} = \begin{bmatrix} 1 & 0 \\ 1 & -3/5 \\ 0 & 0 \\ 0 & -4/5 \\ 0 & 1 \\ 0 & -3/5 \\ 0 & -4/5 \\ 0 & 1 \end{bmatrix} \begin{Bmatrix} X_1 \\ X_2 \end{Bmatrix} + \begin{Bmatrix} 3F/8 \\ 3F/8 \\ -5F/8 \\ F \\ 0 \\ 0 \\ 0 \\ -5F/8 \end{Bmatrix} \tag{5.21}$$

This matrix form is typical of all equilibrium equations obtained in the analysis process; thus, we may represent such equilibrium relations by the general form

$$\mathbf{P} = \mathbf{bX} + \mathbf{P_o} \tag{5.22}$$

where **P** is a vector of bar forces, **b** is an "equilibrium matrix" (the matrix of coefficients of variables **X** in the equilibrium equations), and $\mathbf{P_o}$ is the vector of bar forces associated with applied loads. Note that each column of the **b** matrix defines components of bar forces due to a unit value of one of the redundants, e.g., in this case, the first column of **b** defines bar forces due to a unit value of X_1 and the second column of **b** lists bar forces due to a unit value of X_2. (These are exactly the results contained in Equations 5.14*a* and 5.14*b*, respectively.)

Step 3. In matrix notation, the deformation-force relations given in Equations 5.13 become

$$\begin{Bmatrix} \Delta_{AC} \\ \Delta_{CB} \\ \Delta_{AD} \\ \Delta_{CD} \\ \Delta_{CE} \\ \Delta_{DE} \\ \Delta_{BE} \\ \Delta_{BD} \end{Bmatrix} = \frac{l}{AE} \begin{bmatrix} 3 & 0 & 0 & 0 & 0 & 0 & 0 & 0 \\ 0 & 3 & 0 & 0 & 0 & 0 & 0 & 0 \\ 0 & 0 & 5 & 0 & 0 & 0 & 0 & 0 \\ 0 & 0 & 0 & 4 & 0 & 0 & 0 & 0 \\ 0 & 0 & 0 & 0 & 5 & 0 & 0 & 0 \\ 0 & 0 & 0 & 0 & 0 & 3 & 0 & 0 \\ 0 & 0 & 0 & 0 & 0 & 0 & 4 & 0 \\ 0 & 0 & 0 & 0 & 0 & 0 & 0 & 5 \end{bmatrix} \begin{Bmatrix} P_{AC} \\ P_{CB} \\ P_{AD} \\ P_{CD} \\ P_{CE} \\ P_{DE} \\ P_{BE} \\ P_{BD} \end{Bmatrix} + \begin{Bmatrix} 0 \\ 0 \\ 0 \\ 0 \\ 0 \\ 3l\alpha\Delta T \\ 0 \\ 0 \end{Bmatrix} \tag{5.23}$$

which is also typical of all relations of this type. Note that the matrix of bar flexibilities is diagonal because the elastic behavior of each bar is so simple (i.e., the equations are uncoupled). The constant term that has been factored out, l/AE, is part of the matrix.

We may represent such deformation-force relations by the general form

$$\mathbf{\Delta} = \mathbf{aP} + \hat{\mathbf{\Delta}} \tag{5.24}$$

where $\mathbf{\Delta}$ is a vector of bar deformations (length changes), $\mathbf{a}$ is a diagonal matrix of individual bar flexibilities, $\mathbf{P}$ is the same vector of bar forces as in Equation 5.22, and $\hat{\mathbf{\Delta}}$ is a vector of bar deformations due to temperature or initial effects. The components of $\hat{\mathbf{\Delta}}$ are ''unrestrained'' bar deformations that would occur if the bar forces were equal to zero. (Typical cases of unrestrained bar deformations are initial length changes or length changes due to thermal effects.)

Step 4. For the present purposes, we are interested in the compatibility equations that are expressed in terms of bar deformations, Equations 5.14*c* and 5.14*d*. In matrix notation these equations are

$$\begin{Bmatrix} 0 \\ 0 \end{Bmatrix} = \begin{bmatrix} 1 & 1 & 0 & 0 & 0 & 0 & 0 & 0 \\ 0 & -\frac{3}{5} & 0 & -\frac{4}{5} & 1 & -\frac{3}{5} & -\frac{4}{5} & 1 \end{bmatrix} \begin{Bmatrix} \Delta_{AC} \\ \Delta_{CB} \\ \Delta_{AD} \\ \Delta_{CD} \\ \Delta_{CE} \\ \Delta_{DE} \\ \Delta_{BE} \\ \Delta_{BD} \end{Bmatrix} \tag{5.25}$$

We observe that the matrix of coefficients in this equation is the transpose of the equilibrium matrix in Equation 5.21. In other words,

$$\mathbf{0} = \mathbf{b}^{\mathrm{T}}\mathbf{\Delta} \tag{5.26}$$

Once again, this is not a coincidence but, rather, a reflection of the strong relationship between Steps 2 and 4. We will discuss this point further after completing the formulation.

Step 5. A general form of the solution is obtained by combining Equations 5.22, 5.24 and 5.26. Substituting Equation 5.24 into Equation 5.26 gives

$$\begin{aligned} \mathbf{0} &= \mathbf{b}^{\mathrm{T}}(\mathbf{aP} + \hat{\boldsymbol{\Delta}}) \\ &= \mathbf{b}^{\mathrm{T}}\mathbf{aP} + \mathbf{b}^{\mathrm{T}}\hat{\boldsymbol{\Delta}} \end{aligned} \tag{5.27a}$$

and substituting Equation 5.22 into 5.27*a* leads to

$$\begin{aligned} \mathbf{0} &= \mathbf{b}^{\mathrm{T}}\mathbf{a}(\mathbf{bX} + \mathbf{P}_{\mathrm{o}}) + \mathbf{b}^{\mathrm{T}}\hat{\boldsymbol{\Delta}} \\ &= (\mathbf{b}^{\mathrm{T}}\mathbf{ab})\mathbf{X} + \mathbf{b}^{\mathrm{T}}(\mathbf{aP}_{\mathrm{o}} + \hat{\boldsymbol{\Delta}}) \end{aligned} \tag{5.27b}$$

Finally, we find

$$\boxed{(\mathbf{b}^{\mathrm{T}}\mathbf{ab})\mathbf{X} = -\mathbf{b}^{\mathrm{T}}(\mathbf{aP}_{\mathrm{o}} + \hat{\boldsymbol{\Delta}})} \tag{5.27c}$$

Equation 5.27*c* is the general matrix form of the compatibility equation. It corresponds to Equation 5.18 in the previous example; it follows that the form of Equation 5.27*c* must correspond to Equation 5.19. In other words,

$$\begin{aligned} \mathbf{A} &= \mathbf{b}^{\mathrm{T}}\mathbf{ab} \\ \mathbf{B} &= -\mathbf{b}^{\mathrm{T}}(\mathbf{aP}_{\mathrm{o}} + \hat{\boldsymbol{\Delta}}) \end{aligned} \tag{5.27d}$$

Note that because **a** is a symmetric matrix then **A** must be symmetric because $(\mathbf{b}^{\mathrm{T}}\mathbf{ab})^{\mathrm{T}} = \mathbf{b}^{\mathrm{T}}\mathbf{ab}$, i.e., $\mathbf{A}^{\mathrm{T}} = \mathbf{A}$, the definition of symmetry. This is only one of many ways in which the symmetry property of the flexibility matrix may be demonstrated; we will look at other ways later.

Equation 5.27*c* is the culmination of the entire matrix analysis procedure; it is a single equation that incorporates all five steps of the force method of analysis. For computational purposes, therefore, it is not necessary to explicitly complete each step of the procedure; rather, it is sufficient to generate the specific matrices that appear in Equation 5.27*c*, carry out the algebraic operations indicated, and solve the resulting equation for **X**. We will demonstrate this procedure with some examples, and then present a more formal proof of the validity of the compatibility equation, Equation 5.26.

EXAMPLE 1

We will re-solve the truss using the choice of redundant forces in Figure 5.3*b* and repeated in Figure 5.10*a*; the statical analysis is shown in Figure 5.10*b*.

Step 2. Based on the statical analysis the equilibrium equations are

$$\begin{Bmatrix} P_{AC} \\ P_{CB} \\ P_{AD} \\ P_{CD} \\ P_{CE} \\ P_{DE} \\ P_{BE} \\ P_{BD} \end{Bmatrix} = \begin{bmatrix} 1 & 0 \\ 1 & 3/4 \\ 0 & 0 \\ 0 & 1 \\ 0 & -5/4 \\ 0 & 3/4 \\ 0 & 1 \\ 0 & -5/4 \end{bmatrix} \begin{Bmatrix} X_1 \\ X_2 \end{Bmatrix} + \begin{Bmatrix} 0 \\ -3F/4 \\ -5F/8 \\ 0 \\ 5F/4 \\ -3F/4 \\ -F \\ 5F/8 \end{Bmatrix} \quad \textbf{(5.28a)}$$

which defines the matrices **b** and $\mathbf{P}_o$. Although the order in which the bar forces is arranged is arbitrary, it has been taken to be the same as in Equation 5.21 because this will facilitate a direct comparison with previous results. In comparison to Equation 5.21, the differences in components of the matrices **b** and $\mathbf{P}_o$, which are a consequence of the different choice of redundant forces, should be evident.

Step 3. Because we have preserved the order of the bar forces in Equation 5.28*a*, it follows that the deformation–force relations will be identical to Equation 5.23; this equation therefore defines the **a** and $\widehat{\boldsymbol{\Delta}}$ matrices.

Step 5. We now have all the matrices that appear in Equation 5.27*c*, so there is no need for further analysis to complete Step 4; we substitute directly into Equation 5.27*c*. It is convenient to first evaluate **A** and **B** as defined in Equations 5.27*d*. The first of Equations 5.27*d* is

$$\mathbf{A} = \mathbf{b}^T\mathbf{a}\mathbf{b} = \frac{l}{AE}\begin{bmatrix} 1 & 1 & 0 & 0 & 0 & 0 & 0 & 0 \\ 0 & \frac{3}{4} & 0 & 1 & -\frac{5}{4} & \frac{3}{4} & 1 & -\frac{5}{4} \end{bmatrix} \begin{bmatrix} 3&0&0&0&0&0&0&0 \\ 0&3&0&0&0&0&0&0 \\ 0&0&5&0&0&0&0&0 \\ 0&0&0&4&0&0&0&0 \\ 0&0&0&0&5&0&0&0 \\ 0&0&0&0&0&3&0&0 \\ 0&0&0&0&0&0&4&0 \\ 0&0&0&0&0&0&0&5 \end{bmatrix} \begin{bmatrix} 1 & 0 \\ 1 & 3/4 \\ 0 & 0 \\ 0 & 1 \\ 0 & -5/4 \\ 0 & 3/4 \\ 0 & 1 \\ 0 & -5/4 \end{bmatrix} \quad \textbf{(5.28b)}$$

By performing the matrix multiplications we obtain

$$\mathbf{A} = \frac{l}{AE}\begin{bmatrix} 6 & 9/4 \\ 9/4 & 27 \end{bmatrix} \quad \textbf{(5.28c)}$$

Similarly, the second of Equations 5.27*d* is

$$\mathbf{B} = -\mathbf{b}^T(\mathbf{a}\mathbf{P}_o + \widehat{\boldsymbol{\Delta}})$$

$$= -\begin{bmatrix} 1 & 1 & 0 & 0 & 0 & 0 & 0 & 0 \\ 0 & \frac{3}{4} & 0 & 1 & -\frac{5}{4} & \frac{3}{4} & 1 & -\frac{5}{4} \end{bmatrix} \left\{ \frac{l}{AE} \begin{bmatrix} 3&0&0&0&0&0&0&0 \\ 0&3&0&0&0&0&0&0 \\ 0&0&5&0&0&0&0&0 \\ 0&0&0&4&0&0&0&0 \\ 0&0&0&0&5&0&0&0 \\ 0&0&0&0&0&3&0&0 \\ 0&0&0&0&0&0&4&0 \\ 0&0&0&0&0&0&0&5 \end{bmatrix} \begin{Bmatrix} 0 \\ -3F/4 \\ -5F/8 \\ 0 \\ 5F/4 \\ -3F/4 \\ -F \\ 5F/8 \end{Bmatrix} + \begin{Bmatrix} 0 \\ 0 \\ 0 \\ 0 \\ 0 \\ 3l\alpha\Delta T \\ 0 \\ 0 \end{Bmatrix} \right\}$$

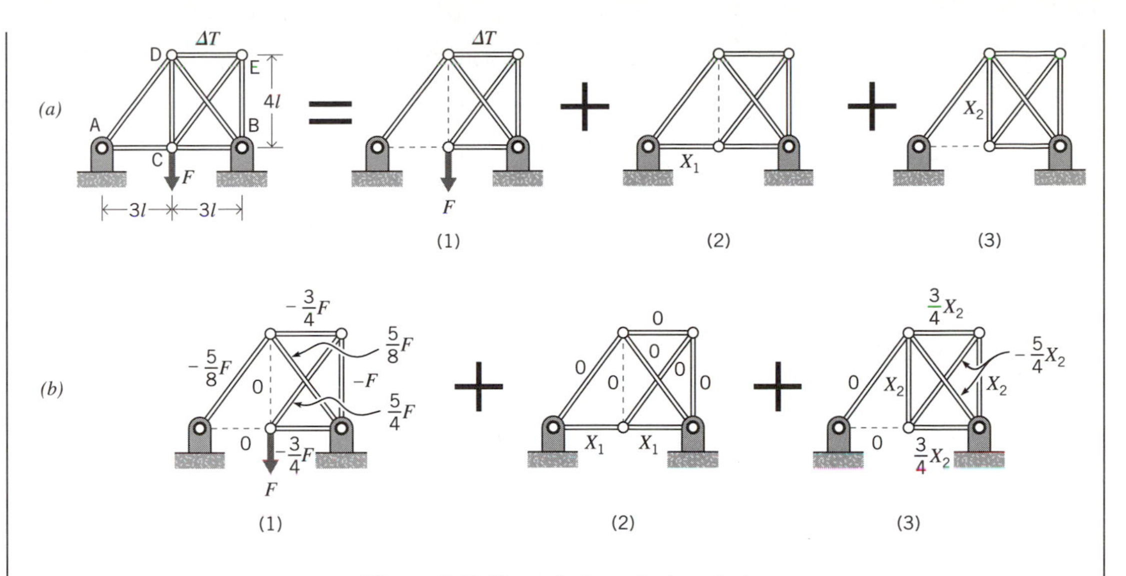

Figure 5.10 Example 1, statical analysis.

or

$$\mathbf{B} = \begin{Bmatrix} \dfrac{9}{4}\dfrac{Fl}{AE} \\ \dfrac{611}{32}\dfrac{Fl}{AE} - \dfrac{9}{4}l\alpha\Delta T \end{Bmatrix} \tag{5.28d}$$

Therefore, Equation 5.27*c* becomes, for this example,

$$\frac{l}{AE}\begin{bmatrix} 6 & 9/4 \\ 9/4 & 27 \end{bmatrix}\begin{Bmatrix} X_1 \\ X_2 \end{Bmatrix} = \begin{Bmatrix} \dfrac{9}{4}\dfrac{Fl}{AE} \\ \dfrac{611}{32}\dfrac{Fl}{AE} - \dfrac{9}{4}l\alpha\Delta T \end{Bmatrix} \tag{5.28e}$$

Inverting the **A** matrix leads to[16]

$$\begin{Bmatrix} X_1 \\ X_2 \end{Bmatrix} = \left(\frac{16AE}{2511l}\begin{bmatrix} 27 & -9/4 \\ -9/4 & 6 \end{bmatrix}\right)\begin{Bmatrix} \dfrac{9}{4}\dfrac{Fl}{AE} \\ \dfrac{611}{32}\dfrac{Fl}{AE} - \dfrac{9}{4}l\alpha\Delta T \end{Bmatrix}$$

[16] Small matrices like this may be inverted explicitly; larger ones require computational inversion.

$$= \begin{Bmatrix} \dfrac{253}{2232}F + \dfrac{1}{31}AE\alpha\Delta T \\ \dfrac{584}{837}F - \dfrac{8}{93}AE\alpha\Delta T \end{Bmatrix} \tag{5.28f}$$

$$= \begin{Bmatrix} 0.113F + 0.032AE\alpha\Delta T \\ 0.698F - 0.086AE\alpha\Delta T \end{Bmatrix}$$

As expected, the values of X_1 and X_2 are identical to the values of P_{AC} and P_{CD}, respectively, in Equations 5.17. Finally, substituting this expression for **X** into Equation 5.28*a* gives

$$\begin{Bmatrix} P_{AC} \\ P_{CB} \\ P_{AD} \\ P_{CD} \\ P_{CE} \\ P_{DE} \\ P_{BE} \\ P_{BD} \end{Bmatrix} = \begin{bmatrix} 1 & 0 \\ 1 & 0.75 \\ 0 & 0 \\ 0 & 1 \\ 0 & -1.25 \\ 0 & 0.75 \\ 0 & 1 \\ 0 & -1.25 \end{bmatrix} \begin{Bmatrix} 0.113F + 0.032AE\alpha\Delta T \\ 0.698F - 0.086AE\alpha\Delta T \end{Bmatrix} + \begin{Bmatrix} 0 \\ -0.75F \\ -0.625F \\ 0 \\ 1.25F \\ -0.75F \\ -F \\ 0.625F \end{Bmatrix} \tag{5.28g}$$

$$= \begin{Bmatrix} 0.113F + 0.032AE\alpha\Delta T \\ -0.113F - 0.032AE\alpha\Delta T \\ -0.625F \\ 0.698F - 0.086AE\alpha\Delta T \\ 0.378F + 0.108AE\alpha\Delta T \\ -0.227F - 0.065AE\alpha\Delta T \\ -0.302F - 0.086AE\alpha\Delta T \\ -0.247F + 0.108AE\alpha\Delta T \end{Bmatrix}$$

which is identical to Equations 5.17.

EXAMPLE 2

Analyze the truss in Figure 5.11. The truss has a pin support at A, and roller supports at B, C, and D. A force F is applied at joint D and a temperature change ΔT occurs in bars AB and DC. Assume that all bars have the same AE.

Step 1. The truss is two-degrees redundant; we arbitrarily select the reactions at D and C as the redundant forces [Figures 5.11(2) and 5.11(3)].

Step 2. The results of the statical analyses are shown in Figures 5.11(1)–5.11(3) and the corresponding equilibrium equations are

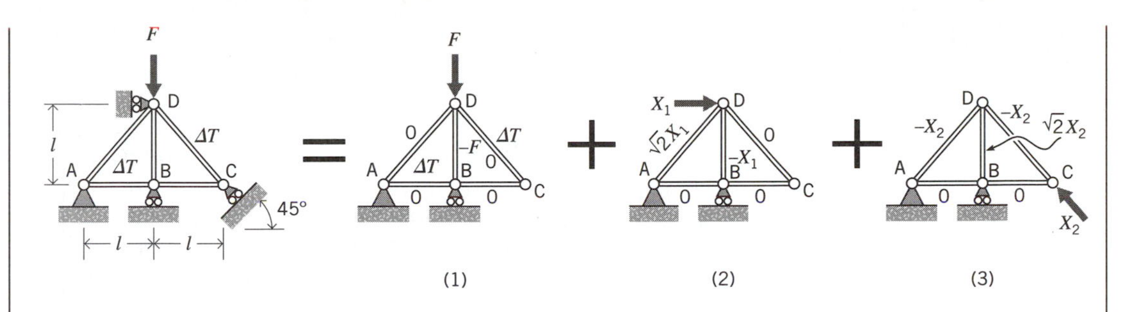

Figure 5.11 Example 2.

$$\begin{Bmatrix} P_{AB} \\ P_{BC} \\ P_{BD} \\ P_{AD} \\ P_{CD} \end{Bmatrix} = \begin{bmatrix} 0 & 0 \\ 0 & 0 \\ -1 & \sqrt{2} \\ \sqrt{2} & -1 \\ 0 & -1 \end{bmatrix} \begin{Bmatrix} X_1 \\ X_2 \end{Bmatrix} + \begin{Bmatrix} 0 \\ 0 \\ -F \\ 0 \\ 0 \end{Bmatrix} \tag{5.29a}$$

which defines the matrices **P**, **b**, and $\mathbf{P_o}$ in the relation

$$\mathbf{P} = \mathbf{bX} + \mathbf{P_o}$$

Step 3. We write the deformation-force relations using the same order of forces selected for **P** above,

$$\begin{Bmatrix} \Delta_{AB} \\ \Delta_{BC} \\ \Delta_{BD} \\ \Delta_{AD} \\ \Delta_{CD} \end{Bmatrix} = \frac{l}{AE} \begin{bmatrix} 1 & 0 & 0 & 0 & 0 \\ 0 & 1 & 0 & 0 & 0 \\ 0 & 0 & 1 & 0 & 0 \\ 0 & 0 & 0 & \sqrt{2} & 0 \\ 0 & 0 & 0 & 0 & \sqrt{2} \end{bmatrix} \begin{Bmatrix} P_{AB} \\ P_{BC} \\ P_{BD} \\ P_{AD} \\ P_{CD} \end{Bmatrix} + \begin{Bmatrix} l\alpha\Delta T \\ 0 \\ 0 \\ 0 \\ \sqrt{2}\, l\alpha\Delta T \end{Bmatrix} \tag{5.29b}$$

which defines the **a** and $\hat{\mathbf{\Delta}}$ matrices in the equation

$$\mathbf{\Delta} = \mathbf{aP} + \hat{\mathbf{\Delta}}$$

We now form Equation 5.27*c*, Step 5.

Step 5. Solving for **A** and **B** gives

$$\mathbf{A} = \mathbf{b}^{\mathrm{T}}\mathbf{ab} = \frac{l}{AE}\begin{bmatrix} 0 & 0 & -1 & \sqrt{2} & 0 \\ 0 & 0 & \sqrt{2} & -1 & -1 \end{bmatrix} \begin{bmatrix} 1 & 0 & 0 & 0 & 0 \\ 0 & 1 & 0 & 0 & 0 \\ 0 & 0 & 1 & 0 & 0 \\ 0 & 0 & 0 & \sqrt{2} & 0 \\ 0 & 0 & 0 & 0 & \sqrt{2} \end{bmatrix} \begin{bmatrix} 0 & 0 \\ 0 & 0 \\ -1 & \sqrt{2} \\ \sqrt{2} & -1 \\ 0 & -1 \end{bmatrix}$$

$$= \frac{l}{AE}\begin{bmatrix} (1 + 2\sqrt{2}) & -(2 + \sqrt{2}) \\ -(2 + \sqrt{2}) & 2(1 + \sqrt{2}) \end{bmatrix}$$

and

$$\mathbf{B} = -\mathbf{b}^T(\mathbf{aP_o} + \hat{\boldsymbol{\Delta}})$$

$$= -\begin{bmatrix} 0 & 0 & -1 & \sqrt{2} & 0 \\ 0 & 0 & \sqrt{2} & -1 & -1 \end{bmatrix} \left\{ \frac{l}{AE} \begin{bmatrix} 1 & 0 & 0 & 0 & 0 \\ 0 & 1 & 0 & 0 & 0 \\ 0 & 0 & 1 & 0 & 0 \\ 0 & 0 & 0 & \sqrt{2} & 0 \\ 0 & 0 & 0 & 0 & \sqrt{2} \end{bmatrix} \begin{Bmatrix} 0 \\ 0 \\ -F \\ 0 \\ 0 \end{Bmatrix} + \begin{Bmatrix} l\alpha\Delta T \\ 0 \\ 0 \\ 0 \\ \sqrt{2}l\alpha\Delta T \end{Bmatrix} \right\}$$

$$= \begin{Bmatrix} -\dfrac{Fl}{AE} \\ \sqrt{2}\dfrac{Fl}{AE} + \sqrt{2}l\alpha\Delta T \end{Bmatrix}$$

So, Equation 5.27*c* becomes

$$\frac{l}{AE}\begin{bmatrix} (1 + 2\sqrt{2}) & -(2 + \sqrt{2}) \\ -(2 + \sqrt{2}) & 2(1 + \sqrt{2}) \end{bmatrix} \begin{Bmatrix} X_1 \\ X_2 \end{Bmatrix} = \begin{Bmatrix} -\dfrac{Fl}{AE} \\ \sqrt{2}\dfrac{Fl}{AE} + \sqrt{2}l\alpha\Delta T \end{Bmatrix} \quad \textbf{(5.29c)}$$

Inverting **A** leads to

$$\begin{Bmatrix} X_1 \\ X_2 \end{Bmatrix} = \frac{AE}{l}\begin{bmatrix} \dfrac{\sqrt{2}}{2} & \dfrac{1}{2} \\ \dfrac{1}{2} & \dfrac{(3\sqrt{2} - 2)}{4} \end{bmatrix} \begin{Bmatrix} -\dfrac{Fl}{AE} \\ \sqrt{2}\dfrac{Fl}{AE} + \sqrt{2}l\alpha\Delta T \end{Bmatrix}$$

$$= \begin{Bmatrix} \dfrac{\sqrt{2}}{2}AE\alpha\Delta T \\ \dfrac{(2 - \sqrt{2})}{2}F + \dfrac{(3 - \sqrt{2})}{2}AE\alpha\Delta T \end{Bmatrix} \quad \textbf{(5.29d)}$$

Substituting Equation 5.29*d* back into Equation 5.29*a* then gives

$$\begin{Bmatrix} P_{AB} \\ P_{BC} \\ P_{BD} \\ P_{AD} \\ P_{CD} \end{Bmatrix} = \begin{Bmatrix} 0 \\ 0 \\ -0.586F + 0.414AE\alpha\Delta T \\ -0.293F + 0.207AE\alpha\Delta T \\ -0.293F - 0.793AE\alpha\Delta T \end{Bmatrix} \quad \textbf{(5.29e)}$$

EXAMPLE 3

The truss shown in Figure 5.12*a* is pinned at A, and has both a roller support and a linear spring at B; the spring is undeformed when there are no loads applied to it. We therefore have a linear relationship between spring force, P_s, and spring deformation, Δ_s, i.e., $P_s = k\Delta_s$ where k is the spring constant. Again, for simplicity, we assume that all bars have the same AE. In this example we shall

(a) Analyze the truss for the special case $AE/kl = 9/8$;

(b) Find the horizontal displacement at B.

To demonstrate concepts, we will carry out the initial phases of the analysis in terms of k (and the other parameters). The spring will be included along with the other components of the structure, i.e., like a member that transmits an axial force; however, the axial flexibility of a spring is $1/k$ (i.e., k^{-1}) as compared to l/AE for a bar.

(a) *Step 1.* This structure is two-degrees redundant because $NJ = 4$, $NR = 3$, and $NB = 7$ (including the spring). We arbitrarily select the internal force in the spring and the internal force in bar AD as the redundant forces, Figures 5.12*a*(2) and 5.12*a*(3), respectively. The temporary removal of the capacity of the spring to transmit an axial force is indicated by the same type of "cut" as for a bar. Note that in this case we cannot select a component of reaction as a redundant force because all three components are required for statical stability.

Step 2. The results of the statical analyses are shown in Figures 5.12*b* and the matrix form of the equilibrium equations is

$$\begin{Bmatrix} P_{AB} \\ P_{AC} \\ P_{AD} \\ P_{BD} \\ P_{BC} \\ P_{CD} \\ P_s \end{Bmatrix} = \begin{bmatrix} 1 & -3/5 \\ 0 & -4/5 \\ 0 & 1 \\ 0 & -4/5 \\ 0 & 1 \\ 0 & -3/5 \\ 1 & 0 \end{bmatrix} \begin{Bmatrix} X_1 \\ X_2 \end{Bmatrix} + \begin{Bmatrix} F \\ 4F/3 \\ 0 \\ 0 \\ -5F/3 \\ F \\ 0 \end{Bmatrix} \qquad \textbf{(5.30a)}$$

Here, both redundant forces are considered to be positive when they are tensile.

Step 3. Again, we write the deformation-force relations using the same order of forces as in the **P** matrix above,

$$\begin{Bmatrix} \Delta_{AB} \\ \Delta_{AC} \\ \Delta_{AD} \\ \Delta_{BD} \\ \Delta_{BC} \\ \Delta_{CD} \\ \Delta_s \end{Bmatrix} = \frac{l}{AE} \begin{bmatrix} 3 & 0 & 0 & 0 & 0 & 0 & 0 \\ 0 & 4 & 0 & 0 & 0 & 0 & 0 \\ 0 & 0 & 5 & 0 & 0 & 0 & 0 \\ 0 & 0 & 0 & 4 & 0 & 0 & 0 \\ 0 & 0 & 0 & 0 & 5 & 0 & 0 \\ 0 & 0 & 0 & 0 & 0 & 3 & 0 \\ 0 & 0 & 0 & 0 & 0 & 0 & \dfrac{AE}{kl} \end{bmatrix} \begin{Bmatrix} P_{AB} \\ P_{AC} \\ P_{AD} \\ P_{BD} \\ P_{BC} \\ P_{CD} \\ P_s \end{Bmatrix} + \begin{Bmatrix} 0 \\ 0 \\ 0 \\ 0 \\ 0 \\ 0 \\ 0 \end{Bmatrix} \qquad \textbf{(5.30b)}$$

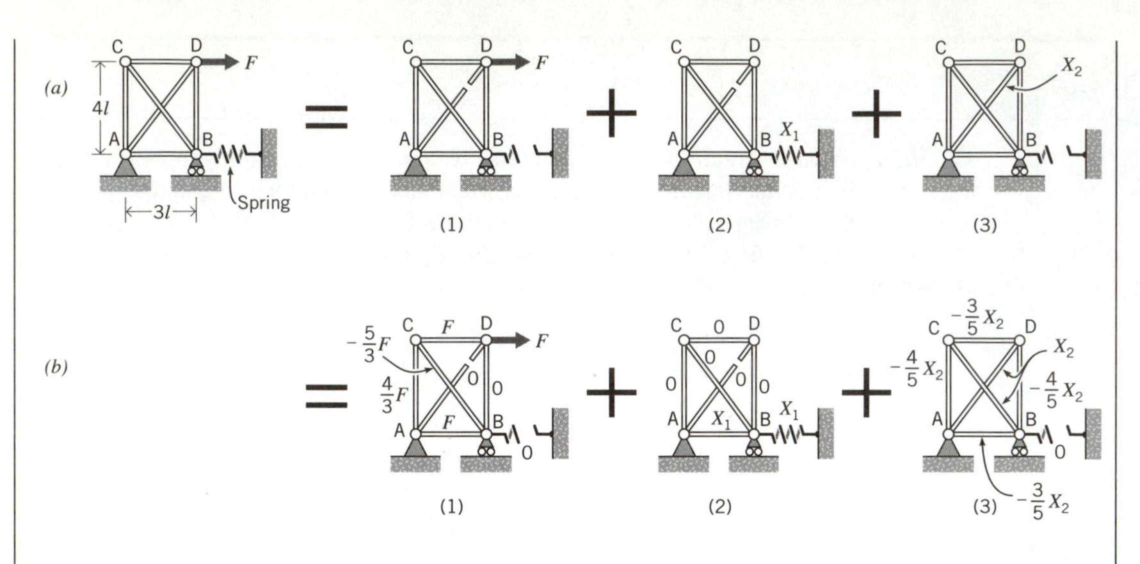

Figure 5.12 Example 3.

Note, in particular, that we have included the deformation-force relation for the spring in the last row, and the fact that $\widehat{\mathbf{\Delta}} = \mathbf{0}$ because there are no thermal or initial effects. We now have all the information required for substitution into Equation 5.27*c*.

Step 5. Solving for **A** and **B** gives

$$\mathbf{A} = \mathbf{b}^{\mathrm{T}}\mathbf{a}\mathbf{b}$$

$$= \frac{l}{AE}\begin{bmatrix} 1 & 0 & 0 & 0 & 0 & 0 & 1 \\ -\frac{3}{5} & -\frac{4}{5} & 1 & -\frac{4}{5} & 1 & -\frac{3}{5} & 0 \end{bmatrix}\begin{bmatrix} 3 & 0 & 0 & 0 & 0 & 0 & 0 \\ 0 & 4 & 0 & 0 & 0 & 0 & 0 \\ 0 & 0 & 5 & 0 & 0 & 0 & 0 \\ 0 & 0 & 0 & 4 & 0 & 0 & 0 \\ 0 & 0 & 0 & 0 & 5 & 0 & 0 \\ 0 & 0 & 0 & 0 & 0 & 3 & 0 \\ 0 & 0 & 0 & 0 & 0 & 0 & \frac{AE}{kl} \end{bmatrix}\begin{bmatrix} 1 & -3/5 \\ 0 & -4/5 \\ 0 & 1 \\ 0 & -4/5 \\ 0 & 1 \\ 0 & -3/5 \\ 1 & 0 \end{bmatrix}$$

$$= \frac{l}{AE}\begin{bmatrix} \left(3 + \frac{AE}{kl}\right) & -\frac{9}{5} \\ -\frac{9}{5} & \frac{432}{25} \end{bmatrix}$$

and

$$\mathbf{B} = -\mathbf{b}^{\mathrm{T}}(\mathbf{a}\mathbf{P}_0 + \widehat{\mathbf{\Delta}})$$

$$= -\begin{bmatrix} 1 & 0 & 0 & 0 & 0 & 0 & 1 \\ -\frac{3}{5} & -\frac{4}{5} & 1 & -\frac{4}{5} & 1 & -\frac{3}{5} & 0 \end{bmatrix}\left\{\frac{l}{AE}\begin{bmatrix} 3 & 0 & 0 & 0 & 0 & 0 & 0 \\ 0 & 4 & 0 & 0 & 0 & 0 & 0 \\ 0 & 0 & 5 & 0 & 0 & 0 & 0 \\ 0 & 0 & 0 & 4 & 0 & 0 & 0 \\ 0 & 0 & 0 & 0 & 5 & 0 & 0 \\ 0 & 0 & 0 & 0 & 0 & 3 & 0 \\ 0 & 0 & 0 & 0 & 0 & 0 & \frac{AE}{kl} \end{bmatrix}\begin{Bmatrix} F \\ 4F/3 \\ 0 \\ 0 \\ -5F/3 \\ F \\ 0 \end{Bmatrix} + \begin{Bmatrix} 0 \\ 0 \\ 0 \\ 0 \\ 0 \\ 0 \\ 0 \end{Bmatrix}\right\}$$

$$= -\frac{Fl}{AE}\begin{Bmatrix} 3 \\ -81/5 \end{Bmatrix}$$

Equation 5.27*c* gives, for this case,

$$\begin{bmatrix} \left(3 + \dfrac{AE}{kl}\right) & -\dfrac{9}{5} \\ -\dfrac{9}{5} & \dfrac{432}{25} \end{bmatrix} \begin{Bmatrix} X_1 \\ X_2 \end{Bmatrix} = F \begin{Bmatrix} -3 \\ \dfrac{81}{5} \end{Bmatrix} \quad \textbf{(5.30c)}$$

Inversion of the equation leads to

$$\begin{Bmatrix} X_1 \\ X_2 \end{Bmatrix} = \frac{5F}{243\left(1 + \dfrac{16}{45}\dfrac{AE}{kl}\right)} \begin{bmatrix} \dfrac{432}{25} & \dfrac{9}{5} \\ \dfrac{9}{5} & \left(3 + \dfrac{AE}{kl}\right) \end{bmatrix} \begin{Bmatrix} -3 \\ \dfrac{81}{5} \end{Bmatrix} = \begin{Bmatrix} -\dfrac{7}{15\left(1 + \dfrac{16}{45}\dfrac{AE}{kl}\right)} F \\ \dfrac{8\left(1 + \dfrac{3}{8}\dfrac{AE}{kl}\right)}{9\left(1 + \dfrac{16}{45}\dfrac{AE}{kl}\right)} F \end{Bmatrix} \quad \textbf{(5.30d)}$$

Because the values of X_1 and X_2 are given in terms of k, it is possible to investigate limiting cases of the value of k. For example, as the spring gets very stiff, i.e., $k \rightarrow \infty$, horizontal motion of the roller approaches zero and the support at B behaves like a pin. In this case the values of the redundant forces are given by

$$\lim_{k\rightarrow\infty} X_1 = -\frac{7}{15}F$$
$$\lim_{k\rightarrow\infty} X_2 = \frac{8}{9}F \quad \textbf{(5.30e)}$$

Similarly, as the spring gets very "soft," i.e., $k \rightarrow 0$, the effect of the spring on horizontal motion at B diminishes; in this case,

$$\lim_{k\rightarrow 0} X_1 = 0$$
$$\lim_{k\rightarrow 0} X_2 = \frac{8}{9}\frac{(3/8)}{(16/45)}F = \frac{15}{16}F \quad \textbf{(5.30f)}$$

The values of X_1 and X_2 in both limiting cases may be verified by reanalyzing the truss without the spring (in the first case, with the roller support replaced by a pin).

Finally, for the specified case, $AE/kl = 9/8$, we obtain

$$\begin{Bmatrix} X_1 \\ X_2 \end{Bmatrix} = \begin{Bmatrix} -1/3 \\ 65/72 \end{Bmatrix} F = \begin{Bmatrix} -0.333 \\ 0.903 \end{Bmatrix} F \tag{5.30g}$$

for which the bar forces and bar deformations, Equations 5.30*a* and 5.30*b*, respectively, are

$$\begin{Bmatrix} P_{AB} \\ P_{AC} \\ P_{AD} \\ P_{BD} \\ P_{BC} \\ P_{CD} \\ P_s \end{Bmatrix} = \begin{Bmatrix} 0.125 \\ 0.611 \\ 0.903 \\ -0.722 \\ -0.764 \\ 0.458 \\ -0.333 \end{Bmatrix} F, \qquad \begin{Bmatrix} \Delta_{AB} \\ \Delta_{AC} \\ \Delta_{AD} \\ \Delta_{BD} \\ \Delta_{BC} \\ \Delta_{CD} \\ \Delta_s \end{Bmatrix} = \begin{Bmatrix} 0.375 \\ 2.444 \\ 4.514 \\ -2.889 \\ -3.819 \\ 1.375 \\ -0.375 \end{Bmatrix} \frac{Fl}{AE} \tag{5.30h}$$

(b) Joint displacements may be evaluated using virtual forces and the real bar deformations in Equation 5.30*h*, as described in Chapter 4. We will demonstrate the procedure again here for the determination of the horizontal displacement of joint B. (We use virtual work even though it may be apparent that the horizontal displacement of joint B must be equal to either the deformation of bar AB or the deformation of the spring; note that these quantities are equal and opposite.)

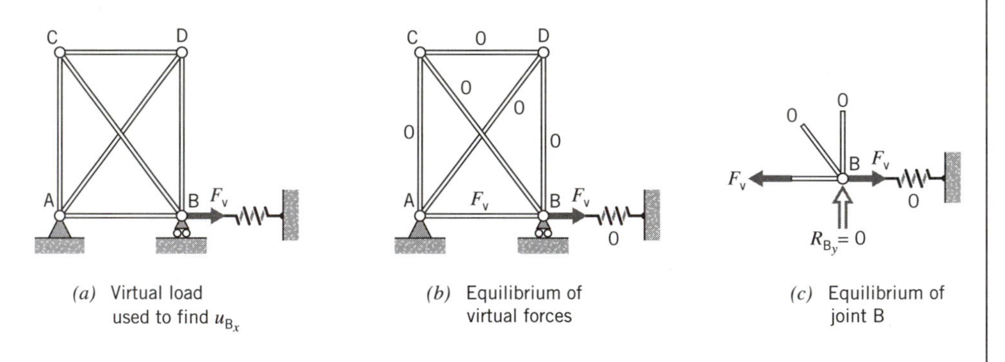

Figure 5.13 Example 3, part (*b*).

We apply an external virtual force F_V at joint B, as shown in Figure 5.13*a*, and find a corresponding set of internal virtual forces that satisfies equilibrium. A simple set of internal virtual forces is shown in Figure 5.13*b*. Figure 5.13*c* is a free-body diagram of joint B, from which we obtain $P_{AB_V} = F_V$; here, P_s and P_{BC} have both arbitrarily been set equal to zero. Therefore, the external virtual work is $u_{BX}F_V$; the only nonzero internal virtual force is P_{AB_V}, so the internal virtual work is $P_{AB_V}\Delta_{AB}$. Equating internal and external virtual work leads to $u_{BX} = \Delta_{AB} = 0.375\,Fl/AE$ (positive in the direction of F_V), as expected. The same result will be obtained using any other internal virtual force state that satisfies equilibrium!

5.2.3 Matrix Formulation for Structures with General Element Flexibility

The deformation-force relationship for a bar is a relationship between a single deformation (length change) and a single axial force that has the simple form

$$\Delta = aP + \hat{\Delta} \tag{5.31a}$$

where $a = l/AE$. In Section 5.3.2 we will show that the corresponding deformation-force relationship for a beam consists of two simultaneous equations in terms of two independent deformation quantities and two independent force quantities. These equations may be expressed in a matrix form, which is similar to Equation 5.31*a*,

$$\begin{Bmatrix} \Delta_1 \\ \Delta_2 \end{Bmatrix} = \begin{bmatrix} a_{11} & a_{12} \\ a_{21} & a_{22} \end{bmatrix} \begin{Bmatrix} M_1 \\ M_2 \end{Bmatrix} + \begin{Bmatrix} \hat{\Delta}_1 \\ \hat{\Delta}_2 \end{Bmatrix} \tag{5.31b}$$

where M_1 and M_2 are moments rather than axial forces, and Δ_1 and Δ_2 are rotational quantities rather than axial deformations. For the current development, we will represent both equations, Equations 5.31*a* and 5.31*b*, by a general form

$$\mathbf{\Delta}^{(e)} = \mathbf{a}^{(e)}\mathbf{P}^{(e)} + \hat{\mathbf{\Delta}}^{(e)} \tag{5.31c}$$

which is similar to Equation 5.24 except that the superscript e denotes a relationship for a single element rather than an entire structure. Here, the matrix $\mathbf{\Delta}^{(e)}$ connotes either the quantity Δ in Equation 5.31*a* or the matrix of Δ's in Equation 5.31*b*, and so on.

By using relations of the type defined by Equation 5.31*c* it is possible to derive the compatibility relation, Equation 5.26, in a general manner that is applicable to both trusses and frames. This will be accomplished by reformulating the procedure for the force method in general terms.

Step 1. Identify redundant forces that will be represented as components of an n-dimensional vector $\mathbf{X}$, i.e., $\mathbf{X}^{\mathrm{T}} = \{X_1 \; . \; . \; . \; X_j \; . \; . \; . \; X_n\}$, where n is the degree of indeterminacy of the structure.

Step 2. Solve equilibrium equations in terms of redundant forces and applied loads, in the form

$$\begin{Bmatrix} \mathbf{P}_1^{(e)} \\ \vdots \\ \mathbf{P}_i^{(e)} \\ \vdots \\ \mathbf{P}_N^{(e)} \end{Bmatrix} = \begin{bmatrix} \mathbf{b}_{11}^{(e)} & \cdots & \mathbf{b}_{1j}^{(e)} & \cdots & \mathbf{b}_{1n}^{(e)} \\ \vdots & \cdots & \vdots & \cdots & \vdots \\ \mathbf{b}_{i1}^{(e)} & \cdots & \mathbf{b}_{ij}^{(e)} & \cdots & \mathbf{b}_{in}^{(e)} \\ \vdots & \cdots & \vdots & \cdots & \vdots \\ \mathbf{b}_{N1}^{(e)} & \cdots & \mathbf{b}_{Nj}^{(e)} & \cdots & \mathbf{b}_{Nn}^{(e)} \end{bmatrix} \begin{Bmatrix} X_1 \\ \vdots \\ X_j \\ \vdots \\ X_n \end{Bmatrix} + \begin{Bmatrix} \mathbf{P}_{\mathbf{o}1}^{(e)} \\ \vdots \\ \mathbf{P}_{\mathbf{o}i}^{(e)} \\ \vdots \\ \mathbf{P}_{\mathbf{o}N}^{(e)} \end{Bmatrix} \tag{5.31d}$$

which is similar to Equation 5.22. Here, subscript N denotes the total number of elements and subscript i denotes the ith element of the structure. Vector $\mathbf{b}_{ij}^{(e)}$ represents a portion of the equilibrium matrix $\mathbf{b}$ corresponding to forces in element i, $\mathbf{P}_i^{(e)}$, due to a unit value of redundant X_j. Equation 5.31*d* can be represented by the equivalent alternate form

$$\mathbf{P} = [\mathbf{b}_1 \ldots \mathbf{b}_j \ldots \mathbf{b}_n]\mathbf{X} + \mathbf{P}_o = \mathbf{b}\mathbf{X} + \mathbf{P}_o \tag{5.31e}$$

where $\mathbf{b}_j$ is the jth column of the $\mathbf{b}$ matrix. It contains all the forces in all the elements produced by a unit value of X_j.

Step 3. The deformation-force relations for the entire structure have the same general form as Equation 5.24,

$$\begin{Bmatrix} \boldsymbol{\Delta}_1^{(e)} \\ \vdots \\ \boldsymbol{\Delta}_i^{(e)} \\ \vdots \\ \boldsymbol{\Delta}_N^{(e)} \end{Bmatrix} = \begin{bmatrix} \mathbf{a}_1^{(e)} & \ldots & \mathbf{0} & \ldots & \mathbf{0} \\ \vdots & \ddots & \vdots & \ldots & \vdots \\ \mathbf{0} & \ldots & \mathbf{a}_i^{(e)} & \ldots & \mathbf{0} \\ \vdots & \ldots & \vdots & \ddots & \vdots \\ \mathbf{0} & \ldots & \mathbf{0} & \ldots & \mathbf{a}_N^{(e)} \end{bmatrix} \begin{Bmatrix} \mathbf{P}_1^{(e)} \\ \vdots \\ \mathbf{P}_i^{(e)} \\ \vdots \\ \mathbf{P}_N^{(e)} \end{Bmatrix} + \begin{Bmatrix} \hat{\boldsymbol{\Delta}}_1^{(e)} \\ \vdots \\ \hat{\boldsymbol{\Delta}}_i^{(e)} \\ \vdots \\ \hat{\boldsymbol{\Delta}}_N^{(e)} \end{Bmatrix} \tag{5.31f}$$

except that the individual components of this equation are the matrices defined in Equation 5.31*c*.

Step 4. In this step, n individual compatibility equations are generated, each by writing a virtual work equation based on a virtual force state that is identical to the equilibrium state associated with a given redundant force. In other words, we choose a virtual force state for which equilibrium is defined by

$$\mathbf{P}_{V_j} = \mathbf{b}_j X_{jV} \tag{5.31g}$$

where F_{jV} is a virtual force ''like'' redundant X_j and, as noted above, $\mathbf{b}_j$ is the jth column of matrix $\mathbf{b}$. Because the real displacement associated with X_{jV} is zero, the external virtual work is zero. The internal virtual work is the product of vectors $\mathbf{P}_{V_j}$ and $\boldsymbol{\Delta}$.[17] Therefore, from the theorem of virtual work,

$$\begin{aligned} 0 &= \mathbf{P}_{V_j}^{\mathrm{T}}\boldsymbol{\Delta} \\ &= X_{jV}\mathbf{b}_j^{\mathrm{T}}\boldsymbol{\Delta} \end{aligned} \tag{5.31h}$$

or, because X_{jV} is arbitrary,

$$0 = \mathbf{b}_j^{\mathrm{T}}\boldsymbol{\Delta} \tag{5.31i}$$

This is a single compatibility equation for the jth redundant force x_j that is typical of all such equations. The other compatibility equations are generated in the same way and involve the other columns of $\mathbf{b}$, i.e., one for each X_j,

---

[17] This will be demonstrated for frames in Section 5.3.3.

$$0 = \mathbf{b}_1^{\mathrm{T}}\boldsymbol{\Delta}$$
$$\vdots$$
$$0 = \mathbf{b}_j^{\mathrm{T}}\boldsymbol{\Delta}$$
$$\vdots$$
$$0 = \mathbf{b}_n^{\mathrm{T}}\boldsymbol{\Delta}$$

These n equations may be combined into a single matrix equation of the form

$$\begin{aligned}\mathbf{0} &= [\mathbf{b}_1 \ldots \mathbf{b}_j \ldots \mathbf{b}_n]^{\mathrm{T}}\boldsymbol{\Delta} \\ &= \mathbf{b}^{\mathrm{T}}\boldsymbol{\Delta}\end{aligned} \tag{5.31j}$$

as was previously stated in Equation 5.26.

Step 5. Equations 5.31*e* and 5.31*f* are substituted into Equation 5.31*j*; the result is the same as previously presented in Equation 5.27*c*, namely

$$(\mathbf{b}^{\mathrm{T}}\mathbf{ab})\mathbf{X} = -\mathbf{b}^{\mathrm{T}}(\mathbf{aP_o} + \boldsymbol{\Delta}) \tag{5.31k}$$

5.3 FRAMES

The force method of analysis of frames follows the same five-step procedure employed for trusses. Again the starting point is the identification of a set of n redundant forces, where n is the degree of indeterminacy of the structure. The degree of indeterminacy may be ascertained using the approach presented in Section 1.7 rather than the "counting" procedure applied to trusses.

As a consequence of the fact that the internal forces are variable functions of position coordinates, every indeterminate frame structure has an infinite number of internal forces (in addition to possible reactions) from which to choose the redundant forces. Nevertheless, the frame structure has only a finite number of redundant forces. Consider, for instance, the structure in Figure 5.14*a*, which we assume may be subjected to arbitrary transverse and axial loads. This structure is one-degree indeterminate. Figures 5.14*b*–5.14*f* show several choices of the redundant force. In Figure 5.14*b* the vertical component of reaction at the left support has been released, in Figure 5.14*c* the moment at the right support has been released, and in Figure 5.14*d* the vertical component (but not the horizontal component) of the right reaction has been released. Alternatively, Figures 5.14*e* and 5.14*f* show internal forces that may be selected as redundant forces. The internal moment at an arbitrary cross section has been released in Figure 5.14*e*; this is represented by the inclusion of a hinge at the designated cross section. (The resulting structure is the same as the beam in Example 3 of Chapter 3.) Similarly, the internal shear at some cross section has been released in Figure 5.14*f*.[18] In each of the last two cases, the internal force may be released at any cross section. Finally, Figure 5.14*g* shows an incorrect choice of redundant force; release of the horizontal component

---

[18] Physical representations of an associated structural modification are possible but are really not necessary; it is generally sufficient to specify the force component that is temporarily removed.

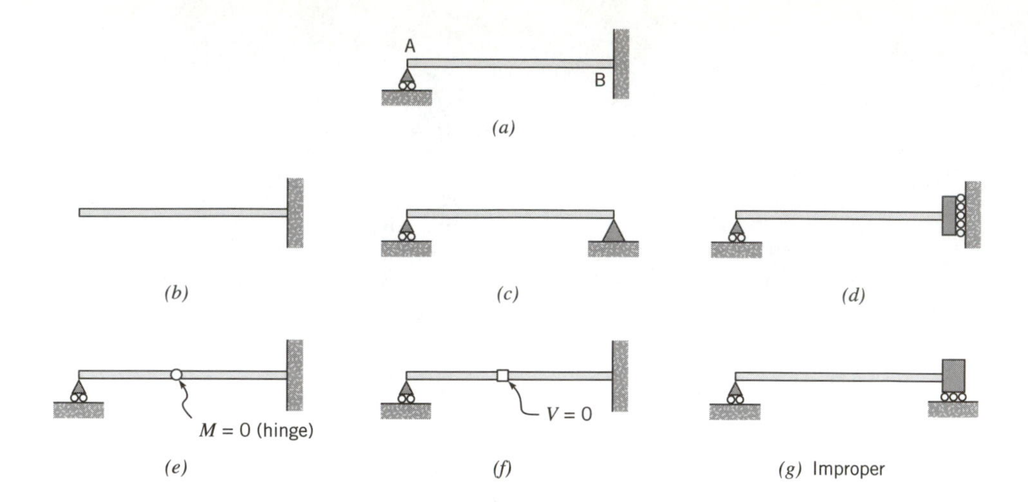

Figure 5.14 (*a*) One-degree indeterminate beam; (*b*)–(*f*) acceptable choices of redundant force; (*g*) improper choice of redundant force.

of reaction renders the structure statically unstable. Likewise, release of an internal axial force would also be incorrect in this case.

The structure in Figure 5.15*a* is two-degrees indeterminate. Recall that the support at C is a ''fixed support with roller,'' which, as shown, is assumed capable of developing a moment and vertical force but no horizontal force. Figures 5.15*b*–5.15*i* show some valid choices of redundant forces. Some of these choices may appear odd (e.g., Figures 5.15*h* and 5.15*i*), but all are statically stable. In Figures 5.15*d* and 5.15*i* the vertical component of reaction at C has been released, a condition represented by a ''fixed support with two rollers.'' In Figure 5.15*d* the second redundant force has been chosen as the moment at support A and in Figure 5.15*i* the second redundant force has been chosen as the internal (bending) moment at B.[19] The structure in Figure 5.15*d* is the same as the one in Example 4, Chapter 3. The choice of redundant forces in Figure 5.15*j* is incorrect because the structure cannot carry applied horizontal forces.

The equilibrium portion of the analysis, Step 2, again is conducted using superposition of forces based on any suitable choice of redundant forces. For instance, Figures 5.16*a*–5.16*c* show the superposition of forces for the beam in Figure 5.14*a* based on the choices of redundant forces in Figures 5.14*b*, 5.14*d*, and 5.14*e*, respectively. (The continuous load is representative of any arbitrary load.) Similarly, Figures 5.17*a* and 5.17*b* show the superposition of forces for the beam in Figure 5.15*a* based on the choices of redundant forces in Figures 5.15*b* and 5.15*f*, respectively. Note that a redundant internal moment, Figures 5.16*c*(2) and 5.17*b*(3), must act in an equal and opposite manner on contiguous cross sections similar to an internal redundant axial force in the case of a truss bar, Figure 5.4*c*. If the beam were truly cut, an arbitrary redundant internal moment would cause a relative rotation between the contiguous

[19] In the subsequent matrix formulation (Section 5.3.3), the choice of an internal force as a redundant force will always be taken at the end of a beam; consequently, we shall usually choose the redundant force at the end of an existing beam (e.g., joint B) rather than at some arbitrary intermediate location.

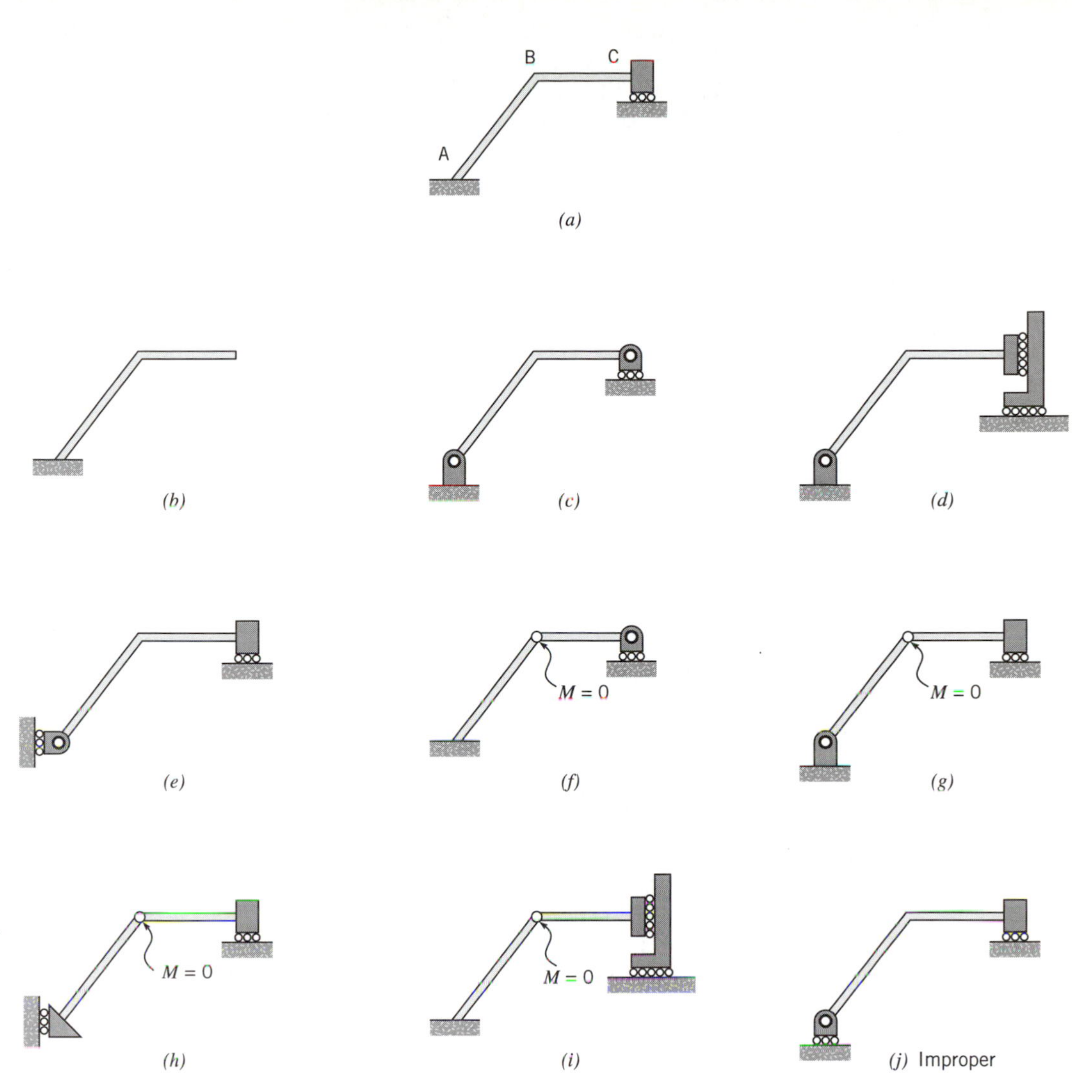

Figure 5.15 (*a*) Two-degree indeterminate beam; (*b*)–(*i*) acceptable choices of redundant forces; (*j*) improper choice of redundant forces.

beam axes, in a manner analogous to the relative displacement between ''cut'' ends of a truss bar. The imposition of zero relative rotation across the ''cut'' will lead directly to the compatibility equation for the redundant moment.

5.3.1 Basic Formulation

To demonstrate the force method for frames, we will begin with a consideration of the same beam as in Figure 5.16*a*, again shown in Figure 5.18*a*. In addition to a uniform continuous load, we assume a temperature change that varies linearly through the depth

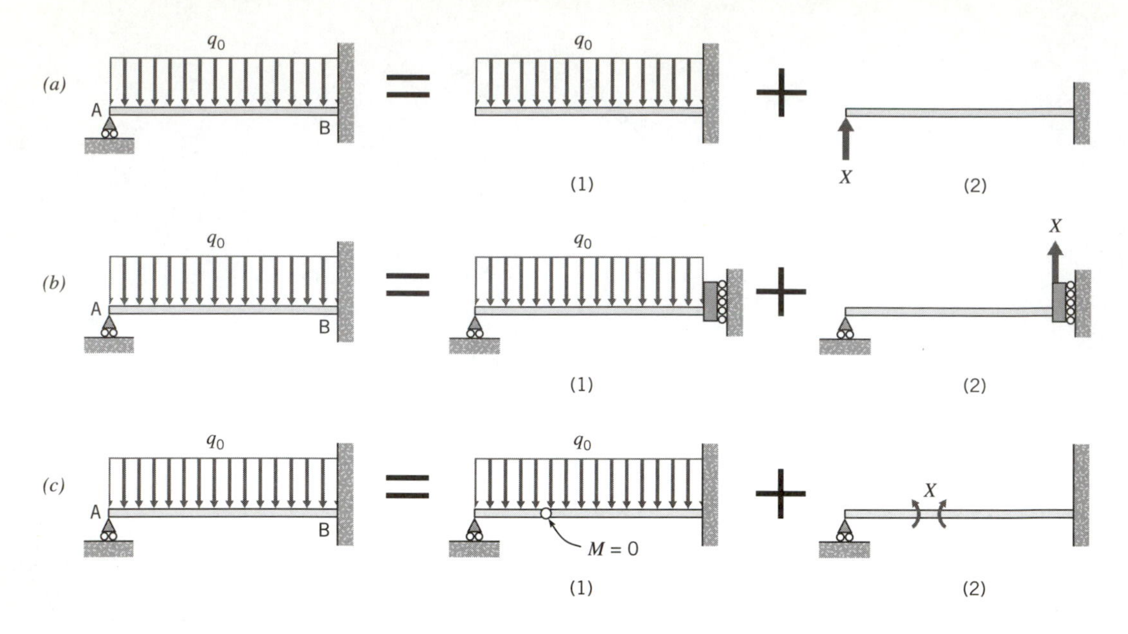

Figure 5.16 Superposition of forces.

and is given by $\Delta T = \Delta T_o y/h$,[20] where ΔT_o is a constant. The beam cross section will also be assumed to be uniform, i.e., EI is constant.

Step 1. We arbitrarily choose the reaction at A as the redundant force and denote it by X [Figure 5.18a(2)].

[20] The quantity h is some characteristic length related to a cross-sectional dimension, e.g., for a rectangular cross section of depth H, $h = H/2$ provides a temperature change that varies linearly from $-\Delta T_o$ at the bottom surface to $+\Delta T_o$ at the top surface.

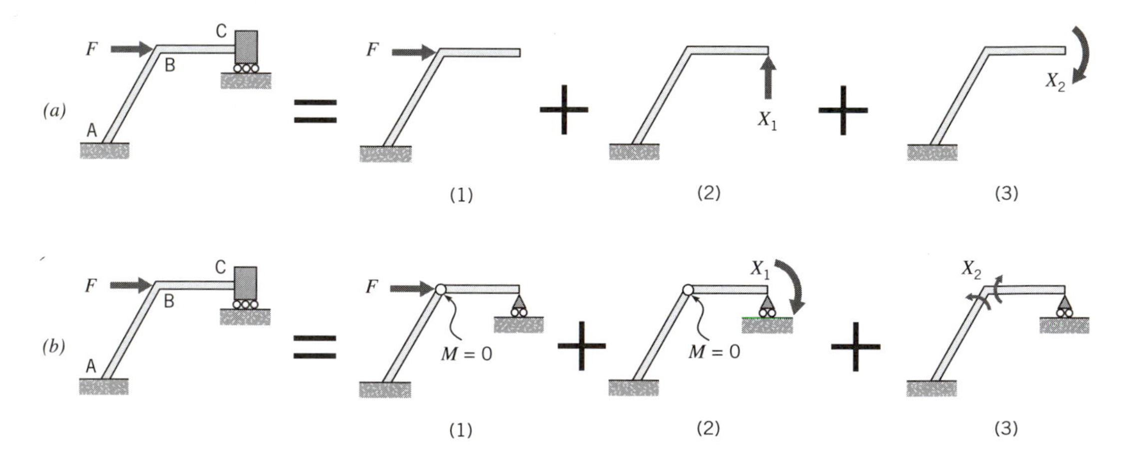

Figure 5.17 Superposition of forces.

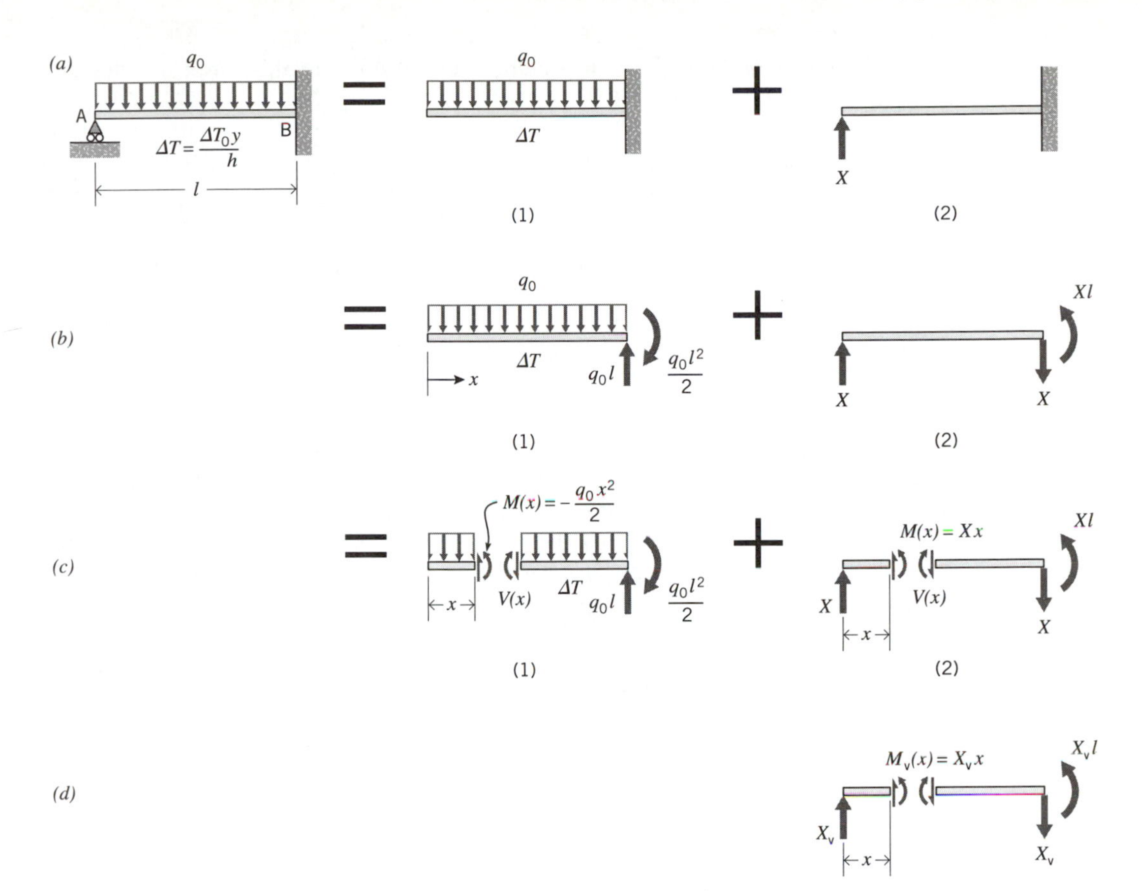

Figure 5.18 Statical analysis of one-degree redundant beam.

Step 2. The statical portion of the analysis is conducted using the superposition of forces indicated in Figures 5.16*a* or 5.18*a*. Free-body diagrams of the structure are shown in Figure 5.18*b*. The internal moment varies with position along the beam so we will have to find the internal bending moment as a function of coordinate x (Figure 5.18*c*). From the free-bodies in Figures 5.18*c*(1) and 5.18*c*(2) we obtain $M(x)$ due to the applied loads and the redundant X, respectively,

$$M(x) = Xx - \frac{q_o x^2}{2} \tag{5.32a}$$

The temperature change cannot induce any forces in the statically determinate structure shown in Figure 5.18*c*(1).

Step 3. For beams, the deformation quantity of interest is the beam curvature, $v''(x)$, Equation 3.16*b*. This curvature is equal to $M(x)$, Equation 5.32*a*, divided by EI, plus the thermal curvature. For the temperature load specified in this case, the thermal curvature is a constant given by[21]

$$v_T''(x) = \frac{M_T(x)}{EI} = -\frac{\alpha}{I}\int_A y\Delta T dA = -\frac{\alpha}{I}\int_A y\left(\frac{\Delta T_o y}{h}\right)dA = -\frac{\alpha\Delta T_o}{h} \quad \textbf{(5.32}b\textbf{)}$$

Thus, the beam curvature is related to forces and temperature changes as follows,

$$v''(x) = \frac{M(x)}{EI} + v_T''(x) = \frac{X}{EI}x - \frac{q_o}{2EI}x^2 - \frac{\alpha\Delta T_o}{h} \quad \textbf{(5.32}c\textbf{)}$$

Step 4. The compatibility equation is obtained from the theorem of virtual work for frames, Equation 4.24. We use a virtual force state ''like'' the real force state associated with X (Figure 5.18*d*); the internal virtual moment for this state is therefore $M_V(x) = X_V x$.[22] The external virtual work is the product of external virtual forces in Figure 5.18*d* and corresponding real displacements in Figure 5.18*a*; this is again zero because no vertical displacement is permitted at A. Equating this to internal virtual work gives

$$\begin{aligned} 0 &= \int_0^l M_V(x)v''(x)dx \\ &= \int_0^l (X_V x)\left(\frac{X}{EI}x - \frac{q_o}{2EI}x^2 - \frac{\alpha\Delta T_o}{h}\right)dx \\ &= \left(\frac{l^3}{3EI}\right)X + \left(-\frac{q_o l^4}{8EI} - \frac{l^2\alpha\Delta T_o}{2h}\right) \end{aligned} \quad \textbf{(5.32}d\textbf{)}$$

which is the desired compatibility equation. We will examine the physical significance of this equation shortly.

Step 5. Solving Equation 5.32*d* provides the value of X,[23]

$$X = \frac{3}{8}q_o l + \frac{3EI\alpha\Delta T_o}{2lh} \quad \textbf{(5.32}e\textbf{)}$$

Other force and displacement quantities may be evaluated by substituting this value into Equations 5.32*a* and 5.32*b*. Reactions may also be obtained by solving the equilibrium equations for the entire structure (Figure 5.18*b*). These reactions are illustrated in Figure 5.19. Note that although the applied load is uniform the two vertical components of reaction are not equal, a reasonable result because the end supports differ.

The physical significance of a compatibility equation similar to Equation 5.32*d* has already been discussed in Example 16, Chapter 4. There, a form of equation similar to

[21] See Example 5, Chapter 3, for an illustration of the computation of a thermal curvature.

[22] Again, because X_V cancels out of the virtual work equation, it would be convenient to let $X_V = 1$; however, we will include X_V for the sake of clarity.

[23] Also see Example 16, Chapter 4.

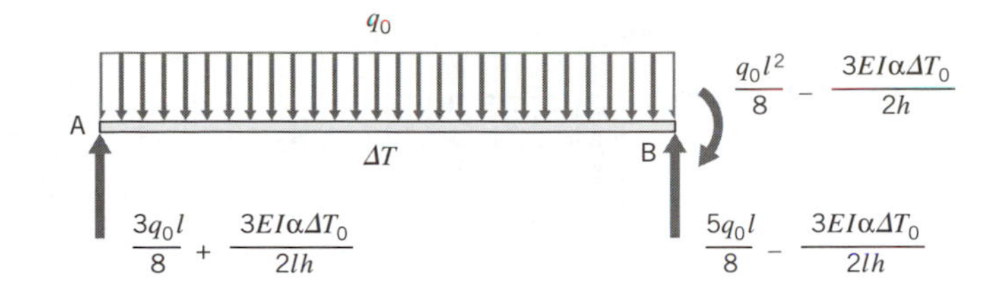

Figure 5.19 Reactions for one-degree redundant beam.

the second line of Equation 5.32*d* was related to the transverse deflections at the free end of the cantilever beams in Figures 5.18*a*(1) and 5.18*a*(2) via the moment-area method.[24] It was shown that the equation indicates that the sum of the tip deflections equals zero, which is consistent with the kinematic constraint condition at the roller support. Similarly, the two terms in the last line of Equation 5.32*d* are the two components of displacement in question, as may be verified from the theorem of virtual work, i.e., the first term is the displacement at A in Figure 5.18*a*(2), and the second term is the displacement at A in Figure 5.18*a*(1). Furthermore, the first term in parentheses is again a flexibility coefficient, in this case the transverse displacement at A due to a unit value of X.

In order to gain additional familiarity with the basic procedure, we will repeat the analysis of this structure using the two alternative choices of redundant forces shown in Figures 5.16*b* and 5.16*c*. Consider first Figure 5.16*b*, for which the redundant is selected as the vertical component of reaction at B. Note that the remaining reactions consist of a vertical force at A and a moment at B. The results of the statical analysis are shown in Figure 5.20. (Here, the horizontal component of reaction at B is identically equal to zero.) The analysis proceeds as follows:

Step 2. The internal bending moment is obtained from Figure 5.20*c*,

$$M(x) = -Xx + q_o lx - \frac{q_o}{2}x^2 \tag{5.33a}$$

Step 3. The deformation-force relation is therefore given by

$$v''(x) = -\frac{X}{EI}x + \frac{q_o l}{EI}x - \frac{q_o}{2EI}x^2 - \frac{\alpha \Delta T_o}{h} \tag{5.33b}$$

Note that the thermal curvature is unchanged from the previous case.

Step 4. The compatibility equation is obtained by using the virtual force state in Figure 5.20*d*, i.e.,

$$0 = \int_0^l (-X_V x)\left(-\frac{X}{EI}x + \frac{q_o l}{EI}x - \frac{q_o}{2EI}x^2 - \frac{\alpha \Delta T_o}{h}\right)dx \tag{5.33c}$$

[24] The moment-area method will also prove useful for identifying the significance of other integrals in subsequent developments; a review of Section 3.4.2 is recommended.

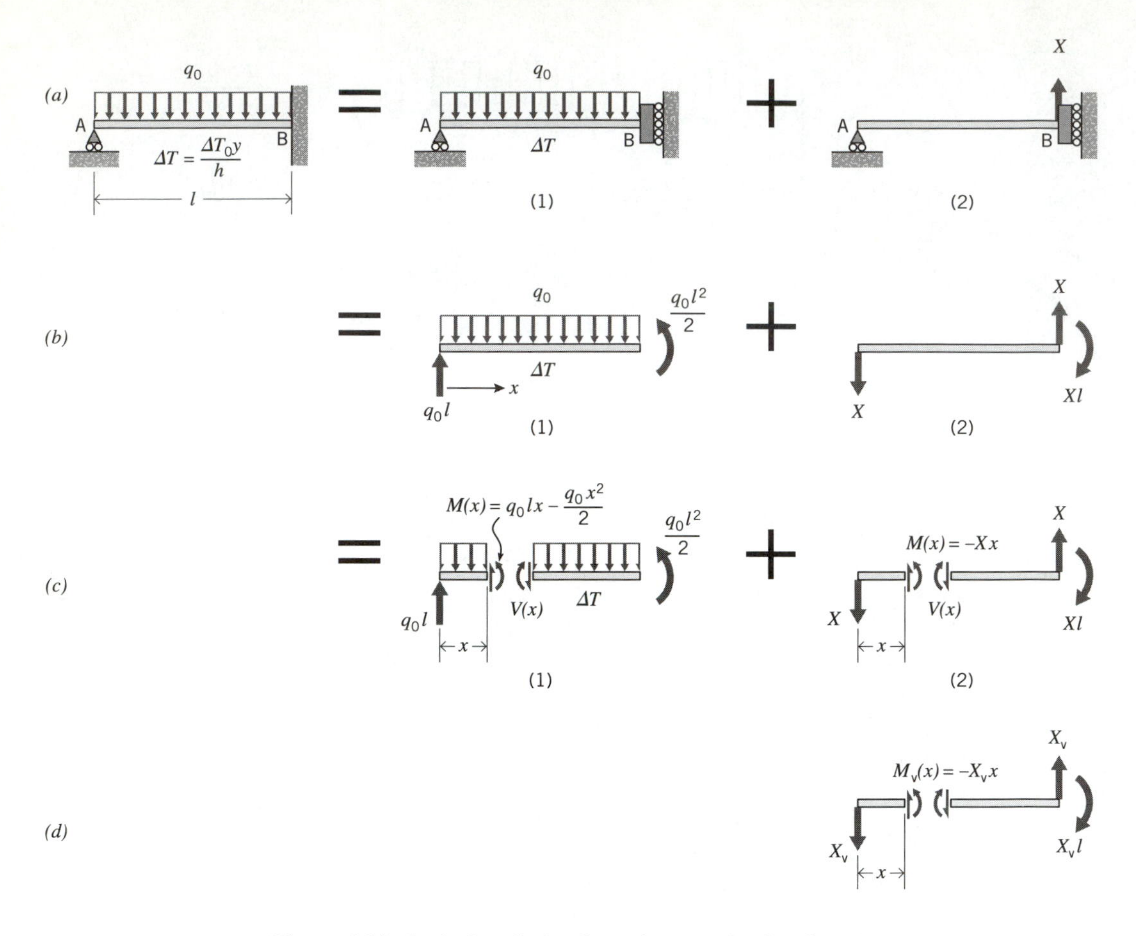

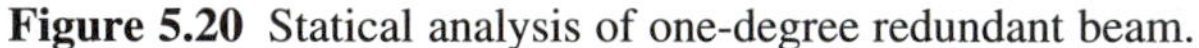

Figure 5.20 Statical analysis of one-degree redundant beam.

$$= \left(\frac{l^3}{3EI}\right)X + \left(-\frac{5q_o l^4}{24EI} + \frac{l^2\alpha \Delta T_o}{2h}\right)$$

In this case, the first term is the transverse displacement at B in Figure 5.20*a*(2), and the second term is the transverse displacement at B in Figure 5.20*a*(1) due to the applied mechanical and thermal loads; the compatibility condition is that the sum of these displacements is zero.

Step 5. The solution of Equation 5.33*c* gives

$$X = \frac{5}{8}q_o l - \frac{3EI\alpha\Delta T_o}{2lh} \qquad \textbf{(5.33}\textbf{\textit{d}}\textbf{)}$$

which agrees with the result previously obtained for the reaction at B (Figure 5.19).

Finally, we consider the case in Figure 5.16*c* for which the redundant force is selected as the internal bending moment at an arbitrary cross section; we will take this cross

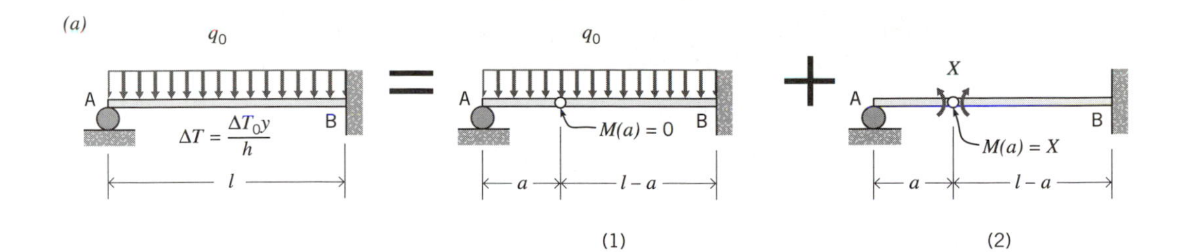

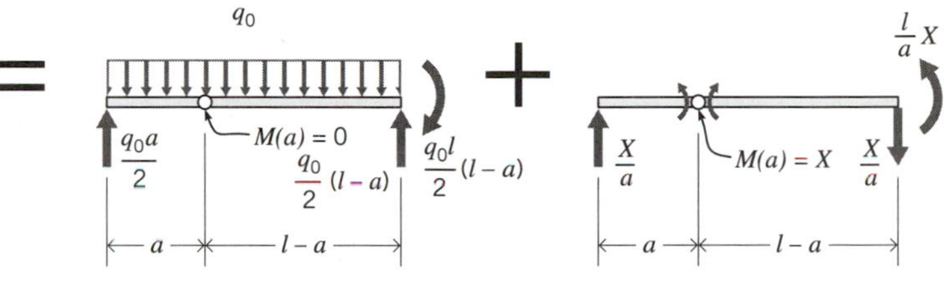

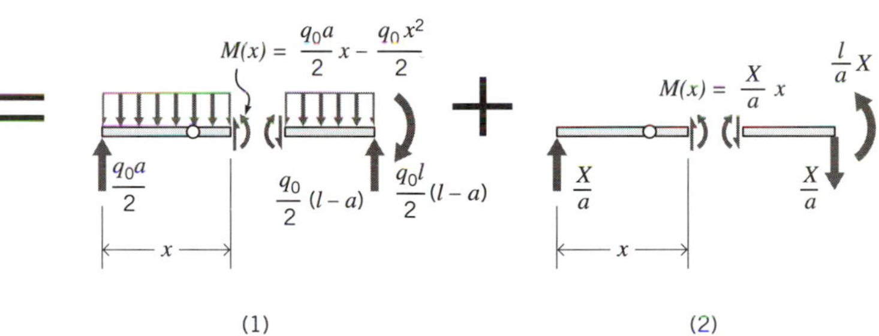

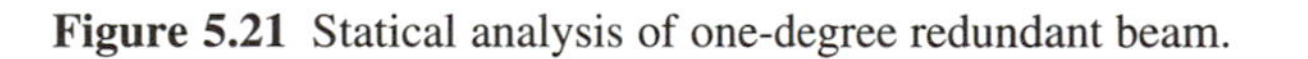

Figure 5.21 Statical analysis of one-degree redundant beam.

section to be located a distance a from the left end of the beam (Figure 5.21). We proceed as follows:

Step 2. The free-body diagrams for the statical portion of the analysis are shown in Figure 5.21. For the structure in Figure 5.21a(1) the internal bending moment at $x = a$ is zero, i.e., M(a) $= 0$; for the structure in Figure 5.21a(2) the internal bending moment at $x = a$ is X, i.e., $M(a) = X$. With this in mind, reactions and internal forces are determined as in Example 3, Chapter 3; the results are shown in Figures 5.21b and

5.21*c*. It should be noted that the bending moments are continuous functions of coordinate x for $0 < x < l$ despite the presence of the ''hinge.'' Thus, we obtain

$$M(x) = \frac{X}{a}x + \frac{q_o a}{2}x - \frac{q_o}{2}x^2, \quad 0 < x < l \tag{5.34a}$$

Step 3. The deformation-force relation is therefore

$$v''(x) = \frac{X}{aEI}x + \frac{q_o a}{2EI}x - \frac{q_o}{2EI}x^2 - \frac{\alpha \Delta T_o}{h}, \quad 0 < x < l \tag{5.34b}$$

(Because this is an expression for curvature of the beam in Figure 5.21*a* there cannot be a discontinuity at $x = a$, i.e., $v''(x)$ is continuous for $0 < x < l$.)

Step 4. The compatibility equation is obtained by using the virtual force state in Figure 5.21*d*, which is ''like'' the real force state in Figure 5.21*c*(2). Because X_V in this case is the internal moment at $x = a$, and because there are no displacements at A or B in the real structure, it follows that the external virtual work is again zero. The virtual work equation becomes

$$\begin{aligned} 0 &= \int_0^l \left(\frac{X_V}{a}x\right)\left(\frac{X}{aEI}x + \frac{q_o a}{2EI}x - \frac{q_o}{2EI}x^2 - \frac{\alpha \Delta T_o}{h}\right)dx \\ &= \left(\frac{l^3}{3aEI}\right)X + \left[\frac{q_o l^3}{24EI}(4a - 3l) - \frac{l^2 \alpha \Delta T_o}{2h}\right] \end{aligned} \tag{5.34c}$$

Here, the first term is the relative rotation between contiguous beam axes on the left and right sides of the beam at $x = a$ [see Figure 5.21*a*(2)], and the second term is the relative rotation between the axes at $x = a$ in Figure 5.21*a*(1);[25] the physical significance of the compatibility condition is that the sum of these relative rotations must equal zero.

Step 5. The solution of Equation 5.34*c* gives

$$X = \frac{q_o a}{8}(3l - 4a) + \frac{3aEI\alpha \Delta T_o}{2lh} \tag{5.34d}$$

which is the correct expression for $M(a)$, see Figure 5.19.

It is worth noting that internal forces (e.g., bending moments, Equation 5.34*a*) and/or deformations (e.g., v'', Equation 5.34*b*) will not always be continuous functions of x for $0 < x < l$; however, the integration in Step 4 must still be conducted over the entire length of the structure using proper analytic expressions for forces or deformations over appropriate subintervals. This will become clearer in the next case, which involves a slightly more complex structure composed of two beam elements.

Consider the two-degree redundant frame in Figure 5.22*a* which is loaded as shown,

---

[25] This may be verified by using virtual work to compute rotations (or *relative* rotations); see, for instance, Example 13, Chapter 4.

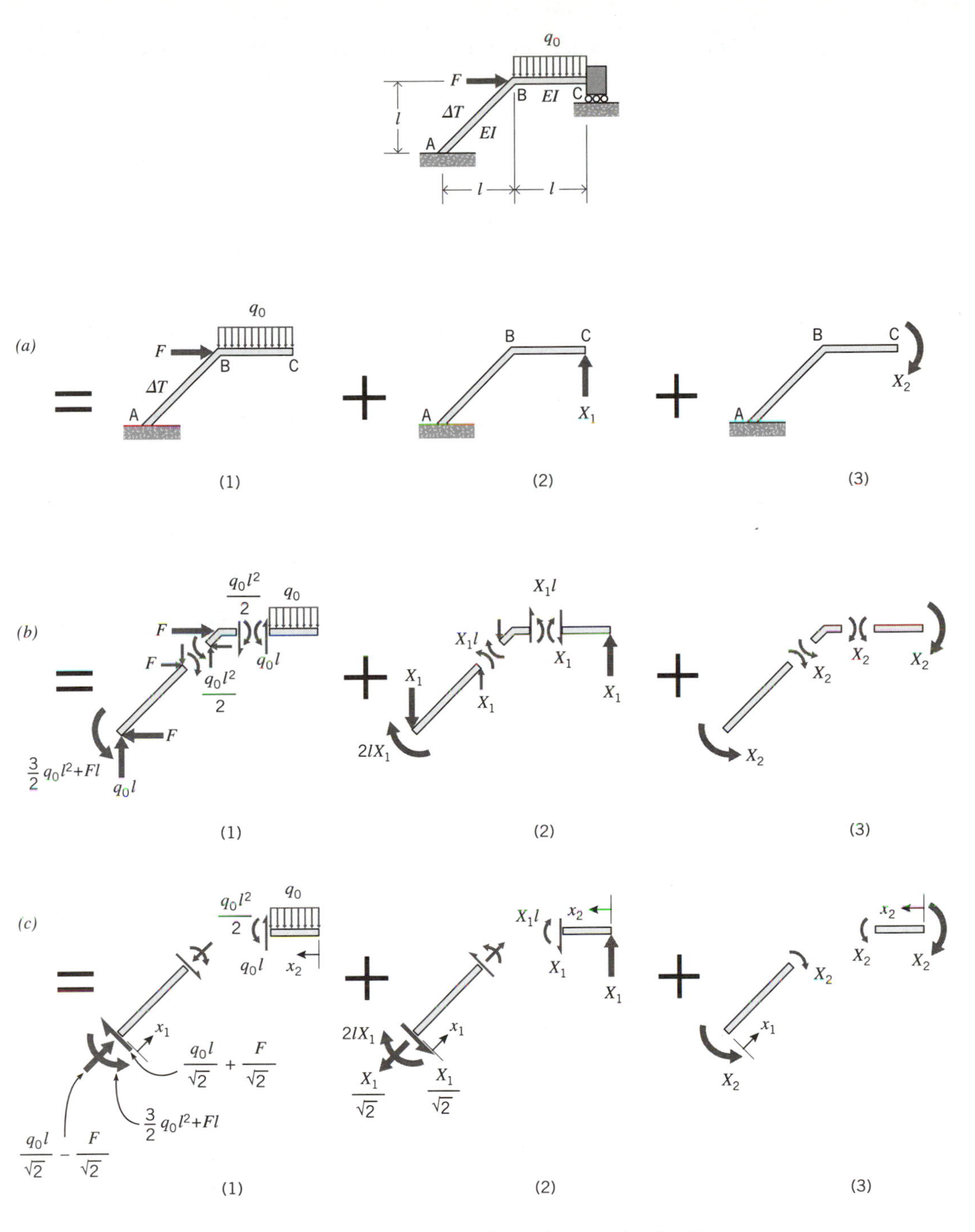

Figure 5.22 Statical analysis of two-degree redundant frame.

i.e., a uniform continuous load on beam BC, a concentrated horizontal load at joint B, and a thermal load $\Delta T = \Delta T_o y/h$ in beam AB. The dimensions are given in the figure, and both beams are assumed to have the same value of EI.

Step 1. The structure will be analyzed taking the two components of reaction at C as redundant forces. Denote the vertical component of reaction at C by X_1 and the moment by X_2, as shown in Figures 5.22*a*(2) and 5.22*a*(3), respectively. Note that X_1 has the dimension of force, while X_2 has the dimension of force $\times$ length. We will see that there is no conceptual problem in dealing with redundant forces with dissimilar dimensions; however, we must be careful to keep all equations dimensionally correct.

Step 2. The statical portion of the analysis is conducted using the basic superposition indicated in Figures 5.22*a*. Figures 5.22*b*(1)–5.22*b*(3) show free-body diagrams[26] corresponding to the three cases in Figures 5.22*a*(1)–5.22*a*(3). The free-body diagrams of beam AB show vertical and horizontal components of forces; these must be converted to transverse and axial components because we need to compute internal bending moments. Figures 5.22*c*(1)–5.22*c*(3) show the appropriate force components. The expressions for internal bending moments in beams AB and BC are now obtained relatively easily. They are written in terms of coordinates x_1 and x_2 identified in the figures; for beam AB

$$M(x_1) = \left(\frac{q_o l}{\sqrt{2}} + \frac{F}{\sqrt{2}} - \frac{X_1}{\sqrt{2}}\right)x_1 + \left(2lX_1 - X_2 - \frac{3}{2}q_o l^2 - Fl\right), \quad 0 < x_1 < \sqrt{2}l \tag{5.35a}$$

and for beam BC

$$M(x_2) = -\frac{q_o}{2}x_2^2 + X_1 x_2 - X_2, \quad 0 < x_2 < l \tag{5.35b}$$

Step 3. The deformation-force relations for beams AB and BC, respectively, are

$$v''(x_1) = \frac{1}{EI}\left(\frac{q_o l}{\sqrt{2}} + \frac{F}{\sqrt{2}} - \frac{X_1}{\sqrt{2}}\right)x_1 + \frac{1}{EI}\left(2lX_1 - X_2 - \frac{3}{2}q_o l^2 - Fl\right) - \frac{\alpha \Delta T_o}{h}, \quad 0 < x_1 < \sqrt{2}l \tag{5.35c}$$

$$v''(x_2) = -\frac{q_o}{2EI}x_2^2 + \frac{X_1}{EI}x_2 - \frac{X_2}{EI}, \quad 0 < x_2 < l \tag{5.35d}$$

Note the inclusion of the thermal curvature in Equation 5.35*c*.[27] (This term is identical to the one in the previous cases because ΔT is the same.)

---

[26] A structure of similar geometry (but with different supports) was analyzed in Example 4, Chapter 3.

[27] Also note that although X_1, X_2, q_o, F, and ΔT_o all have their individual dimensional bases, every term in Equations 5.35*a*–*d* has consistent dimensions.

Step 4. We require two compatibility equations. These are obtained from two virtual work equations, one based on virtual forces identical to those associated with X_1 in Figure 5.22*c*(2), and one based on virtual forces identical to those associated with X_2 in Figure 5.22*c*(3). The internal virtual moments associated with X_{V_1} are

$$
\begin{aligned}
M_V(x_1) &= -\frac{X_{V1}}{\sqrt{2}}x_1 + 2lX_{V1}, \quad 0 < x_1 < \sqrt{2}l \\
M_V(x_2) &= X_{V1}x_2, \quad 0 < x_2 < l
\end{aligned} \tag{5.35e}
$$

External virtual work is zero and the virtual work equation, Equation 4.24, is

$$
\begin{aligned}
0 = \int_0^{\sqrt{2}l} &\left(-\frac{X_{V1}}{\sqrt{2}}x_1 + 2lX_{V1}\right)\left[\frac{1}{EI}\left(\frac{q_o l}{\sqrt{2}} + \frac{F}{\sqrt{2}} - \frac{X_1}{\sqrt{2}}\right)x_1\right. \\
&\left. + \frac{1}{EI}\left(2lX_1 - X_2 - \frac{3}{2}q_o l^2 - Fl\right) - \frac{\alpha \Delta T_o}{h}\right]dx_1 \\
&+ \int_0^l (X_{V1}x_2)\left(-\frac{q_o}{2EI}x_2^2 + \frac{X_1}{EI}x_2 - \frac{X_2}{EI}\right)dx_2
\end{aligned} \tag{5.35f}
$$

Performing the integrations in Equation 5.35*f* leads to

$$
\begin{aligned}
0 = \frac{l^3}{3EI}(1 + 7\sqrt{2})X_1 - \frac{l^2}{2EI}(1 + 3\sqrt{2})X_2 - \frac{1}{EI}&\left[\left(\frac{3 + 38\sqrt{2}}{24}\right)q_o l^4\right. \\
&\left. + \frac{5\sqrt{2}}{6}Fl^3 + \frac{3\sqrt{2}}{2}\frac{EIl^2\,\alpha \Delta T_o}{h}\right]
\end{aligned} \tag{5.35g}
$$

which is the first compatibility equation. As might be anticipated at this point, the first term in this equation is the transverse displacement at C due to X_1, the second term is the transverse displacement at C due to X_2, and the third term is the transverse displacement at C due to applied loads (including thermal loads). The compatibility condition requires that the sum of these displacements must equal zero.

The second compatibility equation is obtained from a virtual force state "like" the force state in Figure 5.22*c*(3), i.e., $X_{V_2} \equiv X_2$. In this case the internal virtual moments are

$$
\begin{aligned}
M_V(x_1) &= -X_{V_2}, \quad 0 < x_1 < \sqrt{2}l \\
M_V(x_2) &= -X_{V_2}, \quad 0 < x_2 < l
\end{aligned} \tag{5.35h}
$$

and the virtual work equation becomes

$$
0 = \int_0^{\sqrt{2}l} (-X_{V2})\left[\frac{1}{EI}\left(\frac{q_o l}{\sqrt{2}} + \frac{F}{\sqrt{2}} - \frac{X_1}{\sqrt{2}}\right)x_1\right.
$$

$$+ \frac{1}{EI}\left(2lX_1 - X_2 - \frac{3}{2}q_o l^2 - Fl\right) - \frac{\alpha \Delta T_o}{h}\Bigg] dx_1$$
$$+ \int_0^l (-X_{V_2})\left(-\frac{q_o}{2EI}x_2^2 + \frac{X_1}{EI}x_2 - \frac{X_2}{EI}\right)dx_2 \tag{5.35i}$$

Integration leads to

$$0 = -\frac{l^2}{2EI}(1 + 3\sqrt{2})X_1 + \frac{l}{EI}(1 + \sqrt{2})X_2$$
$$+ \frac{1}{EI}\left[(1 + 6\sqrt{2})\frac{q_o l^3}{6} + \frac{\sqrt{2}}{2}Fl^2 + \frac{\sqrt{2}EIl\alpha\Delta T_o}{h}\right] \tag{5.35j}$$

which is the second compatibility equation. The first term in this equation is the rotation at C due to X_1, the second term is the rotation at C due to X_2, and the third term is the rotation at C due to applied loads (including thermal loads). The compatibility condition is that the sum of these rotations must equal zero. Note that the coefficient of X_1 in Equation 5.35*j* is equal to the coefficient of X_2 in Equation 5.35*g*, i.e., these influence coefficients are equal. This becomes even more apparent when Equations 5.35*g* and 5.35*j* are written in matrix notation

$$\frac{1}{EI}\begin{bmatrix} \frac{l^3}{3}(1 + 7\sqrt{2}) & -\frac{l^2}{2}(1 + 3\sqrt{2}) \\ -\frac{l^2}{2}(1 + 3\sqrt{2}) & l(1 + \sqrt{2}) \end{bmatrix}\begin{Bmatrix} X_1 \\ X_2 \end{Bmatrix} =$$
$$\frac{1}{EI}\begin{Bmatrix} \left(\frac{3 + 38\sqrt{2}}{24}\right)q_o l^4 + \frac{5\sqrt{2}}{6}Fl^3 + \frac{3\sqrt{2}}{2}\frac{EIl^2\alpha\Delta T_o}{h} \\ -(1 + 6\sqrt{2})\frac{q_o l^3}{6} - \frac{\sqrt{2}}{2}Fl^2 - \frac{\sqrt{2}EIl\alpha\Delta T_o}{h} \end{Bmatrix} \tag{5.36}$$

In other words, the flexibility matrix is symmetric, a result we observed in connection with truss analysis. Note, however, that the dimensions of the coefficients in the flexibility matrix are different. This is a consequence of our choice of redundant forces with different dimensional units.[28]

---

[28] A computational problem can occur as a result of the difference in dimensions if l is evaluated in a system of units where it is either very large or small numerically. For example, a 30-ft ($\approx$10 m) beam will produce flexibility coefficients with l = 30 ft (10 m), l^2 = 900 ft^2 (100 m^2) and l^3 = 27,000 ft^3 (1000 m^3), and the difference between the largest and smallest coefficients is a factor of approximately 30^2 = 900 (10^2 = 100); if the same beam is measured in inches (centimeters), l = 360 in. ($\approx 10^3$ cm), l^2 = 129,600 in.^{2} (10^6 cm^2) and l^3 = 46,656,000 in.^{3} (10^9 cm^3), and the difference between the largest and smallest coefficients is a factor of approximately 360^2 = 129,600 (10^6). When the coefficients differ numerically by large amounts, computer programs for solving the equations may give very inaccurate results.

Solving Equation 5.36 leads to

$$X_1 = \left(\frac{2 + 7\sqrt{2}}{3 + 14\sqrt{2}}\right)F + \left(\frac{5 + 23\sqrt{2}}{3 + 14\sqrt{2}}\right)\frac{q_o l}{2} + \left(\frac{12\sqrt{2}}{3 + 14\sqrt{2}}\right)\frac{EI\alpha\Delta T_o}{lh}$$

$$= 0.522F + 0.823q_o l + 0.744\frac{EI\alpha\Delta T_o}{lh} \tag{5.37}$$

$$X_2 = \left(\frac{2 + 3\sqrt{2}}{3 + 14\sqrt{2}}\right)Fl + \left(\frac{13 + 37\sqrt{2}}{3 + 14\sqrt{2}}\right)\frac{q_o l^2}{12} - \left(\frac{2 - 5\sqrt{2}}{3 + 14\sqrt{2}}\right)\frac{EI\alpha\Delta T_o}{h}$$

$$= 0.274Fl + 0.239q_o l^2 + 0.222\frac{EI\alpha\Delta T_o}{h}$$

These values of X_1 and X_2 may be substituted into the preceding equations, and the expressions for bending moments and curvatures may be determined for beams AB and BC. Displacements may then be evaluated either by integrating the expressions for curvatures, subject to appropriate boundary conditions (see Section 3.4), or by application of virtual work.

Except for writing the matrix form of the compatibility equations (Equation 5.36), at this point it is not possible to directly express the steps we followed in the foregoing procedure in matrix notation. The basic analysis procedure for frames involves integrals rather than the sets of simultaneous equations we had for trusses and, as we have seen, is fairly cumbersome to apply even for a relatively small frame structure. To develop a matrix formulation that is analogous to that for trusses, we must develop a more formal, systematic matrix form of the preceding formulation. The end result, however, will be a matrix formulation with an overall simplicity comparable to that of the matrix formulation for trusses.

5.3.2 Flexibility Matrix for Beams

As we noted in the previous illustration, the continuous variation in the beam curvature forces us to perform integration over each beam to generate the necessary compatibility equations. Because there is an infinite variability within each beam, at first glance it might seem as if there would be no simple matrix forms that could be devised.

In fact, nearly all of the complexity can be eliminated and the overall behavior of the beam can be represented by a simple 2×2 matrix of the form described in Equation 5.31*b*. However, to accomplish this we will restrict ourselves slightly. In particular, we will write out expressions for moments and deformations on the ends of beam elements. If we choose to do so, we can use more than one beam element to represent a single physical beam, although in most cases this will not be necessary. We will not have to limit the types of mechanical forces (concentrated or continuous) applied between ends of the element, nor will we in any way be forced to limit the behavior of the element with regard to thermal or initial deformations. The end moments and end deformations for individual beam elements will become the primary quantities of interest. These end moments assume the same role as bar forces in the matrix truss analysis procedure and, similarly, the end deformations of the beam are analogous to axial deformations of truss bars. Let's first explain the nature of these end moments and end deformations and develop the 2 × 2 matrix form of the flexibility equations for beam elements.

Consider a typical uniform beam element of a frame structure, such as beam IJ in Figure 5.23*a*.[29] The positive direction of the reference axis (x axis) of the beam element is assumed to be in the direction from I to J, and the length of the element will be denoted by l.[30] Let the beam element be subjected to a general set of intermediate transverse applied loads (including continuous and concentrated loads, applied moments, and thermal and initial loads), and also to a complete set of end forces denoted by M_{IJ}, M_{JI}, V_{IJ}, and V_{JI}.[31] Here, we use the order of the subscripts to specify the end of the beam on which the force component acts, i.e., the first subscript identifies the end. Thus, for example, M_{IJ} denotes a moment at end I of beam IJ, and V_{JI} denotes a transverse force at end J of beam IJ. The end forces usually are internal forces in the frame, which appear as end forces only on the free-body diagram of the isolated beam element. Also, Figure 5.23*a* defines positive directions of all end forces.[32] In particular, clockwise end moments are considered to be positive, and transverse end forces that act in the positive y direction are considered to be positive.

We are going to choose the *end moments* in Figure 5.23*a* as independent force quantities in the formulation that follows.[33] For this reason, we proceed by representing the total force state as the superposition of two particular sets of forces, i.e., Figure 5.23*a*(1), end moments equilibrated by transverse end forces, and Figure 5.23*a*(2)], other applied forces equilibrated by transverse end forces. [Applied thermal and initial "loads" do not produce forces in the statically determinate beam of Figure 5.23*a*(2).] Thus, each transverse end force in Figure 5.23*a* is the sum of a transverse end force in Figure 5.23*a*(1) and a transverse end force in Figure 5.23*a*(2); in other words, $V_{IJ} = V_{IJ_1} + V_{IJ_2}$ and $V_{JI} = V_{JI_1} + V_{JI_2}$. The end moments, on the other hand, appear only in Figure 5.23*a*(1) and not in Figure 5.23*a*(2); therefore, the end moments in Figures 5.23*a*(1) are the actual end moments on the beam (Figure 5.23*a*).

The two transverse end forces in Figure 5.23*a*(1) are related to the end moments by two independent equations of equilibrium

$$V_{IJ_1} = -V_{JI_1} = -\frac{M_{IJ} + M_{JI}}{l} \tag{5.38a}$$

[29] We have designated one end of the element as I and the other end as J, with the understanding that this is representative of any beam such as beam AB or beam BC.

[30] The lengths of beam elements may differ; thus, we could represent length of a general element by l_{IJ}, but this would unnecessarily complicate the appearance of equations. It should henceforth be understood that l is the length of element IJ. Similarly, the quantity EI will represent a particular beam stiffness for element IJ.

[31] The term "forces" is used to denote both transverse forces and moments. Also, axial forces may be present but their effects on deformations are not included at this time; these effects will be included later.

[32] This sign convention for positive beam end forces is consistent with the sign conventions used in Chapters 3 and 4, and is separate and distinct from the sign convention for positive internal bending moment or shear.

[33] Other selections of independent end force quantities are possible (e.g., M_{JI} and V_{JI}), but the choice of end moments is most common. (Some combinations of end forces are not independent quantities, e.g., V_{IJ} and V_{JI}.)

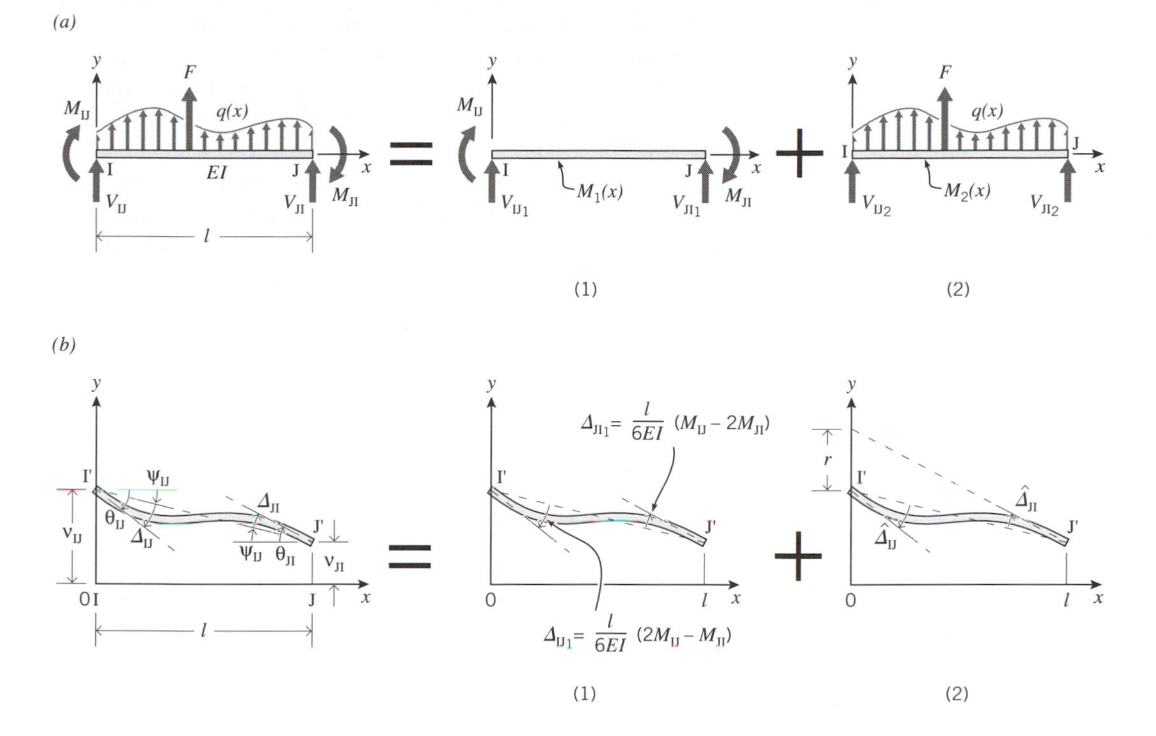

Figure 5.23 (*a*) Forces on beam element; (*b*) Displacements/deformations.

The transverse forces are functions of, i.e., dependent on the end moments in this formulation.

The internal bending moment in beam IJ in Figure 5.23*a*(1) may now be expressed as a function of end moments only,

$$M_1(x) = M_{IJ} + V_{IJ_1}x = M_{IJ}\left(1 - \frac{x}{l}\right) - M_{JI}\left(\frac{x}{l}\right) \tag{5.38b}$$

The internal bending moment throughout beam IJ in Figure 5.23*a*(2) depends on the particular applied mechanical loads; we will denote this bending moment by a general function $M_o(x)$

$$M_2(x) = M_o(x) \tag{5.38c}$$

Because the mechanical loads in Figure 5.23*a*(2) are equilibrated by transverse end forces only, it follows that $M_o(x)$ is exactly the internal bending moment in a simply supported beam. Thus, we expect to be able to determine $M_o(x)$ for very general loadings once they are specified. Again, note that $M_o(x)$ is independent of the end moments.

The total internal bending moment in beam IJ is

$$M(x) = M_1(x) + M_2(x) = M_{IJ}\left(1 - \frac{x}{l}\right) - M_{JI}\left(\frac{x}{l}\right) + M_o(x) \tag{5.38d}$$

The beam in Figure 5.23*a* deforms and displaces as a result of the applied loads.[34] Figure 5.23*b* shows a general representation of the transverse components of displacements[35] and deformations of typical beam IJ. In general, there will be transverse displacements at the I and J ends, denoted by v_{IJ} and v_{JI}, respectively, end rotations θ_{IJ} and θ_{JI}, and a curvature, $v''(x)$. The rotations θ_{IJ} and θ_{JI} are measured with respect to the original orientation of the beam axis. We are also interested in the *end deformations* Δ_{IJ} and Δ_{JI}, which are rotations relative to the displaced beam axis (axis I′J′ in Figure 5.23*b*). As with end moments, we define clockwise end deformations as positive. The end deformations are related to end displacements and end rotations by

$$\Delta_{IJ} = \theta_{IJ} - \left(\frac{v_{IJ} - v_{JI}}{l}\right) = \theta_{IJ} - \psi_{IJ}$$

$$\Delta_{JI} = \theta_{JI} - \left(\frac{v_{IJ} - v_{JI}}{l}\right) = \theta_{JI} - \psi_{IJ} \tag{5.39a}$$

where

$$\psi_{IJ} \equiv \frac{v_{IJ} - v_{JI}}{l} \tag{5.39b}$$

This quantity ψ_{IJ} is a measure of the rigid-body rotation of beam IJ, Figure 5.23*b*.[36]

The beam curvature is given by Equation 3.16*b*; substituting Equation 5.38*d* leads to

$$v''(x) = \frac{1}{EI}[M(x) + M_T(x) + M_I(x)]$$

$$= \frac{M_{IJ}}{EI}\left(1 - \frac{x}{l}\right) - \frac{M_{JI}}{EI}\left(\frac{x}{l}\right) + \frac{1}{EI}[M_o(x) + M_T(x) + M_I(x)] \tag{5.40a}$$

or

$$\boxed{v''(x) = \frac{M_{IJ}}{EI}\left(1 - \frac{x}{l}\right) - \frac{M_{JI}}{EI}\left(\frac{x}{l}\right) + \frac{\widehat{M}(x)}{EI}} \tag{5.40b}$$

where

$$\widehat{M}(x) \equiv M_o(x) + M_T(x) + M_I(x) \tag{5.40c}$$

and where $M_T(x)$ and $M_I(x)$ are thermal and initial pseudomoments, respectively. As is indicated by Equation 5.40*b*, curvature has been expressed explicitly as a function of

---

[34] Recall, for example, the displacements and deformations of beams AB and BC of Example 8, Chapter 3.

[35] Again, any possible axial component of displacement is not shown but may be included by superposition.

[36] We will encounter this terminology again in Chapter 8. Sometimes this quantity is called the side-sway angle.

end moments M_{IJ} and M_{JI}. The remaining quantities in Equation 5.40*b* are the moment $M_o(x)$ due to physical loads applied on the interior of the beam, see Figure 5.23*a*(2), and pseudomoments $M_T(x)$ and $M_I(x)$. The only unknown quantities in Equation 5.40*b* are M_{IJ} and M_{JI}.

To relate the end moments to end deformations, we use virtual work. To compute the end deformation Δ_{IJ}, we choose the virtual force state in Figure 5.24*a*; this virtual force state consists of a virtual moment applied at end I of the beam and equilibrated by transverse forces at each end. The internal virtual moment in the beam is obtained from statics

$$M_V(x) = M_V - \frac{M_V}{l}x = M_V\left(1 - \frac{x}{l}\right) \tag{5.41a}$$

Because the virtual work equation for the beam, Equation 4.17*h*, will be rather lengthy in this case, we will begin with separate expressions for the external and internal virtual work. Corresponding to each of the external virtual forces in Figure 5.24*a* is a real displacement component in Figure 5.23*b*; thus, external virtual work is

$$\begin{aligned} W_{E_V} &= M_V\theta_{IJ} - \left(\frac{M_V}{l}\right)v_{IJ} + \left(\frac{M_V}{l}\right)v_{JI} \\ &= M_V\left[\theta_{IJ} - \left(\frac{v_{IJ} - v_{JI}}{l}\right)\right] \\ &= M_V(\theta_{IJ} - \psi_{IJ}) \\ &= M_V\Delta_{IJ} \end{aligned} \tag{5.41b}$$

The simplification achieved in this expression is somewhat remarkable; in essence, the rigid-body portion of the beam displacement does not contribute to the external virtual work.[37]

[37] Note, also, that the rigid-body portion of the displacement is a linear function of x; therefore, beam curvature, $v''(x)$, will be independent of rigid-body displacement. It follows that internal virtual work, which is a function of beam curvature, must also be independent of rigid-body motion. Thus, rigid-body motion cancels out of the virtual work equation.

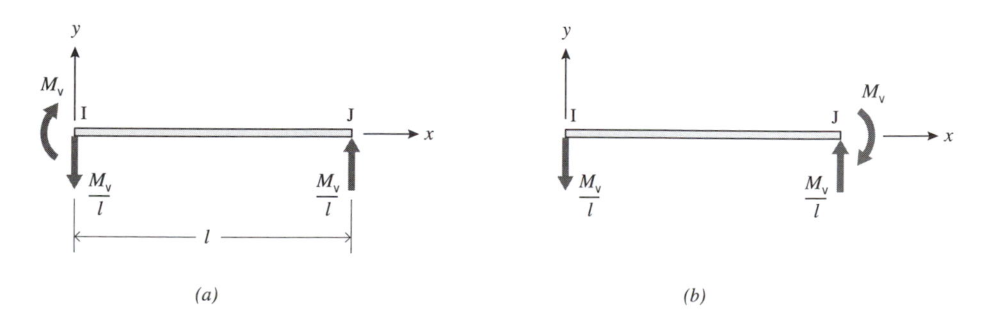

Figure 5.24 (*a*) Virtual forces used to find Δ_{IJ}; (*b*) Virtual forces used to find Δ_{JI}.

The internal virtual work is

$$
\begin{aligned}
W_{I_V} &= \int_0^l \left[M_V\left(1 - \frac{x}{l}\right) \right] v''(x)dx \\
&= \int_0^l \left[M_V\left(1 - \frac{x}{l}\right) \right]\left[\frac{M_{IJ}}{EI}\left(1 - \frac{x}{l}\right) - \frac{M_{JI}}{EI}\left(\frac{x}{l}\right) + \frac{\widehat{M}(x)}{EI} \right]dx \\
&= \frac{M_V}{EI}\int_0^l \left[M_{IJ}\left(1 - \frac{x}{l}\right)^2 - M_{JI}\left(1 - \frac{x}{l}\right)\left(\frac{x}{l}\right) + \widehat{M}(x)\left(1 - \frac{x}{l}\right) \right]dx \qquad \textbf{(5.41c)} \\
&= M_V\left[\frac{l}{3EI}M_{IJ} - \frac{l}{6EI}M_{JI} + \frac{1}{EI}\int_0^l \widehat{M}(x)\left(1 - \frac{x}{l}\right)dx \right] \\
&= M_V\left[\frac{l}{3EI}M_{IJ} - \frac{l}{6EI}M_{JI} + \widehat{\Delta}_{IJ} \right]
\end{aligned}
$$

where

$$\widehat{\Delta}_{IJ} \equiv \frac{1}{EI}\int_0^l \widehat{M}(x)\left(1 - \frac{x}{l}\right)dx \qquad \textbf{(5.41}\boldsymbol{d}\textbf{)}$$

Equating external virtual work, Equation 5.41*b*, and internal virtual work, Equation 5.41*c*, gives

$$\Delta_{IJ} = \frac{l}{3EI}M_{IJ} - \frac{l}{6EI}M_{JI} + \widehat{\Delta}_{IJ} \qquad \textbf{(5.41}\boldsymbol{e}\textbf{)}$$

which is the first of the two desired deformation-moment equations. The quantity $\widehat{\Delta}_{IJ}$ is an unrestrained end deformation that would occur at the I end if the end moments M_{IJ} and M_{JI} were equal to zero; it is analogous to unrestrained bar deformations in trusses when the axial force P is zero.

The second deformation-moment equation is obtained using the virtual force state in Figure 5.24*b*. In this case, the internal virtual moment is

$$M_V(x) = -\frac{M_V}{l}x \qquad \textbf{(5.41}\boldsymbol{f}\textbf{)}$$

We achieve the same type of simplification of the expression for external virtual work as above, and the virtual work equation leads to

$$\Delta_{JI} = -\frac{l}{6EI}M_{IJ} + \frac{l}{3EI}M_{JI} + \widehat{\Delta}_{JI} \qquad \textbf{(5.41}\boldsymbol{g}\textbf{)}$$

where

$$\widehat{\Delta}_{JI} = -\frac{1}{EI}\int_0^l \widehat{M}(x)\frac{x}{l}dx \qquad \textbf{(5.41}\boldsymbol{h}\textbf{)}$$

Equation 5.41g is the second deformation-moment equation; here, again, $\hat{\Delta}_{JI}$ is an unrestrained end deformation.

In matrix notation, the two deformation-moment equations, Equations 5.41e and 5.41g, become

$$\begin{Bmatrix} \Delta_{IJ} \\ \Delta_{JI} \end{Bmatrix} = \frac{l}{6EI}\begin{bmatrix} 2 & -1 \\ -1 & 2 \end{bmatrix}\begin{Bmatrix} M_{IJ} \\ M_{JI} \end{Bmatrix} + \begin{Bmatrix} \hat{\Delta}_{IJ} \\ \hat{\Delta}_{JI} \end{Bmatrix} \tag{5.42a}$$

where

$$\hat{\Delta}_{IJ} \equiv \frac{1}{EI}\int_0^l \hat{M}(x)\left(1 - \frac{x}{l}\right)dx$$
$$\hat{\Delta}_{JI} \equiv -\frac{1}{EI}\int_0^l \hat{M}(x)\frac{x}{l}\,dx \tag{5.42b}$$

and

$$\hat{M}(x) \equiv M_o(x) + M_T(x) + M_I(x) \tag{5.42c}$$

Equation 5.42a is the exact form of the matrix relation previously referred to in general terms in Equation 5.31b. The 2 × 2 matrix, which includes the constant term $l/(6EI)$, is the flexibility matrix for a beam element. Again we observe that the flexibility matrix is symmetric. It should be reemphasized that the only end forces that are explicitly carried in Equation 5.42a are the end moments M_{IJ} and M_{JI}, and that all of the effects of other applied loads (including thermal and initial effects) are contained in the integrals for $\hat{\Delta}_{IJ}$ and $\hat{\Delta}_{JI}$. These integrals are relatively easy to work out, and once we get the $\hat{\Delta}_{IJ}$ and $\hat{\Delta}_{JI}$ for a particular type of loading the result will be applicable to all comparable cases.

The physical significance of $\hat{\Delta}_{IJ}$ and $\hat{\Delta}_{JI}$ is fairly evident from Equation 5.42a. These unrestrained end deformation quantities are the end deformations when M_{IJ} and M_{JI} are zero, i.e., $\hat{\Delta}_{IJ}$ and $\hat{\Delta}_{JI}$ are deformations that are not "restrained" by the independent end moments. In other words, $\hat{\Delta}_{IJ}$ and $\hat{\Delta}_{JI}$ are end deformations due to the applied mechanical loads on the interior of the beam, see Figure 5.23a(2), plus thermal and initial effects.

The unrestrained end deformations are therefore equivalent to rotations at the ends of a simply supported beam due to all loads except M_{IJ} and M_{JI}. The fact that Equations 5.42b define end deformations due to applied loads is also confirmed by the moment-area method, Section 3.4.2. For example, the second part of Equation 5.42b may be rewritten as

$$\hat{\Delta}_{JI} = -\frac{1}{l}\int_0^l \frac{\hat{M}(x)}{EI}x\,dx \tag{5.43}$$

The integral in Equation 5.43 is identical in form to Equation 3.24b. The curvature term $\hat{M}(x)/EI$ is the overall effect of all applied loads on the beam in Figure 5.23a(2).

From the moment-area formulation, we conclude that the integral defines the vertical distance r between tangent lines drawn at the ends of the beam, as shown in Figure 5.23b(2). The distance r divided by the length l is the angle $\widehat{\Delta}_{JI}$ (for small angles); this is exactly the quantity in Equation 5.43. (The negative sign in Equation 5.43 indicates that positive $\widehat{\Delta}_{JI}$ corresponds to a value of r that is measured in the positive y direction.) A similar interpretation applies to the first of Equations 5.42b.

The deformation-force relations in Equations 5.42 enable us to formulate the analysis of frames in a matrix notation that eliminates most (or all) of the integrations inherent in the basic procedure. The only place where integration may be required is in the evaluation of the unrestrained end deformations, Equations 5.42b; even here, however, integration may be simplified using special techniques such as the ones employed in the moment-area method.

5.3.3 Matrix Formulation

The matrix force method for frames utilizes the general sets of equations described in Section 5.2.3. The nature of these equations will be demonstrated by reanalyzing the frame structure that was previously solved using the basic procedure (Figure 5.22).

Step 1. We will choose the same two redundant forces as in Figure 5.22, i.e., the vertical component of reaction (X_1) and the moment (X_2) at C.

Step 2. The equilibrium portion of the matrix analysis procedure requires that the independent end moments be expressed in terms of the redundant forces and the applied loads (just as bar forces were expressed in terms of redundant forces and applied loads for trusses). The independent end moments are identified in Figure 5.25; it is important to keep in mind that the clockwise directions shown for end moments in Figure 5.25 are *defined as positive.* Figures 5.22b contain the equilibrium information for this structure; by comparing Figure 5.25 to Figures 5.22b we find

$$
\begin{aligned}
M_{AB} &= 2lX_1 - X_2 + \left(-Fl - \frac{3q_o l^2}{2}\right) \\
M_{BA} &= -lX_1 + X_2 + \frac{q_o l^2}{2} \\
M_{BC} &= lX_1 - X_2 - \frac{q_o l^2}{2} \\
M_{CB} &= X_2
\end{aligned}
\tag{5.44a}
$$

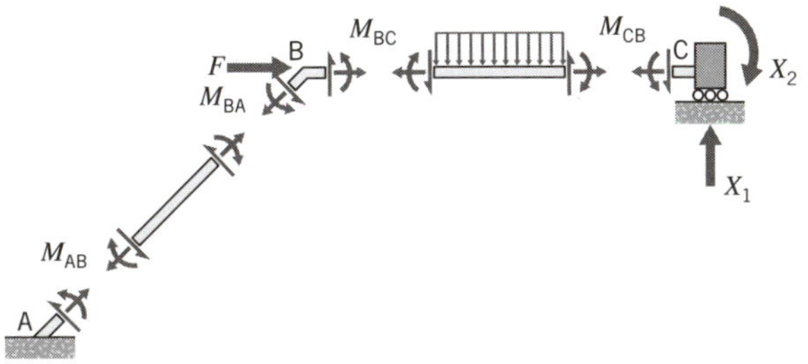

Figure 5.25 End moments in frame.

These equilibrium equations may be expressed in matrix notation as

$$\begin{Bmatrix} M_{AB} \\ M_{BA} \\ M_{BC} \\ M_{CB} \end{Bmatrix} = \begin{bmatrix} 2l & -1 \\ -l & 1 \\ l & -1 \\ 0 & 1 \end{bmatrix} \begin{Bmatrix} X_1 \\ X_2 \end{Bmatrix} + \begin{Bmatrix} -Fl - 3q_o l^2/2 \\ q_o l^2/2 \\ -q_o l^2/2 \\ 0 \end{Bmatrix} \quad \textbf{(5.44}b\textbf{)}$$

which has the basic matrix form (see Equation 5.31*d* or 5.31*e*) for equilibrium equations,

$$\mathbf{P} = \mathbf{bX} + \mathbf{P_o}$$

Note that in Equation 5.44*b* the two independent end moments for each element are grouped together, i.e., the first two rows of Equation 5.44*b* are the end moments for beam AB and the next two rows are the end moments for beam BC.[38] The order of the end moments in each pair of forces is consistent with the order in the deformation-moment relations, Equation 5.42*a*, i.e., for beam AB, $M_{IJ} \leftarrow M_{AB}$ and $M_{JI} \leftarrow M_{BA}$; for beam BC, $M_{IJ} \leftarrow M_{BC}$ and $M_{JI} \leftarrow M_{CB}$. It is also significant to note that it was not necessary to deal with any transverse components of force in writing Equation 5.44*b* because we no longer require the internal bending moments in the beams. (General expressions for the internal bending moments have already been incorporated in the derivation of Equations 5.42.)

As a consequence of our matrix formulation, the equilibrium step in the force method is greatly simplified. All we do now is record the two end moments on each beam element rather than complicated moment expressions.

Step 3. The matrix form of the deformation-moments relations for the structure is obtained by ‘‘assembling’’ the deformation-moments relations for the individual elements, exactly as in Equation 5.31*f*. Let's begin by evaluating the unrestrained end deformations for beam AB and beam BC. (These unrestrained end deformations are the components of vector $\hat{\mathbf{\Delta}}$ in Equation 5.31*f*.)

Beam AB carries only a uniform thermal load $\Delta T = \Delta T_o y/h$; thus, $M_o(x) = M_I(x) = 0$. The thermal load causes a constant curvature $M_T(x)/EI = -\alpha\Delta T_o/h$ (Equation 5.32*b*). We will temporarily denote the length of beam AB by l_{AB}; therefore, from the first of Equations 5.42*b* we obtain

$$\begin{aligned} \hat{\Delta}_{AB} &= \frac{1}{EI}\int_0^{l_{AB}} M_T(x)\left(1 - \frac{x}{l_{AB}}\right)dx \\ &= -\frac{\alpha\Delta T_o}{h}\int_0^{l_{AB}}\left(1 - \frac{x}{l_{AB}}\right)dx \qquad \textbf{(5.44}c\textbf{)} \\ &= -\frac{\alpha\Delta T_o}{2h}l_{AB} \end{aligned}$$

[38] This order is arbitrary, i.e., the first two rows could contain the end moments for beam BC, and the next two rows could contain the end moments for beam AB. It is only necessary to use consistent orders in all equations.

and the second of Equations 5.42*b* gives

$$\widehat{\Delta}_{BA} = -\frac{1}{EI}\int_0^{l_{AB}} M_T(x)\frac{x}{l_{AB}}\,dx$$

$$= \frac{\alpha\Delta T_o}{h}\int_0^{l_{AB}} \frac{x}{l_{AB}}\,dx \tag{5.44d}$$

$$= \frac{\alpha\Delta T_o}{2h}l_{AB}$$

Equations 5.44*c* and 5.44*d* are general in the sense that they are applicable to any uniform beam subjected to this specific thermal load. It is only necessary to substitute the appropriate value of length. For this particular frame $l_{AB} = \sqrt{2}l$, so it follows that

$$\widehat{\Delta}_{AB} = -\sqrt{2}l\alpha\Delta T_o/2h \tag{5.44e}$$

$$\widehat{\Delta}_{BA} = \sqrt{2}l\alpha\Delta T_o/2h$$

The fact that $\widehat{\Delta}_{AB}$ is negative and $\widehat{\Delta}_{BA}$ is positive is consistent with the constant curvature resulting from the thermal load, i.e., the ''unrestrained'' beam deforms as shown in Figure 5.26*a*.

The results in Equations 5.44*e* could have been obtained directly from the moment-area method. For example, Figure 5.26*b* shows the $\widehat{M}/EI$ diagram for the unrestrained beam in Figure 5.26*a*. The first moment of the area of this diagram about $x = l_{AB} = \sqrt{2}l$ gives the vertical distance s; s divided by l_{AB} defines $\widehat{\Delta}_{AB}$.[39]

The unrestrained end deformations for beam BC are determined in a similar manner. In this case, however, there is no thermal (or initial) load; rather, there is a uniform continuous load of magnitude q_o. In accordance with the preceding developments, we regard this continuous load as being equilibrated by transverse end forces only, as in Figure 5.26*c* [or Figure 5.23*a*(2)]. The internal moment in the beam in Figure 5.26*c* is

$$M_o(x) = \frac{q_o l}{2}x - \frac{q_o}{2}x^2 = \frac{q_o}{2}x(l - x) \tag{5.44f}$$

Because $M_T(x) = M_I(x) = 0$, and because $l_{BC} = l$, it follows that Equations 5.42*b* now give

$$\widehat{\Delta}_{BC} = \frac{1}{EI}\int_0^{l}\left[\frac{q_o}{2}x(l - x)\right]\left(1 - \frac{x}{l}\right)dx$$

$$= \frac{q_o l^3}{24EI} \tag{5.44g}$$

[39] Note that both s and r in Figure 5.26*a* are negative quantities, which indicates that the tangents are located above the beam axis, as shown. The end deformations are defined in terms of s and r in the figure.

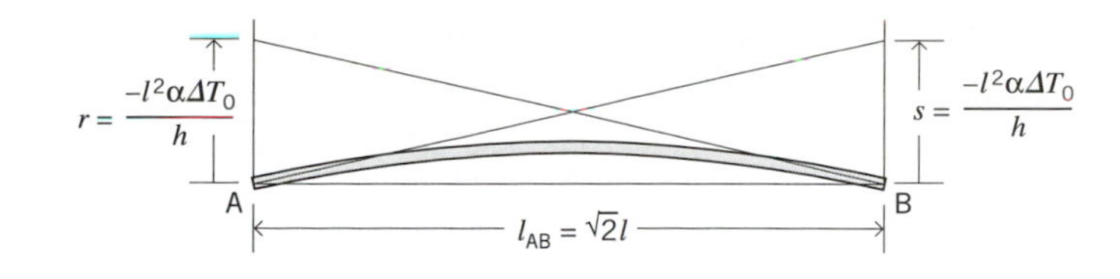

(a) Curvature of beam AB due to thermal load

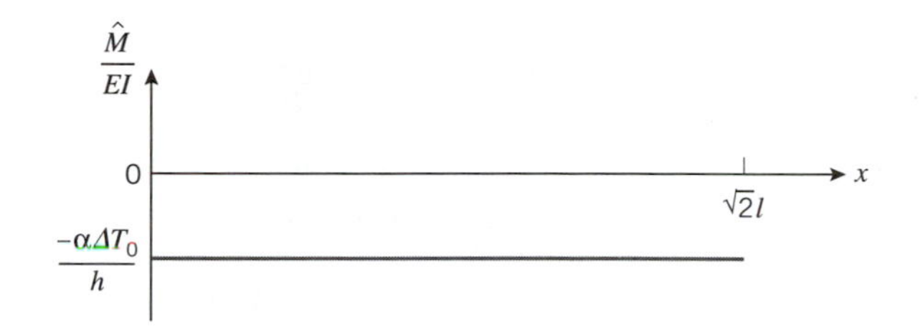

(b) $\hat{M}/EI$ diagram for beam AB

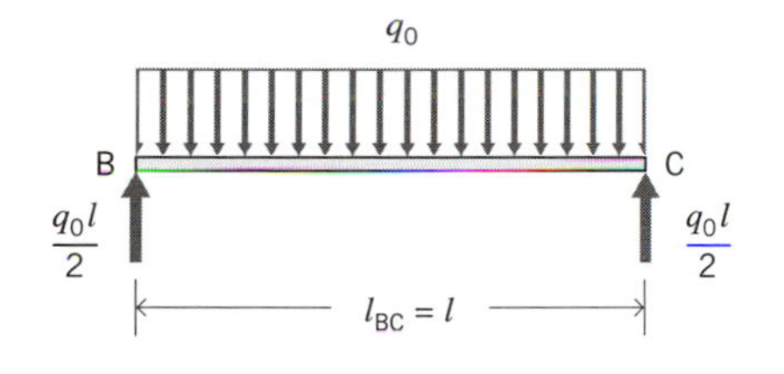

(c) Forces on beam BC for calculation of $\hat{\Delta}_{BC}$ and $\hat{\Delta}_{CB}$

Figure 5.26 Computation of unrestrained end deformations.

$$\widehat{\Delta}_{CB} = -\frac{1}{EI}\int_0^l \left[\frac{q_o}{2}x(l - x)\right]\frac{x}{l}dx$$

$$= -\frac{q_o l^3}{24EI}$$

which, except for the signs, are the same results as in Figure 3.15*c*.[40] These values could also be obtained directly from the moment-area method.

Finally, note that force F is applied directly to a joint, and is therefore not an intermediate load on either beam. Consequently, F does not appear in the $\widehat{\mathbf{\Delta}}$ vector. The matrix form of the deformation-moment relations for the frame can now be written as

[40] The direction of the load in Figure 3.15*c* is opposite to the load here, and there is a distinction between the sign convention for positive end deformations and that for positive slopes.

$$\begin{Bmatrix} \Delta_{AB} \\ \Delta_{BA} \\ \Delta_{BC} \\ \Delta_{CB} \end{Bmatrix} = \frac{l}{6EI} \begin{bmatrix} 2\sqrt{2} & -\sqrt{2} & 0 & 0 \\ -\sqrt{2} & 2\sqrt{2} & 0 & 0 \\ 0 & 0 & 2 & -1 \\ 0 & 0 & -1 & 2 \end{bmatrix} \begin{Bmatrix} M_{AB} \\ M_{BA} \\ M_{BC} \\ M_{CB} \end{Bmatrix} + \begin{Bmatrix} -\sqrt{2}\alpha l \Delta T_o/2h \\ \sqrt{2}\alpha l \Delta T_o/2h \\ q_o l^3/24EI \\ -q_o l^3/24EI \end{Bmatrix} \qquad \textbf{(5.44}h\textbf{)}$$

which has the general matrix form,

$$\mathbf{\Delta} = \mathbf{aP} + \widehat{\mathbf{\Delta}}$$

The first two rows in Equation 5.44h are the deformation-moment relations for beam AB, and the next two rows are the deformation-moment relations for beam BC. The **a** matrix in Equation 5.44h is block-diagonal and symmetric, i.e., the 2×2 symmetric flexibility matrices for the individual beam elements lie along the diagonal of the **a** matrix. Note that the individual beam lengths, $l_{AB} = \sqrt{2}l$ and $l_{BC} = l$, have been incorporated in the **a** matrix. Equation 5.44h is typical of the type of matrix relation defined in Equation 5.31f.

Step 4. The basic matrix form of the compatibility equations is

$$\mathbf{0} = \mathbf{b}^{\mathrm{T}}\mathbf{\Delta}$$

For the frame under consideration, the above equation becomes

$$\begin{Bmatrix} 0 \\ 0 \end{Bmatrix} = \begin{bmatrix} 2l & -l & l & 0 \\ -1 & 1 & -1 & 1 \end{bmatrix} \begin{Bmatrix} \Delta_{AB} \\ \Delta_{BA} \\ \Delta_{BC} \\ \Delta_{CB} \end{Bmatrix} \qquad \textbf{(5.44}i\textbf{)}$$

We write the two equations separately as

$$\begin{aligned} 0 &= 2l\Delta_{AB} - l\Delta_{BA} + l\Delta_{BC} \\ &= l\Delta_{AB} + l(\Delta_{AB} - \Delta_{BA}) + l\Delta_{BC} \end{aligned} \qquad \textbf{(5.44}j\textbf{)}$$

and

$$\begin{aligned} 0 &= -\Delta_{AB} + \Delta_{BA} - \Delta_{BC} + \Delta_{CB} \\ &= (\Delta_{BA} - \Delta_{AB}) + (\Delta_{CB} - \Delta_{BC}) \end{aligned} \qquad \textbf{(5.44}k\textbf{)}$$

In these forms, the compatibility equations are relations that the various end deformations must satisfy to assure that frame displacements are consistent with support conditions. In this regard, the equations are analogous to the relations that must be satisfied by bar deformations in trusses. As with trusses, the significance of these equations may be demonstrated by a direct geometric analysis of the effects of end deformations on the motions of the structure, in this case motions at support C.

Figure 5.27 is a sketch of typical deformations that could occur if the two kinematic constraints at C are removed [as in Figure 5.22a(1)], i.e., if the frame is permitted to undergo a transverse displacement and a rotation at C. Joint B is assumed to displace

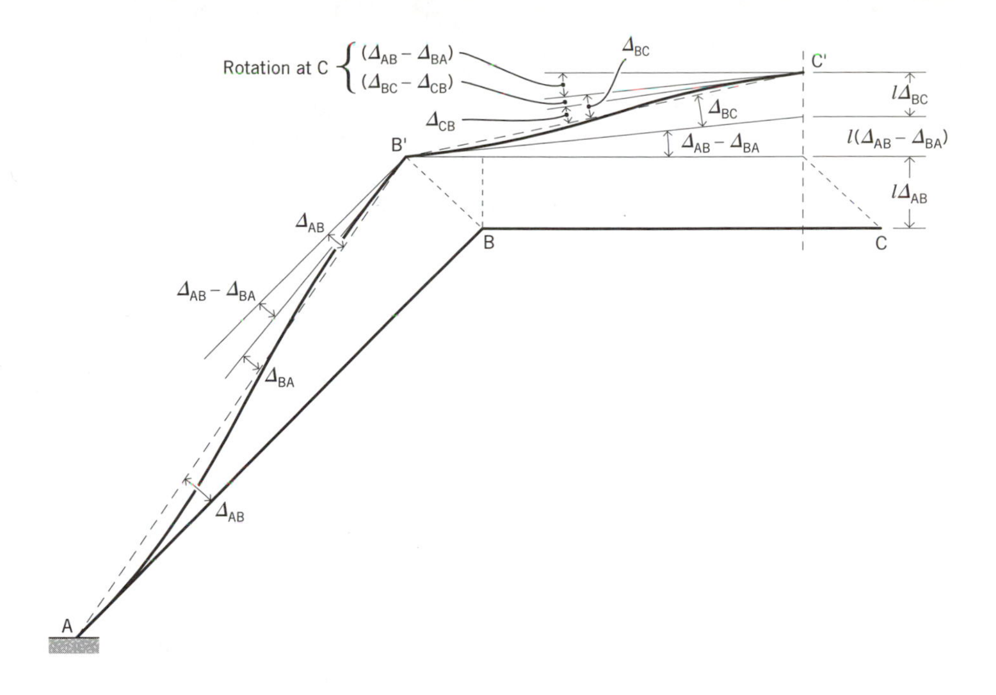

Figure 5.27 Compatibility for frame.

to B′ along a line perpendicular to AB, and end C of beam BC will therefore displace to a point such as C′. The dotted lines A–B′ and B′–C′ show the corresponding rigid-body displacements of the two beam axes. The various end deformations are identified in the figure, as well as several angles and distances that are expressed as functions of these end deformations.

For instance, the length B–B′ is $\Delta_{AB}\sqrt{2}\,l$; it follows that the vertical component of the displacement at B is $\Delta_{AB}l$. Also, because the joint at B is rigid, the rotation at end B of beam AB must be equal to the rotation at end B of beam BC; this particular rotation is $\Delta_{AB} - \Delta_{BA}$ in the figure.

As is illustrated in Figure 5.27, the transverse displacement at C is given by $l\Delta_{AB} + l(\Delta_{AB} - \Delta_{BA}) + l\Delta_{BC}$; this is exactly the quantity that must equal zero according to Equation 5.44*j*. Similarly, the rotation at C is $(\Delta_{AB} - \Delta_{BA}) + (\Delta_{BC} - \Delta_{CB})$, which is exactly the quantity equal to zero in Equation 5.44*k*. In other words, the compatibility equations, even in this form, still provide the requirement that the displacement and rotation at C must be zero.

In subsequent examples we will not need to form the compatibility equations in this step; in general, we will proceed directly to Step 5.

Step 5. The final matrix form of the compatibility equation is given by Equation 5.31*k*,

$$(\mathbf{b}^{\mathrm{T}}\mathbf{a}\mathbf{b})\mathbf{X} = -\mathbf{b}^{\mathrm{T}}(\mathbf{a}\mathbf{P}_{\mathrm{o}} + \hat{\boldsymbol{\Delta}})$$

or

$$\mathbf{A}\mathbf{X} = \mathbf{B}$$

where

$$\mathbf{A} = \mathbf{b}^{\mathrm{T}}\mathbf{a}\mathbf{b}$$
$$\mathbf{B} = -\mathbf{b}^{\mathrm{T}}(\mathbf{a}\mathbf{P}_{\mathbf{o}} + \hat{\boldsymbol{\Delta}})$$

As in the previous truss examples, we will demonstrate the evaluation of the **A** and **B** matrices. We obtain

$$\mathbf{A} = \mathbf{b}^{\mathrm{T}}\mathbf{a}\mathbf{b} = \frac{l}{6EI}\begin{bmatrix} 2l & -l & l & 0 \\ -1 & 1 & -1 & 1 \end{bmatrix}\begin{bmatrix} 2\sqrt{2} & -\sqrt{2} & 0 & 0 \\ -\sqrt{2} & 2\sqrt{2} & 0 & 0 \\ 0 & 0 & 2 & -1 \\ 0 & 0 & -1 & 2 \end{bmatrix}\begin{bmatrix} 2l & -1 \\ -l & 1 \\ l & -1 \\ 0 & 1 \end{bmatrix}$$

$$= \frac{1}{EI}\begin{bmatrix} \frac{l^3}{3}(1 + 7\sqrt{2}) & -\frac{l^2}{2}(1 + 3\sqrt{2}) \\ -\frac{l^2}{2}(1 + 3\sqrt{2}) & l(1 + \sqrt{2}) \end{bmatrix}$$

and

$$\mathbf{B} = -\mathbf{b}^{\mathrm{T}}(\mathbf{a}\mathbf{P}_{\mathrm{o}} + \hat{\boldsymbol{\Delta}})$$

$$= -\begin{bmatrix} 2l & -l & l & 0 \\ -1 & 1 & -1 & 1 \end{bmatrix}\left\{\frac{l}{6EI}\begin{bmatrix} 2\sqrt{2} & -\sqrt{2} & 0 & 0 \\ -\sqrt{2} & 2\sqrt{2} & 0 & 0 \\ 0 & 0 & 2 & -1 \\ 0 & 0 & -1 & 2 \end{bmatrix}\begin{Bmatrix} -Fl - 3q_o l^2/2 \\ q_o l^2/2 \\ -q_o l^2/2 \\ 0 \end{Bmatrix} + \begin{Bmatrix} -\sqrt{2}\, l\alpha\Delta T_o/2h \\ \sqrt{2}\, l\alpha\Delta T_o/2h \\ q_o l^3/24EI \\ -q_o l^3/24EI \end{Bmatrix}\right\}$$

$$= \frac{1}{EI}\begin{Bmatrix} \left(\frac{3 + 38\sqrt{2}}{24}\right)q_o l^4 + \frac{5\sqrt{2}}{6}Fl^3 + \frac{3\sqrt{2}\,EIl^2\alpha\Delta T_o}{2h} \\ -\left(\frac{1 + 6\sqrt{2}}{6}\right)q_o l^3 - \frac{\sqrt{2}}{2}Fl^2 - \frac{\sqrt{2}\,EIl\alpha\Delta T_o}{h} \end{Bmatrix}$$

These **A** and **B** matrices are identical to those in Equation 5.36. We will repeat the results here but, for the sake of conciseness, we show numerical values of the algebraic constants. Thus, the compatibility equation is

$$\frac{1}{EI}\begin{bmatrix} 3.633\,l^3 & -2.621\,l^2 \\ -2.621\,l^2 & 2.414\,l \end{bmatrix}\begin{Bmatrix} X_1 \\ X_2 \end{Bmatrix}$$
$$= \frac{1}{EI}\begin{Bmatrix} 2.364\,q_o l^4 + 1.179\,Fl^3 + 2.121\,EIl^2\alpha\Delta T_o/h \\ -1.581\,q_o l^3 - 0.707\,Fl^2 - 1.414\,EIl\alpha\Delta T_o/h \end{Bmatrix} \qquad \textbf{(5.44}\boldsymbol{l}\textbf{)}$$

Inverting leads to

$$\begin{Bmatrix} X_1 \\ X_2 \end{Bmatrix} = \begin{bmatrix} 1.271/l^3 & 1.380/l^2 \\ 1.380/l^2 & 1.912/l \end{bmatrix} \begin{Bmatrix} 2.364q_ol^4 + 1.179Fl^3 + 2.121EIl^2\alpha\Delta T_o/h \\ -1.581q_ol^3 - 0.707Fl^2 - 1.414EIl\alpha\Delta T_o/h \end{Bmatrix}$$

$$= \begin{Bmatrix} 0.823q_ol + 0.522F + 0.744EI\alpha\Delta T_o/lh \\ 0.239q_ol^2 + 0.274Fl + 0.222EI\alpha\Delta T_o/h \end{Bmatrix} \qquad \textbf{(5.44}\boldsymbol{m}\textbf{)}$$

Substituting the results back into Equations 5.44*b* and 5.44*h* provides

$$\begin{Bmatrix} M_{AB} \\ M_{BA} \\ M_{BC} \\ M_{CB} \end{Bmatrix} = \begin{Bmatrix} -0.093q_ol^2 - 0.230Fl + 1.266EI\alpha\Delta T_o/h \\ -0.084q_ol^2 - 0.248Fl - 0.522EI\alpha\Delta T_o/h \\ 0.084q_ol^2 + 0.248Fl + 0.522EI\alpha\Delta T_o/h \\ 0.239q_ol^2 + 0.274Fl + 0.222EI\alpha\Delta T_o/h \end{Bmatrix} \qquad \textbf{(5.44}\boldsymbol{n}\textbf{)}$$

and

$$\begin{Bmatrix} \Delta_{AB} \\ \Delta_{BA} \\ \Delta_{BC} \\ \Delta_{CB} \end{Bmatrix} = \frac{1}{EI} \begin{Bmatrix} -0.024q_ol^3 - 0.050Fl^2 + 0.013EIl\alpha\Delta T_o/h \\ -0.018q_ol^3 - 0.063Fl^2 + 0.163EIl\alpha\Delta T_o/h \\ 0.030q_ol^3 + 0.037Fl^2 + 0.137EIl\alpha\Delta T_o/h \\ 0.024q_ol^3 + 0.050Fl^2 - 0.013EIl\alpha\Delta T_o/h \end{Bmatrix} \qquad \textbf{(5.44}\boldsymbol{o}\textbf{)}$$

Once the end moments are known, the analysis is essentially complete; the internal forces can be evaluated everywhere in the structure (e.g., Equation 5.38*d*), and the displacements and deformations can be determined using the procedures presented in Chapters 3 or 4.

The computational effort involved in obtaining this matrix solution was considerably less than that required by the basic procedure presented in Equations 5.35. Both approaches require statical analyses of the structure, but the matrix method circumvents the extensive integrations required by the basic procedure. The major computational burden in this approach consists of performing the matrix multiplications to obtain the **A** and **B** matrices, a process that can be automated once the **a**, **b**, $\mathbf{P}_o$ and $\widehat{\boldsymbol{\Delta}}$ matrices have been established.

EXAMPLE 4

We will reanalyze the preceding frame using the choice of redundant forces shown in Figure 5.17*b*. Loads and dimensions will be assumed to be the same as in the previous illustration.

Step 1. We will call the moment at support C redundant force X_1 and the internal moment at joint B redundant force X_2, as shown in Figure 5.28*a*. As usual, the directions in which the redundant forces are assumed to act are chosen arbitrarily.

Step 2. The results of the statical portion of the analysis are shown in Figure 5.28*b*.

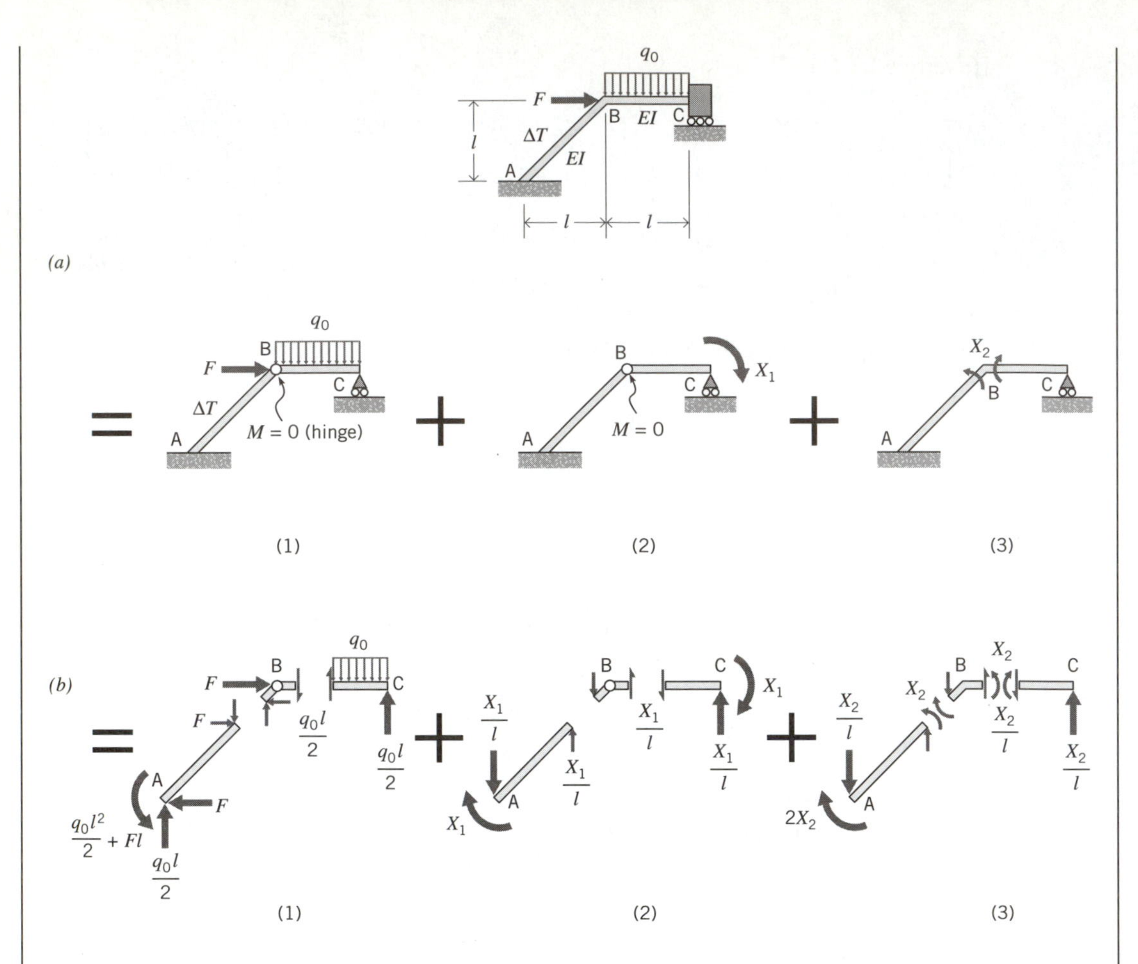

Figure 5.28 Statical analysis of two-degree redundant frame.

Note, in particular, that the redundant moment at end B of beam AB is equal and opposite to the redundant moment at end B of beam BC [Figure 5.28*b*(3)]. As a consequence, one of these end moments is clockwise and the other is counterclockwise, i.e., one is a positive end moment and the other is a negative end moment according to our sign convention in which clockwise end moments are positive. Positive directions of all independent end moments are the same as previously shown in Figure 5.25.

By comparing Figure 5.25 to Figures 5.28*b*(1)–(3), we are able to write the matrix form of the equilibrium equations

$$\begin{Bmatrix} M_{AB} \\ M_{BA} \\ M_{BC} \\ M_{CB} \end{Bmatrix} = \begin{bmatrix} 1 & 2 \\ 0 & -1 \\ 0 & 1 \\ 1 & 0 \end{bmatrix} \begin{Bmatrix} X_1 \\ X_2 \end{Bmatrix} + \begin{Bmatrix} -q_o l^2/2 - Fl \\ 0 \\ 0 \\ 0 \end{Bmatrix} \tag{5.45a}$$

Step 3. Because the order of the end forces in the **P** vector is the same as in the previous illustration we can use the same deformation-moment relations, Equation 5.44*h*,

$$\begin{Bmatrix} \Delta_{AB} \\ \Delta_{BA} \\ \Delta_{BC} \\ \Delta_{CB} \end{Bmatrix} = \frac{l}{6EI}\begin{bmatrix} 2\sqrt{2} & -\sqrt{2} & 0 & 0 \\ -\sqrt{2} & 2\sqrt{2} & 0 & 0 \\ 0 & 0 & 2 & -1 \\ 0 & 0 & -1 & 2 \end{bmatrix}\begin{Bmatrix} M_{AB} \\ M_{BA} \\ M_{BC} \\ M_{CB} \end{Bmatrix} + \begin{Bmatrix} -\sqrt{2}\alpha l\Delta T_o/2h \\ \sqrt{2}\alpha l\Delta T_o/2h \\ q_o l^3/24EI \\ -q_o l^3/24EI \end{Bmatrix} \tag{5.45b}$$

We now have all the matrices that appear in the compatibility equation, so there is no need for any further analysis to complete Step 4; we proceed directly to Step 5.

Step 5. The **A** and **B** matrices are

$$\mathbf{A} = \mathbf{b}^{\mathrm{T}}\mathbf{ab} = \frac{l}{6EI}\begin{bmatrix} 1 & 0 & 0 & 1 \\ 2 & -1 & 1 & 0 \end{bmatrix}\begin{bmatrix} 2\sqrt{2} & -\sqrt{2} & 0 & 0 \\ -\sqrt{2} & 2\sqrt{2} & 0 & 0 \\ 0 & 0 & 2 & -1 \\ 0 & 0 & -1 & 2 \end{bmatrix}\begin{bmatrix} 1 & 2 \\ 0 & -1 \\ 0 & 1 \\ 1 & 0 \end{bmatrix} \tag{5.45c}$$

$$= \frac{l}{6EI}\begin{bmatrix} 2(1+\sqrt{2}) & (5\sqrt{2}-1) \\ (5\sqrt{2}-1) & 2(1+7\sqrt{2}) \end{bmatrix}$$

$$= \frac{l}{EI}\begin{bmatrix} 0.805 & 1.012 \\ 1.012 & 3.633 \end{bmatrix}$$

$$\mathbf{B} = -\mathbf{b}^{\mathrm{T}}(\mathbf{aP_o} + \hat{\mathbf{\Delta}})$$

$$= -\begin{bmatrix} 1 & 0 & 0 & 1 \\ 2 & -1 & 1 & 0 \end{bmatrix}\left[\frac{l}{6EI}\begin{bmatrix} 2\sqrt{2} & -\sqrt{2} & 0 & 0 \\ -\sqrt{2} & 2\sqrt{2} & 0 & 0 \\ 0 & 0 & 2 & -1 \\ 0 & 0 & -1 & 2 \end{bmatrix}\begin{Bmatrix} -q_o l^2/2 - Fl \\ 0 \\ 0 \\ 0 \end{Bmatrix} + \begin{Bmatrix} -\sqrt{2}l\alpha\Delta T_o/2h \\ \sqrt{2}l\alpha\Delta T_o/2h \\ q_o l^3/24EI \\ -q_o l^3/24EI \end{Bmatrix}\right]$$

$$= \frac{l}{EI}\begin{Bmatrix} \left(\dfrac{1+4\sqrt{2}}{24}\right)q_o l^2 + \dfrac{\sqrt{2}}{3}Fl + \dfrac{\sqrt{2}EI\alpha\Delta T_o}{2h} \\ -\left(\dfrac{1-10\sqrt{2}}{24}\right)q_o l^2 + \dfrac{5\sqrt{2}}{6}Fl + \dfrac{3\sqrt{2}EI\alpha\Delta T_o}{2h} \end{Bmatrix} \tag{5.45d}$$

$$= \frac{l}{EI}\begin{Bmatrix} 0.277q_o l^2 + 0.471Fl + 0.707EI\alpha\Delta T_o/h \\ 0.548q_o l^2 + 1.179Fl + 2.121EI\alpha\Delta T_o/h \end{Bmatrix}$$

Therefore, the compatibility equation is

$$\frac{l}{EI}\begin{bmatrix} 0.805 & 1.012 \\ 1.012 & 3.633 \end{bmatrix}\begin{Bmatrix} X_1 \\ X_2 \end{Bmatrix} = \frac{l}{EI}\begin{Bmatrix} 0.277q_o l^2 + 0.471Fl + 0.707EI\alpha\Delta T_o/h \\ 0.548q_o l^2 + 1.179Fl + 2.121EI\alpha\Delta T_o/h \end{Bmatrix} \tag{5.45e}$$

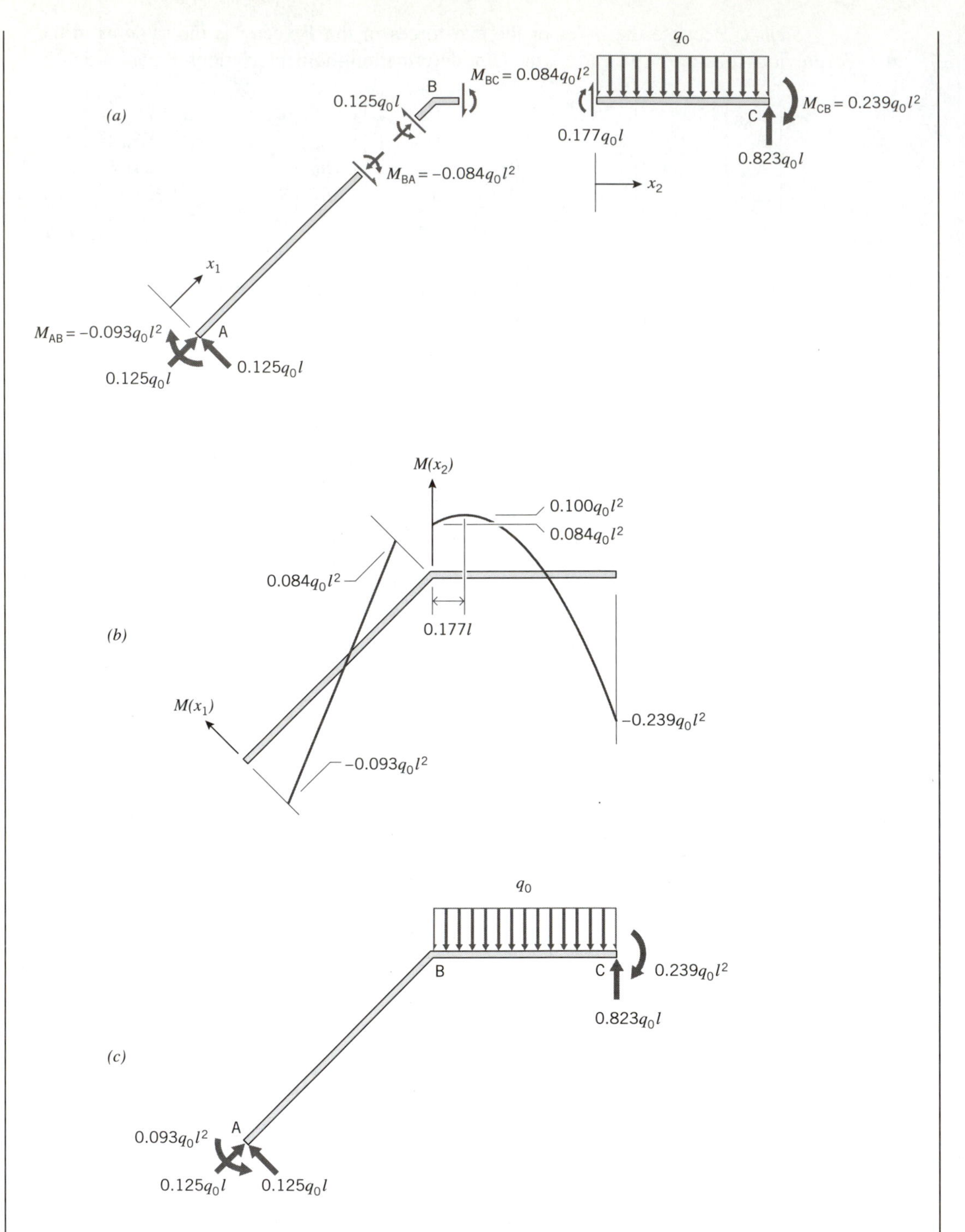

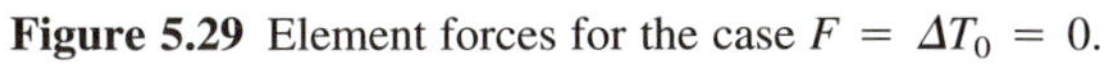
Figure 5.29 Element forces for the case $F = \Delta T_0 = 0$.

Inverting leads to

$$\begin{Bmatrix} X_1 \\ X_2 \end{Bmatrix} = \begin{bmatrix} 1.912 & -0.533 \\ -0.533 & 0.424 \end{bmatrix} \begin{Bmatrix} 0.277q_o l^2 + 0.471Fl + 0.707EI\alpha\Delta T_o/h \\ 0.548q_o l^2 + 1.179Fl + 2.121EI\alpha\Delta T_o/h \end{Bmatrix}$$
$$= \begin{Bmatrix} 0.239q_o l^2 + 0.274Fl + 0.222EI\alpha\Delta T_o/h \\ 0.084q_o l^2 + 0.248Fl + 0.522EI\alpha\Delta T_o/h \end{Bmatrix} \qquad \textbf{(5.45}f\textbf{)}$$

As expected, the value of X_1 in Equation 5.45*f* is identical to the value of X_2 in Equation 5.44*m*, and the value of X_2 in Equation 5.45*f* is the same as $-M_{BA}$ or $+M_{BC}$ in Equation 5.44*n*. Back-substitution would lead to exactly the same results for all other quantities in Equations 5.44*n* and 5.44*o*.

At the completion of any analysis of this type, it is important to determine internal forces in the various components of the structure. To accomplish this, we first sketch the free-body diagrams of the components. The end forces acting on the components can be obtained in either of two equivalent ways: either by superposition of the quantities shown in Figures 5.28*b*(1)–5.28*b*(3) or directly from the end moments given in Equation 5.44*n* combined with equations of statics. We will demonstrate the latter procedure.

To simplify the subsequent presentation of results in this demonstration, we will consider the special case for which the only load on the structure is q_o applied to beam BC, i.e., we set $F = \Delta T_o = 0$. Figure 5.29*a* shows the resulting element end forces for this case. As stated, the end moments are obtained from the solution in Equation 5.44*n*. The transverse end forces on beams AB and BC are computed from equilibrium equations for each free body. We note that transverse forces that are equal and opposite to those at the B ends of elements AB and BC act on joint B. The axial component of force in beam AB is determined from equilibrium of forces on joint B. (The axial force in beam BC is zero because of the roller support at C.)

It is also useful to sketch bending moment diagrams for the elements of the structure as an aid in determining extreme values. The moment diagrams for beams AB and BC are sketched in Figure 5.29*b* directly on a drawing of the structure; this type of presentation is common practice for complex structural configurations, but it is important to specify clearly the individual element coordinate systems to properly interpret sign conventions. The sign convention employed in Figure 5.29*b* is consistent with the one established in Chapter 3. We see from Figure 5.29*b* that the maximum positive bending moment in beam BC occurs at $x_2 = 0.177l$ and is equal to $0.100\ q_o l^2$.

As a final check of the results, we sketch the free-body diagram for the entire structure, Figure 5.29*c*. It may be verified that the forces acting on the structure satisfy all equations of equilibrium.

5.3.4 Combined Bending and Axial Behavior

Up to this point we have not considered length changes caused by axial forces in beam elements of frame structures, i.e., we have tacitly assumed that beam elements are axially rigid. For example, beam AB in the frame in Figure 5.28 transmits an axial force in addition to transverse forces and bending moments, but we have ignored the deformation caused by this axial force in our analysis. In fact, an axial deformation of this type will have an effect on the distribution of forces in an indeterminate structure but, for most realistic frames, the effect will be small. In the following example we will examine the effect of axial deformations on the frame in Figure 5.28. We will see that the analysis requires no new concepts and is quite straightforward.

EXAMPLE 5

The representation of combined bending and axial behavior is based on superposition of our deformation-force relations for beams and bars.[41] Recall that the deformation-force relation for truss bars, Equation 5.31*a*, assumes constant axial force; this relation may therefore be used directly in a deformation-force relation for a "combined" element as long as the axial force in the element is constant.[42] The matrix form of the resulting deformation-force relation for a combined element is

$$\begin{Bmatrix} \Delta_{IJ} \\ \Delta_{JI} \\ \Delta_{IJ}^{(a)} \end{Bmatrix} = \frac{l}{6EI}\begin{bmatrix} 2 & -1 & 0 \\ -1 & 2 & 0 \\ 0 & 0 & 6I/A \end{bmatrix}\begin{Bmatrix} M_{IJ} \\ M_{JI} \\ P_{IJ} \end{Bmatrix} + \begin{Bmatrix} \widehat{\Delta}_{IJ} \\ \widehat{\Delta}_{IJ} \\ \widehat{\Delta}_{IJ}^{(a)} \end{Bmatrix} \tag{5.46}$$

The first two rows of Equation 5.46 are identical to Equation 5.42*a* and the last row is identical to Equation 5.31*a*. To avoid confusion between beam bending deformations and axial deformations, the latter has been redesignated $\Delta_{IJ}^{(a)}$. Thus, the combined deformation-force relations involve three independent deformations (two bending deformations and an axial length change) and three independent forces (two end moments and an axial force). All force and deformation matrices used in the analysis procedure must contain the same sets of quantities for each beam-truss element in the structure. We will now demonstrate the procedure for the frame in Figure 5.28. We assume equal cross-sectional areas for both elements.

Step 1. An interesting fact is that the degree of indeterminacy of the frame is unaffected by our consideration of axial deformations. (The degree of indeterminacy depends on the number of forces only.) Thus, we will use the two redundant forces identified in Figure 5.28*a*.

Step 2. Figure 5.30 shows the same forces as Figure 5.28*b* except that the horizontal and vertical components of forces have been resolved into axial and transverse components. (We require only the axial components.) Note that the axial force in element

---

[41] Such superposition assumes that the two types of behavior are uncoupled, i.e., we assume that an axial force has no effect on the bending moments.

[42] A general deformation-force relation for a bar with variable axial forces will be developed in the next section. The relation will have the same form as Equation 5.31*a*, but P will be an independent force at one of the ends and $\widehat{\Delta}$ will be defined by an integral.

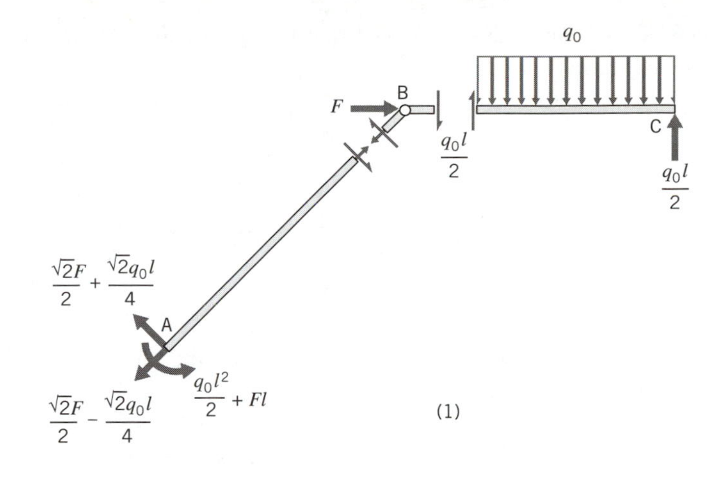

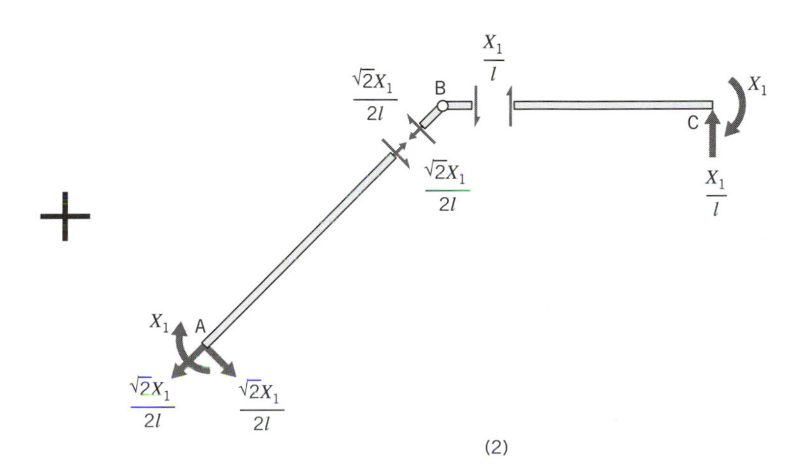

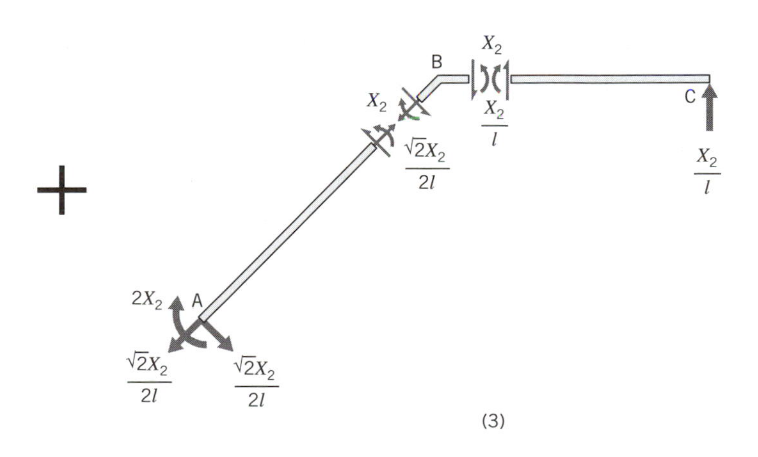

Figure 5.30 Statical analysis of two-degree redundant frame.

AB is constant in this case. The matrix form of the equilibrium equations, including the axial forces, is

$$\begin{Bmatrix} M_{AB} \\ M_{BA} \\ P_{AB} \\ M_{BC} \\ M_{CB} \\ P_{BC} \end{Bmatrix} = \begin{bmatrix} 1 & 2 \\ 0 & -1 \\ \sqrt{2}/2l & \sqrt{2}/2l \\ 0 & 1 \\ 1 & 0 \\ 0 & 0 \end{bmatrix} \begin{Bmatrix} X_1 \\ X_2 \end{Bmatrix} + \begin{Bmatrix} -q_o l^2/2 - Fl \\ 0 \\ -\sqrt{2}q_o l/4 + \sqrt{2}F/2 \\ 0 \\ 0 \\ 0 \end{Bmatrix} \quad \textbf{(5.47a)}$$

Here, the first three terms in the **P** vector are the independent forces in element AB, and the last three terms are the independent forces in element BC. As with truss bars, axial tensile forces are denoted as positive. We have included an axial force term for element BC even though we observe that this axial force is zero.[43]

Step 3. The deformation-force relation for the structure is similar to Equation 5.35*b* except that axial terms are added

$$\begin{Bmatrix} \Delta_{AB} \\ \Delta_{BA} \\ \Delta_{AB}^{(a)} \\ \Delta_{BC} \\ \Delta_{CB} \\ \Delta_{BC}^{(a)} \end{Bmatrix} = \frac{l}{6EI} \begin{bmatrix} 2\sqrt{2} & -\sqrt{2} & & & & \\ -\sqrt{2} & 2\sqrt{2} & & & & \\ & & 6\sqrt{2}I/A & & & \\ & & & 2 & -1 & \\ & & & -1 & 2 & \\ & & & & & 6I/A \end{bmatrix} \begin{Bmatrix} M_{AB} \\ M_{BA} \\ P_{AB} \\ M_{BC} \\ M_{CB} \\ P_{BC} \end{Bmatrix} + \begin{Bmatrix} -\sqrt{2}\alpha l \Delta T_o/2h \\ \sqrt{2}\alpha l \Delta T_o/2h \\ 0 \\ q_o l^3/24EI \\ -q_o l^3/24EI \\ 0 \end{Bmatrix} \quad \textbf{(5.47b)}$$

Here, for the sake of appearance, we have omitted all zeroes in the **a** matrix with the understanding that the matrix is block-diagonal.[44] Again, the first three rows contain the deformation-force relation for element AB in the same order as the terms in Equation 5.46, and the next three rows contain the deformation-force relation for element BC. Note that the thermal load in element AB does not cause any length change ($P_T = 0$, Equation 2.36*c*); consequently, the unrestrained end deformations in the $\widehat{\mathbf{\Delta}}$ matrix have values identical to those in the previous case.

Now that we have all the required matrices we may proceed to Step 5.

Step 5. For this structure with our choice of redundants the **A** matrix (flexibility matrix) is

$$\mathbf{A} = \mathbf{b}^{\mathrm{T}}\mathbf{a}\mathbf{b} = \frac{l}{6EI} \begin{bmatrix} 2(1 + \sqrt{2}) + 3\sqrt{2}\dfrac{I}{l^2 A} & (5\sqrt{2} - 1) + 3\sqrt{2}\dfrac{I}{l^2 A} \\ (5\sqrt{2} - 1) + 3\sqrt{2}\dfrac{I}{l^2 A} & 2(1 + 7\sqrt{2}) + 3\sqrt{2}\dfrac{I}{l^2 A} \end{bmatrix} \quad \textbf{(5.47c)}$$

---

[43] Because the axial force is zero, there will be no axial deformation of element BC; thus, we could have chosen to model element BC as an axially rigid beam.

[44] We will follow this practice in all subsequent examples involving large block-diagonal matrices containing many zeroes.

and the **B** matrix is

$$\mathbf{B} = -\mathbf{b}^{\mathrm{T}}(\mathbf{aP_o} + \hat{\boldsymbol{\Delta}})$$

$$= \frac{l}{EI}\begin{Bmatrix} \dfrac{q_o l^2}{24}\left(1 + 4\sqrt{2} + 6\sqrt{2}\dfrac{I}{l^2A}\right) - \sqrt{2}Fl\left(-\dfrac{1}{3} + \dfrac{I}{2l^2A}\right) + \dfrac{\sqrt{2}EI\alpha\Delta T_o}{2h} \\ -\dfrac{q_o l^2}{24}\left(1 - 10\sqrt{2} - 6\sqrt{2}\dfrac{I}{l^2A}\right) + \sqrt{2}Fl\left(\dfrac{5}{6} - \dfrac{I}{2l^2A}\right) + \dfrac{3\sqrt{2}EI\alpha\Delta T_o}{2h} \end{Bmatrix} \quad \textbf{(5.47}\textit{d}\textbf{)}$$

We observe that these matrices are identical to the previous expressions in Equations 5.45*c* and 5.45*d* except for some additional terms involving the nondimensional parameter I/l^2A. Evidently, this parameter reflects the effects of the axial deformation of element AB on the behavior of the structure. For real structures I/l^2A will usually be no greater than (and probably much less than) approximately 0.001, and the terms involving I/l^2A can therefore be expected to be small in comparison with other terms in the matrices. (The matrices approach the previous expressions as the magnitude of I/l^2A approaches zero.) Thus, axial deformations will generally not have a major impact on behavior. We will demonstrate this point by completing the solution.

For convenience, let $K = I/l^2A$; the compatibility equation now is

$$\frac{l}{EI}\begin{bmatrix} (0.805 + 0.707K) & (1.012 + 0.707K) \\ (1.012 + 0.707K) & (3.633 + 0.707K) \end{bmatrix}\begin{Bmatrix} X_1 \\ X_2 \end{Bmatrix} =$$

$$\frac{l}{EI}\begin{Bmatrix} (0.277 + 0.354K)q_o l^2 + (0.471 - 0.707K)Fl + 0.707\,EI\alpha\Delta T_o/h \\ (0.548 + 0.354K)q_o l^2 + (1.179 - 0.707K)Fl + 2.121\,EI\alpha\Delta T_o/h \end{Bmatrix} \quad \textbf{(5.47}\textit{e}\textbf{)}$$

For illustrative purposes, we will solve this equation for a relatively large value of K, specifically $K = 0.01$, which is approximately ten times the maximum expected value. The result is

$$\begin{Bmatrix} X_1 \\ X_2 \end{Bmatrix} = \begin{bmatrix} 1.899 & -0.532 \\ -0.532 & 0.423 \end{bmatrix}\begin{Bmatrix} 0.281\,q_o l^2 + 0.464\,Fl + 0.707\,EI\alpha\Delta T_o/h \\ 0.551\,q_o l^2 + 1.171\,Fl + 2.121\,EI\alpha\Delta T_o/h \end{Bmatrix} \quad \textbf{(5.47}\textit{f}\textbf{)}$$

$$= \begin{Bmatrix} 0.240\,q_o l^2 + 0.259\,Fl + 0.215\,EI\alpha\Delta T_o/h \\ 0.084\,q_o l^2 + 0.249\,Fl + 0.522\,EI\alpha\Delta T_o/h \end{Bmatrix}$$

Back-substituting leads to

$$\begin{Bmatrix} M_{AB} \\ M_{BA} \\ P_{AB} \\ M_{BC} \\ M_{CB} \\ P_{BC} \end{Bmatrix} = \begin{Bmatrix} -0.091\,q_o l^2 - 0.242\,Fl + 1.260\,EI\alpha\Delta T_o/h \\ -0.084\,q_o l^2 - 0.249\,Fl - 0.522\,EI\alpha\Delta T_o/h \\ -0.124\,q_o l + 1.067\,F + 0.522\,EI\alpha\Delta T_o/lh \\ 0.084\,q_o l^2 + 0.249\,Fl + 0.522\,EI\alpha\Delta T_o/h \\ 0.240\,q_o l^2 + 0.259\,Fl + 0.215\,EI\alpha\Delta T_o/h \\ 0 \end{Bmatrix} \quad \textbf{(5.47}\textit{g}\textbf{)}$$

and

$$\begin{Bmatrix} \Delta_{AB} \\ \Delta_{BA} \\ \Delta_{AB}^{(a)} \\ \Delta_{BC} \\ \Delta_{CB} \\ \Delta_{BC}^{(a)} \end{Bmatrix} = \frac{1}{EI} \begin{Bmatrix} -0.023 q_o l^3 - 0.055 Fl^2 + 0.010 EIl\alpha\Delta T_o/h \\ -0.018 q_o l^3 - 0.060 Fl^2 + 0.164 EIl\alpha\Delta T_o/h \\ -0.002 q_o l^4 + 0.015 Fl^3 + 0.007 EIl^2\alpha\Delta T_o/h \\ 0.030 q_o l^3 + 0.040 Fl^2 + 0.138 EIl\alpha\Delta T_o/h \\ 0.024 q_o l^3 + 0.045 Fl^2 - 0.015 EIl\alpha\Delta T_o/h \\ 0 \end{Bmatrix} \quad \textbf{(5.47h)}$$

Even for this relatively large value of K the maximum difference between the forces in Equation 5.47g and those in Equation 5.44n is only about 5%. Some of the deformations differ from 10%–15%, but this is usually a matter of less concern.[45]

5.3.5 General Axial Behavior

As has been noted, the preceding example involved elements with constant axial forces, the effects of which were included exactly as for idealized truss bars. There are many situations, however, in which axial forces are not constant. The ability to deal with these situations requires a deformation-force relation for bars subjected to variable axial forces. The development is similar to that of the deformation-force relation for beams.

Consider the uniform bar IJ of length l in Figure 5.31a that is subjected to a general set of axial loads, including a continuous load $p(x)$, and end forces P_{IJ} and P_{JI}, which are defined as positive in the directions shown.

We must choose one of the end forces, say P_{JI}, as an independent end force in the formulation. We then represent the total force state as the superposition of two sets of forces, i.e., Figure 5.31a(1), the independent end force in equilibrium with the force P_{IJ_1} at the other (I) end, and Figure 5.31a(2), other applied forces balanced by a force P_{IJ_2} at the I end.

The force at end I in Figure 5.31a(1) is related to the independent force at end J by the equation of equilibrium,

$$P_{IJ_1} = -P_{JI} \quad \textbf{(5.48a)}$$

The internal resultant axial force in the bar in Figure 5.31a(1) is

$$P_1(x) = P_{JI} \quad \textbf{(5.48b)}$$

In other words, $P_1(x)$ = constant. As usual, we define internal tensile forces as positive.

The internal resultant axial force in the bar in Figure 5.31a(2) depends on the

[45] This study is interesting when viewed as a sensitivity analysis, i.e., an investigation of how much effect a small change in some parameter (here, the axial flexibility) has on the overall results. For this example we see that a change in the flexibility coefficients of ~1% due to $K = 0.01$ produces changes in internal forces of ~5% and changes in deformations of ~15%.

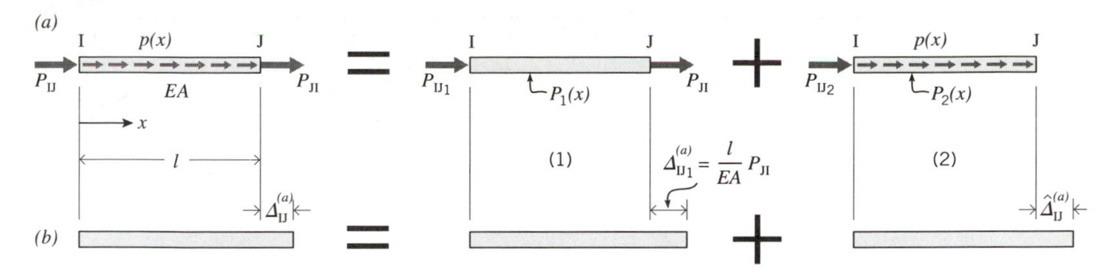

Figure 5.31 (*a*) Forces on bar element; (*b*) Axial deformations.

particular applied mechanical axial forces; we will denote this axial force by a general function $P_o(x)$,

$$P_2(x) = P_o(x) \tag{5.48c}$$

Note that $P_o(x)$ is exactly the axial force in a bar with the I end fixed, and the other end, the J end, free. Thus, we expect to be able to determine $P_o(x)$ for very general axial loadings once they are specified.

The total internal axial force in bar IJ is

$$P(x) = P_{JI} + P_o(x) \tag{5.48d}$$

The bar in Figure 5.31*a* deforms and displaces as a result of the applied loads. Figure 5.31*b* shows a general representation of the axial deformation, i.e., length change. This length change is evaluated by integrating the expression for axial strain, Equation 2.36*e*; the axial strain is given by

$$u'(x) = \frac{P(x) + P_T(x) + P_I(x)}{EA} \tag{5.48e}$$

Substituting Equation 5.48*d* leads to

$$\begin{aligned} u'(x) &= \frac{P_{JI}}{EA} + \frac{1}{EA}[P_o(x) + P_T(x) + P_I(x)] \\ &= \frac{P_{JI}}{EA} + \frac{\widehat{P}(x)}{EA} \end{aligned} \tag{5.48f}$$

where

$$\widehat{P}(x) = P_o(x) + P_T(x) + P_I(x) \tag{5.48g}$$

Equation 5.48*f* is analogous to Equation 5.40*b* for beams.

Integrating Equation 5.48*f* gives the axial deformation,[46]

[46] See, for example, Equation 2.38*b*.

$$\Delta_{IJ}^{(a)} = \frac{l}{EA}P_{JI} + \frac{1}{EA}\int_0^l \widehat{P}(x)dx$$
$$= \frac{l}{EA}P_{JI} + \widehat{\Delta}_{IJ}^{(a)} \qquad \textbf{(5.48}\textbf{\textit{h}}\textbf{)}$$

where

$$\widehat{\Delta}_{IJ}^{(a)} = \frac{1}{EA}\int_0^l \widehat{P}(x)dx \qquad \textbf{(5.48}\textbf{\textit{i}}\textbf{)}$$

Equation 5.48*h* is the exact form of the deformation-force relation for a uniform bar subjected to general axial loads and is analogous to Equation 5.42*a* for a beam. This equation indicates that the total deformation, Δ_{IJ} in Figure 5.31*b*, is the sum of the deformation due to the constant axial force in Figure 5.31*a*(1) and the deformation $\widehat{\Delta}_{IJ}^{(a)}$ in Figure 5.31*b*(2). The quantity $\widehat{\Delta}_{IJ}^{(a)}$ is the unrestrained axial deformation when P_{JI} is zero, i.e., $\widehat{\Delta}_{IJ}^{(a)}$ is the length change that occurs when the bar is not ''restrained'' by the independent axial force P_{JI}. Thus, $\widehat{\Delta}_{IJ}^{(a)}$ is a deformation due to the applied forces in Figure 5.31*a*(2) plus thermal and initial effects.

Because the form of Equation 5.48*h* is still the same as that of Equation 5.31*c*, there is no change in the analysis procedure other than for the possible need to evaluate $\widehat{\Delta}_{IJ}^{(a)}$ for cases involving more general types of loadings than we have considered at this point. This type of calculation will be demonstrated in the next example.

EXAMPLE 6

Find the forces in the structure of Figure 5.32*a* assuming that the only load is due to a constant weight per unit length of the elements, q_o. The frame is otherwise identical to the one in the previous examples.

Step 1. We will use the same choice of redundant forces as in the previous case.

Step 2. The appropriate superposition of forces is shown in Figures 5.32*a*(1)–5.32*a*(3). The statical analyses of the forces in Figures 5.32*a*(2) and 5.32*a*(3) are identical to those in Figures 5.29(2)–5.29(3). The statical analysis of the continuous forces in Figure 5.32*a*(1) is shown in the remainder of Figure 5.32, i.e., Figures 5.32*b* and 5.32*c*. In verifying the forces shown in these figures, note that the vertical force per unit of beam length on beam AB generates a total downward force of $\sqrt{2}q_o l$, and this effect must be carried throughout the statical analysis.

The analysis of the effects of the continuous vertical force on element AB is similar to the analysis conducted in Example 9, Chapter 1. By following the procedures described in that example, we obtain the end forces shown in Figures 5.32*b* and 5.32*c*.

In this case, because element AB transmits a varying axial force, we must designate one of the axial end forces as the independent end force; we will arbitrarily choose P_{BA}. Also, because there are no axial forces in element BC, this time we will consider

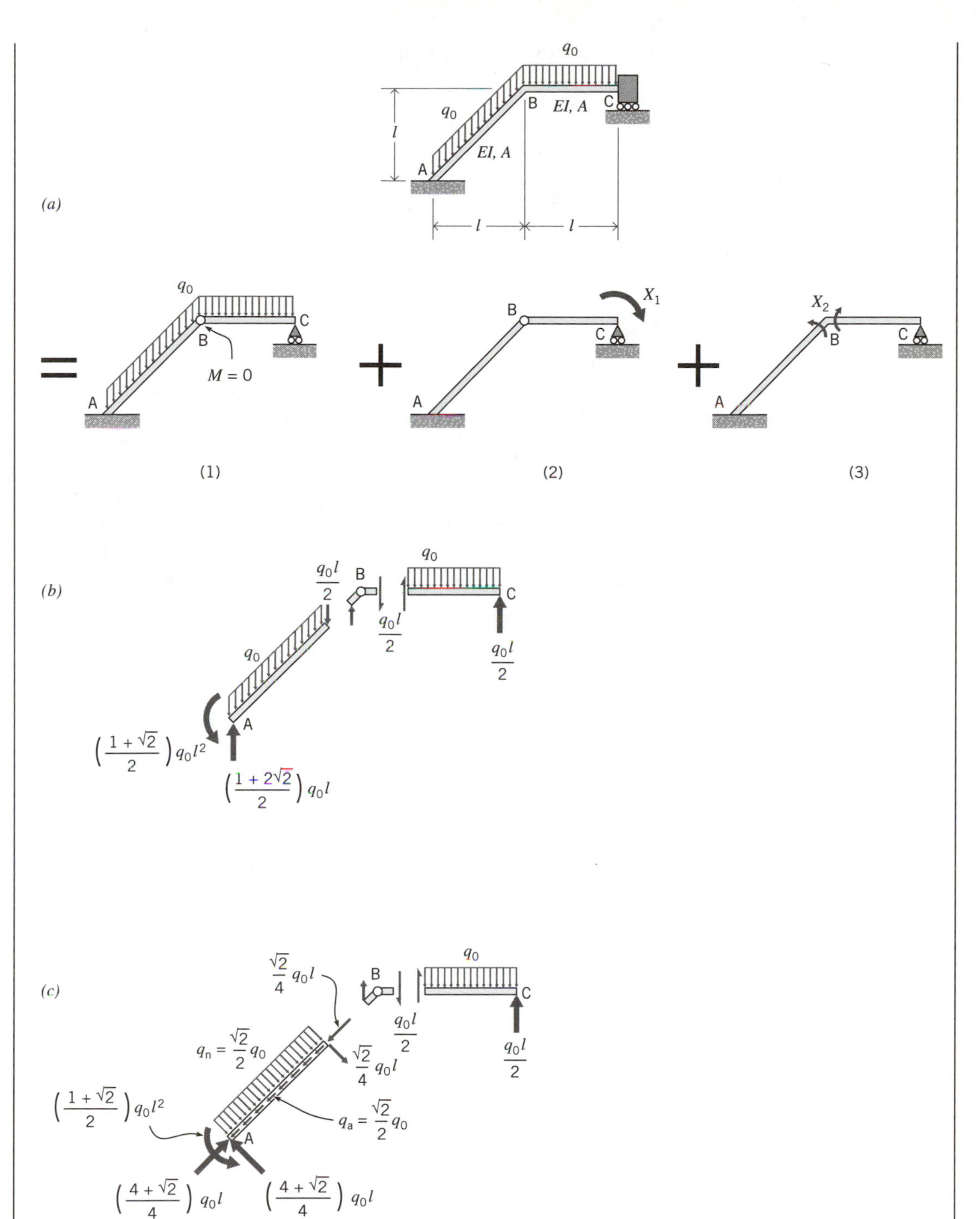

Figure 5.32 Statical analysis of frame subjected to uniform load per unit length of elements.

only the bending effects for this element. Thus, the matrix form of the equilibrium equations is

$$\begin{Bmatrix} M_{AB} \\ M_{BA} \\ P_{BA} \\ M_{BC} \\ M_{CB} \end{Bmatrix} = \begin{bmatrix} 1 & 2 \\ 0 & -1 \\ \sqrt{2}/2l & \sqrt{2}/2l \\ 0 & 1 \\ 1 & 0 \end{bmatrix} \begin{Bmatrix} X_1 \\ X_2 \end{Bmatrix} + \begin{Bmatrix} -(1+\sqrt{2})q_o l^2/2 \\ 0 \\ -\sqrt{2}q_o l/4 \\ 0 \\ 0 \end{Bmatrix} \quad \textbf{(5.49a)}$$

The **b** matrix is identical to the one in Equation 5.47*a* except for the deletion of the last row. Once again, the first three components of the **P** vector are the independent end forces for element AB and the last two components are the independent end forces for element BC. Note, in particular, that the component of P_{BA} in Figure 5.32*c* is a compressive force, and is therefore entered as a negative quantity in the $\mathbf{P_o}$ vector.

Step 3. The bending components of the unrestrained end deformations are computed just as in the previous examples. For beam BC these are exactly the same as in Equation 5.47*b*. In the present example, there is also a transverse continuous load on element AB (but no thermal loads); the unrestrained end deformations due to this transverse load are given by the same equations as for beam BC except that we substitute $\sqrt{2}l$ for l and $q_o/\sqrt{2}$ for q_o.

The axial component of the unrestrained end deformation for element AB is computed from Equation 5.48*i*. Figure 5.33*a* is a free-body of the type defined in Figure 5.31*a*(2), i.e., a free-body of the applied axial forces balanced only by the dependent end force at the A end. Figure 5.33*b* shows $P_o(x)$, which is the resultant internal force in the bar of Figure 5.33*a*

$$P_o(x) = -q_o l + \frac{\sqrt{2}}{2} q_o x \quad \textbf{(5.49b)}$$

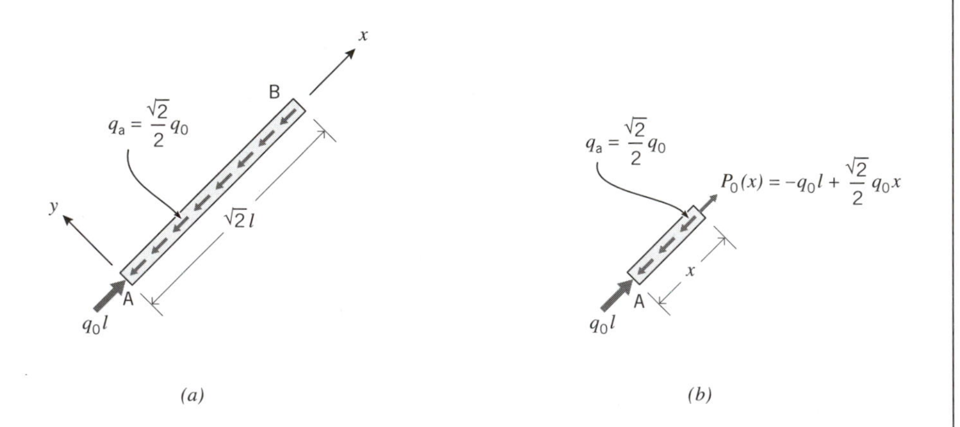

Figure 5.33 Forces on element AB for calculation of $\hat{\Delta}^{(a)}_{AB}$.

Substituting this expression for $P_o(x)$ into Equation 5.48*i* leads to

$$\hat{\Delta}_{AB}^{(a)} = \frac{1}{EA}\int_0^{\sqrt{2}l} \hat{P}(x)dx = \frac{1}{EA}\int_0^{\sqrt{2}l} P_o(x)dx$$

$$= \frac{1}{EA}\int_0^{\sqrt{2}l}\left(-q_o l + \frac{\sqrt{2}}{2}q_o x\right)dx \tag{5.49c}$$

$$= -\frac{\sqrt{2}}{2}\frac{q_o l^2}{EA}$$

The matrix form of the deformation-force equations for the structure now becomes

$$\begin{Bmatrix} \Delta_{AB} \\ \Delta_{BA} \\ \Delta_{AB}^{(a)} \\ \Delta_{BC} \\ \Delta_{CB} \end{Bmatrix} = \frac{l}{6EI}\begin{bmatrix} 2\sqrt{2} & -\sqrt{2} & & & \\ -\sqrt{2} & 2\sqrt{2} & & & \\ & & 6\sqrt{2}I/A & & \\ & & & 2 & -1 \\ & & & -1 & 2 \end{bmatrix}\begin{Bmatrix} M_{AB} \\ M_{BA} \\ P_{BA} \\ M_{BC} \\ M_{CB} \end{Bmatrix} + \begin{Bmatrix} q_o l^3/12EI \\ -q_o l^3/12EI \\ -\sqrt{2}q_o l^2/2EA \\ q_o l^3/24EI \\ -q_o l^3/24EI \end{Bmatrix} \tag{5.49d}$$

which constitutes the remainder of the information required for the analysis; we proceed to Step 5.

Step 5. The matrix operations are similar to those in the preceding examples, so we will present only the basic results. The compatibility equation is

$$\frac{l}{EI}\begin{bmatrix} (0.805 + 0.707K) & (1.012 + 0.707K) \\ (1.012 + 0.707K) & (3.633 + 0.707K) \end{bmatrix}\begin{Bmatrix} X_1 \\ X_2 \end{Bmatrix} = \frac{l}{EI}\begin{Bmatrix} (0.527 + 0.854K)q_o l^2 \\ (1.131 + 0.854K)q_o l^2 \end{Bmatrix} \tag{5.49e}$$

where $K = I/l^2A$. For $K = 0$ we obtain

$$\begin{Bmatrix} X_1 \\ X_2 \end{Bmatrix} = \begin{bmatrix} 1.912 & -0.533 \\ -0.533 & 0.424 \end{bmatrix}\begin{Bmatrix} 0.527q_o l^2 \\ 1.131q_o l^2 \end{Bmatrix} = \begin{Bmatrix} 0.406q_o l^2 \\ 0.198q_o l^2 \end{Bmatrix} \tag{5.49f}$$

and

$$\begin{Bmatrix} M_{AB} \\ M_{BA} \\ P_{BA} \\ M_{BC} \\ M_{CB} \end{Bmatrix} = \begin{Bmatrix} -0.405q_o l^2 \\ -0.198q_o l^2 \\ 0.074q_o l \\ 0.198q_o l^2 \\ 0.406q_o l^2 \end{Bmatrix} \tag{5.49g}$$

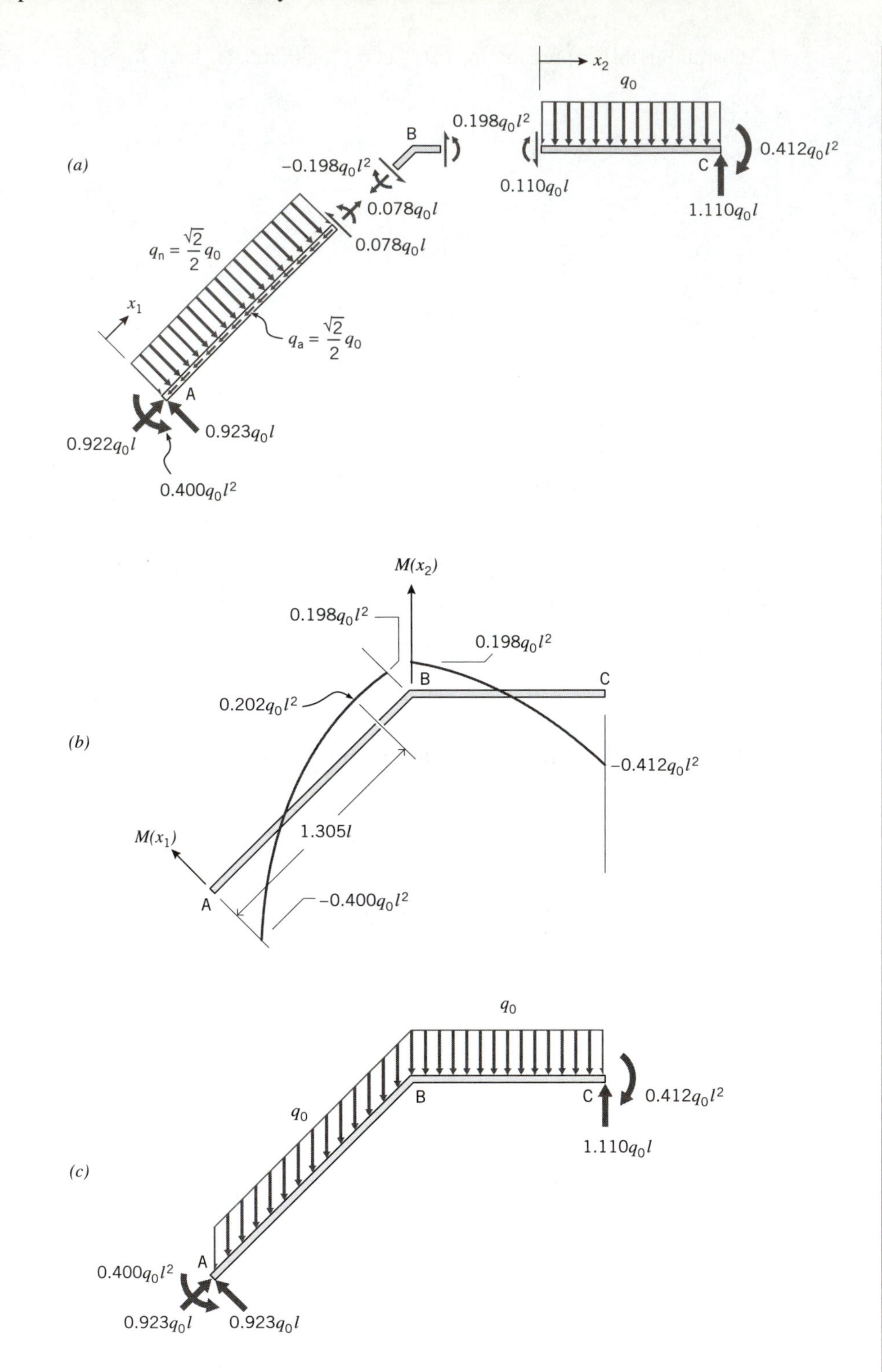

Figure 5.34 Force diagrams.

For $K = 0.01$ the solution is

$$\begin{Bmatrix} X_1 \\ X_2 \end{Bmatrix} = \begin{bmatrix} 1.899 & -0.532 \\ -0.532 & 0.423 \end{bmatrix} \begin{Bmatrix} 0.536q_o l^2 \\ 1.139q_o l^2 \end{Bmatrix} = \begin{Bmatrix} 0.412q_o l^2 \\ 0.198q_o l^2 \end{Bmatrix} \quad \textbf{(5.49}\textbf{\textit{h}}\textbf{)}$$

and

$$\begin{Bmatrix} M_{AB} \\ M_{BA} \\ P_{BA} \\ M_{BC} \\ M_{CB} \end{Bmatrix} = \begin{Bmatrix} -0.400q_o l^2 \\ -0.198q_o l^2 \\ 0.078q_o l \\ 0.198q_o l^2 \\ 0.412q_o l^2 \end{Bmatrix} \quad \textbf{(5.49}\textbf{\textit{i}}\textbf{)}$$

which again differs only slightly from the solution for $K = 0$.

Figure 5.34*a* shows the internal forces corresponding to the solution obtained for $K = 0.01$, Equation 5.49*i*. As in Example 4, the transverse forces have been computed from equilibrium conditions. The bending moment diagrams are sketched in Figure 5.34*b*. Note that the maximum positive bending moment occurs at $x_1 = 1.3l$, and has the value $0.202q_o l^2$. Finally, as in Example 4, we may verify that the entire structure, Figure 5.34*c*, is in equilibrium.

5.4 MISCELLANEOUS CONSIDERATIONS

At this point, our formulation of the force method is essentially complete because we have developed all the procedures required for the analysis of truss and frame structures. Although we have dealt exclusively with two-dimensional structures for the sake of simplicity, the analyses carry over to three-dimensional structures without any conceptual changes. However, three-dimensional analyses involve a great deal more data, and the problem of visualization makes the analyses much more difficult in practice.

Even for the simpler case of two-dimensional structures, there are some additional considerations that deserve further comment and, in this section, we will briefly demonstrate how to deal with concerns such as support settlement, structures composed of different types of elements, joint flexibility, initial length changes, and structures with more redundant forces than those examined thus far. The latter will be explored by examining the behavior of a truss-like structure with rigid joints. This should also enable us to gain some insight into the distinctions between truss and frame behavior.

5.4.1 Support Settlement

Consider the question of how to deal with the effects of support settlement. In all cases, we assume that the magnitude of the settlement is both small and known. We must also recognize that support settlement produces forces only for structures that are externally indeterminate, i.e., structures for which the external reactions cannot be determined from equations of equilibrium alone. This may be visualized by considering the unloaded structures in Figure 5.35. The reactions for the determinate structure in Figure 5.35*a* will be zero even if one of the supports, such as that at A, moves in the vertical direction. On the other hand, if the support at A for the structure in Figure

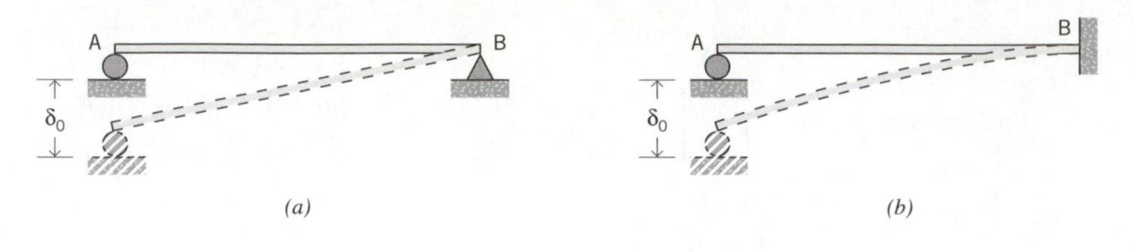

Figure 5.35 Support settlement.

5.35*b* moves vertically, the beam will become distorted, and internal forces and support reactions will develop, even in the absence of applied loads. If loads are applied to these structures, the reactions are determined by superposition of the effects of the loads and of the settlement.

Recall that compatibility equations are derived from virtual work equations and that in all previous examples the external virtual work in these equations was zero; if support movement is present, however, external virtual work may not be zero. For instance, consider the structure in Figure 5.16*a*, and suppose that the support at the left end moves down by a given amount δ_o. The compatibility equation for $\delta_o = 0$ was derived in Equation 5.32*d* (based on a virtual force X_V ''like'' redundant X in Figure 5.16*a*). If we now add the support settlement, the external virtual work becomes $-\delta_o X_V$ rather than 0. The negative sign is a consequence of the fact that F_V is assumed to act vertically upward whereas δ_o is vertically downward. The virtual force X_V cancels out of the equation, which now becomes

$$-\delta_o = \left(\frac{l^3}{3EI}\right)X + \left(-\frac{q_o l^4}{8EI} - \frac{l^2 \alpha \Delta T_o}{2h}\right) \qquad \textbf{(5.50a)}$$

In this case we obtain

$$X = -\frac{3EI}{l^3}\delta_o + \frac{3}{8}q_o l + \frac{3EI\alpha\Delta T_o}{2lh} \qquad \textbf{(5.50b)}$$

Note that for the special case $q_o = \Delta T_o = 0$, Equation 5.50*b* reduces to a familiar relationship between end force X and end displacement δ_o for a simple cantilever beam.

In a matrix formulation, the compatibility equations have been expressed in the form given by Equation 5.31*j*; this was, however, also based on the presumption that external virtual work (the left side of the equation) was zero. In general, we may replace the zero vector on left side of Equation 5.31*j* by another vector, the components of which are the external virtual work terms associated with unit values of the redundants (i.e., X_V taken as $X_j = 1, j = 1, \ldots, n$).[47] We will denote this known vector by $\mathbf{W_e}$ and replace Equation 5.31*j* by

$$\mathbf{W_e} = \mathbf{b}^{\mathrm{T}}\boldsymbol{\Delta} \qquad \textbf{(5.51a)}$$

[47] This is equivalent to canceling X_V out of the equation.

We proceed as in Section 5.2.3, i.e., we substitute Equations 5.31*e* and 5.31*f* into Equation 5.51*a* to obtain

$$(\mathbf{b}^T\mathbf{a}\mathbf{b})\mathbf{X} = \mathbf{W_e} - \mathbf{b}^T(\mathbf{a}\mathbf{P_o} + \hat{\mathbf{\Delta}}) \quad \textbf{(5.51}\textbf{\textit{b}}\textbf{)}$$

This equation is still of the form

$$\mathbf{A}\mathbf{X} = \overline{\mathbf{B}} \quad \textbf{(5.51}\textbf{\textit{c}}\textbf{)}$$

where

$$\overline{\mathbf{B}} = \mathbf{W_e} - \mathbf{b}^T(\mathbf{a}\mathbf{P_o} + \hat{\mathbf{\Delta}}) \quad \textbf{(5.51}\textbf{\textit{d}}\textbf{)}$$

EXAMPLE 7

Find the forces in the structure in Figure 5.36*a* due to a vertical support settlement δ_o at C. The structure is the same frame as in previous examples. We will also include a load consisting of a concentrated moment, M_o, applied at joint B because the equilibrium analysis associated with this load can be instructive. For the sake of conciseness we will take the beams to be axially rigid.

Step 1. The equilibrium analysis becomes interesting and requires some care when one of the redundant forces is chosen as the internal moment at joint B. To illustrate this we will use the same two redundant forces as in Examples 4–6, namely the reaction moment at C and the internal moment at B.

Step 2. In the previous cases, it was not necessary to be very specific about the exact location of the point at B where the internal moment was "released." Here, because there is an external moment load applied at B the equilibrium analysis requires that we be precise about the location of the point of release. We have two choices. Figure 5.36*b* shows the results of the statical analysis of the loaded structure when the internal moment at end B of beam AB is released, and Figure 5.36*c* shows the results when the internal moment at end B of beam BC is released. The force distributions are different in these two figures because there must be an internal moment in the vicinity of the joint that equilibrates the applied load M_o. In Figure 5.36*b* that internal moment is at the B end of beam BC, and in Figure 5.36*c* the internal moment is at the B end of beam AB.

The force distributions associated with the two redundant forces are the same as shown in Figure 5.28*b*(2) and 5.28*b*(3); the forces in Figure 5.28*b*(3) are the same for either location of the release at B. Thus, the **b** matrix is the same as in Equation 5.45*a*.

We will take the $\mathbf{P_o}$ matrix in the equilibrium equations as the one that corresponds to the forces in Figure 5.36*c*. (The end result for values of internal forces, **P**, will be the same regardless of which of the two possible force distributions, Figure 5.36*b* or 5.36*c*, is used. You should verify this yourself.) In this case the equilibrium equations are

$$\begin{Bmatrix} M_{AB} \\ M_{BA} \\ M_{BC} \\ M_{CB} \end{Bmatrix} = \begin{bmatrix} 1 & 2 \\ 0 & -1 \\ 0 & 1 \\ 1 & 0 \end{bmatrix} \begin{Bmatrix} X_1 \\ X_2 \end{Bmatrix} + \begin{Bmatrix} -M_o \\ M_o \\ 0 \\ 0 \end{Bmatrix} \quad \textbf{(5.52}\textbf{\textit{a}}\textbf{)}$$

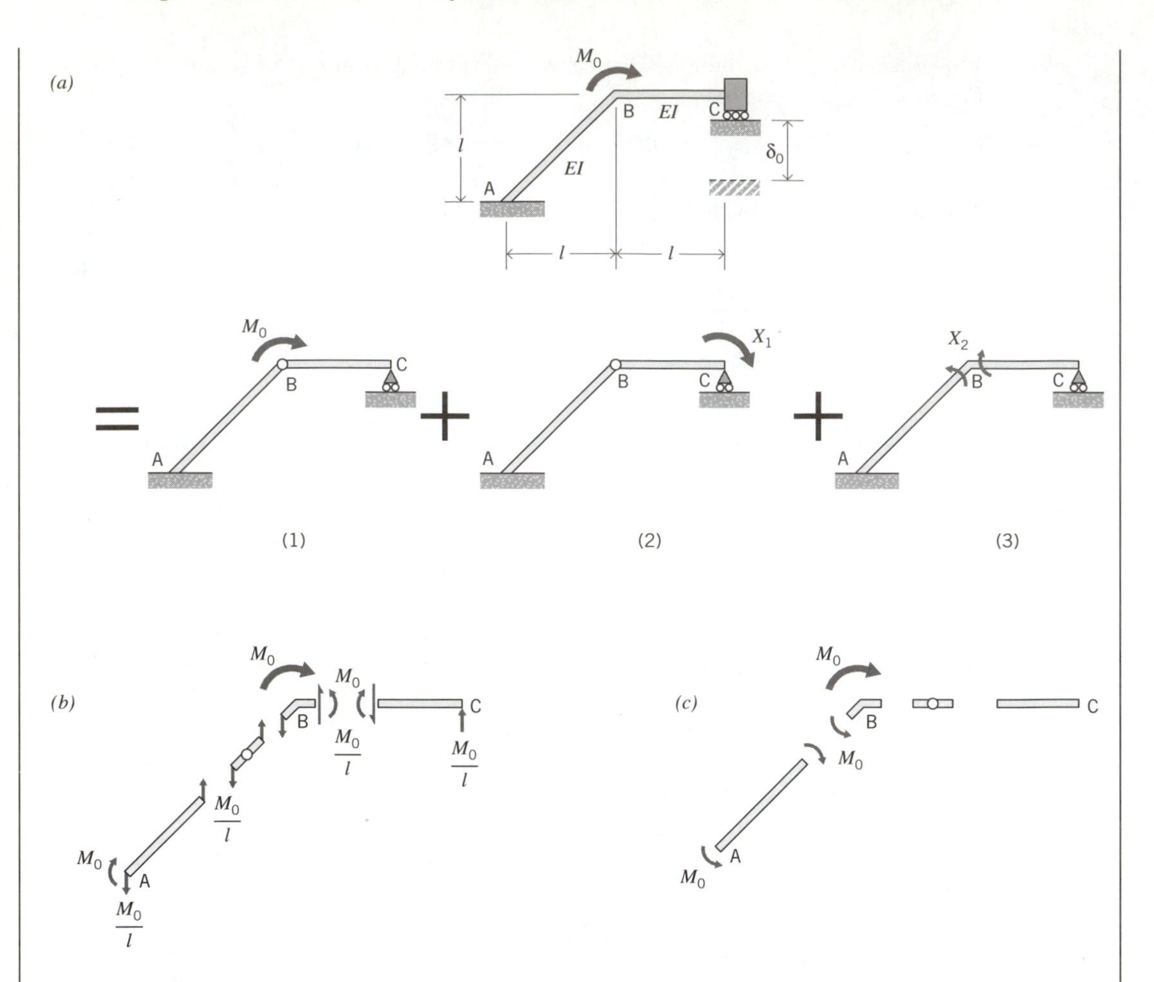

Figure 5.36 Example 7.

Step 3. The **a** matrix is the same as in Equation 5.45*b*. Also, because there are no intermediate loads on either beam, we have $\hat{\boldsymbol{\Delta}} = \mathbf{0}$.

Step 4. We must now define the $\mathbf{W_e}$ vector. The components of this vector are the external virtual work terms associated with the real displacements in Figure 5.36*a*, and virtual force distributions based on unit values of the redundant forces in Figures 5.36*a*(2) and 5.36*a*(3), respectively. The only real displacement that can contribute to the external virtual work is the known vertical settlement, δ_o at C. The vertical reactions at C in both virtual force distributions have magnitude $1/l$. Because the real displacement is opposite in direction to these virtual forces, both virtual work terms are negative. Thus

$$\mathbf{W_e} = \begin{Bmatrix} -\delta_o/l \\ -\delta_o/l \end{Bmatrix} \qquad \textbf{(5.52}\boldsymbol{b}\textbf{)}$$

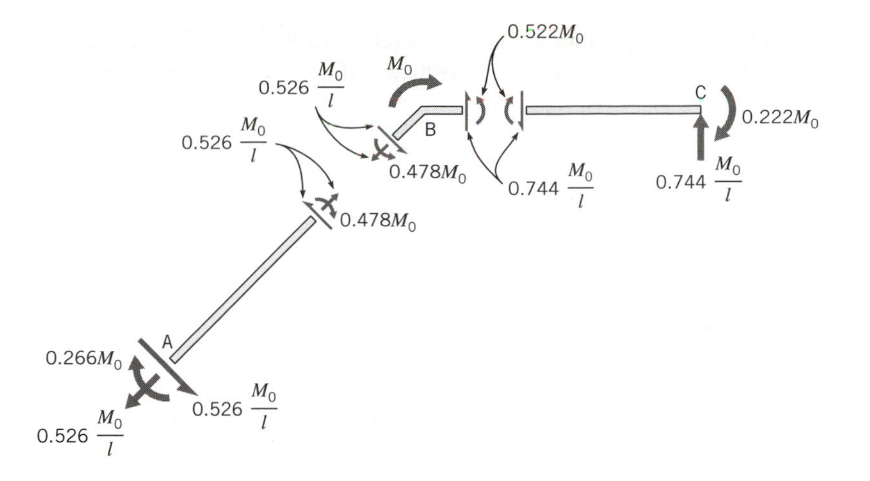

Figure 5.37 Free-body diagrams (forces due to M_0 only).

Step 5. Substituting the matrices defined above into Equation 5.51*b* leads to

$$\frac{l}{EI}\begin{bmatrix} 0.805 & 1.012 \\ 1.012 & 3.633 \end{bmatrix}\begin{Bmatrix} X_1 \\ X_2 \end{Bmatrix} = \frac{l}{EI}\begin{Bmatrix} -EI\delta_o/l^2 + 0.707M_o \\ -EI\delta_o/l^2 + 2.121M_o \end{Bmatrix} \tag{5.52c}$$

The flexibility matrix, **A**, is the same as in Example 4 because **b** and **a** are unchanged. Solving Equation 5.52*c* for X_1 and X_2 gives

$$\begin{Bmatrix} X_1 \\ X_2 \end{Bmatrix} = \begin{Bmatrix} -1.380EI\delta_o/l^2 + 0.222M_o \\ 0.109EI\delta_o/l^2 + 0.522M_o \end{Bmatrix} \tag{5.52d}$$

and

$$\begin{Bmatrix} M_{AB} \\ M_{BA} \\ M_{BC} \\ M_{CB} \end{Bmatrix} = \begin{Bmatrix} -1.162EI\delta_o/l^2 + 0.266M_o \\ -0.109EI\delta_o/l^2 + 0.478M_o \\ 0.109EI\delta_o/l^2 + 0.522M_o \\ -1.380EI\delta_o/l^2 + 0.222M_o \end{Bmatrix} \tag{5.52e}$$

As we expected, the settlement at C produces internal forces and will change the maximum load-carrying capacity of the structure. Also, note that the free-body of joint B, shown in Figure 5.37, is in equilibrium under the action of moments M_{BA}, M_{BC}, and M_o.

5.4.2 Structures Composed of Several Types of Elements

In practice, structures are rarely formed of only beam elements or truss elements. Often both types of elements appear in the same structure together with other types of elements, such as simple springs (axial, torsional, shear) that may be used to represent joint flexibility, and even rigid elements that transmit force but have no flexibility. The next example demonstrates how easily the analysis of structures composed of several types of elements can be performed using the matrix force method.

EXAMPLE 8

Consider the truss-stiffened beam shown in Figure 5.38*a*. Such a truss-stiffened assembly is often used to temporarily strengthen an existing bridge.

The beam elements, AB, BC, and CD, are taken as axially rigid, and the truss is assumed to be composed of ordinary axial bar elements except for bars BE and CF, which are also taken to be axially rigid. This is equivalent to zero flexibility, which means that the elements BE and CF have no contribution to the overall flexibility $\mathbf{A} = \mathbf{b}^T\mathbf{a}\mathbf{b}$. We will assume that all ordinary truss bars have the same cross-sectional area A;[48] furthermore, we assume identical modulus of elasticity E for bars and beams.

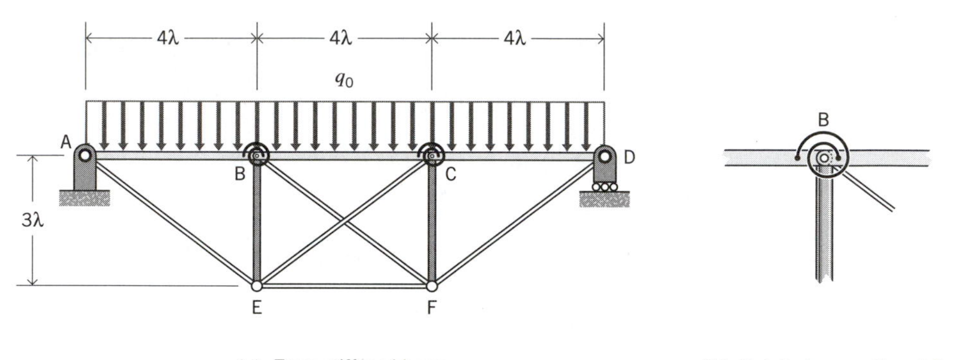

(*a*) Truss-stiffened beam

(*b*) Detail of connection at B

Figure 5.38 Example 8.

We are also going to model the connections between beam elements AB and BC, and BC and CD, as hinges with torsional springs connecting contiguous beams, as shown schematically in Figures 5.38*a* and 5.38*b*; we assume a linear relationship between moment applied to a spring and angular distortion, i.e., $M^{(s)} = k\theta^{(s)}$ where $\theta^{(s)}$ is the angular distortion measured in radians. The quantity k is the spring constant, which we take as the same for both springs. If the springs at joints B and C become rigid ($k \rightarrow \infty$), then a continuous beam ABCD results; otherwise the deflected shape of the initially straight line between A and D will have discontinuities in the slopes, i.e., kinks, at the connections.

We also assume that the rigid truss bars, BE and CF, and bars BF and CE, are

[48] The assumption of axial rigidity for beams AB, BC, and CD, and for bars BE and CF, is equivalent to an assumption that the cross-sectional areas of these elements are much greater than A.

connected to the hinges at B and C but not to the torsional springs, as shown in Figure 5.38*b*; thus, these bars transmit only axial forces.

Step 1. This structure is surprisingly easy to analyze once we establish its degree of indeterminacy. One way to do this is to begin with a statically determinate (and stable) substructure and add to it additional elements until indeterminacy results. Recall that statically determinate structures do not develop internal forces due to initial length changes or thermal effects, but do so only when forces are applied to the structure.

Figure 5.39*a* shows a substructure that is statically determinate. Note that the presence of the torsional springs has no influence on the determinacy of this structure.[49] Even if the beams were initially curved, or assembly produced initial ''kinks'' at B or C, the substructure would be statically determinate.

Now consider adding bars AE and BE and CF and CD (Figure 5.39*b*). This structure is still statically determinate. Note that these elements may be added to the existing subassembly without producing any internal forces even if the elements have initial imperfections (the dashed portion of Figure 5.39*b*).

Next, consider adding element EF to the assembly (Figure 5.39*c*). This element must fit between joints E and F, which are points on a rigid assembly. If bar EF is exactly the right length it can be added without difficulty; however, if the element has any imperfection (or is heated or cooled) it will not fit properly. In this case, additional forces must be used to ''force'' the bar into place; but this means that the bar is redundant and that a redundant force X_1 will exist, which therefore must be defined and considered in the analysis. A similar situation exists for bars BF and CE. Thus the structure is three-degrees redundant (Figure 5.39*d*).

We may achieve some simplification at this point by recognizing the obvious symmetry in the structure and in the loads applied to it. In particular, we are going to make an educated guess that $X_2 = X_3$ and thereby reduce the number of unknown redundant forces to two.

---

[49] If, however, additional torsional springs were added at A or D the structure would be indeterminate. Why?

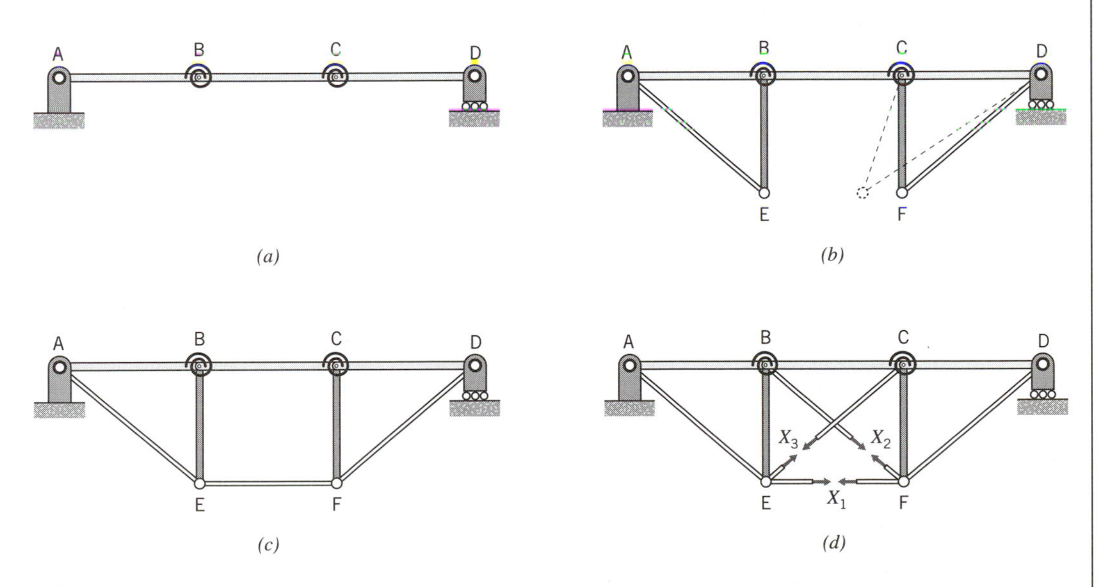

Figure 5.39 Determination of degree of redundancy.

Step 2. Figure 5.40 shows the equilibrium analysis for the structure. Note the representation of the two bars with identical redundant force X_2 in Figure 5.40*a*(3). The internal forces in the structure are shown in Figures 5.40*b*(1)–(3); in particular, Figure 5.40*b*(1) shows the internal forces in the structure in Figure 5.40*a*(1), Figure 5.40*b*(2) shows the internal forces in the structure in Figure 5.40*a*(2), and Figure 5.40*b*(3) shows the internal forces in the structure in Figure 5.40*a*(3). The equilibrium relations are

$$\begin{Bmatrix} M_{AB} \\ M_{BA} \\ M_{BC} \\ M_{CB} \\ M_{CD} \\ M_{DC} \\ M_B^{(s)} \\ M_C^{(s)} \\ P_{AE} \\ P_{CE} \\ P_{BF} \\ P_{DF} \\ P_{EF} \end{Bmatrix} = \begin{bmatrix} 0 & 0 \\ 3 & 12/5 \\ -3 & -12/5 \\ 3 & 12/5 \\ -3 & -12/5 \\ 0 & 0 \\ -3 & -12/5 \\ -3 & -12/5 \\ 5/4\lambda & 1/\lambda \\ 0 & 1/\lambda \\ 0 & 1/\lambda \\ 5/4\lambda & 1/\lambda \\ 1/\lambda & 0 \end{bmatrix} \begin{Bmatrix} X_1\lambda \\ X_2\lambda \end{Bmatrix} + 16q_o\lambda^2 \begin{Bmatrix} 0 \\ -1 \\ 1 \\ -1 \\ 1 \\ 0 \\ 1 \\ 1 \\ 0 \\ 0 \\ 0 \\ 0 \\ 0 \end{Bmatrix} \quad \textbf{(5.53a)}$$

where λ is a gauge length[50] (see Figure 5.38). Note that the axial forces in the bending elements and in the rigid bar elements BE and CF are not recorded because they do not produce element flexibilities. Note also that we are recording data with $X_1\lambda$ and $X_2\lambda$ in the **X** vector. This will make the algebra simpler later in the analysis. (If you are uncomfortable with this, use $X_1' = X_1\lambda$, etc.)

Step 3. The element deformation-force (flexibility) relations are recorded next,

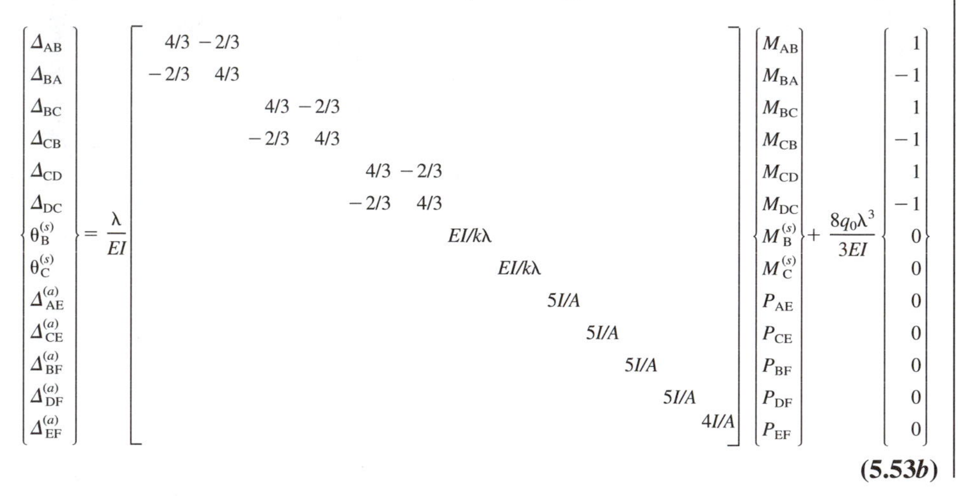

$$\begin{Bmatrix} \Delta_{AB} \\ \Delta_{BA} \\ \Delta_{BC} \\ \Delta_{CB} \\ \Delta_{CD} \\ \Delta_{DC} \\ \theta_B^{(s)} \\ \theta_C^{(s)} \\ \Delta_{AE}^{(a)} \\ \Delta_{CE}^{(a)} \\ \Delta_{BF}^{(a)} \\ \Delta_{DF}^{(a)} \\ \Delta_{EF}^{(a)} \end{Bmatrix} = \frac{\lambda}{EI} \begin{bmatrix} 4/3 & -2/3 & & & & & & & & & & & \\ -2/3 & 4/3 & & & & & & & & & & & \\ & & 4/3 & -2/3 & & & & & & & & & \\ & & -2/3 & 4/3 & & & & & & & & & \\ & & & & 4/3 & -2/3 & & & & & & & \\ & & & & -2/3 & 4/3 & & & & & & & \\ & & & & & & EI/k\lambda & & & & & & \\ & & & & & & & EI/k\lambda & & & & & \\ & & & & & & & & 5I/A & & & & \\ & & & & & & & & & 5I/A & & & \\ & & & & & & & & & & 5I/A & & \\ & & & & & & & & & & & 5I/A & \\ & & & & & & & & & & & & 4I/A \end{bmatrix} \begin{Bmatrix} M_{AB} \\ M_{BA} \\ M_{BC} \\ M_{CB} \\ M_{CD} \\ M_{DC} \\ M_B^{(s)} \\ M_C^{(s)} \\ P_{AE} \\ P_{CE} \\ P_{BF} \\ P_{DF} \\ P_{EF} \end{Bmatrix} + \frac{8q_0\lambda^3}{3EI} \begin{Bmatrix} 1 \\ -1 \\ 1 \\ -1 \\ 1 \\ -1 \\ 0 \\ 0 \\ 0 \\ 0 \\ 0 \\ 0 \\ 0 \end{Bmatrix} \quad \textbf{(5.53b)}$$

[50] We use λ rather than l to avoid possible confusion in the subsequent discussion; λ is referred to as a gauge length because it is not the actual length of any element in this structure.

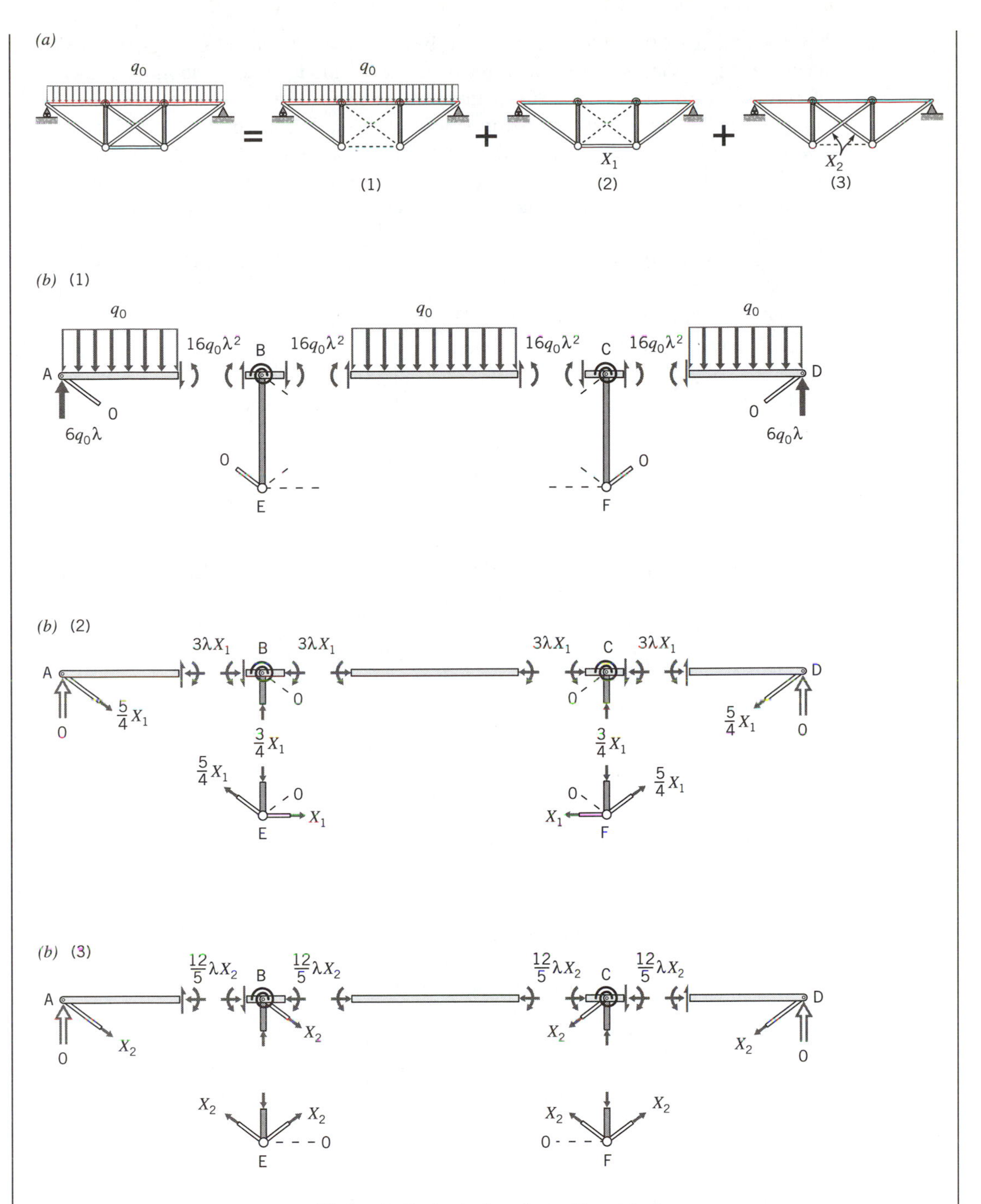

Figure 5.40 Statical analysis, Example 8.

Note that the $\mathbf{\Delta}$ vector contains the various beam, bar, and torsional spring deformation quantities in the same order as the force quantities in the $\mathbf{P}$ vector. The first three pairs of components of $\mathbf{\Delta}$ are the deformations of the three beam elements; the next two components, $\theta_B^{(s)}$ and $\theta_C^{(s)}$, are the deformations of the springs at joints B and C, respectively; the last five components are the bar axial deformations. There are ''unrestrained end deformations'' for the three beam elements. These are computed from Equations 5.44*g* by substituting 4λ for the beam length, and are listed in the first six rows of the $\hat{\mathbf{\Delta}}$ vector.

We now have all the matrices required for solution, and we proceed directly to Step 5.

Step 5. Substituting into Equation 5.31*k* leads to

$$\begin{bmatrix} 60 + 18\alpha + \frac{157}{8}\beta & 48 + \frac{72}{5}\alpha + \frac{25}{2}\beta \\ 48 + \frac{72}{5}\alpha + \frac{25}{2}\beta & \frac{192}{5} + \frac{288}{25}\alpha + 20\beta \end{bmatrix} \begin{Bmatrix} X_1\lambda \\ X_2\lambda \end{Bmatrix} = q_o\lambda^2 \begin{Bmatrix} 352 + 96\alpha \\ \frac{1408}{5} + \frac{384}{5}\alpha \end{Bmatrix} \tag{5.53c}$$

Here, in order to simplify the appearance of the equation, we have let $\alpha = EI/k\lambda$ and $\beta = I/A\lambda^2$; thus, both α and β are pure (dimensionless) constants. The constant β has a different meaning than the similar-appearing constant K in Examples 5 and 6 because I and A are now parameters for different elements. The parameter β can be related to K by

$$\beta = \frac{I}{A\lambda^2} = 16\frac{I}{A'(4\lambda)^2}\left(\frac{A'}{A}\right) = 16K\left(\frac{A'}{A}\right) \tag{5.53d}$$

where A' denotes the cross-sectional area of the beam elements (which have length 4λ).

We may now solve the equation for specified values of the constants α and β. Because the truss bars (tension members, especially) usually have much smaller cross-sectional areas than the beam elements for bridge structures of this type, it is possible to have values of β that are much greater than typical values of K discussed in the previous examples, e.g., $\beta \geq 1.0$ rather than $K = 0.001$. Similarly, α may have any value between 0 and ∞.[51]

Consider a case for which $\beta = 1.0$. This corresponds to a wide flange beam with a length (here 4λ) 13 times its depth (giving $K \approx 0.001$) and a ratio of wide flange beam

[51] A value $\alpha = 0$ corresponds to $k \to \infty$ (rigid torsional springs) and a value $\alpha \to \infty$ corresponds to $k = 0$ (no torsional springs). A computational solution for $\alpha = 0$ can be easily obtained from this formulation, but a computational solution for $\alpha \to \infty$ cannot be obtained explicitly using the current formulation. This is because the structure in Figure 5.39*a* is now a two-hinge mechanism. We can analyze the structure without the springs (Figure 5.39*d*) as a truss, and delete only X_1 as our redundant force. Beams AB, BC, and CD will not have any end moments, and behave like simply supported beams. (See, for example, Problem 3.P15, Chapter 3.)

cross-sectional area over truss bar area of about 60. Let $\alpha = 0.1$, a value typical of riveted connections between beam elements. Inverting Equation 5.53*c* leads to

$$\begin{Bmatrix} X_1\lambda \\ X_2\lambda \end{Bmatrix} = q_o\lambda^2 \begin{Bmatrix} 3.572 \\ 1.143 \end{Bmatrix} \quad or \quad \begin{Bmatrix} X_1 \\ X_2 \end{Bmatrix} = q_o\lambda \begin{Bmatrix} 3.572 \\ 1.143 \end{Bmatrix} \tag{5.53e}$$

Back-substituting provides

$$\mathbf{P} = q_o\lambda \begin{Bmatrix} 0 \\ -2.543\lambda \\ 2.543\lambda \\ -2.543\lambda \\ 2.543\lambda \\ 0 \\ 2.543\lambda \\ 2.543\lambda \\ 5.607 \\ 1.143 \\ 1.143 \\ 5.607 \\ 3.572 \end{Bmatrix}, \qquad \boldsymbol{\Delta} = \frac{q_o\lambda^2}{EI} \begin{Bmatrix} 4.362 \\ -6.057 \\ 7.752 \\ -7.752 \\ 6.057 \\ -4.362 \\ 0.254 \\ 0.254 \\ 28.036\lambda \\ 5.714\lambda \\ 5.714\lambda \\ 28.036\lambda \\ 14.286\lambda \end{Bmatrix} \tag{5.53f}$$

These results show a dramatic change in the moments carried by the main beam ABCD. If no truss support is present ($X_1 = X_2 = 0$) then the moments at points B and C are $\pm 16 q_o\lambda^2$ from Equation 5.53*a*, i.e., more than six times greater. This means that the deflections of the bridge deck are reduced by the same amount due to the additional truss support.

Note that the bending stresses in the bridge deck are much less than the axial stresses in the supporting truss. For example,

$$|\sigma_{bending}|_{max} = \frac{2.543\, q_o\lambda^2 H}{2I}$$
$$|\sigma_{truss}|_{max} = \frac{5.607\, q_o\lambda}{A} \tag{5.53g}$$

so

$$\frac{|\sigma_{bending}|_{max}}{|\sigma_{truss}|_{max}} = \left(\frac{1.272}{5.607}\right)\left(\frac{H}{\lambda}\right)\left(\frac{1}{\beta}\right) = (0.2268)\left(4\frac{H}{4\lambda}\right)\left(\frac{1}{\beta}\right) \tag{5.53h}$$

Using $H/(4\lambda) = 1/13$, we find the ratio is 0.07. Thus, the bending stresses in the bridge deck are only about 1/14 of the stresses in the truss bars. In other words, the bridge structure now carries the applied loading primarily through truss action rather than as a beam in bending. This means that the stresses in the supporting truss are the critical values that govern the behavior of the structure.

The solutions for $\alpha = 0$ and $\alpha \rightarrow \infty$ are left as exercises. Be careful in conducting the $\alpha \rightarrow \infty$ analysis to keep α large but finite; otherwise very poor numerical accuracy may result.

5.4.3 Initial Deformations

The inclusion of initial deformations in the analysis process is quite straightforward. The following example demonstrates the intentional use of initial deformations to improve stress distributions in a structure. (The converse is that unanticipated or unrecognized initial deformations can have a significant detrimental effect on stress distributions.)

EXAMPLE 9

The ratio of bending stresses in the beams compared to worst-case stresses in the supporting truss of the structure in Example 8 can be controlled even further if we introduce initial deformations into the lower truss-stiffener system. Suppose, for example we increase the length of bar EF by the amount $\Delta_{\text{EF}}^{(a)} = \delta_i$.

This effect is relatively easy to include in the analysis. We will simply superpose the effect of the initial deformation on the previous solution, i.e., we will first solve for the effect of the initial deformation alone. In this case, we have a $\hat{\mathbf{\Delta}}$ matrix given by

$$\hat{\mathbf{\Delta}} = \begin{Bmatrix} 0 \\ 0 \\ 0 \\ 0 \\ 0 \\ 0 \\ 0 \\ 0 \\ 0 \\ 0 \\ 0 \\ 0 \\ \delta_i \end{Bmatrix} \tag{5.54a}$$

We take $\mathbf{P_o} = \mathbf{0}$, and recognize that the $\mathbf{b}$ and $\mathbf{a}$ matrices are unchanged from the preceding example. Solving leads to

$$\begin{Bmatrix} X_1 \\ X_2 \end{Bmatrix} = \frac{\delta_i EI}{\lambda^3} \begin{Bmatrix} -5.882 \times 10^{-2} \\ 6.118 \times 10^{-2} \end{Bmatrix} \tag{5.54b}$$

and

$$\mathbf{P} = \frac{\delta_i EI}{\lambda^3}\begin{Bmatrix} 0 \\ -2.963\lambda \\ 2.963\lambda \\ -2.963\lambda \\ 2.963\lambda \\ 0 \\ 2.963\lambda \\ 2.963\lambda \\ -1.235 \\ 6.118 \\ 6.118 \\ -1.235 \\ -5.882 \end{Bmatrix} \times 10^{-2}, \qquad \boldsymbol{\Delta} = \frac{\delta_i}{\lambda}\begin{Bmatrix} 1.975 \\ -3.951 \\ 5.926 \\ -5.926 \\ 3.951 \\ -1.975 \\ 0.296 \\ 0.296 \\ -6.173 \\ 30.589\lambda \\ 30.589\lambda \\ -6.173\lambda \\ 76.472\lambda \end{Bmatrix} \times 10^{-2} \tag{5.54c}$$

By superposing the effect of q_o and the effect of δ_i we obtain

$$\mathbf{P} = q_o\lambda\begin{Bmatrix} 0 \\ -2.543\lambda \\ 2.543\lambda \\ -2.543\lambda \\ 2.543\lambda \\ 0 \\ 2.543\lambda \\ 2.543\lambda \\ 5.607 \\ 1.143 \\ 1.143 \\ 5.607 \\ 3.572 \end{Bmatrix} + \frac{\delta_i EI}{\lambda^3}\begin{Bmatrix} 0 \\ -2.963\lambda \\ 2.963\lambda \\ -2.963\lambda \\ 2.963\lambda \\ 0 \\ 2.963\lambda \\ 2.963\lambda \\ -1.235 \\ 6.118 \\ 6.118 \\ -1.235 \\ -5.882 \end{Bmatrix} \times 10^{-2} \tag{5.54d}$$

Suppose we select $|\sigma_{bending}|_{max} = 0.15|\sigma_{truss}|_{max}$.[52] The choice of which truss element will govern depends on the value of δ_i. For elements AE and DF as "control" elements in the supporting truss we have

$$\left(2.543\,q_o\lambda^2 + 0.02963\frac{\delta_i EI}{\lambda^2}\right)\frac{H}{2I} = 0.15\left(5.607\,q_o\lambda - 0.01235\frac{\delta_i EI}{\lambda^3}\right)\frac{1}{A} \tag{5.54e}$$

or

$$\left(2.543 + 0.02963\frac{\delta_i EI}{q_o\lambda^4}\right) = 0.30\frac{I}{H\lambda A}\left(5.607 - 0.01235\frac{\delta_i EI}{q_o\lambda^4}\right) \tag{5.54f}$$

---

[52] This is a reasonable figure if we use cables for the tension members in the supporting truss. Typically, steel I-beams are designed to have not more than 60% of the yield stress in compression flanges, i.e., about 30 ksi (206 MPa) in a 50 ksi (344 Mpa) steel. Cables (bridge strand) can sustain high stress levels on the order of 200 ksi (1,379 MPa).

Using $4\lambda/H = 13$ and $\beta = I/A\lambda^2 = 1$, we find $0.30I/H\lambda A = 0.0975$. Solving Equation 5.54*f* for δ_i gives $\delta_i = 70.16q_o\lambda^4/EI$. This value gives the following forces in the truss elements,

$$P_{AE} = P_{DF} = 4.740q_o\lambda$$
$$P_{CE} = P_{BF} = 5.436q_o\lambda \qquad \textbf{(5.54g)}$$
$$P_{EF} = -0.555q_o\lambda$$

Thus, if bar AE were stressed to its maximum permissible level for this δ_i, it would follow that bars CE and BF would be overstressed by about 15% and, further, that P_{EF} would be compressive, a loading that cables cannot sustain.

If P_{CE} and P_{BF} were to be used as control forces in the equation to balance stress, we would find that bars AE and DF would then be overstressed. The situation is shown in the sketch in Figure 5.41. Our best compromise is to seek a condition where $P_{AE} = P_{CE}$ (and, therefore, from statics, P_{EF} vanishes). Then δ_i must satisfy

$$5.607 - 0.01235\frac{\delta_i EI}{q_o\lambda^4} = 1.143 + 0.06118\frac{\delta_i EI}{q_o\lambda^4} \qquad \textbf{(5.54h)}$$

or $\delta_i = 60.71q_o\lambda^4/EI$. In this case

$$P_{AE} = P_{DF} = P_{CE} = P_{BF} = 4.857q_o\lambda$$
$$P_{EF} = 0 \qquad \textbf{(5.54i)}$$

Now the maximum bending stress is no longer 15% of the cable stress; the ratio of the two is found by solving for the scale factor α relating the bending stress to the stresses in the cables,

$$4.342q_o\lambda^2\frac{H}{2I} = \alpha\left(\frac{13}{4}\right)\left(\frac{1}{\beta}\right)(4.857q_o\lambda^2)$$
$$\alpha = 0.138 \qquad \textbf{(5.54j)}$$

which is close to the value of 0.15, which we set as our original goal.

So, using a balanced design without any center cable ($P_{EF} = 0$), we can design a cable-stiffened beam in which both the beam and the truss are operating at appropriate stress levels.

Note, finally, that the moment carried by the beam has been reduced from $16q_o\lambda^2$ (no cable stiffening at all) to $4.34q_o\lambda^2$ if the cable stiffeners are employed (no center cable EF), with stress levels appropriate to each element. This means that the truss-stiffened bridge is more than $3\frac{1}{2}$ times stronger than the basic beam.

This small example shows the great improvements one can bring to design (or design modification, as in this case) if it is possible to stiffen the structure and use initial deformations to adjust the behavior.

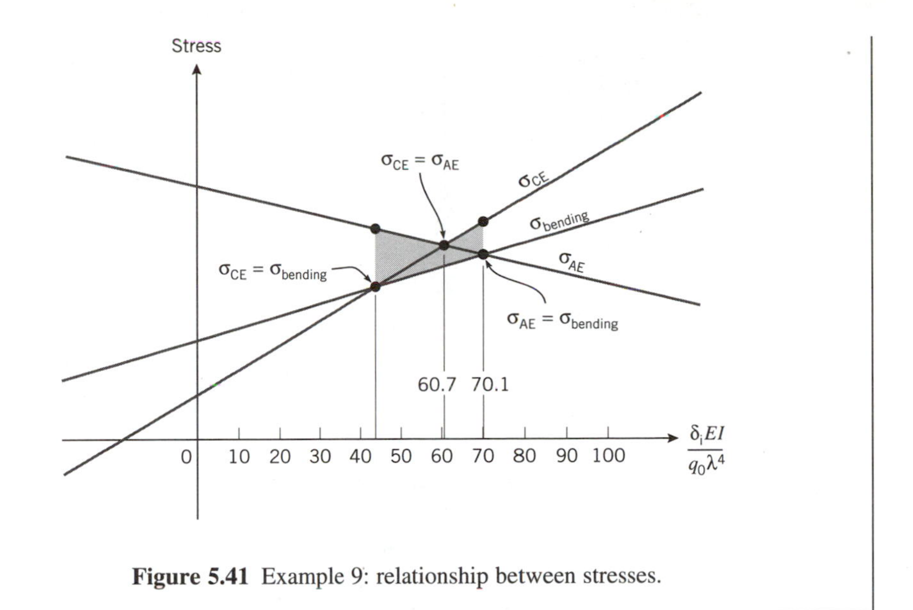

Figure 5.41 Example 9: relationship between stresses.

5.4.4 Effect of Moment-Bearing Joints on Truss Behavior

The preceding examples have demonstrated the manner in which joint flexibility may be included in an analytic model of a structure. Joints between any two elements may be modeled in an identical fashion to the way in which we modeled the joints between beams provided we assign the elements beam-like characteristics. Thus, truss bars, which are after all beams with small bending stiffness, may be joined at moment-resisting connections. While we often ignore moment-resisting connections and treat the structure as if it had pin joints, we can consider the effect of moment-bearing connections if we are willing to upgrade our elements from truss bars to truss-beam elements. These elements are more elaborate and costly to process. Also, the explicit representation of each moment-bearing joint (by torsional springs) will usually require the addition of a very large number of torsional springs to the structure because the connection between each pair of contiguous bars will require a separate spring.

Fortunately, the effect of joint flexibility in trusses composed of slender elements is relatively minor. We will demonstrate this by examining the behavior of a truss-like structure having completely rigid joints rather than pinned or flexible connections. Even without the inclusion of torsional springs, such problems involve much larger matrices than those for ordinary trusses because more redundant forces are involved, and because the elements themselves are more complex.

EXAMPLE 10

Consider the structure in Figure 5.42*a* that is similar to the truss in Figure 5.5*a* except that all connections between contiguous bars are rigid rather than pinned; this implies that all components of this structure must be taken to be axially flexible beam elements. We assume that there are pin connections to the supports at A and B, but that bars AC

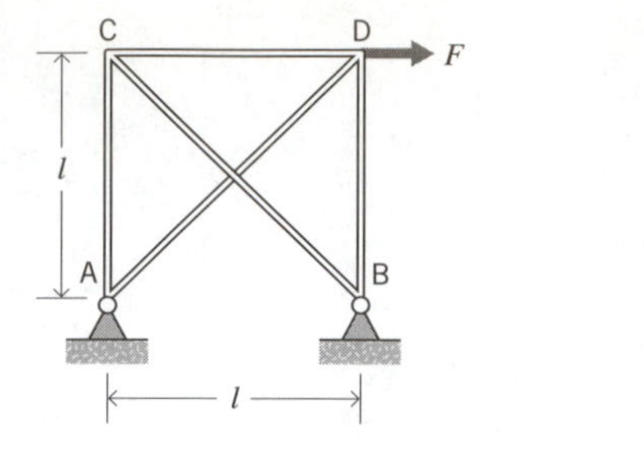

(a) Truss-like structure with rigid joints

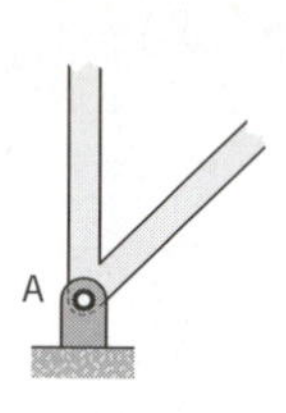

(b) Detail of connection at A

Figure 5.42 Example 10.

and AD, and BC and BD, are rigidly connected to each other, as illustrated in Figure 5.42*b*. Elements AD and BC are still not connected at their midpoints. We will analyze the behavior of this structure when it is subjected to the single load F shown in Figure 5.42*a* (rather than the symmetric loads in Figure 5.5*a*). The thermal load will also be deleted in this example. The values of E, I, and A are taken as identical for all elements.

Step 1. To determine the degree of redundancy of this structure we may begin with the knowledge that if all the joints were pinned we would have a one-degree redundant truss. For the structure in Figure 5.42*a*, we can obtain the equivalent of this one-degree redundant truss by releasing six internal moments, two at joints C and D, and one at joints A and B, as represented by the ''hinges'' in Figure 5.43*b*.[53] Thus, this structure is seven-degrees redundant. We will take the seven redundant forces as the six internal moments noted plus the horizontal component of reaction at B (Figure 5.43*b*).

[53] These are also the locations at which we would add torsional springs if we were modeling joint flexibility. Note also that when three bars are attached to a joint, we can take only two joint releases. If the third one were inserted, the joint would be unstable because it would be unable to resist an applied external moment.

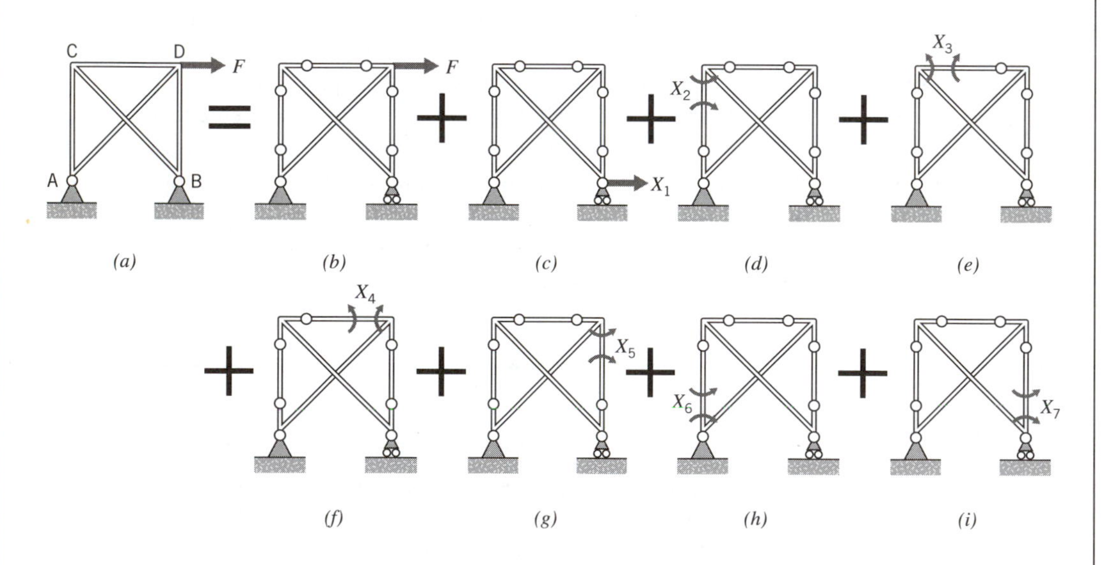

Figure 5.43 Redundant forces.

Step 2. Figure 5.43 illustrates the nature of the superposition of forces for this formulation. The analysis of the forces in Figure 5.43*b* is similar to that in Figure 5.5*c*, and the analysis of the forces in Figure 5.43*c* is identical to that in Figure 5.5*d*. For the sake of brevity, we will discuss the statical analysis of only one of the remaining force states, e.g., the one in Figure 5.43*d*. The appropriate free-body diagrams are shown in Figure 5.44.

The force state in Figure 5.44 results from the application of an internal moment X_2 at end C of component AC. The moment at end A of component AC is zero because the internal moment has been ''released'' at that point; it follows that there must be transverse shears of magnitude X_2/l at both ends of component AC, in the directions shown in the figure. Moment equilibrium of joint C requires that the moment at end C of component BC be equal and opposite to X_2 (because the moment at end C of component CD has also been ''released''). This, in turn, leads to transverse shears of magnitude $X_2/\sqrt{2}l$ at both ends of component BC (again, because the moment at end B of component BC is zero). Because there are no external reactions at joint B, it then follows from an equilibrium analysis of this joint that there must be axial forces in components BC and BD, as shown in Figure 5.44. Analyses of joints D and A provide the remaining forces.

The equilibrium analyses of the forces states in Figures 5.43*e*–5.43*i* are similar to that shown in Figure 5.44. (You should verify this yourself.) The resulting equilibrium equations are

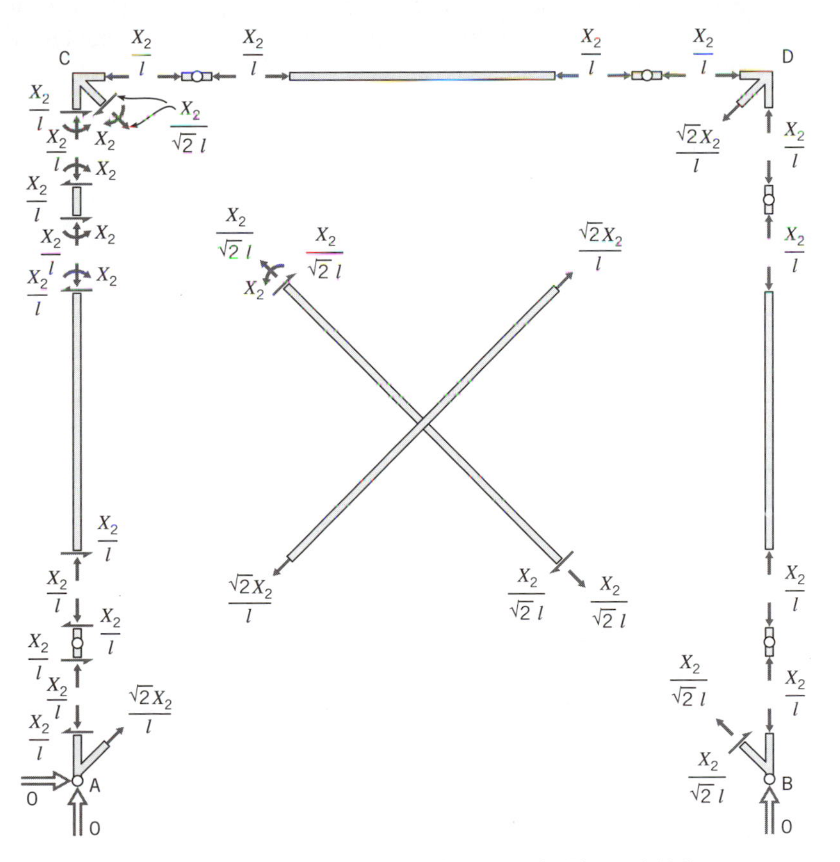

Figure 5.44 Statical analysis of structure in Figure 5.43*d*.

$$\begin{Bmatrix} M_{AC} \\ M_{CA} \\ P_{AC} \\ M_{AD} \\ M_{DA} \\ P_{AD} \\ M_{BC} \\ M_{CB} \\ P_{BC} \\ M_{BD} \\ M_{DB} \\ P_{BD} \\ M_{CD} \\ M_{DC} \\ P_{CD} \end{Bmatrix} = \begin{bmatrix} 0 & 0 & 0 & 0 & 0 & -1 & 0 \\ 0 & 1 & 0 & 0 & 0 & 0 & 0 \\ -1/l & -1/l & 0 & -1/l & 1/l & 0 & 0 \\ 0 & 0 & 0 & 0 & 0 & 1 & 0 \\ 0 & 0 & 0 & 1 & -1 & 0 & 0 \\ \sqrt{2}/l & \sqrt{2}/l & 0 & 1/\sqrt{2}l & -1/\sqrt{2}l & -1/\sqrt{2}l & 0 \\ 0 & 0 & 0 & 0 & 0 & 0 & 1 \\ 0 & -1 & -1 & 0 & 0 & 0 & 0 \\ \sqrt{2}/l & 1/\sqrt{2}l & 1/\sqrt{2}l & 0 & -\sqrt{2}/l & 0 & 1/\sqrt{2}l \\ 0 & 0 & 0 & 0 & 0 & 0 & -1 \\ 0 & 0 & 0 & 0 & 1 & 0 & 0 \\ -1/l & -1/l & -1/l & 0 & 1/l & 0 & 0 \\ 0 & 0 & 1 & 0 & 0 & 0 & 0 \\ 0 & 0 & 0 & -1 & 0 & 0 & 0 \\ -1/l & -1/l & 0 & 0 & 1/l & 1/l & -1/l \end{bmatrix} \begin{Bmatrix} X_1 l \\ X_2 \\ X_3 \\ X_4 \\ X_5 \\ X_6 \\ X_7 \end{Bmatrix} + \begin{Bmatrix} 0 \\ 0 \\ 0 \\ 0 \\ 0 \\ \sqrt{2}F \\ 0 \\ 0 \\ 0 \\ 0 \\ 0 \\ -F \\ 0 \\ 0 \\ 0 \end{Bmatrix} \tag{5.55a}$$

Note that we are recording three forces (two end moments and a constant axial force) for each of the five elements of this structure. Also, we have written $X_1 l$ as the first component of the **X** vector because it will simplify the algebra in the subsequent matrices. (This makes the dimensions of all components of **X** the same.)

Step 3. The deformation-force relations follow directly because there are no unrestrained end deformations to evaluate (i.e., $\widehat{\boldsymbol{\Delta}} = \mathbf{0}$),

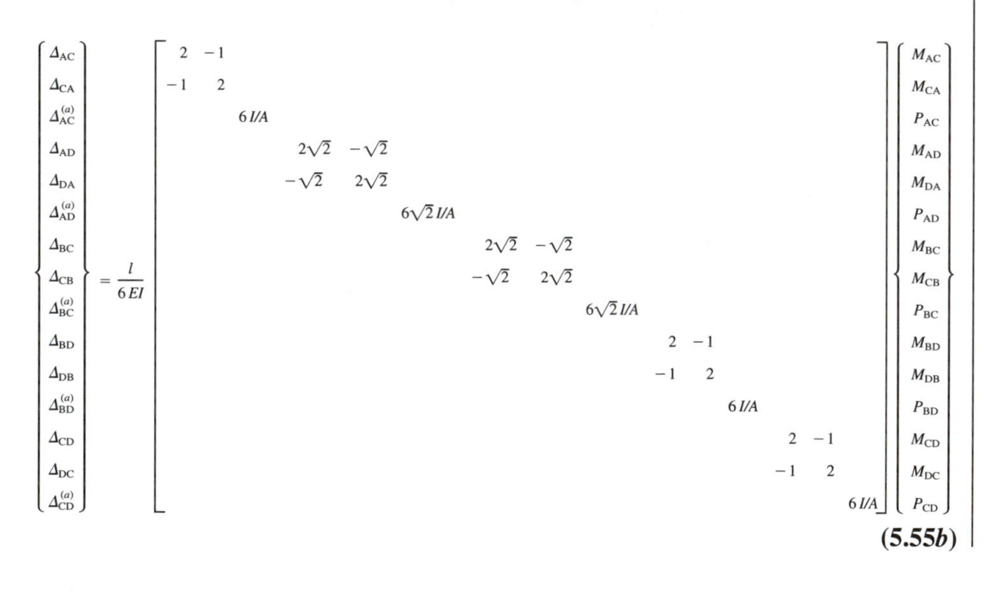

(5.55b)

We now have all the required matrices, and proceed directly to solving the compatibility equations.

Step 5. The flexibility matrix ($\mathbf{b}^T\mathbf{ab}$) for this structure is

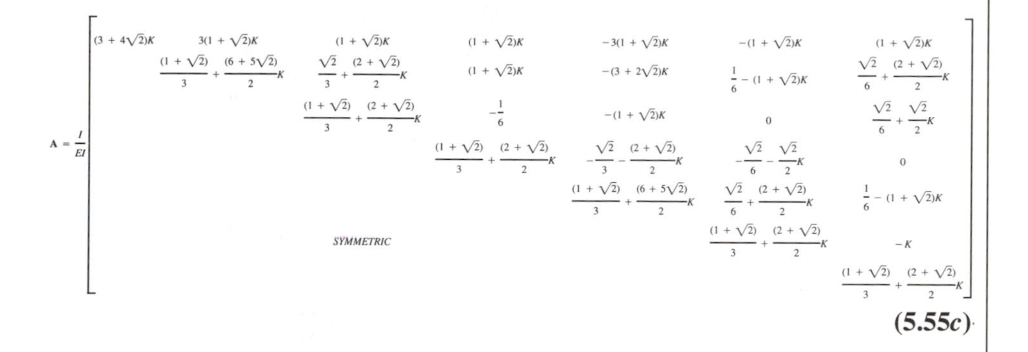

(5.55c)

Here, we have shown only the upper-triangular portion of the symmetric **A** matrix. The quantity K is the same dimensionless parameter we used in Examples 5 and 6, i.e., $K = I/Al^2$.

The **B** matrix ($-\mathbf{b}^T\mathbf{aP_o}$) is

$$\mathbf{B} = \frac{KFl^2}{EI}\begin{Bmatrix} -(1 + 2\sqrt{2}) \\ -(1 + 2\sqrt{2}) \\ -1 \\ -\sqrt{2} \\ (1 + \sqrt{2}) \\ \sqrt{2} \\ 0 \end{Bmatrix} \tag{5.55d}$$

The redundant forces are evaluated by inversion, i.e., $\mathbf{X} = \mathbf{A}^{-1}\mathbf{B}$. For the realistic case of slender elements for which $K = 0.001$ we obtain

$$\begin{Bmatrix} X_1 l \\ X_2 \\ X_3 \\ X_4 \\ X_5 \\ X_6 \\ X_7 \end{Bmatrix} = Fl\begin{Bmatrix} -0.443 \\ -0.002 \\ 0.001 \\ -0.002 \\ -0.003 \\ 0.001 \\ 0.002 \end{Bmatrix} \quad or \quad \begin{Bmatrix} X_1 \\ X_2 \\ X_3 \\ X_4 \\ X_5 \\ X_6 \\ X_7 \end{Bmatrix} = \begin{Bmatrix} -0.443F \\ -0.002Fl \\ 0.001Fl \\ -0.002Fl \\ -0.003Fl \\ 0.001Fl \\ 0.002Fl \end{Bmatrix} \tag{5.55e}$$

Back-substitution gives the internal forces,

$$\mathbf{P} \equiv \begin{Bmatrix} M_{AC} \\ M_{CA} \\ P_{AC} \\ M_{AD} \\ M_{DA} \\ P_{AD} \\ M_{BC} \\ M_{CB} \\ P_{BC} \\ M_{BD} \\ M_{DB} \\ P_{BD} \\ M_{CD} \\ M_{DC} \\ P_{CD} \end{Bmatrix} = \begin{Bmatrix} -0.001\,Fl \\ -0.002\,Fl \\ 0.444\,F \\ 0.001\,Fl \\ 0.000\,Fl \\ 0.784\,F \\ 0.002\,Fl \\ 0.001\,Fl \\ -0.621\,F \\ -0.002\,Fl \\ -0.003\,Fl \\ -0.559\,F \\ 0.001\,Fl \\ 0.002\,Fl \\ 0.441\,F \end{Bmatrix} \qquad \textbf{(5.55f)}$$

We will examine the relative significance of the axial forces and bending moments shortly. First, however, it is instructive to compare this solution to that for a similar structure with pinned joints, i.e., a truss. If all the joints of the structure in Figure 5.42*a* were pinned, it may be shown that the exact solution for internal forces is

$$\mathbf{P} \equiv \begin{Bmatrix} P_{AC} \\ P_{AD} \\ P_{BC} \\ P_{BD} \\ P_{CD} \end{Bmatrix} = F \begin{Bmatrix} 0.442 \\ 0.789 \\ -0.625 \\ -0.558 \\ 0.442 \end{Bmatrix} \qquad \textbf{(5.55g)}$$

These axial forces differ from the corresponding values in Equation 5.55*f* by less than 1%. Thus, this constituent of the solution is virtually identical for the truss and for the structure with rigid joints.

It remains to determine what effect the bending moments may have on the stresses in the elements of this structure. We will examine this aspect of behavior by comparing relative values of stresses due to axial forces and stresses due to bending moments.

Recall that stresses due to axial forces are given by P/A, and maximum stresses due to bending moments are given by $MH/2I$ where H is the depth of the beam cross section.[54] From Equation 5.55*f* we observe that, for this particular solution, typical values of M may be related to typical values of P by a relation of the type $M \approx Pl/200$. If we substitute this expression into the equation for bending stress, we obtain $PlH/(400I)$. The value $K = 0.001$ upon which the solution in Equation 5.55*f* is based

[54] In this discussion we are assuming the cross-sectional shape is symmetric about the middle reference axis so that the distance to the extreme fiber is the same above or below the reference axis. In this case, maximum bending stress is given by Mc/I where $c = H/2$; otherwise the distance $c \neq H/2$ but rather $c = \alpha H$ where $0 < \alpha < 1$.

corresponds to $l \approx 13H$[55] (which is close to the approximate lower limit for the definition of ''slender'' elements); thus, maximum bending stresses for the elements in this structure are given by the approximate expression $PH^2/(30I)$.

We compare stresses due to axial forces with maximum stresses due to bending moments, i.e., compare P/A to $PH^2/(30I)$. It follows that the stresses due to axial forces will be greater than the maximum stresses due to bending provided $30I/(AH^2) > 1$. For typical structural elements, values of the dimensionless parameter $30I/(AH^2)$ are usually greater than 5. Thus, the stresses due to axial forces can be expected to be at least five times greater than the maximum bending stresses.

If we increase ''slenderness'' such that K is reduced to $K \approx 0.0003$ (which corresponds to $l \approx 25H$) then it can be shown that the solution would yield a relation between bending moments and axial forces given by $M \approx Pl/600$. A comparison of stresses would then reveal that axial stresses would be approximately eight times greater than the maximum bending stresses. We can also expect joints that are not completely rigid to cause even lower moments and, hence, lower bending stresses.

Thus, we see that the primary mode of force transmission in truss-like structures composed of slender components is by axial forces regardless of the degree of joint fixity. It should be borne in mind, however, that there is no clear demarcation between significant and insignificant effects of bending, e.g., bending stresses that are 12%–20% of the stresses due to axial forces (as above) may or may not be negligible depending on the margins of safety used to design the structure.

It is the responsibility of the analyst to assess all potentially important factors; for example, if an analyst chooses to model a structure as a truss, the choice should be made only after determining that it is safe to neglect the effects of joint flexibility and bending stresses in the analysis. This is no small responsibility.

5.5 COMPUTING DEFLECTIONS

After completing the analysis of internal forces in an indeterminate structure, it is a relatively straightforward matter to compute any desired deflection. The procedure, using virtual forces, has been discussed in detail in Chapter 4.

EXAMPLE 11

Find the horizontal component of displacement at D, u_{D_x}, for the structure in Figure 5.42*a* (Example 10). We will utilize the solution for internal forces given in Equation 5.55*f* (which is based on $K \equiv I/Al^2 = 0.001$). The internal forces are shown in Figure 5.45.[56]

---

[55] Recall the definition $K \equiv I/Al^2$. For common structural steel shapes, values of I, A, and depth H are available in references such as the *Manual of Steel Construction* published by the American Institute of Steel Construction, Inc. (AISC). For a wide variety of typical shapes we obtain $K \approx 0.001$ for $l \approx 13H$.

[56] It is necessary to add significant figures to the numerical values of moments to obtain reasonable accuracy in the subsequent computations; thus, $M_{AC} = -0.00120Fl$ rather than $-0.001Fl$, etc.

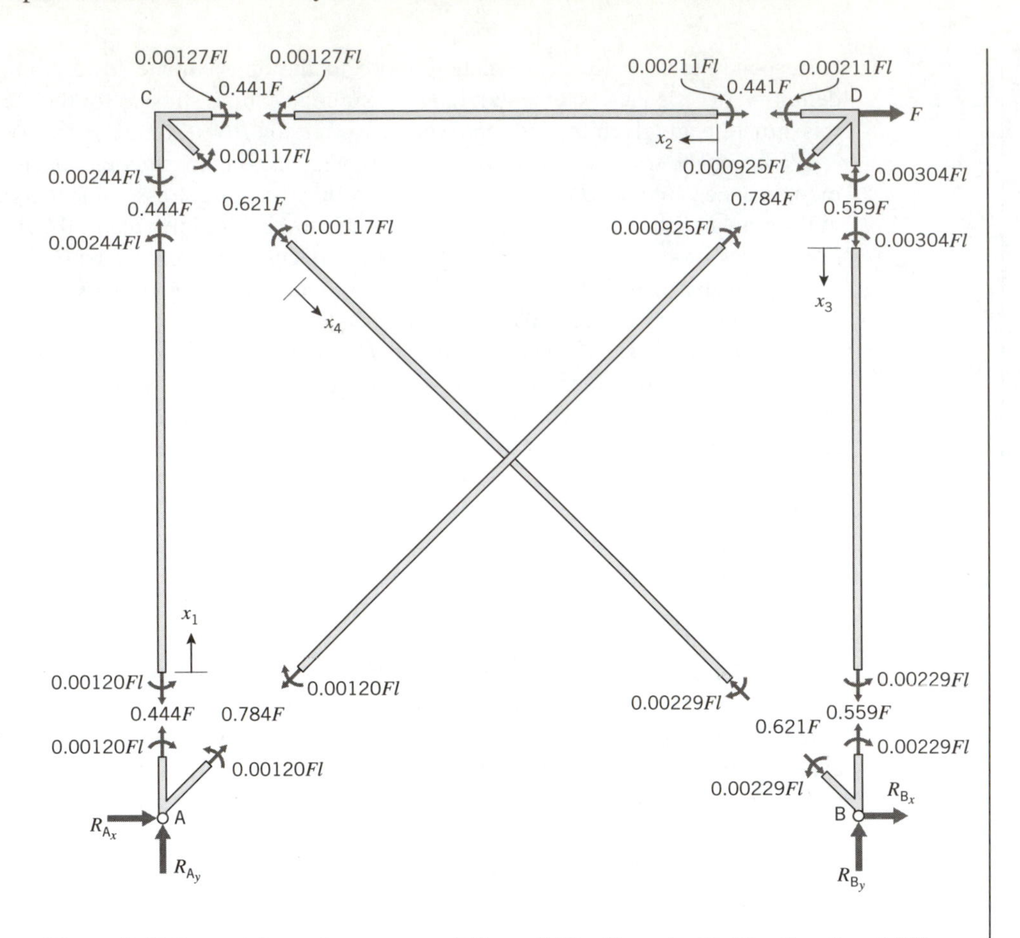

Figure 5.45 Internal forces in structure of Figure 5.39*a*, Example 10. (See Equation 5.55*f*; note that transverse shear components are not shown but are required for equilibrium.)

The internal forces cause deformations of the individual components of the structure, i.e., the axial forces cause length changes and the bending moments cause curvatures. (Transverse shears have a negligible effect on the deformations of slender components; these effects are ignored throughout.) Virtual work for a structure of this type requires superposition of Equation 4.8 for axial forces/deformations and Equation 4.24 for bending moments/curvatures.

We begin by applying a virtual force, F_V, at D in the direction of u_{Dx}. (This gives an external virtual work equal to $F_V u_{Dx}$.) We now require a set of virtual internal forces in equilibrium with F_V. As was noted in Chapter 4, there are many different sets of virtual forces that can satisfy equilibrium in an indeterminate structure. One such set is shown in Figure 5.45 with F_V substituted for the real force F. Three other sets of virtual forces are shown in Figure 5.46. In fact, it will be much easier to perform the desired calculation using any of the sets of forces in Figure 5.46 than it would be using the forces in Figure 5.45. We will demonstrate using all three sets of forces in Figure 5.46.

Force set (*a*). The internal virtual forces in Figure 5.46*a* consist only of axial forces

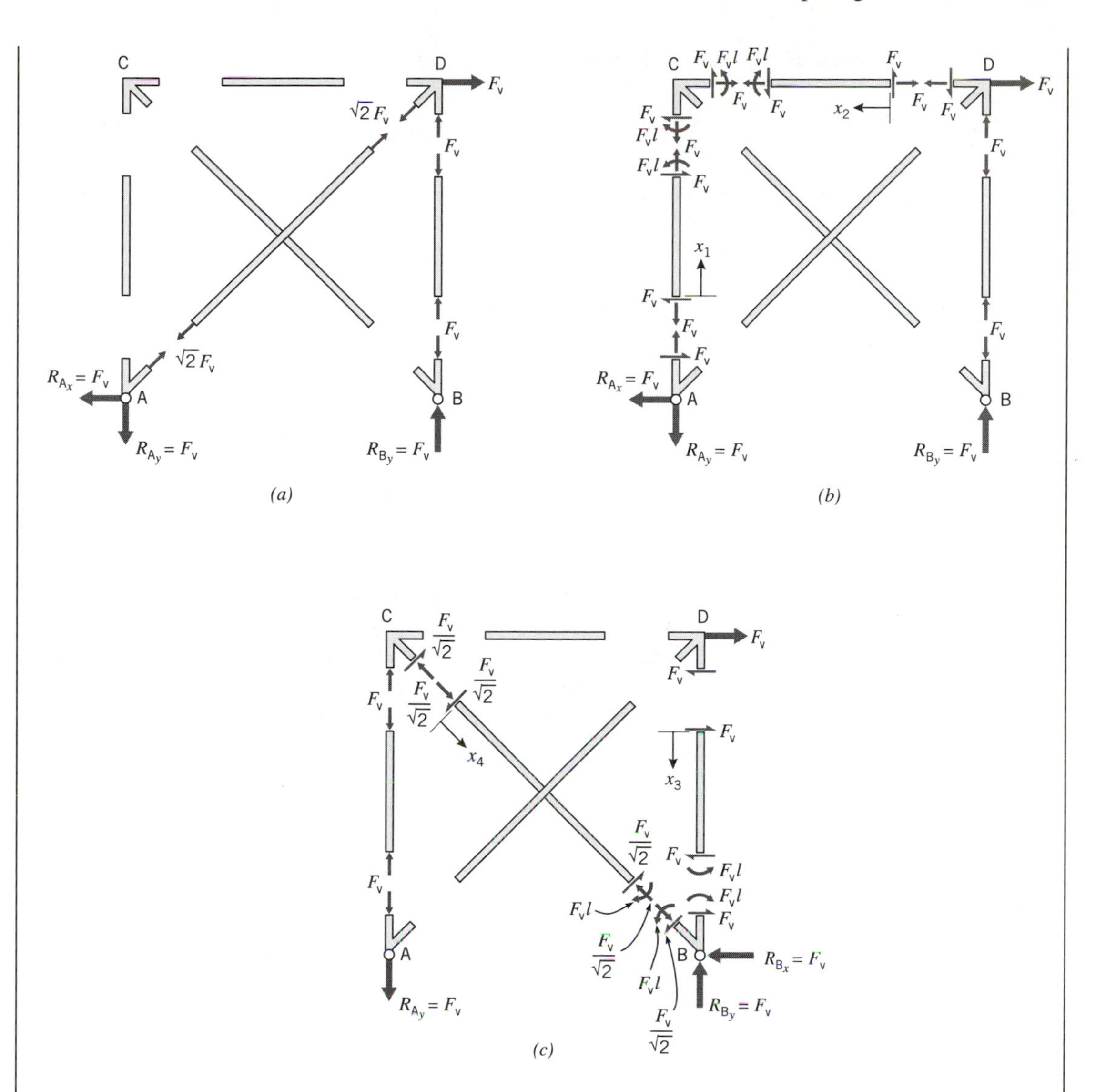

Figure 46 Virtual forces for calculation of u_{D_x}.

in elements AD and BD given by $P_{AD_V} = \sqrt{2}F_V$, $P_{BD_V} = -F_V$. There are no bending moments in any of the elements in this virtual force state.

Consequently, the only internal virtual work will be that associated with the virtual axial forces and the real deformations (length changes) of elements AD and BD. These real deformations are

$$\begin{aligned}\Delta_{AD}^{(a)} &= \frac{P_{AD}l_{AD}}{AE} = \frac{0.784F(\sqrt{2}l)}{AE} = 1.109\frac{Fl}{AE} \\ \Delta_{BD}^{(a)} &= \frac{P_{BD}l_{BD}}{AE} = -0.559\frac{Fl}{AE}\end{aligned} \tag{5.56a}$$

The virtual work equation, Equation 4.8, becomes

$$F_V u_{D_x} = P_{AD_V}\Delta_{AD}^{(a)} + P_{BD_V}\Delta_{BD}^{(a)} \tag{5.56b}$$

Substituting the expressions for P_{AD_V} and P_{BD_V}, and canceling F_V in each term, leads to

$$\begin{aligned} u_{D_x} &= \sqrt{2}\left(1.109\frac{Fl}{AE}\right) + (-1)\left(-0.559\frac{Fl}{AE}\right) \\ &= 2.127\frac{Fl}{AE} \end{aligned} \tag{5.56c}$$

which is the desired displacement. Note that from the definition $K \equiv I/Al^2 = 0.001$ we have $A = I/0.001l^2$, so we may also write $u_{D_x} = 0.00213\,Fl^3/EI$, which is equivalent to Equation 5.56*c*.

Force set (*b*). Once we introduce bending moments into the virtual force state, the computation becomes more involved. Consider the virtual force state in Figure 5.46*b*, which consists of axial forces in elements AC, CD, and BD, and bending moments in elements AC and CD. The virtual axial forces are given by $P_{AC_V} = P_{CD_V} = F_V$, $P_{BD_V} = -F_V$. We express the virtual bending moments in AC and CD as functions of coordinates x_1 and x_2, respectively, in Figure 5.46*b*

$$\begin{aligned} M_V(x_1) &= F_V x_1 \\ M_V(x_2) &= F_V x_2 \end{aligned} \tag{5.56d}$$

Now, we require the real deformations, which consist of length changes $\Delta_{AC}^{(a)}$, $\Delta_{CD}^{(a)}$, and $\Delta_{BD}^{(a)}$, and curvatures $v''(x_1)$ and $v''(x_2)$ for elements AC and CD, respectively. Using the information in Figure 5.45 we obtain

$$\begin{aligned} \Delta_{AC}^{(a)} &= 0.444\frac{Fl}{AE} \\ \Delta_{CD}^{(a)} &= 0.441\frac{Fl}{AE} \\ \Delta_{BD}^{(a)} &= -0.559\frac{Fl}{AE} \\ v''(x_1) &= \frac{1}{EI}(0.00364Fx_1 - 0.00120Fl) = \frac{F}{l^2AE}(3.64x_1 - 1.20l) \\ v''(x_2) &= \frac{1}{EI}(0.00338Fx_2 - 0.00211Fl) = \frac{F}{l^2AE}(3.38x_2 - 2.11l) \end{aligned} \tag{5.56e}$$

Virtual work, Equations 4.8 and 4.24, gives

$$F_V u_{D_x} = P_{AC_V}\Delta_{AC}^{(a)} + P_{CD_V}\Delta_{CD}^{(a)} + P_{BD_V}\Delta_{BD}^{(a)} + \int_0^l M_V(x_1)v''(x_1)dx_1 + \int_0^l M_V(x_2)v''(x_2)dx_2 \quad \textbf{(5.56f)}$$

Substituting Equations 5.56*e* and the values of P_{AC_V}, P_{CD_V}, and P_{BD_V} leads to

$$u_{D_x} = (1)\left(0.444\frac{Fl}{AE}\right) + (1)\left(0.441\frac{Fl}{AE}\right) + (-1)\left(-0.559\frac{Fl}{AE}\right) + \frac{F}{l^2AE}\int_0^l (x_1)(3.64x_1 - 1.20l)dx_1 + \frac{F}{l^2AE}\int_0^l (x_2)(3.38x_2 - 2.11l)dx_2$$

$$= 2.129\frac{Fl}{AE} \quad \textbf{(5.56g)}$$

which is the same result obtained in Equation 5.56*c* except for a small round-off error in the numerical calculations.

Force set (*c*). The virtual force state shown in Figure 5.46*c* is similar in concept to that in Figure 5.46*b*, but constitutes a completely different distribution of forces. Here, elements AC and BC transmit virtual axial forces ($P_{AC_V} = F_V$, $P_{BC_V} = -F_V/\sqrt{2}$), and elements BC and BD transmit virtual bending moments.

In this case we have

$$\Delta_{AC}^{(a)} = 0.444\frac{Fl}{AE}$$

$$\Delta_{BC}^{(a)} = -0.621\frac{F(\sqrt{2}l)}{AE} = -0.878\frac{Fl}{AE}$$

$$v''(x_3) = \frac{F}{l^2AE}(5.33x_3 - 3.04l) \quad \textbf{(5.56h)}$$

$$v''(x_4) = \frac{F}{l^2AE}(-2.45x_4 + 1.17l)$$

and the virtual work equation becomes

$$u_{D_x} = (1)\left(0.444\frac{Fl}{AE}\right) + \left(-\frac{1}{\sqrt{2}}\right)\left(-0.878\frac{Fl}{AE}\right) + \frac{F}{l^2AE}\int_0^l (x_3)(5.33x_3 - 3.04l)dx_3 + \frac{F}{l^2AE}\int_0^{\sqrt{2}l}\left(-\frac{x_4}{\sqrt{2}}\right)(-2.45x_4 + 1.17l)dx_4 = 2.128\frac{Fl}{AE} \quad \textbf{(5.56i)}$$

Again, the deflection is essentially the same as the previous two results. (Note the use of coordinates x_3 and x_4, Figures 5.42 and 5.43*c*, for the computation of virtual moments and real curvatures in elements BC and BD.)

The result is even easier to obtain if we use matrix algebra. To demonstrate, assume we express the internal virtual forces by a vector, $\mathbf{P}_V$, where

$$\mathbf{P}_V \equiv \begin{Bmatrix} M_{\mathrm{AC}_V} \\ M_{\mathrm{CA}_V} \\ P_{\mathrm{AC}_V} \\ M_{\mathrm{AD}_V} \\ M_{\mathrm{DA}_V} \\ P_{\mathrm{AD}_V} \\ M_{\mathrm{BC}_V} \\ M_{\mathrm{CB}_V} \\ P_{\mathrm{BC}_V} \\ M_{\mathrm{BD}_V} \\ M_{\mathrm{DB}_V} \\ P_{\mathrm{BD}_V} \\ M_{\mathrm{CD}_V} \\ M_{\mathrm{DC}_V} \\ P_{\mathrm{CD}_V} \end{Bmatrix} \qquad \mathbf{(5.56\textit{j})}$$

For example, the virtual force states in Figures 5.46*a*–5.46*c*, respectively, have the vector forms

$$\mathbf{P}_{a_V} = \begin{Bmatrix} 0 \\ 0 \\ 0 \\ 0 \\ 0 \\ \sqrt{2} \\ 0 \\ 0 \\ 0 \\ 0 \\ 0 \\ -1 \\ 0 \\ 0 \\ 0 \end{Bmatrix} F_V \ , \quad \mathbf{P}_{b_V} = \begin{Bmatrix} 0 \\ -l \\ 1 \\ 0 \\ 0 \\ 0 \\ 0 \\ 0 \\ 0 \\ 0 \\ 0 \\ -1 \\ l \\ 0 \\ 1 \end{Bmatrix} F_V \ , \quad \mathbf{P}_{c_V} = \begin{Bmatrix} 0 \\ 0 \\ 1 \\ 0 \\ 0 \\ 0 \\ l \\ 0 \\ -1/\sqrt{2} \\ -l \\ 0 \\ 0 \\ 0 \\ 0 \\ 0 \end{Bmatrix} F_V \qquad \mathbf{(5.56\textit{k})}$$

We have seen previously (Section 5.2.3, Equation 5.31*h*) that internal virtual work is given by $\mathbf{P}_V^T\boldsymbol{\Delta}$. In Equation 5.31*h*, external virtual work was zero for every virtual force, X_V, which was chosen "like" a redundant force. Here, however, F_V is an external virtual force unrelated to the redundants, and external virtual work is $F_V u_{D_x}$.[57] Thus, the virtual work equation is

$$F_V u_{D_x} = \mathbf{P}_V^T\boldsymbol{\Delta} \tag{5.56l}$$

We compute $\boldsymbol{\Delta} = \mathbf{aP}$ for this structure from the **a** matrix in Equation 5.55*b* and the **P** vector in Equation 5.55*f*; we obtain

$$\begin{Bmatrix} \Delta_{AC} \\ \Delta_{CA} \\ \Delta_{AC}^{(a)} \\ \Delta_{AD} \\ \Delta_{DA} \\ \Delta_{AD}^{(a)} \\ \Delta_{BC} \\ \Delta_{CB} \\ \Delta_{BC}^{(a)} \\ \Delta_{BD} \\ \Delta_{DB} \\ \Delta_{BD}^{(a)} \\ \Delta_{CD} \\ \Delta_{DC} \\ \Delta_{CD}^{(a)} \end{Bmatrix} = \frac{1}{EI}\begin{Bmatrix} 5.373 \times 10^{-6}Fl^2 \\ -6.118 \times 10^{-4}Fl^2 \\ 4.443 \times 10^{-4}Fl^3 \\ 3.486 \times 10^{-4}Fl^2 \\ 1.529 \times 10^{-4}Fl^2 \\ 1.109 \times 10^{-3}Fl^3 \\ 8.053 \times 10^{-4}Fl^2 \\ 9.315 \times 10^{-6}Fl^2 \\ -8.784 \times 10^{-4}Fl^3 \\ -2.570 \times 10^{-4}Fl^2 \\ -6.315 \times 10^{-4}Fl^2 \\ -5.591 \times 10^{-4}Fl^3 \\ 7.140 \times 10^{-5}Fl^2 \\ 4.929 \times 10^{-4}Fl^2 \\ 4.411 \times 10^{-4}Fl^3 \end{Bmatrix} = \frac{1}{EA}\begin{Bmatrix} 5.373 \times 10^{-3}F \\ -6.118 \times 10^{-1}F \\ 4.443 \times 10^{-1}Fl \\ 3.486 \times 10^{-1}F \\ 1.529 \times 10^{-1}F \\ 1.109Fl \\ 8.053 \times 10^{-1}F \\ 9.315 \times 10^{-3}F \\ -8.784 \times 10^{-1}Fl \\ -2.570 \times 10^{-1}F \\ -6.315 \times 10^{-1}F \\ -5.591 \times 10^{-1}Fl \\ 7.140 \times 10^{-2}F \\ 4.929 \times 10^{-1}F \\ 4.411 \times 10^{-1}Fl \end{Bmatrix} \tag{5.56m}$$

Substituting Equation 5.56*m and any of the vectors in Equation 5.56k* into Equation 5.56*l* gives

$$u_{D_x} = 2.128\frac{Fl}{AE} \tag{5.56n}$$

exactly as expected.

[57] As with redundant forces, F_V is assumed to act at ends of elements.

PROBLEMS

5.P1. Find the bar forces. Assume that all bars have the same *EA*. Note that bar AB has an initial length change Δ_I.

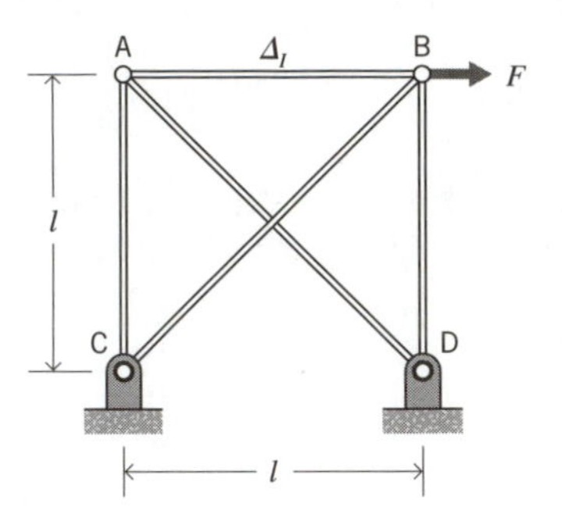

5.P2. Find the bar forces and the reactions. Choose the redundant force as
(a) horizontal component of reaction at B;
(b) force in bar AC.
Assume that all bars have the same *EA*. Note the uniform temperature change in bar CD.

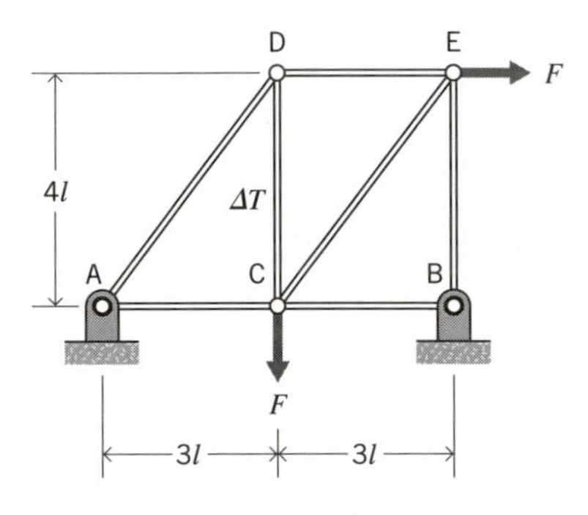

5.P3. Solve for the bar forces. Assume that all bars have the same *EA*.

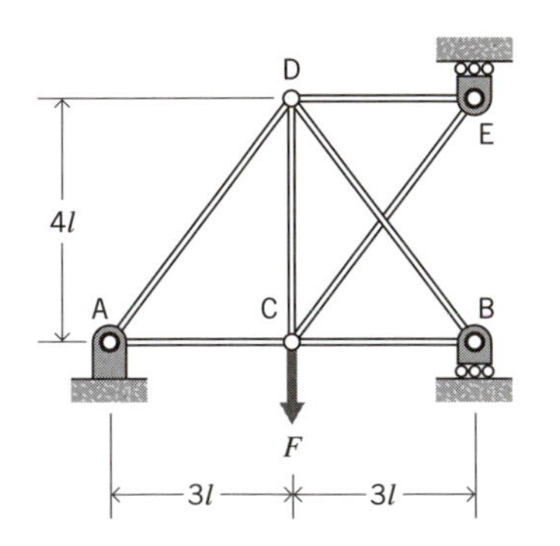

5.P4. Find the bar forces. Assume that all bars have the same EA.

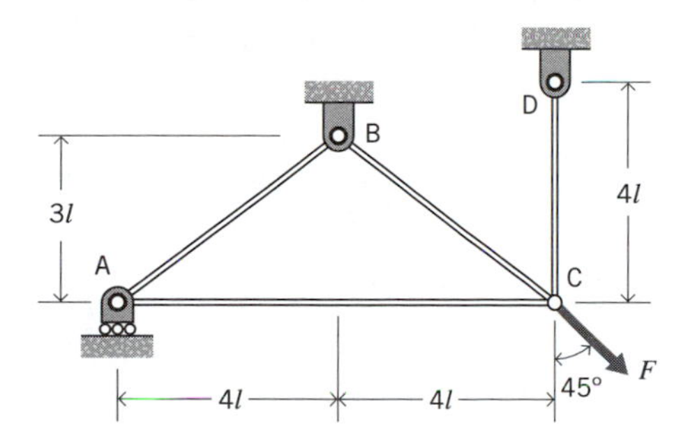

5.P5. Find the bar forces. Assume that all bars have the same EA. The only load is a uniform temperature change ΔT in bar EG.

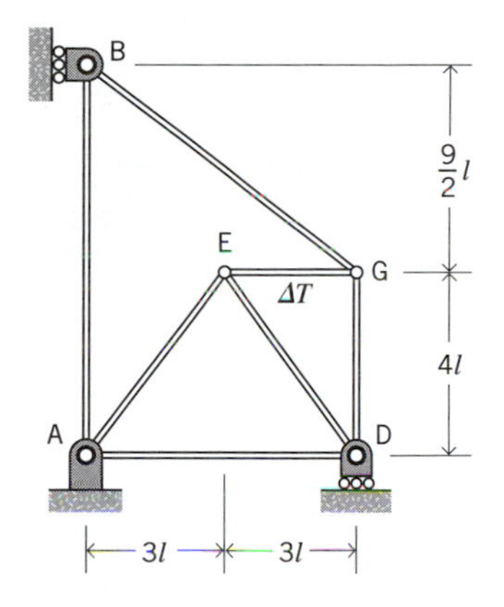

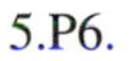

5.P6. Find the bar forces. Assume that all bars have the same EA.

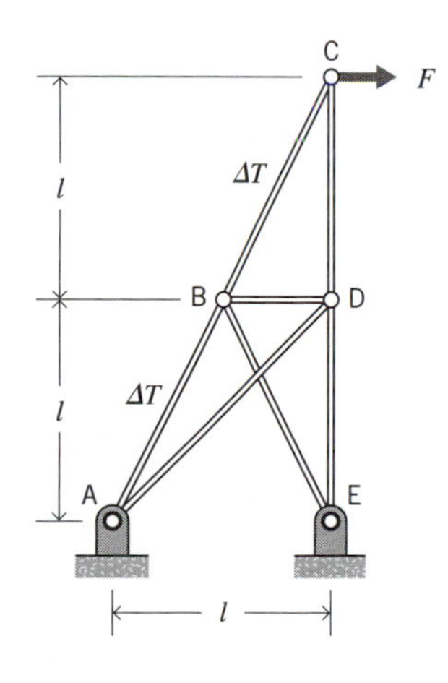

5.P7. A heavy concrete block of weight W is supported by a truss structure and spring, as shown. Assume the block is rigid. All bars have the same EA.
(a) Find the bar forces and the force in the spring.
(b) Find the value of spring stiffness k for which $\theta = 0$.

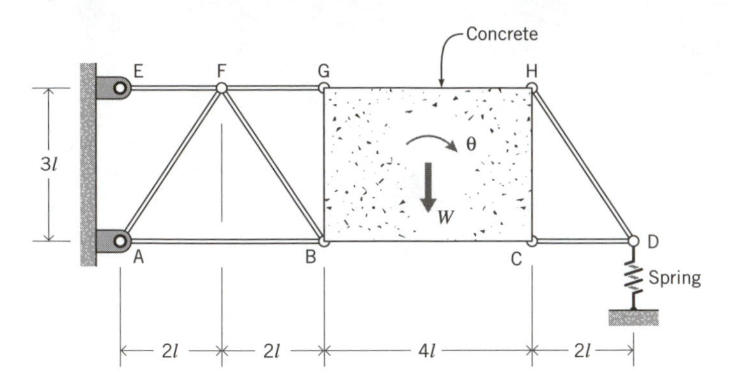

5.P8. Find the bar forces for the truss in Figure 5.2*a* using the choice of redundant forces shown in Figure 5.2*g*. Assume that all bars have the same EA.

5.P9. Find all the bar forces. Assume that all bars have the same EA.

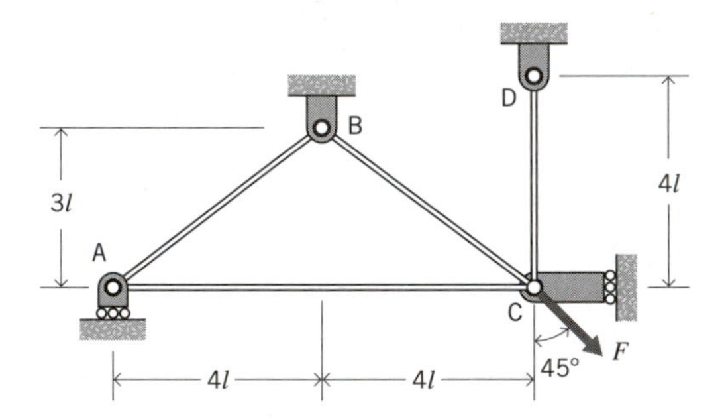

5.P10. The truss is similar to the one in Problem 5.P5 except that bar AG has been added. Find the bar forces if member EG only is subjected to a uniform temperature change ΔT. Assume that all bars have the same EA.

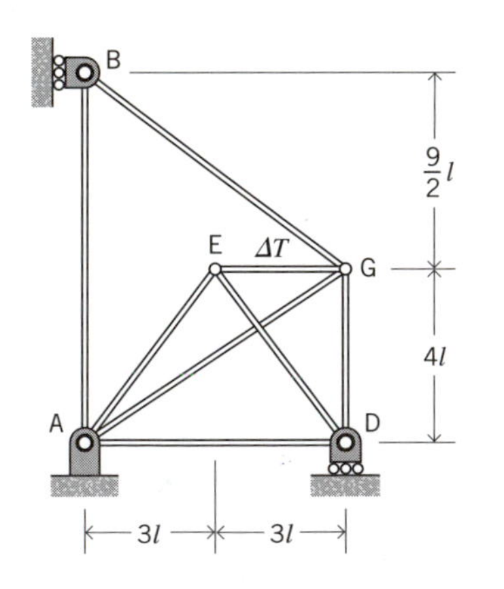

5.P11. Find the bar forces. Assume that all bars have the same EA. Use a different set of redundant forces than in Example 2.

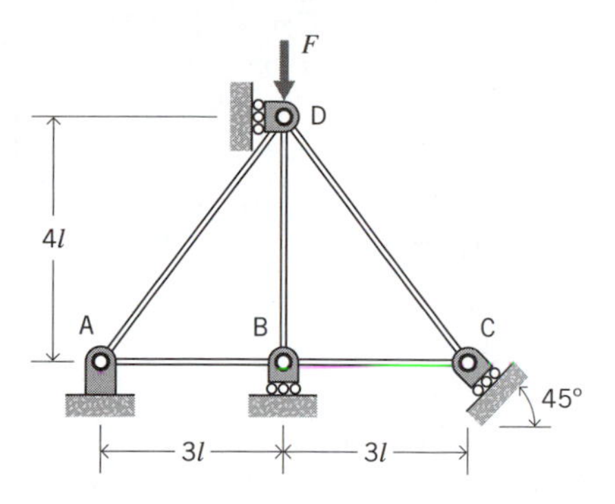

5.P12. Find the bar forces. Assume that all bars have the same EA. The only load is a uniform temperature change in bar CD.

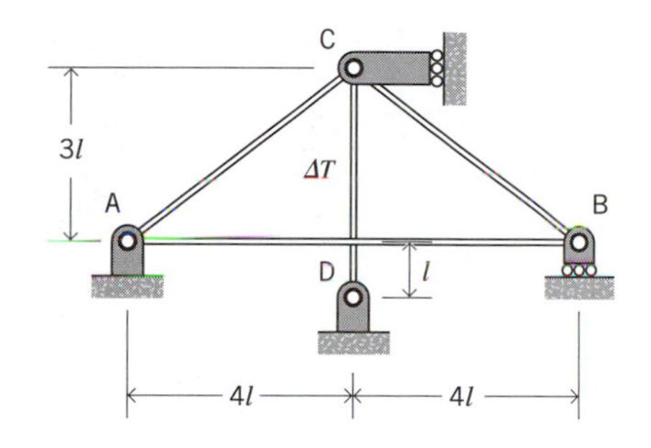

5.P13. Find the bar forces. Assume that all bars have the same EA.

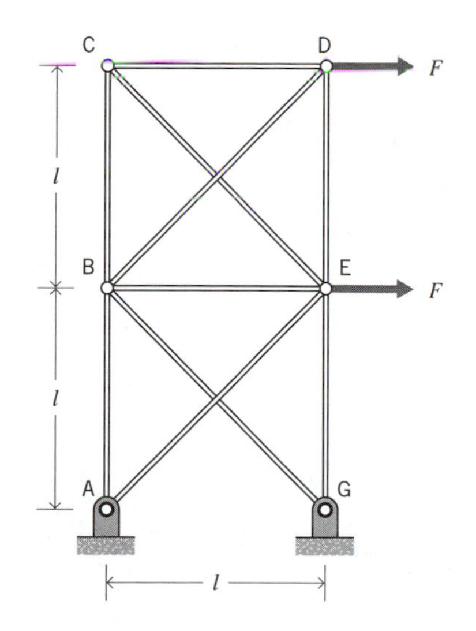

5.P14. Solve for the bar forces. Assume that all bars have the same EA.

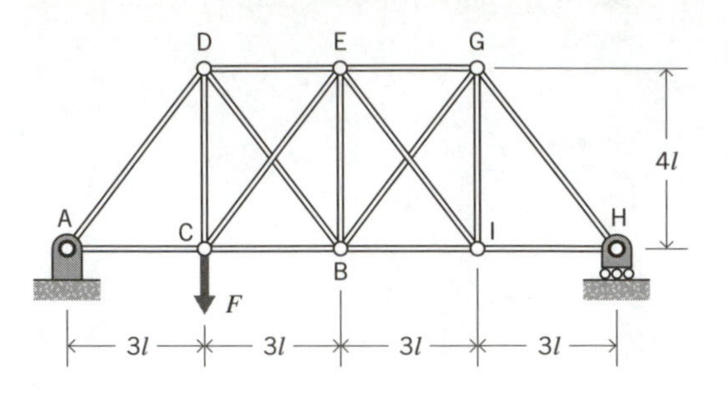

5.P15. Find the bar forces in the power transmission tower shown. Forces F represent weight of the wires; the horizontal component of the forces is due to an earthquake producing a 1g horizontal acceleration. Assume that all bars have the same EA.

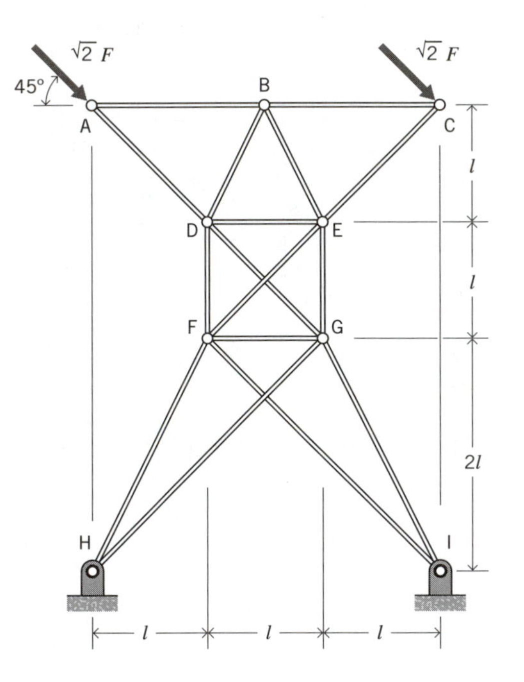

5.P16. Find the bar forces. Assume that all bars have the same EA.

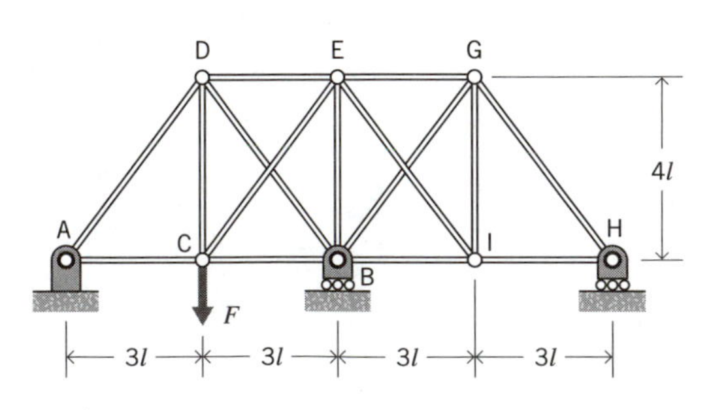

5.P17. The truss is supported by three linear springs with spring constant k. Assume that all bars have the same EA. Find the forces in the springs. Solve the equations for two cases: (*a*) $EA/kl \ll 1$ and (*b*) $EA/kl \gg 1$. Hint: because the geometry and loading are symmetric about axis BE, it is possible to deal with only two redundant forces.

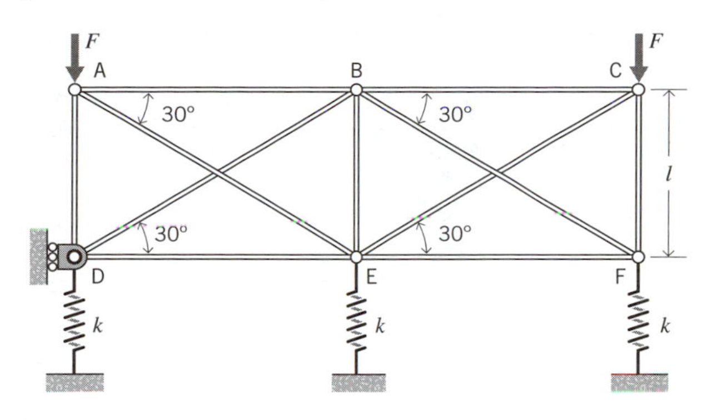

5.P18. Analyze the beam using (*a*) the moment at B and (*b*) the internal shear at $x = a$, as the redundant force.

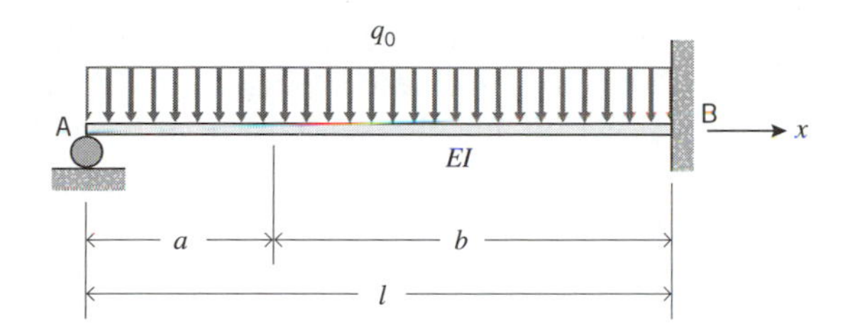

5.P19. The structure consists of a beam AB and a cable BC. The support at B can displace vertically but cannot rotate or displace horizontally. The cable stiffness is such that $EA = 3\sqrt{2}EI/l^2$. Find the force in the cable, and sketch the moment diagram for the beam.

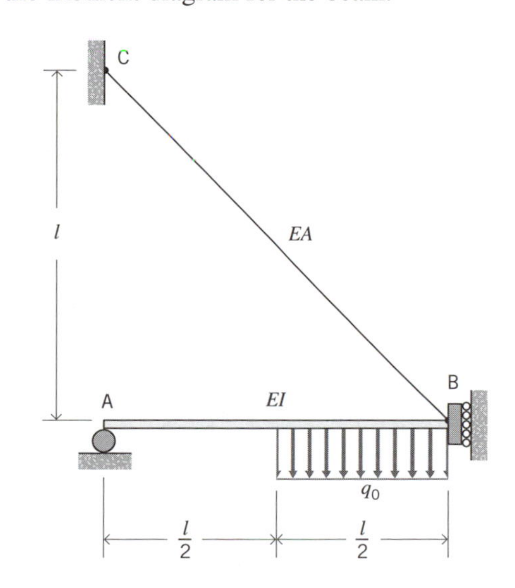

5.P20. The beam is fixed at B and supported by a linear spring with spring constant k at A.

(a) Find the spring force as a function of M_o, l, EI, and k.

(b) Sketch the moment diagram for the special case $k = 3EI/2l^3$.

(c) From (a), find the spring force as $k \rightarrow \infty$.

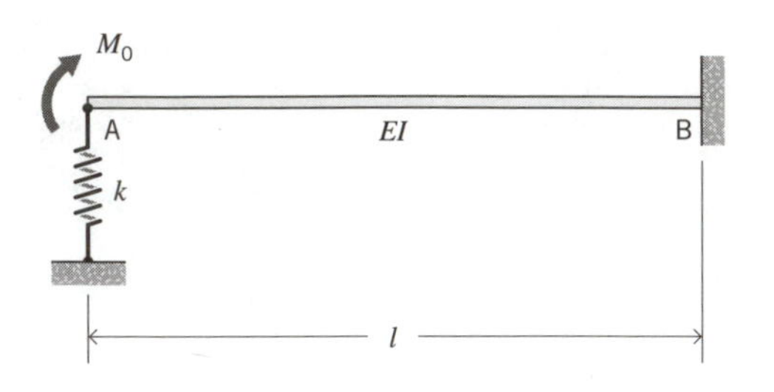

5.P21. The beam is fixed at B and supported by a linear spring with spring constant k at A. The structure is loaded by a concentrated moment M_o applied at the midpoint of the beam.

(a) Find the spring force as a function of M_o, l, EI, and k.

(b) Sketch the moment diagram for the special case $k = 3EI/2l^3$.

(c) From (a), find the spring force as $k \rightarrow \infty$.

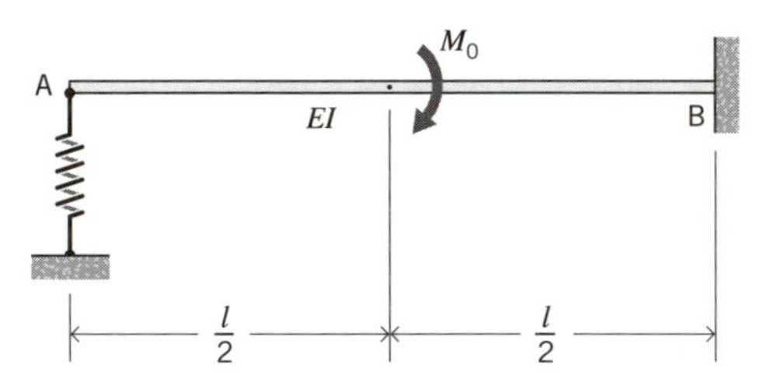

5.P22. Sketch the moment diagram for beam ABC. Assume that beam ABC is axially rigid. The axial stiffness of bars AD, BD, and CD is EA. Assume $I = l^2A$.

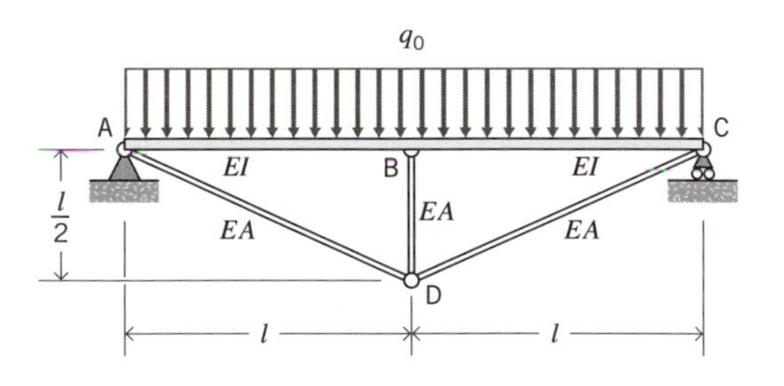

5.P23. Sketch bending moment diagrams for beams AB and BC. Assume that the beams are axially rigid.

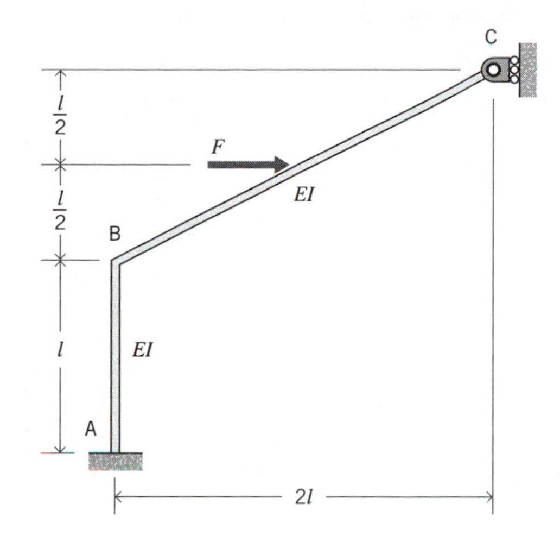

5.P24. The structure is composed of four bars (AC, AD, BC, and BE) and one axially rigid beam (AB). All connections between elements are pinned. The stiffness of all bars is EA and the bending stiffness of the beam is $EI = EAl^2/3$. Find the internal forces.

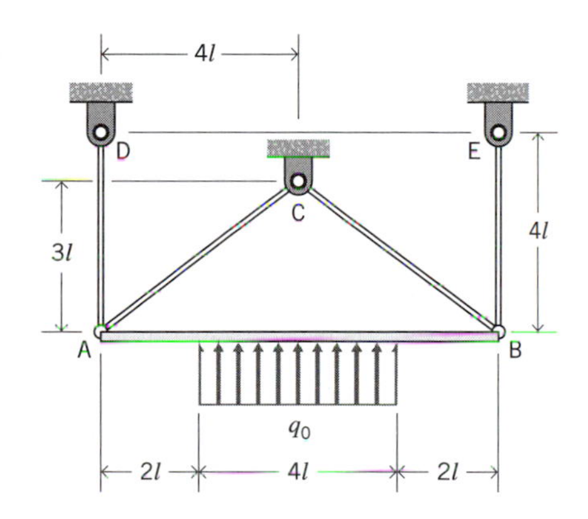

5.P25. The beam is fixed at both ends. Solve for the end forces.

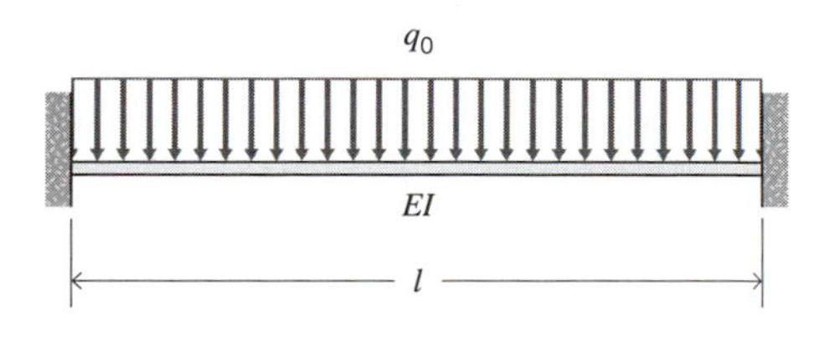

5.P26. The structure is fixed at A and C. There is a hinge connecting the beams at B. Joint B is supported by a linearly elastic spring of stiffness k. Find the internal forces in terms of EI and k. Solve using the internal shear in the beam just to the left of B as one redundant force.

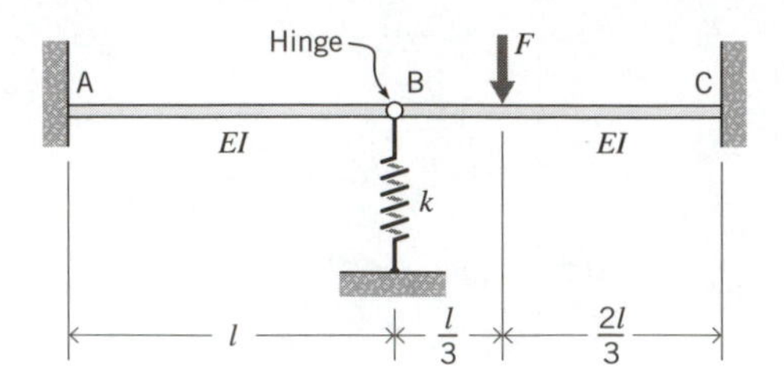

5.P27. The structure has a support at A that is free to roll horizontally but cannot rotate; the support at B can move vertically but also cannot rotate. End B is supported by a linear axial spring. Find the force in the spring and all reactions. Let $k = 3EI/2l^3$.

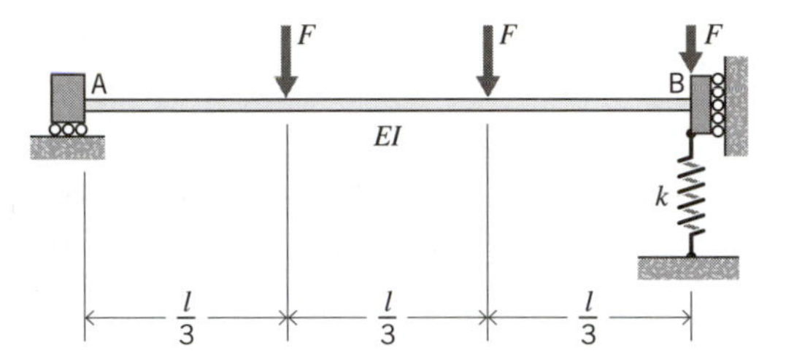

5.P28. Find the end forces on beams AB and BC.

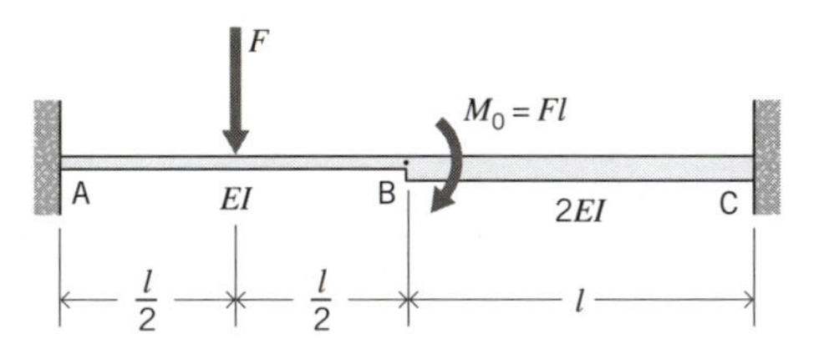

5.P29. The support at C can move vertically but cannot rotate. Find the end forces on each beam.

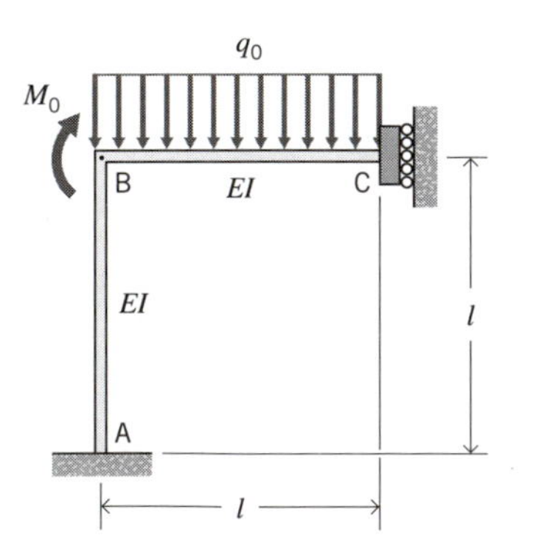

5.P30. The beam is fixed at B and pinned to a roller support at A. There is a linear torsional spring connecting the beam to the support at A. The relationship between moment and deformation for the spring is $M_s = k\Delta_s$ where k is the spring constant and Δ_s is the deformation (angle change). Sketch the moment diagram for the case $k = EI/2l$.

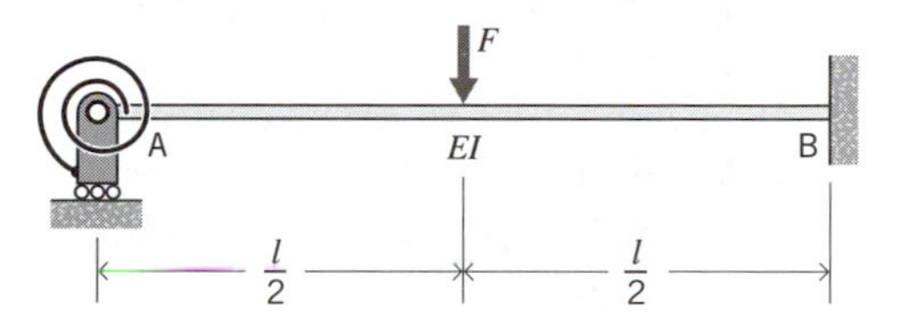

5.P31. The structure consists of a uniform axially rigid beam ABC and a bar BD. There is a fixed support at C that can translate but cannot rotate. Assume that the axial stiffness of the bar is $EA = 3\sqrt{2}EI/l^2$. Find the internal forces.

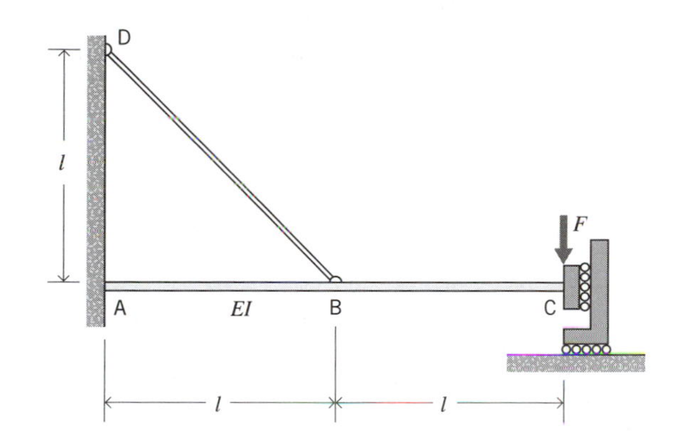

5.P32. The load F in Problem 5.P31 is replaced by a moment M_o (clockwise) applied at the midspan of BC. As in Problem 5.P31, let $EA = 3\sqrt{2}EI/l^2$ and find the internal forces.

5.P33. Beams AB and BC are joined by a hinge at B; there is a torsional spring connecting the ends of the beams at B. The relationship between moment and deformation for the spring is $M_s = k\Delta_s$ where k is the spring constant and Δ_s is the deformation (angle change). Sketch free body diagrams of all components showing the magnitudes of all forces. Let $k = 2EI/l$.

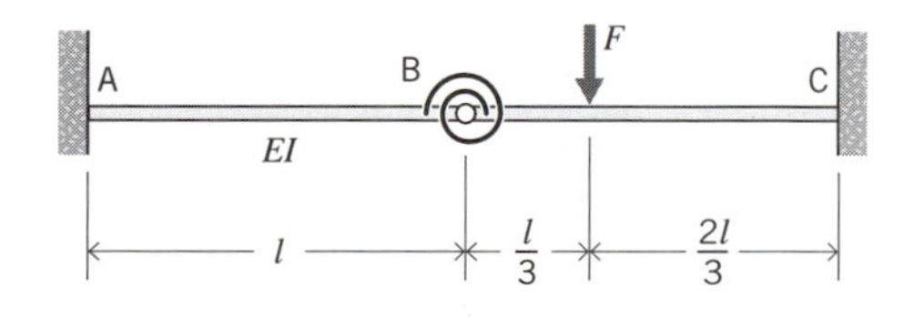

5.P34. For the structure in Problem 5.P33, explain whether or not the following sets of forces can be chosen as redundant forces:

(a) internal shear at end B of beam AB and internal shear at end B of beam BC;
(b) internal shear at end B of beam AB and internal moment at end A of beam AB;
(c) internal shear at end B of beam AB and internal moment at end C of beam BC;
(d) internal shear at end B of beam AB and internal shear at end C of beam BC;
(e) internal shear at end B of beam AB and internal moment at end B of beam BC.

5.P35. Find reactions and sketch bending moment diagrams for all elements. Use internal moments at B and C as redundant forces.

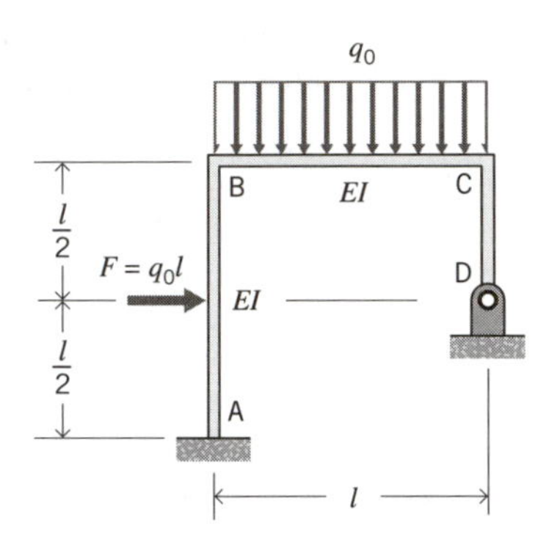

5.P36. The structure has a pin support at A and a fixed support with roller at C. The joints at ends A and B of beam AB are modeled as hinges with torsional springs for which $M_s = k\Delta_s$ where k is the spring constant and Δ_s is the deformation (angle change). There is a uniform load of magnitude q_o applied to half of beam BC, and a moment $M_o = 5q_o l^2/8$ applied to joint B. Find the internal forces. Assume $k = 2EI/l$.

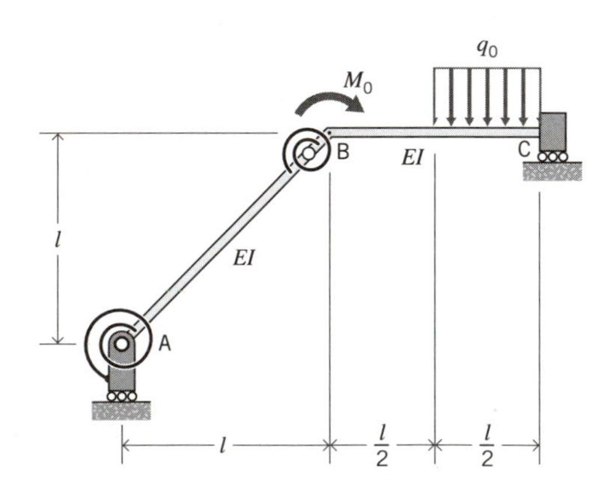

5.P37. The structure is composed of three beams. The connections between beams at A and B are rigid, and there is a hinge at C. All elements have bending stiffness EI. The structure is loaded by a moment M_o applied at the mid-point of beam AB. Find the internal forces.

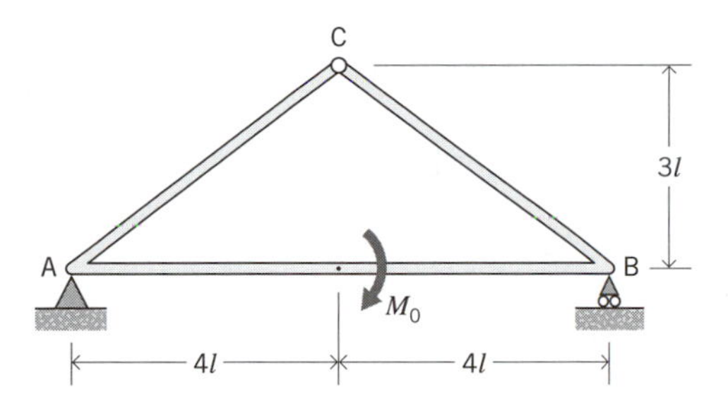

5.P38. Beam ABC is supported at its midpoint by a linear spring with spring constant k. Find internal forces for the case $k = 8EI/l^3$.

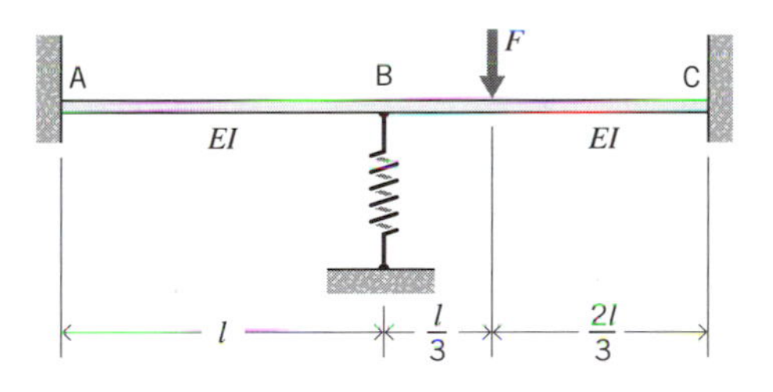

5.P39. Find end forces on all beams. Let $F_1 = 3q_ol$, $F_2 = 2q_ol$.

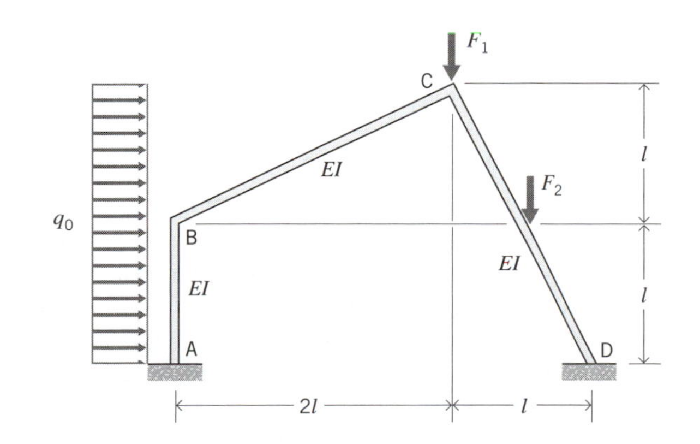

5.P40–P41. The structure is a closed frame. Solve for the end forces on all the elements and sketch free-body diagrams.

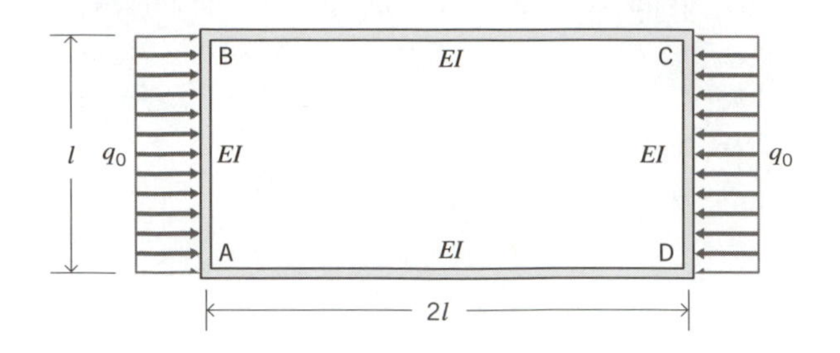

5.P40

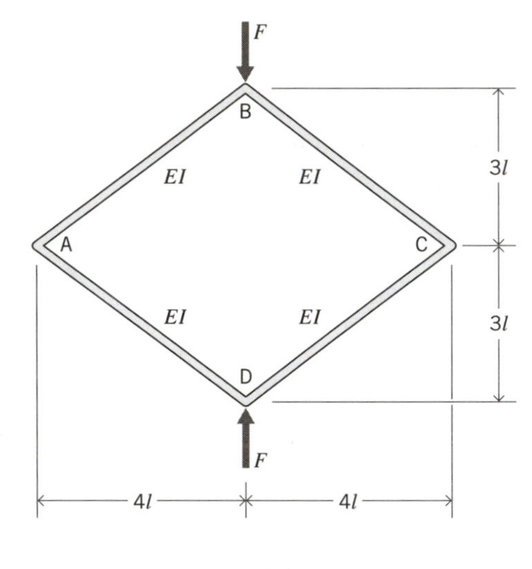

5.P41

5.P42. Assume the elements in Problem 5.P39 are axially deformable rather than axially rigid, and let $I/l^2A = 0.001$ for all elements. Find the end forces using the same values of F_1 and F_2 given in Problem 5.P39.

5.P43. Force F is removed from the truss in Problem 5.P4, but the roller support at A displaces vertically downward a small distance δ_o due to settlement.
(a) Find the bar forces.
(b) Find the horizontal displacement at A.

5.P44. Force F is removed from the structure in Problem 5.P31 but the support at A (only) undergoes a settlement consisting of a small counterclockwise rotation θ_o. Find the element forces and the transverse displacement at C.

5.P45. The truss in Problem 5.P1 is to be designed so that bar BD remains vertical under the application of load F. Find the value of Δ_l in bar BD for which this can be accomplished.

5.P46. Find the horizontal displacement of joint D. Assume that all bars have the same EA and that bar AB undergoes a uniform temperature change ΔT.

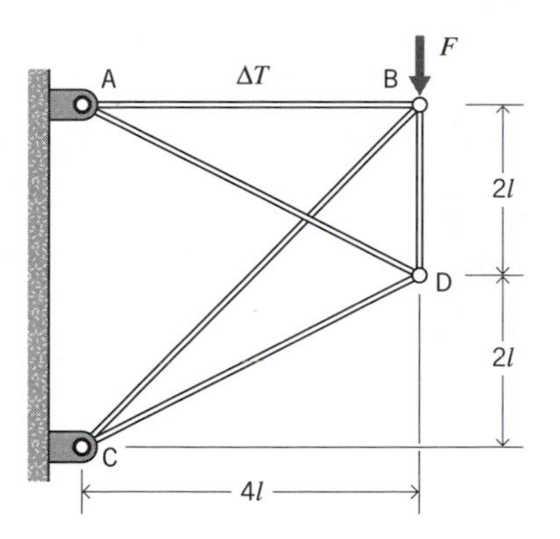

5.P47. The vertical displacement at B in the structure of Problem 5.P22 is too large. It is decided to shorten AD and DC by equal amounts until there is no vertical displacement at B.

(a) What are the forces in AD, DC, and BD when the displacement at B is zero?

(b) Sketch the moment diagram for beam ABC.

(c) How much shortening was required in AD and DC to make the displacement at B zero.

Chapter 6

Symmetry

6.0 INTRODUCTION

Concepts of symmetry are both inherent and beneficial in structural analysis. There are two types of symmetry we encounter. The first may be characterized as behavioral symmetry and the second as physical symmetry. By *behavioral symmetry* we mean mathematical symmetry of the type we observed in the element or structure flexibility matrices; *physical symmetry*, on the other hand, refers to symmetry of the structural configuration and load.

6.1 BEHAVIORAL SYMMETRY FOR BEAMS

In Chapter 5, the development using matrix algebra showed that the flexibility matrices for the linearly elastic structures under consideration would be symmetric, i.e., $\mathbf{A} = \mathbf{b}^{\mathrm{T}}\mathbf{a}\mathbf{b} = (\mathbf{b}^{\mathrm{T}}\mathbf{a}\mathbf{b})^{\mathrm{T}} = \mathbf{A}^{\mathrm{T}}$, if and only if $\mathbf{a} = \mathbf{a}^{\mathrm{T}}$. In other words, $\mathbf{A}$ is symmetric provided $\mathbf{a}$ is symmetric. Matrix $\mathbf{a}$ was observed to be symmetric because it is block-diagonal with symmetric element flexibility matrices, $\mathbf{a}^{(e)}$, situated on the diagonal. Matrix $\mathbf{a}^{(e)}$ for truss bars was, in essence, a 1×1 matrix (i.e., a scalar quantity) and thus symmetric by definition; the 2×2 matrix $\mathbf{a}^{(e)}$ for beams turned out to be symmetric upon derivation. It was not clear whether or not this was mere coincidence. One of the first things we will do in this chapter is prove that the flexibility matrix for any linearly elastic Bernoulli-Euler beam element must be symmetric. The theorem that will be derived in connection with this proof, Betti's theorem, is basic in structural analysis and will be useful for other purposes in subsequent developments.

The fact that all flexibility matrices (element and structure) are symmetric provides important computational advantages, e.g., we need save and/or compute only the upper triangular portion of symmetric matrices, and inversion of symmetric matrices can be accomplished more efficiently than for unsymmetric matrices. Perhaps an even more important result is that structures with this type of matrix can be shown to have unique solutions and, further, to have solutions that are physically reasonable. We shall explore this in detail later.

Just as we have previously derived specialized forms of the theorem of virtual work, we will now derive a specialized form of Betti's theorem that applies to linearly elastic Bernoulli-Euler beams only. This will be sufficient for the purpose of proving that flexibility matrices for such beams are symmetric. As we shall see, this form of Betti's theorem is obtained directly from the theorem of virtual work, Equation 4.18.

6.1.1 Betti's Theorem for Linearly Elastic Bernoulli-Euler Beams

Consider a linearly elastic Bernoulli-Euler beam, such as the one in Figure 5.23*a*, in a state of equilibrium under the action of each of two different sets of external loads and internal forces. We will designate the loads and forces in the first set by attaching a superscript (1) and those in the second set by attaching a superscript (2). Thus, the first set of loads will be denoted by $M_{IJ}^{(1)}$, $M_{JI}^{(1)}$, $q^{(1)}(x)$, etc., and the internal bending moment in the beam when it is subjected to this set of loads will be denoted by $M^{(1)}(x)$. Similarly, the second set of loads will be denoted by $M_{IJ}^{(2)}$, $M_{JI}^{(2)}$, $q^{(2)}(x)$, etc., and the internal bending moment in the beam when it is subjected to this set of loads will be denoted by $M^{(2)}(x)$.

Each set of loads and forces has associated with it a corresponding set of deformations and displacements (e.g., Figure 5.23*b*). These will be denoted in a similar manner, i.e., the deformations and displacements associated with the first set of loads and forces are denoted by $v^{(1)}(x)$, $v'^{(1)}(x)$, $v''^{(1)}(x)$, $v_{IJ}^{(1)}$, etc., and those associated with the second set are denoted by $v^{(2)}(x)$, $v'^{(2)}(x)$, $v''^{(2)}(x)$, $v_{IJ}^{(2)}$, etc. The manner in which beam forces are related to deformations is contained in Equation 3.16*b*, i.e., internal bending moment in each force state is related to the curvature by

$$\begin{aligned} EIv''^{(1)}(x) &= M^{(1)}(x) + M_T^{(1)}(x) + M_I^{(1)}(x) \\ EIv''^{(2)}(x) &= M^{(2)}(x) + M_T^{(2)}(x) + M_I^{(2)}(x) \end{aligned} \qquad \textbf{(6.1a)}$$

The "EI" is the same for both sets of loads because we are dealing with the same beam.

Consider, now, the theorem of virtual work, Equation 4.18. A major feature of this theorem is that the "force state" (state I) and the "displacement state" (state II) in Equation 4.18 are not necessarily related. In all applications thus far, one of these two states has always been "real" and the other has always been "virtual"; however, this is not a requirement of the theorem. In fact, neither state need be virtual and both may be real but unrelated. For instance, state I in Equation 4.18 may be chosen as the *first* set of loads and forces defined above, and state II may be chosen as the unrelated *second* set of displacements and deformations defined above. In this case, the internal virtual work[1] (given by the right side of Equation 4.18) becomes

$$\begin{aligned} W_{I_V} &= \int_0^l M^{(1)}(x)v''^{(2)}(x)dx \\ &= \int_0^l M^{(1)}(x)\left[\frac{M^{(2)}(x) + M_T^{(2)}(x) + M_I^{(2)}(x)}{EI}\right]dx \end{aligned} \tag{6.1b}$$

Here, we have substituted the expression for $v''^{(2)}(x)$ from Equation 6.1a.

Alternately, we may substitute the *second* set of forces for state I and the *first* set of displacements and deformations for state II. We then obtain for internal virtual work

$$\begin{aligned} W_{I_V} &= \int_0^l M^{(2)}(x)v''^{(1)}(x)dx \\ &= \int_0^l M^{(2)}(x)\left[\frac{M^{(1)}(x) + M_T^{(1)}(x) + M_I^{(1)}(x)}{EI}\right]dx \end{aligned} \tag{6.1c}$$

We observe that in the special case $M_T^{(1)}(x) = M_T^{(2)}(x) = M_I^{(1)}(x) = M_I^{(2)}(x) = 0$, these two expressions for internal virtual work both reduce to

$$W_{I_V} = \int_0^l \frac{M^{(1)}(x)M^{(2)}(x)}{EI}dx \tag{6.1d}$$

Thus, when we omit thermal or initial effects and consider only internal forces due to physical loads, the two expressions for internal virtual work, Equations 6.1b and 6.1c, are identical.

Now, associated with the internal virtual work in Equation 6.1b is an external virtual work given by the left side of Equation 4.18. This external virtual work must equal the internal virtual work in Equation 6.1d. Similarly, associated with the internal virtual work in Equation 6.1c is an external virtual work given by the left side of Equation 4.18, which must also equal the internal virtual work in Equation 6.1d. Thus, these two expressions for external virtual work must be equal,

$$\begin{aligned} &V_{\mathrm{JI}}^{(1)}v_{\mathrm{JI}}^{(2)} + V_{\mathrm{IJ}}^{(1)}v_{\mathrm{IJ}}^{(2)} + M_{\mathrm{JI}}^{(1)}\phi_{\mathrm{JI}}^{(2)} + M_{\mathrm{IJ}}^{(1)}\phi_{\mathrm{IJ}}^{(2)} + \\ &\quad \sum_{j=1}^{n(\mathrm{i})} F_j^{(1)}v^{(2)}(x_j) - \sum_{i=1}^{m(\mathrm{i})} M_i^{(1)}v'^{(2)}(x_i) + \int_0^l q^{(1)}(x)v^{(2)}(x)dx = \\ &V_{\mathrm{JI}}^{(2)}v_{\mathrm{JI}}^{(1)} + V_{\mathrm{IJ}}^{(2)}v_{\mathrm{IJ}}^{(1)} + M_{\mathrm{JI}}^{(2)}\phi_{\mathrm{JI}}^{(1)} + M_{\mathrm{IJ}}^{(2)}\phi_{\mathrm{IJ}}^{(1)} + \\ &\quad \sum_{j=1}^{n(2)} F_j^{(2)}v^{(1)}(x_j) - \sum_{i=1}^{m(2)} M_i^{(2)}v'^{(1)}(x_i) + \int_0^l q^{(2)}(x)v^{(1)}(x)dx \end{aligned} \tag{6.1e}$$

[1] We continue to refer to this as ''virtual'' work simply because it is not real work; real work would involve forces and related deformations.

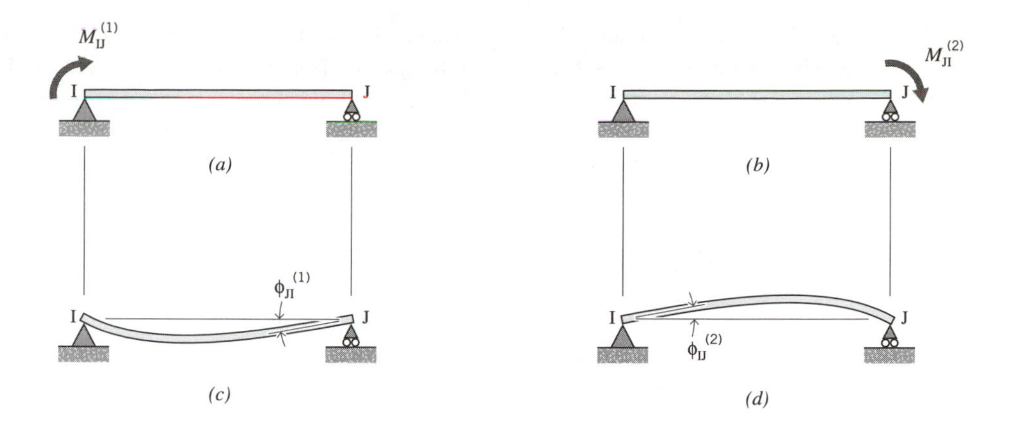

Figure 6.1 Illustration of Maxwell's reciprocal relationship for beam.

This is the desired statement of Betti's theorem for linearly elastic Bernoulli-Euler beams. Essentially, the theorem states that, for a given beam subjected to two distinct sets of loads and corresponding deformations, the external virtual work associated with force set (1) going through displacement set (2) is equal to the external virtual work associated with force set (2) going through displacement set (1). Note that, as opposed to the original form of the theorem of virtual work, Betti's theorem implicitly contains a scalar constitutive relation, in this case a linear relation between axial stress and strain introduced by way of Equations 6.1*a*.

Applications of this new theorem will be less formidable then may appear from Equation 6.1*e*. We will see this in the next section where we will use the theorem to derive Maxwell's reciprocal relationship;[2] the latter proves directly the symmetry of element flexibility matrices. We will also encounter additional applications of Betti's theorem in Chapters 7 and 8.

6.1.2 Maxwell's Reciprocal Relationship

Consider a beam with any given supports that is subjected to two distinct physical load sets, each consisting of a single applied force or moment that can act at any specified point. For example, let's assume a simply-supported beam for which the first load is a moment at the left end, as in Figure 6.1*a*, and the second load is a moment at the right end, as in Figure 6.1*b*. We denote these moments by $M_{IJ}^{(1)}$ and $M_{JI}^{(2)}$, respectively.

Figure 6.1*c* shows the deflections of the beam due to the forces in Figure 6.1*a*, and Figure 6.1*d* shows the deflections of the beam due to the forces in Figure 6.1*b*. We denote the displacement at the right end (the J end) in Figure 6.1*c* in the direction of $M_{JI}^{(2)}$ (i.e., the slope at the J end) by $\phi_{JI}^{(1)}$. Similarly, we denote the displacement at the left end (the I end) in Figure 6.1*d* in the direction of $M_{IJ}^{(1)}$ (i.e., the slope at the I end) by $\phi_{IJ}^{(2)}$.

We apply Betti's theorem, Equation 6.1*e*, to this beam. The left side of Equation

[2] Actually, J. C. Maxwell derived the reciprocal relationship prior to (1864) E. Betti's form of generalization (1872), but it is easier for us to present the derivations in the reverse order.

6.1*e* is the external virtual work associated with the forces in Figure 6.1*a* and the displacements in Figure 6.1*d*; the right side of Equation 6.1*e* is the external virtual work associated with the forces in Figure 6.1*b* and the displacements in Figure 6.1*c*. The former is $-M_{IJ}^{(1)}\phi_{IJ}^{(2)}$, and the latter is $-M_{JI}^{(2)}\phi_{JI}^{(1)}$. (The external virtual work associated with support reactions and displacements is zero in both cases.) Thus, Equation 6.1*e* reduces to

$$M_{IJ}^{(1)}\phi_{IJ}^{(2)} = M_{JI}^{(2)}\phi_{JI}^{(1)} \qquad \textbf{(6.1}f\textbf{)}$$

Consider, now a special case wherein the applied loads are assigned unit values, i.e., $M_{IJ}^{(1)} = 1$, $M_{JI}^{(2)} = 1$. Equation 6.1*f* now reduces to

$$\phi_{IJ}^{(2)} = \phi_{JI}^{(1)} \qquad \textbf{(6.1}g\textbf{)}$$

In other words, for the given beam, the equation states that the slope at the I end due to a unit value of moment applied at the J end is equal to the slope at the J end due to a unit value of moment applied at the I end.

At this point, we recall from previous developments (Equation 5.31*b* or 5.42*a*, Chapter 5) that the flexibility relationship for linear elastic beams has the form

$$\begin{Bmatrix} \Delta_{IJ} \\ \Delta_{JI} \end{Bmatrix} = \begin{bmatrix} a_{11} & a_{12} \\ a_{21} & a_{22} \end{bmatrix} \begin{Bmatrix} M_{IJ} \\ M_{JI} \end{Bmatrix} \qquad \textbf{(6.1}h\textbf{)}$$

where $\Delta_{IJ} = -\phi_{IJ}$ and $\Delta_{JI} = -\phi_{JI}$. We recognize that $-\phi_{JI}^{(1)}$ is the negative of a_{21} in Equation 6.1*h*, i.e., it is the clockwise rotation at the J end due to a unit value of M_{IJ}; similarly, we recognize that $-\phi_{IJ}^{(2)}$ is the negative of a_{12}, i.e., it is the clockwise rotation at the I end due to a unit value of M_{JI}. Therefore, it follows that Equation 6.1*g* is equivalent to

$$a_{21} = a_{12} \qquad \textbf{(6.1}i\textbf{)}$$

Thus, Equation 6.1*g* or 6.1*i* proves the symmetry of the flexibility matrix for a beam. These equations are a particular example of Maxwell's reciprocal relationship. This result may appear obvious based on the apparent physical symmetry of the structure and loads in Figure 6.1, but it is actually quite general.[3] Furthermore, an equality like the one in Equation 6.1*g* holds regardless of the nature or location of either unit load as long as the ''displacement'' quantity at a point is defined as being ''in the direction of'' the unit load applied at that point.

Because Maxwell's relationship does little more than confirm what we have observed previously, it will have no further direct applications.

6.2 PHYSICAL SYMMETRY

A different type of computational simplification results from physical symmetry. When a loaded structure possesses physical symmetry, it is possible to conduct the analysis

[3] In fact, the equality holds even if the beam is not symmetric, e.g., even if the beam has a nonuniform cross section. (Note the *EI* term under the integral in Equation 6.1*d*, which could be a variable function of *x*.)

using a representative substructure suggested by the symmetry of the structural configuration and the loads applied to it. This makes it possible for the structural analyst to reduce the number of elements that must be considered, and will usually also reduce the number of unknowns (e.g., redundant forces) in the analysis. For large structures, consideration of physical symmetry can lead to major reductions in the sizes of all matrices. In fact, the savings in terms of human effort and computer cost are so great that structures that are not truly symmetric are often approximated by symmetric configurations simply to reduce the cost of analysis. (This may be risky unless the analyst can demonstrate that the approximation has little effect on the overall results.)

We will examine two-dimensional structures that possess either of two distinct types of symmetry, i.e., symmetry with respect to a line, or symmetry with respect to a point.

6.2.1 Structures Symmetric With Respect to a Line

Let the structure in Figure 6.2*a* be representative of the general class of structures that are symmetric with respect to a line *L*. This implies that for any point on the axis of an element of the structure, e.g., point *a*, there exists another point *b*, which lies on the axis of an element where the physical and mechanical properties are identical to those at *a*, such that the line segment *ab* is bisected at right angles by *L*. The line *L* is called the *axis of symmetry*. In simplified terms, if the structure is rotated 180° about *L*, the resulting configuration is indistinguishable from the original configuration. Thus, the rotated structure DCBA in Figure 6.2*b* is identical to structure ABCD in Figure 6.2*a*. It should be noted that a structure may possess more than one axis of symmetry, in which case the discussion that follows would apply with respect to all such axes.

The structure in Figure 6.2 is symmetric regardless of the loads to which it is subjected; however, in order to utilize this symmetry to simplify the analysis, it is necessary for the loads to also possess some type of symmetry. There are two basic categories of loads to consider, so-called ''symmetric'' and ''antisymmetric'' loads, Figures 6.3*a* and 6.3*b*, respectively. Here, the definitions of symmetric loads and antisymmetric loads may be related to the results of rotations about the axis of symmetry. Specifically, if a 180° rotation of the loaded structure results in an identical loaded configuration we refer to the loads as symmetric (e.g., Figure 6.3*a*). On the

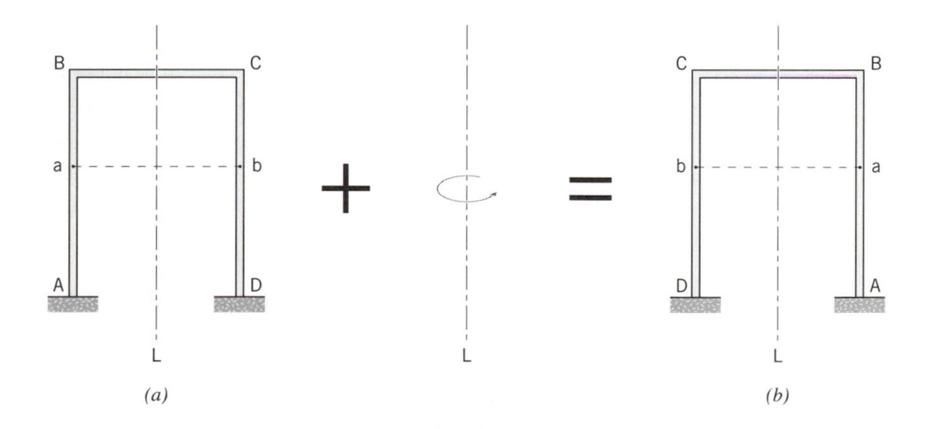

Figure 6.2 Structure that is symmetric with respect to a line *L*.

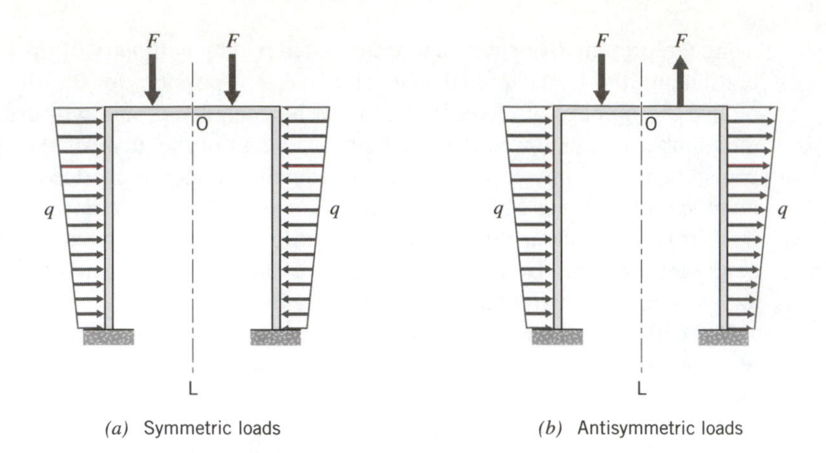

Figure 6.3 Loads with symmetry properties.

other hand, if a 180° rotation of the loaded structure results in a configuration in which the loads are exactly opposite in sign, we refer to these loads as antisymmetric (e.g., Figure 6.3*b*). We may also note that deformations and internal forces must be either symmetric or antisymmetric, depending on whether the applied loads are symmetric or antisymmetric.

As we will see, a symmetric structure that is subjected to either of these categories of loads may be fully analyzed by examining only the portion that lies to one side or the other of *L*. This portion, or substructure, consists of no more than one-half of the original structure.[4]

The requirement that the loads possess symmetry properties is actually not restrictive because any set of arbitrary loads applied to a symmetric linear elastic structure can be decomposed into symmetric and antisymmetric components. Consider, for example, the loaded structure in Figure 6.4*a*. The arbitrary nonsymmetric loads in Figure 6.4*a* are equivalent to the sum of the symmetric loads in Figures 6.4*b* and the antisymmetric loads in Figure 6.4c. Thus, superposition of the results of the analyses of the structures in Figures 6.4*b* and 6.4*c* yields the behavior of the structure in Figure 6.4*a*. From a computational standpoint, it will almost always be preferable to analyze a substructure subjected to the two different sets of loads rather than the full structure subjected to the single set of arbitrary loads.

The main issue remaining to be resolved is the nature of the kinematic and static boundary conditions to be imposed on the substructure at the points where it is to be detached from the full structure, i.e., at points such as O in Figures 6.2 or 6.3, which lie on the axis of symmetry. We will ascertain the appropriate boundary conditions by considering the possible displacements and internal forces that occur or act at these points.

We begin by considering the entire structure. Figure 6.5*a*, for example, shows displacements and internal forces in the vicinity of point O for the representative

[4] Note, however, that this does not necessarily mean one-half the number of elements because, for example, one-half of the structure in Figures 6.2 or 6.3 contains two beams compared to three in the full structure.

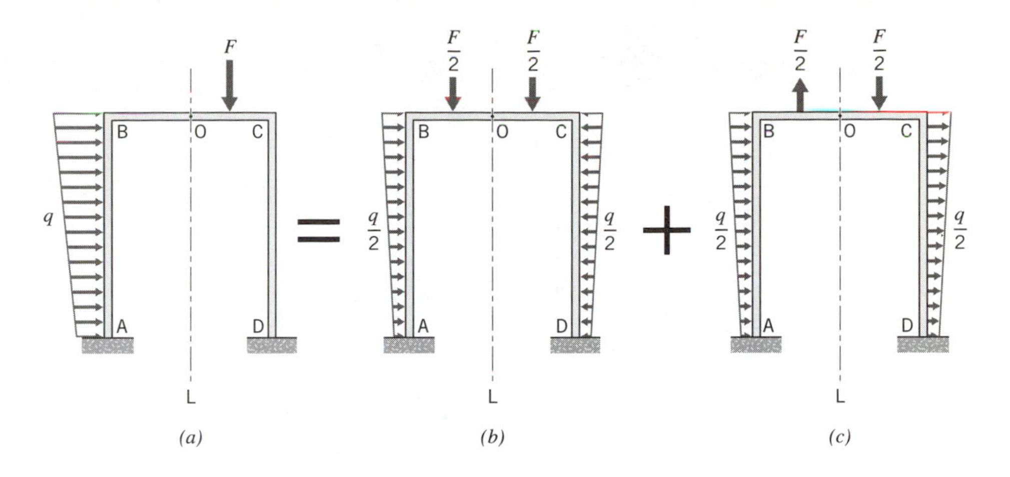

Figure 6.4 Superposition of symmetric and antisymmetric loads.

structure of Figure 6.2*a*. We assume point O displaces to O′, and that the displacement components at point O are u, v, and ϕ, as illustrated. The directions shown for the displacement components are entirely arbitrary. Similarly, the possible internal forces in the vicinity of point O consist of an axial force, P, a transverse force, V, and a moment, M. The directions of these forces are also arbitrary in the figure, but they are equal and opposite on ends of contiguous free-bodies.

If the structure in Figure 6.5*a* is rotated 180° about L, we obtain the structure in Figure 6.5*b*. Now, if the loads on the structure are symmetric, as in Figure 6.3*a*, this rotation must result in a configuration in which displacements and forces are identical to the original configuration; if the loads on the structure are antisymmetric, as in Figure 6.3*b*, this rotation must result in a configuration in which displacements and forces are equal and opposite to those in the original configuration. In each case, this is possible only if some of the displacement and force components are zero.

In particular, if the loads are *symmetric*, displacements and forces in Figures 6.5*a*

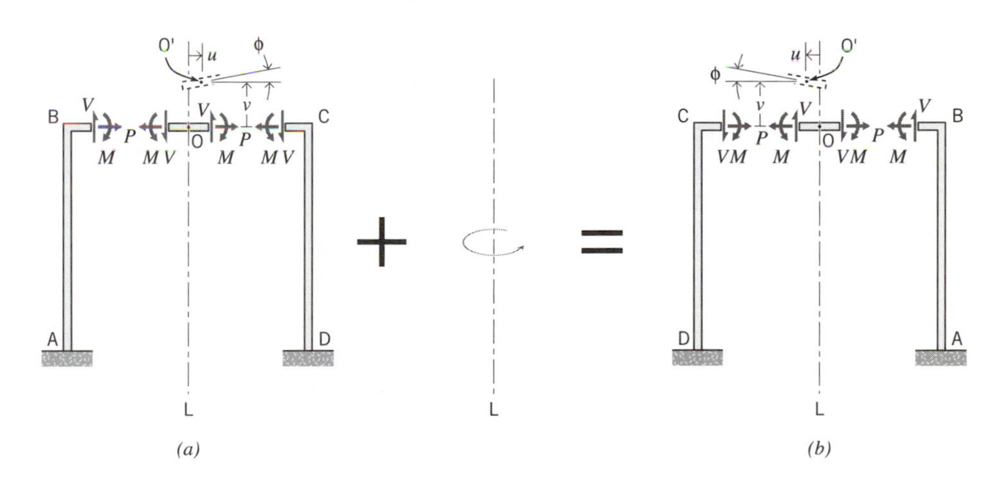

Figure 6.5 Displacements and internal forces at point O.

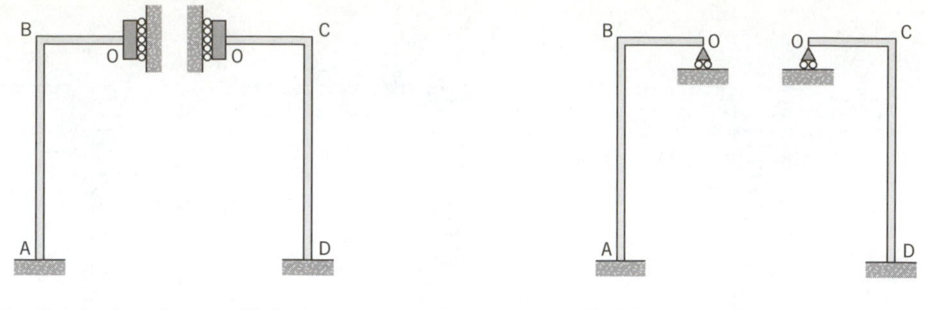

(*a*) Substructures for symmetric loads

(*b*) Substructures for antisymmetric loads

Figure 6.6 Substructure.

and 6.5*b* will be identical only if $u = \phi = V = 0$ (which implies that P, M, and v may be nonzero); if the loads are *antisymmetric*, displacements and forces in Figures 6.5*a* and 6.5*b* will be opposite only if $P = M = v = 0$ (which implies that u, ϕ, and V may be nonzero).[5]

In the first case (symmetric load), the kinematic and static conditions at point O ($u = \phi = V = 0$) are exactly equivalent to those imposed by a fixed support with roller parallel to L, Figure 6.6*a*;[6] in the second case (antisymmetric load), the kinematic and static conditions at point O ($P = M = v = 0$) are exactly equivalent to those imposed by a pin support with roller perpendicular to L (Figure 6.6*b*).

Thus, for example, if we wish to analyze the structure in Figure 6.4*b*, it is sufficient to analyze either of the two halves shown in Figure 6.7*a*; if we wish to analyze the structure in Figure 6.4*c*, it is sufficient to analyze either of the two halves in Figure 6.7*b*. The solution for the other half is symmetric in the case of Figure 6.7*a*, and antisymmetric in the case of Figure 6.7*b*.

In some cases, structural elements will lie on an axis of symmetry and/or concentrated forces will be applied at points on the axis of symmetry, as in Figure 6.8*a*. The nature of the internal forces in an element lying on the axis (such as GH in Figure 6.8*a*) is determined exactly as in Figure 6.5. Specifically, for symmetric loads, we find that such an element can transmit only axial forces, and no bending or shear; for antisymmetric loads, we find that such an element can transmit only bending and shear, and no axial loads. In either case, the substructure is relatively easy to define. We assume half of the element and/or load lies on each side of the axis, as in Figure 6.8*b*. By ''half'' of an element, we mean an element with one-half the bending stiffness or axial stiffness of the original.[7] (In effect, because there is only one element, we assume the

[5] Note the correspondence between displacement components and ''associated'' force quantities in each of these cases: either the axial component of displacement (u) or the axial force (P) is zero; either the transverse component of displacement (v) or the transverse shear (V) is zero; either the rotation (ϕ) or the moment (M) is zero. In no case are both associated quantities simultaneously equal to zero.

[6] Thus, we see that a support of this type does serve a theoretical function even though it is not likely to ever exist in a physical structure.

[7] For symmetric loads, it would not even be necessary to include element GH in the substructure if it were modeled as an axially rigid element. In such a case, we could analyze the substructure without element GH but with a fixed support (without vertical roller) at G and delete the vertical displacement degree of freedom. The vertical reaction at G would be transmitted directly to element GH.

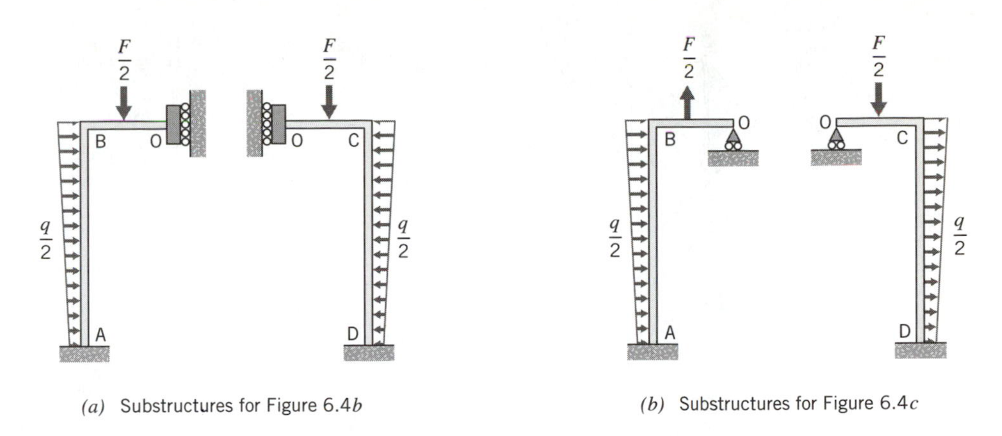

(a) Substructures for Figure 6.4*b*

(b) Substructures for Figure 6.4*c*

Figure 6.7 Use of substructures.

behavior of both halves is identical.) Supports that lie on an axis of symmetry (e.g., at H in Figure 6.8) may also require modification to reflect the kinematic and static conditions consistent with the symmetry.

The preceding illustrations used a representation of a frame structure that transmits a general set of internal forces. For symmetric trusses, the supports defined at points like point O in Figure 6.6*a* must be modified. We will illustrate this straightforward simplification in the following example.

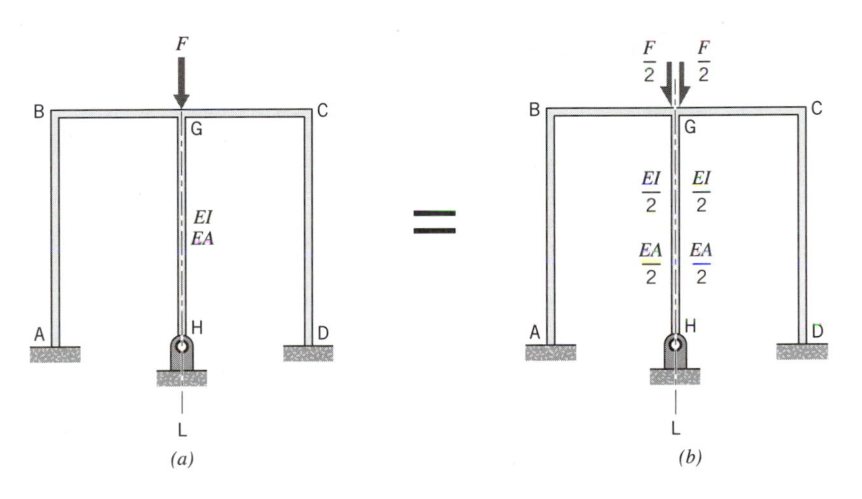

(a)

(b)

Figure 6.8 Structure with element and load on axis of symmetry.

EXAMPLE 1

Find the internal forces in the truss in Figure 6.9*a*. All elements have the same *EA*. The complete truss has 15 bars and is two-degrees redundant. If we utilize symmetry, we will only have to analyze a truss with eight bars and one redundant force.

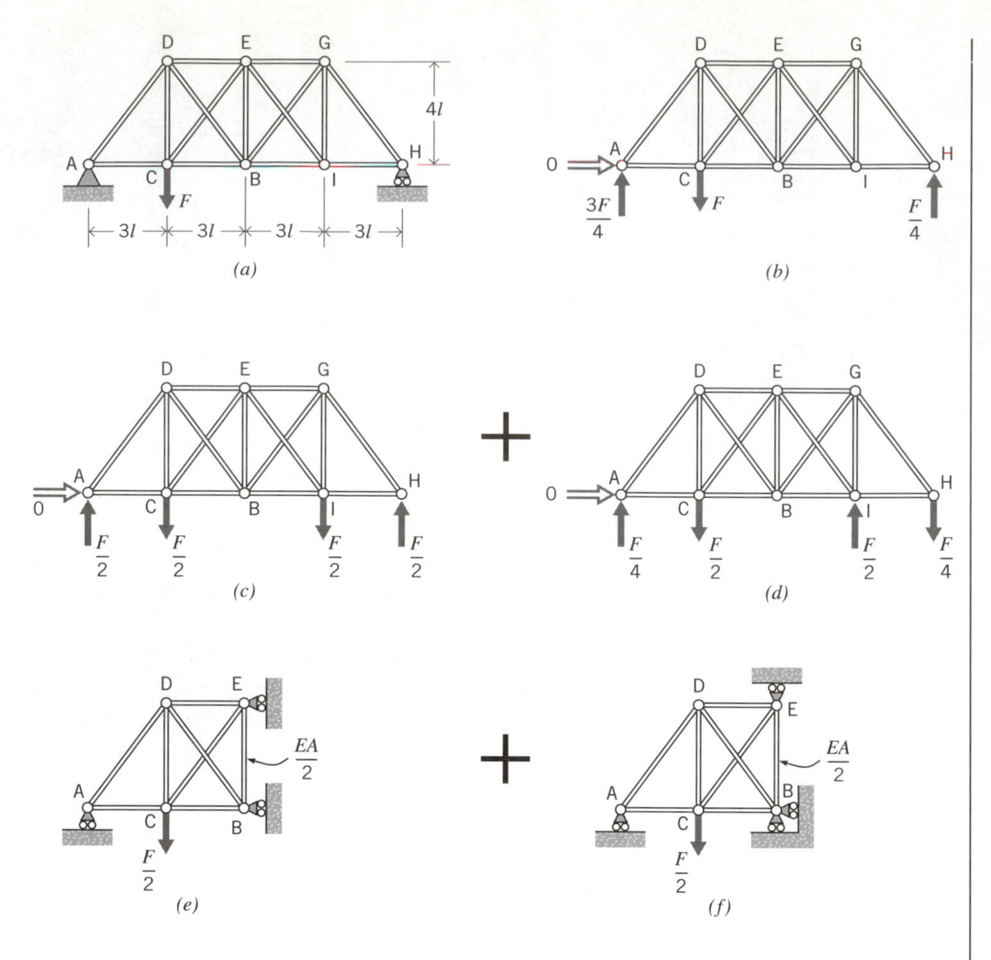

Figure 6.9 Example 1.

We begin by evaluating the determinate reactions at A and H, which are shown in Figure 6.9*b*. We recognize line EB as an axis of symmetry for the truss in Figure 6.9*b*, and we decompose the external loads into the symmetric and antisymmetric components shown in Figures 6.9*c* and 6.9*d*, respectively. Note that both trusses (Figures 6.9*c* and 6.9*d*) are in a state of equilibrium under the loads shown. Also note that the kinematic support conditions at A and H of the truss in Figure 6.9*a* are not entirely symmetric because H can move horizontally but A cannot move. We can impose kinematic symmetry and preserve geometric stability by replacing the pin at A by a roller and, instead, temporarily assume that joint B is pinned to prevent horizontal (but not vertical) motion. Horizontal components of joint displacement can then be computed relative to point B. However, the force state in Figure 6.9*b* should include the actual horizontal reaction at A (zero in this case).

The internal forces in the truss in Figure 6.9*c* can be completely determined from the analysis of the substructure in Figure 6.9*e*. Here, the supports at B and E must be taken as pinned supports (with vertical rollers) rather than fixed supports (with vertical rollers) because we do not permit moments to be applied at truss joints. Note the roller support at A, which provides geometric stability and develops the same reaction force as shown in Figure 6.9*c*. Similarly, the internal forces in the truss in Figure 6.9*d* can be completely determined from the analysis of the substructure in Figure 6.9*f*. In this case, the horizontal roller supports at B and E are identical to those shown at point O in Figure 6.6*b*. The vertical roller support at B provides geometric stability. In each substructure, Figures 6.9*e* and 6.9*f*, the axial stiffness of bar BE is *EA/2* rather than *EA*.

We may now proceed with the analyses of the two substructures. For the sake of clarity, we will describe each analysis separately.

(a) Symmetric Load Case

Step 1. The truss in Figure 6.9*e* is one-degree redundant; we choose the redundant as the internal force in bar CE, and denote it by *X*.

Step 2. The statical analysis is shown in Figure 6.10*a*; Figure 6.10*a*(1) shows the internal forces with the redundant released, and Figure 6.10*a*(2) shows the internal forces when the force in bar CE is *X*. Note that Figure 6.10*a*(2) is identical to Figure 5.9*b*(3).

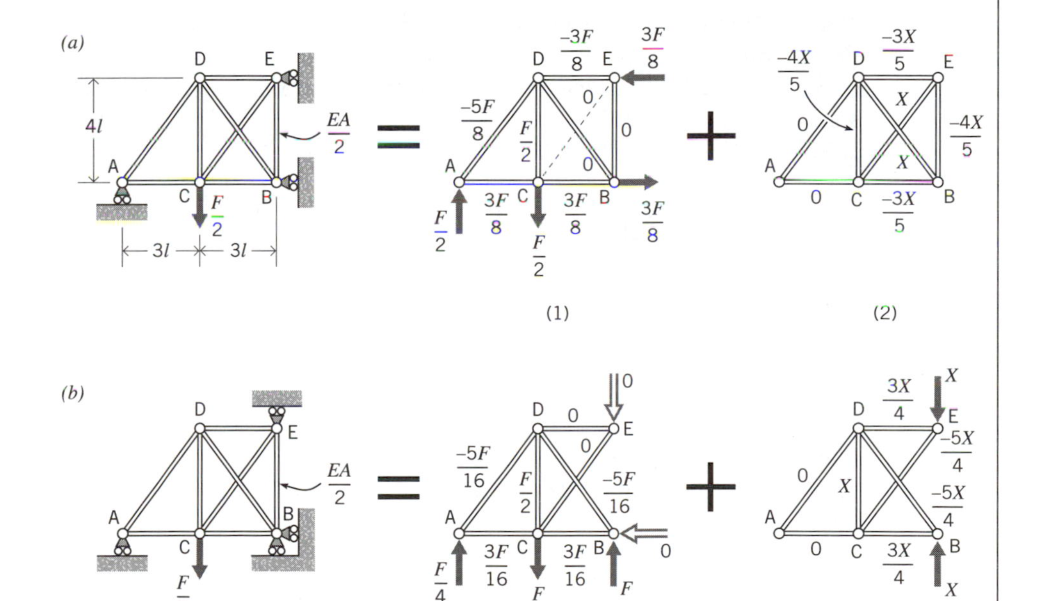

Figure 6.10 Statical analyses: (*a*) Symmetric loads; (*b*) Antisymmetric loads.

The equilibrium equation ($\mathbf{P} = \mathbf{bX} + \mathbf{P_o}$) is

$$\begin{Bmatrix} P_{AC} \\ P_{CB} \\ P_{AD} \\ P_{CD} \\ P_{CE} \\ P_{DE} \\ P_{BE} \\ P_{BD} \end{Bmatrix} = \begin{bmatrix} 0 \\ -3/5 \\ 0 \\ -4/5 \\ 1 \\ -3/5 \\ -4/5 \\ 1 \end{bmatrix} \{X\} + \begin{Bmatrix} 3/8 \\ 3/8 \\ -5/8 \\ 1/2 \\ 0 \\ -3/8 \\ 0 \\ 0 \end{Bmatrix} F \qquad \textbf{(6.2}\textbf{\textit{a}}\textbf{)}$$

Step 3. The deformation-force relation ($\boldsymbol{\Delta} = \mathbf{aP} + \hat{\boldsymbol{\Delta}}$) is almost identical to Equation 5.23. The only differences are that $\hat{\boldsymbol{\Delta}} = \mathbf{0}$, and the axial stiffness of bar BE is $EA/2$. We require only the **a** matrix

$$\mathbf{a} \equiv \frac{l}{AE}\begin{bmatrix} 3 & & & & & & & \\ & 3 & & & & & & \\ & & 5 & & & & & \\ & & & 4 & & & & \\ & & & & 5 & & & \\ & & & & & 3 & & \\ & & & & & & 8 & \\ & & & & & & & 5 \end{bmatrix} \qquad \textbf{(6.2}\textbf{\textit{b}}\textbf{)}$$

Note that the seventh element of the matrix has the value $8l/AE$, which corresponds to the flexibility of bar BE with axial stiffness $EA/2$ and length $4l$.

Because we have all the necessary matrices we skip Step 4 and proceed to the solution, Equation 5.31*k*.

Step 5. We evaluate $\mathbf{A} = \mathbf{b}^T\mathbf{ab}$ and $\mathbf{B} = -\mathbf{b}^T\mathbf{aP_o}$,

$$\mathbf{A} = \frac{l}{AE}\left[\frac{496}{25}\right] = \frac{l}{AE}[19.84], \qquad \mathbf{B} = \frac{Fl}{AE}\left\{\frac{8}{5}\right\} = \frac{Fl}{AE}\{1.6\} \qquad \textbf{(6.2}\textbf{\textit{c}}\textbf{)}$$

Solving for $\mathbf{X} = \mathbf{A}^{-1}\mathbf{B}$ leads to

$$\mathbf{X} = X = \frac{5}{62}F = 0.0806F \qquad \textbf{(6.2}\textbf{\textit{d}}\textbf{)}$$

Back-substituting into Equation 6.2*a* gives the bar forces. For future reference, we will designate the resulting vector of symmetric forces by $\mathbf{P}_s$ rather than $\mathbf{P}$,

$$\mathbf{P}_S = \begin{Bmatrix} P_{AC} \\ P_{CB} \\ P_{AD} \\ P_{CD} \\ P_{CE} \\ P_{DE} \\ P_{BE} \\ P_{BD} \end{Bmatrix} = \begin{Bmatrix} 0.375 \\ 0.327 \\ -0.625 \\ 0.435 \\ 0.081 \\ -0.423 \\ -0.065 \\ 0.081 \end{Bmatrix} F \qquad \textbf{(6.2}\textbf{\textit{e}}\textbf{)}$$

This constitutes the desired solution for the symmetric loads in Figure 6.10*a*.

(b) Antisymmetric Load Case

Step 1. The truss in Figure 6.9*f* is also one-degree redundant, but in this case it is externally redundant;[8] we will choose the redundant force as the reaction at E, again denoted by X.

Step 2. The statical analysis is shown in Figure 6.10*b* and the equilibrium equation is

$$\begin{Bmatrix} P_{AC} \\ P_{CB} \\ P_{AD} \\ P_{CD} \\ P_{CE} \\ P_{DE} \\ P_{BE} \\ P_{BD} \end{Bmatrix} = \begin{bmatrix} 0 \\ 3/4 \\ 0 \\ 1 \\ -5/4 \\ 3/4 \\ 0 \\ -5/4 \end{bmatrix} \{X\} + \begin{Bmatrix} 3/16 \\ 3/16 \\ -5/16 \\ 1/2 \\ 0 \\ 0 \\ 0 \\ -5/16 \end{Bmatrix} F \tag{6.2f}$$

Here, we have set $P_{BE} = 0$ in the **P** vector rather than delete it entirely; this leaves the vector the same dimension as in the previous case. Note that the **b** and $\mathbf{P_o}$ vectors are different from those in Equation 6.2*a*.

Step 3. The **a** matrix is identical to Equation 6.2*b* because the truss is unchanged. We proceed to solve the compatibility equations.

Step 5. We now obtain

$$\mathbf{A} = \frac{l}{AE}[23], \qquad \mathbf{B} = -\frac{Fl}{AE}\left\{\frac{35}{8}\right\} = -\frac{Fl}{AE}\{4.375\} \tag{6.2g}$$

and

$$\mathbf{X} = X = -\frac{35}{184}F = -0.190F \tag{6.2h}$$

We will denote the new vector of antisymmetric bar forces by $\mathbf{P}_A$,

$$\mathbf{P}_A = \begin{Bmatrix} P_{AC} \\ P_{CB} \\ P_{AD} \\ P_{CD} \\ P_{CE} \\ P_{DE} \\ P_{BE} \\ P_{BD} \end{Bmatrix} = \begin{Bmatrix} 0.188 \\ 0.045 \\ -0.313 \\ 0.310 \\ 0.238 \\ -0.143 \\ 0 \\ -0.075 \end{Bmatrix} F \tag{6.2i}$$

[8] We recognize that bar BE cannot change length because the supports at B and E cannot move vertically; thus no force can be developed in bar BE and we may delete it from the structure (see Figure 6.10*b*).

which is the desired solution for internal forces due to the antisymmetric loads in Figure 6.10*b*.

(c) Superposition

To obtain the bar forces due to the loads in Figure 6.9*a* we superpose the two solutions, Equations 6.2*e* and 6.2*i*.

Equation 6.2*e* gives the bar forces for the half of the structure shown in Figure 6.9*e*; the bar forces for the other half are symmetric. In other words, for the symmetric load case, $P_{HG} = P_{AD}$, $P_{HI} = P_{AC}$, and so on. (Note that P_{BE} in Equation 6.2*e* is the force in one-half of bar BE, which implies that the force in bar BE is actually twice the indicated value, i.e., $P_{BE} = -0.129F$.) Similarly, Equation 6.2*i* gives the bar forces for the half of the structure shown in Figure 6.9*f*, but the bar forces for the other half are antisymmetric, i.e., for the antisymmetric load case $P_{HG} = -P_{AD}$, $P_{HI} = -P_{AC}$, etc. (Again, note that P_{BE} in Equation 6.2*i* is the force in one-half of bar BE, and that the force in the whole bar is zero for this case.) Thus, except for bar BE, forces for bars in the left half of the truss are given by $\mathbf{P}_S + \mathbf{P}_A$, and forces for bars in the right half of the truss are given by $\mathbf{P}_S - \mathbf{P}_A$; the final bar forces are

$$\begin{Bmatrix} P_{AC} \\ P_{CB} \\ P_{AD} \\ P_{CD} \\ P_{CE} \\ P_{DE} \\ P_{BE} \\ P_{BD} \end{Bmatrix} = \begin{Bmatrix} 0.563 \\ 0.371 \\ -0.938 \\ 0.745 \\ 0.318 \\ -0.566 \\ -0.129 \\ 0.006 \end{Bmatrix} F, \qquad \begin{Bmatrix} P_{HI} \\ P_{IB} \\ P_{HG} \\ P_{IG} \\ P_{IE} \\ P_{GE} \\ P_{BE} \\ P_{BG} \end{Bmatrix} = \begin{Bmatrix} 0.188 \\ 0.282 \\ -0.313 \\ 0.126 \\ -0.157 \\ -0.281 \\ -0.129 \\ 0.155 \end{Bmatrix} F \tag{6.2j}$$

These forces are shown in Figure 6.11. This completes the analysis.

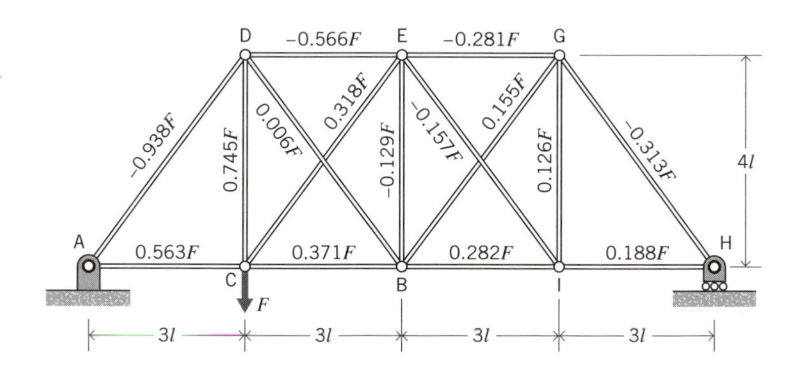

Figure 6.11 Internal forces.

It is important to recognize both the advantages and disadvantages of the use of symmetry. For this simple example there appears to be little computational (numerical) advantage because we have to set up and solve two problems. In fact, the only advantage at this level is that half as many elements are considered in each ''half''problem so that data entry is much easier. The real advantage occurs when we deal with large structures with thousands of variables. Then the reduction in problem size may be the difference between whether or not a structural analysis is economically justified. This is because the cost of computing the results for the complete structural model may be much greater than twice that for the two half models. When matrices exceed the memory capacity of a high-speed computer, and only portions of the matrix can be processed by the computer, a great deal of data swapping occurs between the computer's CPU and external memory. This slows the solution down greatly and, at the same time, causes major increases in solution cost in terms of time and money.

EXAMPLE 2

As a further demonstration of procedures, we will consider another relatively small structure shown in Figure 6.12*a*. This structure consists of a uniform beam, ABC, that has both ends fixed, and that is supported in the middle by a linear spring. The structure is subjected to a single concentrated transverse force F in span BC. We will assume that the beam is axially rigid.

The structure in Figure 6.12*a* is 3-degrees redundant and we cannot evaluate reactions. Because line BD is an axis of symmetry, we can decompose the load into the symmetric and antisymmetric components shown in Figures 6.12*b* and 6.12*c*. We will therefore analyze only the two halves shown in Figures 6.12*d* and 6.12*e*. Note that the former is 2-degrees redundant and the latter is 1-degree redundant. The axial stiffness of the spring is $k/2$ in both figures. For the sake of clarity in the subsequent calculations, we will let $F' = F/2$.

(a) Symmetric Load Case

Step 1. We will analyze the structure in Figure 6.12*d* using the choice of redundant forces shown in Figure 6.13*a*, i.e., the internal force in the spring (X_1) and the moment at B (X_2).

Step 2. The statical analysis is shown in Figure 6.13*b* and the equilibrium equation is

$$\begin{Bmatrix} M_{AB} \\ M_{BA} \\ P_{BD} \end{Bmatrix} = \begin{bmatrix} -l & -1 \\ 0 & 1 \\ 1 & 0 \end{bmatrix} \begin{Bmatrix} X_1 \\ X_2 \end{Bmatrix} + \begin{Bmatrix} -2F'l/3 \\ 0 \\ 0 \end{Bmatrix} \qquad \textbf{(6.3a)}$$

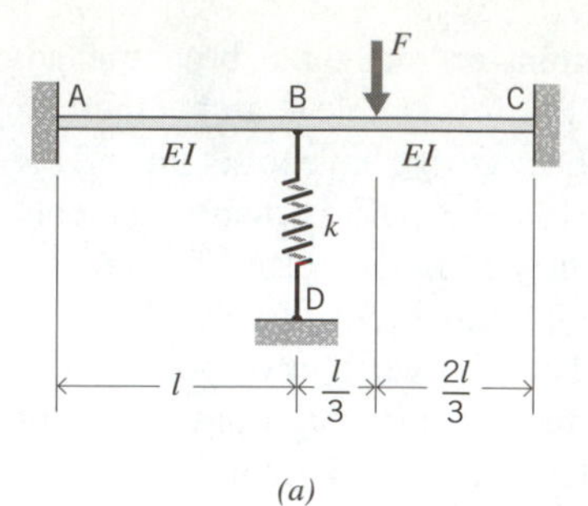

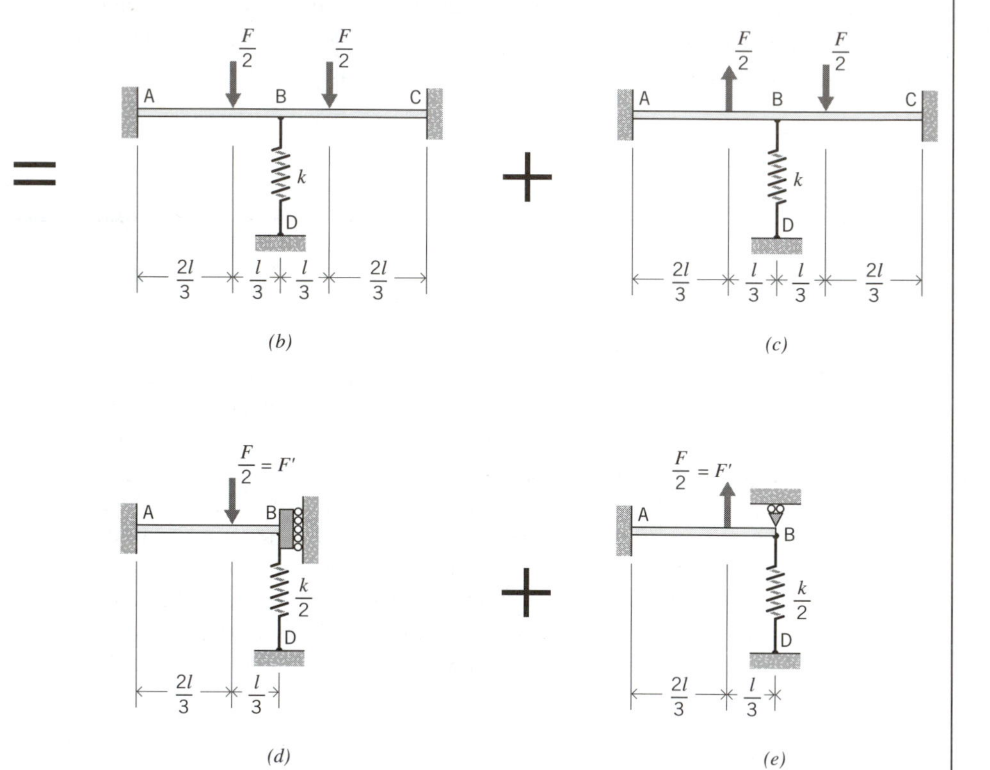

Figure 6.12 Example 2.

Note that X_1 has been taken as positive in the direction shown in Figure 6.13*b*(2), i.e., when it is a tensile force in the spring.

Step 3. The deformation-force relations for the structure in Figure 6.13*a* have the matrix form $\mathbf{\Delta} = \mathbf{aP} + \widehat{\mathbf{\Delta}}$. In this case, the components of $\widehat{\mathbf{\Delta}}$, the vector of unrestrained end deformations, are not all zero; there are unrestrained deformations (rotations) at the ends of beam AB due to the applied force. There are many ways to evaluate these unrestrained end deformations. Here, for illustrative purposes, we will use the definitions in Equations 5.42*b*.

The free-body diagram we require for this computation is shown in Figure 6.13*c*. As

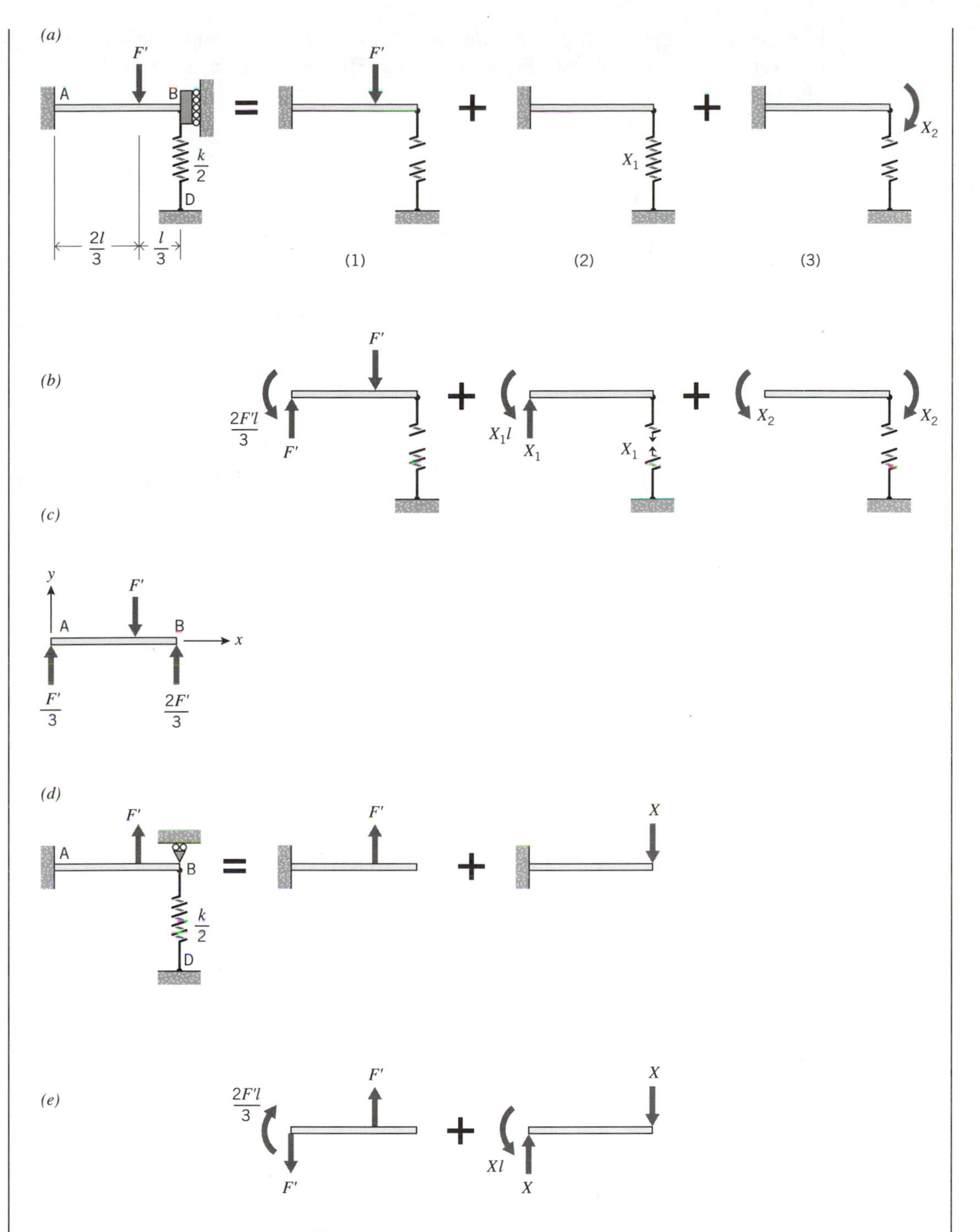

Figure 6.13 Statical analyses, Example 2.

in previous examples (Chapter 5), we regard the load F' as being equilibrated by transverse end forces only [see Figure 5.23*a*(2)]. The internal moment in the beam in Figure 6.13*c* is

$$\widehat{M}(x) = M_o(x) = \begin{cases} \dfrac{F'}{3}x, & 0 < x < \dfrac{2}{3}l \\ \dfrac{2F'}{3}(l - x), & \dfrac{2}{3}l < x < l \end{cases} \tag{6.3b}$$

Substituting relations 6.3*b* into the first of Equation 5.42*b* gives

$$\begin{aligned} \widehat{\Delta}_{AB} &= \frac{1}{EI}\int_0^l \widehat{M}(x)\left(1 - \frac{x}{l}\right)dx \\ &= \frac{1}{EI}\int_0^{2l/3}\left(\frac{F'}{3}x\right)\left(1 - \frac{x}{l}\right)dx + \frac{1}{EI}\int_{2l/3}^{l}\left[\frac{2F'}{3}(l - x)\right]\left(1 - \frac{x}{l}\right)dx \\ &= \frac{4F'l^2}{81EI} = \frac{4Fl^2}{162EI} \end{aligned} \tag{6.3c}$$

while the second of Equations 5.42*b* provides

$$\begin{aligned} \widehat{\Delta}_{BA} &= -\frac{1}{EI}\int_0^l \widehat{M}(x)\left(\frac{x}{l}\right)dx \\ &= -\frac{1}{EI}\int_0^{2l/3}\left(\frac{F'}{3}x\right)\left(\frac{x}{l}\right)dx - \frac{1}{EI}\int_{2l/3}^{l}\left[\frac{2F'}{3}(l - x)\right]\left(\frac{x}{l}\right)dx \\ &= -\frac{5F'l^2}{81EI} = -\frac{5Fl^2}{162EI} \end{aligned} \tag{6.3d}$$

Consequently, the deformation-force relations are

$$\begin{Bmatrix} \Delta_{AB} \\ \Delta_{BA} \\ \Delta_{BD}^{(a)} \end{Bmatrix} = \frac{l}{6EI}\begin{bmatrix} 2 & -1 & \\ -1 & 2 & \\ & & 12EI/kl \end{bmatrix}\begin{Bmatrix} M_{AB} \\ M_{BA} \\ P_{BD} \end{Bmatrix} + \begin{Bmatrix} 4F'l^2/81EI \\ -5F'l^2/81EI \\ 0 \end{Bmatrix} \tag{6.3e}$$

Note that the last term of the **a** matrix has the value $2/k$, which corresponds to the flexibility of the spring in Figure 6.13*a*. We now have all the required matrices and proceed to the solution.

Step 5. We obtain

$$\mathbf{A} = \frac{l}{6EI}\begin{bmatrix} 2l^2(1 + 6EI/kl^3) & 3l \\ 3l & 6 \end{bmatrix}, \qquad \mathbf{B} = -\frac{F'l^2}{81EI}\begin{Bmatrix} 14l \\ 18 \end{Bmatrix} \tag{6.3f}$$

Thus, the compatibility equations are

$$\begin{bmatrix} 2l^2(1 + 6\gamma) & 3l \\ 3l & 6 \end{bmatrix} \begin{Bmatrix} X_1 \\ X_2 \end{Bmatrix} = -\frac{2F'l}{27EI} \begin{Bmatrix} 14l \\ 18 \end{Bmatrix} \tag{6.3g}$$

Here, to simplify this and subsequent equations, we have let $\gamma = EI/kl^3$.
Inverting leads to

$$\begin{Bmatrix} X_1 \\ X_2 \end{Bmatrix} = -\frac{2F'}{81l(1 + 24\gamma)} \begin{bmatrix} 6 & -3l \\ -3l & 2l^2(1 + 6\gamma) \end{bmatrix} \begin{Bmatrix} 14l \\ 18 \end{Bmatrix} \tag{6.3h}$$

from which we obtain

$$\begin{aligned} X_1 &= -\frac{20F'}{27(1 + 24\gamma)} = -\frac{10F}{27(1 + 24\gamma)} \\ X_2 &= \frac{4F'l}{27} \frac{(1 - 36\gamma)}{(1 + 24\gamma)} = \frac{2Fl}{27} \frac{(1 - 36\gamma)}{(1 + 24\gamma)} \end{aligned} \tag{6.3i}$$

Substituting Equation 6.3*h* into Equation 6.3*a* gives

$$\begin{aligned} M_{AB} &= \frac{2F'l}{27} \frac{(1 + 144\gamma)}{(1 + 24\gamma)} = -\frac{Fl}{27} \frac{(1 + 144\gamma)}{(1 + 24\gamma)} \\ M_{BA} &= \frac{4F'l}{27} \frac{(1 - 36\gamma)}{(1 + 24\gamma)} = \frac{2Fl}{27} \frac{(1 - 36\gamma)}{(1 + 24\gamma)} \\ P_{BD} &= -\frac{20F'}{27(1 + 24\gamma)} = -\frac{10F}{27(1 + 24\gamma)} \end{aligned} \tag{6.3j}$$

For the original structure in Figure 6.12*b*, symmetry implies that M_{BC} must be opposite in direction to M_{BA}, and M_{CB} must be opposite in direction to M_{AB}, i.e., $M_{BC} = -M_{BA}$ and $M_{CB} = -M_{AB}$. Also, note that the force in the original spring must be twice the value computed above (because the force in Equation 6.3*j* is essentially the force in ''half'' the spring).

(b) Antisymmetric Load Case

Step 1. The structure in Figure 6.12*e* is 1-degree redundant; we will choose the redundant force as the vertical reaction at B. Observe that the nature of the support at B precludes transverse displacement at B; therefore, there can be no deformation or force in the spring, so it need not be included in the analysis that follows.

Step 2. The statical analysis is shown in Figures 6.13*d*–6.13*e*. We will continue with a matrix formulation even though there is no particular advantage in this case. The equilibrium equation is

$$\begin{Bmatrix} M_{AB} \\ M_{BA} \end{Bmatrix} = \begin{bmatrix} -l \\ 0 \end{bmatrix} \{X\} + \begin{Bmatrix} 2F'l/3 \\ 0 \end{Bmatrix} \tag{6.3k}$$

Step 3. The components of unrestrained end deformations are opposite to the values presented in Equations 6.3*c* and 6.3*d* (because the load is opposite). So the deformation-force relations are

$$\begin{Bmatrix} \Delta_{AB} \\ \Delta_{BA} \end{Bmatrix} = \frac{l}{6EI} \begin{bmatrix} 2 & -1 \\ -1 & 2 \end{bmatrix} \begin{Bmatrix} M_{AB} \\ M_{BA} \end{Bmatrix} + \frac{F'l^2}{81EI} \begin{Bmatrix} -4 \\ 5 \end{Bmatrix} \tag{6.3l}$$

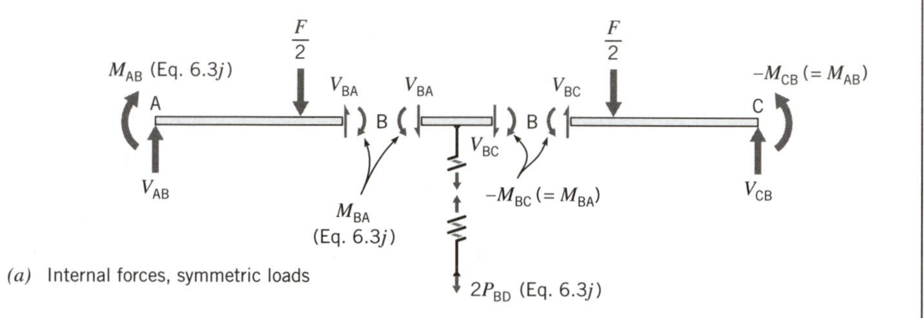

(*a*) Internal forces, symmetric loads

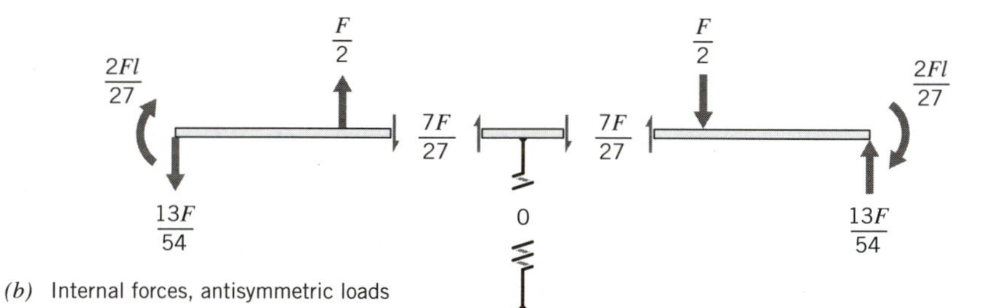

(*b*) Internal forces, antisymmetric loads

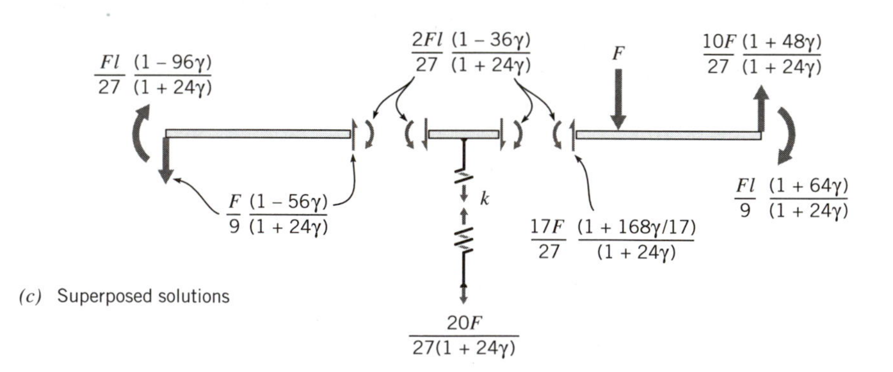

(*c*) Superposed solutions

Figure 6.14 Internal forces, Example 2.

Step 5. We now obtain

$$\mathbf{A} = \frac{l}{6EI}[2l^2], \qquad \mathbf{B} = \frac{14F'l^3}{81EI} \tag{6.3m}$$

which leads to

$$X = \frac{14}{27}F' = \frac{7}{27}F \tag{6.3n}$$

Back-substituting provides

$$\begin{aligned} M_{AB} &= \frac{4}{27}F'l = \frac{2}{27}Fl \\ M_{BA} &= 0 \end{aligned} \tag{6.3o}$$

For the original structure in Figure 6.12*c*, antisymmetry implies that M_{BC} must be in the same direction as M_{BA}, and M_{CB} must be in the same direction as M_{AB}, i.e., $M_{BC} = M_{BA} = 0$ and $M_{CB} = M_{AB} = 2Fl/27$.

(c) Superposition

Figures 6.14*a* and 6.14*b* show the internal forces for the symmetric and antisymmetric cases, respectively. Figure 6.14*c* shows the superposition of these solutions, which constitutes the desired solution for the original structure.

EXAMPLE 3

As a final example, we will examine the closed rectangular frame[9] ABCD (composed of beams AB, BC, CD, and DA) in Figure 6.15*a*. This structure has two axes of symmetry, designated L_1 and L_2 in the figure. Because the loads shown are symmetric with respect to both axes, we have only this one load case. We will assume that the beams are axially rigid.

Step 1. The structure in Figure 6.15*a* is 3-degrees redundant. If we utilize the symmetry, we need only consider one-quarter of the structure, for example the substructure shown in Figure 6.15*b*. This segment is only 1-degree redundant. We will choose the redundant force as the moment at G.

Step 2. The statical analysis is shown in Figures 6.15*b* and 6.15*c* and the matrix equilibrium equation is

[9] Closed frames are common in aircraft and ship structures.

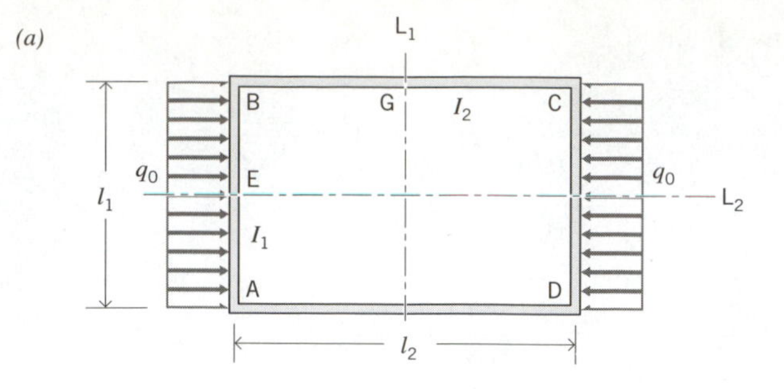

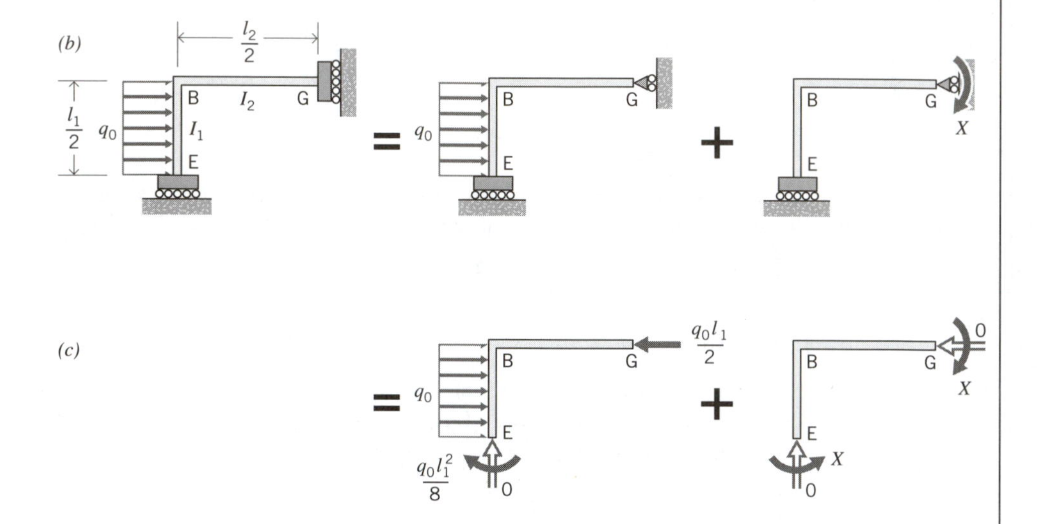

Figure 6.15 Example 3.

$$
\begin{Bmatrix} M_{EB} \\ M_{BE} \\ M_{BG} \\ M_{GB} \end{Bmatrix} = \begin{bmatrix} -1 \\ 1 \\ -1 \\ 1 \end{bmatrix} \{X\} + \begin{Bmatrix} q_o l^2/8 \\ 0 \\ 0 \\ 0 \end{Bmatrix} \tag{6.4a}
$$

Step 3. There are unrestrained end deformations due to the uniform continuous load on beam EB. These have been previously computed in Equations 5.44*g*; here, we obtain the desired unrestrained end deformations by substituting $l_1/2$ for l in Equations 5.44*g*. Thus, the deformation-force relations are

$$
\begin{Bmatrix} \Delta_{EB} \\ \Delta_{BE} \\ \Delta_{BG} \\ \Delta_{GB} \end{Bmatrix} = \frac{1}{12E} \begin{bmatrix} 2l_1/I_1 & -l_1/I_1 & & \\ -l_1/I_1 & 2l_1/I_1 & & \\ & & 2l_2/I_2 & -l_2/I_2 \\ & & -l_2/I_2 & 2l_2/I_2 \end{bmatrix} \begin{Bmatrix} M_{EB} \\ M_{BE} \\ M_{BG} \\ M_{GB} \end{Bmatrix} + \frac{q_o l^3}{192 EI_1} \begin{Bmatrix} 1 \\ -1 \\ 0 \\ 0 \end{Bmatrix} \tag{6.4b}
$$

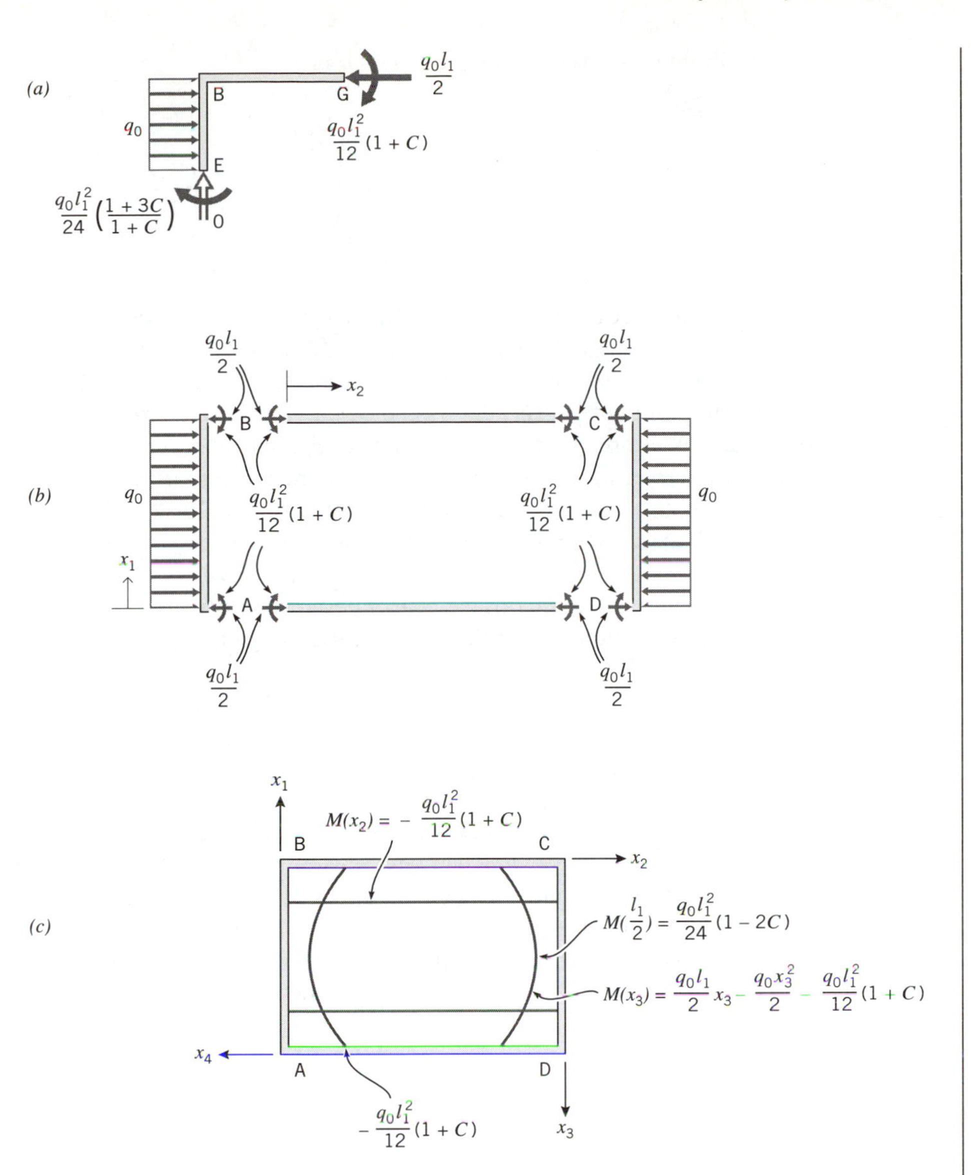

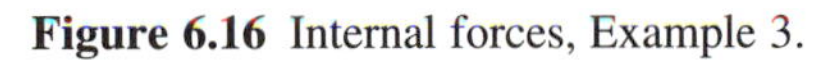
Figure 6.16 Internal forces, Example 3.

Step 5. We obtain

$$\mathbf{A} = \frac{l_1}{2EI_1}\left[1 + \left(\frac{I_1}{I_2}\right)\left(\frac{l_2}{l_1}\right)\right], \qquad \mathbf{B} = \frac{q_0 l_1^3}{24EI_1} \tag{6.4c}$$

and

$$X = \frac{q_0 l_1^2}{12\left[1 + \left(\frac{I_1}{I_2}\right)\left(\frac{l_2}{l_1}\right)\right]} \tag{6.4d}$$

Substituting this result into Equation 6.4*a* provides

$$M_{\mathrm{EB}} = \frac{q_o l_1^2}{24} \frac{\left[1 + 3\left(\frac{I_1}{I_2}\right)\left(\frac{l_2}{l_1}\right)\right]}{\left[1 + \left(\frac{I_1}{I_2}\right)\left(\frac{l_2}{l_1}\right)\right]} = \frac{q_o l_1^2}{24}\left(\frac{1 + 3C}{1 + C}\right) \quad \textbf{(6.4e)}$$

where we have set $C = (I_1/I_2)(l_2/l_1)$; note that C is a dimensionless constant. Because Equation 6.4*a* also gives $M_{\mathrm{BE}} = -M_{\mathrm{BG}} = M_{\mathrm{GB}} = X$, the final result is as shown in Figures 6.16*a* and 6.16*b*. The moment diagrams are sketched directly on a drawing of the structure in Figure 6.16*c*. As was noted previously (Example 4 of Chapter 5), it is important to clearly specify the individual element coordinate systems in order to properly interpret sign conventions in these moment diagrams; it should be understood, for instance, that x_1 is an axial coordinate that is zero at the A end of element AB, and that coordinate axes are right handed (which implies that y_1 is in the same direction as x_4 in the figure).

Even though the original frame constitutes a relatively small structure, the computational simplification achieved through the use of symmetry has been quite significant in this case.

6.2.2 Structures Symmetric With Respect to a Point

Let the structure in Figure 6.17*a* be representative of the general class of structures that are symmetric with respect to a point, O. This implies that for any point on the axis of an element of the structure, e.g., point *a*, there exists another point *b*, which lies on the axis of an element where the physical and mechanical properties are identical to those at *a*, such that point O is the midpoint of the line segment *ab*. The point O is called the *center of symmetry*. In simplified terms, if the structure is rotated 180° about O, the

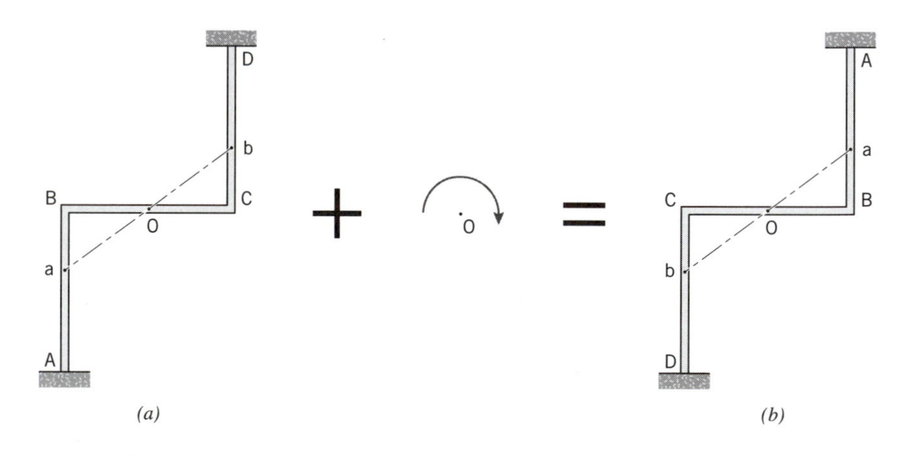

Figure 6.17 Structure that is symmetric with respect to a point O.

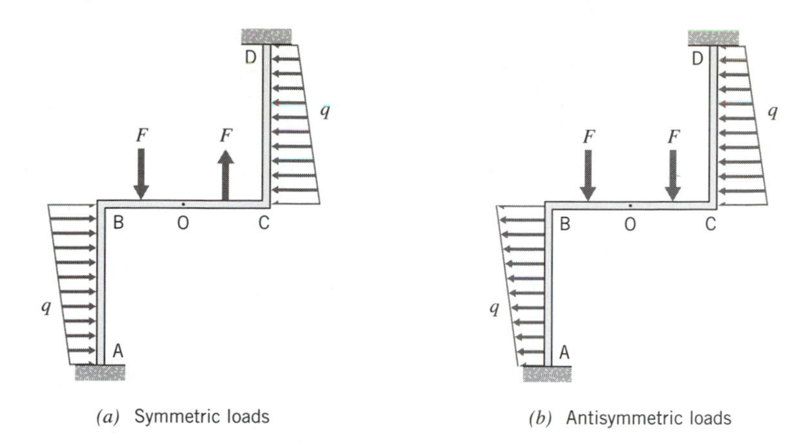

(a) Symmetric loads

(b) Antisymmetric loads

Figure 6.18 Loads with symmetry properties.

resulting configuration is indistinguishable from the original configuration. Thus, the rotated structure DCBA in Figure 6.17*b* is identical to structure ABCD in Figure 6.17*a*.

To utilize this symmetry, it is again necessary to deal with loads that also possess some type of symmetry. We will continue to categorize loads as either symmetric or antisymmetric, based on the results of rotation about O. Specifically, if a 180° rotation of the loaded structure results in an identical loaded configuration, we refer to the loads as symmetric (e.g., Figure 6.18*a*); if a 180° rotation results in a configuration in which the loads are exactly opposite in sign, we refer to the loads as antisymmetric (e.g., Figure 6.18*b*). Again, note that deformations and internal forces must also be either symmetric or antisymmetric depending on whether the applied loads are symmetric or antisymmetric.

To determine the nature of the kinematic and static boundary conditions at point O, we consider the displacements and internal forces at point O (Figure 6.19*a*). If the structure in Figure 6.19*a* is rotated 180° about O, we obtain the structure in Figure 6.19*b*. If the loads on the structure are symmetric, as in Figure 6.18*a*, this rotation must result in a configuration in which displacements and forces are identical to the original

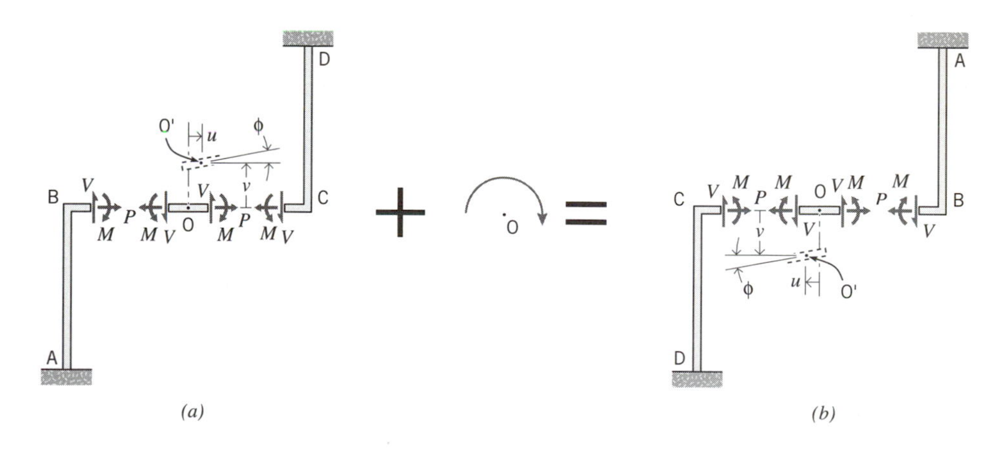

(a)

(b)

Figure 6.19 Displacements and internal forces at point O.

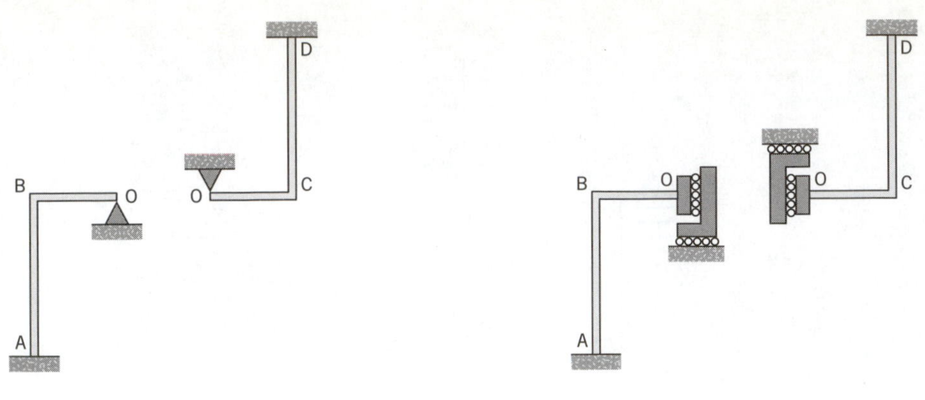

Figure 6.20 Substructures.

configuration; if the loads on the structure are antisymmetric, as in Figure 6.18*b*, this rotation must result in a configuration in which displacements and forces are opposite to those in the original configuration.

Thus, we see that if the loads are *symmetric*, displacements and forces in Figures 6.19*a* and 6.19*b* will be identical only if $u = v = M = 0$ (which implies that P, V, and ϕ may be nonzero); if the loads are *antisymmetric*, displacements and forces in Figures 6.19*a* and 6.19*b* will be opposite only if $P = V = \phi = 0$ (which implies that u, v, and M may be nonzero).[10]

In the first case (symmetric load), the kinematic and static conditions at point O ($u = v = M = 0$) are exactly equivalent to those imposed by a pin support (Figure 6.20*a*); in the second case (antisymmetric load), the kinematic and static conditions at point O ($P = V = \phi = 0$) are exactly equivalent to those imposed by a fixed support with two rollers (Figure 6.20*b*).[11]

Thus, we see a direct parallel between the formulations and procedures for analysis of structures that are symmetric with respect to a line and structures that are symmetric with respect to a point.

---

[10] Again note the correspondence between displacement components and associated force quantities in each of these cases.

[11] Actually, rollers in any two nonparallel directions will suffice. (Once again, we see that a hypothetical support serves a useful function.)

PROBLEMS

6.P1. The truss has pin supports at A and D and is loaded by a vertical force F at C. All bars have the same EA.

(a) Find the bar forces by analyzing one-half of the truss.

(b) Find the vertical displacement at B.

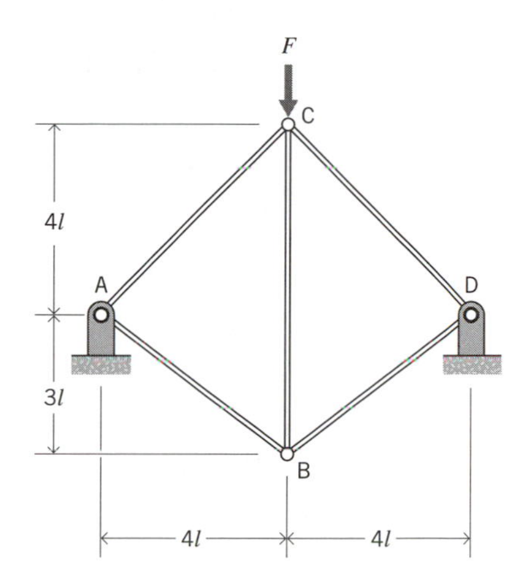

6.P2. The truss loaded as shown has been analyzed in Section 5.2.1. (All bars have the same EA.) Reanalyze the truss taking advantage of the symmetry.

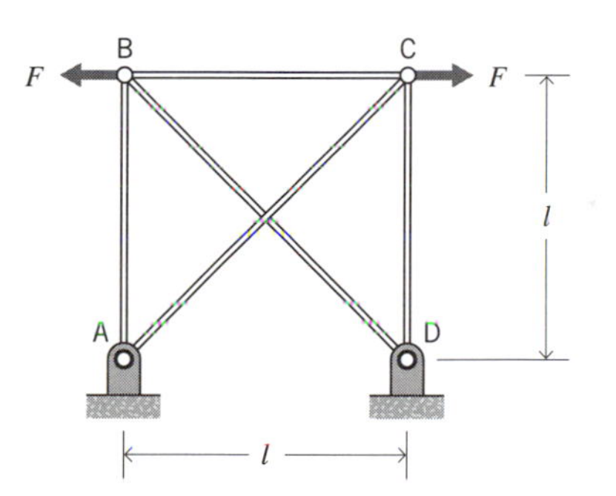

6.P3. In addition to the force F applied to the truss, bar AB has an initial length change Δ_I. Find the bar forces by analyzing one-half of the truss. All bars have the same EA.

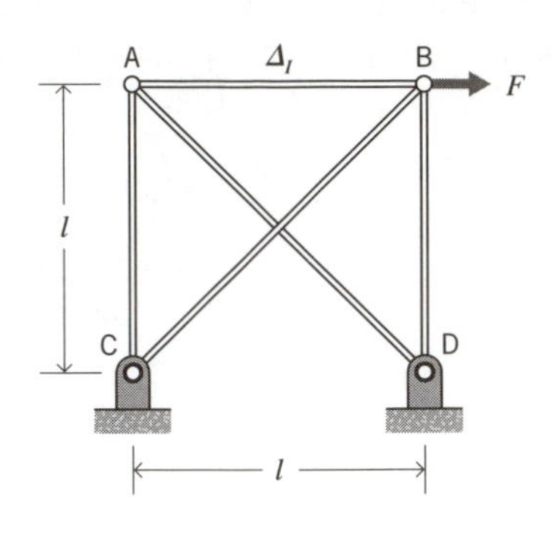

6.P4. Find the bar forces by analyzing one-half of the truss. All bars have the same EA.

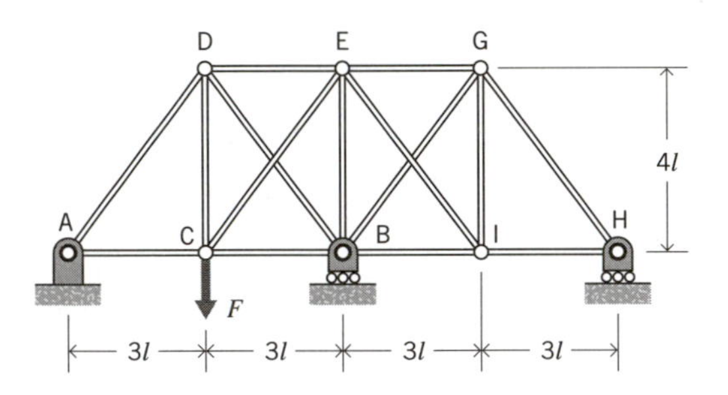

6P.5. The truss is supported by three linear springs with spring constant k. Find the spring forces by analyzing one-half of the structure. All bars have the same EA. Solve the equations for two cases: (a) $EA/kl \ll 1$, (b) $EA/kl \gg 1$.

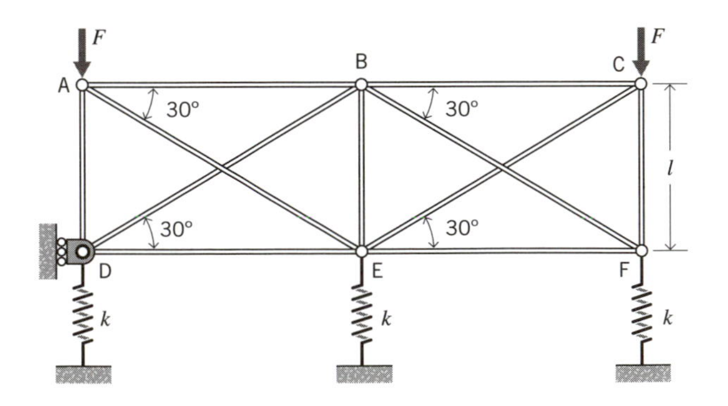

6.P6. The structure is composed of four truss bars (AC, BD, AG, and BG) and one axially rigid beam (AB). All connections between elements are pinned. The stiffness of all bars is EA, and the bending stiffness of the beam is $EI = EAl^2/3$. Find the internal forces by analyzing one-half of the structure.

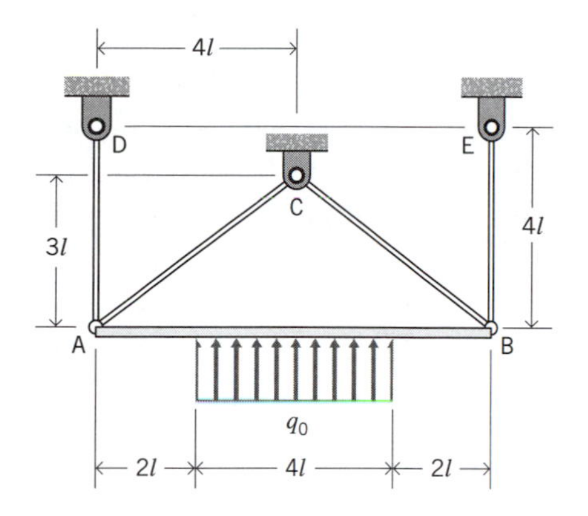

6.P7. The axial stiffness of truss bars AD, BD, and CD is EA. For beam ABC $I = l^2A$. Find the internal forces by analyzing one-half of the structure.

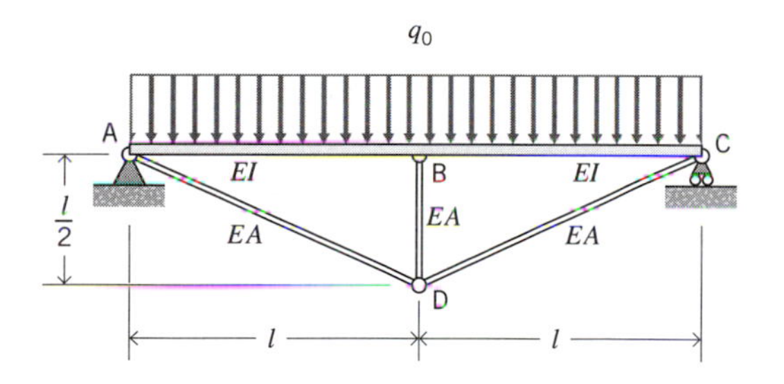

6.P8. Find the reactions at A and B by analyzing one-half of the beam.

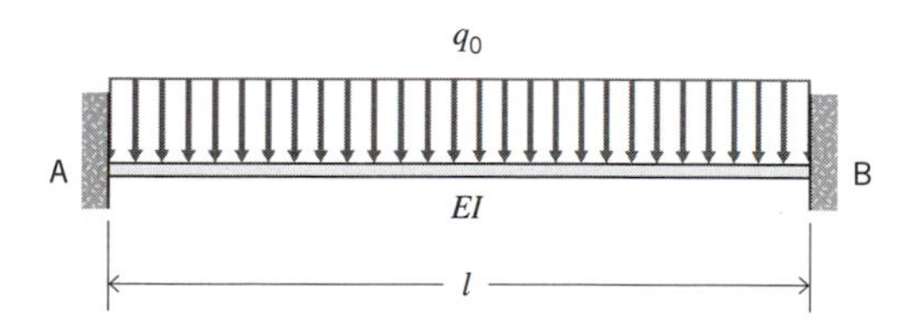

6.P9. Find the reactions at A and B by analyzing one-half of the beam.

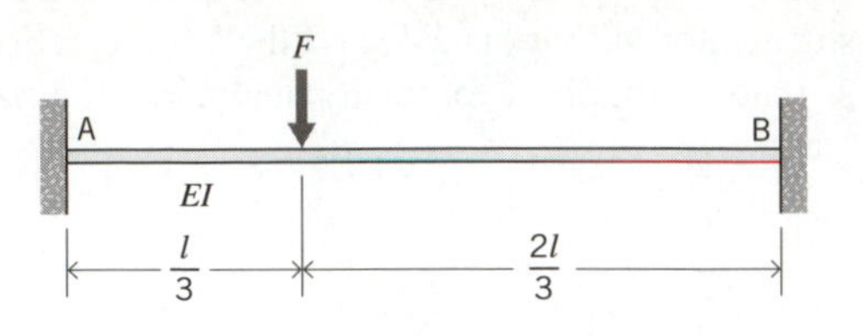

6.P10. The structure is fixed at A and C. There is a hinge connecting the beams at B. Joint B is supported by a linearly elastic spring of stiffness k. Analyze one-half of the structure and find the internal forces in terms of EI and k.

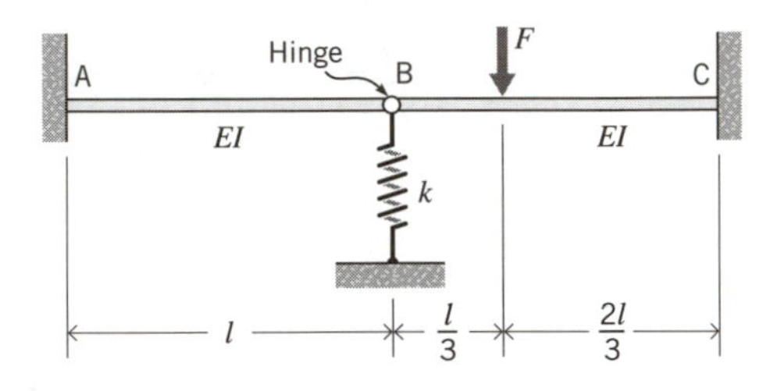

6.P11. The structure is fixed at A and C and hinged at B. There is a linear torsional spring of stiffness k connecting the ends of the beams at B. The relationship between moment and deformation for the spring is $M_s = k\Delta_s$ where Δ_s is the deformation (angle change). Find the internal forces by analyzing one-half of the structure. Let $k = 2EI/l$. Hint: use $2k$ rather than $k/2$. Why?

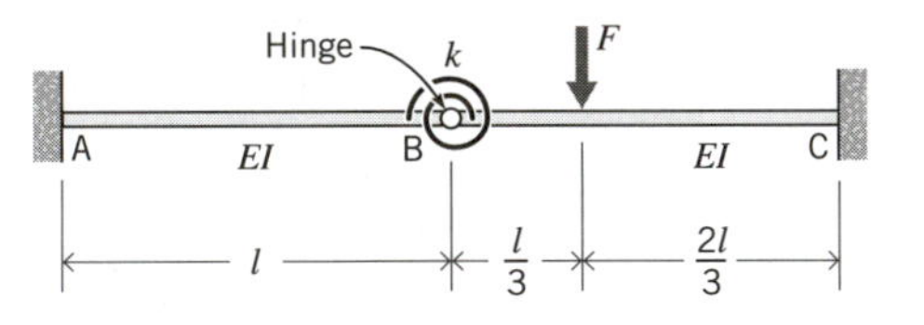

6.P12. Find the internal forces by analyzing one-half of the structure. Sketch free-body diagrams for elements AB, BC, and CD.

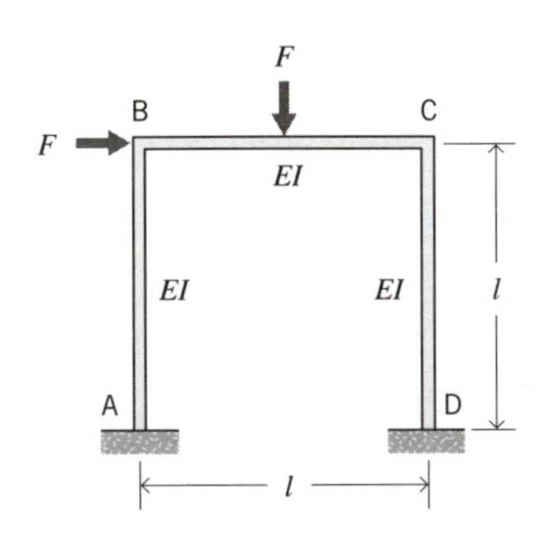

6.P13. The structure is a closed frame with two axes of symmetry. Solve for the internal forces by analyzing beam AB only. Show results by sketching free-body diagrams of all beams and joints.

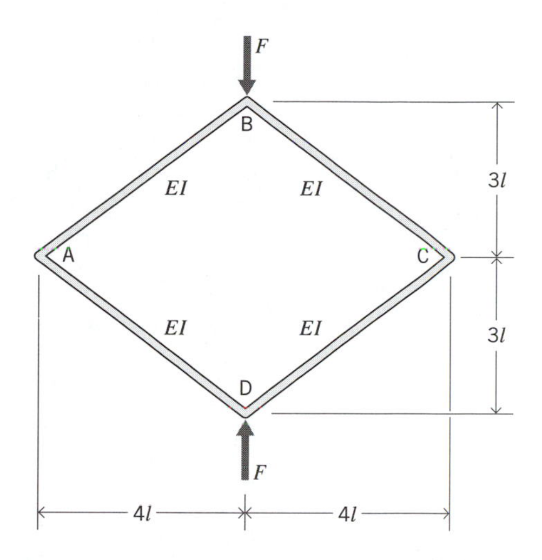

6.P14. The structure is a closed frame with two axes of symmetry. The frame is loaded by external moments applied at A and C. Solve for the internal forces by analyzing beam AB only. Show results by sketching free-body diagrams of all beams and joints. Note: pay careful attention to symmetry conditions for the loading.

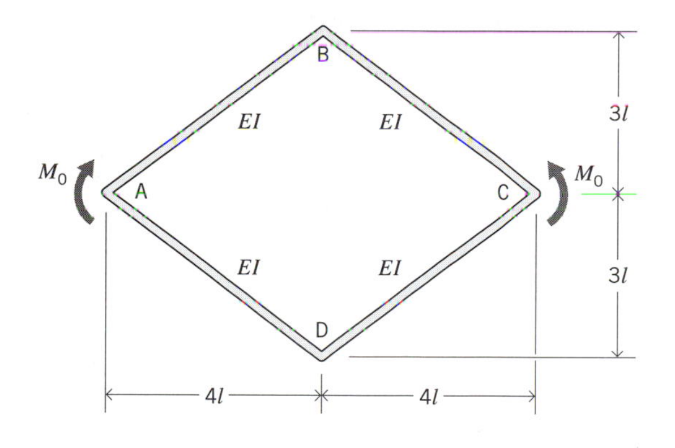

6.P15. The closed frame having two axes of symmetry is simply supported at A and C. Solve for the internal forces by analyzing beam AB only. Show results by sketching free-body diagrams of all beams and joints.

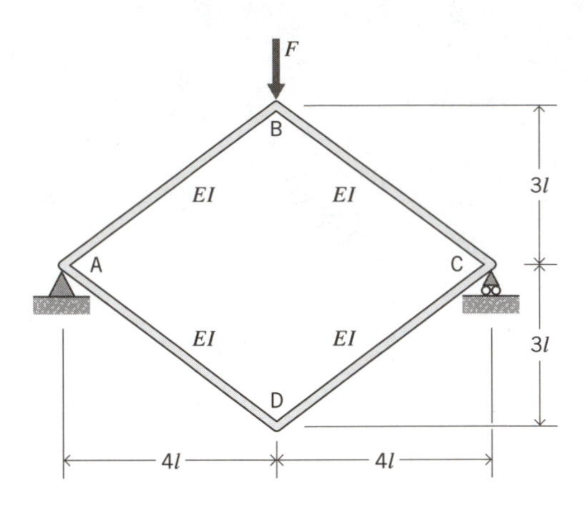

6.P16. A closed frame having two axes of symmetry is simply supported at A and C. Joints between elements at A and C are rigid, but joints between elements at B and D are pinned. Elements at joints B and D are connected by linear torsional springs; the relationship between moment and deformation for the springs is $M_s = k\Delta_s$ where k is the spring constant and Δ_s is the deformation (angle change). Assume $k = 2EI/5l$. Vertical forces F are applied at the midpoints of elements AB and BC. Solve for the internal forces by analyzing beam AB only. Show your results by sketching free-body diagrams of all beams and joints. Hint: use $2k$ rather than $k/2$. Why?

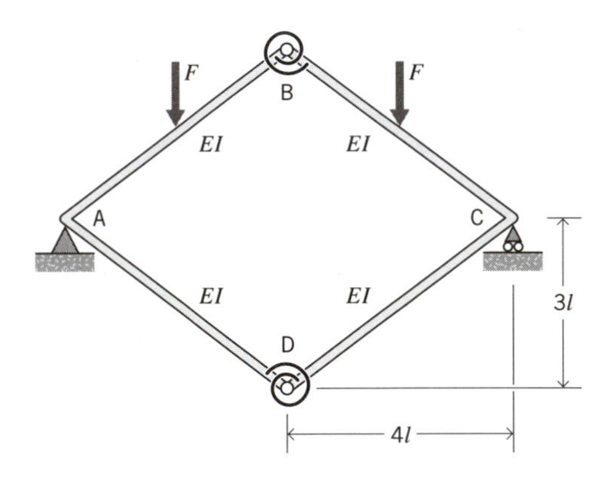

6.P17. The structure is composed of three axially rigid beam elements. The connections between elements at joints A and B are rigid, and there is a hinge at joint C. All elements have a bending stiffness *EI*. The structure is loaded by a concentrated moment M_o applied at point D, which is the midpoint of beam AB. Note that CD is an axis of symmetry for the structure. Find the internal forces by analyzing one-half of the structure. Show your results by sketching free-body diagrams of all beams and joints. Sketch the bending moment diagram for beam AB.

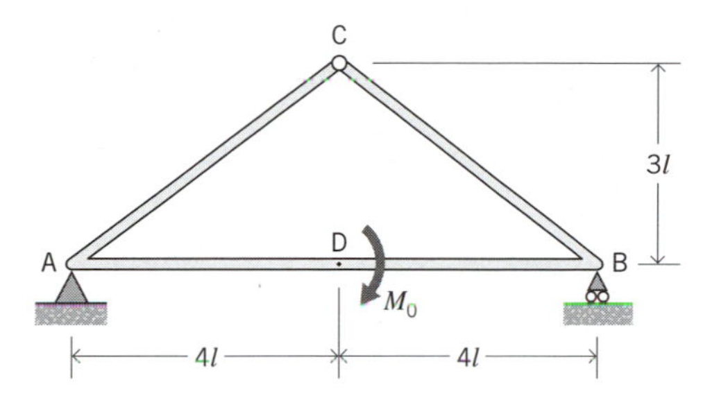

Chapter 7

Influence Lines

7.0 INTRODUCTION

In the preceding chapters, we have developed procedures for the analysis of structures that are subjected to physical loads applied at specified locations. Many structures are actually subjected to physical loads that can have variable positions, for instance vehicle loads on bridges. Furthermore, these so-called ''live loads'' can be applied in various combinations as well as at various locations.

It is often important for the analyst to determine the positions and/or combinations of loads that will produce the maximum structural response, i.e., the maximum magnitude of force, stress, or displacement. In some cases, it may be feasible and convenient to obtain this type of information simply by repeating structural analyses for different locations of loads. There are, however, many cases for which this repetitive approach will not readily reveal the most significant structural behavior. This is particularly true for cases where the analyst must determine where to apply various loads to elicit the desired response.

In these latter cases, it may be possible to investigate only a limited number of response quantities at specific locations in the structure where the analyst thinks the maximum (or near-maximum) responses may occur. For example, in a simple beam subject to vehicle loads, the analyst may restrict attention to the bending moment at midspan even though the maximum may actually occur elsewhere (see Example 1 following). The bending moment at such a specified point can be determined relatively easily even for a complex set of moving loads by using the concept of influence lines.

An *influence line* is a graphical representation of the variation of a particular response quantity (at a given point in the structure) as a function of position of a concentrated unit load. This type of graphical representation can simplify the determination of load positions that lead to maximum responses. The characteristics and shapes of influence lines can be determined from basic analysis procedures or, in many cases, from virtual work considerations.

7.1 MOVING LOADS AND INFLUENCE LINES

Before beginning a detailed examination of influence lines, it will be helpful to understand the types of loads and responses for which applications of the concept are most useful. We will, therefore, start with a simple illustrative example.

EXAMPLE 1

A two-axle truck moves across a simply supported beam bridge of length l, Figure 7.1*a*. The front axle load is F_1 and the rear axle load is F_2, where $F_2 > F_1$. We will denote the distance between axles by d. For illustrative purposes in the computations that follow we will set $F_1 = F_2/2$, and $d = l/5$. We wish to find the maximum bending moment in the beam. The moving loads will be regarded as pseudostatic, i.e., possible dynamic effects will not be considered.

In a case such as this one, the maximum bending moment in the beam can be determined either approximately by trial and error or exactly by mathematical analysis. The trial-and-error procedure consists of constructing bending moment diagrams for various positions of the axle loads, and identifying the maximum bending moment

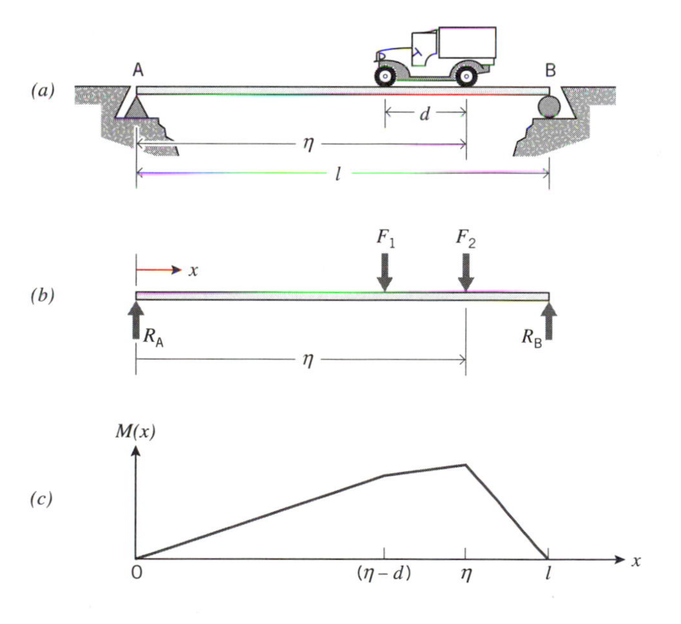

Figure 7.1 Example 1.

from among all diagrams. A typical bending moment diagram will have the shape shown in Figure 7.1*c*; note that this particular bending moment diagram assumes that both axles are on the beam. This procedure involves no new concepts and will receive no further consideration here.

An exact solution can be obtained by finding an expression for maximum bending moment as a function of axle position, e.g., as a function of coordinate η in Figure 7.1*b*. We will assume that the maximum bending moment occurs at the location of F_2, as in Figure 7.1*c*. Again, assuming that both axles are on the beam, for the special case $F_1 = F_2/2$ and $d = l/5$, it may be shown that

$$R_B = \frac{F_2(15\eta - l)}{10l}, \quad \frac{l}{5} < \eta < l \tag{7.1a}$$

and

$$M_{max}(\eta) = R_B(l - \eta) = \frac{F_2(15\eta - l)(l - \eta)}{10l} \tag{7.1b}$$

where R_B is the reaction at end B (Figure 7.1*b*) and $M_{max}(\eta)$ is the bending moment in the beam at the location of load F_2.

Differentiating Equation 7.1*b* with respect to η and setting the result equal to zero leads to

$$\overline{\eta} = \frac{8}{15}l = 0.5333l \tag{7.1c}$$

where $\overline{\eta}$ is the value of η that maximizes $M_{max}(\eta)$. Substituting Equation 7.1*c* into Equation 7.1*b* provides the maximum bending moment due to the moving vehicle,[1]

$$M_{max} = M_{max}(\overline{\eta}) = \frac{49}{150}F_2 l = 0.3267F_2 l \tag{7.1d}$$

We note from Equation 7.1*c* that $\overline{\eta} > l/2$, i.e., that the maximum bending moment does not occur at the middle of the beam. This may be verified by substituting $\eta = l/2$ into Equation. 7.1*b*, for which we obtain

$$M_{max}(l/2) = \frac{13}{40}F_2 l = 0.3250F_2 l \tag{7.2}$$

The difference between the results in Equations 7.1*d* and 7.2 in the preceding example is relatively small. If an analyst has confidence at the outset that the difference will be small, then it is reasonable to restrict attention to the bending moment at the midspan

[1] It may be verified that this value of M_{max} is greater than the maximum bending moment at any position of F_1.

because computations will be considerably simplified, especially for complicated loading cases. The next step is to determine the effect on the selected response of loads applied at arbitrary locations throughout the structure. We accomplish this by introducing the concept of an influence coefficient.

An *influence coefficient* is the value of a particular response at a specific location, say $\bar{x}$, due to a concentrated unit load applied at some point in the structure, say x.[2] The magnitude of the influence coefficient will change if the location of the unit load changes. An *influence line* is a mathematical plot of the magnitude of the influence coefficient as a function of position of the unit load. Thus, the ordinate to the influence line at any point x represents the magnitude of the response quantity at $\bar{x}$ for the unit load applied at point x. This concept will be demonstrated in the following example. As we will show subsequently, the specific characteristics of many influence lines can be obtained from virtual work.

EXAMPLE 2

Define the influence line for bending moment at midspan, $M(l/2)$ (i.e., $\bar{x} = l/2$), for the beam in Figure 7.1*a*; use the influence line to find the maximum value of $M(l/2)$ due to the vehicle loads from Example 1.

In order to construct the influence line for the bending moment at midspan, consider the beam subjected to a positive unit load,[3] as shown in Figure 7.2*a*. This unit load can be applied at any position $0 < x < l$. We seek $M(l/2)$ as a function of x. The reactions at A and B are shown in Figure 7.2*b*. It follows that

$$M(l/2) = R_B\frac{l}{2} = -\frac{x}{2}, \qquad 0 \leq x \leq \frac{l}{2} \tag{7.3a}$$

and

$$M(l/2) = R_A\frac{l}{2} = -\frac{(l-x)}{2}, \qquad \frac{l}{2} \leq x \leq l \tag{7.3b}$$

Equations 7.3*a* and 7.3*b* are plotted in Figure 7.2*c*. Figure 7.2*c* is defined as the influence line for the bending moment at midspan; as noted previously, it is a plot of the value of $M(l/2)$ as a function of position of the unit load.[4]

The influence line in Figure 7.2*c* can be used to find the value of $M(l/2)$ due to any combination of loads. It is only necessary to superpose the effects of the various individual loads, including continuous loads. The value of $M(l/2)$ due to a single

[2] Thus, a flexibility coefficient (Chapter 5) is a special case of an "influence coefficient."

[3] We continue to define positive loads as loads that act in the positive y direction.

[4] This should not be confused with a bending moment diagram, which is a plot of moment at every cross section due to a specified set of loads applied at fixed positions. Thus, even though this particular influence line has the same shape as a bending moment diagram due to a concentrated load applied at the midpoint of the beam, the two plots are separate and distinct.

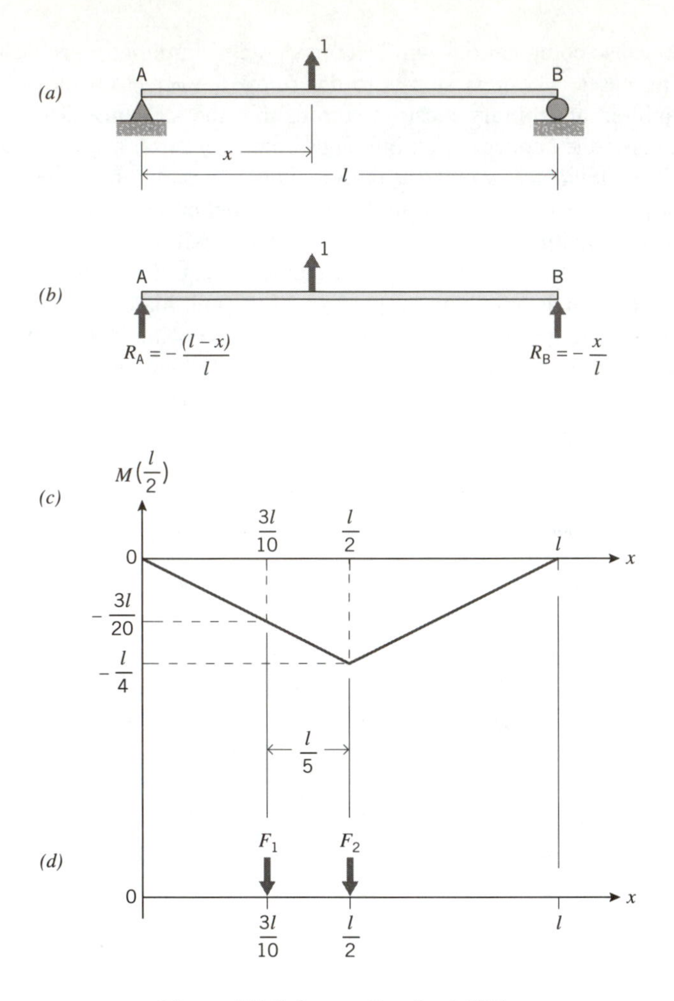

Figure 7.2 Influence line for $M(l/2)$.

concentrated load is equal to the magnitude of the concentrated load multiplied by the ordinate to the influence line at the point of application of the load. For vehicle loads, it is still necessary to find the positions of the loads that will produce the maximum value of $M(l/2)$. In this case, the maximum is obtained with the loads in the position shown in Figure 7.2*d*,

$$M(l/2) = \left(-\frac{3l}{20}\right)(-F_1) + \left(-\frac{l}{4}\right)(-F_2) = \left(\frac{3l}{20}\right)\frac{F_2}{2} + \left(\frac{l}{4}\right)F_2 = \frac{13}{40}F_2 l \quad \textbf{(7.4)}$$

which is the same solution as in Equation 7.2. In more complex cases, trial and error may be required to find the positions and combinations of loads that produce the maximum response at a given position. In general, such trial-and-error procedures will require much less computational effort than multiple reanalyses for various load conditions.

EXAMPLE 3

A bulldozer crosses a simply supported bridge of length l, Figure 7.3*a*. Assume the tracks of the bulldozer apply a uniform distributed load q_o over a distance d, and let $d = l/5$. (*a*) Find the maximum moment at midspan using the influence function defined in Equations 7.3. (*b*) Check the result obtained in part (*a*) by sketching the bending moment diagram for the beam with the bulldozer in the position that produces the maximum value of $M(l/2)$.

(a) In this case it should be apparent that the maximum moment at midspan is obtained with the uniform load exactly centered on the bridge ($\eta = 2\,l/5$), as in Figure 7.3*b*. For this position of the load, we obtain the value of $M(l/2)$ from the influence function in Equation 7.3 by integration,

$$M_{max}(l/2) = \int_{2l/5}^{l/2} (-q_o)\left(-\frac{x}{2}\right)dx + \int_{l/2}^{3l/5} (-q_o)\left(-\frac{l-x}{2}\right)dx \qquad \textbf{(7.5a)}$$

This is a direct extension of the procedure employed in Equation 7.4 for concentrated loads.

Performing the integrations in Equation 7.5*a* leads to

$$M_{max}(l/2) = \frac{9}{200}q_o l^2 \qquad \textbf{(7.5b)}$$

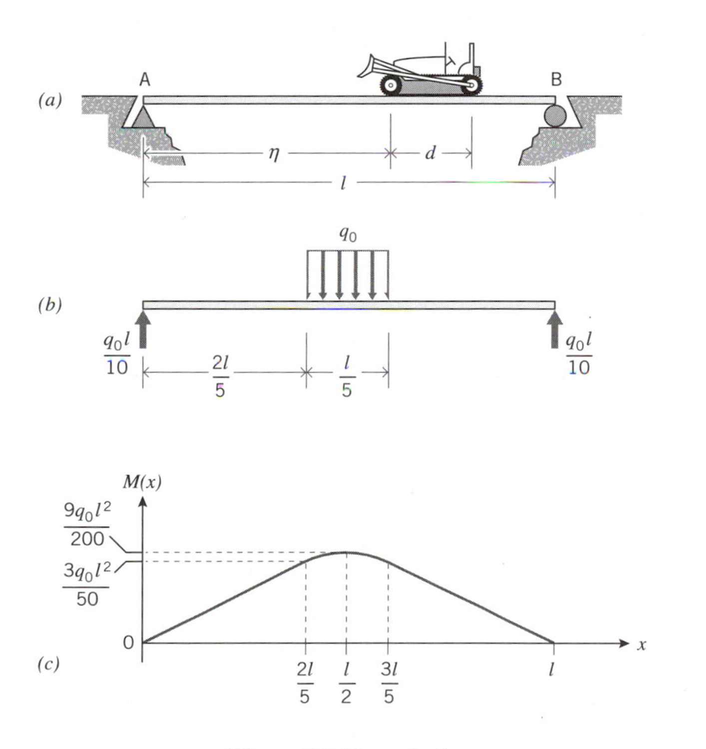

Figure 7.3 Example 3.

(b) The bending moment diagram for the beam in Figure 7.3*b* is shown in Figure 7.3*c*. We see that the value of $M(l/2)$ is identical to the value in Equation 7.5*b*. (Note the difference between the bending moment diagram in Figure 7.3*c* and the influence line in Figure 7.2*c*.)

7.2 VIRTUAL WORK AND INFLUENCE LINES

The procedure used in Example 2 to construct the influence line consisted essentially of the direct calculation of the desired response quantity for various positions of the unit load. This approach is relatively easy to apply for statically determinate structures, but can become burdensome for indeterminate structures. For the latter, virtual work provides a useful alternative for determining properties of influence lines. We will demonstrate the application of virtual work for a statically determinate case similar to the preceding examples and then generalize the result to statically indeterminate structures.

7.2.1 Influence Lines for Statically Determinate Structures

Consider the statically determinate beam in Figure 7.4*a* that is subjected to any general load, denoted by $q(x)$ and that may be continuous or discontinuous. It should be understood that $q(x)$ encompasses concentrated loads. Let $a(x)$ be some "influence function," which defines the influence line for a particular response quantity, say Q, in the beam. For example, if $Q \equiv M(l/2)$ then $a(x)$ would be defined as the function on the right side of Equations 7.3*a* and 7.3*b*. Then, the response Q due to $q(x)$ would, by definition, be given by

$$Q = \int_0^l q(x)\, a(x)\, dx \tag{7.6}$$

Equation 7.6 may be regarded as a generalization of Equation 7.5*a*.

We wish to demonstrate that virtual work can be used to obtain the influence function for any desired force quantity Q in the beam in Figure 7.4*a*. Suppose, for instance, that the response quantity of interest actually is $Q \equiv M(l/2)$. Figure 7.4*b* shows free-body diagrams of the beam in Figure 7.4*a*. The free body of the incremental segment of the midportion of the beam indicates the positive direction of $M(l/2)$ as previously defined in Chapter 3; however, it should be understood that $M(l/2)$ is the internal moment at the midspan of the beam in Figure 7.4*a*.

Let us apply the theorem of virtual work for the real forces in Figure 7.4*a* and a virtual displacement, which is chosen in a very specific way so as to provide the desired information about the particular response quantity $Q \equiv M(l/2)$. Consider the virtual displacement pattern in Figure 7.4*c*, which is obtained by inserting a "force-releasing device" in the beam in Figure 7.4*a* and introducing a unit motion in the device. For a moment such as $M(l/2)$ we introduce a hinge device and apply a unit of angular offset $\Delta\theta = 1$, i.e., a rotation, in the hinge. Note that the direction of $+Q$ in Figure 7.4*b* is the same as the direction of hinge rotation in Figure 7.4*c*. The insertion of the device reduces the statically determinate structure to a mechanism, the components of which

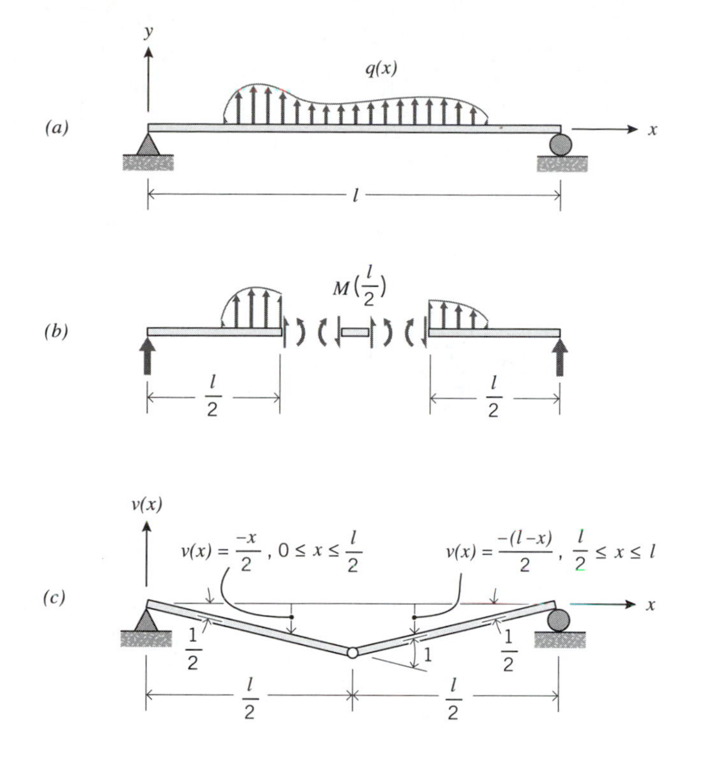

Figure 7.4 Influence line from virtual displacement.

undergo rigid-body rotations; thus, the stipulated virtual displacement is completely defined, as indicated in Figure 7.4*c*.

For the real external and internal forces in Figure 7.4*a*, and the virtual displacements in Figure 7.4*c*, the theorem of virtual work gives[5]

$$M(l/2) \times 1 = \int_0^l q(x)\, v(x)\, dx \tag{7.7}$$

The internal virtual work on the left side of Equation 7.7 is positive because, as stated, the virtual rotation at $x = l/2$ was chosen in the same direction as a real positive bending moment $M(l/2)$. The notable feature of the virtual displacement pattern in Figure 7.4*c* is that $v(x)$ is identical to $a(x)$ in Equation 7.3; this leads to the conclusion that Equations 7.7 and 7.6 are equivalent. In other words, rather than computing the influence function directly from equilibrium considerations (as in Example 2), the same information is obtained from the analytic properties of the particular virtual displacement pattern in Figure 7.4*c*. This type of relationship between influence functions and displacement patterns is actually quite general. It is a useful relationship because it is

[5] See Example 9, Chapter 4.

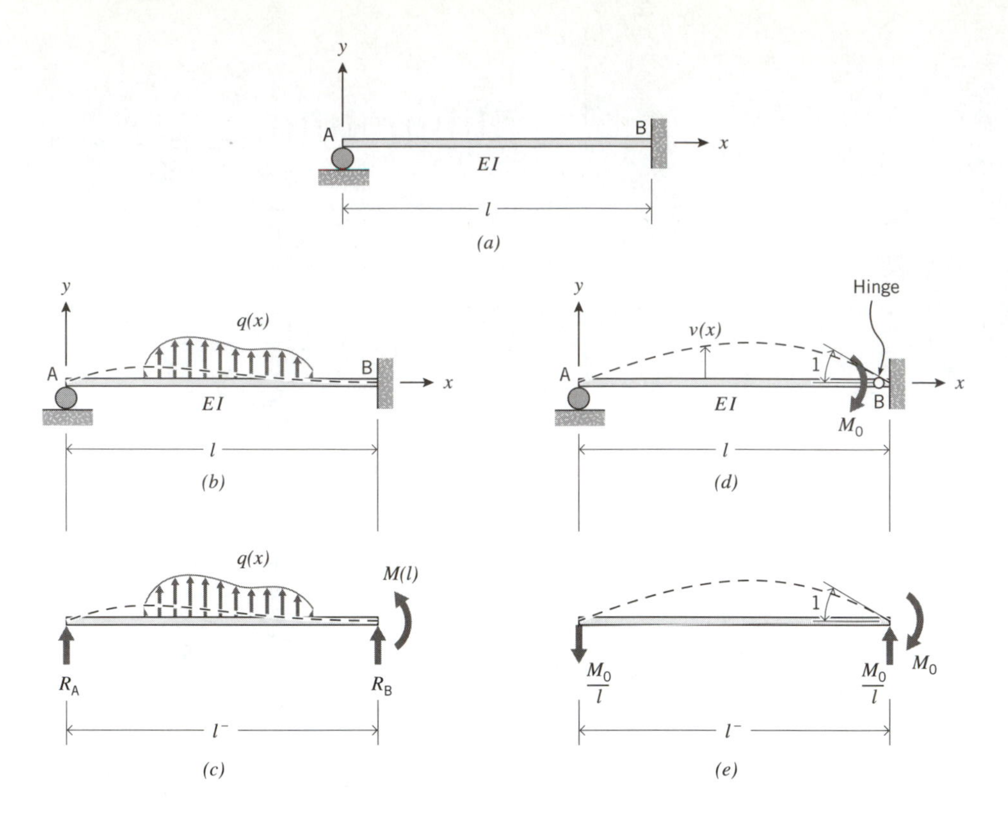

Figure 7.5 Influence line from Betti's theorem.

often easier to define the virtual displacement pattern than it is to directly determine the influence functions.

7.2.2 Influence Lines for Statically Indeterminate Structures: The Müller-Breslau Principle

The type of relationship observed in the preceding section between influence functions and displacements can be shown to exist for statically indeterminate structures. We may demonstrate the relationship by using Betti's theorem (Section 6.1.1). To illustrate, consider the structure in Figure 7.5*a*, which is one-degree indeterminate, and assume that we are interested in determining the influence function for the internal bending moment at $\bar{x} = l^-$, i.e., $Q \equiv M(l^-)$.[6]

To proceed with the application of Betti's theorem, we assume that the structure is subjected to the same type of general load $q(x)$ as was applied to the beam in Figure 7.4*a*. The loaded structure is shown in Figure 7.5*b*, and the free-body diagram of the beam is shown in Figure 7.5*c*. The displacements produced by the loads are indicated schematically by the dotted curve in Figures 7.5*b* and 7.5*c*.

Now, assume that a hinge device is inserted in the structure at $x = l^-$ and that an

[6] As in Chapter 3, the notation l^- denotes a location an infinitesimal distance to the left of $x = l$.

external force, in this case M_o, is applied just to the left of the hinge to cause a unit rotation at the hinge, as shown in Figure 7.5*d*. Note, again, that the direction of this rotation at l^- is the same as that due to a positive value of $Q \equiv M(l^-)$. The corresponding free-body diagram for the beam is shown in Figure 7.5*e*.[7] The displacements produced by M_o are shown by the dotted curve in Figures 7.5*d* and 7.5*e*.

Figures 7.5*c* and 7.5*e* are free-body diagrams of the same beam subjected to two different sets of forces. The forces shown in these figures are external to the free bodies; consequently, Betti's theorem is directly applicable to the beams in these figures.

Betti's theorem equates the external virtual work done by the forces in Figure 7.5*c* going through the displacements in Figure 7.5*e* to the external virtual work done by the forces in Figure 7.5*e* going through the displacements in Figure 7.5*c*. The latter is zero, so we obtain

$$-M(l) \times 1 + \int_0^l q(x)\, v(x)\, dx = 0 \qquad \textbf{(7.8a)}$$

or

$$M(l) = \int_0^l q(x)\, v(x)\, dx \qquad \textbf{(7.8b)}$$

where $v(x)$ is the displacement in Figure 7.5*d* or Figure 7.5*e* and where, for simplicity, we have replaced l^- by l. We observe, again, that Equations 7.8*b* and 7.6 are equivalent.

An expression like Equation 7.8*b* can be obtained for any general force response Q for any statically indeterminate structure. The right-hand sides of all equations like Equation 7.8*b* must be equivalent to the right-hand side of Equation 7.6. Therefore $v(x)$ and $a(x)$ must be identical. This relationship may be stated as follows: the influence function $a(x)$ for a particular force response Q is the same as a specific displacement function $v(x)$, which is obtained by inserting a (hypothetical) device that has the effect of removing from the original structure the capacity to develop the force Q, and applying loads that produce a unit incremental displacement in the same sense as $+Q$ at the point where the device is inserted. This concept is known as the *Müller-Breslau principle*. The principle reflects the same limitations as Betti's theorem from which it is derived, i.e., it is applicable only to linearly elastic structures subjected to mechanical loads.

The Müller-Breslau principle can be demonstrated to hold for a wide range of applications and is, therefore, a general procedure for the determination of influence functions. For instance, we may demonstrate the principle for a more complex case than the preceding one by considering the influence function for internal bending moment at $x = \eta$, $0 < \eta < l$, for the beam in Figure 7.5*a*, i.e., we will take $Q \equiv M(\eta)$. The beam is also shown in Figure 7.6*a* with an applied load $q(x)$ that, as in Figure 7.5*b*, is included for use in the subsequent application of Betti's theorem.

According to the Müller-Breslau principle, the desired influence line corresponds to the schematic dotted deflection curve in Figure 7.6*b*. Specifically, this is the deflection curve for the beam with a hinge introduced at $x = \eta$, and equal and opposite external

[7] The forces in this beam are very similar to those in the beam of Figure 4.18*c*.

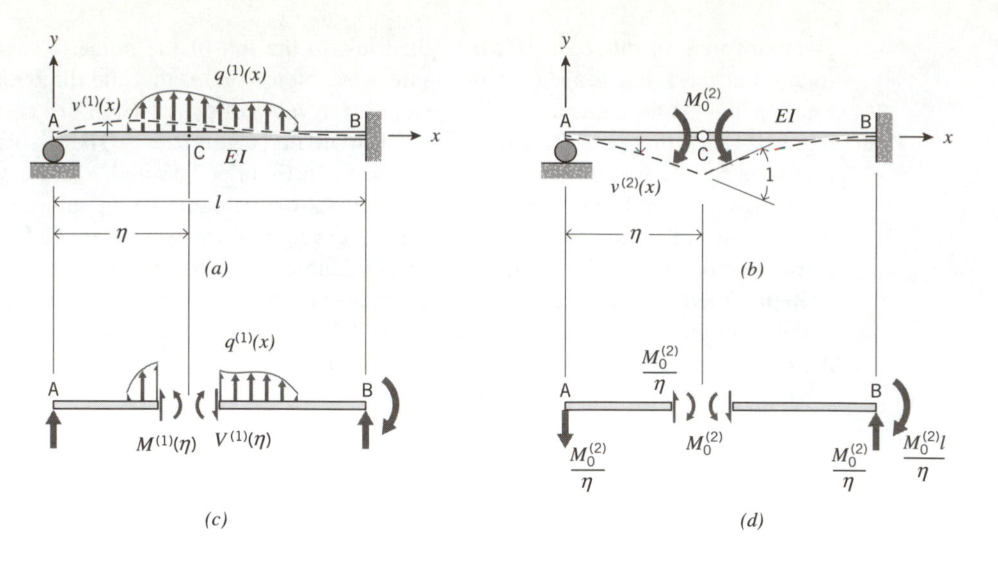

Figure 7.6 Influence line for internal moment in indeterminate beam.

moments, M_o, applied adjacent to the hinge. The applied moments are of such magnitude and sense as to produce a unit rotation at $x = \eta$ in the same direction as positive $M(\eta)$. For the sake of clarity in the subsequent application of Betti's theorem, we may regard the structure in Figure 7.6*b* as being composed of two beams joined by the hinge at $x = \eta$; we will denote the location of the hinge in this structure as joint C.

To demonstrate that the influence function for $M(\eta)$ is the same as the equation for $v(x)$ in Figure 7.6*b*, we proceed as follows. Figures 7.6*c* and 7.6*d* show free-body diagrams for the structures of Figures 7.6*a* and 7.6*b*, respectively. Note that the moments in the vicinity of joint C in Figure 7.6*d* are equal to M_o.[8] We will denote the displacements and forces in Figures 7.6*a* and 7.6*c* by superscripts (1), and those in Figures 7.6*b* and 7.6*d* by superscripts (2), exactly as was done in Section 6.1.1.

We consider segments AC and CB separately and equate the external virtual work done by the forces in Figure 7.6*c* and the displacements in Figure 7.6*b* to the external virtual work done by the forces in Figure 7.6*d* and the displacements in Figure 7.6*a*. For beam AC we obtain

$$M^{(1)}(\eta)\, v'^{(2)}(\eta^-) + V^{(1)}(\eta)\, v^{(2)}(\eta^-) + \int_0^{\eta^-} q^{(1)}(x)\, v^{(2)}(x)\, dx$$

$$= -M_o^{(2)}\, v'^{(1)}(\eta^-) + \frac{M_o^{(2)}}{\eta}\, v^{(1)}(\eta^-) \qquad \textbf{(7.9a)}$$

[8] *ibid*; the internal moments in this case can be obtained by superposing the effects of each of the two applied moments M_o.

Similarly, for beam CB we obtain

$$-M^{(1)}(\eta)\,v'^{(2)}(\eta^{+}) - V^{(1)}(\eta)\,v^{(2)}(\eta^{+}) + \int_{\eta^{+}}^{l} q^{(1)}(x)\,v^{(2)}(x)\,dx = M_o^{(2)}\,v'^{(1)}(\eta^{+}) - \frac{M_o^{(2)}}{\eta}\,v^{(1)}(\eta^{+}) \tag{7.9b}$$

It should be noted that the signs of terms in Equations 7.9*a* and 7.9*b* are based on the actual directions of forces shown in Figure 7.6 and positive analytic values of displacements and slopes as defined in Chapter 3 (Figure 3.14*a*).

The displacements at $x = \eta$ are continuous in Figure 7.6*a*, which means that

$$\begin{aligned} v^{(1)}(\eta^{-}) &= v^{(1)}(\eta^{+}) \\ v'^{(1)}(\eta^{-}) &= v'^{(1)}(\eta^{+}) \end{aligned} \tag{7.9c}$$

Therefore, adding Equations 7.9*a* and 7.9*b* leads to

$$M^{(1)}(\eta)[v'^{(2)}(\eta^{+}) - v'^{(2)}(\eta^{-})] = \int_0^l q^{(1)}(x)\,v^{(2)}(x)\,dx \tag{7.9d}$$

If we stipulate that the term in brackets above is equal to unity, we obtain

$$M(\eta) \times 1 = \int_0^l q(x)\,v(x)\,dx \tag{7.9e}$$

where the superscripts have been deleted. In other words, we have obtained an equation that has the same form as Equation 7.8*b*. We conclude that the Müller-Breslau principle can be demonstrated in a similar manner for any linearly elastic structure.

EXAMPLE 4

Use an influence line to find the end moment M_{BA} for the uniformly loaded beam in Figure 7.7*a*. The moment M_{BA} is the negative of $M(l)$ in Figure 7.5*c*.

As we have noted in the preceding text, the influence line for $M(l)$ or $-M_{BA}$ is defined by the displacement function in Figure 7.5*d*. The beam in Figure 7.5*d* is statically determinate, and the displacement can be determined by integrating the expression for curvature twice. It may be shown that this displacement function is

$$v(x) = \frac{x}{2} - \frac{x^3}{2l^2} \tag{7.10a}$$

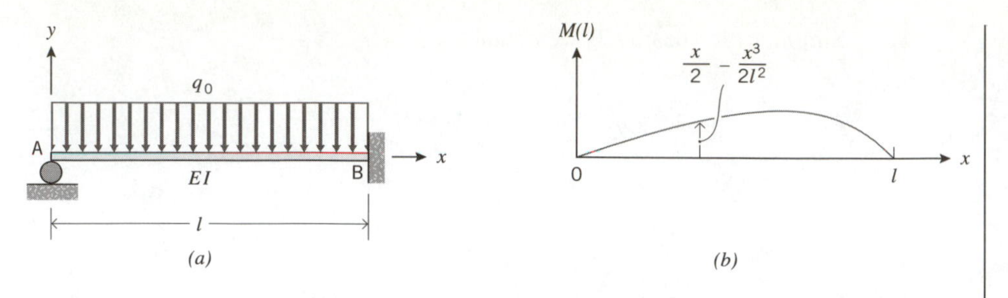

Figure 7.7 Example 4.

The influence line is redrawn in Figure 7.7*b*. Thus, from Equation 7.8*b*, we obtain for the uniform load in Figure 7.7*a*

$$M_{BA} = -\int_0^l (-q_o)\left(\frac{x}{2} - \frac{x^3}{2l^2}\right)dx = \frac{q_o l^2}{8} \qquad \textbf{(7.10b)}$$

which agrees with previous results (cf. Figure 5.19).

EXAMPLE 5

The influence line shown in Figure 7.7*b* and defined in Equation 7.10*a* can be used to find the value of M_{BA} for any load applied to beam AB in Figure 7.5*a*. For example, if the vehicle loads from Example 1 (Figure 7.1*b*) are applied to this beam, we find by trial and error that the maximum value of M_{BA} is obtained with F_2 at approximately $x = 0.636l$ and F_1 at $x = 0.636l - d = 0.436l$. The approximate maximum value of M_{BA} for this case is

$$M_{BA} = -\left[(0.189l)(-F_2) + (0.177l)\left(-\frac{F_2}{2}\right)\right] = 0.278F_2 l \qquad \textbf{(7.11)}$$

where the quantity $0.189l$ is the ordinate to the influence line at $x = 0.636l$ and the quantity $0.177l$ is the ordinate to the influence line at $x = 0.436l$.

Note that $v(x)$ in Equation 7.10*a* assumes its maximum value at $x = l/\sqrt{3} = 0.577l$, but that neither load acts exactly at this point. Also, it should be determined whether or not a greater value of M_{BA} is obtained with F_1 and F_2 reversed, i.e., with the vehicle moving in the opposite direction.

EXAMPLE 6

Derive the analytic expression for the influence line for internal bending moment at $x = \eta$ for the beam in Figure 7.5*a*.

The nature of this influence line has been discussed above. The influence line is defined by $v(x)$ for the statically determinate structure in Figure 7.6*b*.

The derivation of the expressions for $v(x)$ as a function of M_o follows procedures similar to those employed in Example 7, Chapter 3. We note that $v'(x)$ is discontinuous at $x = \eta$, so it is necessary to integrate curvature equations over two distinct regions $0 \le x < \eta$ and $\eta < x \le l$. The curvature in both regions is $M(x)/EI = -M_o x/\eta EI$ (Figure 7.6*d*), and the boundary conditions are $v(0) = v(l) = v'(l) = 0$ and $v(\eta^-) = v(\eta^+)$. The result of the integration is

$$v(x) = -\frac{M_o}{6\eta EI}x^3 + \left(\frac{M_o l^2}{2\eta EI} - \frac{M_o l^3}{3\eta^2 EI}\right)x, \qquad 0 \le x < \eta \qquad \textbf{(7.12}\textbf{\textit{a}}\textbf{)}$$

and

$$v(x) = -\frac{M_o}{6\eta EI}x^3 + \frac{M_o l^2}{2\eta EI}x - \frac{M_o l^3}{3\eta EI}, \qquad \eta < x \le l \qquad \textbf{(7.12}\textbf{\textit{b}}\textbf{)}$$

A unit incremental rotation at $x = \eta$ corresponds to the condition $v'(\eta^+) - v'(\eta^-) = 1$. The value of M_o that produces this unit rotation is found to be

$$M_o = \frac{3\eta^2 EI}{l^3} \qquad \textbf{(7.13)}$$

Substituting Equation 7.13 into Equations 7.12*a* and 7.12*b* leads to

$$v(x) = -\frac{\eta}{2l^3}x^3 + \left(\frac{3\eta}{2l} - 1\right)x, \qquad 0 \le x < \eta \qquad \textbf{(7.14}\textbf{\textit{a}}\textbf{)}$$

and

$$v(x) = -\frac{\eta}{2l^3}x^3 + \frac{3\eta}{2l}x - \eta, \qquad \eta < x \le l \qquad \textbf{(7.14}\textbf{\textit{b}}\textbf{)}$$

Equations 7.14*a* and 7.14*b* define the desired influence line for $M(\eta)$.

It is worth noting that as η n l, Equation 7.14*a* approaches the solution given in Equation 7.10*a*, and Equation 7.14*b* approaches zero. (η n l also implies that x n l in Equation 7.14*b*.) Thus, we see that the influence line in Example 4 is a special case of this solution.

EXAMPLE 7

Derive the analytic expression for the influence line for internal transverse *shear* at $x = \eta$, $V(\eta)$, for the beam in Figure 7.8*a*.

In this case, the Müller-Breslau principle requires that the capacity to develop $V(\eta)$ be removed from the beam, and that a unit incremental displacement be introduced in the direction of $+V(\eta)$. This displacement is illustrated by the dotted curve in Figure 7.8*b*. Note that equal and opposite external forces F are applied in the vicinity of $x = \eta$ to produce the displacement. The internal forces and reactions are found in a manner similar to those in Figure 7.6*d* and are shown in Figure 7.8*c*.

It is also important to recognize that the displacement curve must exhibit the property $v'(\eta^+) = v'(\eta^-)$ so that there would be no net virtual work done by internal *moments* at $x = \eta$ in the beam of Figure 7.6*c* going through the displacements in Figure 7.8*b* in the application of Betti's theorem. The only virtual work is that due to the internal shear $V(\eta)$ in the beam of Figure 7.6*c* undergoing the unit transverse incremental displacement shown in Figure 7.8*b*.

The derivation of the expression for $v(x)$ proceeds as in Example 6. It is again necessary to integrate curvature equations over the two distinct regions $0 \leq x < \eta$ and $\eta < x \leq l$. The curvature in both regions is $M(x)/EI = Fx/EI$, and the boundary

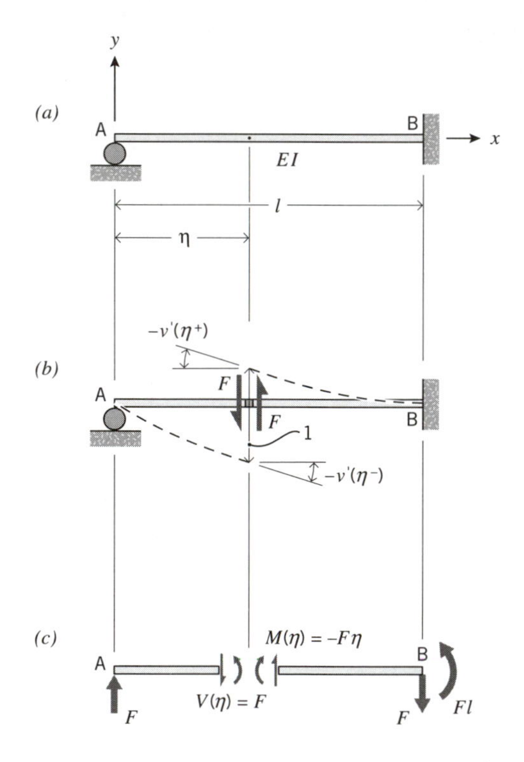

Figure 7.8 Example 7.

conditions are $v(0) = v(l) = v'(l) = 0$ and $v'(\eta^-) = v'(\eta^+)$. The result of the integration is

$$v(x) = \frac{F}{6EI}x^3 - \frac{Fl^2}{2EI}x, \qquad 0 \le x < \eta \tag{7.15a}$$

and

$$v(x) = \frac{F}{6EI}x^3 - \frac{Fl^2}{2EI}x + \frac{Fl^3}{3EI}, \qquad \eta < x \le l \tag{7.15b}$$

The unit incremental displacement corresponds to the condition $v(\eta^+) - v(\eta^-) = 1$. After setting the difference equal to unity, we find

$$F = \frac{3EI}{l^3} \tag{7.16}$$

Substituting Equation 7.16 into Equations 7.15*a* and 7.15*b* leads to

$$v(x) = \frac{x^3}{2l^3} - \frac{3x}{2l}, \qquad 0 \le x < \eta \tag{7.17a}$$

and

$$v(x) = \frac{x^3}{2l^3} - \frac{3x}{2l} + 1, \qquad \eta < x \le l \tag{7.17b}$$

which define the influence line for $V(\eta)$, Figure 7.8*b*.

Although Equations 7.17*a* and 7.17*b* do not explicitly contain the coordinate η, the equations are valid only over the indicated ranges of x. As in the previous example, this solution can be used to obtain influence lines for $V(\eta)$ as $\eta \to 0$ or $\eta \to l$, which is equivalent to providing the influence lines for transverse reactions at A or B.

EXAMPLE 8

Figure 7.9*a* shows a continuous three-span beam. Find the influence line for the internal moment at joint B. Note that the bending stiffness of beam BC is $2EI$.

The beam in Figure 7.9*a* is two-degrees indeterminate. The desired influence line is the dotted displacement curve shown in Figure 7.9*b*, which is obtained by inserting a hinge in the beam at B and introducing a unit increment of rotation at the hinge. The beam with the hinge at B is one-degree indeterminate. Thus, we seek the equation of the displacement curve for the one-degree indeterminate structure in Figure 7.9*b* subject to the loads shown.

The analysis is accomplished in accordance with the concepts developed in Chapter 5.

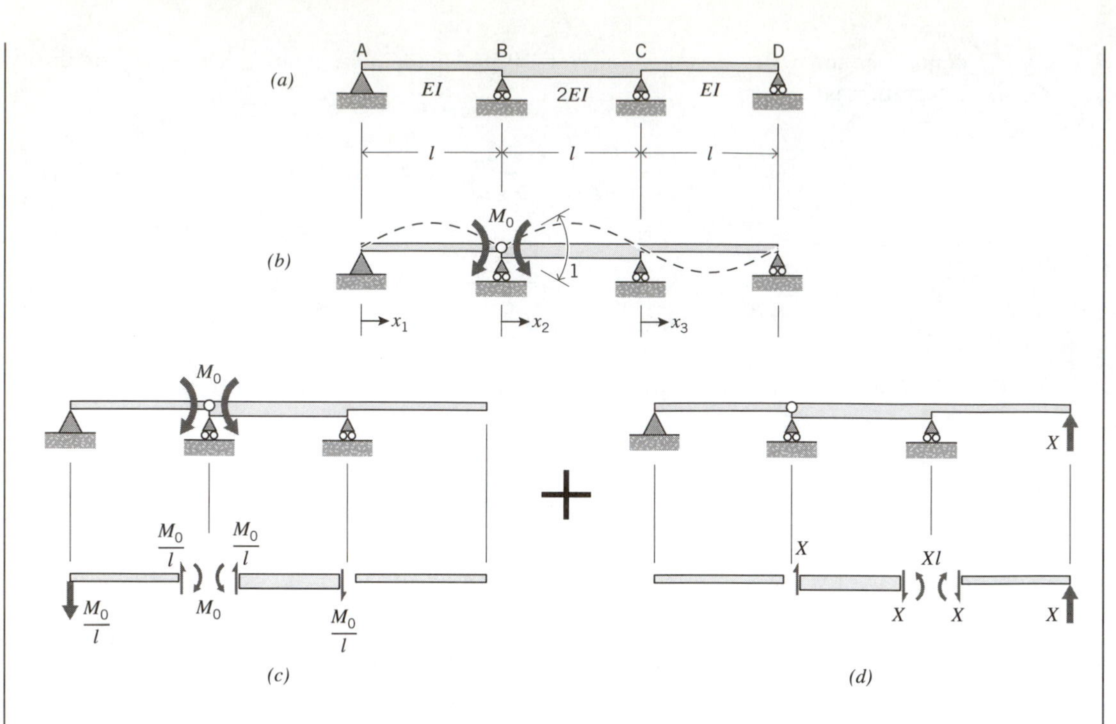

Figure 7.9 Example 8.

We will choose as the redundant force the reaction at D, and superpose the force states in Figures 7.8*c* and 7.8*d*. The matrix equilibrium equation is

$$\begin{Bmatrix} M_{AB} \\ M_{BA} \\ M_{BC} \\ M_{CB} \\ M_{CD} \\ M_{DC} \end{Bmatrix} = \begin{bmatrix} 0 \\ 0 \\ 0 \\ -l \\ l \\ 0 \end{bmatrix} \{X\} + \begin{Bmatrix} 0 \\ M_o \\ -M_o \\ 0 \\ 0 \\ 0 \end{Bmatrix} \tag{7.18a}$$

and the deformation-force equation has the form

$$\begin{Bmatrix} \Delta_{AB} \\ \Delta_{BA} \\ \Delta_{BC} \\ \Delta_{CB} \\ \Delta_{CD} \\ \Delta_{DC} \end{Bmatrix} = \frac{l}{6EI} \begin{bmatrix} 2 & -1 & & & & \\ -1 & 2 & & & & \\ & & 1 & -1/2 & & \\ & & -1/2 & 1 & & \\ & & & & 2 & -1 \\ & & & & -1 & 2 \end{bmatrix} \begin{Bmatrix} M_{AB} \\ M_{BA} \\ M_{BC} \\ M_{CB} \\ M_{CD} \\ M_{DC} \end{Bmatrix} \tag{7.18b}$$

The vector of unrestrained end deformations is zero in Equation 7.18*b* because there are no intermediate loads between the ends of any of the beams.

Solving the resulting compatibility equation leads to

$$X = \frac{M_o}{6l} \tag{7.18c}$$

Back-substituting provides the values of end forces and end deformations,

$$\mathbf{P} \equiv \begin{Bmatrix} M_{AB} \\ M_{BA} \\ M_{BC} \\ M_{CB} \\ M_{CD} \\ M_{DC} \end{Bmatrix} = \begin{Bmatrix} 0 \\ M_o \\ -M_o \\ -M_o/6 \\ M_o/6 \\ 0 \end{Bmatrix} \tag{7.18d}$$

$$\boldsymbol{\Delta} \equiv \begin{Bmatrix} \Delta_{AB} \\ \Delta_{BA} \\ \Delta_{BC} \\ \Delta_{CB} \\ \Delta_{CD} \\ \Delta_{DC} \end{Bmatrix} = \frac{M_o l}{6EI} \begin{Bmatrix} -1 \\ 2 \\ -11/12 \\ 1/3 \\ 1/3 \\ -1/6 \end{Bmatrix} \tag{7.18e}$$

As in Example 6, we must set the difference in slopes between the left end of beam BC and the right end of beam AB equal to unity. Recall that a positive slope corresponds to a negative deformation; therefore, the unit incremental rotation at joint B is obtained from the condition[9]

$$-\Delta_{BC} + \Delta_{BA} = 1 \tag{7.19a}$$

From Equation 7.18*e* we obtain

$$\frac{11 M_o l}{72 EI} + \frac{M_o l}{3 EI} = 1 \tag{7.19b}$$

or

$$M_o = \frac{72 EI}{35 l} \tag{7.19c}$$

Substituting the value of M_o from Equation 7.19*c* into Equation 7.18*d* gives the end moments in the structure when the incremental rotation at B is unity. The curvature for each beam is then obtained from equilibrium considerations. Integration leads to

[9] Note that the side-sways are zero; hence, slopes are directly related to deformations. This result could also be obtained using the theorem of virtual forces by applying equal and opposite virtual moments of unity on each side of the hinge at B and computing **P** for the AB and BC spans.

$$v(x_1) = -\frac{12}{35l^2}x_1^3 + \frac{12}{35}x_1 \tag{7.20a}$$

$$v(x_2) = \frac{7}{35l^2}x_2^3 - \frac{18}{35l}x_2^2 + \frac{11}{35}x_2 \tag{7.20b}$$

and

$$v(x_3) = -\frac{2}{35l^2}x_3^3 + \frac{6}{35l}x_3^2 - \frac{4}{35}x_3 \tag{7.20c}$$

where the coordinates x_1, x_2, and x_3 are defined in Figure 7.9*b*. Equation 7.20 specifies the shape of the influence line in this figure.

An important feature of this influence line is the fact that the sign of the displacement in beam CD is opposite to that in beams AB and BC. This implies that loads applied anywhere in beam CD will cause bending moments at B that are opposite in sign to the bending moments at B that are caused by loads applied anywhere in beams AB and BC. This is the type of information that influence lines are particularly useful in conveying to the analyst. By using the influence line, the analyst can position various live loads in locations where the effects on the response quantity are cumulative rather than subtractive.

7.2.3 Influence Lines for Trusses

The concept of influence lines is also applicable to trusses that are subjected to live loads. In the case of trusses, influence lines are usually most useful for bridges where live loads are confined to a roadway or deck surface. Influence lines are typically constructed for values of bar forces as functions of position of a unit load that moves across the bridge deck.

In our previous analyses of trusses, it has been assumed that loads consist of concentrated forces applied only at joints. Loads on the decks of truss bridges, on the other hand, can be applied anywhere and can consist of continuous as well as concentrated forces. This is not inconsistent with our analytic model of trusses because it is assumed that all applied forces are distributed to nodes through a deck system consisting of longitudinal beams (''stringers'') and transverse beams (''floor beams''). The longitudinal elements of the deck system comprise a series of simply supported beams spanning between adjacent nodes (Figures 7.10*a* and 7.10*b*); the transverse beams connect identical nodes of parallel trusses. The manner in which an arbitrarily located force is assumed to be distributed to nodes is illustrated in Figure 7.10*c*.

As was the case with frame structures, influence lines for trusses can be constructed either by direct calculation of a response quantity for various positions of a unit load, or by virtual work concepts. Applications of virtual displacements have previously been demonstrated in Chapter 4, and Betti's theorem is applicable to the special case of structures composed of truss bars. Consequently, all preceding concepts pertaining to the derivation of influence lines, including the Müller-Breslau principle, are applicable to truss structures.

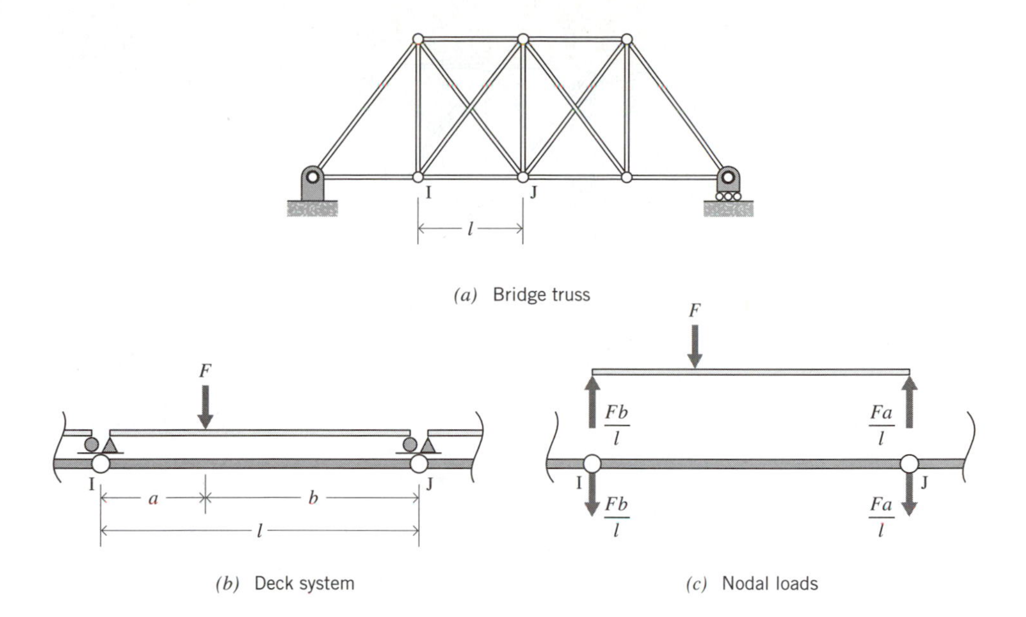

Figure 7.10 Distribution of live loads to truss nodes.

EXAMPLE 9

Find the influence line for the force in bar DE of the bridge truss in Figure 7.11*a*. Assume that all bars have the same axial stiffness *EA*.

The ordinate of the influence line for the bridge deck at any point x (Figure 7.11*a*) is the value of the bar force P_{DE} due to a unit load applied at x along the bridge deck, line ACBIH. Because the truss is two-degrees indeterminate, the Müller-Breslau principle indicates that this influence line is identical to the displaced shape of line ACBIH when the force in bar DE is removed from the truss and a unit incremental displacement is introduced in the direction of positive bar force, i.e., a unit of elongation is inserted between D and E by application of a force F.

Removing the capacity of bar DE to transmit an axial force is not equivalent to removing the bar completely because the relative motion at the ''cut'' in the bar is the primary quantity we are evaluating.[10] When the force in bar DE is removed, the truss is one-degree indeterminate. We will choose the force in bar EG as the redundant force (X). The resulting determinate truss, with the equal and opposite external forces F, which produce a displacement in the direction of bar DE, is shown in Figure 7.11*b*. The corresponding internal forces are also shown in this figure, and the internal forces due to the redundant force are shown in Figure 7.11*c*.

---

[10] See Figure 5.4.

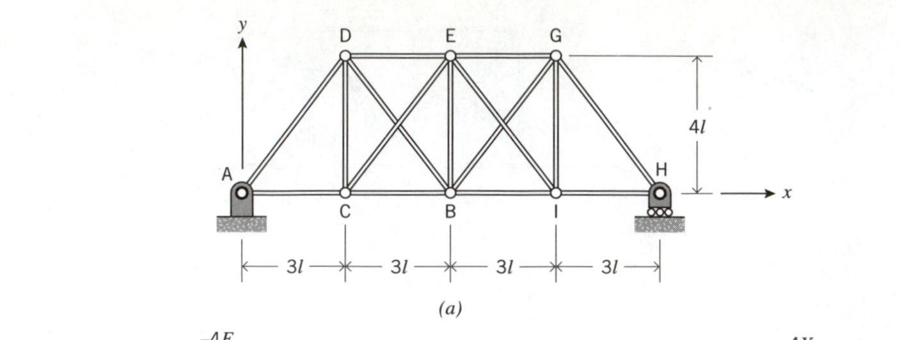

(a)

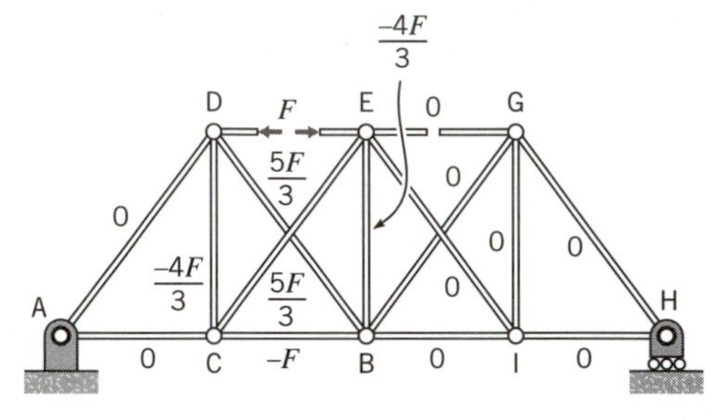

(b)

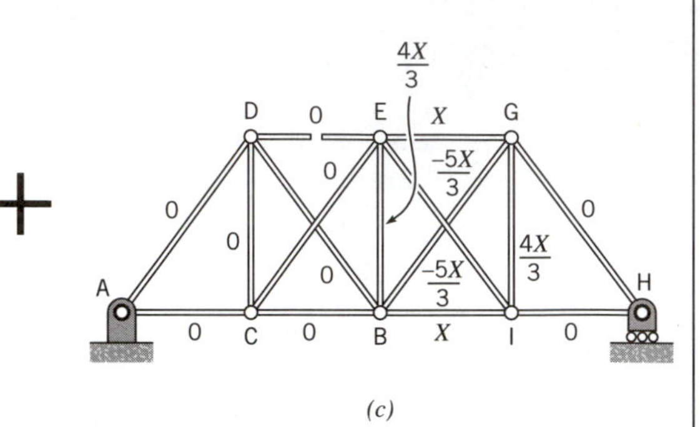

(c)

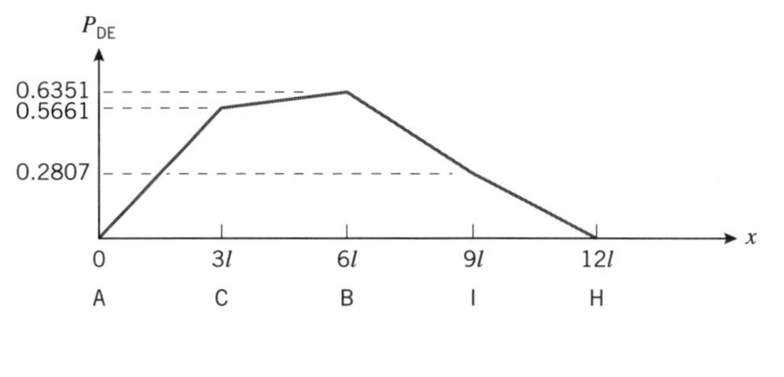

(d)

Figure 7.11 Example 9.

The analysis proceeds as in Chapter 5, and the details will be omitted here. We obtain

$$X = \frac{4}{27}F \tag{7.21a}$$

and

$$\boldsymbol{\Delta} \equiv \begin{Bmatrix} \Delta_{AC} \\ \Delta_{CB} \\ \Delta_{AD} \\ \Delta_{CD} \\ \Delta_{CE} \\ \Delta_{DE} \\ \Delta_{BE} \\ \Delta_{BD} \\ \Delta_{HI} \\ \Delta_{IB} \\ \Delta_{HG} \\ \Delta_{IG} \\ \Delta_{IE} \\ \Delta_{GE} \\ \Delta_{BG} \end{Bmatrix} = \frac{Fl}{EA} \begin{Bmatrix} 0 \\ -3 \\ 0 \\ -16/3 \\ 25/3 \\ 3 \\ -368/81 \\ 25/3 \\ 0 \\ 4/9 \\ 0 \\ 64/81 \\ -100/81 \\ 4/9 \\ -100/81 \end{Bmatrix} \tag{7.21b}$$

where $\boldsymbol{\Delta}$ is the vector of bar deformations.

The incremental axial displacement across the cut in bar DE, u_{DE}, can be determined using virtual forces exactly like the forces in Figure 7.11*b* (see Section 5.5). The result is

$$u_{DE} = \frac{11408\,Fl}{243\,EA} = 46.95\,\frac{Fl}{EA} \tag{7.22a}$$

Setting $u_{DE} = 1$ leads to

$$F = \frac{243\,EA}{11408\,l} = 0.02130\,\frac{EA}{l} \tag{7.22b}$$

Substituting Equation 7.22b into Equation 7.21b gives the bar deformations corresponding to the unit incremental displacement

$$\boldsymbol{\Delta} \equiv \begin{Bmatrix} \Delta_{\mathrm{AC}} \\ \Delta_{\mathrm{CB}} \\ \Delta_{\mathrm{AD}} \\ \Delta_{\mathrm{CD}} \\ \Delta_{\mathrm{CE}} \\ \Delta_{\mathrm{DE}} \\ \Delta_{\mathrm{BE}} \\ \Delta_{\mathrm{BD}} \\ \Delta_{\mathrm{HI}} \\ \Delta_{\mathrm{IB}} \\ \Delta_{\mathrm{HG}} \\ \Delta_{\mathrm{IG}} \\ \Delta_{\mathrm{IE}} \\ \Delta_{\mathrm{GE}} \\ \Delta_{\mathrm{BG}} \end{Bmatrix} = \begin{Bmatrix} 0 \\ -0.06390 \\ 0 \\ -0.1136 \\ 0.1775 \\ -0.06390 \\ -0.09677 \\ 0.1775 \\ 0 \\ 0.00947 \\ 0 \\ 0.01683 \\ -0.02630 \\ 0.00947 \\ -0.02630 \end{Bmatrix} \tag{7.23}$$

As noted above, the desired influence line is the displaced shape of line ACBIH when the bar deformations are as given in Equation 7.23. To obtain the displaced shape of line ACBIH, it is sufficient to obtain the vertical displacements of joints C, B, and I because bars AC, CB, BI, and IH must remain straight. To obtain the displacement of a joint, we apply a vertical external virtual force at the joint and determine a set of internal virtual forces that is in equilibrium with the joint force. The joint displacement is then computed using the procedure in Section 5.5. The three sets of computations produce the following results.

$$\begin{aligned} u_{\mathrm{C}} &= 0.566 \\ u_{\mathrm{B}} &= 0.635 \\ u_{\mathrm{I}} &= 0.281 \end{aligned} \tag{7.24}$$

The displacements in Equation 7.24 are plotted in Figure 7.11d, which constitutes the influence line for the force in bar DE. The sign conventions employed in this case are consistent with previous sign conventions for influence lines for frames, i.e., positive displacements and live loads are directed in the positive y direction, and positive bar forces are tension forces.

It should be noted that the ordinates to the influence line in Figure 7.11d are consistent with the data presented in Figure 6.11. Because the ordinate to the influence line at any point is the value of P_{DE} due to a unit load applied at that point on the bridge deck, then a concentrated force F (directed in the negative y direction) applied at joint C (i.e., at $x = 3l$) results in $P_{\mathrm{DE}} = -0.566F$, exactly as shown in Figure 6.11. Similarly, when the force F acts at joint I (i.e., at $x = 9l$) we obtain $P_{\mathrm{DE}} = -0.281F$ that, by symmetry, must be the same as the force in bar EG when the force is at joint C.

We will end with the observation that if a structure is n-degrees indeterminate, then once we have computed influence lines for each of n redundant forces, all remaining forces may be determined from the information contained in those influence lines combined with basic statics.

PROBLEMS

7.P1. Use the influence function defined in Equation 7.3 to find the moment at midspan of the beam in Figure 7.3a when the bulldozer is positioned at $\eta = 3l/10$. As in Example 3, assume the bulldozer applies a uniform load q_o over a distance $d = l/5$. Sketch the bending moment diagram for the beam with the bulldozer in this position.

7.P2. For the bulldozer crossing the bridge in Figure 7.3a and Example 3, use influence lines to
(a) find the maximum transverse shear at the midpoint of the bridge;
(b) find the maximum transverse shear at the end of the bridge.

7.P3. A military half-track crosses a simply supported bridge of length l. The front axle load is F and the tracks apply a uniform load q_o over the length d. Let $d = l/5$ and $F = q_o d/2$. Use influence lines to
(a) find the maximum bending moment at the midpoint of the bridge;
(b) find the maximum transverse shear at end A of the bridge.

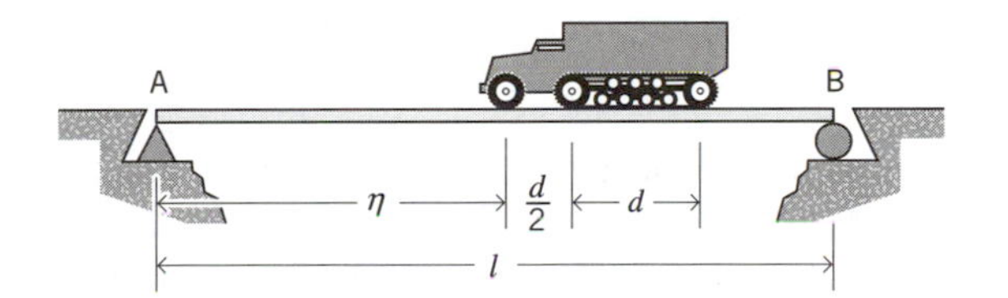

7.P4. Derive the displacement function in Equation 7.10a.

7.P5. Derive the analytic expression for the influence function for internal transverse shear at end A of the beam in Figure 7.5a; compare to the limiting case of Equation 7.17. Use the solution to find the support reaction at end A due to a uniform load q_o applied along the entire length of the beam.

7.P6. Derive the influence function for
(a) internal bending moment at B; and
(b) reaction at C.
Assume a uniform load q_o can be applied over any desired portion of beam ABC. What is the maximum magnitude of the reaction at C?

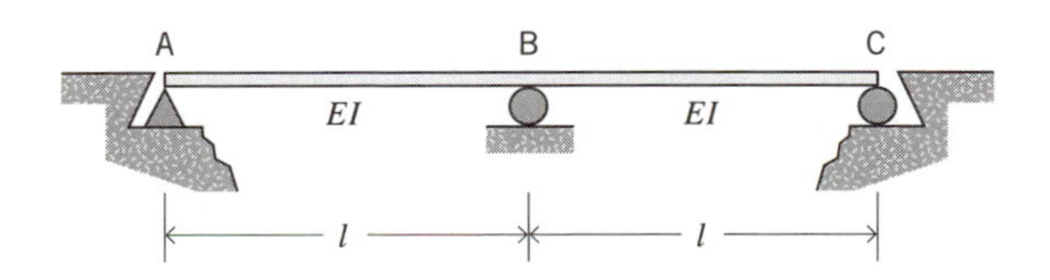

7.P7. (*a*) Derive the influence function for the internal moment at end B of the beam with fixed ends shown.

(*b*) Find the maximum moment at end B due to the vehicle load in Example 1.

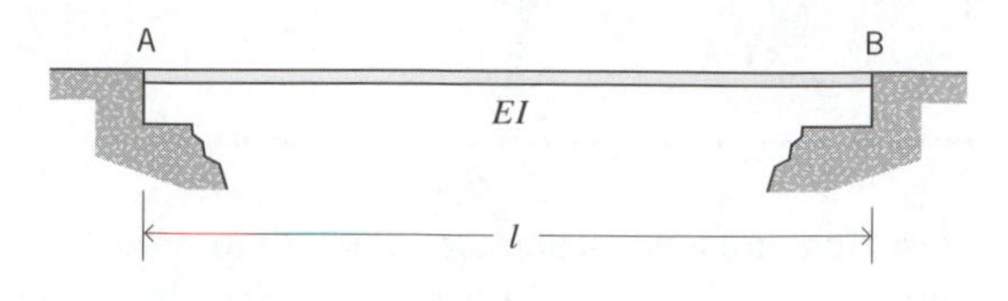

7.P8. For the structure in Figure 7.9*a*, find the magnitudes of the maximum positive and maximum negative moment at B due to a uniform live load q_o which can be applied to one or more segments of the structure, e.g., AB and/or BC and/or CD. Use the influence function defined in Equation 7.20.

7.P9. The beam is built in at end A, simply supported at B and C. Sketch the influence line for the bending moment at A. Find the maximum positive and negative values of this bending moment due to a uniform live load q_o.

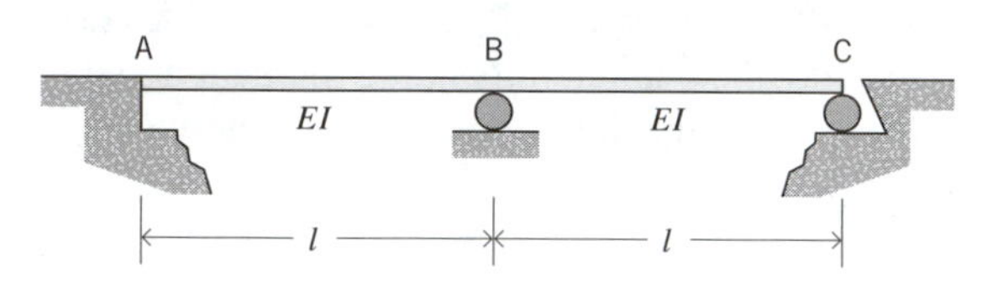

7.P10. Find the maximum values of the positive and negative bending moments at A in the beam in Problem 7.P9 due to the vehicle load in Example 1.

7.P11. Sketch and define the shape of the influence line for the bending moment at B. Find the bending moment at B due to a uniform live load q_o. Let $EA = EI/l^2$.

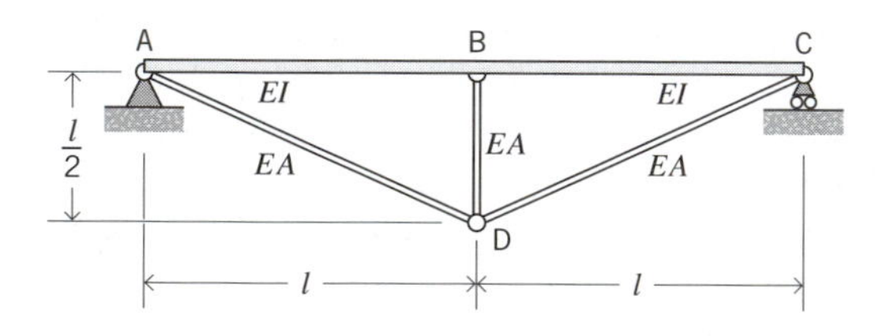

7.P12. (*a*) Define the influence line for the force in bar DE;
(*b*) Define the influence line for the force in bar CD;
(*c*) Define the influence line for the force in bar BD;
(*d*) Find the maximum value of each of the above bar forces due to a uniform live load q_o.

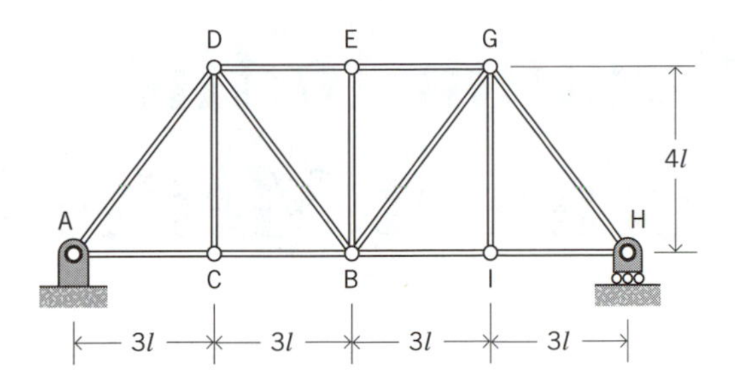

7.P13. Define the influence line for the force in bar CD of the truss in Figure 7.11*a*. Find the maximum value of the force in bar CB due to a uniform live load q_o.

7.P14. Define the influence line for the force in bar BD of the truss in Figure 7.11*a*. Find the maximum value of the force in bar BD due to a uniform live load q_o.

Chapter 8

Introduction to the Displacement Method

8.0 INTRODUCTION

In the preceding study of the force method of analysis, it was shown (Section 5.5) that we can compute deflections at connections of a structure once the forces acting on the ends of all the elements are known. In the displacement method we reverse the order of the procedure, i.e., we first introduce deflections at connections and then compute forces. These deflections, referred to as ''system displacements,'' replace the redundant forces as the unknown quantities in the analysis. Compatibility is imposed at the outset by relating element motions to the system displacements.

At first glance, the idea of attempting to impose compatibility at the beginning of the analysis might seem to be a poor way to formulate the problem since compatibility was not a particularly easy concept to contend with in the force method. The difficulty we encountered previously was due to the fact that, in essence, we began with deformations at the element level, and then constructed equations for displacements at the system level.[1] As we shall see, if we begin with displacements at the system level and then compute deformations of the elements, the procedure will actually become quite simple.

The definition of allowable joint motions, and the imposition of compatibility by means of relating element motions to the system displacements, constitute the first two steps in the five-step process that characterizes the displacement method. An overview of the process may be summarized as follows:

---

[1] See, for example, Figure 5.6, which illustrates the manner in which displacement of a truss support is related to element deformations.

Step 1. Identify a set of m independent system displacements.
Step 2. Impose compatibility.
Step 3. Write element force-deformation relations.
Step 4. Find m equilibrium equations.
Step 5. Solve equilibrium equations for system displacements.

Step 2 involves only kinematics relating the elements and their connections. Step 3 is the statement relating a set of element forces to the element motion. The equilibrium equations, Step 4, are obtained by using the theorem of virtual displacements and, as we saw with the force method, employing the data in Step 2. Virtual work yields m equations in terms of the m independent system displacements, which are solved in Step 5. We back-substitute the results into Step 2 to find element motions and into Step 3 to get element forces.

There are two slightly different approaches that can be taken in formulating Steps 2 and 3 above. The first approach involves the use of the same element deformation quantities with which we dealt in the force method, i.e., axial deformations (length changes) of truss bars, and end deformations of beam elements. The second approach uses element displacement data directly rather than deformations. (This is possible because each of the deformations is a function of end displacement quantities.)

We will begin our examination of the displacement method using the first approach involving element deformations. This approach is, in some respects, closest to an inverse of the force method developed in Chapter 5. The second approach involving element end displacements requires some additional development and will be presented in Chapter 9. The approach presented in this Chapter is the more ''compact'' of the two but requires some care on the part of the analyst. The approach in Chapter 9 employs larger element matrices but requires less knowledge of the details of the method by the analyst, and is better suited to computer implementation. The second method is the basis of most modern computational procedures.

8.1 SYSTEM DISPLACEMENTS

As was noted in Sections 1.6.1 and 1.6.2, the motions at a connection generally consist of displacements and element rotations. Each component of the motion at the connection is referred to as a kinematic degree of freedom or, sometimes, a displacement degree of freedom where the word ''displacement'' is generalized to include rotations of the connections.

A structure containing many connections and elements can have a very large number of independent kinematic degrees of freedom. For example, in a typical planar truss the x and y components of displacement at any joint are independent of those of any other joint.[2] In many structures support displacements also introduce kinematic degrees of freedom. Knowledge of all displacements at joints and supports completely defines the deformed shape of a loaded structure. Given these displacements, we can determine all element deformations and then find the element forces. In other words, a knowledge of these displacements completely defines the behavior of the structure. The indepen-

---

[2] In other words, a single bar of a planar truss can have, at most, two independent components of displacement at each end.

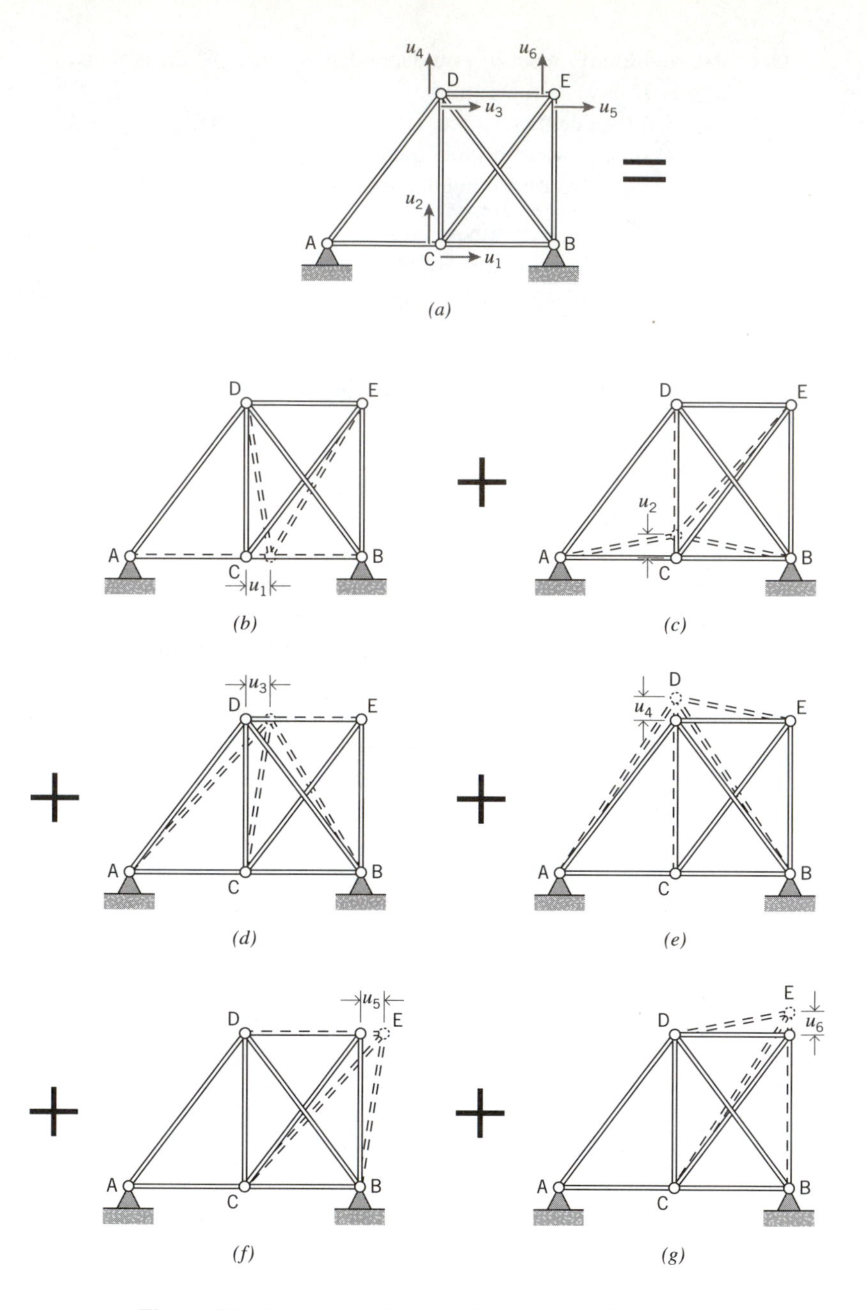

Figure 8.1 Illustration of system displacements for a truss.

dent displacements, from which the behavior of a structure may be completely determined, are referred to as the system displacements.

To illustrate, consider the truss in Figure 5.9*a*, which is also shown in Figure 8.1*a*. This truss has pin supports at A and B and joints at C, D, and E. At each of the joints, the system displacements are taken to be the components of displacement in the coordinate directions, i.e., as the x and y components of displacement. Because there are three joints (and no displacements at the supports), it follows that there are six system displacements for this truss.[3] These are numbered u_1 through u_6 in Figure 8.1*a* (rather than u_{C_x}, u_{C_y}, etc.). In Figure 8.1*a*, each of the independent displacement components is denoted by an arrow, which indicates positive direction of the designated component.

The physical significance of these system displacements is illustrated in Figures 8.1*b*–8.1*g*; there, each displacement is shown separately. The dotted lines in these figures indicate the motions of the bars that develop due to each displacement. For example, in Figure 8.1*b*, joint C (only) is shown displaced to the right a distance u_1. This means that the right end of bar AC, the left end of bar CB, and also the C ends of bars DC and EC, must move to the right by this amount. The other figures are similar. Since each of Figures 8.1*b*–8.1*g* shows the motions associated with only one of the system displacements, it follows that the overall motions of the bars must be given by the sum of the motions due to each of the system displacements.

As we will see shortly, it is relatively easy to determine individual bar deformations due to each of the system displacements in these figures. The procedure involves only a simple geometric analysis, and is similar to that employed in Example 7 of Chapter 4.[4] As noted previously, this relationship is much simpler to derive than the inverse compatibility relation obtained in the force method.

Figures 5.11 and 8.2*a* show another truss for which all connections are at supports; the support at A is a pin, while those at B, C, and D are rollers with different orientations. In this case, there are only three system displacements, denoted u_1, u_2, and u_3 in Figure 8.2*a*. Note that each of the system displacements corresponds to a physically realizable motion that would result from some set of applied forces. The motions associated with each of the system displacements is shown in Figures 8.2*b*–8.2*d*.

Coupling between displacements at different connections in a structure can occur as a consequence of restrictions imposed by the analytic model of the structure. In particular, the assumption of axial rigidity for elements may lead to coupling of displacements at connections. For example, suppose one bar, such as DE, of the truss in Figure 8.1*a* is assumed to be axially rigid (see Figure 8.3).[5] In this case, since bar DE cannot change length, the horizontal components of displacement of joints D and E are coupled, i.e., horizontal motion of joint D requires an identical horizontal motion of joint E. The system displacements are illustrated in Figure 8.3. Vertical motions at D and E remain independent because they do not involve any length changes of bar DE. Thus, there are only five independent system displacements. Note that the motions of bar DE illustrated in Figures 8.3*d*–8.3*f* consist of a rigid-body translation and two rigid-body rotations, i.e., three independent rigid-body motions. Also, we observe that the simultaneous horizontal displacements of joints D and E (Figure 8.3*d*) involve more bars than any of the other system displacements.

The situation would be similar but somewhat more complicated if a diagonal bar, such as CE in Figure 8.4*a*, were assumed to be axially rigid. In this case some motions

---

[3] Thus, for this case, there are six unknown system displacements but only two unknown redundant forces, so analysis by the force method would require solution of a much smaller number of simultaneous equations.

[4] The fact that this example involved virtual displacements rather than real displacements does not alter the concept; in fact, as in this example, we will be using virtual displacements that are just ''like'' the real system displacements to obtain the required equilibrium equations. The application of virtual work will involve deformation information that is identical to that generated for each of the real system displacements.

[5] See, for example, bars BE and CF in Example 8, Chapter 5.

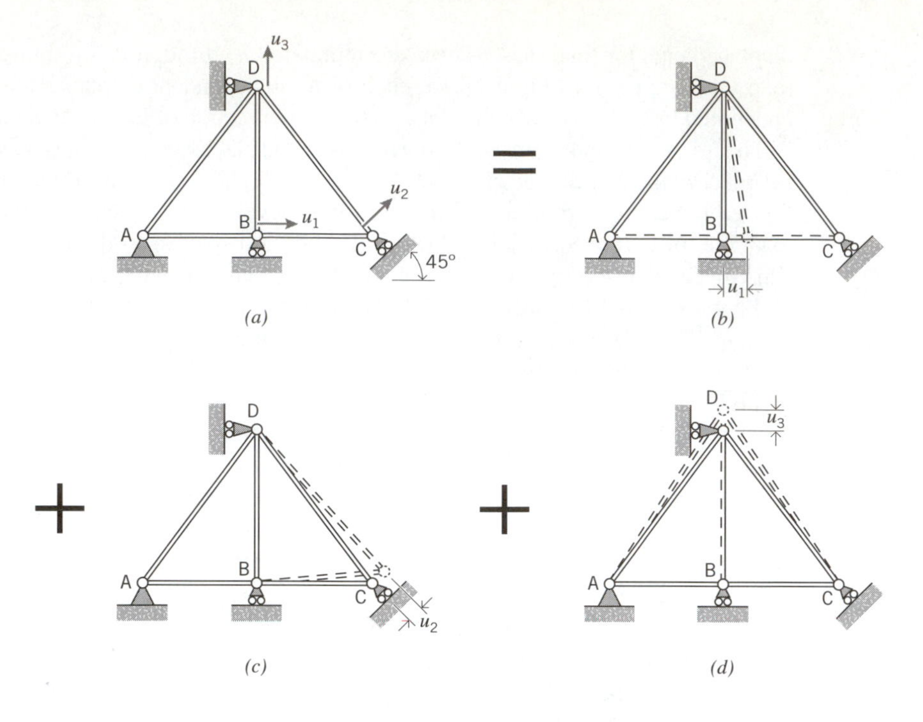

Figure 8.2 Illustration of system displacements for a truss.

of joints C and E are coupled. As with bar DE in Figure 8.3, there must be three independent rigid-body motions for bar CE. For the sake of convenience and clarity, we choose to represent these rigid-body motions as in Figures 8.4*b*, 8.4*c*, and 8.4*f*. In Figure 8.4*b*, we show equal horizontal displacements of joints C and E, denoted by u_1, which corresponds to a horizontal rigid-body translation of bar CE. Similarly, the vertical displacement u_2 of joints C and E in Figure 8.4*c* corresponds to a vertical translation of bar CE. The third independent rigid-body motion, a rotation of bar CE, u_5, is shown in Figure 8.4*f*; here, $l_{CE}u_5$ is a displacement at E in the direction perpendicular to bar CE. The remaining two system displacements, u_3 and u_4 shown in Figures 8.4*d* and 8.4*e*, are the same as in Figures 8.1 and 8.3.

Although we do not usually consider truss bars to be axially rigid, this has been a common assumption for beams and, consequently, can be a complicating factor in the analysis of frame structures (just as it can be for trusses). The somewhat surprising corollary is that the introduction of axial flexibility in beams actually simplifies those aspects of the displacement formulation that involve information about system and element motions.

Just as for trusses, the system displacements for a frame are a set of independent motions at connections from which all element deformations and forces can be completely determined. The end deformations for beams are defined in terms of rotations and transverse displacements (Equation 5.39*a*), so it also follows that the independent system motions generally consist of joint rotations as well as displacements.

We will begin our discussion of system displacements for simple frame structures with beams that are axially rigid. Consider, for instance, the beam in Figure 8.5*a* (Figure 5.14*a*), which is fixed at the right end and simply supported at the left end. For

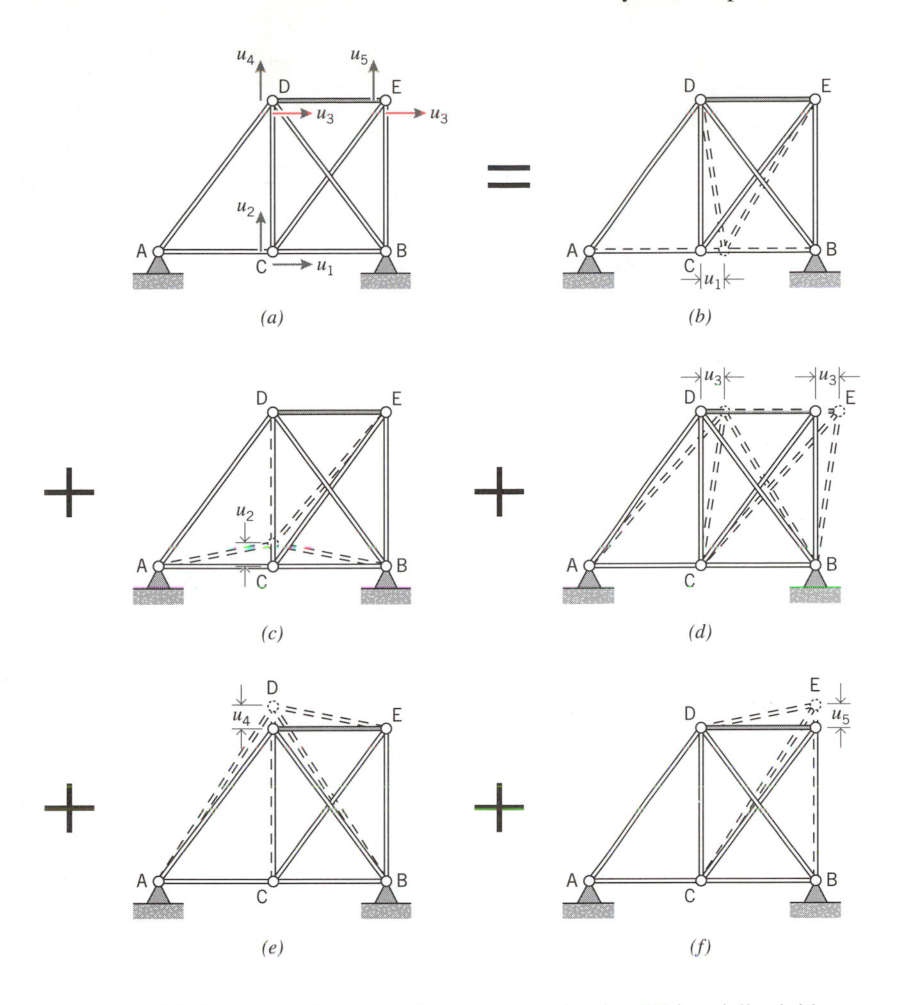

Figure 8.3 System displacements for truss assuming bar DE is axially rigid.

this structure there is only one possible end motion, namely a rotation at connection A. Once the magnitude of this rotation is determined for a specified set of loads, the deflections (Figure 8.5*b*) and the internal forces are also uniquely determined. The rotation at A is therefore the only system displacement for this structure. We will henceforth designate this system displacement by the symbol u (Figure 8.5*c*). As with redundant forces, we arbitrarily specify a positive direction for this unknown displacement to be clockwise rotation positive as indicated by the arrow in the figure. Figure 8.5*b* is thus a schematic representation of the motion of the entire structure due to the system displacement u.

Consider next the structure in Figure 8.6*a*, which is composed of three beams with rigid joints at B and C. In this case the connections at B and C can both rotate. The rotation at connection B involves a rotation of end B of beam AB and a rotation of end B of beam BC; similarly, the rotation at connection C involves a rotation of end C of beam BC and a rotation of end C of beam CD. Since each of these two rotations may be imposed separately, these rotations are independent system displacements, designated as u_1 and u_2 in Figure 8.6*a*. Figures 8.6*b* and 8.6*c* show the motions associated

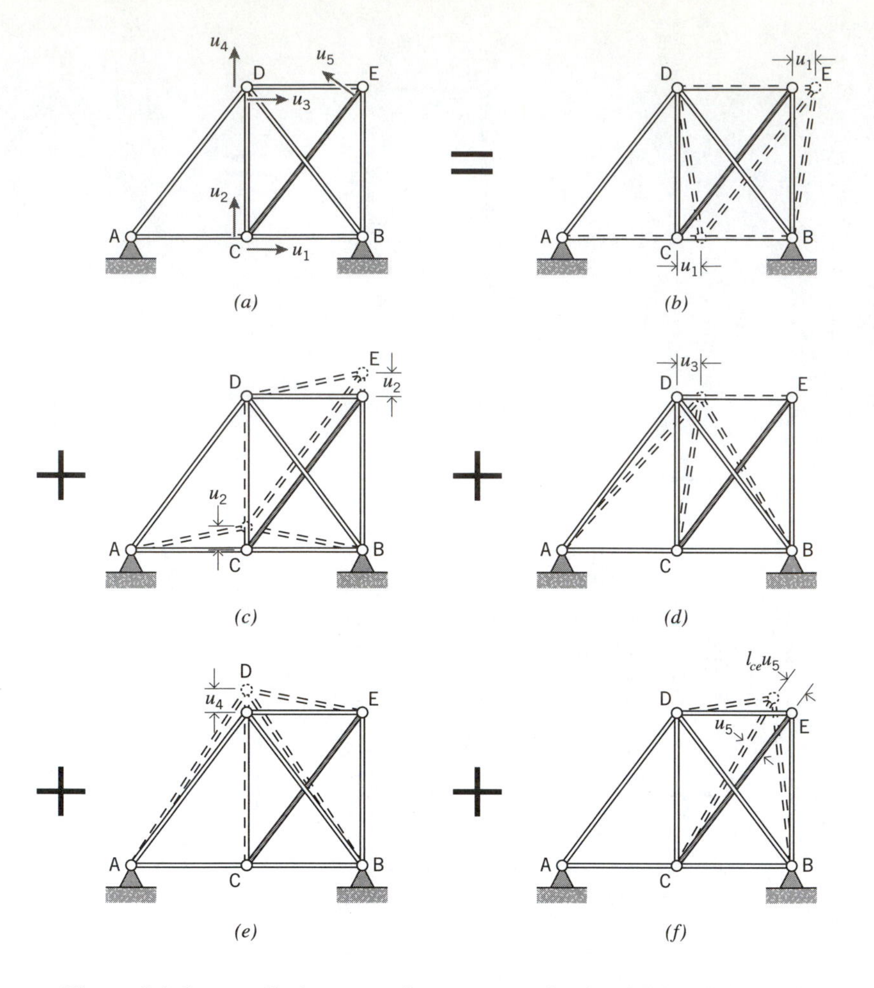

Figure 8.4 System displacements for truss assuming bar CE is axially rigid.

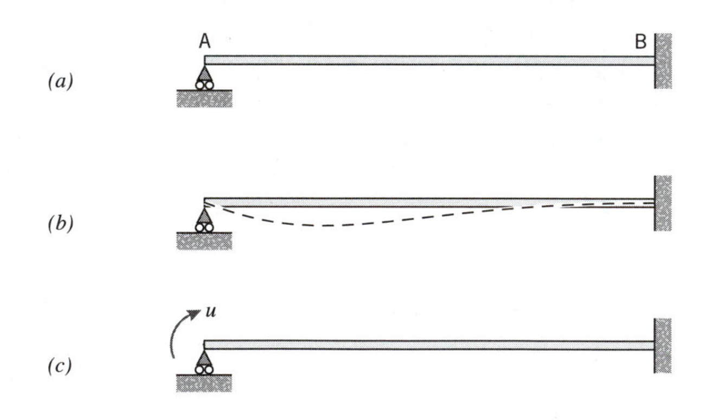

Figure 8.5 Illustration of system displacement for a particular beam structure.

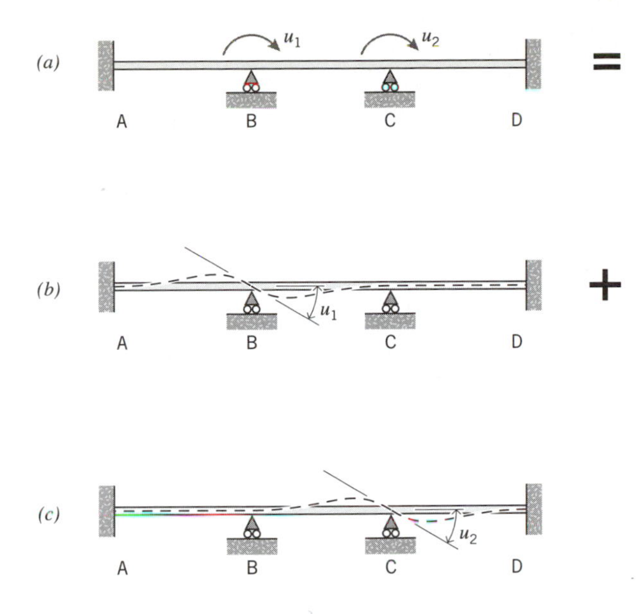

Figure 8.6 System displacements for beam structure.

with u_1 and u_2, e.g., in Figure 8.6*b* we show the motion due to u_1 only. Figure 8.6*c* is similar, but shows the motion due to u_2 only. Note that each of these figures shows the motions associated with one of the system displacements while assuming the other is restrained or prevented from occurring. These figures are analogous to the previous sketches of truss motion due to a single component of system displacement.

All of the system displacements for the two preceding structures consisted of rotations at connections; there were no transverse displacements at the ends of any beam elements. The situation is different for the structure shown in Figure 8.7*a*. Here, there is a fixed support with vertical roller at C, so it is possible for the support at C to move in the vertical direction. Such vertical movement involves a transverse displacement at end C of beam BC. Therefore, in this case, the system displacements consist of a rotation, u_1, at B, and a vertical displacement, u_2, at C.[6] The motions associated with u_1 and u_2, respectively, are illustrated in Figures 8.7*b* and 8.7*c*.

The system displacement u_2 for the structure in Figure 8.7*a* does not give rise to any motions at connections other than the vertical displacement at C. This is not the case for the frame structure in Figure 8.8*a* (Figure 5.15*a*). Here, the joint at B can rotate and displace in both the horizontal and vertical directions; similarly, the connection at C can displace horizontally (but cannot rotate). However, the horizontal and vertical components of displacement at B and C are not independent motions because beam BC is assumed to be axially rigid. Therefore a horizontal displacement of end B of beam BC requires an identical horizontal displacement of end C; this is similar to the motion of bar DE in Figure 8.3*d*. Furthermore, since beam AB is also axially rigid, end B of beam AB can only move along a line perpendicular to AB; therefore, end B

---

[6] We are using the symbol u to denote both rotations and displacements, just as we previously used the symbol X to denote both redundant forces and moments.

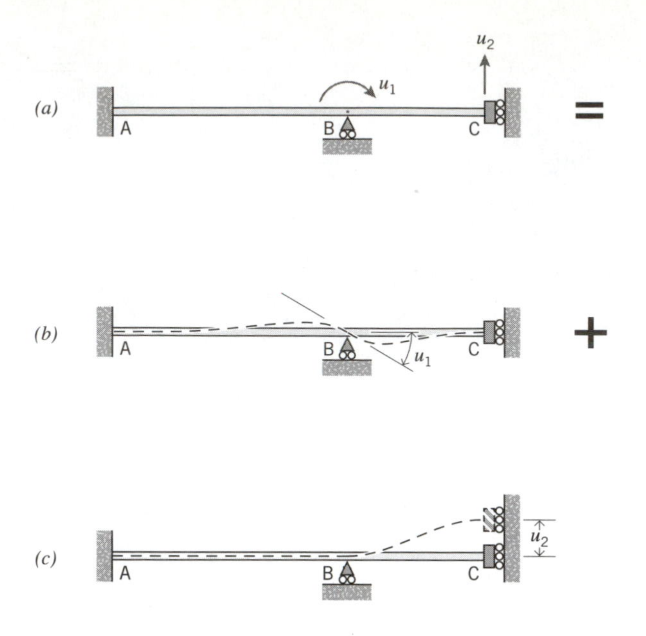

Figure 8.7 System displacements that include transverse displacement.

of beam BC, which must move along a line perpendicular to BC, will displace to the point of intersection of the perpendicular lines. In other words, the horizontal and vertical motions at B are coupled. In summary, there are only two system displacements, i.e., two independent joint motions, for the structure in Figure 8.8*a*.[7]

The coupling of displacements illustrated for the structure in Figure 8.8 is typical of situations that can arise in the analysis of frames composed of axially rigid beams. On the other hand, if the beams are assumed to be axially flexible, there is a major change in the nature of the system displacements, i.e., the displacements become uncoupled. For the frame in Figure 8.8*a*, for example, each of the possible components of joint displacement would now be an independent motion, as illustrated in Figure 8.9. The rotational degree of freedom in Figure 8.9*b* is unaltered; however, now associated with each of the system displacements u_2–u_4 (Figures 8.9*c*–8.9*e*) are length changes of one or both of the beams. In Figure 8.9*c*, the horizontal displacement of joint B requires a shortening of beam BC and a lengthening of beam AB; in Figure 8.9*d*, the vertical displacement of joint B requires only a lengthening of beam AB, and in Figure 8.9*e*, the horizontal displacement of the roller at C requires only a lengthening of beam BC. The relationships between these length changes and the system displacements will be computed for the beams of a frame structure in exactly the same manner as for bars in trusses.

The motions in Figures 8.9*c*–8.9*e* are generally easier to analyze than the motion in Figure 8.8*c*. This is a major advantage derived from the introduction of axial flexibility. The difference, however, is one of detail rather than fundamental concept. With the

[7] Since u_2 is, at the same time, horizontal displacement and vertical displacement, we could have just as easily and correctly drawn Figure 8.8*a* with u_2 shown as a downward deflection. Figure 8.8*c* would have been the same!

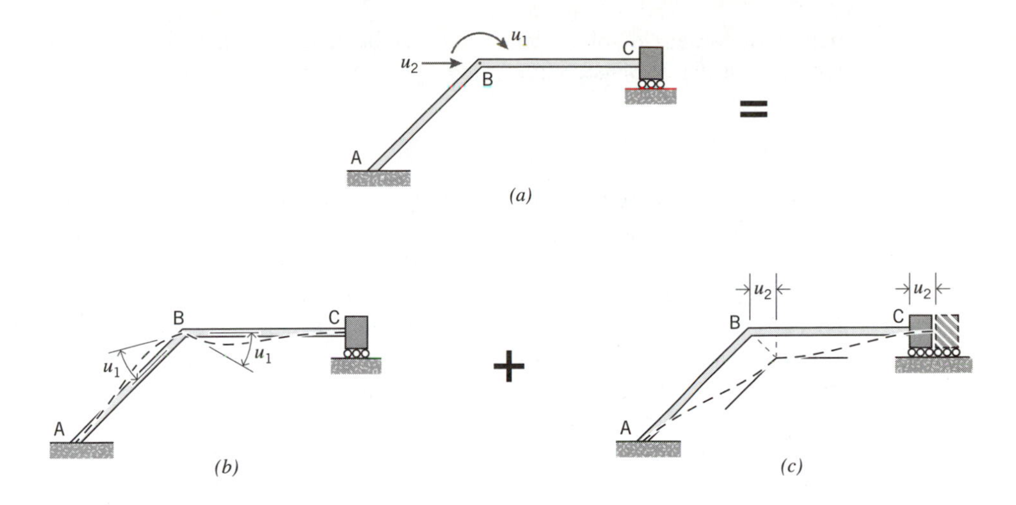

Figure 8.8 System displacements for frame (axially rigid beam elements).

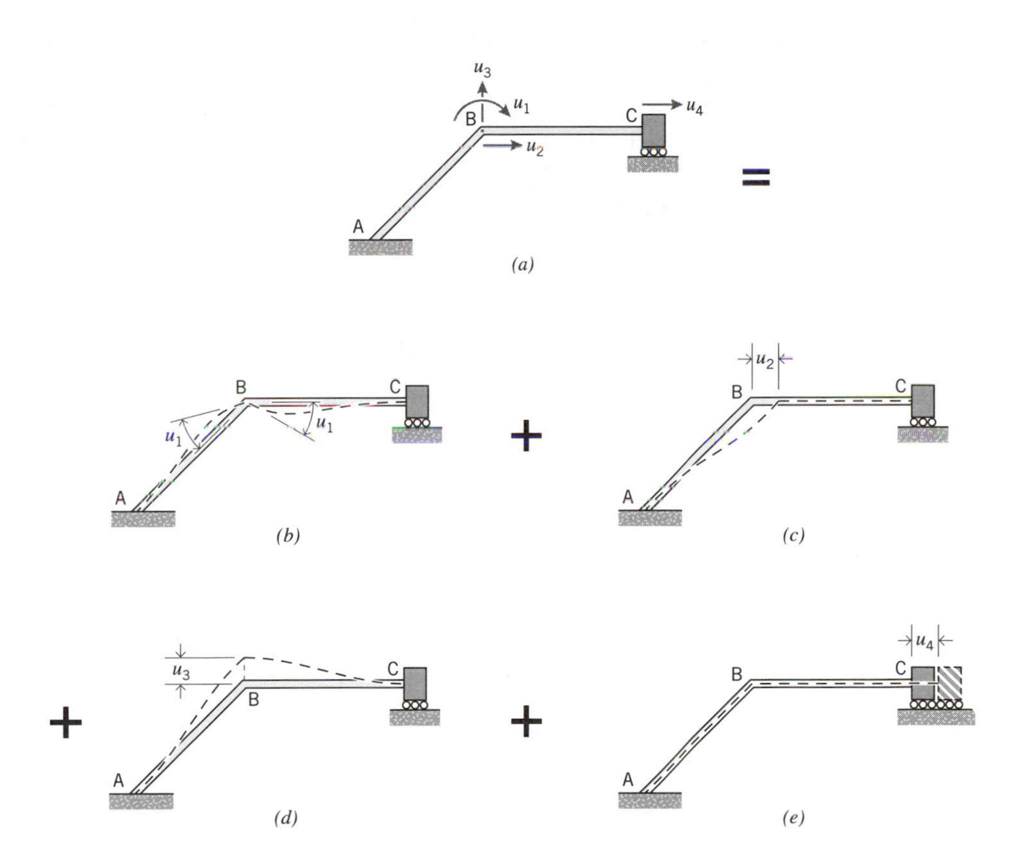

Figure 8.9 System displacements for frame (axially flexible beam elements).

nature of system displacements now established, we will proceed to illustrate the specifics of the displacement method.

8.2 ANALYSIS OF TRUSSES

Since the displacement method generally leads to fairly large systems of equations, it is convenient to proceed directly with a matrix formulation. For illustrative purposes we will begin with the truss in Figure 8.10*a*. (We analyzed this truss in Chapter 5 using the force method.) The truss is subjected to a vertical force F at node D and temperature changes ΔT in bars AB and CD; we assume all bars have the same EA. The three system displacements for this truss were identified in Figure 8.2, so we may move directly to Step 2, the imposition of compatibility.

Step 2. In the displacement method we seek to relate bar deformations, i.e., length changes, to the system displacements. We do this by determining bar deformations in each of Figures 8.2*b*–8.2*d*. For instance, Figure 8.10*b* (Figure 8.2*b*) shows the bar motions (dotted lines) associated with the system displacement u_1 (horizontal displacement of support B only). We assume that all bars remain connected to joint B when it displaces to B′, i.e., we assume that the motions in the figure satisfy compatibility. It follows that bar AB in Figure 8.10*b* must elongate by an amount u_1, and bar BC must shorten by the same amount, u_1, i.e., $\Delta_{AB} = u_1$ and $\Delta_{BC} = -u_1$. End B of bar BD displaces to the right a distance u_1 but, since this displacement is in a direction perpendicular to the bar, this (small) motion does not involve any length change. Furthermore, since supports A, C, and D do not displace in Figure 8.10*b*, it also follows that bars AD and CD undergo no length changes due to system displacement u_1.

Figure 8.10*c* shows the motions associated with system displacement u_2. Here, joint

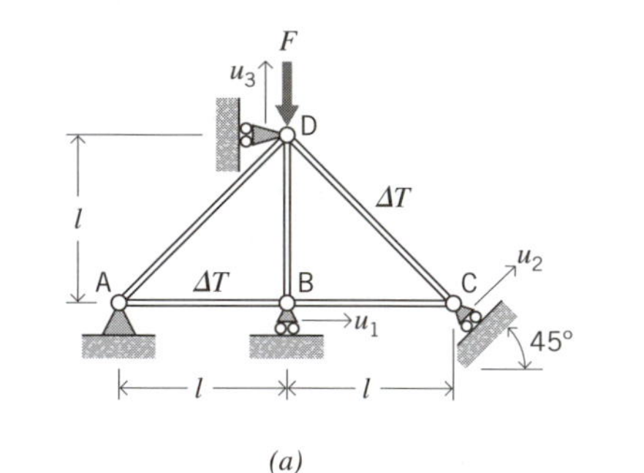

(a)

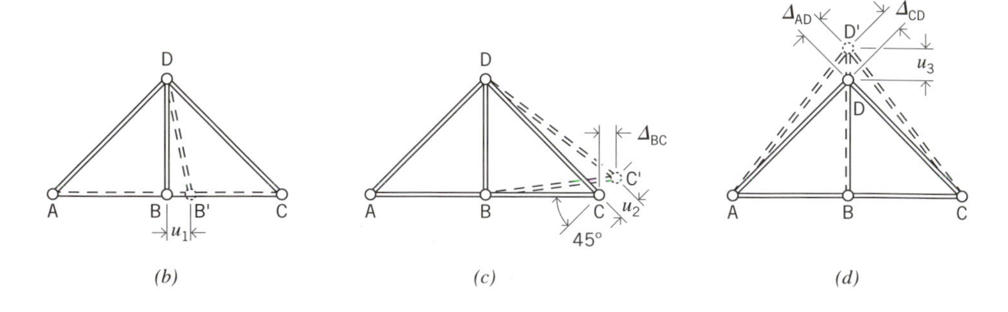

(b) *(c)* *(d)*

Figure 8.10 Illustration of truss compatibility.

C displaces in the direction of the support surface (which is oriented at 45° from the horizontal) to C′. This results in an elongation of bar BC, $\Delta_{BC} = u_2/\sqrt{2}$, which is determined from the geometry of the right triangle with hypotenuse CC′ in the figure. This is the only bar deformation in Figure 8.10*c*; there is no deformation of bar CD for this particular truss geometry because the direction of u_2 is perpendicular to bar CD.

Finally, as is shown in Figure 8.10*d*, there are three bar deformations associated with system displacement u_3, i.e., the displacement of joint D to D′ requires elongations of bars AD, BD, and CD. In this case we have $\Delta_{BD} = u_3$, $\Delta_{AD} = u_3/\sqrt{2}$ and $\Delta_{CD} = u_3/\sqrt{2}$. As noted previously, the total deformation of a bar is obtained by summing the deformations due to each of the system displacements. Thus, for example, $\Delta_{BC} = -u_1 + u_2/\sqrt{2} + 0u_3$.

It is important to recognize that the individual bar deformations in Figures 8.10*b*–8.10*d* were obtained solely from analyses of the geometries of the deformed trusses; the procedure is usually quite straightforward. Once the relations between bar deformations and system displacements are obtained, the matrix form of these compatibility equations follows directly. In the current instance, we have

$$\begin{Bmatrix} \Delta_{AB} \\ \Delta_{BC} \\ \Delta_{BD} \\ \Delta_{AD} \\ \Delta_{CD} \end{Bmatrix} = \begin{bmatrix} 1 & 0 & 0 \\ -1 & 1/\sqrt{2} & 0 \\ 0 & 0 & 1 \\ 0 & 0 & 1/\sqrt{2} \\ 0 & 0 & 1/\sqrt{2} \end{bmatrix} \begin{Bmatrix} u_1 \\ u_2 \\ u_3 \end{Bmatrix} \tag{8.1}$$

Note that each column of the 5 × 3 matrix represents the deformations associated with a unit value of one system displacement, e.g., the second column contains the bar deformations due to a unit value of u_2, etc. We will represent equations such as Equation 8.1 by the general form

$$\boldsymbol{\Delta} = \boldsymbol{\beta}\mathbf{u} \tag{8.2}$$

where $\boldsymbol{\Delta}$ is a vector of bar deformations (as in the force formulation, Chapter 5), $\mathbf{u}$ is a vector of system displacements, and $\boldsymbol{\beta}$ is a "compatibility matrix."

Step 3. The force-deformation relation for a uniform truss bar is the inverse of the deformation-force relation in Equation 5.31*a*, i.e., $P_i = (EA/l)_i\Delta_i + \widehat{P}_i$, where $\widehat{P}_i = -(EA/l)_i\widehat{\Delta}_i$ is interpreted as a "fixed-end" axial force, the axial force due to thermal or initial effects when $\Delta_i = 0$. For the case of a uniform thermal load ΔT, for example, $\widehat{\Delta}$ in Equation 5.31*a* is equal to $\alpha l\Delta T$ so $\widehat{P}_i = -(EA\alpha\Delta T)_i$.[8]

For the truss in Figure 8.10*a*, the matrix form of the force-deformation equations is

$$\begin{Bmatrix} P_{AB} \\ P_{BC} \\ P_{BD} \\ P_{AD} \\ P_{CD} \end{Bmatrix} = \frac{EA}{l} \begin{bmatrix} 1 & & & & \\ & 1 & & & \\ & & 1 & & \\ & & & 1/\sqrt{2} & \\ & & & & 1/\sqrt{2} \end{bmatrix} \begin{Bmatrix} \Delta_{AB} \\ \Delta_{BC} \\ \Delta_{BD} \\ \Delta_{AD} \\ \Delta_{CD} \end{Bmatrix} + \begin{Bmatrix} -EA\alpha\Delta T \\ 0 \\ 0 \\ 0 \\ -EA\alpha\Delta T \end{Bmatrix} \tag{8.3}$$

[8] An alternate derivation was presented in Example 9, Chapter 2.

Note the manner in which individual bar lengths have been incorporated into the matrix, i.e., the lengths of bars AB, BC, and BD are all equal to l, while the lengths of bars AD and CD are equal to $\sqrt{2}l$.

We will represent equations such as Equation 8.3 by the general form

$$\mathbf{P} = \mathbf{K}\boldsymbol{\Delta} + \widehat{\mathbf{P}} \tag{8.4}$$

where $\mathbf{P}$ is the vector of bar forces, $\mathbf{K}$ is a diagonal matrix of individual bar stiffnesses, and $\widehat{\mathbf{P}}$ is a vector of fixed-end forces. In this illustration, the term EA/l is a common factor of all components of $\mathbf{K}$.

Step 4. The equilibrium equations are obtained from the theorem of virtual work, specifically by using virtual displacements "like" the system displacements. (This is analogous to the procedure used in the force method, where compatibility equations were obtained from virtual forces "like" the redundant forces.) In this case, we will obtain three equations that can be solved for the unknown system displacements.

To illustrate, consider the theorem of virtual work for the truss in Figure 8.10*a* with virtual displacements chosen like the motions in Figure 8.10*b*. Denote the virtual displacement of joint B by u_{1_V}. It follows that the virtual bar deformations are $\Delta_{AB_V} = u_{1_V}$ and $\Delta_{BC_V} = -u_{1_V}$, and $\Delta_{BD} = \Delta_{AD} = \Delta_{CD} = 0$. External virtual work is zero because there are no virtual displacements at joints in the directions of any real reactions or of force F in Figure 8.10*a*. Internal virtual work is equal to $P_{AB}\Delta_{AB_V} + P_{BC}\Delta_{BC_V}$, or $W_{I_V} = P_{AB}u_{1_V} - P_{BC}u_{1_V}$. Equating external and internal virtual work leads to

$$0 = P_{AB} - P_{BC} \tag{8.5a}$$

This is the equation of horizontal force equilibrium for joint B of the truss in Figure 8.10*a*.

Equation 8.5*a* can also be written in the form

$$0 = [1 \quad -1 \quad 0 \quad 0 \quad 0]\begin{Bmatrix} P_{AB} \\ P_{BC} \\ P_{BD} \\ P_{AD} \\ P_{CD} \end{Bmatrix} \tag{8.5b}$$

or

$$0 = \boldsymbol{\beta}_1^T\mathbf{P} \tag{8.5c}$$

where $\boldsymbol{\beta}_1$ is the first column of the $\boldsymbol{\beta}$ matrix. As in previous formulations (e.g., Equation 5.31*i*), this follows from the close relationship between Step 2 and Step 4; the information is identical to the information on real bar deformations contained in the $\boldsymbol{\beta}$ matrix.

The right side of Equation 8.5*c* is the internal virtual work; as noted for this case, the external virtual work, the left side of the equation, is zero because there is no applied horizontal force at B. In general, the external virtual work is the product of the virtual displacement at a joint and any real force in the direction of the virtual displacement at that joint. The external virtual work is frequently nonzero. For example, consider the equilibrium equation obtained from virtual work using virtual displacements like the u_3 motion in Figure 8.10*d*. Here, external virtual work is equal to $-Fu_{3_V}$, where u_{3_V} is a virtual displacement "like" real system displacement u_3 and F is the vertical force at joint D in Figure 8.10*a*. The negative sign is a consequence of the fact that force F is opposite in direction to u_{3_V}. Thus, we obtain

$$-F = \boldsymbol{\beta}_3^{\mathrm{T}}\mathbf{P} = [0 \quad 0 \quad 1 \quad 1/\sqrt{2} \quad 1/\sqrt{2}]\begin{Bmatrix} P_{\mathrm{AB}} \\ P_{\mathrm{BC}} \\ P_{\mathrm{BD}} \\ P_{\mathrm{AD}} \\ P_{\mathrm{CD}} \end{Bmatrix} = P_{\mathrm{BD}} + \frac{P_{\mathrm{AD}}}{\sqrt{2}} + \frac{P_{\mathrm{CD}}}{\sqrt{2}} \qquad \mathbf{(8.5}\textbf{\textit{d}}\mathbf{)}$$

where $\boldsymbol{\beta}_3$ is the third column of the $\boldsymbol{\beta}$ matrix and u_{3_V} has been cancelled from both sides of the equation. This is the equation of vertical force equilibrium for joint D of the truss in Figure 8.10*a*.

The equation of equilibrium associated with u_2 is obtained in a similar manner, i.e., by choosing virtual displacements like the real u_2 motion in Figure 8.10*c*. The resulting equation and the previous two equations may all be expressed in matrix form as

$$\begin{Bmatrix} 0 \\ 0 \\ -F \end{Bmatrix} = \begin{bmatrix} 1 & -1 & 0 & 0 & 0 \\ 0 & 1/\sqrt{2} & 0 & 0 & 0 \\ 0 & 0 & 1 & 1/\sqrt{2} & 1/\sqrt{2} \end{bmatrix}\begin{Bmatrix} P_{\mathrm{AB}} \\ P_{\mathrm{BC}} \\ P_{\mathrm{BD}} \\ P_{\mathrm{AD}} \\ P_{\mathrm{CD}} \end{Bmatrix} \qquad \mathbf{(8.5}\textbf{\textit{e}}\mathbf{)}$$

or

$$\mathbf{F} = \boldsymbol{\beta}^{\mathrm{T}}\mathbf{P} \qquad \mathbf{(8.6)}$$

where $\mathbf{F}$ is a vector of forces obtained from the individual external virtual work terms.

Step 5. In a manner analogous to developments in the force method, the equations generated in Steps 2–4 may be combined into a single matrix equilibrium equation in terms of the unknown system displacements $\mathbf{u}$. Substituting Equations 8.2 and 8.4 into Equation 8.6 gives

$$\begin{aligned} \mathbf{F} &= \boldsymbol{\beta}^{\mathrm{T}}\mathbf{P} \\ &= \boldsymbol{\beta}^{\mathrm{T}}(\mathbf{K}\boldsymbol{\Delta} + \widehat{\mathbf{P}}) \\ &= \boldsymbol{\beta}^{\mathrm{T}}[\mathbf{K}(\boldsymbol{\beta}\mathbf{u}) + \widehat{\mathbf{P}}] \\ &= (\boldsymbol{\beta}^{\mathrm{T}}\mathbf{K}\boldsymbol{\beta})\mathbf{u} + \boldsymbol{\beta}^{\mathrm{T}}\widehat{\mathbf{P}} \end{aligned} \qquad \mathbf{(8.7}\textbf{\textit{a}}\mathbf{)}$$

or

$$\boxed{(\boldsymbol{\beta}^{\mathrm{T}}\mathbf{K}\boldsymbol{\beta})\mathbf{u} = \mathbf{F} - \boldsymbol{\beta}^{\mathrm{T}}\widehat{\mathbf{P}}} \qquad \mathbf{(8.7}\textbf{\textit{b}}\mathbf{)}$$

Equation 8.7*b* is the analog of Equation 5.27*c*. This equation is usually written in the form

$$\mathbf{k}\mathbf{u} = \mathbf{F}^* \tag{8.7c}$$

where $\mathbf{k} \equiv \boldsymbol{\beta}^{\mathrm{T}}\mathbf{K}\boldsymbol{\beta}$ is the system stiffness matrix and $\mathbf{F}^* \equiv \mathbf{F} - \boldsymbol{\beta}^{\mathrm{T}}\hat{\mathbf{P}}$ is a "resultant" force vector. Once again we observe that **k** must be symmetric because **K** is symmetric.[9]

We solve for **u** by inverting **k**, i.e.,

$$\mathbf{u} = \mathbf{k}^{-1}\mathbf{F}^* \tag{8.7d}$$

Back-substituting the result into Equations 8.2 and 8.4 provides the desired solution.

We will demonstrate the procedure by completing the solution for the illustrative truss. We perform the indicated matrix multiplications to get **k** and **F***,

$$\mathbf{k} = \boldsymbol{\beta}^{\mathrm{T}}\mathbf{K}\boldsymbol{\beta}$$

$$= \frac{EA}{l}\begin{bmatrix} 1 & -1 & 0 & 0 & 0 \\ 0 & 1/\sqrt{2} & 0 & 0 & 0 \\ 0 & 0 & 1 & 1/\sqrt{2} & 1/\sqrt{2} \end{bmatrix}\begin{bmatrix} 1 & & & & \\ & 1 & & & \\ & & 1 & & \\ & & & 1/\sqrt{2} & \\ & & & & 1/\sqrt{2} \end{bmatrix}\begin{bmatrix} 1 & 0 & 0 \\ -1 & 1/\sqrt{2} & 0 \\ 0 & 0 & 1 \\ 0 & 0 & 1/\sqrt{2} \\ 0 & 0 & 1/\sqrt{2} \end{bmatrix} \tag{8.8a}$$

$$= \frac{EA}{l}\begin{bmatrix} 2 & -1/\sqrt{2} & 0 \\ -1/\sqrt{2} & 1/2 & 0 \\ 0 & 0 & (2+\sqrt{2})/2 \end{bmatrix}$$

and

$$\mathbf{F}^* = \mathbf{F} - \boldsymbol{\beta}^{\mathrm{T}}\hat{\mathbf{P}} = \begin{Bmatrix} 0 \\ 0 \\ -F \end{Bmatrix} - \begin{bmatrix} 1 & -1 & 0 & 0 & 0 \\ 0 & 1/\sqrt{2} & 0 & 0 & 0 \\ 0 & 0 & 1 & 1/\sqrt{2} & 1/\sqrt{2} \end{bmatrix}\begin{Bmatrix} -EA\alpha\Delta T \\ 0 \\ 0 \\ 0 \\ -EA\alpha\Delta T \end{Bmatrix} \tag{8.8b}$$

$$= \begin{Bmatrix} EA\alpha\Delta T \\ 0 \\ -F + EA\alpha\Delta T/\sqrt{2} \end{Bmatrix}$$

[9] This symmetry of the system stiffness matrix may also be demonstrated from Maxwell's reciprocal relations, Section 6.1.2. For example, the case in which displacements are assigned unit values in Equation 6.1*f* leads to $k_{ij} = k_{ji}$.

Therefore, Equation 8.7*b* or 8.7*c* becomes

$$\frac{EA}{l}\begin{bmatrix} 2 & -1/\sqrt{2} & 0 \\ -1/\sqrt{2} & 1/2 & 0 \\ 0 & 0 & (2+\sqrt{2})/2 \end{bmatrix}\begin{Bmatrix} u_1 \\ u_2 \\ u_3 \end{Bmatrix} = \begin{Bmatrix} EA\alpha\Delta T \\ 0 \\ -F + EA\alpha\Delta T/\sqrt{2} \end{Bmatrix} \qquad \textbf{(8.8c)}$$

Inverting Equation 8.8*c* (as in Equation 8.7*d*) leads to

$$\begin{Bmatrix} u_1 \\ u_2 \\ u_3 \end{Bmatrix} = \frac{l}{EA}\begin{bmatrix} 1 & \sqrt{2} & 0 \\ \sqrt{2} & 4 & 0 \\ 0 & 0 & (2-\sqrt{2}) \end{bmatrix}\begin{Bmatrix} EA\alpha\Delta T \\ 0 \\ -F + EA\alpha\Delta T/\sqrt{2} \end{Bmatrix}$$

$$= \frac{l}{EA}\begin{Bmatrix} EA\alpha\Delta T \\ \sqrt{2}EA\alpha\Delta T \\ (-2F + \sqrt{2}EA\alpha\Delta T)/(2+\sqrt{2}) \end{Bmatrix} \qquad \textbf{(8.8d)}$$

Substituting Equation 8.8*d* into Equations 8.1 and 8.2 and into Equations 8.3 and 8.4 gives

$$\begin{Bmatrix} \Delta_{AB} \\ \Delta_{BC} \\ \Delta_{BD} \\ \Delta_{AD} \\ \Delta_{CD} \end{Bmatrix} = \frac{l}{EA}\begin{bmatrix} 1 & 0 & 0 \\ -1 & 1/\sqrt{2} & 0 \\ 0 & 0 & 1 \\ 0 & 0 & 1/\sqrt{2} \\ 0 & 0 & 1/\sqrt{2} \end{bmatrix}\begin{Bmatrix} EA\alpha\Delta T \\ \sqrt{2}EA\alpha\Delta T \\ (-2F + \sqrt{2}EA\alpha\Delta T)/(2+\sqrt{2}) \end{Bmatrix}$$

$$= \frac{l}{EA}\begin{Bmatrix} EA\alpha\Delta T \\ 0 \\ -0.586F + 0.414EA\alpha\Delta T \\ -0.414F + 0.293EA\alpha\Delta T \\ -0.414F + 0.293EA\alpha\Delta T \end{Bmatrix} \qquad \textbf{(8.8e)}$$

and

$$\begin{Bmatrix} P_{AB} \\ P_{BC} \\ P_{BD} \\ P_{AD} \\ P_{CD} \end{Bmatrix} = \begin{bmatrix} 1 & & & & \\ & 1 & & & \\ & & 1 & & \\ & & & 1/\sqrt{2} & \\ & & & & 1/\sqrt{2} \end{bmatrix}\begin{Bmatrix} EA\alpha\Delta T \\ 0 \\ -0.586F + 0.414EA\alpha\Delta T \\ -0.414F + 0.293EA\alpha\Delta T \\ -0.414F + 0.293EA\alpha\Delta T \end{Bmatrix} + \begin{Bmatrix} -EA\alpha\Delta T \\ 0 \\ 0 \\ 0 \\ -EA\alpha\Delta T \end{Bmatrix} \qquad \textbf{(8.8f)}$$

$$= \begin{Bmatrix} 0 \\ 0 \\ -0.586F + 0.414EA\alpha\Delta T \\ -0.293F + 0.207EA\alpha\Delta T \\ -0.293F - 0.793EA\alpha\Delta T \end{Bmatrix}$$

The latter is identical to the solution obtained by the force method, Equation 5.29*e*.

To reiterate, the solution procedure requires the matrices that appear in Equation 8.7*b*. With the possible exception of the **F** vector, these matrices are generally obtained in a very straightforward manner once the system displacements are identified. The following examples help clarify the concepts.

EXAMPLE 1

Analyze the truss in Figure 8.11*a* using the displacement method. This truss is similar to the one in Figures 5.9 and 5.10, except that there are now two distinct forces, F_1 and F_2, and the thermal load in bar DE has been deleted. Assume that EA is constant for all bars.

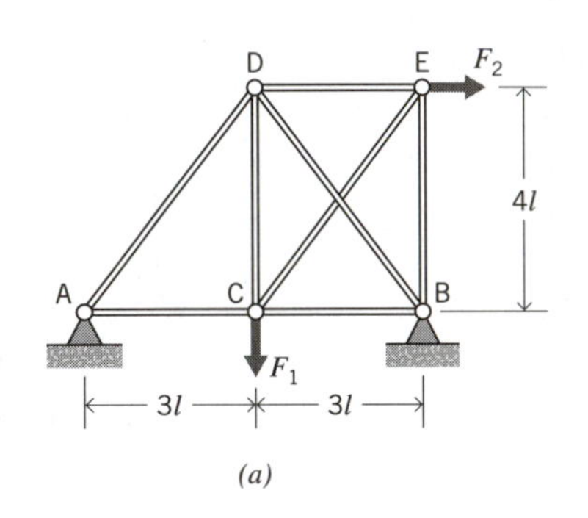

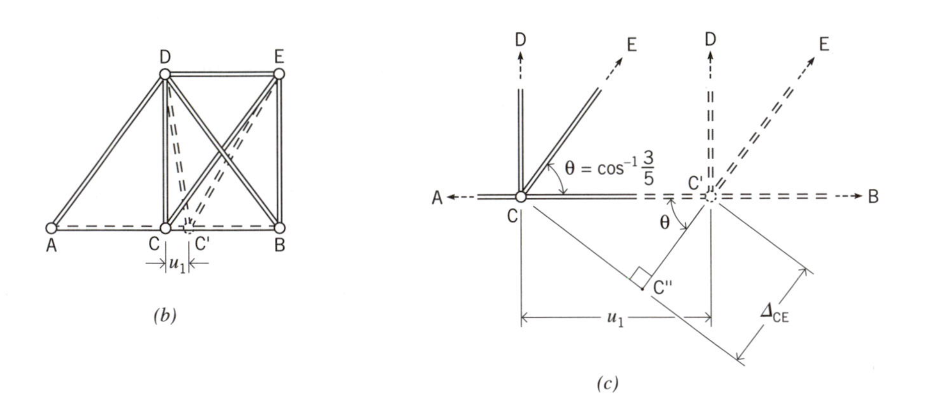

Figure 8.11 Illustration of calculation of bar deformation.

Step 1. The six system displacements for this truss are identified in Figure 8.1; recall that the numbering sequence is arbitrary.

Step 2. We relate bar deformations to system displacements by considering the motions in each of Figures 8.1*b*–8.1*g*. We will illustrate the computations only for system displacement u_1 (Figure 8.1*b*), because the remaining computations are similar and repetitive.

The motions associated with u_1 are shown by the dotted lines in Figure 8.11*b*. As indicated, joint C (only) undergoes a horizontal displacement u_1 to C′. Thus the C ends of bars AC, BC, CD, and CE all move to C′. Consequently, $\Delta_{AC} = u_1$ and $\Delta_{BC} = -u_1$, as in the preceding illustration (Figure 8.10*b*). Also, $\Delta_{CD} = 0$ because the displacement at end C is in a direction perpendicular to the bar. To determine the deformation of bar CE, it is necessary to analyze the geometry in detail; Figure 8.11*c* shows a typical enlarged view of the displacements at C, which we can use for this purpose. Here, Δ_{CE} is determined from the geometry of right triangle CC″C′, of which side CC′ has length u_1. CC″ is the direction perpendicular to bar CE, and is thus the line along which end C of bar CE would move if this bar did not change length; side C′C″ is the amount by which bar CE must shorten for end C to displace to C′.[10] Triangle CC″C′ is similar to triangle CEB and is thus a 3-4-5 right triangle. We conclude that $C'C'' = (3/5)u_1$ and it follows that $\Delta_{CE} = -(3/5)u_1$. Since the remaining bar deformations in Figure 8.11*b* are all zero, we may write

$$\begin{Bmatrix} \Delta_{AC} \\ \Delta_{CB} \\ \Delta_{AD} \\ \Delta_{CD} \\ \Delta_{CE} \\ \Delta_{DE} \\ \Delta_{BE} \\ \Delta_{BD} \end{Bmatrix} = \begin{bmatrix} 1 \\ -1 \\ 0 \\ 0 \\ -3/5 \\ 0 \\ 0 \\ 0 \end{bmatrix} u_1 \tag{8.9a}$$

The matrix on the right side of this equation is the first column of the $\boldsymbol{\beta}$ matrix; the remaining five columns are found from similar considerations of the motions in Figures 8.1*c*–8.1*g*. Therefore, the compatibility equation (Equation 8.2) is

$$\begin{Bmatrix} \Delta_{AC} \\ \Delta_{CB} \\ \Delta_{AD} \\ \Delta_{CD} \\ \Delta_{CE} \\ \Delta_{DE} \\ \Delta_{BE} \\ \Delta_{BD} \end{Bmatrix} = \begin{bmatrix} 1 & 0 & 0 & 0 & 0 & 0 \\ -1 & 0 & 0 & 0 & 0 & 0 \\ 0 & 0 & 3/5 & 4/5 & 0 & 0 \\ 0 & -1 & 0 & 1 & 0 & 0 \\ -3/5 & -4/5 & 0 & 0 & 3/5 & 4/5 \\ 0 & 0 & -1 & 0 & 1 & 0 \\ 0 & 0 & 0 & 0 & 0 & 1 \\ 0 & 0 & -3/5 & 4/5 & 0 & 0 \end{bmatrix} \begin{Bmatrix} u_1 \\ u_2 \\ u_3 \\ u_4 \\ u_5 \\ u_6 \end{Bmatrix} \tag{8.9b}$$

[10] Recall that displacement u_1 is assumed to be very small compared to the lengths of the bars. This implies that the bar rotations are also very small; consequently, in Figure 8.11*c*, lines such as CE and C′E are taken to be nearly parallel for purposes of computation.

Step 3. The force-deformation relations, Equation 8.4, have the form

$$\begin{Bmatrix} P_{AC} \\ P_{CB} \\ P_{AD} \\ P_{CD} \\ P_{CE} \\ P_{DE} \\ P_{BE} \\ P_{BD} \end{Bmatrix} = \frac{EA}{l} \begin{bmatrix} 1/3 &&&&&&& \\ & 1/3 &&&&&& \\ && 1/5 &&&&& \\ &&& 1/4 &&&& \\ &&&& 1/5 &&& \\ &&&&& 1/3 && \\ &&&&&& 1/4 & \\ &&&&&&& 1/5 \end{bmatrix} \begin{Bmatrix} \Delta_{AC} \\ \Delta_{CB} \\ \Delta_{AD} \\ \Delta_{CD} \\ \Delta_{CE} \\ \Delta_{DE} \\ \Delta_{BE} \\ \Delta_{BD} \end{Bmatrix} + \begin{Bmatrix} 0 \\ 0 \\ 0 \\ 0 \\ 0 \\ 0 \\ 0 \\ 0 \end{Bmatrix} \qquad \textbf{(8.9c)}$$

This defines the **K** matrix; in this case, $\hat{\mathbf{P}} = \mathbf{0}$ because there are no thermal or initial effects.

Step 4. The only input matrix that is not yet defined is **F** in Equation 8.6. As explained previously, we may determine the components of **F** by considering external virtual work associated with the real forces in Figure 8.11*a* and virtual displacement patterns like each of the motions in Figures 8.1*b*–8.1*g*. For example, the external virtual work due to virtual displacements like the motions in Figure 8.1*b* (or Figure 8.11*b*) is zero because there are no virtual displacements at any joints that are in the directions of any of the applied forces in Figure 8.11*a*. Since this is a portion of the first equilibrium equation, it follows that the first component of **F** is zero.

On the other hand, the external virtual work due to virtual displacements like the motions in Figure 8.1*c* is not zero because there is a virtual displacement at joint C that is in the directions of F_1 in Figure 8.11*a*. In this case, the external virtual work is $-F_1 u_{2_V}$; the negative sign is a consequence of the fact that the directions of F_1 and u_{2_V} are opposite. Since u_{2_V} cancels from the equilibrium equation, it follows that the second component of **F** is $-F_1$.

Similarly, we find that the only other nonzero component of **F** is the fifth component, i.e., the component of the fifth equilibrium equation that is obtained from virtual displacements like the motions in Figure 8.1*f*. Here, there is a virtual displacement u_{5_V} at joint E that is in the same direction as F_2 in Figure 8.11*a*; hence, the fifth component of **F** is F_2. Therefore, we have

$$\mathbf{F} = \begin{Bmatrix} 0 \\ -F_1 \\ 0 \\ 0 \\ F_2 \\ 0 \end{Bmatrix} \qquad \textbf{(8.9d)}$$

Step 5. Now that all matrices are defined, we perform the multiplications indicated in Equation 8.7*b*. The stiffness matrix is

$$\mathbf{k} = \boldsymbol{\beta}^{\mathrm{T}}\mathbf{K}\boldsymbol{\beta} = \frac{EA}{l}\begin{bmatrix} 0.739 & 0.096 & 0 & 0 & -0.072 & -0.096 \\ & 0.378 & 0 & -0.25 & -0.096 & -0.128 \\ & & 0.477 & 0 & -0.333 & 0 \\ & & & 0.506 & 0 & 0 \\ & \textit{Sym.} & & & 0.405 & 0.096 \\ & & & & & 0.378 \end{bmatrix} \tag{8.9e}$$

Since $\widehat{\mathbf{P}} = \mathbf{0}$ we have $\mathbf{F}^* = \mathbf{F}$, so Equation 8.7*b* becomes

$$\frac{EA}{l}\begin{bmatrix} 0.739 & 0.096 & 0 & 0 & -0.072 & -0.096 \\ & 0.378 & 0 & -0.25 & -0.096 & -0.128 \\ & & 0.477 & 0 & -0.333 & 0 \\ & & & 0.506 & 0 & 0 \\ & \textit{Sym.} & & & 0.405 & 0.096 \\ & & & & & 0.378 \end{bmatrix}\begin{Bmatrix} u_1 \\ u_2 \\ u_3 \\ u_4 \\ u_5 \\ u_6 \end{Bmatrix} = \begin{Bmatrix} 0 \\ -F_1 \\ 0 \\ 0 \\ F_2 \\ 0 \end{Bmatrix} \tag{8.9f}$$

Solving for $\mathbf{u}$ by inverting $\mathbf{k}$ (Equation 8.7*d*) gives

$$\begin{Bmatrix} u_1 \\ u_2 \\ u_3 \\ u_4 \\ u_5 \\ u_6 \end{Bmatrix} = \frac{l}{EA}\begin{bmatrix} 0.340F_1 + 0.321F_2 \\ -5.516F_1 + 2.254F_2 \\ -1.574F_1 + 5.459F_2 \\ -2.726F_1 + 1.114F_2 \\ -2.254F_1 + 7.818F_2 \\ -1.209F_1 - 1.141F_2 \end{bmatrix} \tag{8.9g}$$

Back-substituting $\mathbf{u}$ into Equation 8.9*b* leads to

$$\begin{Bmatrix} \Delta_{AC} \\ \Delta_{CB} \\ \Delta_{AD} \\ \Delta_{CD} \\ \Delta_{CE} \\ \Delta_{DE} \\ \Delta_{BE} \\ \Delta_{BD} \end{Bmatrix} = \frac{l}{EA}\begin{Bmatrix} 0.340F_1 + 0.321F_2 \\ -0.340F_1 - 0.321F_2 \\ -3.125F_1 + 4.167F_2 \\ 2.791F_1 - 1.141F_2 \\ 1.889F_1 + 1.782F_2 \\ -0.680F_1 + 2.358F_2 \\ -1.209F_1 - 1.141F_2 \\ -1.236F_1 - 2.385F_2 \end{Bmatrix} \tag{8.9h}$$

and from Equation 8.9*c* we obtain

$$\begin{Bmatrix} P_{AC} \\ P_{CB} \\ P_{AD} \\ P_{CD} \\ P_{CE} \\ P_{DE} \\ P_{BE} \\ P_{BD} \end{Bmatrix} = \begin{Bmatrix} 0.113F_1 + 0.107F_2 \\ -0.113F_1 - 0.107F_2 \\ -0.625F_1 + 0.833F_2 \\ 0.698F_1 - 0.285F_2 \\ 0.378F_1 + 0.356F_2 \\ -0.227F_1 + 0.786F_2 \\ -0.302F_1 - 0.285F_2 \\ -0.247F_1 - 0.477F_2 \end{Bmatrix} \quad \textbf{(8.9}\textbf{\textit{i}}\textbf{)}$$

For the case in which $F_2 = 0$ this is identical to the result in Equation 5.17.

EXAMPLE 2

Reanalyze the truss in Example 1 (Figure 8.11*a*) assuming that bar CE is axially rigid (as in Figure 8.4). (This example will add to our understanding of the computation of **F**, and will also illustrate complications that arise in the computation of axial forces in axially rigid members.)

Step 1. A set of five system displacements is shown in Figure 8.4, so we may proceed to writing the compatibility equations.

Step 2. As was observed in Section 8.1, the motions in Figures 8.4*b* and 8.4*c* involve rigid-body translations of bar CE; thus, the motions associated with the system displacements u_1 and u_2 do not involve length changes of bar CE, and differ in this respect from the motions in Figures 8.1*b* and 8.1*c*. As a consequence of the rigid-body translations of bar CE, the motions in Figures 8.4*b* and 8.4*c* now include displacements at joint E. In Figure 8.4*b*, the displacement at E is in a direction perpendicular to bar BE and does not result in a deformation of this bar; however, it does result in a deformation of bar DE, i.e., $\Delta_{DE} = u_1$. In Figure 8.4*c* the displacement at E implies that $\Delta_{BE} = u_2$. The deformations of all remaining bars in Figures 8.4*b* and 8.4*c* are identical to those in Figures 8.1*b* and 8.1*c*.

The motions in Figures 8.4*d* and 8.4*e* are identical to the motions in Figures 8.1*d* and 8.1*e*, so the deformations associated with system displacements u_3 and u_4 are also identical to those in Example 1.

System displacements u_5 and u_6 in Figures 8.1*f* and 8.1*g* are now replaced by the single system displacement u_5 in Figure 8.4*f*, where u_5 is the angle of rotation of rigid bar CE. Hence, the displacement at E in the direction perpendicular to bar CE is $l_{CE}u_5 = 5lu_5$, as shown in Figure 8.4*f*. Figure 8.12 shows an enlarged view of the geometry of the motion at joint E; here joint E is shown displaced to F', so $EE' = 5lu_5$. We see that system displacement u_5 involves deformation of bars BE and DE only, and from 3-4-5 right triangle EE″E′ in Figure 8.12 we conclude that $\Delta_{BE} = EE''$ and $\Delta_{DE} = -E'E''$. It follows that $\Delta_{BE} = (3/5)(5lu_5) = 3lu_5$, and $\Delta_{DE} = -(4/5)(5lu_5) = 4lu_5$.

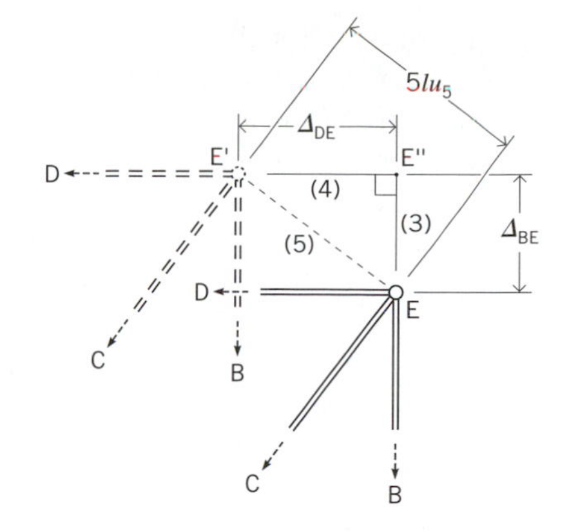

Figure 8.12 Bar deformations for displacement u_5.

The compatibility equations therefore have the form

$$\begin{Bmatrix} \Delta_{AC} \\ \Delta_{CB} \\ \Delta_{AD} \\ \Delta_{CD} \\ \Delta_{CE} \\ \Delta_{DE} \\ \Delta_{BE} \\ \Delta_{BD} \end{Bmatrix} = \begin{bmatrix} 1 & 0 & 0 & 0 & 0 \\ -1 & 0 & 0 & 0 & 0 \\ 0 & 0 & 3/5 & 4/5 & 0 \\ 0 & -1 & 0 & 1 & 0 \\ 0 & 0 & 0 & 0 & 0 \\ 1 & 0 & -1 & 0 & -4l \\ 0 & 1 & 0 & 0 & 3l \\ 0 & 0 & -3/5 & 4/5 & 0 \end{bmatrix} \begin{Bmatrix} u_1 \\ u_2 \\ u_3 \\ u_4 \\ u_5 \end{Bmatrix} \tag{8.10a}$$

To effect a direct comparison with the previous formulation, Δ_{CE} has been retained in this equation even though it is identically equal to zero.

Step 3. $\Delta_{CE} = 0$ implies that the stiffness (EA) of bar CE must be infinite. As we shall see, this also implies (and equilibrium requires) that P_{CE} can have a finite (nonzero) value (even though $\Delta_{CE} = 0$). This force can be calculated once the unknown system displacements are found.

The force-deformation relations have the form

$$\begin{Bmatrix} P_{AC} \\ P_{CB} \\ P_{AD} \\ P_{CD} \\ P_{CE} \\ P_{DE} \\ P_{BE} \\ P_{BD} \end{Bmatrix} = \frac{EA}{l} \begin{bmatrix} 1/3 & & & & & & & \\ & 1/3 & & & & & & \\ & & 1/5 & & & & & \\ & & & 1/4 & & & & \\ & & & & \infty & & & \\ & & & & & 1/3 & & \\ & & & & & & 1/4 & \\ & & & & & & & 1/5 \end{bmatrix} \begin{Bmatrix} \Delta_{AC} \\ \Delta_{CB} \\ \Delta_{AD} \\ \Delta_{CD} \\ \Delta_{CE} \\ \Delta_{DE} \\ \Delta_{BE} \\ \Delta_{BD} \end{Bmatrix} + \begin{Bmatrix} 0 \\ 0 \\ 0 \\ 0 \\ 0 \\ 0 \\ 0 \\ 0 \end{Bmatrix} \tag{8.10b}$$

Step 4. As in the previous example, we determine components of **F** by considering external virtual work associated with the real forces in Figure 8.11*a* and virtual displacement patterns like each of the motions in Figures 8.4*b*–8.4*f*.

In this case, the external virtual work due to virtual displacements like the motions in Figure 8.4*b* is equal to $F_2 u_{1_V}$ because there is a virtual displacement u_{1_V} at joint E, which is in the direction of F_2 in Figure 8.11*a*. The external virtual work due to virtual displacements like the motions in Figure 8.4*c* is the same as in Example 1, i.e., $-F_1 u_{2_V}$ because there is a virtual displacement u_{2_V} at joint C, which is opposite in direction to F_1 in Figure 8.11*a*. The external virtual works associated with virtual displacements like the motions in Figures 8.4*d* and 8.4*e* are again zero, exactly as in Example 1. Finally, virtual displacements like the motions in Figure 8.4*f* give rise to an external virtual work equal to $-4l\ F_2 u_{5_V}$ because joint E has a horizontal component of displacement $4l\ u_{5_V}$, as in Figure 8.12, which is opposite in direction to F_2.

Thus, the **F** matrix now becomes

$$\mathbf{F} = \begin{Bmatrix} F_2 \\ -F_1 \\ 0 \\ 0 \\ -4lF_2 \end{Bmatrix} \tag{8.10c}$$

Note that the coupling of joint displacements makes it conceivable that some components of **F** might involve more than a single force. For instance, if there had been a vertical load F_3 (say upward) at joint E in Figure 8.11*a*, the external virtual work associated with virtual displacements like Figure 8.4*c* would be $(-F_1 + F_3)u_{2_V}$, and so the second component of **F** would be $(-F_1 + F_3)$ rather than $-F_1$.

Step 5. Once again, we have all the matrices required in Equation 8.7*b*. We obtain

$$\mathbf{k} = \boldsymbol{\beta}^{\mathrm{T}}\mathbf{K}\boldsymbol{\beta} = \frac{EA}{l}\begin{bmatrix} 1 & 0 & -0.333 & 0 & -1.333l \\ & 0.5 & 0 & -0.25 & 0.75l \\ & & 0.477 & 0 & 1.333l \\ & & & 0.506 & 0 \\ & Sym. & & & 7.583l^2 \end{bmatrix} \tag{8.10d}$$

We observe that $\mathbf{F}^* = \mathbf{F}$ since $\boldsymbol{\beta}^{\mathrm{T}}\hat{\mathbf{P}} = \mathbf{0}$. Thus Equation 8.7*b* becomes

$$\frac{EA}{l}\begin{bmatrix} 1 & 0 & -0.333 & 0 & -1.333l \\ & 0.5 & 0 & -0.25 & 0.75l \\ & & 0.477 & 0 & 1.333l \\ & & & 0.506 & 0 \\ & Sym. & & & 7.583l^2 \end{bmatrix}\begin{Bmatrix} u_1 \\ u_2 \\ u_3 \\ u_4 \\ u_5 \end{Bmatrix} = \begin{Bmatrix} F_2 \\ -F_1 \\ 0 \\ 0 \\ -4lF_2 \end{Bmatrix} \tag{8.10e}$$

from which we obtain

$$\begin{Bmatrix} u_1 \\ u_2 \\ u_3 \\ u_4 \\ u_5 \end{Bmatrix} = \frac{l}{EA}\begin{bmatrix} 0.485F_1 + 0.457F_2 \\ -4.499F_1 + 3.215F_2 \\ -2.245F_1 + 4.827F_2 \\ -2.223F_1 + 1.588F_2 \\ 0.925F_1/l - 1.614F_2/l \end{bmatrix} \tag{8.10f}$$

Back-substitution leads to

$$\begin{Bmatrix} \Delta_{AC} \\ \Delta_{CB} \\ \Delta_{AD} \\ \Delta_{CD} \\ \Delta_{CE} \\ \Delta_{DE} \\ \Delta_{BE} \\ \Delta_{BD} \end{Bmatrix} = \frac{l}{EA} \begin{Bmatrix} 0.485F_1 + 0.457F_2 \\ -0.485F_1 - 0.457F_2 \\ -3.125F_1 + 4.167F_2 \\ 2.276F_1 - 1.626F_2 \\ 0 \\ -0.970F_1 + 2.085F_2 \\ -1.724F_1 - 1.626F_2 \\ -0.431F_1 - 1.625F_2 \end{Bmatrix} \tag{8.10g}$$

and

$$\begin{Bmatrix} P_{AC} \\ P_{CB} \\ P_{AD} \\ P_{CD} \\ P_{CE} \\ P_{DE} \\ P_{BE} \\ P_{BD} \end{Bmatrix} = \begin{Bmatrix} 0.162F_1 + 0.152F_2 \\ -0.162F_1 - 0.152F_2 \\ -0.625F_1 + 0.833F_2 \\ 0.569F_1 - 0.407F_2 \\ \text{undefined} \\ -0.323F_1 + 0.695F_2 \\ -0.431F_1 - 0.407F_2 \\ -0.086F_1 - 0.325F_2 \end{Bmatrix} \tag{8.10h}$$

Undefined force P_{CE} can now be determined from a consideration of equilibrium of a truss joint, either joint C or joint E in Figure 8.11*a*. To illustrate, consider the free-body diagram of joint E shown in Figure 8.13; summing forces in the y direction gives

$$0 = P_{BE} + \frac{4}{5}P_{CE} \tag{8.11a}$$

or

$$P_{CE} = -\frac{5}{4}P_{BE} \tag{8.11b}$$

Substituting the value of P_{BE} from Equation 8.10*h* leads to

$$\begin{aligned} P_{CE} &= -\frac{5}{4}(-0.431F_1 - 0.407F_2) \\ &= 0.539F_1 + 0.508F_2 \end{aligned} \tag{8.11c}$$

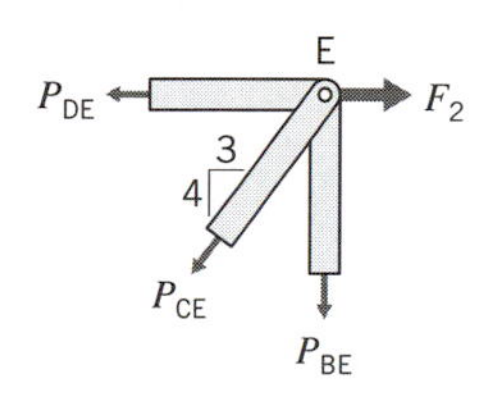

Figure 8.13 Free-body of jont E.

which is the desired value. (The same result will be obtained if we sum forces in the x direction.) Thus, for future reference, the final **P** matrix is

$$\begin{Bmatrix} P_{AC} \\ P_{CB} \\ P_{AD} \\ P_{CD} \\ P_{CE} \\ P_{DE} \\ P_{BE} \\ P_{BD} \end{Bmatrix} = \begin{Bmatrix} 0.162F_1 + 0.152F_2 \\ -0.162F_1 - 0.152F_2 \\ -0.625F_1 + 0.833F_2 \\ 0.569F_1 - 0.407F_2 \\ 0.539F_1 + 0.508F_2 \\ -0.323F_1 + 0.695F_2 \\ -0.431F_1 - 0.407F_2 \\ -0.086F_1 - 0.325F_2 \end{Bmatrix} \tag{8.11d}$$

There is an alternate approach that may be employed to analyze this structure, i.e., we may obtain an approximate solution by using the previous formulation in Example 1 with the axial stiffness of bar CE set to some arbitrarily large numerical value. This approaches the case of a rigid bar. For example, suppose we increase the stiffness of bar CE by a factor of 1000, so that the force-deformation relation for this one bar is $P_{CE} = 1000EA/5l = 200EA/l$. We will assume that the force-deformation relations for the other bars are unchanged, and we will reanalyze the structure using the six system displacements defined in Example 1 (Figure 8.1).

The $\boldsymbol{\beta}$ matrix is not a function of individual bar stiffnesses and remains as defined in Equation 8.9*b*. The modified form of Equation 8.9*c* is

$$\begin{Bmatrix} P_{AC} \\ P_{CB} \\ P_{AD} \\ P_{CD} \\ P_{CE} \\ P_{DE} \\ P_{BE} \\ P_{BD} \end{Bmatrix} = \frac{EA}{l} \begin{bmatrix} 1/3 & & & & & & & \\ & 1/3 & & & & & & \\ & & 1/5 & & & & & \\ & & & 1/4 & & & & \\ & & & & 200 & & & \\ & & & & & 1/3 & & \\ & & & & & & 1/4 & \\ & & & & & & & 1/5 \end{bmatrix} \begin{Bmatrix} \Delta_{AC} \\ \Delta_{CB} \\ \Delta_{AD} \\ \Delta_{CD} \\ \Delta_{CE} \\ \Delta_{DE} \\ \Delta_{BE} \\ \Delta_{BD} \end{Bmatrix} + \begin{Bmatrix} 0 \\ 0 \\ 0 \\ 0 \\ 0 \\ 0 \\ 0 \\ 0 \end{Bmatrix} \tag{8.12a}$$

which defines the new **K** matrix. The new stiffness matrix is

$$\mathbf{k} = \boldsymbol{\beta}^{T}\mathbf{K}\boldsymbol{\beta} = \frac{EA}{l} \begin{bmatrix} 72.667 & 96 & 0 & 0 & -72.0 & -96 \\ & 128.25 & 0 & -0.25 & -96 & -128 \\ & & 0.477 & 0 & -0.333 & 0 \\ & & & 0.506 & 0 & 0 \\ & Sym. & & & 72.333 & 96 \\ & & & & & 128.25 \end{bmatrix} \tag{8.12b}$$

The force matrix is still given by Equation 8.9*d*, so Equation 8.7*b* now becomes

$$\frac{EA}{l}\begin{bmatrix} 72.667 & 96 & 0 & 0 & -72.0 & -96 \\ & 128.25 & 0 & -0.25 & -96 & -128 \\ & & 0.477 & 0 & -0.333 & 0 \\ & & & 0.506 & 0 & 0 \\ & Sym. & & & 72.333 & 96 \\ & & & & & 128.25 \end{bmatrix}\begin{Bmatrix} u_1 \\ u_2 \\ u_3 \\ u_4 \\ u_5 \\ u_6 \end{Bmatrix} = \begin{Bmatrix} 0 \\ -F_1 \\ 0 \\ 0 \\ F_2 \\ 0 \end{Bmatrix} \quad \textbf{(8.12c)}$$

from which we find

$$\begin{Bmatrix} u_1 \\ u_2 \\ u_3 \\ u_4 \\ u_5 \\ u_6 \end{Bmatrix} = \frac{l}{EA}\begin{bmatrix} 0.485F_1 + 0.457F_2 \\ -4.500F_1 + 3.213F_2 \\ -2.244F_1 + 4.828F_2 \\ -2.223F_1 + 1.588F_2 \\ -3.213F_1 + 6.913F_2 \\ -1.723F_1 - 1.626F_2 \end{bmatrix} \quad \textbf{(8.12d)}$$

Back-substituting gives

$$\begin{Bmatrix} \Delta_{AC} \\ \Delta_{CB} \\ \Delta_{AD} \\ \Delta_{CD} \\ \Delta_{CE} \\ \Delta_{DE} \\ \Delta_{BE} \\ \Delta_{BD} \end{Bmatrix} = \frac{l}{EA}\begin{Bmatrix} 0.485F_1 + 0.457F_2 \\ -0.485F_1 - 0.457F_2 \\ -3.125F_1 + 4.167F_2 \\ 2.277F_1 - 1.626F_2 \\ 0.003F_1 + 0.003F_2 \\ -0.969F_1 + 2.086F_2 \\ -1.723F_1 - 1.626F_2 \\ -0.432F_1 - 1.627F_2 \end{Bmatrix} \quad \textbf{(8.12e)}$$

and

$$\begin{Bmatrix} P_{AC} \\ P_{CB} \\ P_{AD} \\ P_{CD} \\ P_{CE} \\ P_{DE} \\ P_{BE} \\ P_{BD} \end{Bmatrix} = \begin{Bmatrix} 0.162F_1 + 0.152F_2 \\ -0.162F_1 - 0.152F_2 \\ -0.625F_1 + 0.833F_2 \\ 0.569F_1 - 0.406F_2 \\ 0.539F_1 + 0.508F_2 \\ -0.323F_1 + 0.695F_2 \\ -0.431F_1 - 0.406F_2 \\ -0.086F_1 - 0.325F_2 \end{Bmatrix} \quad \textbf{(8.12f)}$$

Equations 8.12*e* and 8.12*f* are very nearly indistinguishable from Equations 8.10*g* and 8.11*d*, respectively.

8.3 ANALYSIS OF FRAMES

We turn our attention now to the basic formulation of the analysis procedure for frames composed of axially rigid beams. This study will enable us to establish the basic concepts of the displacement method of analysis for frames, and will serve as the foundation for discussions of the more advanced formulations that follow.

8.3.1 Stiffness Matrix for Beams

Recall the general uniform beam that is subjected to the arbitrary set of forces depicted in Figure 5.23*a*, and that undergoes the displacements and deformations defined in Figure 5.23*b*. The flexibility matrix for this beam, which relates the end deformations to the independent end moments, was originally presented in Equation 5.42*a* and is given by

$$\mathbf{a}^{(e)} = \frac{l}{6EI}\begin{bmatrix} 2 & -1 \\ -1 & 2 \end{bmatrix} \tag{8.13a}$$

The stiffness matrix for a beam element is the inverse of this, i.e.,

$$\mathbf{k}^{(e)} = \mathbf{a}^{(e)-1} = \frac{2EI}{l}\begin{bmatrix} 2 & 1 \\ 1 & 2 \end{bmatrix} \tag{8.13b}$$

The moment-deformation relations for a beam element then follow from Equation 5.42*a*,

$$\begin{aligned} \begin{Bmatrix} M_{\mathrm{IJ}} \\ M_{\mathrm{JI}} \end{Bmatrix} &= \frac{2EI}{l}\begin{bmatrix} 2 & 1 \\ 1 & 2 \end{bmatrix}\left(\begin{Bmatrix} \Delta_{\mathrm{IJ}} \\ \Delta_{\mathrm{JI}} \end{Bmatrix} - \begin{Bmatrix} \widehat{\Delta}_{\mathrm{IJ}} \\ \widehat{\Delta}_{\mathrm{JI}} \end{Bmatrix}\right) \\ &= \frac{2EI}{l}\begin{bmatrix} 2 & 1 \\ 1 & 2 \end{bmatrix}\begin{Bmatrix} \Delta_{\mathrm{IJ}} \\ \Delta_{\mathrm{JI}} \end{Bmatrix} + \begin{Bmatrix} -(4EI/l)\widehat{\Delta}_{\mathrm{IJ}} - (2EI/l)\widehat{\Delta}_{\mathrm{JI}} \\ -(2EI/l)\widehat{\Delta}_{\mathrm{IJ}} - (4EI/l)\widehat{\Delta}_{\mathrm{JI}} \end{Bmatrix} \end{aligned} \tag{8.13c}$$

We identify the components of the last vector as fixed-end moments, i.e., end moments when end deformations Δ_{IJ} and Δ_{JI} are zero; therefore, we rewrite the equation as

$$\begin{Bmatrix} M_{\mathrm{IJ}} \\ M_{\mathrm{JI}} \end{Bmatrix} = \frac{2EI}{l}\begin{bmatrix} 2 & 1 \\ 1 & 2 \end{bmatrix}\begin{Bmatrix} \Delta_{\mathrm{IJ}} \\ \Delta_{\mathrm{JI}} \end{Bmatrix} + \begin{Bmatrix} \widehat{M}_{\mathrm{IJ}} \\ \widehat{M}_{\mathrm{JI}} \end{Bmatrix} \tag{8.13d}$$

which has the same basic form as Equation 8.4. Substituting the expressions for $\widehat{\Delta}_{\mathrm{IJ}}$ and $\widehat{\Delta}_{\mathrm{JI}}$ from Equations 5.42*b* into the last vector in Equation 8.13*c* provides a definition of $\widehat{M}_{\mathrm{IJ}}$ and $\widehat{M}_{\mathrm{JI}}$ which is the analog of Equations 5.42*b*,

$$\begin{aligned} \widehat{M}_{\mathrm{IJ}} &= \frac{2}{l^2}\left(\int_0^l \widehat{M}(x)x\,dx - 2\int_0^l \widehat{M}(x)(l - x)dx\right) \\ \widehat{M}_{\mathrm{JI}} &= \frac{2}{l^2}\left(2\int_0^l \widehat{M}(x)x\,dx - \int_0^l \widehat{M}(x)(l - x)dx\right) \end{aligned} \tag{8.13e}$$

When the equations are written in this form, we recognize the integrals as first moments of areas under the $\widehat{M}(x)$ diagram about $x = 0$ or $x = l$.[11] These equations may also be expressed in slightly more compact form as

$$\widehat{M}_{IJ} = -\int_0^l \widehat{M}(x)\left(\frac{4}{l} - \frac{6x}{l^2}\right)dx$$
$$\widehat{M}_{JI} = -\int_0^l \widehat{M}(x)\left(\frac{2}{l} - \frac{6x}{l^2}\right)dx \tag{8.13f}$$

In the analysis procedure that follows, the deformations Δ_{IJ} and Δ_{JI} in Equation 8.13*d* are related to system displacements consisting of a set of joint rotations and displacements. Consequently, it is convenient to modify Equation 8.13*d* by substituting the definitions of Δ_{IJ} and Δ_{JI} from Equations 5.39*a*. The desired form of the moment-deformation relations is

$$\begin{Bmatrix} M_{IJ} \\ M_{JI} \end{Bmatrix} = \frac{2EI}{l}\begin{bmatrix} 2 & 1 \\ 1 & 2 \end{bmatrix}\begin{Bmatrix} \theta_{IJ} - \psi_{IJ} \\ \theta_{JI} - \psi_{IJ} \end{Bmatrix} + \begin{Bmatrix} \widehat{M}_{IJ} \\ \widehat{M}_{JI} \end{Bmatrix} \tag{8.13g}$$

where

$$\widehat{M}_{IJ} = \frac{2}{l^2}\left(\int_0^l \widehat{M}(x)x\,dx - 2\int_0^l \widehat{M}(x)(l - x)dx\right)$$
$$\widehat{M}_{JI} = \frac{2}{l^2}\left(2\int_0^l \widehat{M}(x)x\,dx - \int_0^l \widehat{M}(x)(l - x)dx\right) \tag{8.13e}$$

and

$$\widehat{M}(x) = M_o(x) + M_T(x) + M_I(x) \tag{8.13h}$$

$$\psi_{IJ} \equiv \frac{v_{IJ} - v_{JI}}{l} \tag{8.13i}$$

Here, ψ is the "side-sway angle" defined in Equation 5.39*b*. Thus, Equation 8.13*g* contains deformations that are expressed in terms of end rotations and rigid-body rotation.[12] As with Equation 5.42*a*, we see that the only end forces that are explicitly carried in Equation 8.13*g* are the end moments M_{IJ} and M_{JI}; all effects of other applied loads (including thermal and initial effects) are contained in the integrals for $\widehat{M}_{IJ}$ and $\widehat{M}_{JI}$. It should be understood that $M_o(x)$ in Equation 8.13*h* is the same as defined in

---

[11] See Equation 3.24*b* or 3.24*c*.

[12] We continue to regard this as a force-deformation relation because it is essentially the inverse of Equation 5.42*a*, and also because each element of the **Δ** vector is a combination of displacement components. Later, we will reformulate the relations in such a way that each element of the **Δ** vector is a single end displacement or rotation.

Figure 5.23*a*(2) and Equation 5.38*c*, i.e., it is the bending moment in the unrestrained beam due to applied interior physical loads. Even though the integrals in Equations 8.13*g* are usually not difficult to work out, there is an important and relatively simple alternate procedure available for evaluating fixed-end moments due to applied physical loads (not including thermal or initial loads). This alternate procedure, which is an application of Betti's theorem (Section 6.1.1), does not require the evaluation of the integrals in Equation 8.13*g* (but may require the evaluation of other integrals). This procedure will be discussed shortly.

Before proceeding to illustrations of the analysis procedure for frames, it is interesting to note the expanded form of Equations 8.13*g*, which is

$$
\begin{aligned}
M_{IJ} &= \frac{2EI}{l}(2\theta_{IJ} + \theta_{JI} - 3\psi_{IJ}) + \widehat{M}_{IJ} \\
M_{JI} &= \frac{2EI}{l}(\theta_{IJ} + 2\theta_{JI} - 3\psi_{IJ}) + \widehat{M}_{JI}
\end{aligned}
\qquad \textbf{(8.14)}
$$

These are referred to historically as the slope-deflection equations, and were used in analysis long before the matrix formulation was developed.[13] The earlier method of analysis utilized these equations in a long-hand format, but is otherwise conceptually identical to the modern formulation. The matrix approach that follows leads to exactly the same system equilibrium equations as the older slope-deflection method. Since the slope-deflection equations are only an alternate expression of Equations 8.13*g* they will not be discussed further here.

8.3.2 Matrix Formulation

For illustrative purposes we will begin with a structure similar to the one in Figure 8.6, specifically the structure in Figure 8.14. This structure is subjected to two identical concentrated forces F in span BC, and a temperature change $\Delta T = \Delta T_o y/h$[14] in spans AB and CD, as shown.

Step 1. The two system displacements for this structure were identified in Figure 8.6, so we may proceed directly to the consideration of compatibility.

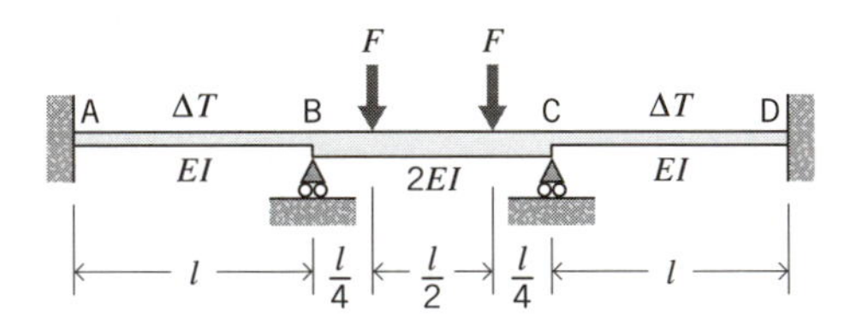

Figure 8.14 Frame structure.

[13] The slope-deflection equations are attributed to Bendixen (1914) and Maney (1915).

[14] See Example 5, Chapter 3, for a review of the significance of terms.

Step 2. The imposition of compatibility requires that we relate element end deformations to the system displacements. This means that we must relate the components of the $\mathbf{\Delta}$ vector in Equation 8.2 to the system displacements; but the components of the $\mathbf{\Delta}$ vector in Equation 8.2 are the quantities $\theta_{IJ} - \psi_{IJ}$ and $\theta_{JI} - \psi_{IJ}$ (Equation 8.13*g*) for each of the beams, i.e.,

$$\mathbf{\Delta} \equiv \begin{Bmatrix} \Delta_{AB} \\ \Delta_{BA} \\ \Delta_{BC} \\ \Delta_{CB} \\ \Delta_{CD} \\ \Delta_{DC} \end{Bmatrix} = \begin{Bmatrix} \theta_{AB} - \psi_{AB} \\ \theta_{BA} - \psi_{AB} \\ \theta_{BC} - \psi_{BC} \\ \theta_{CB} - \psi_{BC} \\ \theta_{CD} - \psi_{CD} \\ \theta_{DC} - \psi_{CD} \end{Bmatrix} \tag{8.15a}$$

To determine $\theta_{IJ} - \psi_{IJ}$ and $\theta_{JI} - \psi_{IJ}$ for the beams, we begin by observing that in this case there are no transverse displacements at the ends of any of the beams, i.e., $v_{AB} = v_{BA} = v_{BC} = v_{CB} = v_{CD} = v_{DC} = 0$. Thus, according to Equation 8.13*i* we have $\psi_{AB} = \psi_{BC} = \psi_{CD} = 0$. In other words, there are no side-sway angles ψ for this frame. It follows from Equation 8.15*a* that beam end deformations (Δ_{IJ} and Δ_{JI}) are identical to end rotations (θ_{IJ} and θ_{JI}). The end rotations are obtained in terms of the system displacements from Figures 8.6*b* and 8.6*c*, e.g., from Figure 8.6*b* we recognize that $\theta_{AB} = 0$, $\theta_{BA} = u_1$, $\theta_{BC} = u_1$, $\theta_{CB} = 0$, $\theta_{CD} = 0$, and $\theta_{DC} = 0$. Similarly, from Figure 8.6*c* we have $\theta_{AB} = 0$, $\theta_{BA} = 0$, $\theta_{BC} = 0$, $\theta_{CB} = u_2$, $\theta_{CD} = u_2$, and $\theta_{DC} = 0$. (Note that we continue to define clockwise end rotations as positive.) Therefore, the compatibility equation is

$$\mathbf{\Delta} \equiv \begin{Bmatrix} \Delta_{AB} \\ \Delta_{BA} \\ \Delta_{BC} \\ \Delta_{CB} \\ \Delta_{CD} \\ \Delta_{DC} \end{Bmatrix} = \begin{Bmatrix} \theta_{AB} - \psi_{AB} \\ \theta_{BA} - \psi_{AB} \\ \theta_{BC} - \psi_{BC} \\ \theta_{CB} - \psi_{BC} \\ \theta_{CD} - \psi_{CD} \\ \theta_{DC} - \psi_{CD} \end{Bmatrix} = \begin{bmatrix} 0 & 0 \\ 1 & 0 \\ 1 & 0 \\ 0 & 1 \\ 0 & 1 \\ 0 & 0 \end{bmatrix} \begin{Bmatrix} u_1 \\ u_2 \end{Bmatrix} \tag{8.15b}$$

This defines the $\boldsymbol{\beta}$ matrix.

Step 3. The matrix moment-deformation equation for this structure consists of an assemblage of Equation 8.13*g* for each of the three beam elements. Since ''fixed-end moments'' are a component of these equations, we will begin with their computation.

Consider first the fixed-end moments due to ΔT in beams AB and CD. There are no applied intermediate physical loads or initial effects, so for both of these beams we have[15]

$$\widehat{M}(x) = M_T(x) = -E\alpha \int_A y\,\Delta T dA = -\frac{E\alpha\Delta T_o}{h}\int_A y^2 dA = -\frac{EI\alpha\Delta T_o}{h} \tag{8.15c}$$

[15] *ibid*; see also Equation 5.32*b*.

Substituting into the first of Equations 8.13e gives[16]

$$\widehat{M}_{AB} = \widehat{M}_{CD} = \frac{2}{l^2}\left(-\frac{EI\alpha\Delta T_o}{h}\int_0^l xdx + \frac{2EI\alpha\Delta T_o}{h}\int_0^l (l - x)dx\right) = \frac{EI\alpha\Delta T_o}{h} \qquad \textbf{(8.15}d\textbf{)}$$

Similarly, from the second of Equations 8.13e (or from symmetry) we find

$$\widehat{M}_{BA} = \widehat{M}_{DC} = -\frac{EI\alpha\Delta T_o}{h} \qquad \textbf{(8.15}e\textbf{)}$$

To calculate the fixed-end moments for beam BC using Equation 8.13e, we must first determine the bending moment $\widehat{M}(x)$ in the ''unrestrained'' beam. In this case, the bending moment is due entirely to the applied forces F (there are no thermal or initial effects), so $\widehat{M}(x) = M_o(x)$. Figures 8.15a and 8.15b show the unrestrained beam, and Figure 8.15c shows the corresponding bending moment diagram. It should be emphasized that this is not the actual bending moment in beam BC in Figure 8.14. Rather, it is the bending moment without the effects of the end moments M_{BC} and M_{CB}; this bending moment is the quantity required in Equation 8.13e for the computation of the fixed-end moments $\widehat{M}_{BC}$ and $\widehat{M}_{CB}$. It may seem odd that bending moments in an ''unrestrained'' beam are used to compute end moments in a ''fully restrained'' beam (i.e., in a beam with end rotations prevented), but that is a consequence of the development of these equations. (As we will see shortly, Betti's theorem provides a direct computational procedure that accentuates the physical significance of fixed-end moments.)

From Figure 8.15c we obtain the desired bending moment as a function of coordinate x,

$$\begin{aligned} M_o(x) &= Fx, && 0 < x < \frac{l}{4} \\ M_o(x) &= \frac{Fl}{4}, && \frac{l}{4} < x < \frac{3l}{4} \\ M_o(x) &= F(l - x), && \frac{3l}{4} < x < l \end{aligned} \qquad \textbf{(8.15}f\textbf{)}$$

We substitute into the first of Equations 8.13e to get

$$\begin{aligned} \widehat{M}_{BC} = \frac{2}{l^2}\Bigg[&\int_0^{l/4} Fx^2dx + \int_{l/4}^{3l/4}\left(\frac{Fl}{4}\right)xdx + \int_{3l/4}^{l} F(l - x)xdx \\ &-2\left(\int_0^{l/4} Fx(l - x)dx + \int_{l/4}^{3l/4}\left(\frac{Fl}{4}\right)(l - x)dx + \int_{3l/4}^{l} F(l - x)^2dx\right)\Bigg] \end{aligned} \qquad \textbf{(8.15}g\textbf{)}$$

---

[16] Alternately, we may substitute the unrestrained end deformations calculated in Equation 5.44c into the expressions for fixed-end moments in Equation 8.13c.

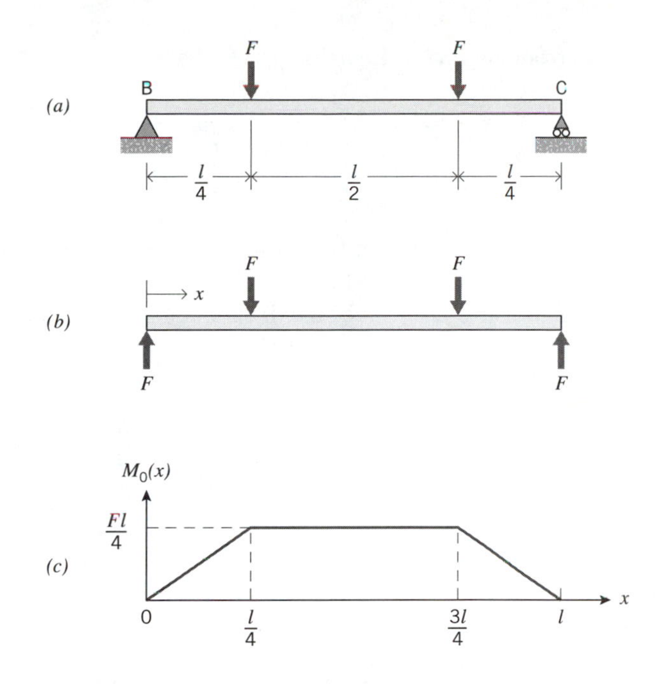

Figure 8.15 Beam BC (unrestrained, i.e., $M_{BC} = M_{CB} = 0$).

Performing the integrations leads to

$$\widehat{M}_{BC} = -\frac{3}{16}Fl \qquad \textbf{(8.15}\boldsymbol{h}\textbf{)}$$

Similarly, from the second of Equations 8.13*e* (or from the symmetry in Figure 8.15) we obtain

$$\widehat{M}_{CB} = \frac{3}{16}Fl \qquad \textbf{(8.15}\boldsymbol{i}\textbf{)}$$

The results in Equations 8.15*h* and 8.15*i* can also be obtained from the moment-area interpretation of Equation 8.13*e* mentioned previously. For example, we note that the first integral in Equation 8.13*e* is the first moment of the area under the $M_o(x)$ diagram in Figure 8.15*c* about $x = 0$, and the second integral is the first moment of the area under the $M_o(x)$ diagram about $x = l$. Denote the first integral by Q_1 and the second integral by Q_2; we observe from the symmetry of the moment diagram that $Q_1 = Q_2$. From Figure 8.15*c* we obtain

$$Q_1 = \frac{1}{2}\left(\frac{l}{4}\right)\left(\frac{Fl}{4}\right)\left[\left(\frac{2}{3}\right)\left(\frac{l}{4}\right)\right] + \left(\frac{l}{2}\right)\left(\frac{Fl}{4}\right)\left(\frac{l}{4} + \frac{l}{4}\right) + \frac{1}{2}\left(\frac{l}{4}\right)\left(\frac{Fl}{4}\right)\left[\frac{3l}{4} + \left(\frac{1}{3}\right)\left(\frac{l}{4}\right)\right] \qquad \textbf{(8.15}\boldsymbol{j}\textbf{)}$$

$$= \frac{3}{32}Fl^3$$

It follows from the first of Equations 8.13*e* that

$$\widehat{M}_{BC} = \frac{2}{l^2}(Q_1 - 2Q_2) = \frac{2}{l^2}\left(\frac{3}{32}Fl - 2\frac{3}{32}Fl\right) = -\frac{3}{16}Fl \qquad \textbf{(8.15}\textbf{\textit{k}}\textbf{)}$$

which agrees with Equation 8.15*h*, as expected. The second of Equations 8.13*e* gives the same result as Equation 8.15*i*.

Now that we have determined the fixed-end moments for all beams of the structure, we can assemble the matrix moment-deformation equation,

$$\begin{Bmatrix} M_{AB} \\ M_{BA} \\ M_{BC} \\ M_{CB} \\ M_{CD} \\ M_{DC} \end{Bmatrix} = \frac{2EI}{l} \begin{bmatrix} 2 & 1 & & & & \\ 1 & 2 & & & & \\ & & 4 & 2 & & \\ & & 2 & 4 & & \\ & & & & 2 & 1 \\ & & & & 1 & 2 \end{bmatrix} \begin{Bmatrix} \Delta_{AB} \\ \Delta_{BA} \\ \Delta_{BC} \\ \Delta_{CB} \\ \Delta_{CD} \\ \Delta_{DC} \end{Bmatrix} + \begin{Bmatrix} EI\alpha\Delta T_o/h \\ -EI\alpha\Delta T_o/h \\ -3Fl/16 \\ 3Fl/16 \\ EI\alpha\Delta T_o/h \\ -EI\alpha\Delta T_o/h \end{Bmatrix} \qquad \textbf{(8.15}\textbf{\textit{l}}\textbf{)}$$

The first two rows of this equation are the force-deformation relations (Equation 8.13*d* or 8.13*g*) for beam AB, the third and fourth rows in the equation are the force-deformation relations for beam BC, which has length l and bending stiffness $2EI$, and the last two rows are the force-deformation relations for beam CD. Equation 8.15*l* defines the **K** matrix and $\widehat{\mathbf{P}}$ vector.

Step 4, Let us temporarily accept without any formal proof the proposition that Equation 8.6 is the matrix form of the equilibrium equations for frame structures. The validity of this proposition will be demonstrated in the next section using virtual work. As with trusses, the main difficulty we encounter with equilibrium equations for frames is the determination of the components of the **F** vector. This is also the case for the frame structure under consideration herein, i.e., the two components of the **F** vector must be determined somehow. In order to demonstrate the computation, we begin by designating these components by f_1 and f_2; therefore, from Equation 8.6 we obtain

$$\begin{Bmatrix} f_1 \\ f_2 \end{Bmatrix} = \begin{bmatrix} 0 & 1 & 1 & 0 & 0 & 0 \\ 0 & 0 & 0 & 1 & 1 & 0 \end{bmatrix} \begin{Bmatrix} M_{AB} \\ M_{BA} \\ M_{BC} \\ M_{CB} \\ M_{CD} \\ M_{DC} \end{Bmatrix} \qquad \textbf{(8.15}\textbf{\textit{m}}\textbf{)}$$

Thus, the first of the two equilibrium equations has the form $f_1 = M_{BA} + M_{BC}$ and the second has the form $f_2 = M_{CB} + M_{CD}$.

Although we will develop a more general procedure in the next section, we can ascertain what the values of f_1 and f_2 must be in an *ad hoc* manner by relating each

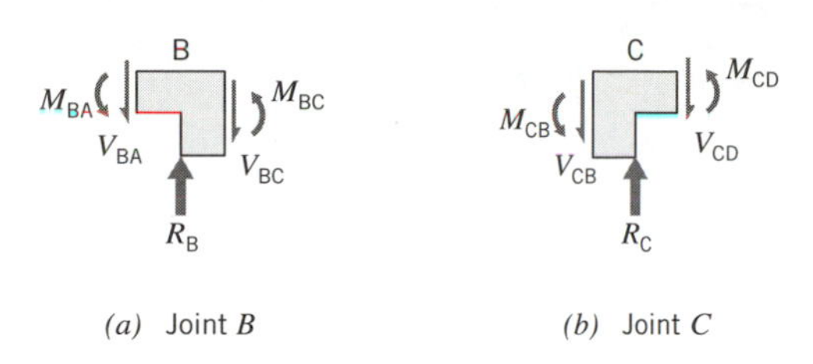

Figure 8.16 Free-body diagrams of joints.

equation to an appropriate free-body.[17] The appropriate free-bodies, in this case, are those of joints B and C, in Figures 8.16*a* and 8.16*b*. Since the right side of each equation involves the moments in one of these free-bodies, it should be apparent that each equation is an expression of moment equilibrium for one of the free-bodies in Figure 8.16, i.e., the first equation must be the equation of moment equilibrium for the free-body in Figure 8.16*a*, and the second equation must be the equation of moment equilibrium for the free-body in Figure 8.16*b*. Since there are no external moments (other than M_{BA}, M_{BC}, M_{CB}, and M_{CD}) applied to either joint in Figure 8.16*a* and 8.16*b*,[18] equilibrium requires that $M_{BA} + M_{BC} = 0$ and $M_{CB} + M_{CD} = 0$, so it follows that $f_1 = f_2 = 0$. Thus, the correct equilibrium equations are given by

$$\begin{Bmatrix} 0 \\ 0 \end{Bmatrix} = \begin{bmatrix} 0 & 1 & 1 & 0 & 0 & 0 \\ 0 & 0 & 0 & 1 & 1 & 0 \end{bmatrix} \begin{Bmatrix} M_{AB} \\ M_{BA} \\ M_{BC} \\ M_{CB} \\ M_{CD} \\ M_{DC} \end{Bmatrix} \tag{8.15n}$$

In other words, $\mathbf{F} = \mathbf{0} = \boldsymbol{\beta}^T \mathbf{P}$; this is the final matrix needed to complete the solution.

Step 5. Performing the usual matrix multiplications gives

$$\mathbf{k} = \boldsymbol{\beta}^T \mathbf{K} \boldsymbol{\beta} = \frac{2EI}{l} \begin{bmatrix} 0 & 1 & 1 & 0 & 0 & 0 \\ 0 & 0 & 0 & 1 & 1 & 0 \end{bmatrix} \begin{bmatrix} 2 & 1 & & & & \\ 1 & 2 & & & & \\ & & 4 & 2 & & \\ & & 2 & 4 & & \\ & & & & 2 & 1 \\ & & & & 1 & 2 \end{bmatrix} \begin{bmatrix} 0 & 0 \\ 1 & 0 \\ 1 & 0 \\ 0 & 1 \\ 0 & 1 \\ 0 & 0 \end{bmatrix} \tag{8.15o}$$

---

[17] This is analogous to writing equilibrium equations for free-bodies of truss joints (rather than using virtual work to obtain the equilibrium equations). In effect, we identify a free-body on which the only end forces are all of the force quantities that appear on the right side of the equilibrium equation.

[18] In other words, there are no applied external moment loads at B or C on the structure in Figure 8.14 that appear in the free-bodies in Figure 8.16. (M_{BA}, M_{BC}, M_{CB}, and M_{CD} are external moments on the free bodies in Figure 8.16, but are internal forces in the structure.)

$$= \frac{EI}{l}\begin{bmatrix} 12 & 4 \\ 4 & 12 \end{bmatrix}$$

and

$$\mathbf{F}^* = \mathbf{F} - \boldsymbol{\beta}^{\mathrm{T}}\widehat{\mathbf{P}} = \begin{Bmatrix} 0 \\ 0 \end{Bmatrix} - \begin{bmatrix} 0 & 1 & 1 & 0 & 0 & 0 \\ 0 & 0 & 0 & 1 & 1 & 0 \end{bmatrix} \begin{Bmatrix} EI\alpha\Delta T_o/h \\ -EI\alpha\Delta T_o/h \\ -3Fl/16 \\ 3Fl/16 \\ EI\alpha\Delta T_o/h \\ -EI\alpha\Delta T_o/h \end{Bmatrix}$$

$$= \begin{Bmatrix} \dfrac{3Fl}{16} + \dfrac{EI\alpha\Delta T_o}{h} \\ -\dfrac{3Fl}{16} - \dfrac{EI\alpha\Delta T_o}{h} \end{Bmatrix} \tag{8.15p}$$

So, Equation 8.7c is

$$\frac{EI}{l}\begin{bmatrix} 12 & 4 \\ 4 & 12 \end{bmatrix}\begin{Bmatrix} u_1 \\ u_2 \end{Bmatrix} = \begin{Bmatrix} \dfrac{3Fl}{16} + \dfrac{EI\alpha\Delta T_o}{h} \\ -\dfrac{3Fl}{16} - \dfrac{EI\alpha\Delta T_o}{h} \end{Bmatrix} \tag{8.15q}$$

Inverting leads to

$$\begin{Bmatrix} u_1 \\ u_2 \end{Bmatrix} = \begin{Bmatrix} \dfrac{3Fl^2}{128EI} + \dfrac{\alpha l\Delta T_o}{8h} \\ -\dfrac{3Fl^2}{128EI} - \dfrac{\alpha l\Delta T_o}{8h} \end{Bmatrix} \tag{8.15r}$$

Back-substituting provides

$$\boldsymbol{\Delta} \equiv \begin{Bmatrix} \Delta_{AB} \\ \Delta_{BA} \\ \Delta_{BC} \\ \Delta_{CB} \\ \Delta_{CD} \\ \Delta_{DC} \end{Bmatrix} = \begin{Bmatrix} 0 \\ \dfrac{3Fl^2}{128EI} + \dfrac{\alpha l\Delta T_0}{8h} \\ \dfrac{3Fl^2}{128EI} + \dfrac{\alpha l\Delta T_o}{8h} \\ -\dfrac{3Fl^2}{128EI} - \dfrac{\alpha l\Delta T_o}{8h} \\ -\dfrac{3Fl^2}{128EI} - \dfrac{\alpha l\Delta T_o}{8h} \\ 0 \end{Bmatrix} \tag{8.15s}$$

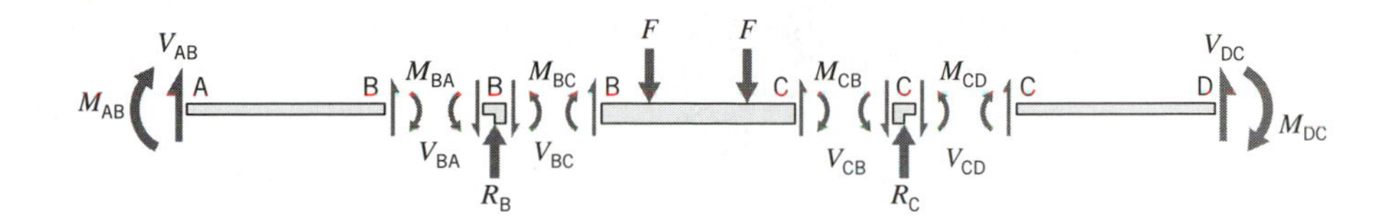

Figure 8.17 Free-body diagrams of structural elements.

and

$$\mathbf{P} \equiv \begin{Bmatrix} M_{AB} \\ M_{BA} \\ M_{BC} \\ M_{CB} \\ M_{CD} \\ M_{DC} \end{Bmatrix} = \begin{Bmatrix} \dfrac{3Fl}{64} + \dfrac{5EI\alpha\Delta T_o}{4h} \\ \dfrac{3Fl}{32} - \dfrac{EI\alpha\Delta T_o}{2h} \\ -\dfrac{3Fl}{32} + \dfrac{EI\alpha\Delta T_o}{2h} \\ \dfrac{3Fl}{32} - \dfrac{EI\alpha\Delta T_o}{2h} \\ -\dfrac{3Fl}{32} + \dfrac{EI\alpha\Delta T_o}{2h} \\ -\dfrac{3Fl}{64} - \dfrac{5EI\alpha\Delta T_o}{4h} \end{Bmatrix} \tag{8.15t}$$

Figure 8.17 shows free-body diagrams of the various beams and joints that comprise the structure. (All forces are shown acting in positive directions in this figure.) Given the values of moments in Equation 8.15*t*, the end shears and reactions R_B and R_C may be computed from equilibrium equations applied to the distinct free-bodies.

In this example, we were able to obtain the equilibrium equations (Equation 8.15*n*) in a straightforward manner from free-bodies that were relatively easy to identify, and it was not necessary to use virtual work. As with trusses, we observe that the components of **F** are the nodal loads (zero in this case) in the directions of the system displacements; however, the situation becomes more complicated when transverse deflections at the ends of individual beams are not zero, i.e., when side-sway angles are present ($\psi_{IJ} \neq 0$). In such circumstances, appropriate free-bodies may be difficult to identify, and virtual work may be the method of choice for obtaining equilibrium equations.

8.3.3 Equilibrium by Virtual Work

To establish a systematic procedure for obtaining equilibrium equations, it is convenient to first develop an expression of virtual work for frames in matrix form. In particular, we seek an expression of virtual work for a frame when we choose virtual displacements that are ''like'' system displacements.

We begin with a single beam of a frame that may be subjected to the general set of

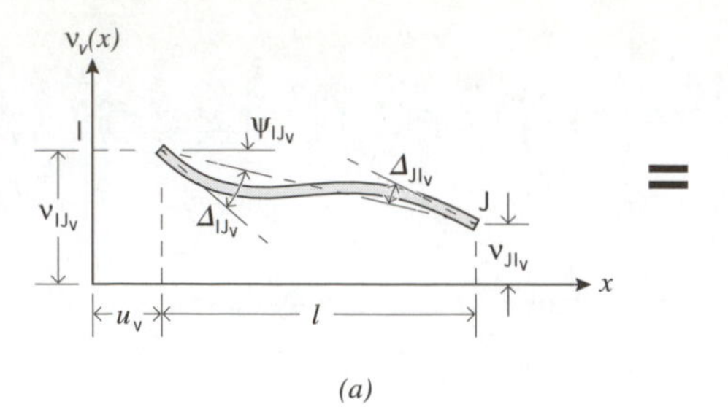

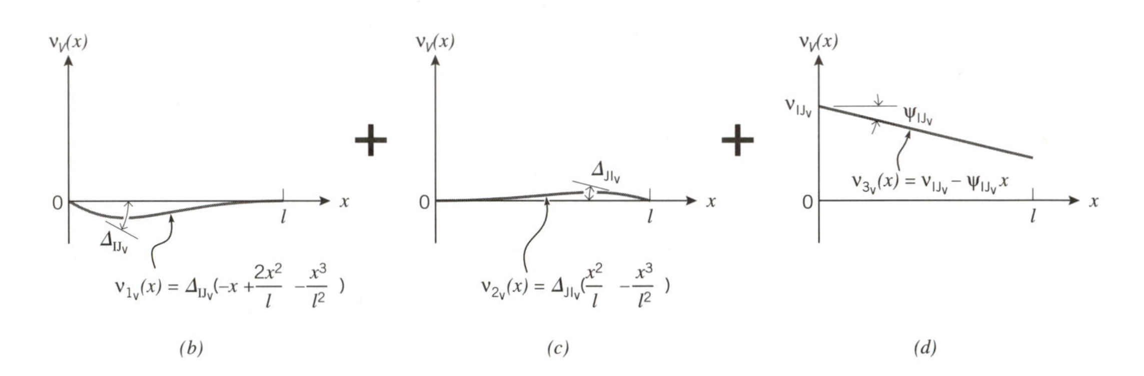

Figure 8.18 Virtual displacements.

real loads and end forces in Figure 5.23*a*. The internal bending moment is given by Equation 5.38*d*,

$$M(x) = M_{IJ}\left(1 - \frac{x}{l}\right) - M_{JI}\left(\frac{x}{l}\right) + M_o(x) \qquad \textbf{(8.16a)}$$

The virtual displacements are chosen like those due to end deformations and side-sway, see Figure 8.18.[19] The general displacement in Figure 8.18*a* may be regarded as the superposition of the three displacement patterns shown in Figures 8.18*b*–8.18*d*: Figure 8.18*b* shows the transverse displacement due to Δ_{IJ_V} only, Figure 8.18*c* shows the transverse displacement due to Δ_{JI_V} only, and Figure 8.18*d* shows the transverse displacement due to ''side-sway'' (ψ_{IJ_V}) only. The displacement functions are determined using the procedure in Section 3.4.1. For example, in Figure 8.18*b* we find $v_{1_V}(x)$ due to an external moment applied at end I for a beam with kinematic boundary conditions $v_{1_V}(0) = v_{1_V}(l) = v'_{1_V}(l) = 0$ and, since $v'_{1_V}(0) \equiv -\Delta_{IJ_V}$, we substitute the relation for the applied moment in terms of Δ_{IJ_V}.[20] (Note that here we are using the subscript V to distinguish these displacements as virtual.)

[19] Figure 8.18 is a representation of the motion of a single beam due to a system displacement (see, for example, beams AB or BC in Figure 8.8*c*).

[20] See, for example, Problem 3.P40 at the end of Chapter 3.

The overall transverse virtual displacement is given by

$$v_V(x) = \left(-x + \frac{2x^2}{l} - \frac{x^3}{l^2}\right)\Delta_{IJ_V} + \left(\frac{x^2}{l} - \frac{x^3}{l^2}\right)\Delta_{JI_V} + v_{IJ_V} - \psi_{IJ_V}x \qquad \textbf{(8.16}\textbf{\textit{b}}\textbf{)}$$

The expression for internal virtual work requires curvature, i.e., the second derivative of Equation 8.16*b*,

$$v_V''(x) = \left(\frac{4}{l} - \frac{6x}{l^2}\right)\Delta_{IJ_V} + \left(\frac{2}{l} - \frac{6x}{l^2}\right)\Delta_{JI_V} \qquad \textbf{(8.16}\textbf{\textit{c}}\textbf{)}$$

Since the beam under consideration is a typical component of a frame structure, we will apply the theorem of virtual work for frames, Equation 4.24. For this beam, the internal virtual work (the integral on the right side of Equation 4.24) is given by

$$W_{I_V} = \int_0^l M(x)v_V''(x)dx = \int_0^l \left[M_{IJ}\left(1 - \frac{x}{l}\right) - M_{JI}\left(\frac{x}{l}\right) + M_o(x)\right]\left[\left(\frac{4}{l} - \frac{6x}{l^2}\right)\Delta_{IJ_V} + \left(\frac{2}{l} - \frac{6x}{l^2}\right)\Delta_{JI_V}\right]dx \qquad \textbf{(8.17}\textbf{\textit{a}}\textbf{)}$$

It may be verified that

$$\int_0^l \left(1 - \frac{x}{l}\right)\left(\frac{4}{l} - \frac{6x}{l^2}\right)dx = \int_0^l \left(-\frac{x}{l}\right)\left(\frac{2}{l} - \frac{6x}{l^2}\right)dx = 1$$
$$\int_0^l \left(1 - \frac{x}{l}\right)\left(\frac{2}{l} - \frac{6x}{l^2}\right)dx = \int_0^l \left(-\frac{x}{l}\right)\left(\frac{4}{l} - \frac{6x}{l^2}\right)dx = 0 \qquad \textbf{(8.17}\textbf{\textit{b}}\textbf{)}$$

Therefore, Equation 8.17*a* reduces to

$$W_{I_V} = \Delta_{IJ_V}M_{IJ} + \Delta_{JI_V}M_{JI} + \Delta_{IJ_V}\int_0^l M_o(x)\left(\frac{4}{l} - \frac{6x}{l^2}\right)dx + \Delta_{JI_V}\int_0^l M_o(x)\left(\frac{2}{l} - \frac{6x}{l^2}\right)dx \qquad \textbf{(8.17}\textbf{\textit{c}}\textbf{)}$$

The integrals in Equation 8.17*c* are the negative of the fixed-end moment expressions in Equations 8.13*f*, with only $M_o(x)$ rather than $\widehat{M}(x)$ appearing in the integrals. Since $M_o(x)$ is one of the components of $\widehat{M}(x)$, it follows from the interpretation of Equations 8.13*f* that each of the integrals in Equation 8.17*c* must be the negative of a fixed-end moment due to applied interior physical loads. In other words, the integrals are the negatives of fixed-end moments due only to the applied physical loads and not to thermal and/or initial effects. Thus, Equation 8.17*c* may be written as

$$W_{I_V} = \Delta_{IJ_V}M_{IJ} + \Delta_{JI_V}M_{JI} - \Delta_{IJ_V}\widehat{M}^o_{IJ} - \Delta_{JI_V}\widehat{M}^o_{JI} \qquad \textbf{(8.17}\textbf{\textit{d}}\textbf{)}$$

where

$$\widehat{M}^o_{\mathrm{IJ}} \equiv -\int_0^l M_o(x)\left(\frac{4}{l} - \frac{6x}{l^2}\right)dx$$
$$\widehat{M}^o_{\mathrm{JI}} \equiv -\int_0^l M_o(x)\left(\frac{2}{l} - \frac{6x}{l^2}\right)dx \qquad \textbf{(8.17e)}$$

denote the fixed-end moments due to the interior physical loads. Equation 8.17*d* has the matrix form

$$W_{I_V} = \begin{Bmatrix} \Delta_{\mathrm{IJ}_V} \\ \Delta_{\mathrm{JI}_V} \end{Bmatrix}^{\mathrm{T}} \left(\begin{Bmatrix} M_{\mathrm{IJ}} \\ M_{\mathrm{JI}} \end{Bmatrix} - \begin{Bmatrix} \widehat{M}^o_{\mathrm{IJ}} \\ \widehat{M}^o_{\mathrm{JI}} \end{Bmatrix} \right) \qquad \textbf{(8.17f)}$$

For purposes of the current development, external virtual work due to the applied loads is expressed by the integral on the left side of Equation 4.24,[21]

$$W_{E_V} = \int_0^l q(x)v_V(x)dx + u_V\int_0^l p(x)dx = \int_0^l q(x)\left[\left(-x + \frac{2x^2}{l} - \frac{x^3}{l^2}\right)\Delta_{\mathrm{IJ}_V}\right.$$
$$\left. + \left(\frac{x^2}{l} - \frac{x^3}{l^2}\right)\Delta_{\mathrm{JI}_V} + v_{\mathrm{IJ}_V} - \psi_{\mathrm{IJ}_V}x\right]dx + u_V\int_0^l p(x)dx \qquad \textbf{(8.17g)}$$

or

$$W_{E_V} = \Delta_{\mathrm{IJ}_V}\int_0^l q(x)\left(-x + \frac{2x^2}{l} - \frac{x^3}{l^2}\right)dx + \Delta_{\mathrm{JI}_V}\int_0^l q(x)\left(\frac{x^2}{l} - \frac{x^3}{l^2}\right)dx$$
$$+ \int_0^l q(x)(v_{\mathrm{IJ}_V} - \psi_{\mathrm{IJ}_V}x)dx + u_V\int_0^l p(x)dx \qquad \textbf{(8.17h)}$$

In equations 8.17*g* and 8.17*h*, note that in addition to the external virtual work due to virtual lateral motion, in this case $v_V(x)$, we also have to include the work done by axial loads going through a virtual axial displacement u_V, which is constant because the beam is axially rigid.

The first integral in Equation 8.17*h* is another expression for the negative of fixed-end moment $\widehat{M}^o_{\mathrm{IJ}}$ due to a general interior load $q(x)$, and the second integral is another expression for the negative of fixed-end moment $\widehat{M}^o_{\mathrm{JI}}$. There are a number of ways to demonstrate this fact. One way is to use the result for the fixed-end moments due to a single concentrated transverse load applied at an arbitrary location (Figure 8.19*a*). The end moments $\widehat{M}^o_{\mathrm{IJ}}$ and $\widehat{M}^o_{\mathrm{JI}}$ acting on the beam in Figure 8.19*a* are determined from the

[21] For the sake of convenience in the development, we will let $q(x)$ represent a general set of lateral physical loads on the beam, and we will omit virtual work terms involving loads F_j and M_i. (In fact, F_j and M_i may be regarded as limiting cases of continuous force distributions.) The conclusions that will be drawn about the nature of the external virtual work terms will be general, and will encompass concentrated loads F_j and M_i.

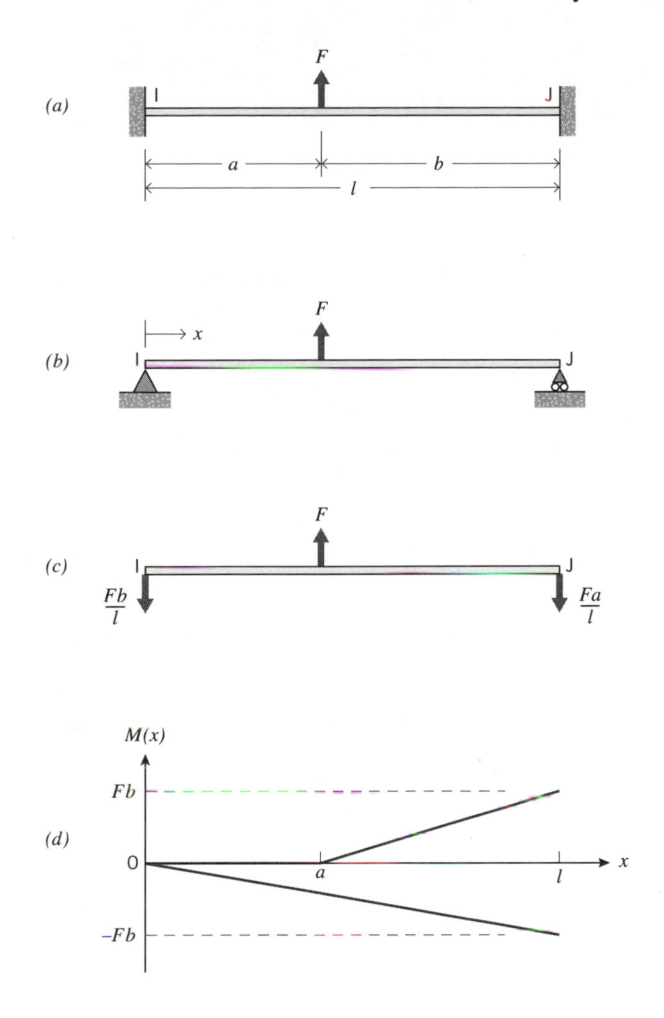

Figure 8.19 Beam with concentrated load.

bending moments in the unrestrained beam in Figures 8.19*b*–8.19*d*; the latter are given by

$$M_o(x) = \frac{Fb}{l}x, \quad 0 < x < a$$

$$M_o(x) = \frac{Fb}{l}x - F(x - a), \quad a < x < l \tag{8.17i}$$

Substituting Equations 8.17*i* into Equations 8.13*e* leads to

$$\widehat{M}^o_{\mathrm{IJ}} = \frac{Fab^2}{l^2}$$

$$\widehat{M}^o_{\mathrm{JI}} = -\frac{Fa^2b}{l^2} \tag{8.17j}$$

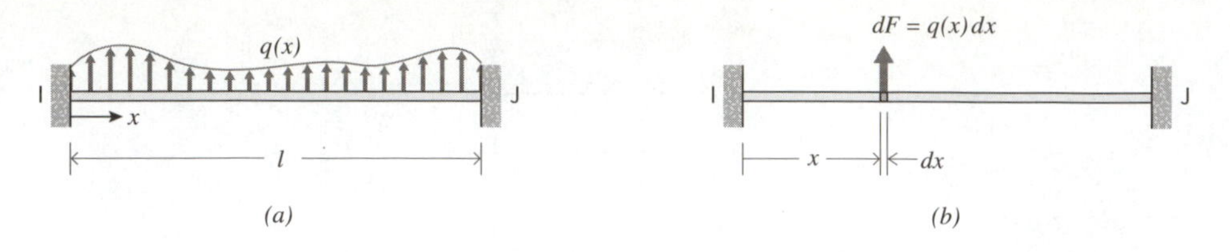

Figure 8.20 Beam with continuous load.

A general continuous force $q(x)$ (Figure 8.20*a*) is the sum of an infinite number of incremental concentrated forces of the type shown in Figure 8.20*b*. The increment of fixed-end moment due to the load in Figure 8.20*b* is obtained from Equation 8.17*j* by substituting dF for F, x for a, and $l - x$ for b,

$$d\widehat{M}^o_{IJ} = \frac{x(l-x)^2}{l^2} dF = \frac{q(x)x(l-x)^2}{l^2} dx$$
$$d\widehat{M}^o_{JI} = -\frac{x^2(l-x)}{l^2} dF = -\frac{q(x)x^2(l-x)}{l^2} dx \qquad \textbf{(8.17}k\textbf{)}$$

It follows that the fixed-end moments in Figure 8.20*a* must be given by

$$\widehat{M}^o_{IJ} = \int_0^l \frac{q(x)x(l-x)^2}{l^2} dx = -\int_0^l q(x)\left(-x + \frac{2x^2}{l} - \frac{x^3}{l^2}\right)dx$$
$$\widehat{M}^o_{JI} = -\int_0^l \frac{q(x)x^2(l-x)}{l^2} dx = -\int_0^l q(x)\left(\frac{x^2}{l} - \frac{x^3}{l^2}\right)dx \qquad \textbf{(8.17}l\textbf{)}$$

The integrals in Equation 8.17*h* are thus shown to be equivalent to fixed-end moments.

The last two integrals in Equation 8.17*h* are the virtual work done by the real physical (interior) loads on the beam going through the virtual displacements *associated only with rigid-body motion*, e.g., the motion in Figure 8.18*d*. We will use $W^o_{E_V}$ to denote this ''rigid-body component'' of virtual work due to physical loads,

$$W^o_{E_V} = \int_0^l q(x)[v_{IJ_V}(x) - \psi_{IJ_V}x]dx + u_V \int_0^l p(x)dx \qquad \textbf{(8.17}m\textbf{)}$$

The side-sway and axial components of virtual work, which had no counterpart in truss analysis, are the major complicating factor in the current formulation. The significance of this virtual work will be demonstrated in subsequent examples.

Equation 8.17*h* may now be rewritten in the form

$$W_{E_V} = -\begin{Bmatrix} \Delta_{IJ_V} \\ \Delta_{JI_V} \end{Bmatrix}^T \begin{Bmatrix} \widehat{M}^o_{IJ} \\ \widehat{M}^o_{JI} \end{Bmatrix} + W^o_{E_V} \qquad \textbf{(8.17}n\textbf{)}$$

We observe that the matrix term above is identical to the second term in Equation 8.17*f*; therefore, we may anticipate that these terms will cancel when external and internal virtual work are equated.

For a complete frame structure, the internal virtual work is obtained by summing the expression in Equation 8.17f over all the beams; the external virtual work is obtained by summing the expression in Equation 8.17n over all the beams and adding the external virtual work terms associated with concentrated loads (F_j and M_i) applied at the joints.

The sum of internal virtual work (Equation 8.17f) over all beams is

$$\sum_{n=1}^{NB}(W_{I_V})_n = \sum_{n=1}^{NB}\begin{Bmatrix}\Delta_{\mathrm{IJ}_V}\\ \Delta_{\mathrm{JI}_V}\end{Bmatrix}_n^{\mathrm{T}}\left(\begin{Bmatrix}M_{\mathrm{IJ}}\\ M_{\mathrm{JI}}\end{Bmatrix}_n - \begin{Bmatrix}\widehat{M}^o_{\mathrm{IJ}}\\ \widehat{M}^o_{\mathrm{JI}}\end{Bmatrix}_n\right)$$

$$= \sum_{n=1}^{NB}\begin{Bmatrix}\Delta_{\mathrm{IJ}_V}\\ \Delta_{\mathrm{JI}_V}\end{Bmatrix}_n^{\mathrm{T}}\begin{Bmatrix}M_{\mathrm{IJ}}\\ M_{\mathrm{JI}}\end{Bmatrix}_n - \sum_{n=1}^{NB}\begin{Bmatrix}\Delta_{\mathrm{IJ}_V}\\ \Delta_{\mathrm{JI}_V}\end{Bmatrix}_n^{\mathrm{T}}\begin{Bmatrix}\widehat{M}^o_{\mathrm{IJ}}\\ \widehat{M}^o_{\mathrm{JI}}\end{Bmatrix}_n \qquad \textbf{(8.18}a\textbf{)}$$

$$= \mathbf{D}_V^{\mathrm{T}}\mathbf{P} - \mathbf{D}_V^{\mathrm{T}}\widehat{\mathbf{P}}^{\mathbf{o}}$$

where $\boldsymbol{\Delta}_V$ is the vector of all element virtual end deformations (similar to Equation 8.15a, for example), $\mathbf{P}$ is the usual vector of all element end moments, $\widehat{\mathbf{P}}^{\mathbf{o}}$ is a vector of all fixed-end moments due only to physical loads, and NB is the number of beam elements.

As has been shown previously for trusses, the external virtual work at the joints must have a form like $\sum_{k=1}^{NDOF} u_{k_V}F_k$, where $NDOF$ is the number of independent displacements, u_{k_V} is a virtual displacement identical to the kth system displacement, and F_k is the corresponding k^{th} generalized force, which is a function of the concentrated forces or moments applied at the joints.[22] As will be demonstrated in the examples that follow, the sum of the side-sway components of virtual work of forces applied on the interior of the beams (Equation 8.17m) may be expressed in a similar form,

$$\sum_{n=1}^{NB}(W^o_{E_V})_n \equiv \sum_{k=1}^{NDOF} u_{k_V}F^o_k \qquad \textbf{(8.18}b\textbf{)}$$

where F^o_k is also a generalized force but is a function of the physical loads applied to the interiors of the various beam elements.[23]

Therefore, the sum of all external virtual work over the entire structure is

$$\sum_{n=1}^{NB}(W_{E_V})_n + u_{k_V}F_k = -\sum_{n=1}^{NB}\begin{Bmatrix}\Delta_{\mathrm{IJ}_V}\\ \Delta_{\mathrm{JI}_V}\end{Bmatrix}_n^{\mathrm{T}}\begin{Bmatrix}\widehat{M}^o_{\mathrm{IJ}}\\ \widehat{M}^o_{\mathrm{JI}}\end{Bmatrix}_n + \sum_{n=1}^{NB}(W_{E_V})_n + \sum_{k=1}^{NDOF} u_{k_V}F_k$$

$$= -\boldsymbol{\Delta}_V^{\mathrm{T}}\widehat{\mathbf{P}}^{\mathbf{o}} + \sum_{k=1}^{NDOF} u_{k_V}F^o_k + \sum_{k=1}^{NDOF} u_{k_V}F_k \qquad \textbf{(8.18}c\textbf{)}$$

$$= -\boldsymbol{\Delta}_V^{\mathrm{T}}\widehat{\mathbf{P}}^{\mathbf{o}} + \sum_{k=1}^{NDOF} u_{k_V}(F^o_k + F_k)$$

---

[22] For example, F_k may involve more than a single load if joint displacements are coupled (see Example 2, Step 4).

[23] Consequently, F^o_k may be a function of continuous loads as well as concentrated loads.

Equating total external and internal virtual work (Equations 8.18*c* and 8.18*a*, respectively) leads to

$$-\mathbf{\Delta}_V^{\mathrm{T}}\widehat{\mathbf{P}}^{\mathbf{o}} + \sum_{k=1}^{NDOF} u_{k_V}(F_k^o + F_k) = \mathbf{\Delta}_V^{\mathrm{T}}\mathbf{P} - \mathbf{\Delta}_V^{\mathrm{T}}\mathbf{P}^{\mathbf{o}} \qquad \textbf{(8.18}\textbf{\textit{d}}\textbf{)}$$

or

$$\sum_{k=1}^{NDOF} u_{k_V}(F_k^o + F_k) = \mathbf{\Delta}_V^{\mathrm{T}}\mathbf{P} \qquad \textbf{(8.18}\textbf{\textit{e}}\textbf{)}$$

Finally, since the virtual deformations, $\mathbf{\Delta}_V$, are obtained from virtual displacements in exactly the same manner as real deformations are related to real system displacements, we can write as a special case when only u_{k_V} is nonzero.

$$\mathbf{\Delta}_V = \boldsymbol{\beta}_k u_{k_V} \qquad \textbf{(8.18}\textbf{\textit{f}}\textbf{)}$$

where vector $\boldsymbol{\beta}_k$ is the kth column of the compatibility matrix (Equation 8.2). Substituting Equation 8.18*f* into Equation 8.18*e* gives

$$u_{k_V}(F_k^o + F_k) = u_{k_V}\boldsymbol{\beta}_k^{\mathrm{T}}\mathbf{P} \qquad \textbf{(8.18}\textbf{\textit{g}}\textbf{)}$$

or

$$(F_k^o + F_k) = \boldsymbol{\beta}_k^{\mathrm{T}}\mathbf{P} \qquad \textbf{(8.18}\textbf{\textit{h}}\textbf{)}$$

which is the desired equilibrium equation. The remaining equilibrium equations are obtained in a similar manner, i.e., from virtual work based on virtual displacement patterns like those due to each of the other system displacements. The combined equilibrium equations can be written as[24]

$$(\mathbf{F}^{\mathbf{o}} + \mathbf{F}) = \boldsymbol{\beta}^{\mathrm{T}}\mathbf{P} \qquad \textbf{(8.18}\textbf{\textit{i}}\textbf{)}$$

where $\mathbf{F}^{\mathbf{o}}$ and $\mathbf{F}$ are force vectors and $\boldsymbol{\beta}$ is the compatibility matrix; this is the same form as Equation 8.6. In other words, we conclude that the equilibrium equations generated in Step 4 of the analysis procedure must have the form previously defined in Equation 8.6 (and used in Equation 8.15*m*); however, it must be remembered that the composite force vector $(\mathbf{F}^{\mathbf{o}} + \mathbf{F})$ may now include terms in $\mathbf{F}^{\mathbf{o}}$ arising from interior applied loads when rigid-body motion is present.

Substituting $\mathbf{F}^{\mathbf{o}} + \mathbf{F}$ for $\mathbf{F}$ in Equation 8.7*a* gives the final form of the equilibrium equations for the frame structure,

$$\boxed{(\boldsymbol{\beta}^{\mathrm{T}}\mathbf{K}\boldsymbol{\beta})\mathbf{u} = (\mathbf{F}^{\mathbf{o}} + \mathbf{F}) - \boldsymbol{\beta}^{\mathrm{T}}\widehat{\mathbf{P}}} \qquad \textbf{(8.19}\textbf{\textit{a}}\textbf{)}$$

[24] Rather than developing the equations of equilibrium one at a time, we can let $\mathbf{u}_V$ be a vector of arbitrary virtual displacements like the system displacements, and $\mathbf{F}^{\mathbf{o}}$ and $\mathbf{F}$ be force vectors in Equations 8.18*b* and 8.18*c*. Then the matrix form, Equation 8.18*i*, would be obtained directly. The current development more closely reflects the analysis procedure that will be followed in subsequent examples.

or

$$\mathbf{ku} = \mathbf{F}^* \qquad \textbf{(8.19}\textbf{\textit{b}}\textbf{)}$$

where, in this case, $\mathbf{F}^* = (\mathbf{F}^\mathbf{o} + \mathbf{F}) - \boldsymbol{\beta}^T\widehat{\mathbf{P}}$. It is essential to understand the virtual work performed by all the loads acting on the elements to properly assemble the composite force vector $(\mathbf{F}^\mathbf{o} + \mathbf{F})$. It is also important to recognize that interior physical loads may appear in both the $\mathbf{F}^\mathbf{o}$ and the $\widehat{\mathbf{P}}$ vectors. On the other hand, the effects of thermal or initial loads appear only in the $\widehat{\mathbf{P}}$ vector. We will use the remainder of this section to illustrate the calculations of the $\mathbf{F}^\mathbf{o}$ and $\mathbf{F}$ vectors necessary to conduct an analysis of frame structures.

EXAMPLE 3

Consider the structure in Figure 8.21, which is one half of the structure in Figure 8.14. Point G in Figure 8.21 is on the axis of symmetry of the structure in Figure 8.14 and, because the loads are symmetric, the appropriate support at G is a vertical roller clamped against rotation, as shown. Beam BG now has a central force F and beam AB has the same thermal load ΔT as the structure in Figure 8.14.

Step 1. The two system displacements for this structure were identified in Figure 8.7, so we proceed to the consideration of compatibility.

Step 2. We must examine each of the system displacements in detail, and superpose each of the displacements to define overall system behavior.

For system displacement u_1 (Figure 8.7*b*) there are no transverse displacements at the ends of either beam element, i.e., $v_{AB} = v_{BA} = v_{BG} = v_{GB} = 0$. It follows from Equations 5.39 that $\psi_{AB} = \psi_{BG} = 0$ and element end deformations are equal to end rotations. We have $\theta_{AB} = 0$, $\theta_{BA} = u_1$, $\theta_{BG} = u_1$, and $\theta_{GB} = 0$; therefore, $\Delta_{AB} = 0$, $\Delta_{BA} = u_1$, $\Delta_{BG} = u_1$, and $\Delta_{GB} = 0$.

From Figure 8.7*c* we observe that the transverse end displacements are related to system displacement u_2 by $v_{AB} = v_{BA} = v_{BG} = 0$ and $v_{GB} = +u_2$;[25] therefore, from Equation 8.13*i*, we have $\psi_{AB} = 0$ and

[25] It is very important to adhere to the sign convention for end displacements. It must be remembered that transverse end displacements (such as v_{BG} and v_{GB}) are defined as positive when acting in the direction shown in Figure 8.18; thus, if u_2 had been defined as positive in the opposite direction (i.e., downward) then we would have $v_{GB} = -u_2$.

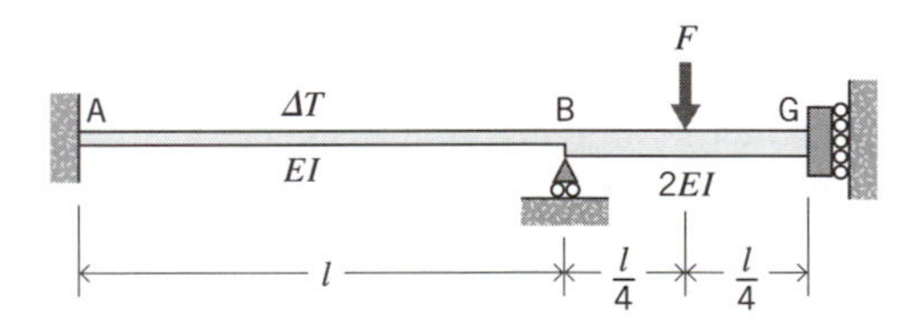

Figure 8.21 Example 3.

$$\psi_{BG} = \frac{v_{BG} - v_{GB}}{l_{BG}} = \frac{0 - u_2}{l/2} = -\frac{2}{l}u_2 \tag{8.20a}$$

The negative sign for ψ_{BG} indicates a counterclockwise rotation of the beam axis, which is consistent with the motion depicted in Figure 8.7*c*.

Since $\theta_{AB} = \theta_{BA} = \theta_{BG} = \theta_{GB} = 0$, it follows that $\Delta_{AB} = \Delta_{BA} = 0$, and

$$\begin{aligned} \Delta_{BG} &= \theta_{BG} - \psi_{BG} = 0 - \left(-\frac{2}{l}u_2\right) = \frac{2}{l}u_2 \\ \Delta_{GB} &= \theta_{GB} - \psi_{BG} = 0 - \left(-\frac{2}{l}u_2\right) = \frac{2}{l}u_2 \end{aligned} \tag{8.20b}$$

Thus, taking into account both u_1 and u_2, the compatibility equation becomes

$$\boldsymbol{\Delta} \equiv \begin{Bmatrix} \Delta_{AB} \\ \Delta_{BA} \\ \Delta_{BG} \\ \Delta_{GB} \end{Bmatrix} = \begin{Bmatrix} \theta_{AB} - \psi_{AB} \\ \theta_{BA} - \psi_{AB} \\ \theta_{BG} - \psi_{BG} \\ \theta_{GB} - \psi_{BG} \end{Bmatrix} = \begin{bmatrix} 0 & 0 \\ 1 & 0 \\ 1 & 2/l \\ 0 & 2/l \end{bmatrix} \begin{Bmatrix} u_1 \\ u_2 \end{Bmatrix} \tag{8.20c}$$

which defines the $\boldsymbol{\beta}$ matrix.

Step 3. The fixed-end moments on beam AB due to ΔT were computed previously in Equations 8.15*c* and 8.15*e*. The fixed-end moments on beam BG due to the central force F are obtained from Equations 8.17*j* by substituting $l/2$ for l and $a = b = l/4$ (since the length of beam BG is $l/2$). In this case, the result is

$$\widehat{M}_{BG} = -\widehat{M}_{GB} = -\frac{Fl}{16} \tag{8.20d}$$

Therefore, the matrix force-deformation equation is

$$\begin{Bmatrix} M_{AB} \\ M_{BA} \\ M_{BG} \\ M_{GB} \end{Bmatrix} = \frac{2EI}{l} \begin{bmatrix} 2 & 1 & & \\ 1 & 2 & & \\ & & 8 & 4 \\ & & 4 & 8 \end{bmatrix} \begin{Bmatrix} \Delta_{AB} \\ \Delta_{BA} \\ \Delta_{BG} \\ \Delta_{GB} \end{Bmatrix} + \begin{Bmatrix} EI\alpha\Delta T_o/h \\ -EI\alpha\Delta T_o/h \\ -Fl/16 \\ Fl/16 \end{Bmatrix} \tag{8.20e}$$

Note that the third and fourth rows (the moment-deformation equations for beam BG) reflect an element length $l/2$ and bending stiffness $2EI$. Equation 8.20*e* defines the **K** matrix and $\widehat{\mathbf{P}}$ vector.

Step 4. We now turn our attention to the determination of the composite force vector $\mathbf{F}^o + \mathbf{F}$, which is required in Equation 8.19*a*. This force vector is part of the equilibrium equation derived in Equation 8.18*i*. For illustrative purposes, we will demonstrate several alternate approaches to the determination of the components of this vector.

First, we may employ the same type of *ad hoc* procedure discussed in the preceding illustration, Equation 8.15*m*. We write the matrix equilibrium equation, Equation 8.18*i*,

$$\begin{Bmatrix} f_1 \\ f_2 \end{Bmatrix} = \begin{bmatrix} 0 & 1 & 1 & 0 \\ 0 & 0 & 2/l & 2/l \end{bmatrix} \begin{Bmatrix} M_{AB} \\ M_{BA} \\ M_{BG} \\ M_{GB} \end{Bmatrix} \quad \textbf{(8.20}\boldsymbol{f}\textbf{)}$$

where f_1 and f_2 are the ''unknown'' components of the composite force vector $\mathbf{F}^o + \mathbf{F}$. Thus, the first of the two equilibrium equations has the form $f_1 = M_{BA} + M_{BG}$ and the second has the form $f_2 = 2M_{BG}/l + 2M_{GB}/l$. As we saw previously, the first equation must be the equation of moment equilibrium for joint B and, because there are no external moments applied at B, we must have $f_1 = 0$. The second equation is the equation of moment equilibrium for beam BG, and is obtained from the free-body in Figure 8.22,

$$\Sigma M_B = 0 = M_{BG} + M_{GB} + F\frac{l}{4} \quad \textbf{(8.20}\boldsymbol{g}\textbf{)}$$

Note that V_{GB} must be zero because of the roller support at G, so $V_{BG} = F$ as shown. Equation 8.20*g* can be rewritten in a form that matches the second equilibrium equation,

$$-\frac{F}{2} = \frac{2}{l}M_{BG} + \frac{2}{l}M_{GB} \quad \textbf{(8.20}\boldsymbol{h}\textbf{)}$$

Consequently, it follows that $f_2 = -F/2$. We conclude that

$$\mathbf{F}^o + \mathbf{F} = \begin{Bmatrix} 0 \\ -F/2 \end{Bmatrix} \quad \textbf{(8.20}\boldsymbol{i}\textbf{)}$$

The *ad hoc* procedure, although applicable here, can become much more abstruse and difficult to conduct as the complexity of the structure increases (see Example 4, which follows, for instance). Therefore, it is important to be able to determine the components of $\mathbf{F}^o + \mathbf{F}$ from a more systematic approach and, in particular, by considering external virtual work done by the applied loads. We will now demonstrate the procedure.

Recall that each component of $\mathbf{F}$ is an external virtual work done by joint loads moving through virtual displacements like the motions associated with one of the system displacements. Specifically, we obtain the components of $\mathbf{F}$ from virtual displacements

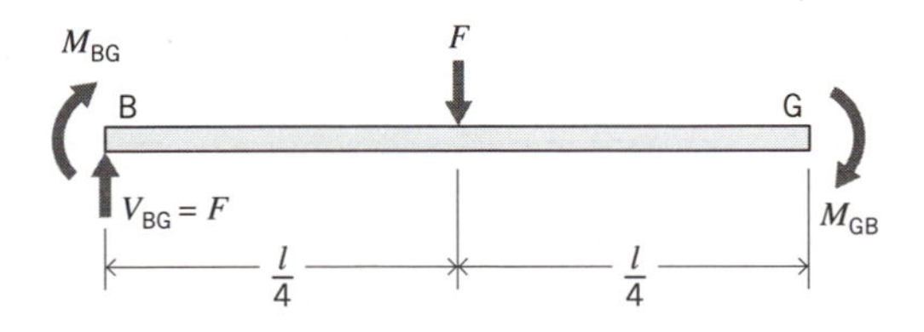

Figure 8.22 Free-body diagram of beam BG.

that are like unit values of system displacements. For example, let the motions in Figure 8.7*b* be regarded as virtual displacements like those associated with $u_1 = 1$. There are no joint loads in Figure 8.21 that cause any virtual work for these motions, so the first component of **F** must equal zero. Similarly, for virtual displacements like system displacement $u_2 = 1$ (Figure 8.7*c*), there are again no real applied forces at the joints in Figure 8.21 that cause any virtual work, so the second component of **F** must also equal zero. Thus,

$$\mathbf{F} = \begin{Bmatrix} 0 \\ 0 \end{Bmatrix} \tag{8.20j}$$

Each component of $\mathbf{F}^{\mathbf{o}}$ is an external virtual work done by interior forces on the beams moving through virtual displacements like the rigid-body motions of the elements produced by the side-sway due to one of the system displacements. Once again, we require unit values of the system displacements. For example, Figure 8.23 shows the side-sway component of a virtual displacement $u_{2_V} = 1$, which is obtained from motions like those in Figure 8.7*c*. Again, this side-sway component of motion is the rigid-body portion of the motion in Figure 8.7*c*. In this case, the load F on the interior of beam BG in Figure 8.21 undergoes a virtual displacement with a magnitude $l/2$. Therefore, the external virtual work is $-F/2$ (negative because the displacement is in a direction opposite to the force). Since this virtual work is associated with u_{2_V}, this defines the second component of $\mathbf{F}^{\mathbf{o}}$.

The first component of $\mathbf{F}^{\mathbf{o}}$ is obtained from the side-sway component of a virtual displacement $u_{1_V} = 1$, which is nonexistent in this case because the motions associated with u_{1_V} do not involve side-sway. Therefore, the first component of $\mathbf{F}^{\mathbf{o}}$ is zero, so we have

$$\mathbf{F}^{\mathbf{o}} = \begin{Bmatrix} 0 \\ -F/2 \end{Bmatrix} \tag{8.20k}$$

and

$$\mathbf{F}^{\mathbf{o}} + \mathbf{F} = \begin{Bmatrix} 0 \\ -F/2 \end{Bmatrix} + \begin{Bmatrix} 0 \\ 0 \end{Bmatrix} = \begin{Bmatrix} 0 \\ -F/2 \end{Bmatrix} \tag{8.20l}$$

which is the same result presented in Equation 8.20*i*.

The virtual work approach to obtaining $\mathbf{F}^{\mathbf{o}} + \mathbf{F}$ is much more straightforward than the *ad hoc* approach because the former deals only with external virtual work, while

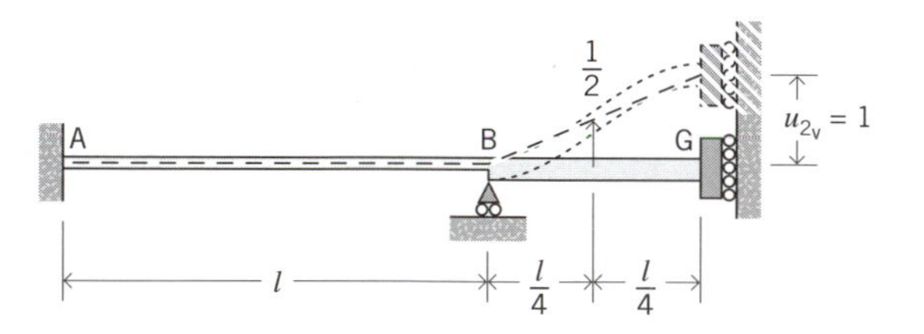

Figure 8.23 ''Side-way component'' of virtual displacement $u_{2_V} = 1$.

the latter requires that we deal with the entire equilibrium equation. In fact, if it is so desired, virtual work can be used to obtain the complete equilibrium equations in a much more direct fashion than is possible with the *ad hoc* approach. This direct procedure, which circumvents the prerequisite of identifying an appropriate free-body, has been introduced previously in Example 17, Chapter 4. (In fact, Equation 4.29*b* is identical to Equation 8.20*g*; rewriting Equation 4.29 in the form of Equation 8.20*h* provides the desired component of $\mathbf{F}^o + \mathbf{F}$.) In essence, the procedure requires that we introduce a virtual concentrated curvature at the location of each end moment that appears in an equilibrium equation; a direct application of virtual work will lead to the complete equilibrium equation. Since the procedure has been discussed in Chapter 4, it will not be repeated here; however, it should be reviewed, and may be used as the basis for an independent check of computations. (The procedure will be demonstrated again in Example 4, which follows.)

Step 5. Now that we have all required matrices, we substitute into Equation 8.19*a*. The stiffness matrix is given by

$$\mathbf{k} = \boldsymbol{\beta}^T\mathbf{K}\boldsymbol{\beta} = \frac{EI}{l}\begin{bmatrix} 0 & 1 & 1 & 0 \\ 0 & 0 & 2/l & 2/l \end{bmatrix}\begin{bmatrix} 4 & 2 & & \\ 2 & 4 & & \\ & & 16 & 8 \\ & & 8 & 16 \end{bmatrix}\begin{bmatrix} 0 & 0 \\ 1 & 0 \\ 1 & 2/l \\ 0 & 2/l \end{bmatrix} = \frac{EI}{l}\begin{bmatrix} 20 & 48/l \\ 48/l & 192/l^2 \end{bmatrix} \tag{8.21a}$$

and $\mathbf{F}^*$ is

$$\mathbf{F}^* = (\mathbf{F}^o + \mathbf{F}) - \boldsymbol{\beta}^T\hat{\mathbf{P}} = \begin{Bmatrix} 0 \\ -F/2 \end{Bmatrix} - \begin{bmatrix} 0 & 1 & 1 & 0 \\ 0 & 0 & 2/l & 2/l \end{bmatrix}\begin{Bmatrix} EI\alpha\Delta T_o/h \\ -EI\alpha\Delta T_o/h \\ -Fl/16 \\ Fl/16 \end{Bmatrix}$$
$$= \begin{Bmatrix} Fl/16 + EI\alpha\Delta T_o/h \\ -F/2 \end{Bmatrix} \tag{8.21b}$$

Solving for $\mathbf{u}$ leads to

$$\mathbf{u} \equiv \begin{Bmatrix} u_1 \\ u_2 \end{Bmatrix} = \mathbf{k}^{-1}\mathbf{F}^* = \frac{l}{EI}\begin{bmatrix} \dfrac{1}{8} & -\dfrac{l}{32} \\ -\dfrac{l}{32} & \dfrac{5l^2}{384} \end{bmatrix}\begin{Bmatrix} \dfrac{Fl}{16} + \dfrac{EI\alpha\Delta T_o}{h} \\ -\dfrac{F}{2} \end{Bmatrix} = \begin{Bmatrix} \dfrac{3Fl^2}{128EI} + \dfrac{\alpha l\Delta T_o}{8h} \\ -\dfrac{13Fl^3}{1536EI} - \dfrac{\alpha l^2\Delta T_o}{32h} \end{Bmatrix} \tag{8.21c}$$

The value of u_1 (the rotation at B) is identical to that in Equation 8.15*r*. Here, however, u_2 has a different meaning than in the previous case, i.e., u_2 is now a transverse displacement at G (Figure 8.21) rather than a rotation at C (Figure 8.14; hence, the difference in dimensionality).

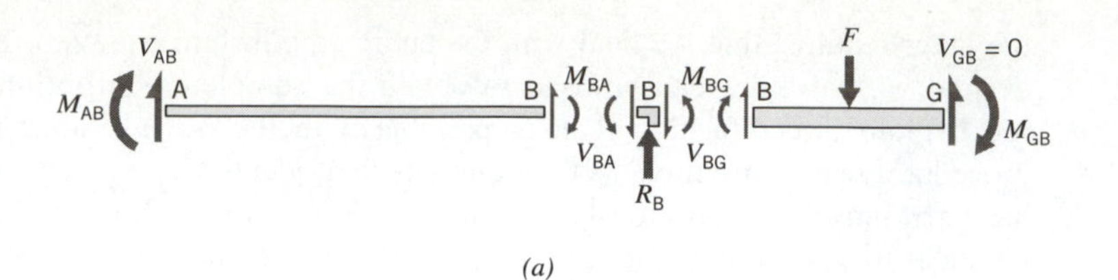

(a)

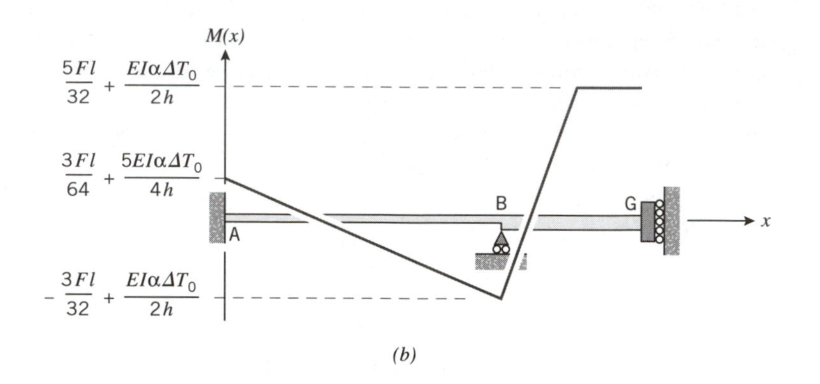

(b)

Figure 8.24 Internal forces, Example 3.

Back substitution results in

$$\boldsymbol{\Delta} \equiv \begin{Bmatrix} \Delta_{AB} \\ \Delta_{BA} \\ \Delta_{BG} \\ \Delta_{GB} \end{Bmatrix} = \boldsymbol{\beta}\mathbf{u} = \begin{Bmatrix} 0 \\ \dfrac{3Fl^2}{128EI} + \dfrac{\alpha l\Delta T_o}{8h} \\ \dfrac{5Fl^2}{768EI} + \dfrac{\alpha l\Delta T_o}{16h} \\ -\dfrac{13Fl^2}{768EI} - \dfrac{\alpha l\Delta T_o}{16h} \end{Bmatrix} \quad \textbf{(8.21}\textbf{\textit{d}}\textbf{)}$$

and

$$\mathbf{P} \equiv \begin{Bmatrix} M_{AB} \\ M_{BA} \\ M_{BG} \\ M_{GB} \end{Bmatrix} = \mathbf{K}\boldsymbol{\Delta} + \widehat{\mathbf{P}} = \begin{Bmatrix} \dfrac{3Fl}{64} + \dfrac{5EI\alpha\Delta T_o}{4h} \\ \dfrac{3Fl}{32} - \dfrac{EI\alpha\Delta T_o}{2h} \\ -\dfrac{3Fl}{32} + \dfrac{EI\alpha\Delta T_o}{2h} \\ -\dfrac{5Fl}{32} - \dfrac{EI\alpha\Delta T_o}{2h} \end{Bmatrix} \quad \textbf{(8.21}\textbf{\textit{e}}\textbf{)}$$

The first three components of the **P** vector have the same meaning as the first three components of the **P** vector in Equation 8.15t and are seen to be identical. The last component, M_{GB}, is different from the fourth component of Equation 8.15*t* because M_{GB} is equivalent to the bending moment at the midspan of beam BC in Figure 8.14 (as may be verified from the previous solution). In other words, the two solutions are identical. Figure 8.24*a* shows the free-body diagrams of the various elements of this structure, and Figure 8.24*b* shows the moment diagrams for beams AB and BG.

EXAMPLE 4

Consider the structure in Figure 8.25, which was analyzed in Chapter 5 (Figure 5.22) using the force method. Beam AB is subjected to a thermal load $\Delta T = \Delta T_o y/h$, a uniform continuous load q_o is applied on beam BC, and a concentrated force F is applied at joint B. We will reanalyze this structure using the displacement method. The beam elements will be assumed to be axially rigid.

Step 1. The two system displacements have been defined in Figure 8.8.

Step 2. For system displacement u_1, which involves no side-sway (Figure 8.8*b*), we have $\Delta_{AB} = \theta_{AB} = 0$, $\Delta_{BA} = \theta_{BA} = u_1$, $\Delta_{BC} = \theta_{BC} = u_1$, $\Delta_{CB} = \theta_{CB} = 0$. This defines the first column of the $\boldsymbol{\beta}$ matrix.

For system displacement u_2, both beams exhibit side-sway (Figure 8.8*c*). Figure 8.26*a* is a detailed sketch of the motion associated with u_2; this sketch is used to determine the transverse components of displacements at the B end of beams AB and BC as functions of u_2.

Since u_2 is defined as a horizontal displacement at B (and C), this system displacement must result in joint B moving to B′, which is the intersection of lines B″B′ (perpendicular to BC) and BB′ (perpendicular to AB), as discussed in Section 8.1. For the particular geometry of this structure, BB″B′ is an isosceles right triangle for which $BB'' = B''B' = u_2$ and $BB' = \sqrt{2}u_2$. Thus, we have $v_{AB} = 0$, $v_{BA} = -\sqrt{2}u_2$, $v_{BC} = -u_2$, and $v_{CB} = 0$. Note the negative signs, which are a consequence of the sign convention for transverse end displacements. (See footnote 25 in the preceding example.) It follows that the side-sways in Figure 8.26 are

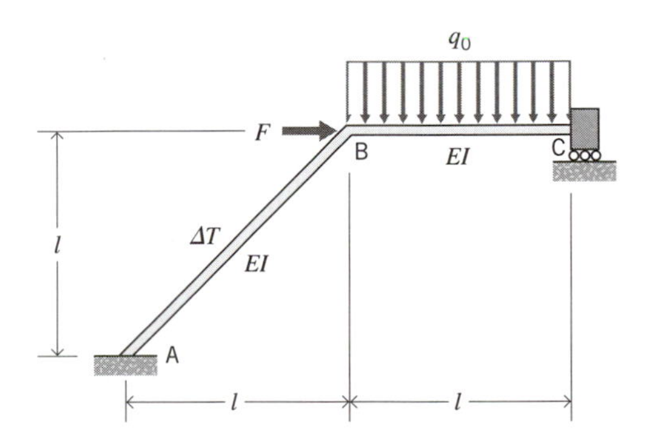

Figure 8.25 Example 4.

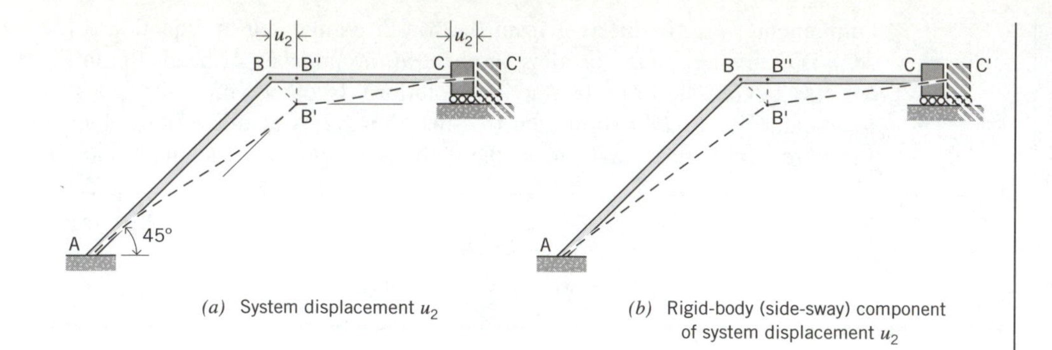

(*a*) System displacement u_2

(*b*) Rigid-body (side-sway) component of system displacement u_2

Figure 8.26 Frame motions.

$$\psi_{AB} = \frac{v_{AB} - v_{BA}}{l_{AB}} = \frac{0 - (-\sqrt{2}u_2)}{\sqrt{2}l} = \frac{u_2}{l}$$

$$\psi_{BC} = \frac{v_{BC} - v_{CB}}{l_{BC}} = \frac{(-u_2) - 0}{l} = -\frac{u_2}{l} \tag{8.22a}$$

From Figure 8.26*a* we have $\theta_{AB} = \theta_{BA} = \theta_{BC} = \theta_{CB} = 0$; therefore, for system displacement u_2,

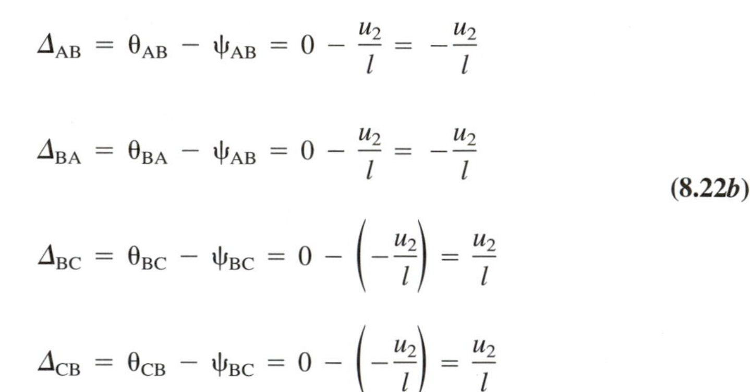

$$\Delta_{AB} = \theta_{AB} - \psi_{AB} = 0 - \frac{u_2}{l} = -\frac{u_2}{l}$$

$$\Delta_{BA} = \theta_{BA} - \psi_{AB} = 0 - \frac{u_2}{l} = -\frac{u_2}{l}$$

$$\Delta_{BC} = \theta_{BC} - \psi_{BC} = 0 - \left(-\frac{u_2}{l}\right) = \frac{u_2}{l} \tag{8.22b}$$

$$\Delta_{CB} = \theta_{CB} - \psi_{BC} = 0 - \left(-\frac{u_2}{l}\right) = \frac{u_2}{l}$$

which defines the second column of the $\boldsymbol{\beta}$ matrix.

Thus, the compatibility equation is

$$\boldsymbol{\Delta} \equiv \begin{Bmatrix} \Delta_{AB} \\ \Delta_{BA} \\ \Delta_{BC} \\ \Delta_{CB} \end{Bmatrix} = \begin{Bmatrix} \theta_{AB} - \psi_{AB} \\ \theta_{BA} - \psi_{AB} \\ \theta_{BC} - \psi_{BC} \\ \theta_{CB} - \psi_{BC} \end{Bmatrix} = \begin{bmatrix} 0 & -1/l \\ 1 & -1/l \\ 1 & 1/l \\ 0 & 1/l \end{bmatrix} \begin{Bmatrix} u_1 \\ u_2 \end{Bmatrix} \tag{8.22c}$$

Step 3. Since the thermal load (ΔT) in beam AB is the same as in the previous cases, we may obtain the fixed-end moments due to ΔT from Equations 8.15*d* and 8.15*e*,

$$\widehat{M}_{AB} = -\widehat{M}_{BA} = \frac{EI\alpha\Delta T_o}{h} \tag{8.22d}$$

For beam BC, there are fixed-end moments due to the uniform continuous load q_o. These may be determined from Equation 8.17l by setting $q(x) = -q_o$ (negative because the load is directed downward),

$$\widehat{M}_{BC} = -\widehat{M}_{CB} = -\frac{q_o l^2}{12} \tag{8.22e}$$

Consequently, the force-deformation relations are

$$\begin{Bmatrix} M_{AB} \\ M_{BA} \\ M_{BC} \\ M_{CB} \end{Bmatrix} = \frac{EI}{\sqrt{2}l}\begin{bmatrix} 4 & 2 & & \\ 2 & 4 & & \\ & & 4\sqrt{2} & 2\sqrt{2} \\ & & 2\sqrt{2} & 4\sqrt{2} \end{bmatrix}\begin{Bmatrix} \Delta_{AB} \\ \Delta_{BA} \\ \Delta_{BC} \\ \Delta_{CB} \end{Bmatrix} + \begin{Bmatrix} EI\alpha\Delta T_o/h \\ -EI\alpha\Delta T_o/h \\ -q_o l^2/12 \\ q_o l^2/12 \end{Bmatrix} \tag{8.22f}$$

Note the fact that the stiffness matrix in Equation 8.22f reflects beam lengths $l_{AB} = \sqrt{2}l$ and $l_{BC} = l$.

Step 4. If we attempt to use the *ad hoc* procedure to obtain the components of $\mathbf{F}^o + \mathbf{F}$ we begin with

$$\begin{Bmatrix} f_1 \\ f_2 \end{Bmatrix} = \begin{bmatrix} 0 & 1 & 1 & 0 \\ -1/l & -1/l & 1/l & 1/l \end{bmatrix}\begin{Bmatrix} M_{AB} \\ M_{BA} \\ M_{BC} \\ M_{CB} \end{Bmatrix} \tag{8.22g}$$

and we note that the right-hand side of the second equilibrium equation contains four end moments, M_{AB}, M_{BA}, M_{BC}, and M_{CB}. It is not possible to identify a single free-body on which these four quantities appear as external forces. Evidently, this equation must contain a combination of equilibrium information obtained from at least two free-bodies. This introduces a complication that can be avoided if we use the virtual work approach.

Recall that each component of $\mathbf{F}$ is the virtual work done by real joint loads moving through virtual motions based on a unit value of one of the system displacements. In particular, the first component of $\mathbf{F}$ is the virtual work done by joint loads undergoing motions like those in Figure 8.8b, which is zero in this case. The second component of $\mathbf{F}$ is the virtual work done by joint loads undergoing motions like those in Figure 8.8c (or Figure 8.26a); this is not zero because horizontal force F at joint B (Figure 8.25) moves through horizontal virtual displacement $u_{2_V} = 1$. Thus

$$\mathbf{F} = \begin{Bmatrix} 0 \\ F \end{Bmatrix} \tag{8.22h}$$

Each component of $\mathbf{F}^o$ is an external virtual work done by interior forces on the beams moving through (virtual) rigid-body displacements associated with the side-sway component of one of the (virtual) system displacements. There is no side-sway component of motion corresponding to virtual system displacement u_{1_V}, so the first component of $\mathbf{F}^o$ is zero. Figure 8.26b shows the side-sway component of motion associated with system displacement u_{2_V}. (For $u_{2_V} = 1$, BB$''$ = B$''$B$'$ = 1 in Figure

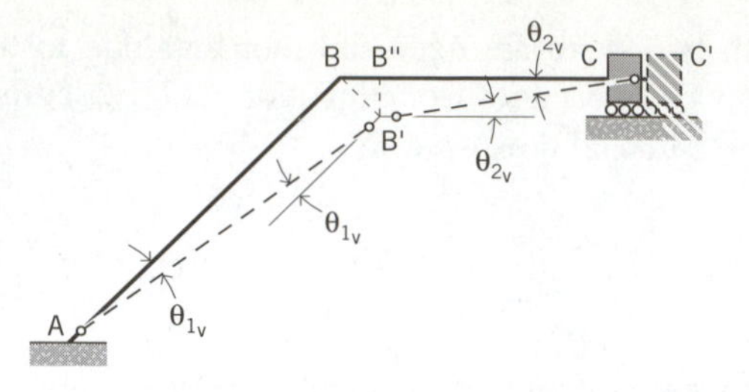

Figure 8.27 Virtual displacements (Solid circles indicate locations of "concentrated curvatures.").

8.26*b*.) It follows that there is a virtual work due to the real interior load q_o on beam BC (Figure 8.25) undergoing a virtual rigid-body displacement due to $u_{2_V} = 1$ (see Figure 8.26*b*). In this case, this virtual work is equal to $q_o l/2$,[26] so

$$\mathbf{F^o} = \begin{Bmatrix} 0 \\ q_o l/2 \end{Bmatrix} \tag{8.22i}$$

The required composite force vector is

$$\mathbf{F^o} + \mathbf{F} = \begin{Bmatrix} 0 \\ F + q_o l/2 \end{Bmatrix} \tag{8.22j}$$

Before completing the analysis, we will review the applicability of the procedure that was presented in Example 17 of Chapter 4 (cited in the previous example) for finding the components of $\mathbf{F^o} + \mathbf{F}$. In particular, we utilize the virtual displacement pattern shown in Figure 8.27. This virtual displacement involves only rigid-body motions and concentrated curvatures (hinges) at the ends of the beam elements. (Note the similarity to Figure 8.26*b*; the locations of the concentrated curvatures coincide with the locations of the real end moments that appear in the second row of Equation 8.22*g*.) Following the procedure described in Chapter 4, we apply the theorem of virtual work using the real loads in Figure 8.25.

We observe from Figure 8.27 that $BB' = \sqrt{2}l\theta_{1_V}$. It follows from geometry that $B''B' = BB' = l\theta_{1_V}$, so $\theta_{2_V} = \theta_{1_V} = \theta_V$. From virtual work we obtain

$$W_{E_V} = Fl\theta_V + \frac{q_o l^2}{2}\theta_V \tag{8.22k}$$

and

$$W_{I_V} = -M_{AB}\theta_V - M_{BA}\theta_V + M_{BC}\theta_V + M_{CB}\theta_V \tag{8.22l}$$

---

[26] See Example 9, Chapter 4, for an illustration of the calculation of external virtual work in a similar case.

Equating leads to

$$F + \frac{q_o l}{2} = -\frac{M_{AB}}{l} - \frac{M_{BA}}{l} + \frac{M_{BC}}{l} + \frac{M_{CB}}{l} \tag{8.22m}$$

which defines f_2 in Equation 8.22*g*. This agrees with the previous result.

Step 5. The solution proceeds in the usual manner,

$$\mathbf{k} = \frac{EI}{\sqrt{2}l}\begin{bmatrix} 0 & 1 & 1 & 0 \\ -1/l & -1/l & 1/l & 1/l \end{bmatrix}\begin{bmatrix} 4 & 2 & & \\ 2 & 4 & & \\ & & 4\sqrt{2} & 2\sqrt{2} \\ & & 2\sqrt{2} & 4\sqrt{2} \end{bmatrix}\begin{bmatrix} 0 & -1/l \\ 1 & -1/l \\ 1 & 1/l \\ 0 & 1/l \end{bmatrix}$$

$$= \frac{EI}{\sqrt{2}l}\begin{bmatrix} 4(1+\sqrt{2}) & \frac{6}{l}(\sqrt{2}-1) \\ \frac{6}{l}(\sqrt{2}-1) & \frac{12}{l^2}(1+\sqrt{2}) \end{bmatrix} \tag{8.23a}$$

and

$$\mathbf{F}^* = \begin{Bmatrix} 0 \\ F + q_o l/2 \end{Bmatrix} - \begin{bmatrix} 0 & 1 & 1 & 0 \\ -1/l & -1/l & 1/l & 1/l \end{bmatrix}\begin{Bmatrix} EI\alpha\Delta T_o/h \\ -EI\alpha\Delta T_o/h \\ -q_o l^2/12 \\ q_o l^2/12 \end{Bmatrix}$$

$$= \begin{Bmatrix} \dfrac{q_o l^2}{12} + \dfrac{EI\alpha\Delta T_o}{h} \\ F + \dfrac{q_o l}{2} \end{Bmatrix} \tag{8.23b}$$

from which we obtain

$$\mathbf{u} = \begin{Bmatrix} -0.0128\dfrac{Fl^2}{EI} + 0.0061\dfrac{q_o l^3}{EI} + 0.1498\dfrac{\alpha l\Delta T_o}{h} \\ 0.0499\dfrac{Fl^3}{EI} + 0.0239\dfrac{q_o l^4}{EI} - 0.0128\dfrac{\alpha l^2\Delta T_o}{h} \end{Bmatrix} \tag{8.23c}$$

Back-substituting gives

$$\begin{Bmatrix} \Delta_{AB} \\ \Delta_{BA} \\ \Delta_{BC} \\ \Delta_{CB} \end{Bmatrix} = \frac{1}{EI}\begin{Bmatrix} -0.024q_o l^3 - 0.050Fl^2 + 0.013EIl\alpha\Delta T_o/h \\ -0.018q_o l^3 - 0.063Fl^2 + 0.163EIl\alpha\Delta T_o/h \\ 0.030q_o l^3 + 0.037Fl^2 + 0.137EIl\alpha\Delta T_o/h \\ 0.024q_o l^3 + 0.050Fl^2 - 0.013EIl\alpha\Delta T_o/h \end{Bmatrix} \tag{8.23d}$$

and

$$\begin{Bmatrix} M_{AB} \\ M_{BA} \\ M_{BC} \\ M_{CB} \end{Bmatrix} = \begin{Bmatrix} -0.093q_o l^2 - 0.230Fl + 1.266EI\alpha\Delta T_o/h \\ -0.084q_o l^2 - 0.248Fl - 0.522EI\alpha\Delta T_o/h \\ 0.084q_o l^2 + 0.248Fl + 0.522EI\alpha\Delta T_o/h \\ 0.239q_o l^2 + 0.274Fl + 0.222EI\alpha\Delta T_o/h \end{Bmatrix} \tag{8.23e}$$

which are identical to Equations 5.44*o* and 5.44*n*, respectively.

EXAMPLE 5

The gabled frame in Figure 8.28 is subjected to a uniform continuous horizontal load q_o (e.g., wind load), and to concentrated vertical forces F_1 at joint C and F_2 at the midpoint of beam CD, as shown. In addition, there is a thermal load $\Delta T = \Delta T_o y/h$ in beam BC. The geometry is such that beams BC and CD are perpendicular. All beams are assumed to be axially rigid.

Step 1. There are three system displacements (Figure 8.29). System displacements u_1 and u_2 are rotations at rigid joints B and C, respectively, and u_3 is defined as the horizontal displacement at B (Figure 8.29*d*).

Step 2. Since u_1 and u_2 do not involve side-sway, the first two columns of the $\boldsymbol{\beta}$ matrix are obtained in a straightforward manner. For u_1 we have $\Delta_{AB} = \theta_{AB} = 0$, $\Delta_{BA} = \theta_{BA} = u_1$, $\Delta_{BC} = \theta_{BC} = u_1$, $\Delta_{CB} = \theta_{CB} = 0$, $\Delta_{CD} = \theta_{CD} = 0$, $\Delta_{DC} = \theta_{DC} = 0$, which defines the first column; for u_2 we have $\Delta_{AB} = \theta_{AB} = 0$, $\Delta_{BA} = \theta_{BA} =$

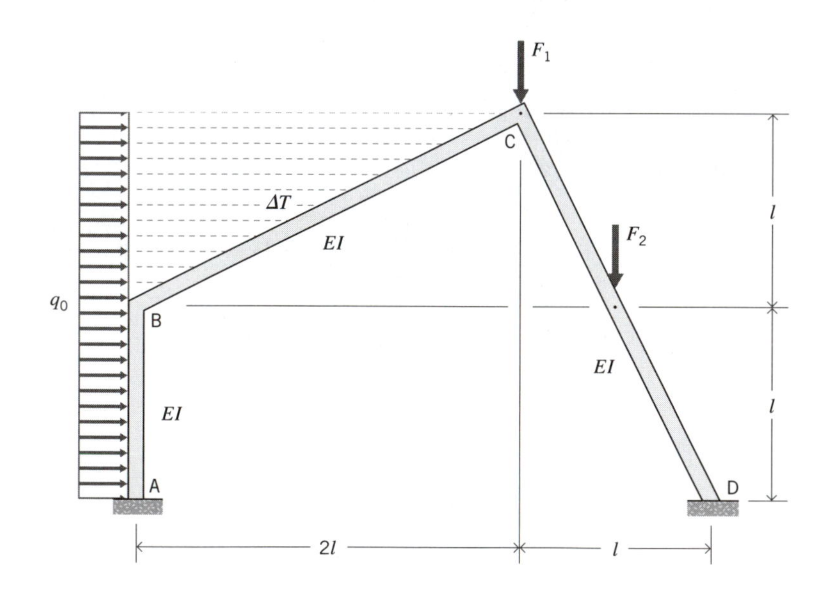

Figure 8.28 Example 5.

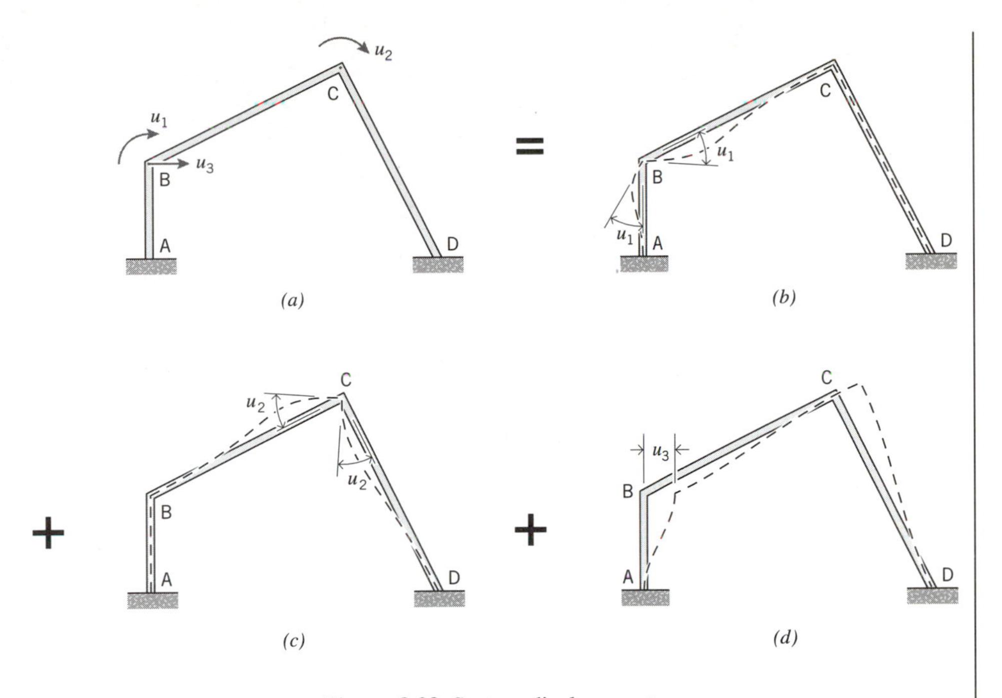

Figure 8.29 System displacements.

0, $\Delta_{BC} = \theta_{BC} = 0$, $\Delta_{CB} = \theta_{CB} = u_2$, $\Delta_{CD} = \theta_{CD} = u_2$, $\Delta_{DC} = \theta_{DC} = 0$, which defines the second column.

Since the beams are axially rigid, displacement u_3 results in side-sway for all three beams (see Figure 8.29*d*). The geometry of the side-sway motion is detailed in Figure 8.30; joint B moves to B′, which lies along a line perpendicular to AB, and joint C moves to C′, which lies along a line perpendicular to CD. (Since CD is perpendicular to BC, C′ lies along the extension of BC.) Distance CC′ is determined from right triangle CC′C″ for which CC″ = BB′ = u_3. Thus, transverse end displacements for the three beams are given by $v_{AB} = 0$, $v_{BA} = -u_3$, $v_{BC} = -u_3/\sqrt{5}$, $v_{CB} = 0$, $v_{CD} = 2u_3/\sqrt{5}$, $v_{DC} = 0$, and side-sways are

$$
\begin{aligned}
\psi_{AB} &= \frac{v_{AB} - v_{BA}}{l_{AB}} = \frac{0 - (-u_3)}{l} = \frac{1}{l}u_3 \\
\psi_{BC} &= \frac{v_{BC} - v_{CB}}{l_{BC}} = \frac{(-u_3/\sqrt{5}) - 0}{\sqrt{5}l} = -\frac{1}{5l}u_3 \\
\psi_{CD} &= \frac{v_{CD} - v_{DC}}{l_{CD}} = \frac{(2u_3/\sqrt{5}) - 0}{\sqrt{5}l} = \frac{2}{5l}u_3
\end{aligned}
\qquad \textbf{(8.24a)}
$$

All end rotations are zero for system displacement u_3, so the end deformations ($\Delta_{IJ} = \theta_{IJ} - \psi_{IJ}$) are

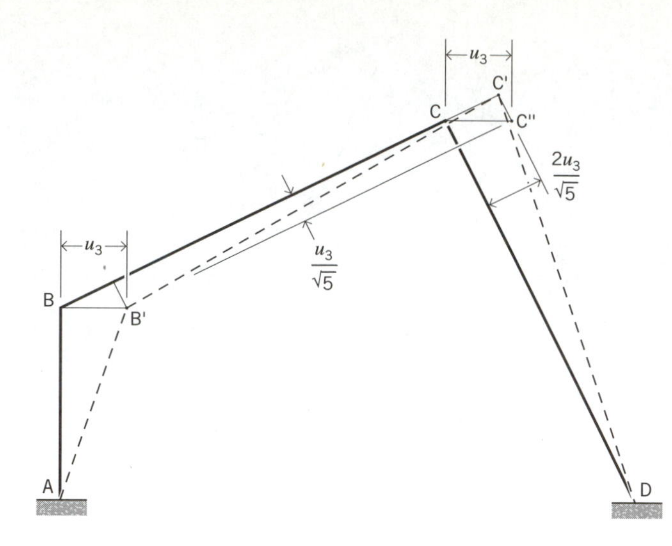

Figure 8.30 Geometry of system displacement u_3; rigid-body (side-sway) component of u_3.

$$\Delta_{AB} = \Delta_{BA} = 0 - \frac{1}{l}u_3 = -\frac{1}{l}u_3$$

$$\Delta_{BC} = \Delta_{CB} = 0 - \left(-\frac{1}{5l}u_3\right) = \frac{1}{5l}u_3 \tag{8.24b}$$

$$\Delta_{CD} = \Delta_{DC} = 0 - \frac{2}{5l}u_3 = -\frac{2}{5l}u_3$$

This defines the third column of the $\boldsymbol{\beta}$ matrix.

Therefore, the compatibility relations are

$$\boldsymbol{\Delta} \equiv \begin{Bmatrix} \Delta_{AB} \\ \Delta_{BA} \\ \Delta_{BC} \\ \Delta_{CB} \\ \Delta_{CD} \\ \Delta_{DC} \end{Bmatrix} = \begin{bmatrix} 0 & 0 & -1/l \\ 1 & 0 & -1/l \\ 1 & 0 & 1/5l \\ 0 & 1 & 1/5l \\ 0 & 1 & -2/5l \\ 0 & 0 & -2/5l \end{bmatrix} \begin{Bmatrix} u_1 \\ u_2 \\ u_3 \end{Bmatrix} \tag{8.24c}$$

Step 3. Fixed-end moments are developed in all three beams. Those in beam AB are due to the transverse load q_o and are given by Equation 8.22*e*,

$$\widehat{M}_{AB} = -\widehat{M}_{BA} = -\frac{q_o l^2}{12} \tag{8.24d}$$

The fixed-end moments in beam BC are due to the thermal load and to the transverse component of load q_o. The transverse component of q_o is determined from Equation

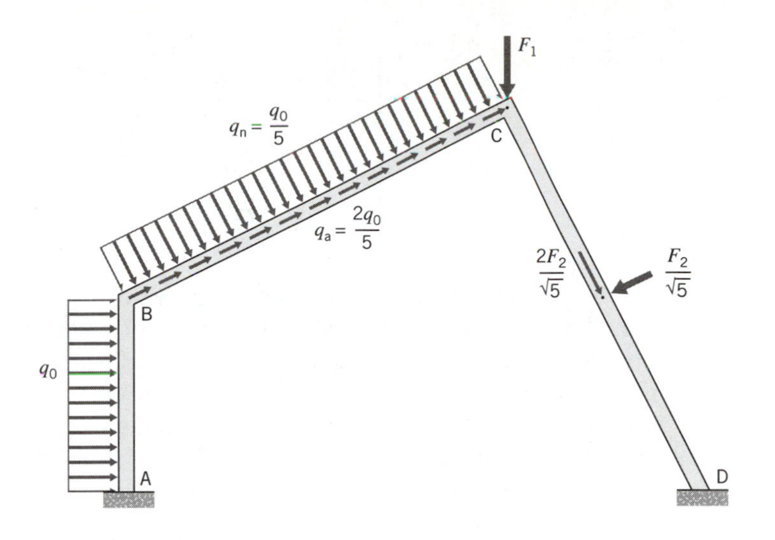

Figure 8.31 Components of physical loads.

1.60, and is equal to $q_o/5$ (Figure 8.31).[27] Thus, the portions of the fixed-end moments due to the distributed load are also obtained from Equation 8.22*e* by substituting $q_o/5$ for q_o and $\sqrt{5}l$ for l. The fixed-end moments due to the thermal load are the same as those in Equation 8.22*d*, so the total fixed-end moments in beam BC are

$$\widehat{M}_{\mathrm{BC}} = -\widehat{M}_{\mathrm{CB}} = -\frac{q_o l^2}{12} + \frac{EI\alpha\Delta T_o}{h} \tag{8.24e}$$

The fixed-end moments in beam CD are due to the transverse component of F_2 (Figure 8.31) and are obtained from Equation 8.17*j* by substituting $-F_2/\sqrt{5}$ for F, $\sqrt{5}l$ for l, and $a = b = \sqrt{5}l/2$; the result is

$$\widehat{M}_{\mathrm{CD}} = -\widehat{M}_{\mathrm{DC}} = \frac{F_2 l}{8} \tag{8.24f}$$

Therefore, the matrix moment-deformation equation is

$$\begin{Bmatrix} M_{\mathrm{AB}} \\ M_{\mathrm{BA}} \\ M_{\mathrm{BC}} \\ M_{\mathrm{CB}} \\ M_{\mathrm{CD}} \\ M_{\mathrm{DC}} \end{Bmatrix} = \frac{EI}{l} \begin{bmatrix} 4 & 2 & & & & \\ 2 & 4 & & & & \\ & & 4/\sqrt{5} & 2/\sqrt{5} & & \\ & & 2/\sqrt{5} & 4/\sqrt{5} & & \\ & & & & 4/\sqrt{5} & 2/\sqrt{5} \\ & & & & 2/\sqrt{5} & 4/\sqrt{5} \end{bmatrix} \begin{Bmatrix} \Delta_{\mathrm{AB}} \\ \Delta_{\mathrm{BA}} \\ \Delta_{\mathrm{BC}} \\ \Delta_{\mathrm{CB}} \\ \Delta_{\mathrm{CD}} \\ \Delta_{\mathrm{DC}} \end{Bmatrix} +$$

[27] Note the definition of angle θ when using Equation 1.60 to determine this component of the continuous load.

$$+\begin{Bmatrix} -q_o l^2/12 \\ q_o l^2/12 \\ -q_o l^2/12 + EI\alpha\Delta T_o/h \\ q_o l^2/12 - EI\alpha\Delta T_o/h \\ -F_2 l/8 \\ F_2 l/8 \end{Bmatrix} \tag{8.24g}$$

Step 4. We will obtain the components of **F** and $\mathbf{F}^o$ by using virtual work. The first component of **F** is the virtual work done by joint forces undergoing motions like those in Figure 8.29*b*, and the second component of **F** is the virtual work done by joint forces undergoing motions like those in Figure 8.29*c*; these are both zero. The third component of **F** is the virtual work done by joint forces undergoing motions like those in Figures 8.29*d* or 8.30. In this case, there is virtual work done by F_1 moving through the vertical motion of joint C. The magnitude of this vertical motion is $2u_{3_V}/5$ (see Figure 8.32). Thus, for a unit displacement $u_{3_V} = 1$ the virtual work is $-2F_1/5$. Consequently,

$$\mathbf{F} = \begin{Bmatrix} 0 \\ 0 \\ -2F_1/5 \end{Bmatrix} \tag{8.24h}$$

Note that although u_3 was initially associated with horizontal motion at joint B, in fact it is also associated with both horizontal and vertical motion at joint C. This means that downward loads at joint C produce horizontal motions at joint B!

Each component of $\mathbf{F}^o$ is an external virtual work done by interior forces on the beams moving through (virtual) rigid-body displacements associated with the side-sway component of one of the (virtual) system displacements. Since no side-sway motion develops due to virtual system displacements u_{1_V} or u_{2_V}, the first two compo-

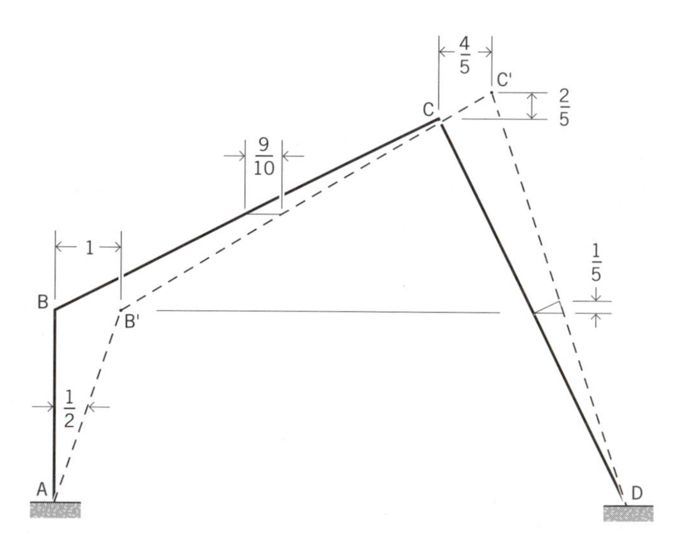

Figure 8.32 Side-sway component of virtual displacement $u3_v = 1$.

nents of $\mathbf{F^o}$ are zero. Figures 8.30 and 8.32 show the side-sway components of motion associated with virtual system displacement u_{3_V}. Figure 8.32, in particular, shows the details of motion due to $u_{3_V} = 1$ needed to compute the virtual work.

The interior physical load on each of the beams contributes to the external virtual work. For beam AB the external virtual work due to the uniform load q_o is equal to its resultant $(q_o l)$ multiplied by the virtual displacement at the centroid of the load $(l/2)$, giving the final result of $q_o l/2$. The virtual work due to the load on beam BC may be determined from the resultant of the horizontal load $(q_o l)$ multiplied by the virtual displacement at its centroid. The latter is the horizontal component of virtual displacement at the midpoint of beam BC, and is the mean of the horizontal components of end displacements, i.e., $(1 + 4/5)/2 = 9/10$, so the virtual work is $9q_o l/10$. For beam CD, the vertical concentrated (downward) force F_2 at the midpoint undergoes a virtual (upward) displacement 1/5, so the virtual work is $-F_2/5$.[28] Thus, all the beam loads contribute to produce the result

$$\mathbf{F^o} = \begin{Bmatrix} 0 \\ 0 \\ q_o l/2 + 9q_o l/10 - F_2/5 \end{Bmatrix} = \begin{Bmatrix} 0 \\ 0 \\ 7q_o l/5 - F_2/5 \end{Bmatrix} \tag{8.24i}$$

Therefore, the total $\mathbf{F^o} + \mathbf{F}$ is

$$\mathbf{F^o} + \mathbf{F} = \begin{Bmatrix} 0 \\ 0 \\ 7q_o l/5 - 2F_1/5 - F_2/5 \end{Bmatrix} \tag{8.24j}$$

Step 5. The stiffness matrix is

$$\mathbf{k} = \boldsymbol{\beta}^{\mathrm{T}}\mathbf{K}\boldsymbol{\beta} = \frac{EI}{l}\begin{bmatrix} 5.789 & 0.894 & -5.463/l \\ 0.894 & 3.578 & -0.537/l \\ -5.463/l & -0.537/l & 13.073/l^2 \end{bmatrix} \tag{8.25a}$$

and $\mathbf{F}^*$ is

$$\mathbf{F}^* = (\mathbf{F^o} + \mathbf{F}) - \boldsymbol{\beta}^{\mathrm{T}}\hat{\mathbf{P}} = \begin{Bmatrix} -EI\alpha\Delta T_o/h \\ F_2 l/8 - q_o l^2/12 + EI\alpha\Delta T_o/h \\ -2F_1/5 - F_2/5 + 7q_o l/5 \end{Bmatrix} \tag{8.25b}$$

[28] The same result can be obtained by using components of forces, e.g., the transverse component of F_2 $(F_2/\sqrt{5})$ undergoes a virtual displacement $1/\sqrt{5}$ (Figure 8.30), and the axial component of F_2 undergoes no virtual displacement. (Therefore, the axial component has no effect on structural response.) Similarly, the resultant of the transverse component of the distributed force on beam BC is $(q_o/5)(\sqrt{5}l) = \sqrt{5}q_o l/5$, and the transverse virtual displacement at the midpoint is $1/(2\sqrt{5})$ (Figure 8.30); the resultant of the axial component of distributed force is $(2q_o/5)(\sqrt{5}l) = 2\sqrt{5}q_o l/5$ and the axial virtual displacement at the midpoint is $2/\sqrt{5}$ (Figure 8.30). Thus, this axial component of force does produce structural response.

The system displacement vector is

$$\mathbf{u} = \mathbf{k}^{-1}\mathbf{F}^*$$

$$= \begin{Bmatrix} -0.049F_1l^2/EI - 0.031F_2l^2/EI + 0.175q_ol^3/EI - 0.352\alpha l\Delta T_o/h \\ 0.005F_1l^2/EI + 0.039F_2l^2/EI - 0.040q_ol^3/EI + 0.348\alpha l\Delta T_o/h \\ -0.051F_1l^3/EI - 0.027F_2l^3/EI + 0.178q_ol^4/EI - 0.133\alpha l^2\Delta T_o/h \end{Bmatrix} \quad \textbf{(8.25c)}$$

Back-substituting gives

$$\boldsymbol{\Delta} = \begin{Bmatrix} 0.051F_1l^2/EI + 0.027F_2l^2/EI - 0.178q_ol^3/EI + 0.133\alpha l\Delta T_o/h \\ 0.002F_1l^2/EI - 0.004F_2l^2/EI - 0.004q_ol^3/EI - 0.219\alpha l\Delta T_o/h \\ -0.059F_1l^2/EI - 0.037F_2l^2/EI + 0.210q_ol^3/EI - 0.378\alpha l\Delta T_o/h \\ -0.006F_1l^2/EI + 0.033F_2l^2/EI - 0.004q_ol^3/EI + 0.321\alpha l\Delta T_o/h \\ 0.025F_1l^2/EI + 0.049F_2l^2/EI - 0.112q_ol^3/EI + 0.401\alpha l\Delta T_o/h \\ 0.020F_1l^2/EI + 0.011F_2l^2/EI - 0.071q_ol^3/EI + 0.053\alpha l\Delta T_o/h \end{Bmatrix} \quad \textbf{(8.25d)}$$

and

$$\mathbf{P} = \begin{Bmatrix} 0.207F_1l + 0.098F_2l - 0.805q_ol^2 + 0.093EI\alpha\Delta T_o/h \\ 0.110F_1l + 0.036F_2l - 0.289q_ol^2 - 0.611EI\alpha\Delta T_o/h \\ -0.110F_1l - 0.036F_2l + 0.289q_ol^2 + 0.611EI\alpha\Delta T_o/h \\ -0.063F_1l + 0.027F_2l + 0.263q_ol^2 - 0.764EI\alpha\Delta T_o/h \\ 0.063F_1l - 0.027F_2l - 0.263q_ol^2 + 0.764EI\alpha\Delta T_o/h \\ 0.058F_1l + 0.188F_2l - 0.227q_ol^2 + 0.453EI\alpha\Delta T_o/h \end{Bmatrix} \quad \textbf{(8.25e)}$$

The preceding examples have illustrated the two major complications that arise in the analysis of frames when side-sway motion is present, namely the coupling of displacements when elements are axially rigid and the difficulties inherent in the computation of $\mathbf{F}^o + \mathbf{F}$. As was noted in Section 8.1, the introduction of axial flexibility will eliminate the former complication entirely and will thereby simplify the computation of $\mathbf{F}$. However, the difficulties associated with the computation of $\mathbf{F}^o$ are not entirely eliminated by the introduction of axial flexibility. This is because most of these difficulties were a consequence of the use of element force-deformation relations (rather than force-displacement relations). Thus, there is no significant computational advantage to be gained from the introduction of axial flexibility as long as we use force-deformation relations as our base. To make further progress beyond this point, our analysis procedure will have to be reformulated using force-displacement relations. This will be deferred to Chapter 9.

In the remainder of this chapter, we will examine a few additional topics of general interest for conducting realistic frame analyses.

8.3.4 Fixed-End Moments From Betti's Theorem

Thus far, we have used Equation 8.13e (or some variation thereof) to obtain fixed-end moments. The computational effort required for these evaluations can become consid-

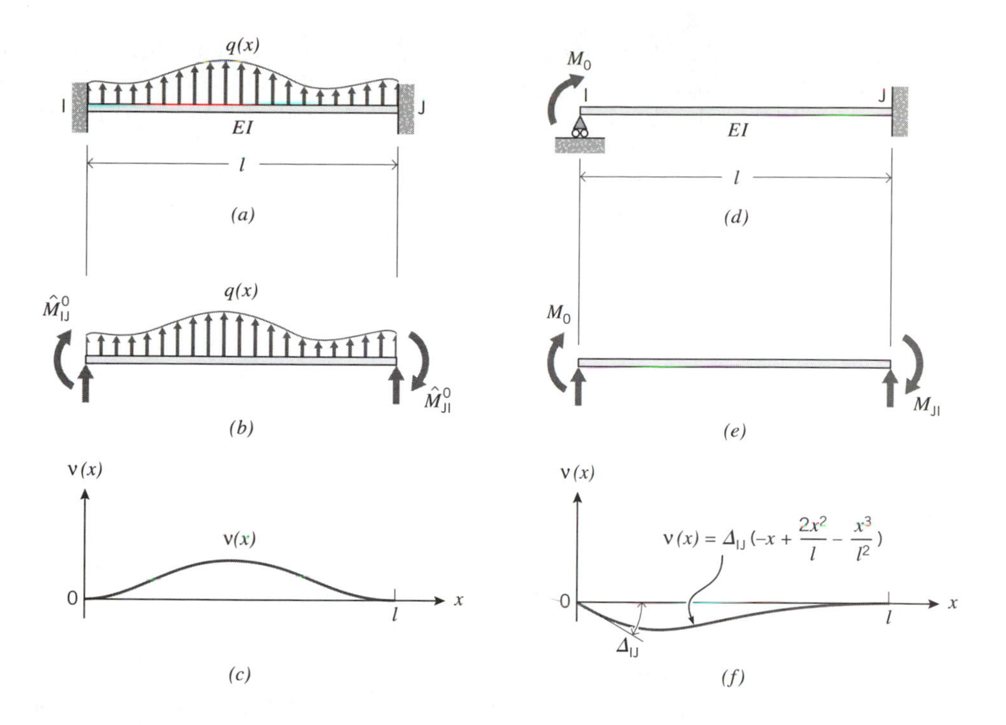

Figure 8.33 *(a–c)*: Beam with fixed-ends subjected to arbitrary load; *(d)–(f)*: Beam subjected to moment load at end.

erable for any but the simplest of load conditions. When the fixed-end moments are due to physical loads, the computational effort required to find them can often be reduced significantly by an application of Betti's theorem. In addition, the theorem gives a useful work-based interpretation of the fixed-end moments.

To illustrate, consider a uniform beam with ends fixed against rotation, which is subjected to some specified set of interior physical loads, such as an arbitrary continuous load (Figure 8.33*a*). Figure 8.33*b* shows a free-body diagram with all of the forces acting on the beam. The quantities $\widehat{M}^o_{IJ}$ and $\widehat{M}^o_{JI}$ in this figure are fixed-end moments by definition. The real (compatible) displacements due to the forces are depicted schematically in Figure 8.33*c*; it is important to note again that the rotations (and transverse displacements) are zero at the ends. We will use Betti's theorem to evaluate $\widehat{M}^o_{IJ}$.

To accomplish this, we consider the same beam with different end supports and applied loads. In particular, we consider the beam to be fixed at the J end, simply supported at the I end, and subjected only to a concentrated moment M_o applied at the I end (Figure 8.33*d*). Figure 8.33*e* shows the free-body diagram for this case, and Figure 8.33*f* shows the real (compatible) displacements due to the forces in Figure 8.33*e*; the displacement as a function of coordinate x is given by the expression previously recorded in Figure 8.18*b*. It is important to recognize that this displacement function is known.

We regard the forces and displacements in Figures 8.33*b* and 8.33*c* as set (1), and the forces and displacements in Figures 8.33*e* and 8.33*f* as set (2). Recall Betti's theorem, Equation 6.1*e*, which states that the external virtual work associated with force set (1) going through displacement set (2) is equal to the external virtual work associated with force set (2) going through displacement set (1). The latter external virtual work is identically zero because all the forces in Figure 8.33*e* are applied at the ends of the beam, but all the end motions in Figure 8.33*c* are zero.

On the other hand, there is external virtual work due to $\widehat{M^o_{IJ}}$ at I in Figure 8.33*b* undergoing the end displacement (rotation) Δ_{IJ} in Figure 8.33*f*, and also due to $q(x)$ in Figure 8.33*b* undergoing the displacement $v(x)$ in Figure 8.33*f*; the remaining forces in Figure 8.33*b* undergo no displacements in Figure 8.33*f*. Thus, Betti's theorem, Equation 6.1*e*, gives[29]

$$\begin{aligned} 0 &= \widehat{M^o_{IJ}}\Delta_{IJ} + \int_0^l q(x)v(x)dx \\ &= \widehat{M^o_{IJ}}\Delta_{IJ} + \Delta_{IJ}\int_0^l q(x)\left(-x + \frac{2}{l}x^2 - \frac{1}{l^2}x^3\right)dx \end{aligned} \qquad \textbf{(8.26a)}$$

from which we obtain

$$\widehat{M^o_{IJ}} = -\int_0^l q(x)\left(-x + \frac{2x^2}{l} - \frac{x^3}{l^2}\right)dx = \int_0^l q(x)\frac{x(l-x)^2}{l^2}dx \qquad \textbf{(8.26b)}$$

which is exactly the same as the first of Equations 8.17*l*!

This approach is completely general; we may substitute any physical loads in Figure 8.33*a* and use the known displacement in Figure 8.33*f* to compute $\widehat{M^o_{IJ}}$. We may also compute $\widehat{M^o_{JI}}$ in an analogous manner by substituting the displacement function in Figure 8.18*c* for that in Figure 8.33*f*. The procedure is demonstrated in the following examples.

EXAMPLE 6

Find $\widehat{M^o_{JI}}$ for the beam in Figure 8.33*a*.

We replace Figures 8.33*d*–8.33*f* with Figures 8.34*a*–8.34*c*, respectively. In this case, end I is taken to be fixed against rotation, and there is a moment M_0 applied at end J, which is simply supported. The known displacement function in Figure 8.34*c* is the same as that in Figure 8.18*c*.

The external virtual work associated with Figures 8.34*b* and 8.33*c* is again identically equal to zero because all the end motions in Figure 8.33*c* are zero. (In fact, this will always be the case in these applications.)

As in the preceding illustration, there is external virtual work due to $\widehat{M^o_{JI}}$ at J in Figure 8.33*b* undergoing the end displacement (rotation) Δ_{JI} in Figure 8.34*c*, and also due to $q(x)$ in Figure 8.33*b* undergoing the displacement $v(x)$ in Figure 8.34*c*.

---

[29] Sign conventions must be rigorously followed. The first term is positive because the moment and the rotation are both clockwise; the second term is positive because both $q(x)$ and $v(x)$ are defined as positive when they are in the positive y direction. [Note that $v'(0) = -\Delta_{IJ}$ in Figure 8.33*f*.]

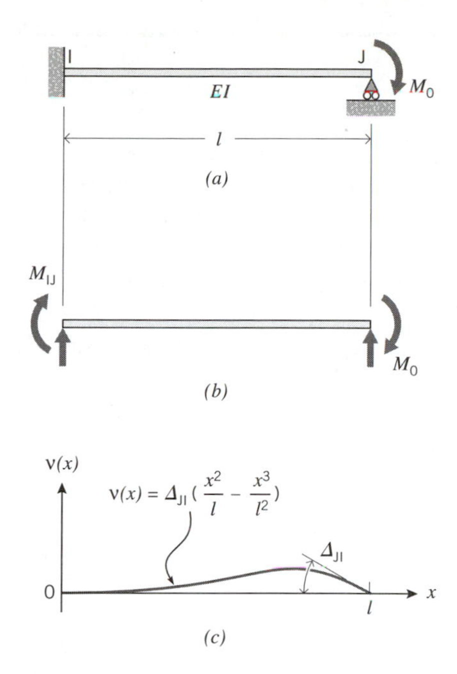

Figure 8.34 Forces and displacements for computing $\widehat{M}_{JI0}$.

Equation 6.1*e* reduces to

$$0 = \widehat{M}^{o}_{JI}\Delta_{JI} + \int_0^l q(x)v(x)dx$$
$$= \widehat{M}^{o}_{JI}\Delta_{JI} + \Delta_{JI}\int_0^l q(x)\left(\frac{x^2}{l} - \frac{x^3}{l^2}\right)dx \quad \textbf{(8.27a)}$$

from which we obtain

$$\widehat{M}^{o}_{JI} = -\int_0^l q(x)\left(\frac{x^2}{l} - \frac{x^3}{l^2}\right)dx = -\int_0^l q(x)\frac{x^2(l-x)}{l^2}dx \quad \textbf{(8.27b)}$$

which is identical to the second of Equations 8.17*l*.

The same result may be obtained from the displacement in Figure 8.33*f* if we "rotate" the beam in Figure 8.33*a* 180° about its vertical axis of symmetry since this reverses the locations of the fixed-end moments. This also reverses the direction of the positive x axis in Figure 8.33*b*, so in this case it is necessary to transform coordinates by substituting $l - x$ for x and $-dx$ for dx in the preceding equations. Under this transformation, the integral in Equation 8.27*b* becomes the integral in Equation 8.26*b*, and vice versa. In effect, this rotation eliminates any need to use the displacement function in Figure 8.34*c*.

EXAMPLE 7

Find fixed-end moments for a uniform beam that is subjected to a single concentrated transverse force applied at an arbitrary location (Figure 8.35*a*).

Figures 8.35*b* and 8.35*c* show the free-body diagram and schematic displacements, respectively, for this beam; these constitute set (1). The forces and displacements in Figures 8.33*e* and 8.33*f* again constitute set (2).

The external virtual work done by force (2) and displacement (1) is identically zero. For force (1) and displacement (2), there is external virtual work done by $\widehat{M}^o_{IJ}$ in Figure 8.35*b* undergoing end rotation Δ_{IJ} in Figure 8.33*f*, and also due to F in Figure 8.35*b* undergoing the transverse displacement at $x = a$ in Figure 8.33*f*. This transverse displacement is given by

$$v(a) = \Delta_{IJ}\left(-a + \frac{2a^2}{l} - \frac{a^3}{l^2}\right) = -\Delta_{IJ}\frac{ab^2}{l^2} \tag{8.28a}$$

Therefore, Betti's theorem gives

$$\begin{aligned} 0 &= \widehat{M}^o_{IJ}\Delta_{IJ} + Fv(a) \\ &= \widehat{M}^o_{IJ}\Delta_{IJ} - F\Delta_{IJ}\frac{ab^2}{l^2} \end{aligned} \tag{8.28b}$$

It follows that

$$\widehat{M}^o_{IJ} = \frac{ab^2}{l^2}F \tag{8.28c}$$

which is identical to the first of Equations 8.17*j*.

To find $\widehat{M}^o_{JI}$, we will rotate the beam in Figure 8.35 about its vertical axis of symmetry. This leads to the free-body diagram in Figure 8.35*d*; note the directions of the end moments, which are still positive with respect to the rotated axes. We follow the same procedure used above, except that Figure 8.35*d* replaces Figure 8.35*b*.

In this case, there is external virtual work done by $\widehat{M}^o_{JI}$ in Figure 8.35*d* undergoing end rotation Δ_{IJ} in Figure 8.33*f*, and also done by F in Figure 8.35*d* undergoing the transverse displacement at $x = b$ in Figure 8.33*f*. This transverse displacement is

$$v(b) = \Delta_{IJ}\left(-b + \frac{2b^2}{l} - \frac{b^3}{l^2}\right) = -\Delta_{IJ}\frac{a^2b}{l^2} \tag{8.28d}$$

and Betti's theorem gives

$$\begin{aligned} 0 &= -\widehat{M}^o_{JI}\Delta_{IJ} + Fv(b) \\ &= -\widehat{M}^o_{JI}\Delta_{IJ} - F\Delta_{IJ}\frac{a^2b}{l^2} \end{aligned} \tag{8.28e}$$

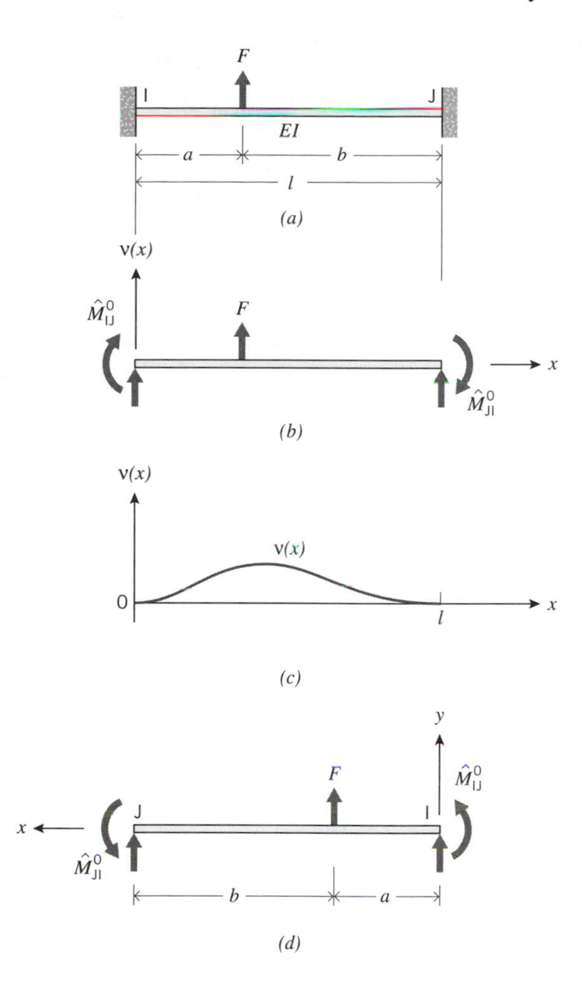

Figure 8.35 Example 7.

The first term above is negative because $\widehat{M}^o_{JI}$ in Figure 8.35*d* is counterclockwise while Δ_{IJ} in Figure 8.33*f* is clockwise. We find

$$\widehat{M}^o_{JI} = -\frac{a^2 b}{l^2} F \tag{8.28f}$$

which is identical to the second of Equations 8.17*j*.

EXAMPLE 8

Find the fixed-end moments for a uniform beam that is subjected to a concentrated moment M_o applied at an arbitrary location (Figure 8.36*a*).[30] Figures 8.36*b* and 8.36*c* show the free-body diagram and schematic displacements, respectively.

[30] This moment is unrelated to M_o in Figure 8.33*d*.

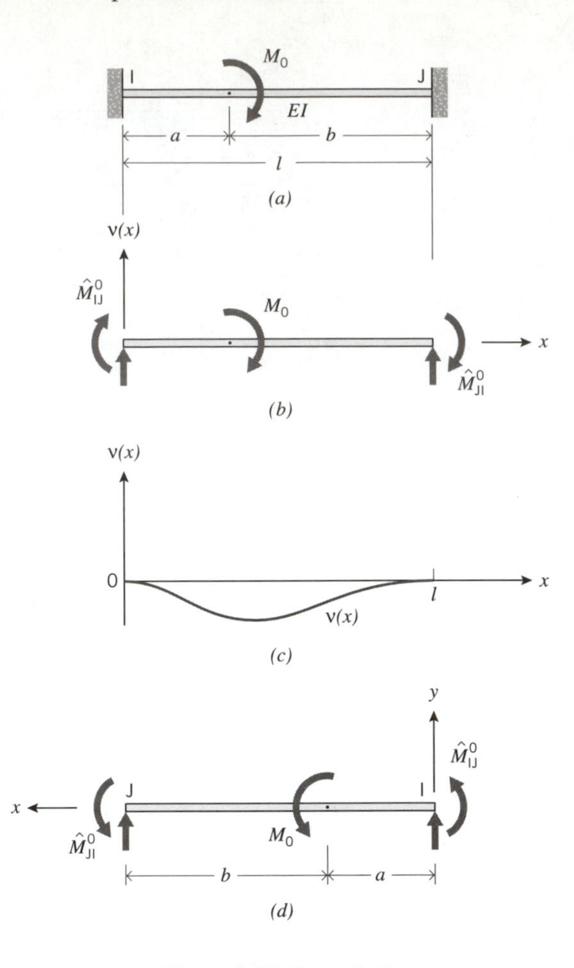

Figure 8.36 Example 8.

The external virtual work done by the forces in Figure 8.33*e* and the displacements in Figure 8.36*c* is identically zero. There is external virtual work done by $\widehat{M}^o_{IJ}$ in Figure 8.36*b* undergoing end rotation Δ_{IJ} in Figure 8.33*f*, and also by M_o in Figure 8.36*b* undergoing the rotation at $x = a$ in Figure 8.33*f*. Since clockwise rotations are defined as positive, it follows that the rotation at $x = a$ is the negative of the slope at $x = a$; the latter is

$$v'(a) = \Delta_{IJ}\left(-1 + \frac{4a}{l} - \frac{3a^2}{l^2}\right) = -\Delta_{IJ}\frac{b(l - 3a)}{l^2} \tag{8.29a}$$

Therefore, Betti's theorem gives

$$\begin{aligned} 0 &= \widehat{M}^o_{IJ}\Delta_{IJ} + M_o[-v'(a)] \\ &= \widehat{M}^o_{IJ}\Delta_{IJ} + M_o\Delta_{IJ}\frac{b(l - 3a)}{l^2} \end{aligned} \tag{8.29b}$$

from which we obtain

$$\widehat{M}^o_{IJ} = -\frac{b(l - 3a)}{l^2} M_o \quad \textbf{(8.29c)}$$

It is interesting to note that $\widehat{M}^o_{IJ}$ changes sign depending on whether a is greater or less than $l/3$.

Figure 8.36*d* shows the rotated beam, which together with the displacements in Figure 8.33*f* will be used to evaluate $\widehat{M}^o_{JI}$. In this case, we require the slope of the beam in Figure 8.33*f* at $x = b$, which is

$$v'(b) = \Delta_{IJ}\left(-1 + \frac{4b}{l} - \frac{3b^2}{l^2}\right) = -\Delta_{IJ}\frac{a(l - 3b)}{l^2} \quad \textbf{(8.29d)}$$

Betti's theorem leads to

$$\begin{aligned} 0 &= -\widehat{M}^o_{JI}\Delta_{IJ} + M_o[v'(b)] \\ &= -\widehat{M}^o_{JI}\Delta_{IJ} - M_o\Delta_{IJ}\frac{a(l - 3b)}{l^2} \end{aligned} \quad \textbf{(8.29e)}$$

or

$$\widehat{M}^o_{JI} = -\frac{a(l - 3b)}{l^2} M_o \quad \textbf{(8.29f)}$$

Appendix A contains a table summarizing the most frequently encountered fixed-end moments.

8.4 MISCELLANEOUS CONSIDERATIONS

Before proceeding to further development of the displacement method in Chapter 9, we will address two familiar structural concerns in the context of the present formulation. In particular, we will briefly demonstrate how to deal with support settlement and with structures composed of some different types of elements.

8.4.1 Support Settlement

As in Section 5.4.1, we assume that the magnitudes of any support settlements are both small and known. We recognize that "support settlements" are merely specific structural motions that occur in addition to the unknown system displacements.[31] Thus,

[31] In other words, we may superpose the effects of support settlements and the behavior in the absence of support settlements.

support settlements may be regarded as prespecified system displacements that may be separated from the unknown system displacements in the analysis.

This may be accomplished by appending to the compatibility equation an expression for element deformations in terms of settlement motions. Thus, Equation 8.2 becomes

$$\mathbf{\Delta} = \boldsymbol{\beta}\mathbf{u} + \boldsymbol{\beta}_s\mathbf{u}_s \tag{8.30a}$$

where $\boldsymbol{\beta}_s$ is a compatibility matrix associated with the vector of support settlements $\mathbf{u}_s$.[32]

The force-deformation equations (Step 3), and the equilibrium equations (Step 4), are unaltered by the occurrence of support settlements. Therefore, the desired form of the matrix equilibrium equation is obtained by substituting Equations 8.30*a* and 8.4 into Equation 8.18*i*,

$$\begin{aligned}(\mathbf{F}^o + \mathbf{F}) &= \boldsymbol{\beta}^T\mathbf{P} \\ &= \boldsymbol{\beta}^T(\mathbf{K}\mathbf{\Delta} + \widehat{\mathbf{P}}) \\ &= \boldsymbol{\beta}^T[\mathbf{K}(\boldsymbol{\beta}\mathbf{u} + \boldsymbol{\beta}_s\mathbf{u}_s) + \widehat{\mathbf{P}}] \\ &= (\boldsymbol{\beta}^T\mathbf{K}\boldsymbol{\beta})\mathbf{u} + \boldsymbol{\beta}^T(\widehat{\mathbf{P}} + \mathbf{K}\boldsymbol{\beta}_s\mathbf{u}_s)\end{aligned} \tag{8.30b}$$

or

$$(\boldsymbol{\beta}^T\mathbf{K}\boldsymbol{\beta})\mathbf{u} = (\mathbf{F}^o + \mathbf{F}) - \boldsymbol{\beta}^T(\widehat{\mathbf{P}} + \mathbf{K}\boldsymbol{\beta}_s\mathbf{u}_s) \tag{8.30c}$$

which we may write in the usual form as

$$\mathbf{k}\mathbf{u} = \overline{\mathbf{F}}^* \tag{8.30d}$$

where

$$\overline{\mathbf{F}}^* \equiv (\mathbf{F}^o + \mathbf{F}) - \boldsymbol{\beta}^T(\widehat{\mathbf{P}} + \mathbf{K}\boldsymbol{\beta}_s\mathbf{u}_s) \equiv \mathbf{F}^* - \boldsymbol{\beta}^T\mathbf{K}\boldsymbol{\beta}_s\mathbf{u}_s \tag{8.30e}$$

For trusses $\mathbf{F}^o = \mathbf{0}$, so $\mathbf{F}^*$ on the right side of Equation 8.30*e* reduces to $\mathbf{F} - \boldsymbol{\beta}^T\widehat{\mathbf{P}}$ (Equation 8.7*c*).

EXAMPLE 9

The frame structure in Figure 8.37*a* is subjected to a moment M_o at joint B and undergoes a vertical settlement δ_o at support C. Find the internal forces. Assume the

[32] Equation 8.30*a* may also be expressed in compact (partioned) matrix form,

$$\mathbf{\Delta} = [\boldsymbol{\beta} \quad \boldsymbol{\beta}_s]\begin{Bmatrix}\mathbf{u} \\ \mathbf{u}_s\end{Bmatrix} = \boldsymbol{\beta}^*\mathbf{u}^*$$

but we will leave it as is for the sake of clarity.

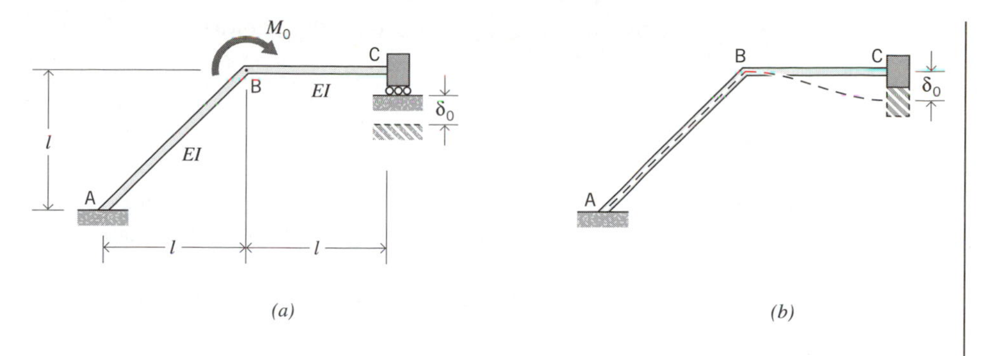

Figure 8.37 Example 9.

beam elements are axially rigid. (This structure has been analyzed for the same load and settlement using the force method; see Example 7 in Chapter 5, and Figure 5.34.)

Step 1. The structure has the two system displacements illustrated in Figure 8.8.

Step 2. The relationship between system displacements and deformations for this structure has been analyzed in Example 4, and the β matrix is exactly the same as in Equation 8.22*c*; however, we must now append to Equation 8.22*c* the beam deformations associated with the support settlement. Figure 8.37*b* shows the motions due to δ_o. In this case, the only end motion is the vertical settlement at C.[33] For the motions in Figure 8.37*b* we have $\theta_{AB} = \theta_{BA} = \theta_{BC} = \theta_{CB} = 0$ and $\psi_{AB} = 0$, but $\psi_{BC} = [(0 - (-\delta_o)]/l = \delta_o/l$, so the end deformations are given by

$$\boldsymbol{\Delta} \equiv \begin{Bmatrix} \Delta_{AB} \\ \Delta_{BA} \\ \Delta_{BC} \\ \Delta_{CB} \end{Bmatrix} = \begin{Bmatrix} \theta_{AB} - \psi_{AB} \\ \theta_{BA} - \psi_{AB} \\ \theta_{BC} - \psi_{BC} \\ \theta_{CB} - \psi_{BC} \end{Bmatrix} = \begin{bmatrix} 0 \\ 0 \\ -1/l \\ -1/l \end{bmatrix} \{\delta_o\} \tag{8.31a}$$

which defines the $\boldsymbol{\beta}_s$ matrix.

Therefore, for this example, Equation 8.30*a* becomes

$$\boldsymbol{\Delta} \equiv \begin{Bmatrix} \Delta_{AB} \\ \Delta_{BA} \\ \Delta_{BC} \\ \Delta_{CB} \end{Bmatrix} = \begin{bmatrix} 0 & -1/l \\ 1 & -1/l \\ 1 & 1/l \\ 0 & 1/l \end{bmatrix} \begin{Bmatrix} u_1 \\ u_2 \end{Bmatrix} + \begin{bmatrix} 0 \\ 0 \\ -1/l \\ -1/l \end{bmatrix} \{\delta_o\} \tag{8.31b}$$

Step 3. The matrix force-deformation relation is given by Equation 8.22*f* with $\widehat{\mathbf{P}} = \mathbf{0}$ since there are no interior loads (including thermal effects) on either beam element,

$$\begin{Bmatrix} M_{AB} \\ M_{BA} \\ M_{BC} \\ M_{CB} \end{Bmatrix} = \frac{\text{EI}}{\sqrt{2}l} \begin{bmatrix} 4 & 2 & & \\ 2 & 4 & & \\ & & 4\sqrt{2} & 2\sqrt{2} \\ & & 2\sqrt{2} & 4\sqrt{2} \end{bmatrix} \begin{Bmatrix} \Delta_{AB} \\ \Delta_{BA} \\ \Delta_{BC} \\ \Delta_{CB} \end{Bmatrix} + \begin{Bmatrix} 0 \\ 0 \\ 0 \\ 0 \end{Bmatrix} \tag{8.31c}$$

[33] In other words, Figure 8.37*b* is exactly the same as a sketch of motions due to a system displacement "like" δ_o; the only distinction is that δ_o is a known rather than unknown quantity.

Step 4. The only load is M_o applied at joint B; therefore, $\mathbf{F^o} = \mathbf{0}$ but

$$\mathbf{F} = \begin{Bmatrix} M_o \\ 0 \end{Bmatrix} \tag{8.31d}$$

since there is external virtual work done by M_o in Figure 8.37*b* going through virtual displacements like the motions in Figure 8.8*b*. This is the last matrix required for the solution.

Step 5. The **k** matrix is the same as computed in Equation 8.23*a*. Equation 8.30*e* leads to

$$\overline{\mathbf{F}}^* = \begin{Bmatrix} M_o \\ 0 \end{Bmatrix} - \frac{EI}{\sqrt{2}l}\begin{bmatrix} 0 & 1 & 1 & 0 \\ -1/l & -1/l & 1/l & 1/l \end{bmatrix}\begin{bmatrix} 4 & 2 & & \\ 2 & 4 & & \\ & & 4\sqrt{2} & 2\sqrt{2} \\ & & 2\sqrt{2} & 4\sqrt{2} \end{bmatrix}\begin{bmatrix} 0 \\ 0 \\ -1/l \\ -1/l \end{bmatrix}\{\delta_o\} \tag{8.31e}$$

$$= \begin{Bmatrix} M_o + 6EI\delta_o/l^2 \\ 12EI\delta_o/l^3 \end{Bmatrix}$$

so we obtain

$$\mathbf{u} = \mathbf{k}^{-1}\overline{\mathbf{F}}^* = \begin{Bmatrix} 0.150lM_o/EI + 0.744\delta_o/l \\ -0.013l^2M_o/EI + 0.522\delta_o \end{Bmatrix} \tag{8.31f}$$

Back-substituting into Equation 8.31*b* gives

$$\begin{Bmatrix} \Delta_{AB} \\ \Delta_{BA} \\ \Delta_{BC} \\ \Delta_{CB} \end{Bmatrix} = \begin{Bmatrix} 0.013M_ol/EI - 0.522\delta_o/l \\ 0.163M_ol/EI + 0.222\delta_o/l \\ 0.137M_ol/EI + 0.266\delta_o/l \\ -0.013M_ol/EI - 0.478\delta_o/l \end{Bmatrix} \tag{8.31g}$$

Substituting Equation 8.31*g* into Equation 8.31*c* leads to exactly the same result as in Equation 5.52*e*.

8.4.2 Structures Composed of Several Types of Elements

The displacement method is particularly well suited to the analysis of structures composed of various types of elements, as the final example of this chapter will demonstrate.

EXAMPLE 10

The structure in Figure 8.38*a* consists of two axially rigid beams, AB and BC, which are joined by a hinge at B. There is a linear torsional spring of stiffness k_2 between the

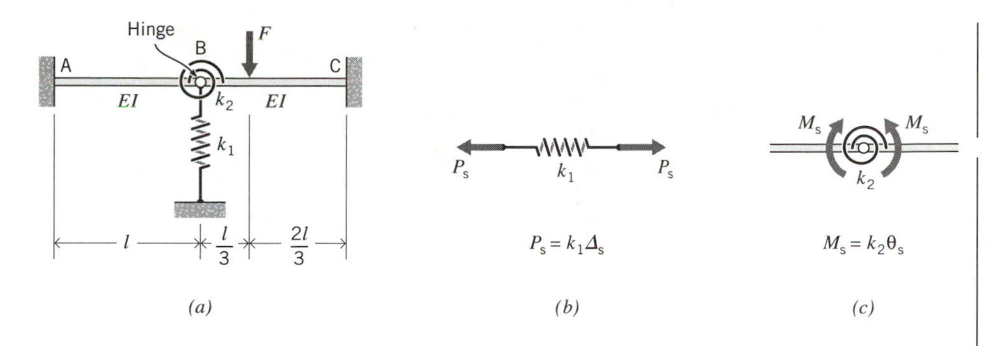

Figure 8.38 Example 10.

adjacent ends of the beams,[34] and also a linear axial spring of stiffness k_1 acting perpendicular to the beams at B. The only load is the force F shown. Find the internal forces.

Step 1. There are three system displacements for this structure, as shown in Figure 8.39. Note that u_1 is the rotation of end B of beam AB, and u_2 is the rotation of end B of beam BC; the two rotations are independent because of the hinge at B. System displacement u_3 is defined as the transverse displacement at B (upward positive).

Step 2. The relationships between beam deformations and system displacements are similar to those in preceding examples and will not be discussed in detail. We note that for system displacement u_3 the side-sways are $\psi_{AB} = -u_3/l$ and $\psi_{BC} = u_3/l$. Also, the axial spring is assumed to have a force-deformation relationship of the type defined in Figure 8.38*b*, i.e., tensile forces and increases in length are positive quantities (as for truss bars).

The torsional spring, however, requires some attention to establish an appropriate sign convention. We will use the force-deformation relationship defined in Figure 8.38*c* as the basis for our convention. This figure arbitrarily defines the torsional moment shown as positive, which implies that angle changes in the same direction are also positive. We must rely on this type of definition because the forces and deformations of a torsional spring do not have the same type of clear physical associations as for an axial spring.

Given the definition in Figure 8.38*c*, it follows that the system displacement in Figure 8.39*b* causes a positive deformation of the torsional spring of magnitude u_1, and the system displacement in Figure 8.39*c* causes a negative deformation of magnitude u_2. Thus, the compatibility equation is

$$\boldsymbol{\Delta} \equiv \begin{Bmatrix} \Delta_{AB} \\ \Delta_{BA} \\ \Delta_{BC} \\ \Delta_{CB} \\ \Delta_s \\ \theta_s \end{Bmatrix} = \begin{Bmatrix} \theta_{AB} - \psi_{AB} \\ \theta_{BA} - \psi_{AB} \\ \theta_{BC} - \psi_{BC} \\ \theta_{CB} - \psi_{BC} \\ \Delta_s \\ \theta_s \end{Bmatrix} = \begin{bmatrix} 0 & 0 & 1/l \\ 1 & 0 & 1/l \\ 0 & 1 & -1/l \\ 0 & 0 & -1/l \\ 0 & 0 & 1 \\ 1 & -1 & 0 \end{bmatrix} \begin{Bmatrix} u_1 \\ u_2 \\ u_3 \end{Bmatrix} \qquad \textbf{(8.32a)}$$

[34] This type of spring was introduced in Example 8 of Chapter 5; see Figure 5.35.

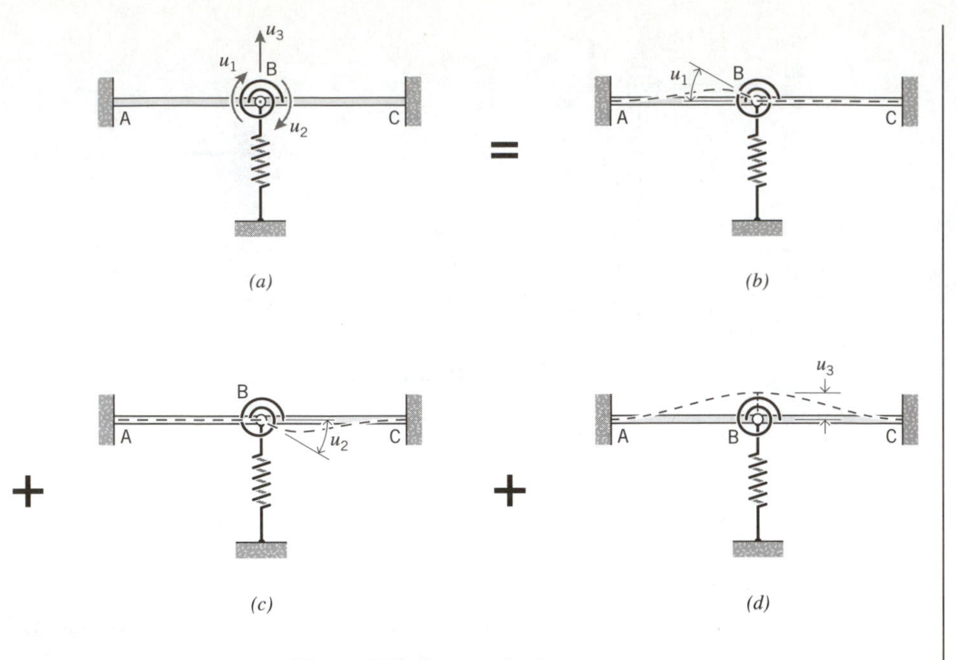

Figure 8.39 System displacements.

Step 3. The matrix force-deformation relation is now fairly straightforward,

$$\begin{Bmatrix} M_{AB} \\ M_{BA} \\ M_{BC} \\ M_{CB} \\ P_s \\ M_s \end{Bmatrix} = \frac{EI}{l} \begin{bmatrix} 4 & 2 & & & & \\ 2 & 4 & & & & \\ & & 4 & 2 & & \\ & & 2 & 4 & & \\ & & & & k_1 l/EI & \\ & & & & & k_2 l/EI \end{bmatrix} \begin{Bmatrix} \Delta_{AB} \\ \Delta_{BA} \\ \Delta_{BC} \\ \Delta_{CB} \\ \Delta_s \\ \theta_s \end{Bmatrix} + \begin{Bmatrix} 0 \\ 0 \\ -4Fl/27 \\ 2Fl/27 \\ 0 \\ 0 \end{Bmatrix} \qquad \textbf{(8.32b)}$$

Note the fixed-end moments for beam BC (Figure 8.38*a*), which are obtained from Equations 8.28*b* and 8.28*c* (with $a = l/3$, $b = 2l/3$).

Step 4. There are no concentrated forces applied in the vicinity of joint B, so all components of **F** are zero. Similarly, since there are no side-sway components of motion associated with (virtual) displacements like Figures 8.39*b* and 8.39*c*, the first two components of $\mathbf{F^o}$ are also zero. The third component of $\mathbf{F^o}$, however, is not zero because there is virtual work done by the real interior force F on beam BC (Figure 8.38*a*) moving through a (virtual) rigid-body side-sway component of motion associated with $u_3 = 1$. The virtual displacement at the point of application of F has magnitude 2/3, and is directed upward; therefore, the virtual work is $-2F/3$, so we have

$$\mathbf{F^o} + \mathbf{F} = \begin{Bmatrix} 0 \\ 0 \\ -2F/3 \end{Bmatrix} + \begin{Bmatrix} 0 \\ 0 \\ 0 \end{Bmatrix} = \begin{Bmatrix} 0 \\ 0 \\ -2F/3 \end{Bmatrix} \qquad \textbf{(8.32c)}$$

Step 5. From the preceding matrices we obtain

$$\mathbf{k} = \boldsymbol{\beta}^{\mathrm{T}}\mathbf{K}\boldsymbol{\beta} = \begin{bmatrix} k_2 + \dfrac{4EI}{l} & -k_2 & \dfrac{6EI}{l^2} \\ -k_2 & k_2 + \dfrac{4EI}{l} & -\dfrac{6EI}{l^2} \\ \dfrac{6EI}{l^2} & -\dfrac{6EI}{l^2} & k_1 + \dfrac{24EI}{l^3} \end{bmatrix} \tag{8.32d}$$

and

$$\mathbf{F}^* = (\mathbf{F}^{\mathrm{o}} + \mathbf{F}) - \boldsymbol{\beta}^{\mathrm{T}}\hat{\mathbf{P}} = \begin{Bmatrix} 0 \\ 4Fl/27 \\ -20F/27 \end{Bmatrix} \tag{8.32e}$$

Solving for **u** leads to

$$\begin{Bmatrix} u_1 \\ u_2 \\ u_3 \end{Bmatrix} = \frac{1}{54EI\left(24 + \dfrac{12EI}{k_2 l} + \dfrac{2k_1 l^2}{k_2} + \dfrac{k_1 l^3}{EI}\right)} \begin{Bmatrix} Fl^2\left(24 + \dfrac{k_1 l^3}{EI} + \dfrac{84EI}{k_2 l}\right) \\ Fl^2\left(24 + \dfrac{k_1 l^3}{EI} + \dfrac{4k_1 l^2}{k_2} - \dfrac{60EI}{k_2 l}\right) \\ -8Fl^3\left(5 + \dfrac{7EI}{k_2 l}\right) \end{Bmatrix} \tag{8.32f}$$

Back-substituting gives

$$\begin{Bmatrix} \Delta_{\mathrm{AB}} \\ \Delta_{\mathrm{BA}} \\ \Delta_{\mathrm{BC}} \\ \Delta_{\mathrm{CB}} \\ \Delta_s \\ \theta_s \end{Bmatrix} = \frac{1}{54EI\left(24 + \dfrac{12EI}{k_2 l} + \dfrac{2k_1 l^2}{k_2} + \dfrac{k_1 l^3}{EI}\right)} \begin{Bmatrix} 8Fl^2\left(5 + \dfrac{7EI}{k_2 l}\right) \\ Fl^2\left(-16 + \dfrac{k_1 l^3}{EI} + \dfrac{28EI}{k_2 l}\right) \\ Fl^2\left(64 + \dfrac{k_1 l^3}{EI} - \dfrac{4EI}{k_2 l} + \dfrac{4k_1 l^2}{k_2}\right) \\ 8Fl^2\left(5 + \dfrac{7EI}{k_2 l}\right) \\ -8Fl^3\left(5 + \dfrac{7EI}{k_2 l}\right) \\ 4Fl^2\left(\dfrac{36EI}{k_2 l} - \dfrac{k_1 l^2}{k_2}\right) \end{Bmatrix} \tag{8.32g}$$

and

$$\begin{Bmatrix} M_{AB} \\ M_{BA} \\ M_{BC} \\ M_{CB} \\ P_s \\ M_s \end{Bmatrix} = \frac{1}{324\left(2 + \dfrac{EI}{k_2 l} + \dfrac{k_1 l^2}{6k_2} + \dfrac{k_1 l^3}{12EI}\right)} \begin{Bmatrix} Fl\left(-96 + \dfrac{k_1 l^3}{EI} - \dfrac{84EI}{k_2 l}\right) \\ 2Fl\left(-36 + \dfrac{k_1 l^3}{EI}\right) \\ -2Fl\left(-36 + \dfrac{k_1 l^3}{EI}\right) \\ Fl\left(192 + \dfrac{3k_1 l^3}{EI} + \dfrac{132EI}{k_2 l} + \dfrac{8k_1 l^2}{k_2}\right) \\ -4F\left(\dfrac{5k_1 l^3}{EI} + \dfrac{7k_1 l^2}{k_2}\right) \\ 2Fl\left(36 - \dfrac{k_1 l^3}{EI}\right) \end{Bmatrix} \tag{8.32h}$$

It is interesting to note that in the limit as $k_2 \to \infty$ the connection at B approaches a rigid joint, and the structure becomes identical to the one in Example 2 of Chapter 6 (Figure 6.12*a*). From Equation 8.32*h* we obtain

$$\underset{\lim k_2 \to \infty}{\mathbf{P}} = \frac{Fl}{27(1 + 24\gamma)} \begin{Bmatrix} (1 - 96\gamma) \\ 2(1 - 36\gamma) \\ -2(1 - 36\gamma) \\ 3(1 + 64\gamma) \\ -20/l \\ -2(1 - 36\gamma) \end{Bmatrix} \tag{8.32i}$$

where $\gamma = EI/k_1 l^3$. This is exactly the same as the solution shown in Figure 6.14*c*.

PROBLEMS

8.P1. Find the bar forces. Assume that all bars have the samed *EA*. (Compare results to the solution obtained in Problem 5.P9, Chapter 5.)

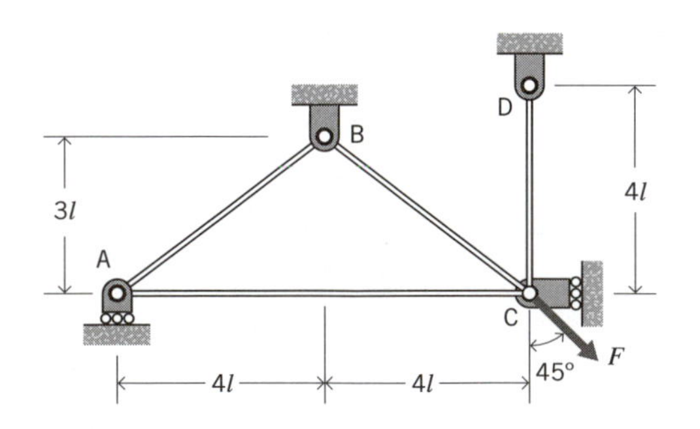

8.P2. Find the bar forces. Assume that all bars have the same EA. The only load is a uniform temperature change in bar CD. (Compare results to the solution obtained in Problem 5.P12, Chapter 5.)

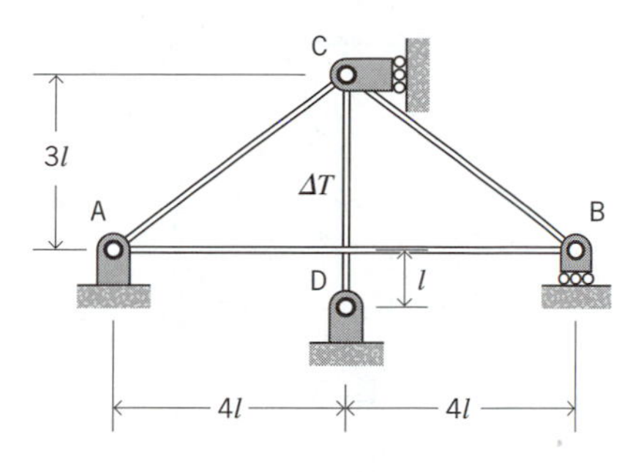

8.P3. Find the bar forces. Assume that all bars have the same EA. (Compare results to the solution obtained in Problem 5.P4, Chapter 5.)

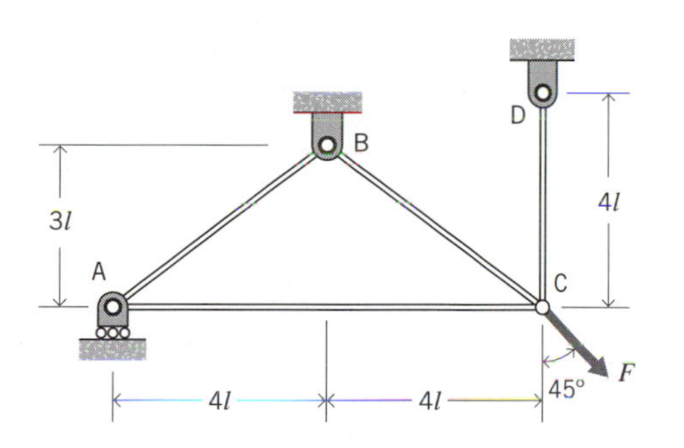

8.P4. Solve for the bar forces. Assume that all bars have the same EA except bar CD, which is axially rigid.

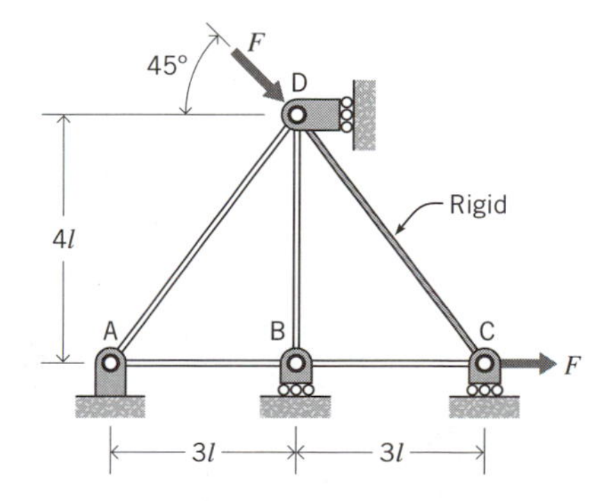

8.P5. Find the bar forces. Assume that all bars have the same EA. (Compare results to the solution obtained in Problem 5.P11, Chapter 5.)

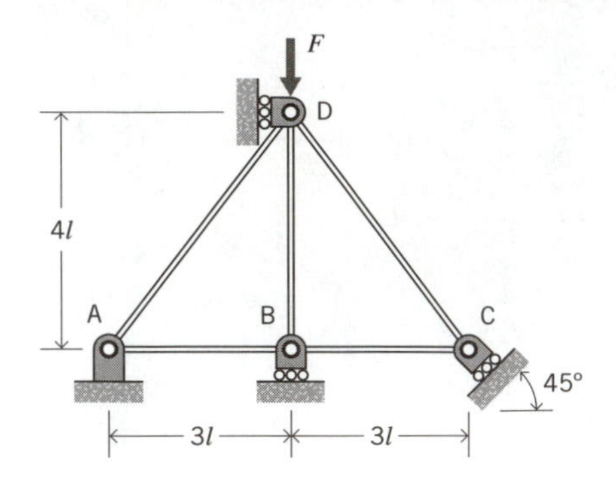

8.P6. Solve for the bar forces in the truss in Problem 8.P5. Assume that all bars have the same EA except bar CD, which is axially rigid.

8.P7. Solve for the bar forces. Assume that all bars have the same EA except bar CD, which is axially rigid.

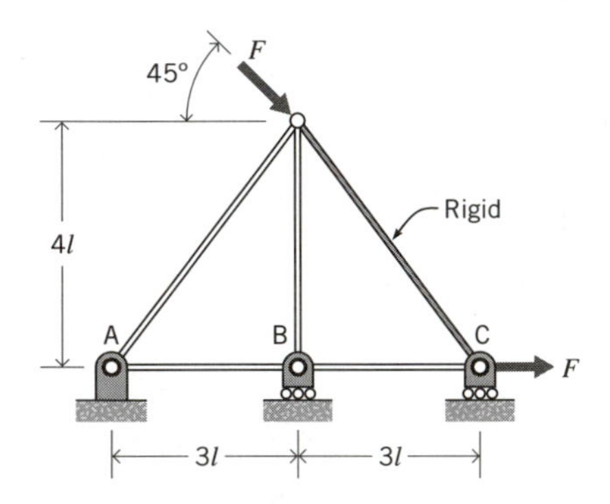

8.P8. Find the bar forces. Assume that all bars have the same EA. (Compare results to the solution obtained in Problem 5.P1, Chapter 5.)

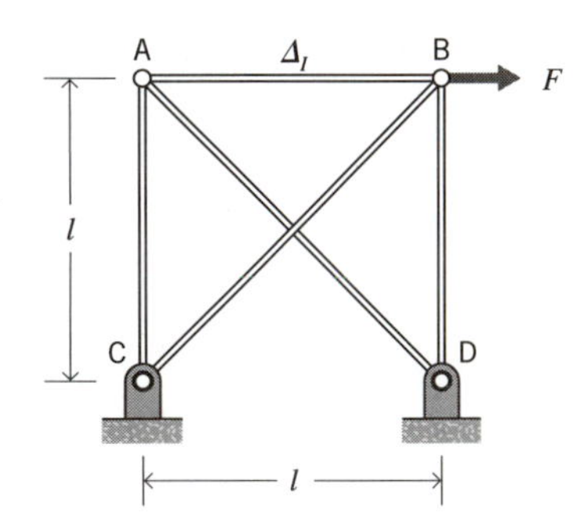

8.P9. Find the bar forces. Assume that all bars have the same EA. (Compare results to the solution obtained in Problem 6.P1, Chapter 6.)

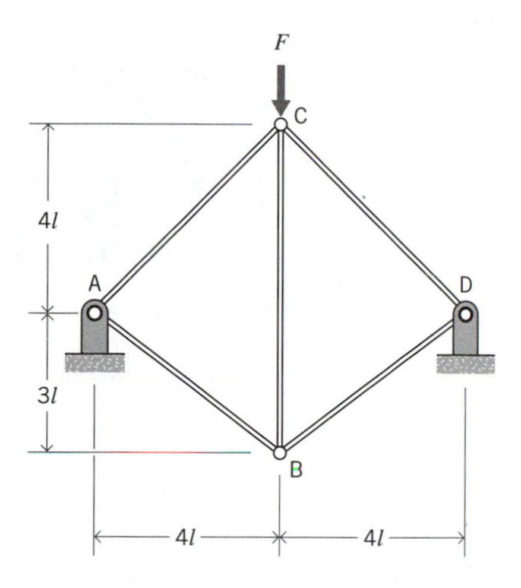

8.P10. Find the bar forces. Assume that all bars have the same EA. (Compare results to the solution obtained in Problem 5.P6, Chapter 5.)

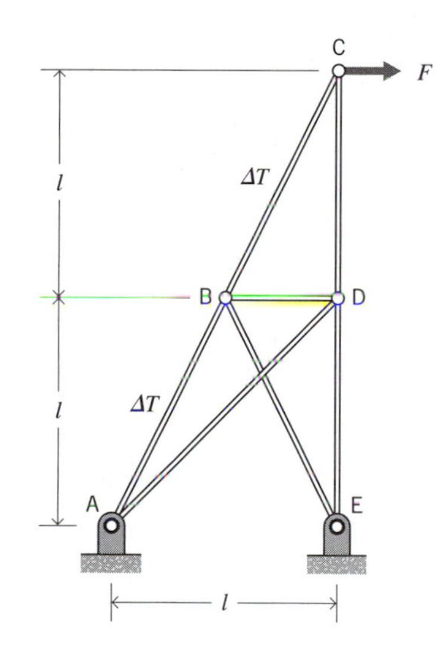

8.P11. Find the bar forces. Assume that all bars have the same EA. (Compare results to the solution obtained in Problem 5.P3, Chapter 5.)

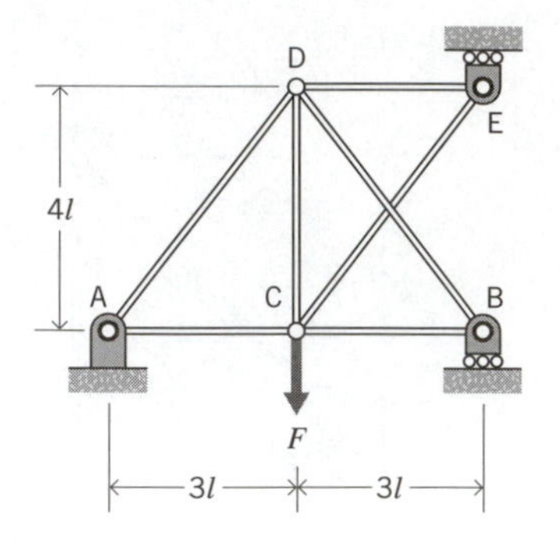

8.P12. A heavy concrete block of weight W is supported by a truss structure and spring, as shown. All bars have the same EA.

(a) Find the bar forces and the force in the spring.

(b) Find the value of spring stiffness k for which $\theta = 0$.

(Compare the results to the solution obtained in Problem 5.P7, Chapter 5.)

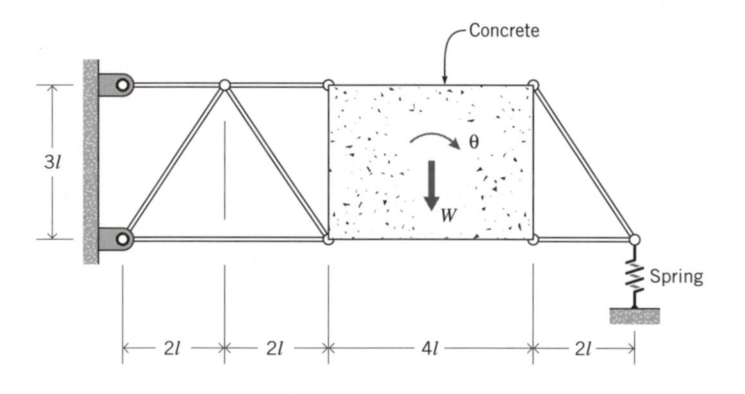

8.P13. Find the bar forces. Assume that all bars have the same EA. Take advantage of symmetry and analyze only one-half of the truss. (Compare the results to the solution obtained in Example 1, Chapter 6.)

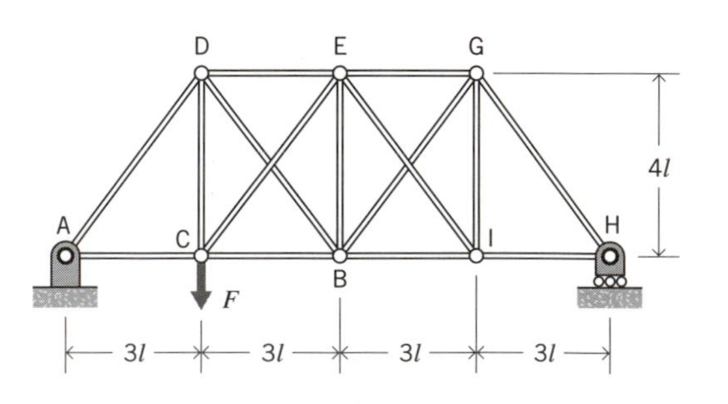

8.P14. Find the bar forces. Assume that all bars have the same EA. Take advantage of symmetry and analyze only one-half of the truss. (Compare the results to the solution obtained in Problem 5.P16, Chapter 5 or Problem 6.P4, Chapter 6.)

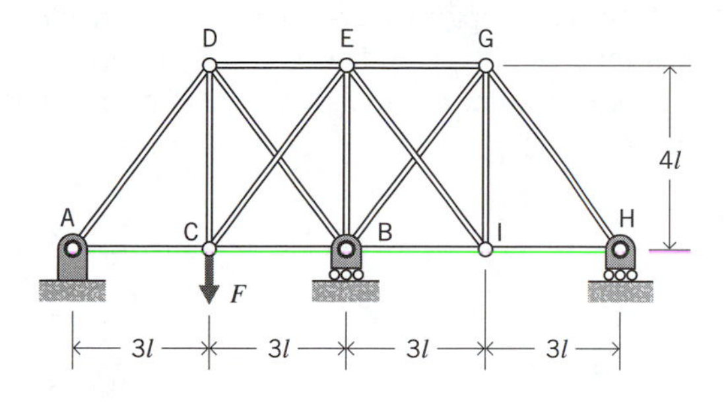

8.P15. The truss is supported by three linear springs with spring constant k. Assume that all bars have the same EA. Find the forces in the springs by analyzing one-half of the structure. Solve the equations for two cases: (a) $EA/kl \ll 1$, (b) $EA/kl \gg 1$. (Compare the results to the solution obtained in Problem 5.P17, Chapter 5 or Problem 6.P5, Chapter 6.)

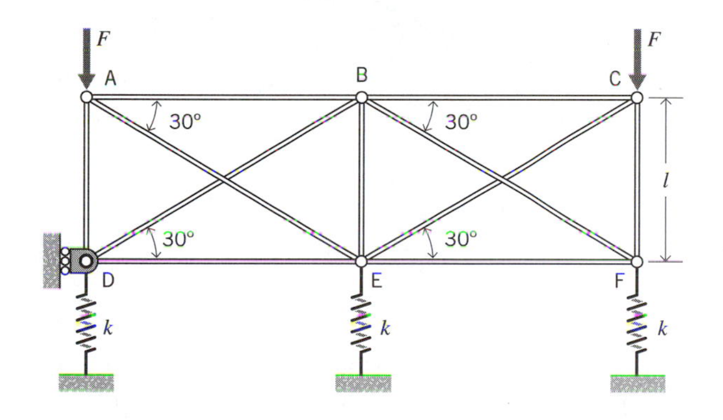

8.P16. A "twisted" beam/truss structure has the configuration shown before any loads are applied. Beam element DEF is rigid, truss bars AD, BE, and CF all have stiffness EA. The dotted lines in the figure lie in the xy plane; beam DEF lies in a plane parallel to the yz plane, bars AD and CF lie in planes parallel to the xz plane. The support at point E permits axial displacement u_x and rotation about the x axis θ_x; all other motions are zero at this support. Loads F_x and M_x are applied at joint E. Find the forces in bars AD, BE, and CF. Find the rotation when $M_x = 0$.

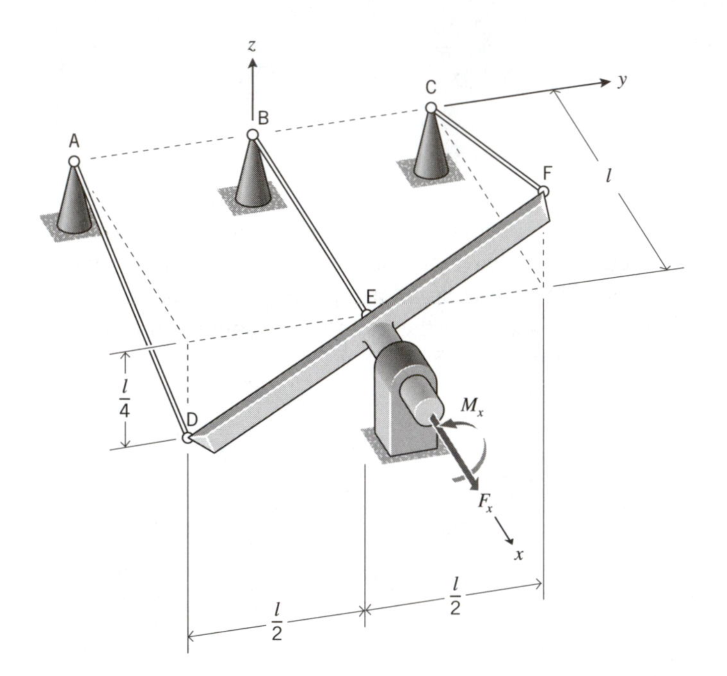

8.P17. Show that the system displacements in Equations 8.10f and 8.12d define identical motions at joint E of the truss in Figure 8.4.

8.P18. The structure has a support at A that is free to roll horizontally but cannot rotate; the support at B can move vertically but also cannot rotate. End B is supported by a linear axial spring. Find the force in the spring and all the reactions. Let $k = 3EI/2l^3$. (Compare the results to the solution obtained in Problem 5.P27, Chapter 5.)

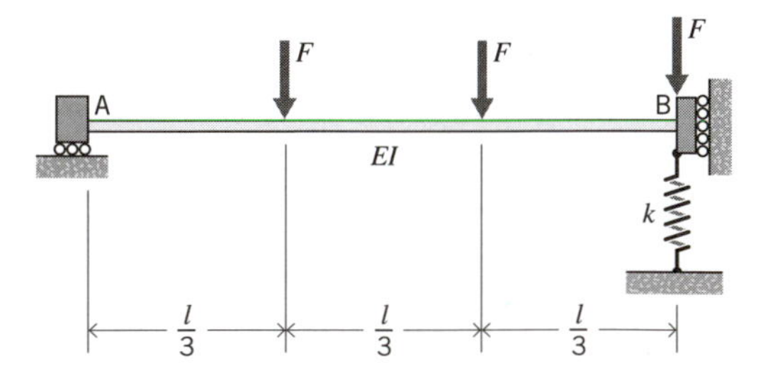

8.P19. The beam is fixed at B and pinned to a roller support at A. There is a linear torsional spring connecting the beam to the support at A. The relationship between moment and deformation for the spring is $M_s = k\Delta_s$ where k is the spring constant and Δ_s is the deformation (angle change). Sketch the moment diagram for the case $k = EI/2l$. (Compare the results to the solution obtained in Problem 5.P30, Chapter 5.)

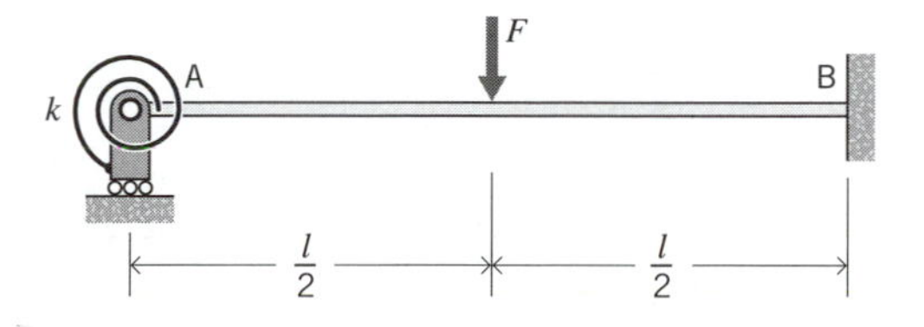

8.P20. Find the internal forces and sketch the moment diagram for the structure.

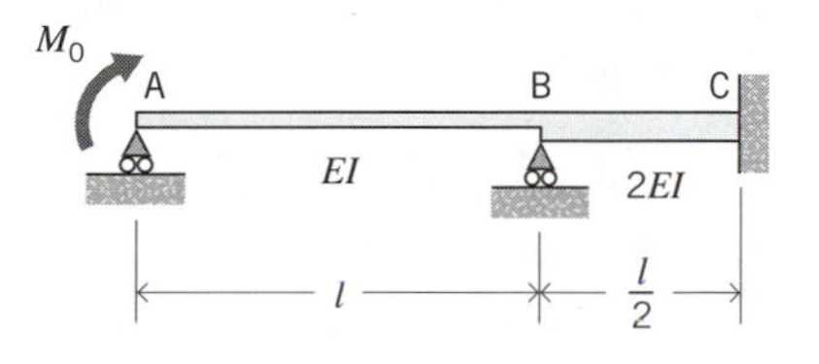

8.P21. Find the internal forces and sketch the moment diagram for the structure.

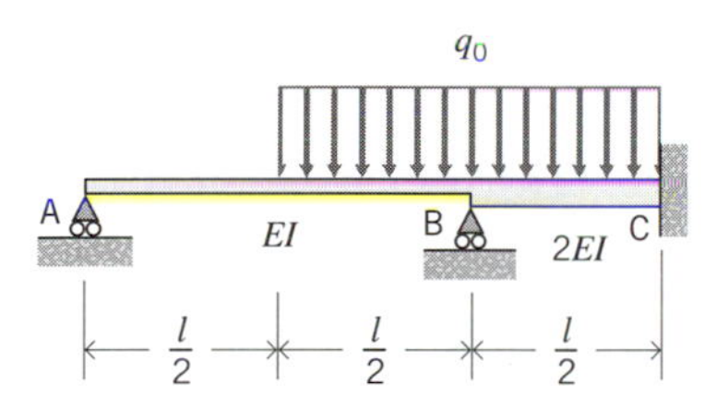

8.P22. The beam is fixed at B and supported by a linear spring with spring constant k at A.

(a) Find the spring force as a function of M_o, l, EI, and k.
(b) Sketch the moment diagram for the special case $k = 3EI/2l^3$.
(c) From (a), find the spring force as $k \to \infty$.
(Compare the results to the solution obtained in Problem 5.P20, Chapter 5.)

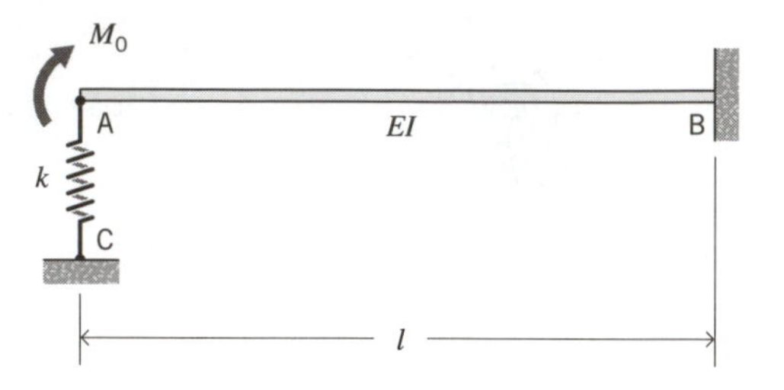

8.P23. The beam is fixed at B and supported by a linear spring with spring constant k at A. The structure is loaded by a moment M_o applied at the midpoint of the beam.

(a) Find the spring force as a function of M_o, l, EI, and k.
(b) Sketch the moment diagram for the special case $k = 3EI/2l^3$.
(c) From (a), find the spring force as $k \to \infty$.
(Compare the results to the solution obtained in Problem 5.P21, Chapter 5.)

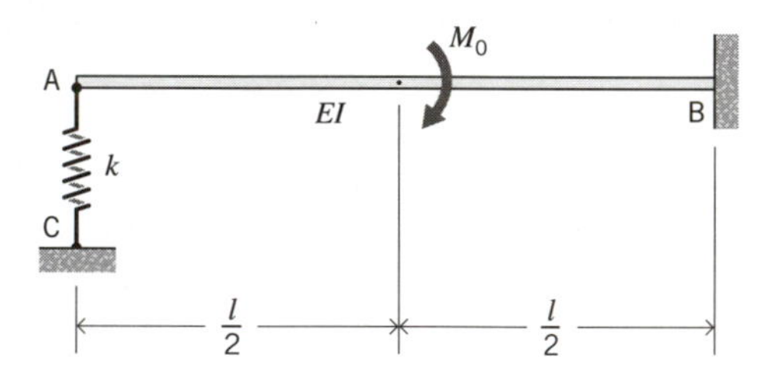

8.P24. The structure consists of a beam AB and a cable BC. The support at B can displace vertically but cannot rotate or displace horizontally. The cable stiffness is such that $EA = 3\sqrt{2}EI/l^2$. Find the force in the cable and sketch the moment diagram for the beam. (Compare the results to the solution obtained in Problem 5.P19, Chapter 5.)

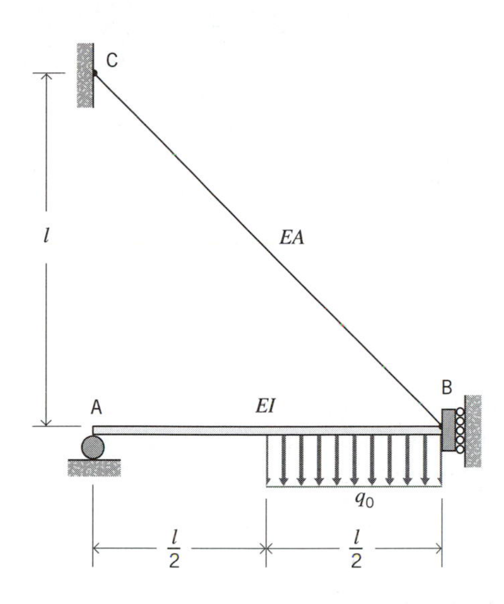

8.P25. For the structure in Figure 8.25, use equilibrium analyses of free-bodies to show directly that $f_2 = F + q_o l/2$ in Equation 8.22g.

8.P26. Find the end forces on beams AB and BC. (Compare the results to the solution obtained in Problem 5.P28, Chapter 5.)

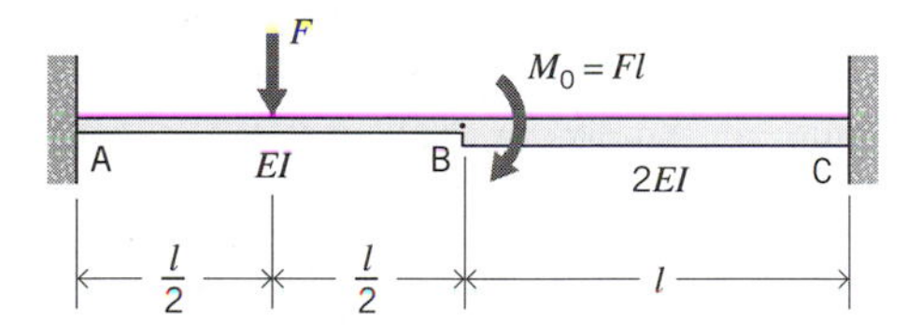

8.P27. The support at C can move vertically but cannot rotate. Find the end forces on each component. (Compare the results to the solution obtained in Problem 5.P29, Chapter 5.)

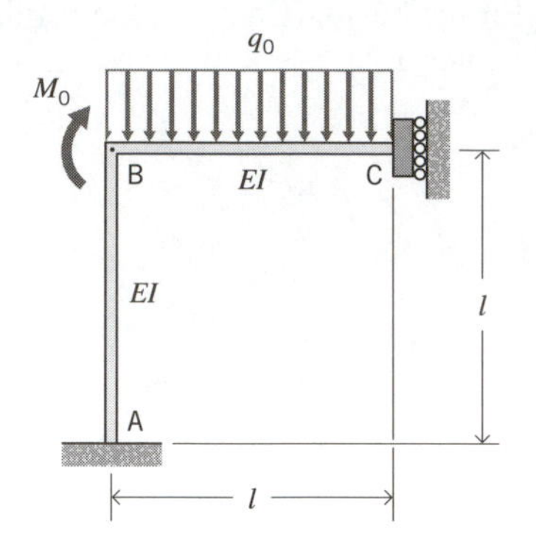

8.P28. The structure shown is composed of axially rigid beams AB and BC rigidly connected at B, and is subjected to a uniform load q_o per unit of horizontal length. Find the internal forces and sketch the moment diagram. Identify the location and magnitude of the maximum moment.

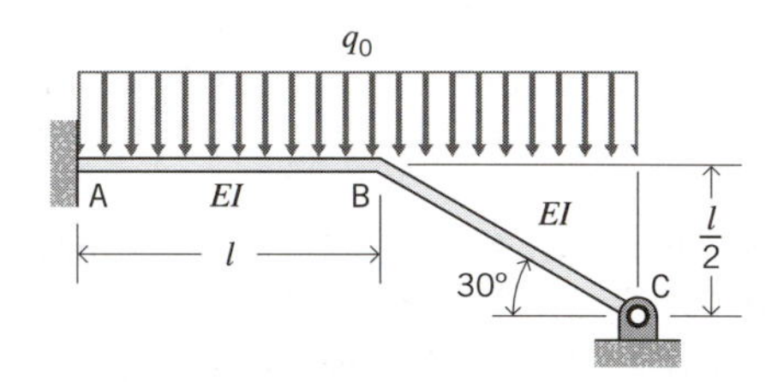

8.P29. Suppose the pin support at C in Problem 8.P28 is replaced by a horizontal roller. Find the internal forces and sketch the moment diagram.

8.P30. Sketch the moment diagrams for beams AB and BC. (Compare the results to the solution obtained in Problem 5.P23, Chapter 5.)

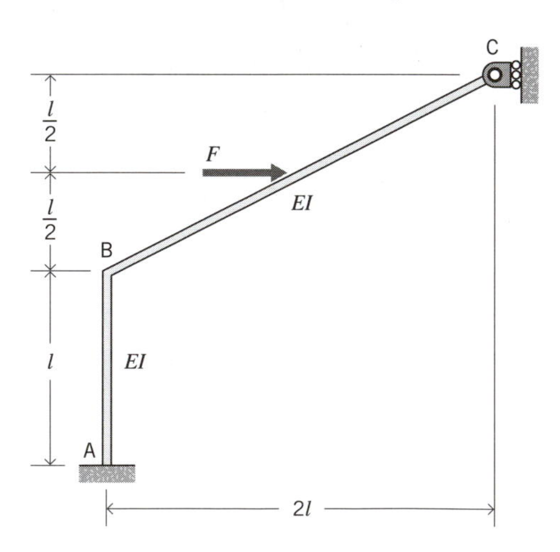

8.P31. The structure consists of a uniform axially rigid beam ABC and a bar BD. There is a fixed support at C, which can translate but cannot rotate. Assume that the axial stiffness of the bar is $EA = 3\sqrt{2}EI/l^2$. Find the internal forces. (Compare the results to the solution obtained in Problem 5.P31, Chapter 5.)

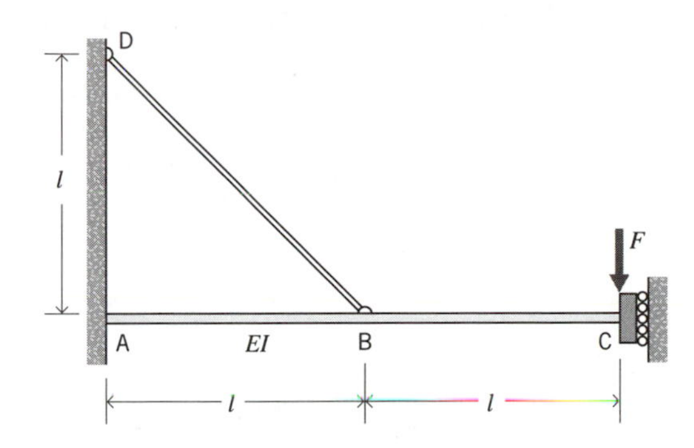

8.P32. Find the reactions and sketch the bending moment diagrams for all elements. (Compare the results to the solution obtained in Problem 5.P35, Chapter 5.)

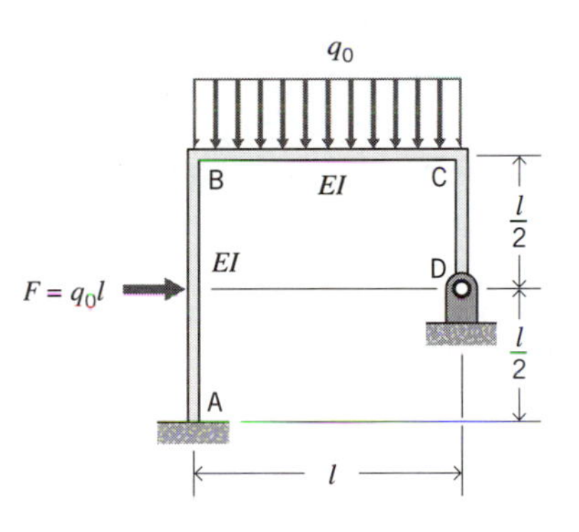

8.P33. The frame structure shown is composed of three uniform axially rigid beams AB, BC, and AC; the bending stiffness of all beams is *EI.* Connections between beams at joint A and C are rigid, but there is a pin connection between beams at joint B. The structure is symmetric about a vertical axis through B. Find the internal forces by analyzing the portion of the structure that lies to one side of the axis of symmetry. Show the results on free-body diagrams of all elements of the original structure.

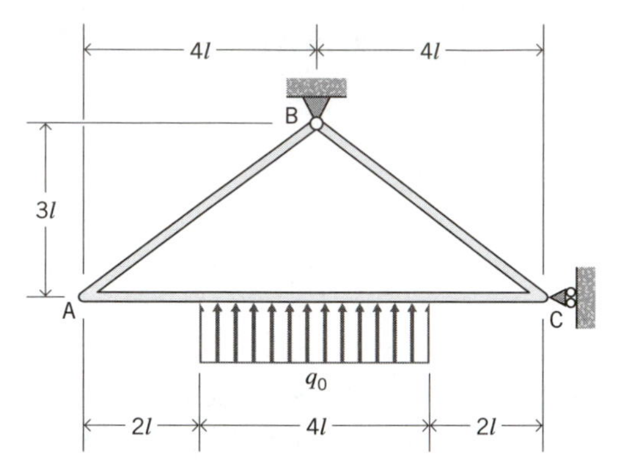

8.P34. The frame structure shown is composed of three uniform axially rigid beams AB, BC, and AC; the bending stiffness of all beams is *EI.* Connections between beams at joint A and C are rigid, but there is a pin connection between beams at joint B. The structure is symmetric about a vertical axis through B and is loaded by a moment applied at the midpoint of beam AC. Find the internal forces by analyzing the portion of the stucture that lies to one side of the axis of symmetry. Show the results on free-body diagrams of all elements of the original structure.

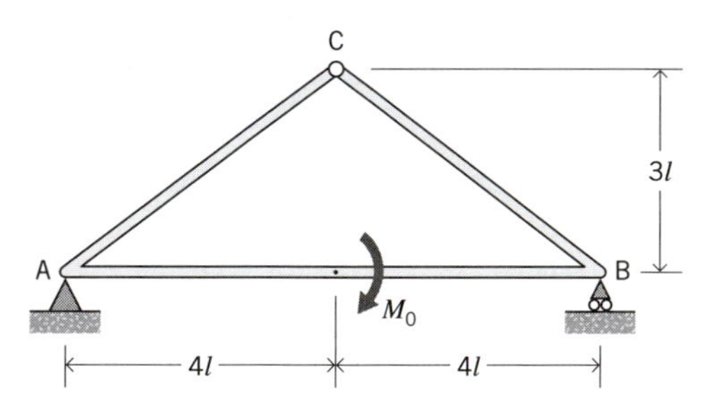

8.P35. The structure has a pin support at A and a fixed support with roller at C. The joints at ends A and B of beam AB are modeled as hinges with torsional springs for which $M_s = k\Delta_s$ where k is the spring constant and Δ_s is the deformation (angle change). There is a uniform load of magnitude q_o applied to half of beam BC, and a moment $M_o = 5q_o l^2/8$ applied to joint B. Find the internal forces. Assume $k = 2EI/l$. (Compare the results to the solution obtained in Problem 5.P36, Chapter 5.)

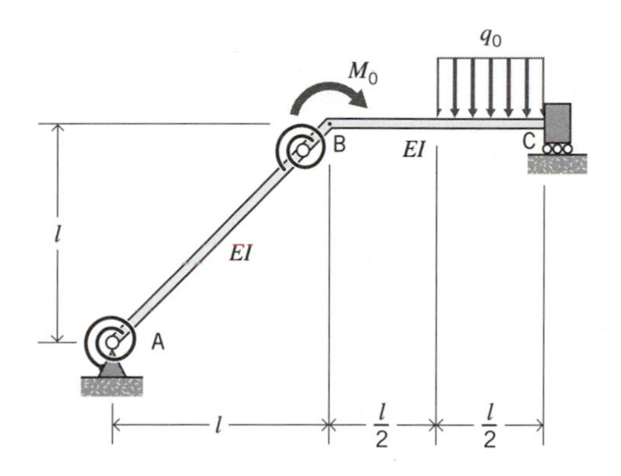

8.P36. The structure is a closed frame with two axes of symmetry. Solve for internal forces by analyzing beam AB only. Assume that the beams are axially rigid. Show the results by sketching free-body diagrams of all beams and joints. (Compare the results to the solution obtained in Problem 5.P41, Chapter 5 or Problem 6.P13, Chapter 6.)

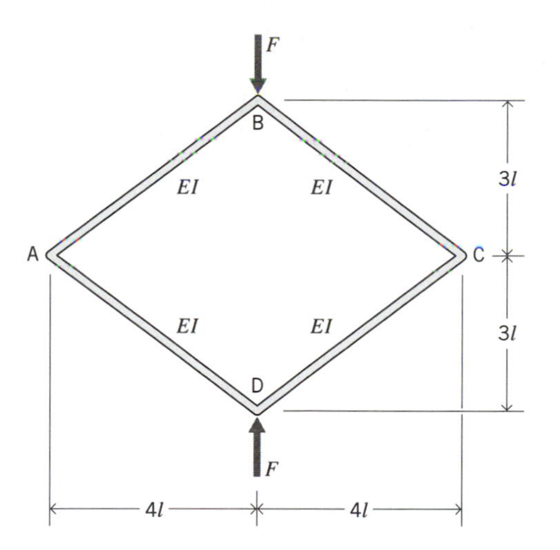

8.P37. The structure is a closed frame with two axes of symmetry. The frame is loaded by external moments applied at A and C. Assume that the beams are axially rigid. Solve for internal forces by analyzing beam AB only. Show the results by sketching free-body diagrams of all beams and joints. (Compare the results to the solution obtained in Problem 6.P14, Chapter 6.)

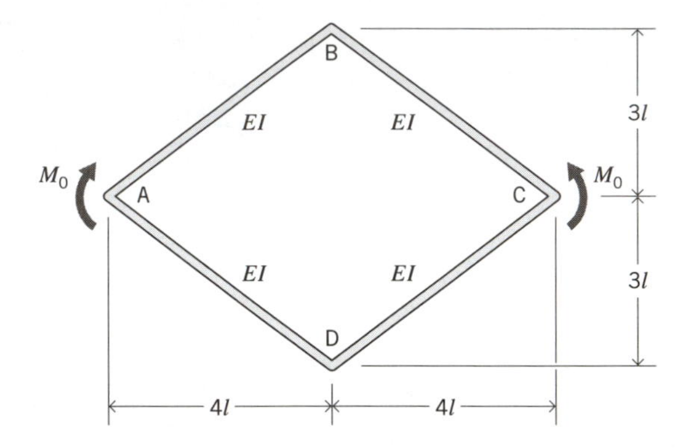

8.P38. A closed frame having two axes of symmetry is simply supported at A and C. Joints between elements at A and C are rigid, but joints between elements at B and D are pinned. Beams at joints B and D are connected by linear torsional springs; the relationship between moment and deformation for the springs is $M_s = k\Delta_s$ where k is the spring constant and Δ_s is the deformation (angle change). Assume $k = 2EI/5l$. Vertical loads F are applied at the midpoints of beams AB and BC. Assume that the beams are axially rigid. Solve for the internal forces by analyzing beam AB only. Show the results by sketching free-body diagrams of all beams and joints. (Compare the results to the solution obtained in Problem 6.P16, Chapter 6.)

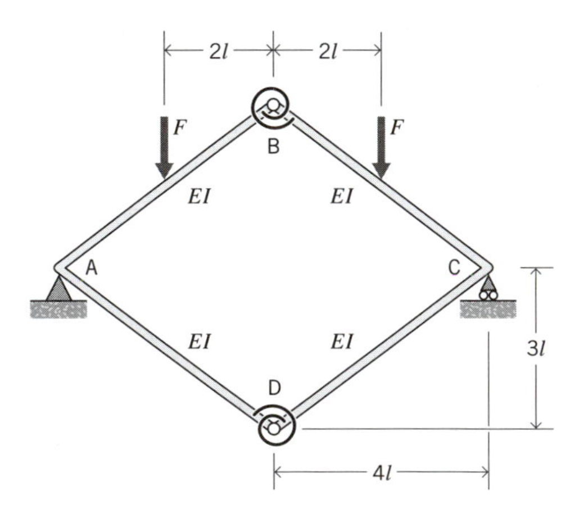

8.P39. (*a*) Consider a uniform beam IJ, which is subjected to a general set of loads and displacements except for M_{JI}, which is known to be zero, a situation that might occur if, for example, the J-end is a pin connection. Show that Equations 8.14 (or Equations 8.13*g*) reduce to

$$M_{IJ} = \frac{3EI}{l}(\theta_{IJ} - \psi_{IJ}) + \left(\widehat{M}_{IJ} - \frac{\widehat{M}_{JI}}{2}\right)$$

This modification of the beam stiffness relations simplifies the analysis somewhat because the rotation at the J-end is a known function of the remaining displacements and the applied loads and, thus, need not be treated as an unknown system displacement. (Also see Section 9.7.4.)

(*b*) Analyze the structure shown using the beam stiffness relation given in (*a*) to characterize the pin-ended members. Assume that all beams are axially rigid. Sketch the moment diagram for the structure.

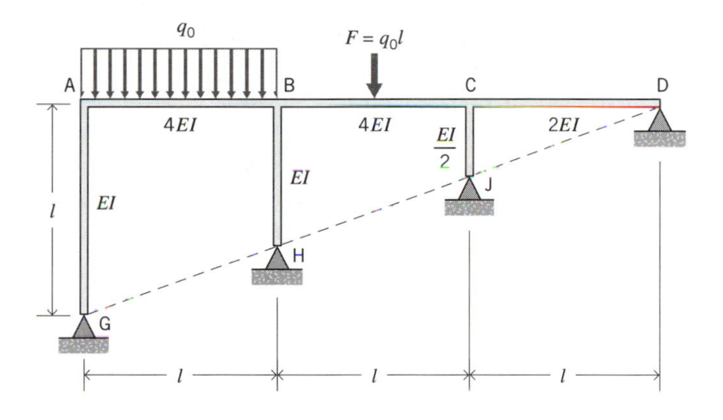

8.P40. Analyze the structure shown using the beam stiffness relation given in Problem 8.P39 where appropriate. Assume that all beams are axially rigid.

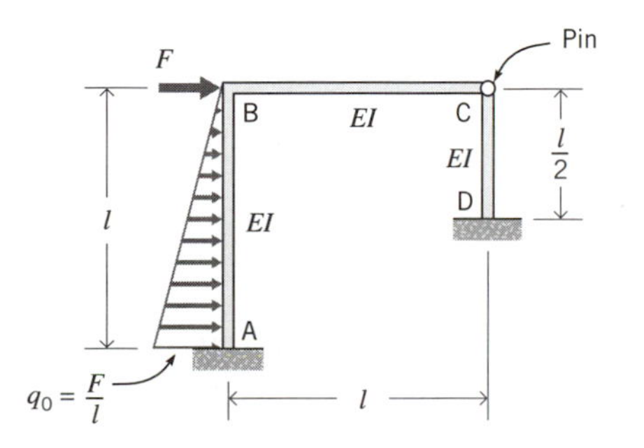

8.P41. The structure shown is an eccentrically braced frame. The structure is symmetric about a vertical axis through the midpoint of CC′. Beams AB and AC are pinned at A, and joints at B and C are rigid. Find the internal forces using the beam stiffness relation given in Problem 8.P39 where appropriate.

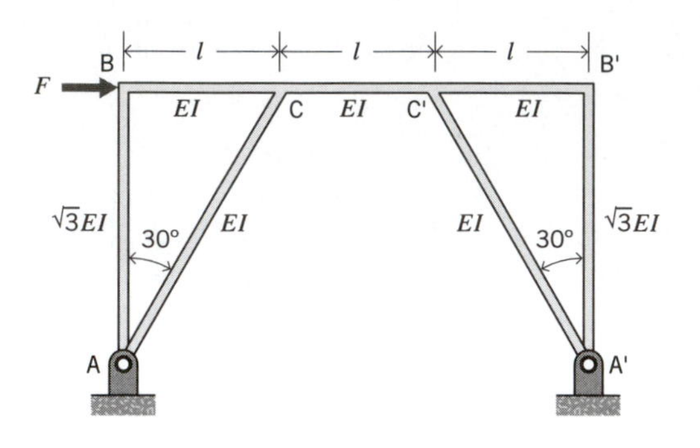

8.P42. Force F is removed from the truss in Problem 8.P3 but the roller support at A displaces vertically downward a small distance δ_o due to settlement.
(a) Find the bar forces.
(b) Find the horizontal displacement at A.
(Compare the results to the solution obtained in Problem 5.P43, Chapter 5.)

8.P43. Force F is removed from the structure in Problem 8.P31 but the support at A (only) undergoes a settlement consisting of a small counterclockwise rotation θ_o. Find the element forces and transverse displacement at C.

8.P44. Find the internal forces in the structure shown in Figure 8.37a due to a vertical settlement δ_o at A and a clockwise rotation θ_o at C.

Chapter 9

Displacement Method: Advanced Formulation

9.0 INTRODUCTION

In Chapter 8 we encountered two significant complications when we used the basic formulation of the displacement method to analyze planar structures. The first was the coupling of joint displacements when elements were modeled as axially rigid. This coupling required solution of some nontrivial geometry problems associated with structural motions. The second complication was in the determination of the proper contributions of applied physical loads to the equilibrium equations. Concentrated loads applied at joints were relatively easy to deal with, but interior loads on beams also appeared in the equilibrium equations whenever they did external virtual work during side-sway motions. Both of these factors made it necessary to exercise considerable care in generating the required matrices and virtually excluded the possibility of developing an efficient algorithm for computer implementation of the procedure.

As was noted previously, the first complication can be eliminated by removing the restrictive assumption of axial rigidity, and the second by utilizing force-displacement rather than force-deformation element stiffness relationships. This will require the use of larger matrices than those employed in Chapter 8, but will result in a procedure ideally suited to computer automation. In fact, as will be shown subsequently, it is possible to generate the system stiffness matrix directly from displacement information without performing any matrix multiplications.

We will begin by developing the element force-displacement relationships. In particular, we will reformulate the procedure for analysis of frame structures assuming axially rigid beams, and then include the effects of axial flexibility. Truss bars will be represented as a special class of elements with zero bending stiffness. Subsequently, the stiffness matrix developed with both bending and axial stiffness will be written in a global rather than element-level coordinate system. This will make it possible to directly correlate element and system displacements and, ultimately, avoid the triple matrix product $\boldsymbol{\beta}^T\mathbf{K}\boldsymbol{\beta}$ now required to find the system stiffness. This procedure is known as the direct stiffness method (see Section 9.6).

9.1 FORCE-DISPLACEMENT STIFFNESS MATRIX FOR BEAMS

For our reformulation, we seek to expand the two-by-two (2×2) moment-deformation relationship given by Equation 8.13*g* into a four-by-four (4×4) relationship between the four end forces in Figure 5.23*a* (M_{IJ}, M_{JI}, V_{IJ}, and V_{JI}) and the four end displacements in Figure 5.23*b* (θ_{IJ}, θ_{JI}, v_{IJ}, and v_{JI}). In other words, we seek four equations rather than the two in Equation 8.13*g*.

The first two of the desired equations are precisely those in Equation 8.13*g* (or Equation 8.14) except that we substitute the definition of ψ_{IJ} in terms of v_{IJ} and v_{JI} from Equation 8.13*i*,

$$
\begin{aligned}
M_{IJ} &= \frac{EI}{l}\left[4\theta_{IJ} + 2\theta_{JI} - 6\left(\frac{v_{IJ} - v_{JI}}{l}\right)\right] + \widehat{M}_{IJ} \\
M_{JI} &= \frac{EI}{l}\left[2\theta_{IJ} + 4\theta_{JI} - 6\left(\frac{v_{IJ} - v_{JI}}{l}\right)\right] + \widehat{M}_{JI}
\end{aligned} \tag{9.1a}
$$

The second two equations, for V_{IJ} and V_{JI}, are obtained directly from equilibrium, i.e., by superposition of the forces shown in Figures 5.23*a*(1) and 5.23*a*(2),

$$
\begin{aligned}
V_{IJ} &= V_{IJ_1} + V_{IJ_2} = -\frac{M_{IJ} + M_{JI}}{l} - \frac{1}{l}\int_0^l q(x)\,(l - x)\,dx \\
V_{JI} &= V_{JI_1} + V_{JI_2} = \frac{M_{IJ} + M_{JI}}{l} - \frac{1}{l}\int_0^l q(x)\,x\,dx
\end{aligned} \tag{9.1b}
$$

Here, we have substituted the expressions for V_{IJ_1} and V_{JI_1} in terms of M_{IJ} and M_{JI}, which were obtained previously from equilibrium considerations in Equations 5.38*a*. The integrals in Equations 9.1*b* are expressions for V_{IJ_2} and V_{JI_2}, respectively, in terms of a general load $q(x)$ in Figure 5.23*a*(2).[1]

Substituting Equations 9.1*a* into Equations 9.1*b* leads to

$$
\begin{aligned}
V_{IJ} = \frac{EI}{l}&\left[-\frac{6}{l}\theta_{IJ} - \frac{6}{l}\theta_{JI} + \frac{12}{l^2}(v_{IJ} - v_{JI})\right] \\
&-\frac{1}{l}(\widehat{M}_{IJ} + \widehat{M}_{JI}) - \frac{1}{l}\int_0^l q(x)\,(l - x)\,dx
\end{aligned} \tag{9.1c}
$$

---

[1] These expressions for V_{IJ_2} and V_{JI_2} are obtained from equations of moment equilibrium about either the J end or the I end of the beam in Figure 5.23*a*(2). Once again, for the sake of simplicity, we assume that the continuous load $q(x)$ also encompasses concentrated loads.

$$V_{JI} = \frac{EI}{l}\left[\frac{6}{l}\theta_{IJ} + \frac{6}{l}\theta_{JI} - \frac{12}{l^2}(v_{IJ} - v_{JI})\right]$$
$$+\frac{1}{l}(\widehat{M}_{IJ} + \widehat{M}_{JI}) - \frac{1}{l}\int_0^l q(x)\,x\,dx$$

Equations 9.1a and 9.1c may now be written in matrix form as

$$\begin{Bmatrix} M_{IJ} \\ M_{JI} \\ V_{IJ} \\ V_{JI} \end{Bmatrix} = \frac{EI}{l}\begin{bmatrix} 4 & 2 & -6/l & 6/l \\ 2 & 4 & -6/l & 6/l \\ -6/l & -6/l & 12/l^2 & -12/l^2 \\ 6/l & 6/l & -12/l^2 & 12/l^2 \end{bmatrix}\begin{Bmatrix} \theta_{IJ} \\ \theta_{JI} \\ v_{IJ} \\ v_{JI} \end{Bmatrix} + \begin{Bmatrix} \widehat{M}_{IJ} \\ \widehat{M}_{JI} \\ \widehat{V}_{IJ} \\ \widehat{V}_{JI} \end{Bmatrix} \quad \textbf{(9.1}\boldsymbol{d}\textbf{)}$$

where

$$\widehat{V}_{IJ} = -\frac{1}{l}(\widehat{M}_{IJ} + \widehat{M}_{JI}) - \frac{1}{l}\int_0^l q(x)\,(l - x)\,dx$$
$$\widehat{V}_{JI} = \frac{1}{l}(\widehat{M}_{IJ} + \widehat{M}_{JI}) - \frac{1}{l}\int_0^l q(x)\,x\,dx \quad \textbf{(9.1}\boldsymbol{e}\textbf{)}$$

Equation 9.1d is the desired matrix force-displacement relation for an axially rigid beam. Equations 9.1e define fixed-end forces in the same way that Equations 8.13e define fixed-end moments; we will refer to $\widehat{V}_{IJ}$ and $\widehat{V}_{JI}$ as "fixed-end shears." Equations 9.1e may be recognized as the results obtained by summing moments of forces about the ends of a fixed beam that is subjected to the applied load $q(x)$ (Figure 9.1).

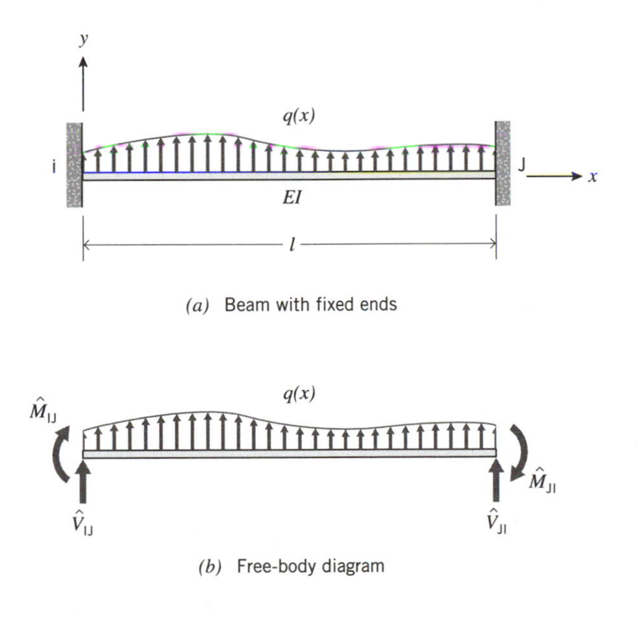

Figure 9.1 Fixed-end forces.

It should be noted that the element stiffness matrix in Equation 9.1*d* is singular because the last two rows are linear combinations of the first two rows, a consequence of Equations 9.1*b*, i.e., that the four end forces must satisfy two equations of equilibrium. Thus, this matrix is of rank 2 exactly like the element stiffness matrix in Equation 8.13*g*. In other words, this matrix contains no more stiffness information than the one in Equation 8.13*g*. However, the new form contains some additional information that is essential for the developments that follow. Before proceeding further we will illustrate an application of the force-displacement relation in Equation 9.1*d* and examine some of the implications of its use.

EXAMPLE 1

Let's reanalyze the structure in Figure 8.21 (Example 3, Chapter 8) using the element force-displacement relations in Equation 9.1*d* rather than Equation 8.13*g*.

Step 1. The system displacements are the same as in Figure 8.7 and Example 3, Chapter 8.

Step 2. In Example 3, Chapter 8, the element end displacements associated with each of the system displacements were combined into end deformations. In the present formulation, the same displacements are used directly in the compatibility equations. Thus, for system displacement u_1 we have $v_{AB} = v_{BA} = v_{BG} = v_{GB} = 0$, $\theta_{AB} = \theta_{GB} = 0$, and $\theta_{BA} = \theta_{BG} = u_1$. For system displacement u_2 we have $\theta_{AB} = \theta_{BA} = \theta_{BG} = \theta_{GB} = 0$, $v_{AB} = v_{BA} = v_{BG} = 0$, and $v_{GB} = +u_2$. Therefore, the matrix form of the compatibility equations now becomes

$$\begin{Bmatrix} \theta_{AB} \\ \theta_{BA} \\ v_{AB} \\ v_{BA} \\ \theta_{BG} \\ \theta_{GB} \\ v_{BG} \\ v_{GB} \end{Bmatrix} = \begin{vmatrix} 0 & 0 \\ 1 & 0 \\ 0 & 0 \\ 0 & 0 \\ 1 & 0 \\ 0 & 0 \\ 0 & 0 \\ 0 & 1 \end{vmatrix} \begin{Bmatrix} u_1 \\ u_2 \end{Bmatrix} \quad \textbf{(9.2}\boldsymbol{a}\textbf{)}$$

Here, the first four rows relate end displacements of beam AB to the system displacements, and the last four rows relate end displacements of beam BG to the system displacements. The ordering of the end displacements for each beam is consistent with that in Equation 9.1*d*. We will denote vectors of element end displacements, such as the one on the left side of Equation 9.2*a*, by the symbol $\boldsymbol{\delta}$ to distinguish it from the vector of element end deformations, $\boldsymbol{\Delta}$, in Equation 8.20*c*; thus, Equation 9.2*a* has the general form $\boldsymbol{\delta} = \boldsymbol{\beta u}$.

Step 3. As part of the force-displacement relations, we require the fixed-end forces (moments and shears) due to the applied loads. The fixed-end moments have been calculated in Chapter 8; the fixed-end shears may be determined from equilibrium.

Consider the fixed-end forces in beam AB (Figure 9.2*a*). Figure 9.2*b* shows a free-body diagram of the beam with the end moments due to the thermal load, computed

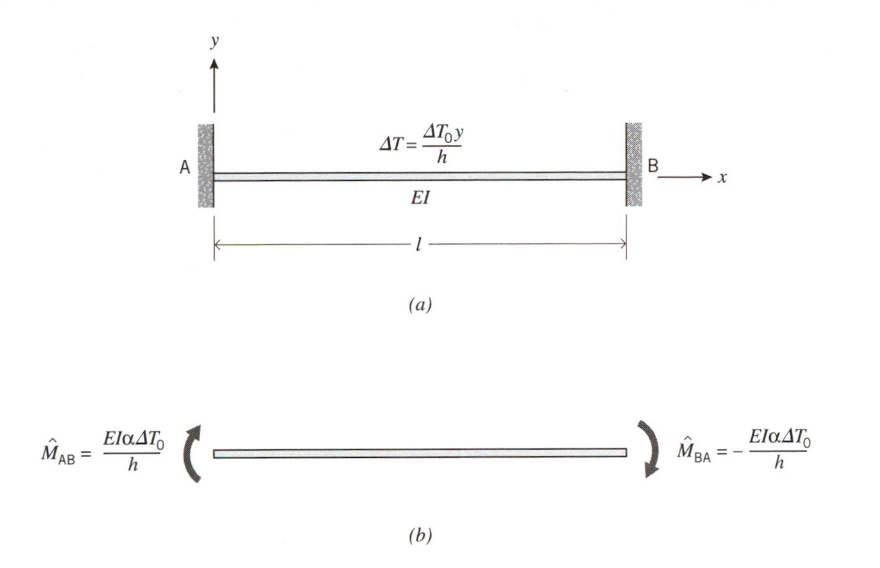

Figure 9.2 Fixed-end forces for beam AB.

previously in Chapter 8. Equilibrium requires that the end shears, $\hat{V}_{AB}$ and $\hat{V}_{BA}$ be zero in this case.

A similar equilibrium analysis of beam BG (Figure 9.3), shows that the fixed-end shears are $\hat{V}_{BG} = \hat{V}_{GB} = F/2$. Note that both of these shears are positive, i.e., they act in the positive direction shown in Figure 5.23*a*.

With all fixed-end forces determined, the force-displacement relations for the structure may be expressed in the matrix form

$$\begin{Bmatrix} M_{AB} \\ M_{BA} \\ V_{AB} \\ V_{BA} \\ M_{BG} \\ M_{GB} \\ V_{BG} \\ V_{GB} \end{Bmatrix} = \frac{EI}{l} \begin{bmatrix} 4 & 2 & -6/l & 6/l & & & & \\ 2 & 4 & -6/l & 6/l & & & & \\ -6/l & -6/l & 12/l^2 & -12/l^2 & & & & \\ 6/l & 6/l & -12/l^2 & 12/l^2 & & & & \\ & & & & 16 & 8 & -48/l & 48/l \\ & & & & 8 & 16 & -48/l & 48/l \\ & & & & -48/l & -48/l & 192/l^2 & -192/l^2 \\ & & & & 48/l & 48/l & -192/l^2 & 192/l^2 \end{bmatrix} \begin{Bmatrix} \theta_{AB} \\ \theta_{BA} \\ v_{AB} \\ v_{BA} \\ \theta_{BG} \\ \theta_{GB} \\ v_{BG} \\ v_{GB} \end{Bmatrix} + \begin{Bmatrix} -EI\alpha\Delta T_o l/h \\ -EI\alpha\Delta T_o l/h \\ 0 \\ 0 \\ -Fl/16 \\ Fl/16 \\ F/2 \\ F/2 \end{Bmatrix} \tag{9.2b}$$

Equation 9.2*b* is analogous to Equation 8.20*e*. Note that the submatrix containing the force-displacement relations for element BG reflects the beam bending stiffness $2EI$

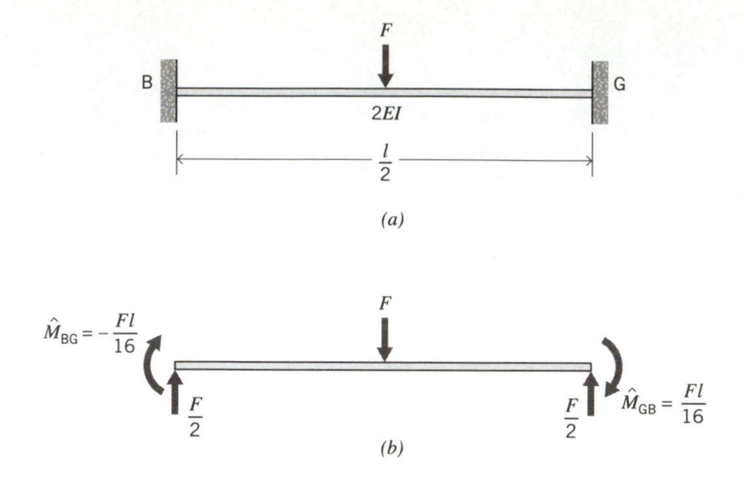

Figure 9.3 Fixed-end forces for beam BG.

and length $l/2$. Our symbolic representation of equations of this type has the general form $\mathbf{P} = \mathbf{K}\boldsymbol{\delta} + \widehat{\mathbf{P}}$.[2]

Step 4. For illustrative purposes, we will return temporarily to the *ad hoc* procedure that we used to generate the system equilibrium equations in Chapter 8, i.e., use $\boldsymbol{\beta}^T$ to write the matrix equilibrium equation in the form

$$\begin{Bmatrix} f_1 \\ f_2 \end{Bmatrix} = \begin{bmatrix} 0 & 1 & 0 & 0 & 1 & 0 & 0 & 0 \\ 0 & 0 & 0 & 0 & 0 & 0 & 0 & 1 \end{bmatrix} \begin{Bmatrix} M_{AB} \\ M_{BA} \\ V_{AB} \\ V_{BA} \\ M_{BG} \\ M_{GB} \\ V_{BG} \\ V_{GB} \end{Bmatrix} \tag{9.2c}$$

In this equation, f_1 and f_2 are ''unknown'' components of a force vector associated with displacements u_1 and u_2, respectively. The first of the two equilibrium equations has the form $f_1 = M_{BA} + M_{BG}$ exactly as in the previous solution, but the second now has the form $f_2 = V_{GB}$ (rather than $f_2 = 2M_{BG}/l + 2M_{GB}/l$). As we saw previously, the first equation is the equation of moment equilibrium for joint B, and in this case we have $f_1 = 0$. The second equation is the equation of transverse force equilibrium for joint G and, since there is a vertical roller at G, we must have $V_{GB} = 0 = f_2$ (see Figure 8.22). Thus, there has been a substantial simplification of the second equilibrium equation. Force f_2 is now zero. The interior force F on beam BG is completely taken into account without considering effects of side-sway as was done previously. Evidently, the use of element force-displacement information produces a fundamental

[2] We continue to use **P** and $\widehat{\mathbf{P}}$ to represent force vectors, **K** to represent the assemblage of element stiffness matrices, and $\boldsymbol{\beta}$ to represent the compatibility matrix, because it will be clear from any context whether the formulation involves matrices based on the use of displacements or deformations.

change in the system equilibrium equations. We will examine this phenomenon more fully in the next section.

Step 5. After substituting the expressions generated in Steps 2 and 3, the system equilibrium equations will again assume the general matrix form $(\boldsymbol{\beta}^T\mathbf{K}\boldsymbol{\beta})\mathbf{u} = \mathbf{F} - \boldsymbol{\beta}^T\widehat{\mathbf{P}}$ (Equation 8.7*b*). For this example, it is a relatively straightforward matter to show that the indicated matrix multiplications lead to exactly the same results as in Equations 8.21*a* and 8.21*b*. Specifically, the product $\boldsymbol{\beta}^T\mathbf{K}\boldsymbol{\beta}$ obtained by using the $\boldsymbol{\beta}$ matrix in Equation 9.2*a* and the $\mathbf{K}$ matrix in Equation 9.2*b* produces exactly the same system stiffness matrix $\mathbf{k}$ previously generated in Equation 8.21*a*. Similarly, $\mathbf{F} - \boldsymbol{\beta}^T\widehat{\mathbf{P}}$ is identical to the quantity computed in Equation 8.21*b*. In other words, we obtain exactly the same solution as in Example 3 of Chapter 8.

In the preceding example we obtained the correct solution in a much more direct manner than was possible using the formulation in Chapter 8. In particular, the effect of an interior transverse force was naturally accounted for when we used the element stiffness relation in Equation 9.1*d*.

Unfortunately, similar simplifications will not be obtained in all cases. Equation 9.1*d* models only axially rigid beams and, by itself, is not sufficient to account for all possible effects of interior axial loads on structural behavior. For instance, consider the structure in Figure 8.28 (Example 5, Chapter 8): if we analyze this structure using the element stiffness relation in Equation 9.1*d*, it is still necessary to assemble an $\mathbf{F}^{\circ}$ vector to properly include the effect of the axial component of the distributed force that is applied to beam BC. This final difficulty will be removed when we add axial stiffness terms to Equation 9.1*d*. First, however, we will examine the nature of the system equilibrium equations that result from the use of Equations 9.1*d*. This will be accomplished by appropriately modifying the developments in Section 8.3.3.

9.2 EQUILIBRIUM EQUATIONS

Consider the previously derived expression for internal virtual work, Equation 8.17*d*, for a single beam element of a structure that is subjected to a virtual displacement like a system displacement,

$$W_{I_V} = \Delta_{\mathrm{IJ}_V}(M_{\mathrm{IJ}} - \widehat{M}^o_{\mathrm{IJ}}) + \Delta_{\mathrm{JI}_V}(M_{\mathrm{JI}} - \widehat{M}^o_{\mathrm{JI}}) \tag{9.3a}$$

We substitute the definitions of the virtual deformations, Δ_{IJ_V} and Δ_{JI_V}, in terms of displacements θ_{IJ_V}, θ_{JI_V}, v_{IJ_V}, and v_{JI_V} from Equation 5.39*a*, and rearrange terms to obtain

$$\begin{aligned} W_{I_V} = {} & \theta_{\mathrm{IJ}_V}(M_{\mathrm{IJ}} - \widehat{M}^o_{\mathrm{IJ}}) + \theta_{\mathrm{JI}_V}(M_{\mathrm{JI}} - \widehat{M}^o_{\mathrm{JI}}) \\ & + v_{\mathrm{IJ}_V}\left[-\frac{(M_{\mathrm{IJ}} + M_{\mathrm{JI}})}{l} + \frac{(\widehat{M}^o_{\mathrm{IJ}} + \widehat{M}^o_{\mathrm{JI}})}{l}\right] \\ & + v_{\mathrm{JI}_V}\left[\frac{(M_{\mathrm{IJ}} + M_{\mathrm{JI}})}{l} - \frac{(\widehat{M}^o_{\mathrm{IJ}} + \widehat{M}^o_{\mathrm{JI}})}{l}\right] \end{aligned} \tag{9.3b}$$

Adding and subtracting integrals like those in Equations 9.1e leads to

$$W_{I_V} = \theta_{IJ_V}(M_{IJ} - \widehat{M}^o_{IJ}) + \theta_{JI_V}(M_{JI} - \widehat{M}^o_{JI})$$

$$+ v_{IJ_V}\left\{\left[-\frac{(M_{IJ} + M_{JI})}{l} - \frac{1}{l}\int_0^l q(x)\,(l - x)\,dx\right]\right.$$

$$\left. + \left[\frac{(\widehat{M}^o_{IJ} + \widehat{M}^o_{JI})}{l} + \frac{1}{l}\int_0^l q(x)\,(l - x)\,dx\right]\right\} \tag{9.3c}$$

$$+ v_{JI_V}\left\{\left[\frac{(M_{IJ} + M_{JI})}{l} - \frac{1}{l}\int_0^l q(x)\,x\,dx\right]\right.$$

$$\left. + \left[-\frac{(\widehat{M}^o_{IJ} + \widehat{M}^o_{JI})}{l} + \frac{1}{l}\int_0^l q(x)\,x\,dx\right]\right\}$$

We now recognize each of the terms in square brackets as either an end force from Equation 9.1b or a fixed-end force (associated with applied physical loads only) from Equation 9.1e. Substituting the appropriate force quantities enables us to rewrite Equation 9.3c as

$$W_{I_V} = \theta_{IJ_V}(M_{IJ} - \widehat{M}^o_{IJ}) + \theta_{JI_V}(M_{JI} - \widehat{M}^o_{JI}) + v_{IJ_V}(V_{IJ} - \widehat{V}^o_{IJ}) + V_{JI_V}(V_{JI} - \widehat{V}^o_{JI}) \tag{9.3d}$$

or, in matrix form,

$$W_{I_V} = \begin{Bmatrix} \theta_{IJ_V} \\ \theta_{JI_V} \\ v_{IJ_V} \\ v_{JI_V} \end{Bmatrix}^T \left\{\begin{Bmatrix} M_{IJ} \\ M_{JI} \\ V_{IJ} \\ V_{JI} \end{Bmatrix} - \begin{Bmatrix} \widehat{M}^o_{IJ} \\ \widehat{M}^o_{JI} \\ \widehat{V}^o_{IJ} \\ \widehat{V}^o_{JI} \end{Bmatrix}\right\} \tag{9.3e}$$

Equation 9.3e is the analog of Equation 8.17f in terms of element end displacements rather than deformations.

The modifications of the expression for external virtual work proceed in a similar manner. We begin with Equation 8.17h,

$$W_{E_V} = \Delta_{IJ_V}(-\widehat{M}^o_{IJ}) + \Delta_{JI_V}(-\widehat{M}^o_{JI}) + \int_0^l q(x)(v_{IJ_V} - \psi_{IJ_V}x)\,dx + u_V\int_0^l p(x)dx \tag{9.3f}$$

Again, the last two terms in Equation 9.3f are due to rigid-body effects, the first one due to lateral rigid-body motion and the second due to axial motion. In the last integral u_V is constant because the beam is axially rigid. Substituting the definitions of the virtual deformations in terms of displacements from Equations 5.39a and 5.39b, and simplifying, leads to

$$W_{E_V} = -\theta_{IJ_V}\widehat{M}^o_{IJ} - \theta_{JI_V}\widehat{M}^o_{JI} - v_{IJ_V}\left[-\frac{(\widehat{M}^o_{IJ} + \widehat{M}^o_{JI})}{l} - \int_0^l q(x)\left(1 - \frac{x}{l}\right)dx\right] +$$

$$- v_{\mathrm{JI}_V}\left[\frac{(\widehat{M}^o_{\mathrm{IJ}} + \widehat{M}^o_{\mathrm{JI}})}{l} - \int_0^l q(x)\frac{x}{l}\,dx\right] + u_V\int_0^l p(x)dx \tag{9.3g}$$

Once again, we recognize the terms in brackets as similar to the expression in Equations 9.1e, i.e., fixed end shears due to physical loads only. Thus, Equation 9.3g becomes

$$W_{E_V} = -\theta_{\mathrm{IJ}_V}\widehat{M}^o_{\mathrm{IJ}} - \theta_{\mathrm{JI}_V}\widehat{M}^o_{\mathrm{JI}} - v_{\mathrm{IJ}_V}\widehat{V}^o_{\mathrm{IJ}} - v_{\mathrm{JI}_V}\widehat{V}^o_{\mathrm{JI}} + W^o_{E_V} \tag{9.3h}$$

or

$$W_{E_V} = -\begin{Bmatrix}\theta_{\mathrm{IJ}_V}\\ \theta_{\mathrm{JI}_V}\\ v_{\mathrm{IJ}_V}\\ v_{\mathrm{JI}_V}\end{Bmatrix}^{\mathrm{T}}\begin{Bmatrix}\widehat{M}^o_{\mathrm{IJ}}\\ \widehat{M}^o_{\mathrm{JI}}\\ \widehat{V}^o_{\mathrm{IJ}}\\ \widehat{V}^o_{\mathrm{JI}}\end{Bmatrix} + W^o_{E_V} \tag{9.3i}$$

Equation 9.3i is analogous to Equation 8.17n except that, now, $W^o_{E_V}$ is the external virtual work due to axial forces only. The virtual work due to lateral motion is now contained in the first vector relation $-\boldsymbol{\delta}_V^{\mathrm{T}}\widehat{\mathbf{P}}^{\mathbf{o}}$.

As in Section 8.3.3, we complete the analysis by finding the sum of the virtual work over the entire structure. The total internal virtual work is

$$\sum_{n=1}^{NB}(W_{I_V})_n = \sum_{n=1}^{NB}\begin{Bmatrix}\theta_{\mathrm{IJ}_V}\\ \theta_{\mathrm{JI}_V}\\ v_{\mathrm{IJ}_V}\\ v_{\mathrm{JI}_V}\end{Bmatrix}_n^{\mathrm{T}}\left[\begin{Bmatrix}M_{\mathrm{IJ}}\\ M_{\mathrm{JI}}\\ V_{\mathrm{IJ}}\\ V_{\mathrm{JI}}\end{Bmatrix}_n - \begin{Bmatrix}\widehat{M}^o_{\mathrm{IJ}}\\ \widehat{M}^o_{\mathrm{JI}}\\ \widehat{V}^o_{\mathrm{IJ}}\\ \widehat{V}^o_{\mathrm{JI}}\end{Bmatrix}_n\right] \tag{9.3j}$$

$$= \boldsymbol{\delta}_V^{\mathrm{T}}(\mathbf{P} - \widehat{\mathbf{P}}^{\mathbf{o}})$$

The total external virtual work, including joint loads, is

$$\sum_{n=1}^{NB}(W_{E_V})_n + \sum_{k=1}^{NDOF}u_{k_V}(F^o_k + F_k) = -\sum_{n=1}^{NB}\begin{Bmatrix}\theta_{\mathrm{IJ}_V}\\ \theta_{\mathrm{JI}_V}\\ v_{\mathrm{IJ}_V}\\ v_{\mathrm{JI}_V}\end{Bmatrix}_n^{\mathrm{T}}\begin{Bmatrix}\widehat{M}^o_{\mathrm{IJ}}\\ \widehat{M}^o_{\mathrm{JI}}\\ \widehat{V}^o_{\mathrm{IJ}}\\ \widehat{V}^o_{\mathrm{JI}}\end{Bmatrix}_n + \sum_{k=1}^{NDOF}u_{k_V}(F^o_k + F_k) \tag{9.3k}$$

$$= -\boldsymbol{\delta}_V^{\mathrm{T}}\widehat{\mathbf{P}}^{\mathbf{o}} + \sum_{k=1}^{NDOF}u_{k_V}(F^o_k + F_k)$$

where, as in Chapter 8, F^o_k represents the external virtual work of axial forces in the beams going through a unit displacement $u_{k_V} = 1$, the F_k are the forces applied directly on the joints, and $NDOF \equiv m$ is the total number of system degrees of freedom.

Equating external and internal virtual work leads to

$$\sum_{k=1}^{NDOF}u_{k_V}(F^o_k + F_k) = \boldsymbol{\delta}_V^{\mathrm{T}}\mathbf{P} \tag{9.3l}$$

As in Equation 8.18f, we substitute $\boldsymbol{\delta}_V = \boldsymbol{\beta}_k u_{k_V}$, which leads to

$$F^o_k + F_k = \boldsymbol{\beta}_k^{\mathrm{T}}\mathbf{P} \tag{9.3m}$$

Equation 9.3*m* is the desired equilibrium equation. It follows that the set of equilibrium equations for the system has the matrix form

$$\mathbf{F}^{o} + \mathbf{F} = \boldsymbol{\beta}^{T}\mathbf{P} \tag{9.3n}$$

9.3 FORCE-DISPLACEMENT STIFFNESS MATRIX FOR AXIALLY LOADED BARS

The deformation-force relation for a uniform bar subjected to a general axial load was presented in Equation 5.48*h* based on the positive directions of forces defined in Figure 5.30,

$$\begin{aligned} \Delta_{IJ}^{(a)} &= \frac{l}{EA} P_{JI} + \frac{1}{EA}\int_0^l \widehat{P}(x)\,dx \\ &= \frac{l}{EA} P_{JI} + \widehat{\Delta}_{IJ}^{(a)} \end{aligned} \tag{5.48h}$$

The axial deformation (length change) is related to end displacements by

$$\Delta_{IJ}^{(a)} = u_{JI} - u_{IJ} \tag{9.4a}$$

Therefore, inverting Equation 5.48*h* and substituting Equation 9.4*a* leads to

$$P_{JI} = \frac{EA}{l}(u_{JI} - u_{IJ}) - \frac{1}{l}\int_0^l \widehat{P}(x)\,dx \tag{9.4b}$$

We find P_{IJ} from equilibrium of the bar in Figure 5.30*a* by summing forces in the *x* direction,

$$\begin{aligned} P_{IJ} &= -P_{JI} - \int_0^l p(x)\,dx \\ &= \frac{EA}{l}(u_{IJ} - u_{JI}) + \frac{1}{l}\int_0^l \widehat{P}(x)\,dx - \int_0^l p(x)\,dx \end{aligned} \tag{9.4c}$$

Equations 9.4*c* and 9.4*b* may be arranged in matrix form,

$$\begin{Bmatrix} P_{IJ} \\ P_{JI} \end{Bmatrix} = \frac{EA}{l}\begin{bmatrix} 1 & -1 \\ -1 & 1 \end{bmatrix}\begin{Bmatrix} u_{IJ} \\ u_{JI} \end{Bmatrix} + \begin{Bmatrix} \widehat{P}_{IJ} \\ \widehat{P}_{JI} \end{Bmatrix} \tag{9.4d}$$

where

$$\begin{aligned} \widehat{P}_{IJ} &= \frac{1}{l}\int_0^l \widehat{P}(x)\,dx - \int_0^l p(x)\,dx \\ \widehat{P}_{JI} &= -\frac{1}{l}\int_0^l \widehat{P}(x)\,dx \end{aligned} \tag{9.4e}$$

Equation 9.4*d* is the desired force-displacement relation for the bar. In this case, we identify the quantities in Equations 9.4*e* as fixed-end axial forces.[3] Just as for the beam, the stiffness matrix in Equation 9.4*d* is singular.

For structures composed of this class of bar elements, use of the stiffness relation in Equation 9.4*d* can be shown to lead to system equilibrium equations that have exactly the same form as Equation 9.3*n*.[4] In other words, the use of force-displacement stiffness matrices for bars directly accounts for all effects of interior axial loads including axial rigid-body motion, and the components of the force vector in the system equilibrium equations arise only from joint loads.

For the analysis of trusses, where element interior loads are not considered, this force-displacement relation offers no advantage over the prior force-deformation relation. Therefore, we will proceed with applications involving elements having both axial and bending stiffness. This will allow us to clearly demonstrate the advantages of the new formulation. We will return to trusses later.

9.4 COMBINED BENDING AND AXIAL BEHAVIOR

Assuming that bending and axial behaviors are uncoupled, the force-displacement element stiffness relation, considering combined bending and axial effects, is obtained from superposition,

$$\begin{Bmatrix} M_{IJ} \\ M_{JI} \\ V_{IJ} \\ V_{JI} \\ P_{IJ} \\ P_{JI} \end{Bmatrix} = \frac{EI}{l} \begin{bmatrix} 4 & 2 & -6/l & 6/l & & \\ 2 & 4 & -6/l & 6/l & & \\ -6/l & -6/l & 12/l^2 & -12/l^2 & & \\ 6/l & 6/l & -12/l^2 & 12/l^2 & & \\ & & & & A/I & -A/I \\ & & & & -A/I & A/I \end{bmatrix} \begin{Bmatrix} \theta_{IJ} \\ \theta_{JI} \\ v_{IJ} \\ v_{JI} \\ u_{IJ} \\ u_{JI} \end{Bmatrix} + \begin{Bmatrix} \widehat{M}_{IJ} \\ \widehat{M}_{JI} \\ \widehat{V}_{IJ} \\ \widehat{V}_{JI} \\ \widehat{P}_{IJ} \\ \widehat{P}_{JI} \end{Bmatrix} \quad \textbf{(9.5)}$$

The fixed-end forces have all been defined previously and require no further discussion here. Once again, we note that it would not be difficult to demonstrate that use of this stiffness relation will lead to system equilibrium equations that have the matrix form given in Equation 9.3*n*. Finally, although the combined axial/bending element involves a larger matrix form, in this case a 6 × 6, all interior applied leads are completely accounted for in the fixed-end force vector. Thus, there is no longer a need for the troublesome $\mathbf{F}^{o}$ vector, and the equilibrium equation is simply $\mathbf{F} = \boldsymbol{\beta}^{T}\mathbf{P}$.

[3] A review of the interpretation of $\widehat{P}(x)$ is recommended; see the derivation in Section 5.3.5, and Example 6 in the same section.

[4] The derivation is analogous to that employed in Section 9.2 and will not be repeated here.

EXAMPLE 2

We will reanalyze the frame in Example 6 of Chapter 5 (Figure 5.31*a*), again shown in Figure 9.4*a*. The load consists of a weight per unit length of the elements, q_o. This load has transverse and axial components in beam AB, as shown in Figure 9.4*b* (and Figure 5.31*c*).

Step 1. Since we are assuming axial and bending stiffness for the elements, we have the four system displacements shown in Figure 8.9*a*. In this case, u_2 and u_3 are horizontal and vertical displacements of joint B, respectively, and u_4 is the horizontal displacement of joint C (the only possible motion of joint C).

Step 2. System displacement u_1 causes only end rotations $\theta_{BA} = \theta_{BC} = u_1$. System displacement u_2 causes the following three distinct components of end displacement: (1) transverse displacement at end B of beam AB, $v_{BA} = -u_2/\sqrt{2}$, (2) axial displacement at end B of beam AB, $u_{BA} = +u_2/\sqrt{2}$, and (3) axial displacement at end B of beam BC, $u_{BC} = +u_2$. These displacements are illustrated in Figure 9.4*c*. (The axial displacements are determined in exactly the same manner as they are for truss bars, *cf.* Figures 8.11*b* and 8.11*c*.)

Similarly, system displacement u_3 also causes three distinct components of end

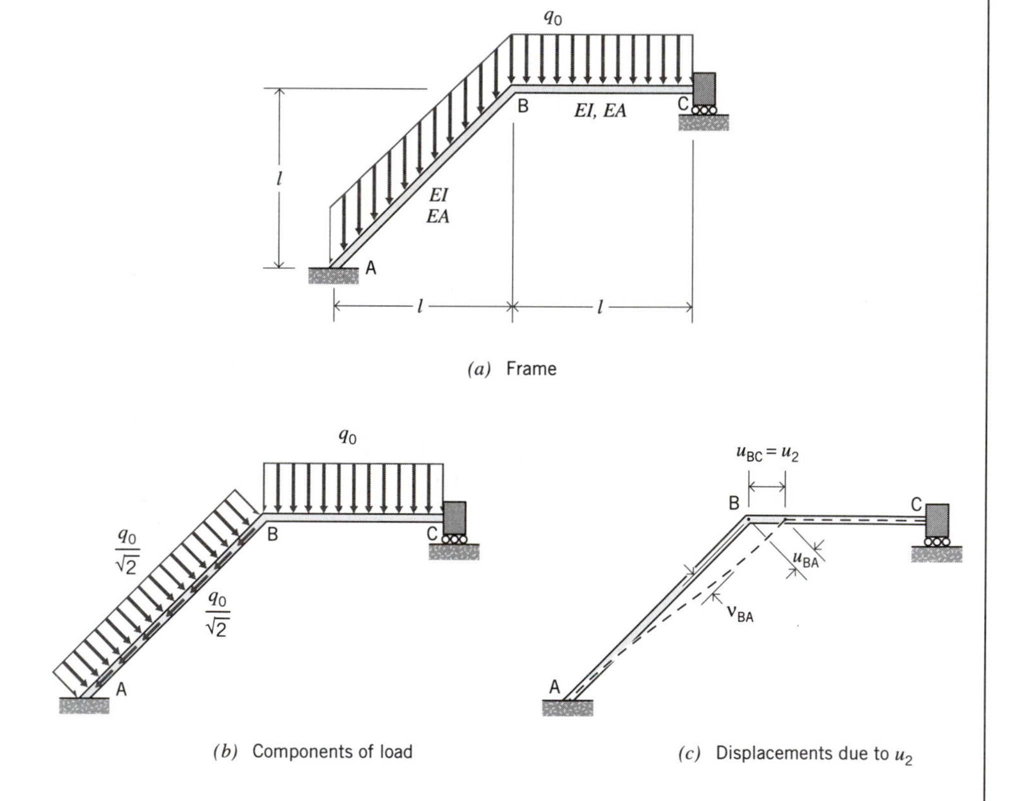

Figure 9.4 Example 2.

displacement: (1) transverse displacement at end B of beam AB, $v_{BA} = +u_3/\sqrt{2}$, (2) axial displacement at end B of beam AB, $u_{BA} = +u_3/\sqrt{2}$, and (3) transverse displacement at end B of beam BC, $v_{BC} = +u_3$. System displacement u_4 causes only an axial displacement at end C of beam BC, $u_{CB} = u_4$. Note that positive directions of the axial displacements are taken to be the same as those of the axial end forces.

Thus, the compatibility equation is

$$\begin{Bmatrix} \theta_{AB} \\ \theta_{BA} \\ v_{AB} \\ v_{BA} \\ u_{AB} \\ u_{BA} \\ \theta_{BC} \\ \theta_{CB} \\ v_{BC} \\ v_{CB} \\ u_{BC} \\ u_{CB} \end{Bmatrix} = \begin{bmatrix} 0 & 0 & 0 & 0 \\ 1 & 0 & 0 & 0 \\ 0 & 0 & 0 & 0 \\ 0 & -1/\sqrt{2} & 1/\sqrt{2} & 0 \\ 0 & 0 & 0 & 0 \\ 0 & 1/\sqrt{2} & 1/\sqrt{2} & 0 \\ 1 & 0 & 0 & 0 \\ 0 & 0 & 0 & 0 \\ 0 & 0 & 1 & 0 \\ 0 & 0 & 0 & 0 \\ 0 & 1 & 0 & 0 \\ 0 & 0 & 0 & 1 \end{bmatrix} \begin{Bmatrix} u_1 \\ u_2 \\ u_3 \\ u_4 \end{Bmatrix} \tag{9.6a}$$

Step 3. To assemble the force-displacement relation for the structure, we require the fixed-end forces for each of the two elements. For element AB, fixed-end moments and shears arise from the transverse component of the uniform load, as shown in Figure 9.5*a* and 9.5*b*;[5] the axial component of the uniform load gives rise to the fixed-end forces shown in Figures 9.5*c* and 9.5*d*. The latter are computed from Equations 9.4*e*. The quantity $\widehat{P}(x)$ in Equations 9.4*e* has been defined in Equation 5.48*g*. In this particular case, we have $\widehat{P}(x) = P_o(x)$, where the quantity $P_o(x)$ has been previously defined in Equation 5.49*b* for the same axial load. Thus, for example, when we substitute the expression for $P_o(x)$ and $\sqrt{2}l$ for l in the second of Equations 9.4*e*, we obtain

$$\widehat{P}_{BA} = -\frac{1}{\sqrt{2}l}\int_0^{\sqrt{2}l} P_o(x)\,dx = -\frac{1}{\sqrt{2}l}\int_0^{\sqrt{2}l}\left(-q_o l + \frac{q_o}{\sqrt{2}}x\right)dx = \frac{q_o l}{2} \tag{9.6b}$$

Substituting $p(x) = -q_o/\sqrt{2}$ into the first of Equations 9.4*e* then gives

$$\widehat{P}_{AB} = -\widehat{P}_{BA} - \int_0^{\sqrt{2}l} p(x)\,dx = -\frac{q_o l}{2} - \int_0^{\sqrt{2}l}\left(-\frac{q_o}{\sqrt{2}}\right)dx = \frac{q_o l}{2} \tag{9.6c}$$

For element BC, the transverse uniform load gives the fixed-end moments and shears shown in Figure 9.6; there are no fixed-end axial forces because there is no interior axial load.

[5] See Section 8.3.4 or Appendix A.

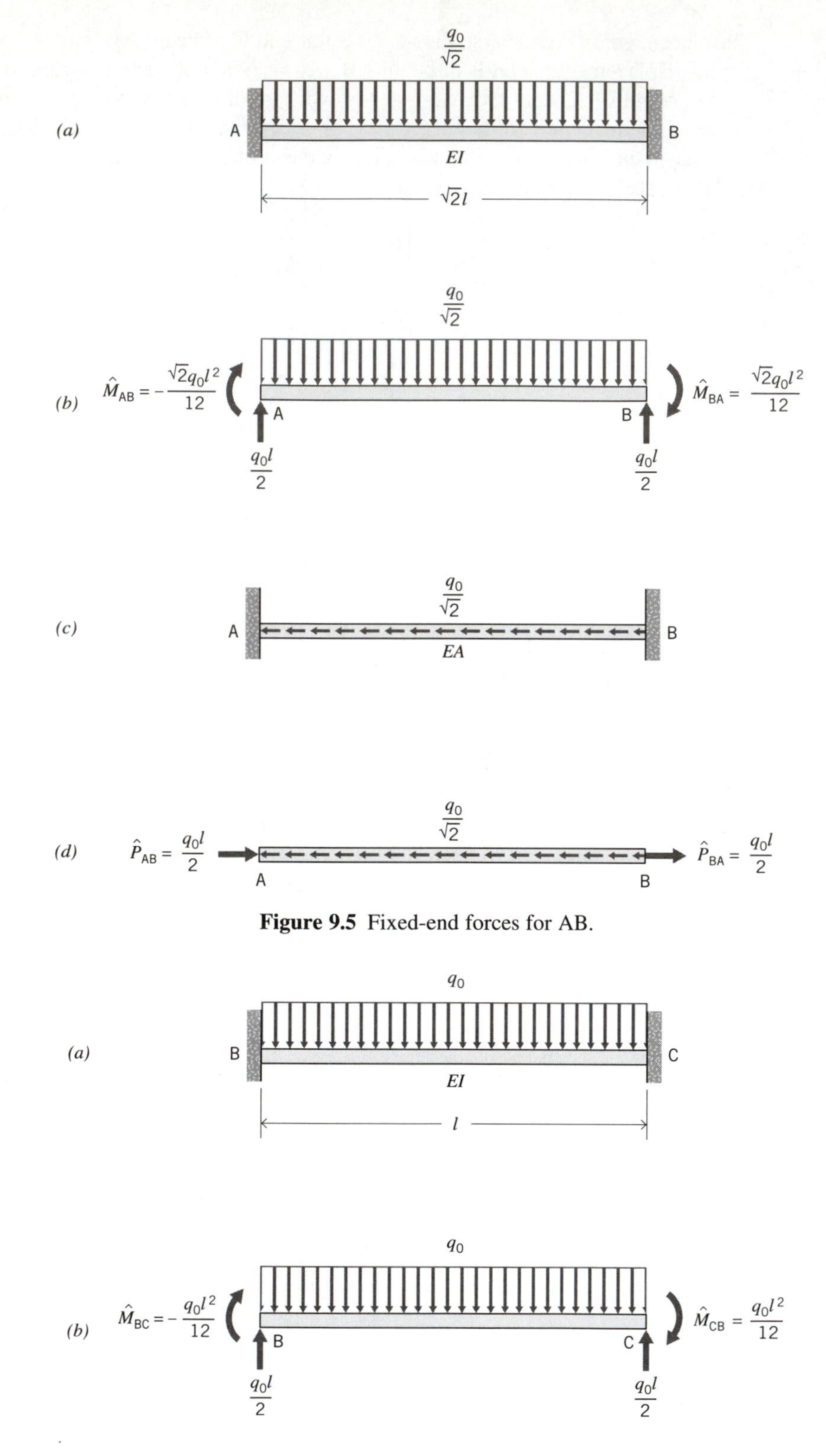

Figure 9.5 Fixed-end forces for AB.

Figure 9.6 Fixed-end forces for BC.

The force-displacement relation, therefore, is

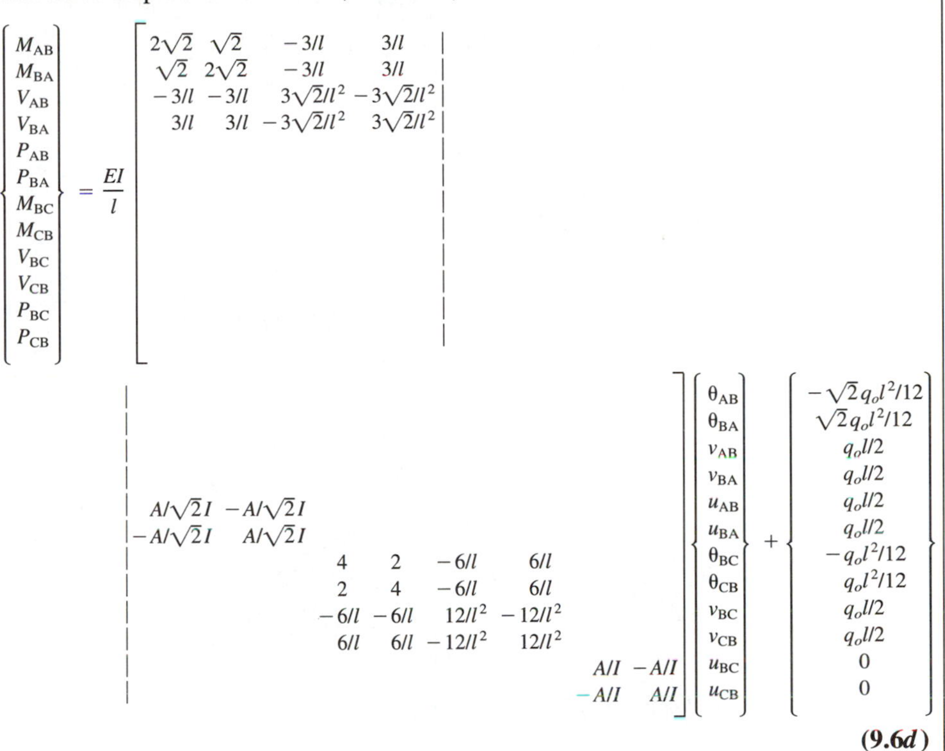

(9.6d)

Note, again, that the stiffness matrix for element AB reflects the length $\sqrt{2}l$.

Step 4. We require only the force vector **F** and, because there are no concentrated loads applied at joints B or C, we have

$$\mathbf{F} = \begin{Bmatrix} 0 \\ 0 \\ 0 \\ 0 \end{Bmatrix} \tag{9.6e}$$

Step 5. We perform the matrix operations $(\boldsymbol{\beta}^T\mathbf{K}\boldsymbol{\beta})\mathbf{u} = \mathbf{F} - \boldsymbol{\beta}^T\widehat{\mathbf{P}}$ and obtain

$$\frac{EI}{l}\begin{bmatrix} (4+2\sqrt{2}) & \left(-\dfrac{3\sqrt{2}}{2l}\right) & \left(\dfrac{-12+3\sqrt{2}}{2l}\right) & 0 \\ & \left(\dfrac{3\sqrt{2}}{2l^2}+\dfrac{4+\sqrt{2}}{4Cl^2}\right) & \left(-\dfrac{3\sqrt{2}}{2l^2}+\dfrac{\sqrt{2}}{4Cl^2}\right) & \left(-\dfrac{1}{Cl^2}\right) \\ & Sym & \left(\dfrac{12}{l^2}+\dfrac{3\sqrt{2}}{2l^2}+\dfrac{\sqrt{2}}{4Cl^2}\right) & 0 \\ & & & \left(\dfrac{1}{Cl^2}\right) \end{bmatrix}\begin{Bmatrix} u_1 \\ u_2 \\ u_3 \\ u_4 \end{Bmatrix} = \begin{Bmatrix} \left(\dfrac{1-\sqrt{2}}{12}\right)q_o l^2 \\ 0 \\ -\left(\dfrac{1+\sqrt{2}}{2}\right)q_o l \\ 0 \end{Bmatrix} \tag{9.6f}$$

in which $C \equiv I/(l^2A)$.

For illustrative purposes, we set $C = 0.01$. Solving for **u** gives

$$\begin{Bmatrix} u_1 \\ u_2 \\ u_3 \\ u_4 \end{Bmatrix} = \begin{Bmatrix} -0.0238\, q_o l^3/EI \\ 0.0543\, q_o l^4/EI \\ -0.0627\, q_o l^4/EI \\ 0.0543\, q_o l^4/EI \end{Bmatrix} \tag{9.6g}$$

Back-substituting into Equation 9.6*a* provides

$$\begin{Bmatrix} \theta_{AB} \\ \theta_{BA} \\ v_{AB} \\ v_{BA} \\ u_{AB} \\ u_{BA} \\ \theta_{BC} \\ \theta_{CB} \\ v_{BC} \\ v_{CB} \\ u_{BC} \\ u_{CB} \end{Bmatrix} = \begin{Bmatrix} 0 \\ -0.0238\, q_o l^3/EI \\ 0 \\ -0.0827\, q_o l^4/EI \\ 0 \\ -0.0060\, q_o l^4/EI \\ -0.0238\, q_o l^3/EI \\ 0 \\ -0.0627\, q_o l^4/EI \\ 0 \\ 0.0543\, q_o l^4/EI \\ 0.0543\, q_o l^4/EI \end{Bmatrix} \tag{9.6h}$$

and from Equation 9.6*d* we obtain

$$\begin{Bmatrix} M_{AB} \\ M_{BA} \\ V_{AB} \\ V_{BA} \\ P_{AB} \\ P_{BA} \\ M_{BC} \\ M_{CB} \\ V_{BC} \\ V_{CB} \\ P_{BC} \\ P_{CB} \end{Bmatrix} = \begin{Bmatrix} -0.3997\, q_o l^2 \\ -0.1977\, q_o l^2 \\ 0.9224\, q_o l \\ 0.0776\, q_o l \\ 0.9224\, q_o l \\ 0.0776\, q_o l \\ 0.1977\, q_o l^2 \\ 0.4120\, q_o l^2 \\ -0.1097\, q_o l \\ 1.1097\, q_o l \\ 0 \\ 0 \end{Bmatrix} \tag{9.6i}$$

The values of M_{AB}, M_{BA}, P_{BA}, M_{BC}, and M_{CB} are seen to be identical to the corresponding values in Equation 5.49*i*.[6]

[6] Choosing a very small value of C, e.g., $C = 0.0001$, will approximate the solution for axially rigid elements ($A \to \infty$) in Equation 5.49*g*.

EXAMPLE 3

Consider the gabled frame in Figure 8.28, which was analyzed in Example 5 of Chapter 8. We will assume that all components of the frame have axial stiffness *EA* as well as bending stiffness *EI*.

Step 1. This structure has six system displacements, which we number as in Figure 9.7*a*.[7] Figures 9.7*b*–9.7*g* show the motions associated with each of these system displacements.

Step 2. Since the motions of joints B and C are now uncoupled, it is a very straightforward matter to determine the element end displacements associated with each of the system displacements. For example, system displacement u_1 (wherein B moves horizontally to B′, as illustrated in Figure 9.8) gives rise to only three end motions: (1) transverse displacement at end B of beam AB (i.e., BB′), $v_{BA} = -u_1$, (2) transverse displacement at end B of beam BC (i.e., B″B′), $v_{BC} = -u_1/\sqrt{5}$, and (3) axial displacement at end B of beam BC (i.e., BB″), $u_{BC} = 2u_1/\sqrt{5}$. End motions due to the other system displacements are determined in a similar manner.

The matrix compatibility equation is, therefore,

$$\begin{Bmatrix} \theta_{AB} \\ \theta_{BA} \\ v_{AB} \\ v_{BA} \\ u_{AB} \\ u_{BA} \\ \theta_{BC} \\ \theta_{CB} \\ v_{BC} \\ v_{CB} \\ u_{BC} \\ u_{CB} \\ \theta_{CD} \\ \theta_{DC} \\ v_{CD} \\ v_{DC} \\ u_{CD} \\ u_{DC} \end{Bmatrix} = \begin{bmatrix} 0 & 0 & 0 & 0 & 0 & 0 \\ 0 & 0 & 1 & 0 & 0 & 0 \\ 0 & 0 & 0 & 0 & 0 & 0 \\ -1 & 0 & 0 & 0 & 0 & 0 \\ 0 & 0 & 0 & 0 & 0 & 0 \\ 0 & 1 & 0 & 0 & 0 & 0 \\ 0 & 0 & 1 & 0 & 0 & 0 \\ 0 & 0 & 0 & 0 & 0 & 1 \\ -1/\sqrt{5} & 2/\sqrt{5} & 0 & 0 & 0 & 0 \\ 0 & 0 & 0 & -1/\sqrt{5} & 2/\sqrt{5} & 0 \\ 2/\sqrt{5} & 1/\sqrt{5} & 0 & 0 & 0 & 0 \\ 0 & 0 & 0 & 2/\sqrt{5} & 1/\sqrt{5} & 0 \\ 0 & 0 & 0 & 0 & 0 & 1 \\ 0 & 0 & 0 & 0 & 0 & 0 \\ 0 & 0 & 0 & 2/\sqrt{5} & 1/\sqrt{5} & 0 \\ 0 & 0 & 0 & 0 & 0 & 0 \\ 0 & 0 & 0 & 1/\sqrt{5} & -2/\sqrt{5} & 0 \\ 0 & 0 & 0 & 0 & 0 & 0 \end{bmatrix} \begin{Bmatrix} u_1 \\ u_2 \\ u_3 \\ u_4 \\ u_5 \\ u_6 \end{Bmatrix} \quad \textbf{(9.7a)}$$

[7] Note that the numbering sequence is different from Figure 8.29*a*, e.g., u_1 is now the horizontal displacement of joint B rather than the rotation, etc.

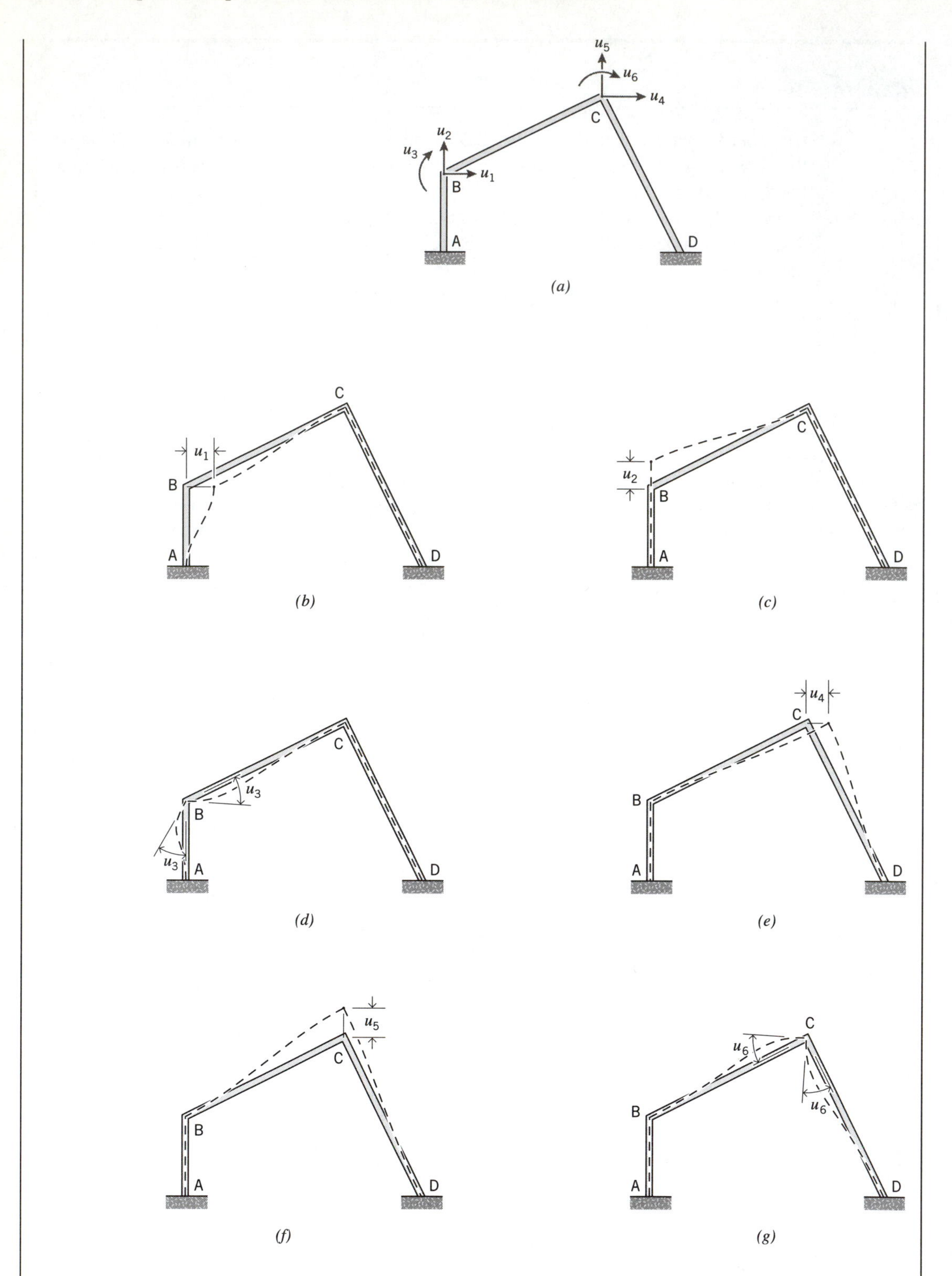

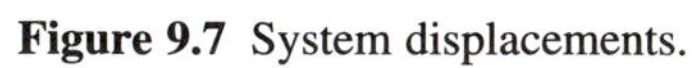

Figure 9.7 System displacements.

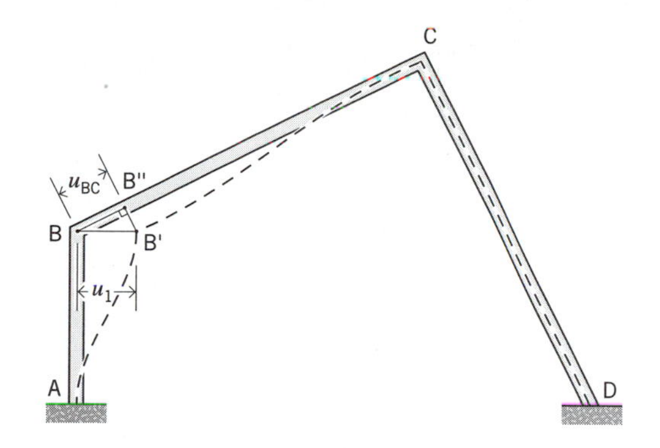

Figure 9.8 End displacements due to u_1.

Step 3. The **K** matrix for this structure is

$$
\mathbf{K} = \frac{EI}{l}\begin{bmatrix}
\begin{matrix} 4 & 2 & -6/l & 6/l \\ 2 & 4 & -6/l & 6/l \\ -6/l & -6/l & 12/l^2 & -12/l^2 \\ 6/l & 6/l & -12/l^2 & 12/l^2 \end{matrix} & & & & & \\
& \begin{matrix} A/I & -A/I \\ -A/I & A/I \end{matrix} & & & & \\
& & \begin{matrix} 4/\sqrt{5} & 2/\sqrt{5} & -6/5l & 6/5l \\ 2/\sqrt{5} & 4/\sqrt{5} & -6/5l & 6/5l \\ -6/5l & -6/5l & 12/5\sqrt{5}\,l^2 & -12/5\sqrt{5}\,l^2 \\ 6/5l & 6/5l & -12/5\sqrt{5}\,l^2 & 12/5\sqrt{5}\,l^2 \end{matrix} & & & \\
& & & \begin{matrix} A/\sqrt{5}\,I & -A/\sqrt{5}\,I \\ -A/\sqrt{5}\,I & A/\sqrt{5}\,I \end{matrix} & & \\
& & & & \begin{matrix} 4/\sqrt{5} & 2/\sqrt{5} & -6/5l & 6/5l \\ 2/\sqrt{5} & 4/\sqrt{5} & -6/5l & 6/5l \\ -6/5l & -6/5l & 12/5\sqrt{5}\,l^2 & -12/5\sqrt{5}\,l^2 \\ 6/5l & 6/5l & -12/5\sqrt{5}\,l^2 & 12/5\sqrt{5}\,l^2 \end{matrix} & \\
& & & & & \begin{matrix} A/\sqrt{5}\,I & -A/\sqrt{5}\,I \\ -A/\sqrt{5}\,I & A/\sqrt{5}\,I \end{matrix}
\end{bmatrix} \tag{9.7b}
$$

Note that the matrix incorporates lengths $\sqrt{5}l$ for elements BC and CD.

To assemble the $\widehat{\mathbf{P}}$ vector, we require the fixed-end forces due to the various components of loads shown in Figure 8.31. Note that element AB is subjected only to a uniform transverse load, and so develops only fixed-end moments and shears. Element BC is subjected to a thermal load and to uniform transverse and axial loads, and develops fixed-end moments, shears, and axial forces. Element CD is subjected to interior concentrated transverse and axial loads, and therefore also develops fixed-end moments, shears, and axial forces.

The determinations of fixed-end forces for elements AB and BC are similar to those of the preceding examples, and will not be detailed here. Similarly, the computation of fixed-end moments and shears for element CD is similar to that for beam BG of Exam-

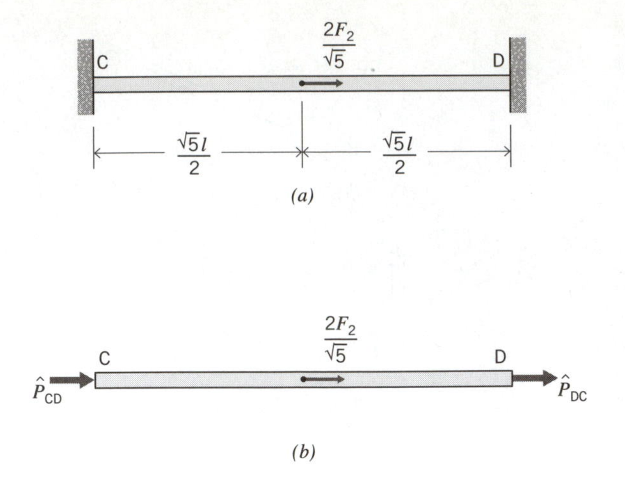

Figure 9.9 Fixed-end axial forces in CD.

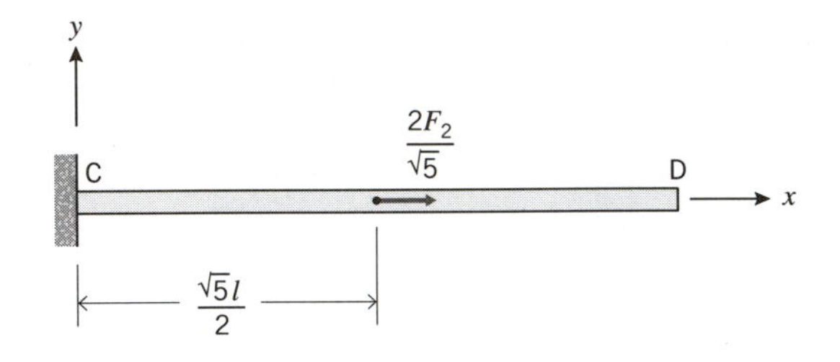

(a) Axial force in bar CD (supported at C only)

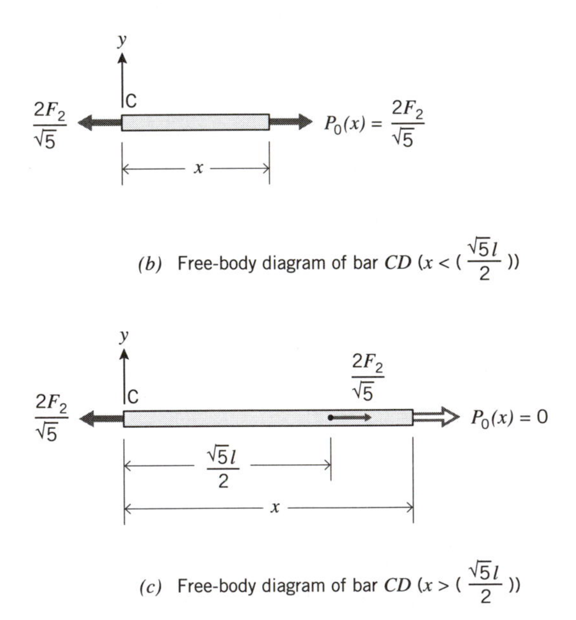

(b) Free-body diagram of bar CD $(x < (\frac{\sqrt{5}l}{2}))$

(c) Free-body diagram of bar CD $(x > (\frac{\sqrt{5}l}{2}))$

Figure 9.10 Axial forces in determinate bar CD.

ple 1, and will also not be illustrated here; however, the computation of the fixed-end axial forces deserves some discussion.

Figure 9.9 shows the component of interior axial load and corresponding fixed-end axial forces of interest. These fixed-end axial forces are determined from Equations 9.4*e* using the internal resultant forces shown in Figure 9.10. (Figure 9.10*a* is equivalent to Figure 5.31*a*(2); the free-body diagrams in Figures 9.10*b* and 9.10*c* follow from Figure 9.10*a* and define the quantity $P_o(x)$ in Equation 5.48*g*.) Note that $P_o(x) = 2F_2/\sqrt{5}$ for $0 < x < \sqrt{5}\,l/2$, and $P_o(x) = 0$ for $\sqrt{5}\,l/2 < x < l$. Since there are no thermal or initial loads, $\widehat{P}(x) = P_o(x)$. Thus, the second of Equations 9.4*e* gives

$$\widehat{P}_{\text{JI}} = -\frac{1}{\sqrt{5}\,l}\int_0^{\sqrt{5}\,l/2}\left(\frac{2F_2}{\sqrt{5}}\right)dx = -\frac{F_2}{\sqrt{5}} \tag{9.7c}$$

From equilibrium of the bar in Figure 9.9*b*, it follows that[8]

$$\widehat{P}_{\text{IJ}} = -\frac{2F_2}{\sqrt{5}} + \frac{F_2}{\sqrt{5}} = -\frac{F_2}{\sqrt{5}} \tag{9.7d}$$

The complete vector of fixed-end forces is

$$\widehat{\mathbf{P}} = \begin{Bmatrix} \widehat{M}_{\text{AB}} \\ \widehat{M}_{\text{BA}} \\ \widehat{V}_{\text{AB}} \\ \widehat{V}_{\text{BA}} \\ \widehat{P}_{\text{AB}} \\ \widehat{P}_{\text{BA}} \\ \widehat{M}_{\text{BC}} \\ \widehat{M}_{\text{CB}} \\ \widehat{V}_{\text{BC}} \\ \widehat{V}_{\text{CB}} \\ \widehat{P}_{\text{BC}} \\ \widehat{P}_{\text{CB}} \\ \widehat{M}_{\text{CD}} \\ \widehat{M}_{\text{DC}} \\ \widehat{V}_{\text{CD}} \\ \widehat{V}_{\text{DC}} \\ \widehat{P}_{\text{CD}} \\ \widehat{P}_{\text{DC}} \end{Bmatrix} = \begin{Bmatrix} -q_o l^2/12 \\ q_o l^2/12 \\ q_o l/2 \\ q_o l/2 \\ 0 \\ 0 \\ -q_o l^2/12 + EI\alpha\Delta T_o/h \\ q_o l^2/12 - EI\alpha\Delta T_o/h \\ q_o l/2\sqrt{5} \\ q_o l/2\sqrt{5} \\ -q_o l/\sqrt{5} \\ -q_o l/\sqrt{5} \\ -F_2 l/8 \\ F_2 l/8 \\ F_2/2\sqrt{5} \\ F_2/2\sqrt{5} \\ -F_2/\sqrt{5} \\ -F_2/\sqrt{5} \end{Bmatrix} \tag{9.7e}$$

The fact that there are no fixed-end axial forces in element BC due to the thermal load is a consequence of the particular load *and not a general result*. The thermal load is $\Delta T = \Delta T_o y/h$, for which $M_T(x) = -EI\alpha\Delta T_o/h$ but $P_T(x) = 0$ (see Equation 2.36*c*).

---

[8] In this case, it is implied that the last integral in the first of Equations 9.4*e* reduces to the concentrated load applied at the midpoint of the element, i.e., to the load $2F_2/\sqrt{5}$ in Figure 9.9*b*.

Other thermal loads can lead to nonzero values of $P_T(x)$ and consequent fixed-end axial forces. For instance, a thermal load of form $\Delta T = \Delta T_o y/h + \Delta T_1$ produces the same value of $M_T(x)$, and $P_T(x) = EA\alpha\Delta T_1$, from which we also obtain $\widehat{P}_{IJ} = -\widehat{P}_{JI} = EA\alpha\Delta T_1$ (Equations 9.4*e*); in other words, there would be fixed-end axial forces in addition to the fixed-end moments in Equation 9.7*e*.

Step 4. The only concentrated load applied at a joint is the vertical load F_1 at C. This load will do external virtual work for a unit virtual displacement like system displacement u_5; thus, the force vector is

$$\mathbf{F} = \begin{Bmatrix} 0 \\ 0 \\ 0 \\ 0 \\ -F_1 \\ 0 \end{Bmatrix} \tag{9.7f}$$

Step 5. We now have all of the matrices required for solution. We obtain

$$\mathbf{k} = \frac{EI}{l}\begin{bmatrix} \left(\frac{12.215}{l^2} + \frac{0.358}{Cl^2}\right) & \left(-\frac{0.429}{l^2} + \frac{0.179}{Cl^2}\right) & \left(-\frac{5.463}{l}\right) & \left(-\frac{0.215}{l^2} - \frac{0.358}{Cl^2}\right) & \left(\frac{0.429}{l^2} - \frac{0.179}{Cl^2}\right) & \left(\frac{0.537}{l}\right) \\ & \left(\frac{0.859}{l^2} + \frac{1.089}{Cl^2}\right) & \left(-\frac{1.073}{l}\right) & \left(\frac{0.429}{l^2} - \frac{0.179}{Cl^2}\right) & \left(-\frac{0.859}{l^2} - \frac{0.089}{Cl^2}\right) & \left(-\frac{1.073}{l}\right) \\ & & (5.789) & \left(-\frac{0.537}{l}\right) & \left(\frac{1.073}{l}\right) & (0.894) \\ & & & \left(\frac{1.073}{l^2} + \frac{0.447}{Cl^2}\right) & 0 & \left(-\frac{1.610}{l}\right) \\ & Sym & & & \left(\frac{1.073}{l^2} + \frac{0.447}{Cl^2}\right) & \left(\frac{0.537}{l}\right) \\ & & & & & (3.578) \end{bmatrix} \tag{9.7g}$$

in which we have again let $C \equiv I/l^2A$. Vector $\mathbf{F}^*$ is

$$\mathbf{F}^* = \mathbf{F} - \boldsymbol{\beta}^T\widehat{\mathbf{P}} = \begin{Bmatrix} q_o l \\ 0 \\ -EI\alpha\Delta T_o/h \\ q_o l/2 \\ -F_1 - F_2/2 \\ F_2 l/8 - q_o l^2/12 + EI\alpha\Delta T_o/h \end{Bmatrix} \tag{9.7h}$$

To compare with the solution in Example 5 of Chapter 8, we arbitrarily set $C = 0.0001$ ($A \to \infty$) and solve for **u**; the result is

$$\mathbf{u} = \begin{Bmatrix} u_1 \\ u_2 \\ u_3 \\ u_4 \\ u_5 \\ u_6 \end{Bmatrix} = \begin{Bmatrix} -0.0507F_1l^3/EI - 0.0267F_2l^3/EI + 0.1784q_ol^4/EI - 0.1327\alpha l^2\Delta T_o/h \\ 0 \\ -0.0485F_1l^2/EI - 0.0312F_2l^2/EI + 0.1746q_ol^3/EI - 0.3517\alpha l\Delta T_o/h \\ -0.0405F_1l^3/EI - 0.0214F_2l^3/EI + 0.1428q_ol^4/EI - 0.1061\alpha l^2\Delta T_o/h \\ -0.0205F_1l^3/EI - 0.0108F_2l^3/EI + 0.0713q_ol^4/EI - 0.0530\alpha l^2\Delta T_o/h \\ 0.0045F_1l^2/EI + 0.0387F_2l^2/EI - 0.0401q_ol^3/EI + 0.3476\alpha l\Delta T_o/h \end{Bmatrix} \tag{9.7i}$$

Note that the values of u_1, u_3, and u_6 in Equation 9.7*i* agree with the values of u_3, u_1, and u_2, respectively, in Equation 8.25*c*; also, the result $u_2 = 0$ in Equation 9.7*i* is consistent with the approximation of axial rigidity for beam AB.[9]

Back-substituting into Equation 9.7*a* provides

$$\boldsymbol{\delta} = \begin{Bmatrix} \theta_{AB} \\ \theta_{BA} \\ v_{AB} \\ v_{BA} \\ u_{AB} \\ u_{BA} \\ \theta_{BC} \\ \theta_{CB} \\ v_{BC} \\ v_{CB} \\ u_{BC} \\ u_{CB} \\ \theta_{CD} \\ \theta_{DC} \\ v_{CD} \\ v_{DC} \\ u_{CD} \\ u_{DC} \end{Bmatrix} = \begin{Bmatrix} 0 \\ -0.0485F_1l^2/EI - 0.0312F_2l^2/EI + 0.1746q_ol^3/EI - 0.3517\alpha\Delta T_o l/h \\ 0 \\ 0.0507F_1l^3/EI + 0.0267F_2l^3/EI - 0.1784q_ol^4/EI + 0.1327\alpha\Delta T_o l^2/h \\ 0 \\ 0 \\ -0.0485F_1l^2/EI - 0.0312F_2l^2/EI + 0.1746q_ol^3/EI - 0.3517\alpha\Delta T_o l/h \\ 0.0045F_1l^2/EI + 0.0387F_2l^2/EI - 0.0401q_ol^3/EI + 0.3476\alpha\Delta T_o l/h \\ 0.0226F_1l^3/EI + 0.0120F_2l^3/EI - 0.0798q_ol^4/EI + 0.0594\alpha\Delta T_o l^2/h \\ -0.0002F_1l^3/EI - 0.0001F_2l^3/EI - 0.0001q_ol^4/EI \\ -0.0453F_1l^3/EI - 0.0239F_2l^3/EI + 0.1596q_ol^4/EI - 0.1187\alpha\Delta T_o l^2/h \\ -0.0454F_1l^3/EI - 0.0240F_2l^3/EI + 0.1596q_ol^4/EI - 0.1186\alpha\Delta T_o l^2/h \\ 0.0045F_1l^2/EI + 0.0387F_2l^2/EI - 0.0401q_ol^3/EI + 0.3476\alpha\Delta T_o l/h \\ 0 \\ -0.0454F_1l^3/EI - 0.0240F_2l^3/EI + 0.1596q_ol^4/EI - 0.1186\alpha\Delta T_o l^2/h \\ 0 \\ 0.0002F_1l^3/EI + 0.0001F_2l^3/EI + 0.0001q_ol^4/EI \\ 0 \end{Bmatrix} \tag{9.7j}$$

[9] All components of u_2 are zero to four decimal places.

and, from the force-displacement equations, we obtain

$$\mathbf{P} = \begin{Bmatrix} M_{AB} \\ M_{BA} \\ V_{AB} \\ V_{BA} \\ P_{AB} \\ P_{BA} \\ M_{BC} \\ M_{CB} \\ V_{BC} \\ V_{CB} \\ P_{BC} \\ P_{CB} \\ M_{CD} \\ M_{DC} \\ V_{CD} \\ V_{DC} \\ P_{CD} \\ P_{DC} \end{Bmatrix} = \begin{Bmatrix} 0.2070F_1l + 0.0980F_2l - 0.8046q_ol^2 + 0.0929EI\alpha\Delta T_o/h \\ 0.1100F_1l + 0.0356F_2l - 0.2887q_ol^2 - 0.6105EI\alpha\Delta T_o/h \\ -0.3170F_1 - 0.1337F_2 + 1.5933q_ol + 0.5176EI\alpha\Delta T_o/lh \\ 0.3170F_1 + 0.1337F_2 - 0.5933q_ol - 0.5176EI\alpha\Delta T_o/lh \\ 0.2448F_1 + 0.0712F_2 - 0.3227q_ol - 0.1820EI\alpha\Delta T_o/lh \\ -0.2448F_1 - 0.0712F_2 + 0.3227q_ol + 0.1820EI\alpha\Delta T_o/lh \\ -0.1100F_1l - 0.0356F_2l + 0.2887q_ol^2 + 0.6105EI\alpha\Delta T_o/h \\ -0.0626F_1l + 0.0269F_2l + 0.2633q_ol^2 - 0.7640EI\alpha\Delta T_o/h \\ 0.0772F_1 + 0.0039F_2 - 0.0233q_ol + 0.0687EI\alpha\Delta T_o/lh \\ -0.0772F_1 - 0.0039F_2 + 0.4705q_ol - 0.0687EI\alpha\Delta T_o/lh \\ 0.3930F_1 + 0.1514F_2 - 0.6749q_ol - 0.5444EI\alpha\Delta T_o/lh \\ -0.3930F_1 - 0.1514F_2 - 0.2195q_ol + 0.5444EI\alpha\Delta T_o/lh \\ 0.0626F_1l - 0.0269F_2l - 0.2633q_ol^2 + 0.7640EI\alpha\Delta T_o/h \\ 0.0585F_1l + 0.1884F_2l - 0.2274q_ol^2 + 0.4532EI\alpha\Delta T_o/h \\ -0.0542F_1 + 0.1514F_2 + 0.2195q_ol - 0.5444EI\alpha\Delta T_o/lh \\ 0.0542F_1 + 0.2958F_2 - 0.2195q_ol + 0.5444EI\alpha\Delta T_o/lh \\ 0.8172F_1 - 0.0039F_2 + 0.4705q_ol - 0.0687EI\alpha\Delta T_o/lh \\ -0.8172F_1 - 0.8905F_2 - 0.4705q_ol + 0.0687EI\alpha\Delta T_o/lh \end{Bmatrix} \quad \textbf{(9.7}\textbf{\textit{k}}\textbf{)}$$

which agrees with corresponding values in Equation 8.25*e*.

9.5 ELEMENT STIFFNESS MATRIX IN GLOBAL COORDINATES

In all of our applications thus far, we have employed two distinct types of coordinate systems, namely local (element) coordinates defined in the axial and transverse directions for each element, and global (system) coordinates defined in the horizontal and vertical directions for the structure. The use of element information defined with reference to the local coordinates permitted us to work out the stiffness relations in a straightforward way by separating the bending effects from the axial effects; the use of system displacements defined with reference to the global coordinates proved to be quite convenient for determining the compatibility relations for the structure.

As may have already been observed, the analysis of compatibility (Step 2) is greatly simplified when the local and global coordinate systems coincide, i.e., for elements that are either horizontal or vertical (e.g., element AB in Figure 9.7*a*). In such circumstances, there is a direct correspondence between element end displacements and system displacements. Specifically, each end displacement is identical to one of the system displacements. As a result, the compatibility matrix contains only 0's and 1's with, at most, one 1 in any row of the matrix. (See, for instance, the portion of the compatibility matrix associated with element AB in Equation 9.7*a*.) Elements that are oriented at some angle (e.g., elements BC and CD in Figure 9.7*a*) have more complicated compatibility matrices, i.e., have end displacements that are algebraic functions of two system displacements. (See, for instance, the portions of the compatibility matrix associated with elements BC and CD in Equation 9.7*a*.)

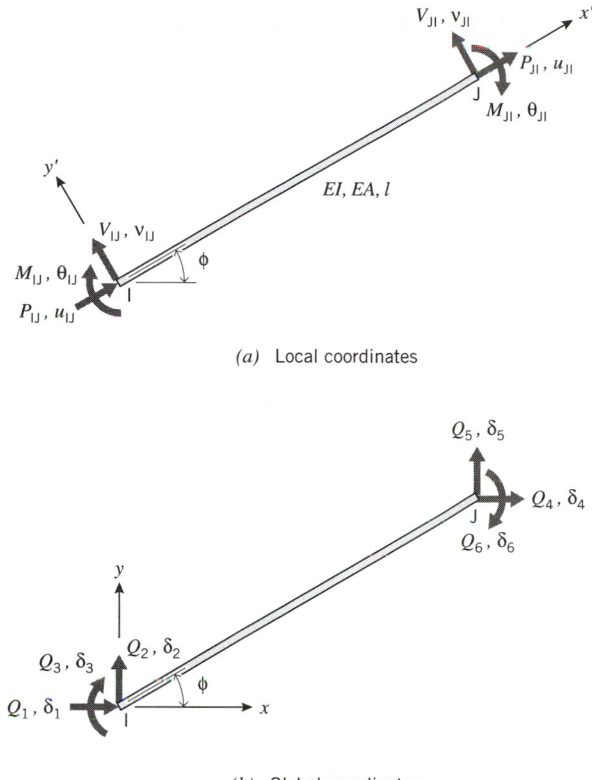

Figure 9.11 Element end forces/displacements.

In addition to the relative simplicity in correlating element and system displacements, there is a second advantage to having global and local coordinates coincide: since the system stiffness matrix, **k**, is obtained from the matrix triple product $\boldsymbol{\beta}^T\mathbf{K}\boldsymbol{\beta}$, the 1's in the $\boldsymbol{\beta}$ matrix have the effect of moving terms from **K** to specific locations in **k** without changing the values. This makes it possible to assemble **k** without performing any multiplications at all, and is the basis for the ''direct stiffness method,'' which will be developed in Section 9.6. To implement this mathematical feature, however, it is necessary to have all element end displacements correspond directly to system displacements regardless of the orientation of the element. Since the elements themselves cannot be rotated, it is necessary to express the element stiffness in terms of horizontal and vertical end displacements. This requires that we reformulate the element stiffness relation, Equation 9.5, in terms of global coordinates, a process that involves a transformation of coordinate axes.

Consider the typical element IJ in Figure 9.11*a*, which is oriented at an arbitrary angle ϕ to the horizontal. (The x'-y'coordinate system in this figure is the local coordinate system.) The element is assumed to be subjected to a general set of interior loads (not shown), and to have end forces and end displacements as given by Equation 9.5.

For the sake of clarity, we will represent the stiffness relation for this element (Equation 9.5) by the matrix form

$$\mathbf{P}_l^{(e)} = \mathbf{K}_l^{(e)} \boldsymbol{\delta}_l^{(e)} + \widehat{\mathbf{P}}_l^{(e)} \tag{9.8a}$$

where the subscript l denotes ''local coordinates'' and the superscript (e) denotes a single element.

Figure 9.11*b* shows the same element with all end forces and end displacements resolved into horizontal and vertical components. (The x-y coordinate system in this figure is the global coordinate system.) We represent the force quantities in Figure 9.11*b* by Q_1–Q_6, and the displacement quantities by δ_1–δ_6.[10] We will denote vectors of the end forces and end displacements in Figure 9.11*b* by $\mathbf{P}_g^{(e)}$ and $\boldsymbol{\delta}_g^{(e)}$, respectively,

$$\mathbf{P}_g^{(e)} \equiv \begin{Bmatrix} Q_1 \\ Q_2 \\ Q_3 \\ Q_4 \\ Q_5 \\ Q_6 \end{Bmatrix}, \qquad \boldsymbol{\delta}_g^{(e)} \equiv \begin{Bmatrix} \delta_1 \\ \delta_2 \\ \delta_3 \\ \delta_4 \\ \delta_5 \\ \delta_6 \end{Bmatrix} \tag{9.8b}$$

We seek to convert Equation 9.8*a* to a relation between the end displacements and end forces in Figure 9.11*b*, i.e., to a relation of form

$$\mathbf{P}_g^{(e)} = \mathbf{K}_g^{(e)} \boldsymbol{\delta}_g^{(e)} + \widehat{\mathbf{P}}_g^{(e)} \tag{9.8c}$$

where $\mathbf{K}_g^{(e)}$ is the stiffness matrix for the element in Figure 9.11*b*.

The quantities in Figure 9.11*a* are related to those in Figure 9.11*b* via basic two-dimensional coordinate rotation equations, namely

$$\begin{aligned}
u_{\mathrm{IJ}} &= \delta_1 \cos\phi + \delta_2 \sin\phi \\
v_{\mathrm{IJ}} &= -\delta_1 \sin\phi + \delta_2 \cos\phi \\
\theta_{\mathrm{IJ}} &= \delta_3 \\
u_{\mathrm{JI}} &= \delta_4 \cos\phi + \delta_5 \sin\phi \\
v_{\mathrm{JI}} &= -\delta_4 \sin\phi + \delta_5 \cos\phi \\
\theta_{\mathrm{JI}} &= \delta_6
\end{aligned} \tag{9.8d}$$

[10] The numbering scheme is: Q_1, Q_2, and Q_3 are the horizontal force, vertical force, and moment, respectively, at the I end, and Q_4, Q_5, and Q_6 are the horizontal force, vertical force, and moment, respectively, at the J end, etc.

and

$$
\begin{aligned}
Q_1 &= -V_{\mathrm{IJ}} \sin\phi + P_{\mathrm{IJ}} \cos\phi \\
Q_2 &= \quad V_{\mathrm{IJ}} \cos\phi + P_{\mathrm{IJ}} \sin\phi \\
Q_3 &= \quad M_{\mathrm{IJ}} \\
Q_4 &= -V_{\mathrm{JI}} \sin\phi + P_{\mathrm{JI}} \cos\phi \\
Q_5 &= \quad V_{\mathrm{JI}} \cos\phi + P_{\mathrm{JI}} \sin\phi \\
Q_6 &= \quad M_{\mathrm{JI}}
\end{aligned}
\tag{9.8e}
$$

By reordering Equations 9.8*d* we obtain a relation between $\boldsymbol{\delta}_g^{(e)}$ and $\boldsymbol{\delta}_l^{(e)}$,

$$
\boldsymbol{\delta}_l^{(e)} \equiv \begin{Bmatrix} \theta_{\mathrm{IJ}} \\ \theta_{\mathrm{JI}} \\ v_{\mathrm{IJ}} \\ v_{\mathrm{JI}} \\ u_{\mathrm{IJ}} \\ u_{\mathrm{JI}} \end{Bmatrix} = \begin{bmatrix} 0 & 0 & 1 & 0 & 0 & 0 \\ 0 & 0 & 0 & 0 & 0 & 1 \\ -S & C & 0 & 0 & 0 & 0 \\ 0 & 0 & 0 & -S & C & 0 \\ C & S & 0 & 0 & 0 & 0 \\ 0 & 0 & 0 & C & S & 0 \end{bmatrix} \begin{Bmatrix} \delta_1 \\ \delta_2 \\ \delta_3 \\ \delta_4 \\ \delta_5 \\ \delta_6 \end{Bmatrix} = \mathbf{T}\boldsymbol{\delta}_g^{(e)} \tag{9.8f}
$$

where, for the sake of conciseness in Equation 9.8*f*, we have employed the notation

$$
\begin{aligned}
S &\equiv \sin\phi \\
C &\equiv \cos\phi
\end{aligned}
\tag{9.8g}
$$

The square matrix in Equation 9.8*f* is a transformation matrix containing only trigonometric functions of ϕ; we denote this matrix by the symbol $\mathbf{T}$.

Similarly, the matrix form of Equations 9.8*e* is

$$
\mathbf{P}_g^{(e)} \equiv \begin{Bmatrix} Q_1 \\ Q_2 \\ Q_3 \\ Q_4 \\ Q_5 \\ Q_6 \end{Bmatrix} = \begin{vmatrix} 0 & 0 & -S & 0 & C & 0 \\ 0 & 0 & C & 0 & S & 0 \\ 1 & 0 & 0 & 0 & 0 & 0 \\ 0 & 0 & 0 & -S & 0 & C \\ 0 & 0 & 0 & C & 0 & S \\ 0 & 1 & 0 & 0 & 0 & 0 \end{vmatrix} \begin{Bmatrix} M_{\mathrm{IJ}} \\ M_{\mathrm{JI}} \\ V_{\mathrm{IJ}} \\ V_{\mathrm{JI}} \\ P_{\mathrm{IJ}} \\ P_{\mathrm{JI}} \end{Bmatrix} = \mathbf{T}^{\mathrm{T}}\mathbf{P}_l^{(e)} \tag{9.8h}
$$

We observe that the transformation matrix in Equation 9.8*h* is the transpose of $\mathbf{T}$. There is a comparable matrix equation relating the fixed-end forces $\widehat{\mathbf{P}}_l^{(e)}$ and $\widehat{\mathbf{P}}_g^{(e)}$, namely

$$
\widehat{\mathbf{P}}_g^{(e)} = \mathbf{T}^{\mathrm{T}}\widehat{\mathbf{P}}_l^{(e)} \tag{9.8i}
$$

We obtain the desired force-displacement relation, Equation 9.8*c*, by transformation of Equation 9.8*a*; premultiplying Equation 9.8*a* by $\mathbf{T}^{\mathrm{T}}$ gives

$$\mathbf{T}^{\mathrm{T}}\mathbf{P}_l^{(e)} = \mathbf{T}^{\mathrm{T}}\mathbf{K}_l^{(e)}\boldsymbol{\delta}_l^{(e)} + \mathbf{T}^{\mathrm{T}}\widehat{\mathbf{P}}_l^{(e)} \tag{9.8j}$$

Substituting Equations 9.8*f*, 9.8*h*, and 9.8*i* into Equation 9.8*j* leads to

$$\begin{aligned}\mathbf{P}_g^{(e)} &= \mathbf{T}^{\mathrm{T}}\mathbf{K}_l^{(e)}(\mathbf{T}\boldsymbol{\delta}_g^{(e)}) + \widehat{\mathbf{P}}_g^{(e)} \\ &= (\mathbf{T}^{\mathrm{T}}\mathbf{K}_l^{(e)}\mathbf{T})\,\boldsymbol{\delta}_g^{(e)} + \widehat{\mathbf{P}}_g^{(e)} \\ &= \mathbf{K}_g^{(e)}\boldsymbol{\delta}_g^{(e)} + \widehat{\mathbf{P}}_g^{(e)}\end{aligned} \tag{9.8k}$$

Here, we have recognized that the matrix triple product (in parentheses in the second equation) must be equivalent to the stiffness matrix in Equation 9.8*c*,

$$\mathbf{K}_g^{(e)} = \mathbf{T}^{\mathrm{T}}\mathbf{K}_l^{(e)}\mathbf{T} \tag{9.8l}$$

Carrying out the matrix multiplications in Equation 9.8*l* (using the definition of $\mathbf{T}$ in Equation 9.8*f* and the definition of $\mathbf{K}_l^{(e)}$ in Equation 9.5) provides the explicit form of $\mathbf{K}_g^{(e)}$ in Equation 9.8*c*,

$$\mathbf{K}_g^{(e)} = \frac{EI}{l}\begin{bmatrix} \left(\frac{12S^2}{l^2} + \frac{A}{I}C^2\right) & \left(-\frac{12SC}{l^2} + \frac{A}{I}SC\right) & \left(\frac{6S}{l}\right) & \left(-\frac{12S^2}{l^2} - \frac{A}{I}C^2\right) & \left(\frac{12SC}{l^2} - \frac{A}{I}SC\right) & \left(\frac{6S}{l}\right) \\ & \left(\frac{12C^2}{l^2} + \frac{A}{I}S^2\right) & \left(-\frac{6C}{l}\right) & \left(\frac{12SC}{l^2} - \frac{A}{I}SC\right) & \left(-\frac{12C^2}{l^2} - \frac{A}{I}S^2\right) & \left(-\frac{6C}{l}\right) \\ & & (4) & \left(-\frac{6S}{l}\right) & \left(\frac{6C}{l}\right) & (2) \\ & \mathit{Sym.} & & \left(\frac{12S^2}{l^2} + \frac{A}{I}C^2\right) & \left(-\frac{12SC}{l^2} + \frac{A}{I}SC\right) & \left(-\frac{6S}{l}\right) \\ & & & & \left(\frac{12C^2}{l^2} + \frac{A}{I}S^2\right) & \left(\frac{6C}{l}\right) \\ & & & & & (4) \end{bmatrix} \tag{9.8m}$$

Similarly, from Equation 9.8*i* we obtain the explicit form of $\widehat{\mathbf{P}}_g^{(e)}$,

$$\widehat{\mathbf{P}}_g^{(e)} \equiv \begin{Bmatrix} \widehat{Q}_1 \\ \widehat{Q}_2 \\ \widehat{Q}_3 \\ \widehat{Q}_4 \\ \widehat{Q}_5 \\ \widehat{Q}_6 \end{Bmatrix} = \begin{Bmatrix} -S\widehat{V}_{\mathrm{IJ}} + C\widehat{P}_{\mathrm{IJ}} \\ C\widehat{V}_{\mathrm{IJ}} + S\widehat{P}_{\mathrm{IJ}} \\ \widehat{M}_{\mathrm{IJ}} \\ -S\widehat{V}_{\mathrm{JI}} + C\widehat{P}_{\mathrm{JI}} \\ C\widehat{V}_{\mathrm{JI}} + S\widehat{P}_{\mathrm{JI}} \\ \widehat{M}_{\mathrm{JI}} \end{Bmatrix} \tag{9.8n}$$

We may use Equation 9.8*c* in exactly the same way that Equation 9.8*a* was used in previous applications. To illustrate, we will reanalyze the structure in Example 3. This will serve to demonstrate the simplified compatibility matrix that results when all element end displacements are defined in terms of the global coordinates.

EXAMPLE 4

We will reconsider the structure shown in Figure 9.7 and follow the same five-step procedure as in Example 3. As in Example 3, we will assume that the three elements have axial stiffness EA as well as bending stiffness EI.

Step 1. The gabled frame has the six system displacements shown in Figure 9.7. Note, in particular, that the system displacements at each joint have been numbered in the same sequence as the element end displacements in Figure 9.11*b*, i.e., horizontal displacement is u_i, vertical displacement is u_{i+1}, and rotation is u_{i+2}.

Step 2. For the sake of convenience (and with a view toward subsequent illustration of the developments in Section 9.6), we will discuss the compatibility matrices for each element separately.

Consider, for example, element AB, which is a special case of the element in Figure 9.11*b* when $\phi = \pi/2$;[11] the length of AB is l. As in Figure 9.11*b*, we will denote the end displacements by $\delta_1^{(AB)}$–$\delta_6^{(AB)}$; as in the figure, $\delta_1^{(AB)}$ is the horizontal displacement at end A, etc. (We add the superscripts to clearly distinguish the particular element under consideration.) Since there is a fixed support at A in Figure 9.7*a*, we must have $\delta_1^{(AB)} = \delta_2^{(AB)} = \delta_3^{(AB)} = 0$. At end B, $\delta_4^{(AB)}$ coincides with system displacement u_1, $\delta_5^{(AB)}$ coincides with system displacement u_2, and $\delta_6^{(AB)}$coincides with system displacement u_3, i.e., $\delta_4^{(AB)} = u_1$, $\delta_5^{(AB)} = u_2$, and $\delta_6^{(AB)} = u_3$. In other words, for AB, we have

$$\boldsymbol{\delta}_g^{(AB)} \equiv \begin{Bmatrix} \delta_1^{(AB)} \\ \delta_2^{(AB)} \\ \delta_3^{(AB)} \\ \delta_4^{(AB)} \\ \delta_5^{(AB)} \\ \delta_6^{(AB)} \end{Bmatrix} = \begin{bmatrix} 0 & 0 & 0 & 0 & 0 & 0 \\ 0 & 0 & 0 & 0 & 0 & 0 \\ 0 & 0 & 0 & 0 & 0 & 0 \\ 1 & 0 & 0 & 0 & 0 & 0 \\ 0 & 1 & 0 & 0 & 0 & 0 \\ 0 & 0 & 1 & 0 & 0 & 0 \end{bmatrix} \begin{Bmatrix} u_1 \\ u_2 \\ u_3 \\ u_4 \\ u_5 \\ u_6 \end{Bmatrix} = \boldsymbol{\beta}^{(AB)}\mathbf{u} \qquad \textbf{(9.9}\textbf{\textit{a}}\textbf{)}$$

where $\boldsymbol{\delta}_g^{(AB)}$ is the vector of end displacements and $\boldsymbol{\beta}^{(AB)}$ is the compatibility matrix for the element.

The compatibility relations for elements BC and CD are obtained in a similar manner. Element BC is a special case of the element in Figure 9.11*b* for which $\phi = \tan^{-1}(1/2) = 26.6°$ and length is equal to $\sqrt{5}l$. At end B of BC, $\delta_1^{(BC)}$ coincides with system displacement u_1, $\delta_2^{(BC)}$ coincides with system displacement u_2, and $\delta_3^{(BC)}$ coincides with system displacement u_3. At end C of BC, $\delta_4^{(BC)}$ coincides with system displacement u_4,

[11] By referring to this element as AB rather than BA, we are implying that end A coincides with the I end of element IJ in Figure 9.11*b*; this also specifies that $\phi = +\pi/2$ rather than $-\pi/2$.

$\delta_5^{(BC)}$ coincides with system displacement u_5, and $\delta_6^{(BC)}$ coincides with system displacement u_6. In other words,

$$\boldsymbol{\delta}_g^{(BC)} \equiv \begin{Bmatrix} \delta_1^{(BC)} \\ \delta_2^{(BC)} \\ \delta_3^{(BC)} \\ \delta_4^{(BC)} \\ \delta_5^{(BC)} \\ \delta_6^{(BC)} \end{Bmatrix} = \begin{bmatrix} 1 & 0 & 0 & 0 & 0 & 0 \\ 0 & 1 & 0 & 0 & 0 & 0 \\ 0 & 0 & 1 & 0 & 0 & 0 \\ 0 & 0 & 0 & 1 & 0 & 0 \\ 0 & 0 & 0 & 0 & 1 & 0 \\ 0 & 0 & 0 & 0 & 0 & 1 \end{bmatrix} \begin{Bmatrix} u_1 \\ u_2 \\ u_3 \\ u_4 \\ u_5 \\ u_6 \end{Bmatrix} = \boldsymbol{\beta}^{(BC)}\mathbf{u} \tag{9.9b}$$

Element CD is obtained from the element in Figure 9.11*b* when $\phi = \tan^{-1}(-2) = -63.4°$ and length is $\sqrt{5}l$. At end C of CD, $\delta_1^{(CD)}$ coincides with system displacement u_4, $\delta_2^{(CD)}$ coincides with system displacement u_5, and $\delta_3^{(CD)}$ coincides with system displacement u_6; at end D of CD, $\delta_4^{(CD)} = \delta_5^{(CD)} = \delta_6^{(CD)} = 0$, so that

$$\boldsymbol{\delta}_g^{(CD)} \equiv \begin{Bmatrix} \delta_1^{(CD)} \\ \delta_2^{(CD)} \\ \delta_3^{(CD)} \\ \delta_4^{(CD)} \\ \delta_5^{(CD)} \\ \delta_6^{(CD)} \end{Bmatrix} = \begin{bmatrix} 0 & 0 & 0 & 1 & 0 & 0 \\ 0 & 0 & 0 & 0 & 1 & 0 \\ 0 & 0 & 0 & 0 & 0 & 1 \\ 0 & 0 & 0 & 0 & 0 & 0 \\ 0 & 0 & 0 & 0 & 0 & 0 \\ 0 & 0 & 0 & 0 & 0 & 0 \end{bmatrix} \begin{Bmatrix} u_1 \\ u_2 \\ u_3 \\ u_4 \\ u_5 \\ u_6 \end{Bmatrix} = \boldsymbol{\beta}^{(CD)}\mathbf{u} \tag{9.9c}$$

The complete matrix compatibility equation is obtained by assembling Equations 9.9*a*–9.9*c*; the form is

$$\boldsymbol{\delta}_g \equiv \begin{Bmatrix} \boldsymbol{\delta}_g^{(AB)} \\ \boldsymbol{\delta}_g^{(BC)} \\ \boldsymbol{\delta}_g^{(CD)} \end{Bmatrix} = \begin{bmatrix} \boldsymbol{\beta}^{(AB)} \\ \boldsymbol{\beta}^{(BC)} \\ \boldsymbol{\beta}^{(CD)} \end{bmatrix} \begin{Bmatrix} u_1 \\ u_2 \\ u_3 \\ u_4 \\ u_5 \\ u_6 \end{Bmatrix} = \boldsymbol{\beta}\mathbf{u} \tag{9.9d}$$

Equation 9.9*d* replaces the previous Equation 9.7*a*, and relates end displacements in global coordinates to system displacements. As may be observed, the $\boldsymbol{\beta}$ matrix in Equation 9.9*d* contains only 0's and 1's.

Step 3. The individual element stiffness matrices are obtained from Equation 9.8*m* by substituting the specific values of ϕ and length noted above. For element AB, $\phi = \pi/2$ and length is l, so we obtain

$$\mathbf{K}_g^{(AB)} = \frac{EI}{l} \begin{bmatrix} 12/l^2 & 0 & 6/l & -12/l^2 & 0 & 6/l \\ & A/I & 0 & 0 & -A/I & 0 \\ & & 4 & -6/l & 0 & 2 \\ & Sym. & & 12/l^2 & 0 & -6/l \\ & & & & A/I & 0 \\ & & & & & 4 \end{bmatrix} \tag{9.10a}$$

Since the axis of element AB is parallel to a global coordinate direction, the individual terms in this matrix are identical to terms in the first submatrix of **K** in Equation 9.7*b*; only the order is different, a consequence of the reordering of the components of the matrix of element end displacements.

For element BC ($\phi = 26.6°$, length $= \sqrt{5}l$) Equation 9.8*m* becomes

$$\mathbf{K}_g^{(BC)} = \frac{EI}{l}\begin{bmatrix} \left(\frac{0.2147}{l^2}+\frac{0.3578A}{I}\right) & \left(-\frac{0.4293}{l^2}+\frac{0.1789A}{I}\right) & \left(\frac{0.5367}{l}\right) & \left(-\frac{0.2147}{l^2}-\frac{0.3578A}{I}\right) & \left(\frac{0.4293}{l^2}-\frac{0.1789A}{I}\right) & \left(\frac{0.5367}{l}\right) \\ & \left(\frac{0.8587}{l^2}+\frac{0.0894A}{I}\right) & \left(-\frac{1.0733}{l}\right) & \left(\frac{0.4293}{l^2}-\frac{0.1789A}{I}\right) & \left(-\frac{0.8587}{l^2}-\frac{0.0894A}{I}\right) & \left(-\frac{1.0733}{l}\right) \\ & & (1.7889) & \left(-\frac{0.5367}{l}\right) & \left(\frac{1.0733}{l}\right) & (0.8944) \\ & \mathit{Sym.} & & \left(\frac{0.2147}{l^2}+\frac{0.3578A}{I}\right) & \left(-\frac{0.4293}{l^2}+\frac{0.1789A}{I}\right) & \left(-\frac{0.5367}{l}\right) \\ & & & & \left(\frac{0.8587}{l^2}+\frac{0.0894A}{I}\right) & \left(\frac{1.0733}{l}\right) \\ & & & & & (1.7889) \end{bmatrix} \quad \textbf{(9.10}\boldsymbol{b}\textbf{)}$$

and for element CD ($\phi = -63.4°$, length $= \sqrt{5}l$) we have

$$\mathbf{K}_g^{(CD)} = \frac{EI}{l}\begin{bmatrix} \left(\frac{0.8587}{l^2}+\frac{0.0894A}{I}\right) & \left(\frac{0.4293}{l^2}-\frac{0.1789A}{I}\right) & \left(-\frac{1.0733}{l}\right) & \left(-\frac{0.8587}{l^2}-\frac{0.0894A}{I}\right) & \left(-\frac{0.4293}{l^2}+\frac{0.1789A}{I}\right) & \left(-\frac{1.0733}{l}\right) \\ & \left(\frac{0.2147}{l^2}+\frac{0.3578A}{I}\right) & \left(-\frac{0.5367}{l}\right) & \left(-\frac{0.4293}{l^2}+\frac{0.1789A}{I}\right) & \left(-\frac{0.2147}{l^2}-\frac{0.3578A}{I}\right) & \left(-\frac{0.5367}{l}\right) \\ & & (1.7889) & \left(\frac{1.0733}{l}\right) & \left(\frac{0.5367}{l}\right) & (0.8944) \\ & \mathit{Sym.} & & \left(\frac{0.8587}{l^2}+\frac{0.0894A}{I}\right) & \left(\frac{0.4293}{l^2}-\frac{0.1789A}{I}\right) & \left(\frac{1.0733}{l}\right) \\ & & & & \left(\frac{0.2147}{l^2}+\frac{0.3578A}{I}\right) & \left(\frac{0.5367}{l}\right) \\ & & & & & (1.7889) \end{bmatrix} \quad \textbf{(9.10}\boldsymbol{c}\textbf{)}$$

The element stiffness matrices in Equations 9.10*b* and 9.10*c* may appear to be more complicated than the corresponding submatrices in Equation 9.7*b*, but the computation using Equation 9.8*m* is very straightforward, particularly when automated.

The assemblage of the three element stiffness matrices is

$$\mathbf{K}_g = \begin{bmatrix} \mathbf{K}_g^{(AB)} & & \\ & \mathbf{K}_g^{(BC)} & \\ & & \mathbf{K}_g^{(CD)} \end{bmatrix} \quad \textbf{(9.10}\boldsymbol{d}\textbf{)}$$

The matrix $\mathbf{K}_g$ defined in Equation 9.10*d* replaces **K** in Equation 9.7*b*.

The individual vectors of global fixed-end forces are obtained in a similar manner from Equation 9.8*n*. It is important to note, however, that the element-level fixed-end forces, Equation 9.7*e*, are required in Equation 9.8*n*; using the values in Equation 9.7*e* we obtain,

$$\widehat{\mathbf{P}}_g^{(AB)} \equiv \begin{Bmatrix} \widehat{Q}_1^{(AB)} \\ \widehat{Q}_2^{(AB)} \\ \widehat{Q}_3^{(AB)} \\ \widehat{Q}_4^{(AB)} \\ \widehat{Q}_5^{(AB)} \\ \widehat{Q}_6^{(AB)} \end{Bmatrix} = \begin{Bmatrix} -q_o l/2 \\ 0 \\ -q_o l^2/12 \\ -q_o l/2 \\ 0 \\ q_o l^2/12 \end{Bmatrix} \tag{9.10e}$$

$$\widehat{\mathbf{P}}_g^{(BC)} \equiv \begin{Bmatrix} \widehat{Q}_1^{(BC)} \\ \widehat{Q}_2^{(BC)} \\ \widehat{Q}_3^{(BC)} \\ \widehat{Q}_4^{(BC)} \\ \widehat{Q}_5^{(BC)} \\ \widehat{Q}_6^{(BC)} \end{Bmatrix} = \begin{Bmatrix} -q_o l/2 \\ 0 \\ -q_o l^2/12 + EI\alpha\Delta T_o/h \\ -q_o l/2 \\ 0 \\ q_o l^2/12 - EI\alpha\Delta T_o/h \end{Bmatrix} \tag{9.10f}$$

and

$$\widehat{\mathbf{P}}_g^{(CD)} \equiv \begin{Bmatrix} \widehat{Q}_1^{(CD)} \\ \widehat{Q}_2^{(CD)} \\ \widehat{Q}_3^{(CD)} \\ \widehat{Q}_4^{(CD)} \\ \widehat{Q}_5^{(CD)} \\ \widehat{Q}_6^{(CD)} \end{Bmatrix} = \begin{Bmatrix} 0 \\ F_2/2 \\ -F_2 l/8 \\ 0 \\ F_2/2 \\ F_2 l/8 \end{Bmatrix} \tag{9.10g}$$

The composite vector of global fixed-end forces is given by

$$\widehat{\mathbf{P}}_g = \begin{Bmatrix} \widehat{\mathbf{P}}_g^{(AB)} \\ \widehat{\mathbf{P}}_g^{(BC)} \\ \widehat{\mathbf{P}}_g^{(CD)} \end{Bmatrix} \tag{9.10h}$$

Vector $\widehat{\mathbf{P}}_g$ replaces $\widehat{\mathbf{P}}$ in Equation 9.7*e*. The final form of the force-displacement relation for the entire structure is

$$\begin{Bmatrix} \mathbf{P}_g^{(AB)} \\ \mathbf{P}_g^{(BC)} \\ \mathbf{P}_g^{(CD)} \end{Bmatrix} = \begin{bmatrix} \mathbf{K}_g^{(AB)} & & \\ & \mathbf{K}_g^{(BC)} & \\ & & \mathbf{K}_g^{(CD)} \end{bmatrix} \begin{Bmatrix} \boldsymbol{\delta}_g^{(AB)} \\ \boldsymbol{\delta}_g^{(BC)} \\ \boldsymbol{\delta}_g^{(CD)} \end{Bmatrix} + \begin{Bmatrix} \widehat{\mathbf{P}}_g^{(AB)} \\ \widehat{\mathbf{P}}_g^{(BC)} \\ \widehat{\mathbf{P}}_g^{(CD)} \end{Bmatrix} \tag{9.10i}$$

which we will denote by the usual form

$$\mathbf{P}_g = \mathbf{K}_g \boldsymbol{\delta}_g + \widehat{\mathbf{P}}_g \tag{9.10j}$$

Step 4. Recall that a major feature of formulations involving element displacements (rather than element deformations) is that the system equilibrium equations have the form of Equation 9.3n, i.e., the components of the force vector $\mathbf{F}$ are associated only with concentrated joint loads. In the present case, there is only one joint load to consider, namely F_1 at C. This force acts in the (opposite) direction of system displacement u_5; thus, the force vector is

$$\mathbf{F} = \begin{Bmatrix} 0 \\ 0 \\ 0 \\ 0 \\ -F_1 \\ 0 \end{Bmatrix} \qquad \textbf{(9.11}\boldsymbol{a}\textbf{)}$$

Step 5. We now have all matrices required for solution. The final form of the system equilibrium equations is

$$(\boldsymbol{\beta}^{\mathrm{T}}\mathbf{K}_g\boldsymbol{\beta})\mathbf{u} = \mathbf{F} - \boldsymbol{\beta}^{\mathrm{T}}\widehat{\mathbf{P}}_g \qquad \textbf{(9.11}\boldsymbol{b}\textbf{)}$$

or

$$\mathbf{ku} = \mathbf{F}^* \qquad \textbf{(9.11}\boldsymbol{c}\textbf{)}$$

The matrices in Equation 9.11c must be exactly the same as the matrices in Equations 9.7g and 9.7h because the equilibrium relation between applied loads and system displacements is unique! (In this particular example, the equivalence of the $\mathbf{k}$ matrices may be verified by substituting $\boldsymbol{\beta}$ defined in Equation 9.9d and $\mathbf{K}_g$ defined in Equation 9.10d and evaluating $\boldsymbol{\beta}^{\mathrm{T}}\mathbf{K}_g\boldsymbol{\beta}$; similarly, the equivalence of the $\mathbf{F}^*$ vectors may be verified by substituting $\mathbf{F}$ from Equation 9.11a and $\widehat{\mathbf{P}}_g$ from Equation 9.10h on the right side of Equation 9.11b.) Since the two solutions are identical, the computations will not be illustrated here.

The formulation and computational procedures employed in Example 4 were essentially the same as those in Example 3. From this point on, however, the computational procedures will be modified to take advantage of the simplified compatibility matrices that result from the use of Equation 9.8c. In particular, it will be possible to assemble the system stiffness matrix without performing any matrix multiplications; in fact, this "direct stiffness method" does not even require explicit compatibility matrices!

9.6 DIRECT STIFFNESS METHOD

Let us consider Equation 9.11b in greater detail. We substitute the expressions for $\boldsymbol{\beta}$, $\mathbf{K}_g$, and $\widehat{\mathbf{P}}_g$ from Equations 9.9d, 9.10d, and 9.10h, respectively,

$$\left([\boldsymbol{\beta}^{(AB)^{\mathrm{T}}} \quad \boldsymbol{\beta}^{(BC)^{\mathrm{T}}} \quad \boldsymbol{\beta}^{(CD)^{\mathrm{T}}}] \begin{bmatrix} \mathbf{K}_g^{(AB)} & & \\ & \mathbf{K}_g^{(BC)} & \\ & & \mathbf{K}_g^{(CD)} \end{bmatrix} \begin{bmatrix} \boldsymbol{\beta}^{(AB)} \\ \boldsymbol{\beta}^{(BC)} \\ \boldsymbol{\beta}^{(CD)} \end{bmatrix} \right) \mathbf{u} \qquad \textbf{(9.12}\boldsymbol{a}\textbf{)}$$

$$= \mathbf{F} - [\boldsymbol{\beta}^{(AB)^T} \quad \boldsymbol{\beta}^{(BC)^T} \quad \boldsymbol{\beta}^{(CD)^T}] \begin{Bmatrix} \widehat{\mathbf{P}}_g^{(AB)} \\ \widehat{\mathbf{P}}_g^{(BC)} \\ \widehat{\mathbf{P}}_g^{(CD)} \end{Bmatrix}$$

The expanded form of Equation 9.12*a* is

$$\begin{aligned} &[(\boldsymbol{\beta}^{(AB)^T}\mathbf{K}_g^{(AB)}\boldsymbol{\beta}^{(AB)}) + (\boldsymbol{\beta}^{(BC)^T}\mathbf{K}_g^{(BC)}\boldsymbol{\beta}^{(BC)}) + (\boldsymbol{\beta}^{(CD)^T}\mathbf{K}_g^{(CD)}\boldsymbol{\beta}^{(CD)})]\mathbf{u} \\ &= \mathbf{F} - \boldsymbol{\beta}^{(AB)^T}\widehat{\mathbf{P}}_g^{(AB)} - \boldsymbol{\beta}^{(BC)^T}\widehat{\mathbf{P}}_g^{(BC)} - \boldsymbol{\beta}^{(CD)^T}\widehat{\mathbf{P}}_g^{(CD)} \end{aligned} \qquad \textbf{(9.12}\boldsymbol{b}\textbf{)}$$

By comparing Equation 9.12*b* to Equation 9.11*c*, we see that the term in square brackets on the left side of Equation 9.12*b* is the system stiffness matrix **k**; thus, Equation 9.12*b* shows that **k** may be obtained as the sum of a set of *NB* individual stiffness matrices, where *NB* is the number of elements in the structure. Furthermore, each of the *NB* stiffness matrices in Equation 9.12*b* has the same dimension as **k**, namely $m \times m$, where m is the number of system displacements,[12] and each is associated with only a single element of the structure. In other words,

$$\mathbf{k} = \mathbf{k}^{(AB)} + \mathbf{k}^{(BC)} + \mathbf{k}^{(CD)} \qquad \textbf{(9.12}\boldsymbol{c}\textbf{)}$$

where

$$\begin{aligned} \mathbf{k}^{(AB)} &\equiv \boldsymbol{\beta}^{(AB)^T}\mathbf{K}_g^{(AB)}\boldsymbol{\beta}^{(AB)} \\ \mathbf{k}^{(BC)} &\equiv \boldsymbol{\beta}^{(BC)^T}\mathbf{K}_g^{(BC)}\boldsymbol{\beta}^{(BC)} \\ \mathbf{k}^{(CD)} &\equiv \boldsymbol{\beta}^{(CD)^T}\mathbf{K}_g^{(CD)}\boldsymbol{\beta}^{(CD)} \end{aligned} \qquad \textbf{(9.12}\boldsymbol{d}\textbf{)}$$

Similarly, each of the matrix products on the right side of Equation 9.12*b* has the same dimension as **F**, namely $m \times 1$, and each is also associated with a single element.

Now, let's examine a typical matrix triple product in Equation 9.12*d*, say $\boldsymbol{\beta}^{(CD)^T}\mathbf{K}_g^{(CD)}\boldsymbol{\beta}^{(CD)}$. For the sake of convenience, we will denote $\mathbf{K}_g^{(CD)}$, Equation 9.10*c*, by the symbolic form

$$\mathbf{K}_g^{(CD)} = \begin{bmatrix} K_{11}^{(CD)} & K_{12}^{(CD)} & K_{13}^{(CD)} & K_{14}^{(CD)} & K_{15}^{(CD)} & K_{16}^{(CD)} \\ & K_{22}^{(CD)} & K_{23}^{(CD)} & K_{24}^{(CD)} & K_{25}^{(CD)} & K_{26}^{(CD)} \\ & & K_{33}^{(CD)} & K_{34}^{(CD)} & K_{35}^{(CD)} & K_{36}^{(CD)} \\ & & & K_{44}^{(CD)} & K_{45}^{(CD)} & K_{46}^{(CD)} \\ & \mathit{Sym.} & & & K_{55}^{(CD)} & K_{56}^{(CD)} \\ & & & & & K_{66}^{(CD)} \end{bmatrix} \qquad \textbf{(9.12}\boldsymbol{e}\textbf{)}$$

[12] The fact that **k** in Equation 9.11*c* is a 6×6 matrix just like $\mathbf{K}_g^{(AB)}$, $\mathbf{K}_g^{(BC)}$, or $\mathbf{K}_g^{(CD)}$ is entirely coincidental, and is due to the fact that $m = 6$ in Figure 9.7*a*. In general, m may be any integer, so the $\boldsymbol{\beta}^{(e)}$ matrix associated with each element will be $6 \times m$ and **k** will be $m \times m$.

In this case $\boldsymbol{\beta}^{(CD)}$ is given in Equation 9.9*c*, so the triple product is

$$\mathbf{k}^{(CD)} = \begin{bmatrix} 0 & 0 & 0 & 0 & 0 & 0 \\ 0 & 0 & 0 & 0 & 0 & 0 \\ 0 & 0 & 0 & 0 & 0 & 0 \\ 1 & 0 & 0 & 0 & 0 & 0 \\ 0 & 1 & 0 & 0 & 0 & 0 \\ 0 & 0 & 1 & 0 & 0 & 0 \end{bmatrix} \begin{bmatrix} K_{11}^{(CD)} & K_{12}^{(CD)} & K_{13}^{(CD)} & K_{14}^{(CD)} & K_{15}^{(CD)} & K_{16}^{(CD)} \\ & K_{22}^{(CD)} & K_{23}^{(CD)} & K_{24}^{(CD)} & K_{25}^{(CD)} & K_{26}^{(CD)} \\ & & K_{33}^{(CD)} & K_{34}^{(CD)} & K_{35}^{(CD)} & K_{36}^{(CD)} \\ & & & K_{44}^{(CD)} & K_{45}^{(CD)} & K_{46}^{(CD)} \\ & \textit{Sym.} & & & K_{55}^{(CD)} & K_{56}^{(CD)} \\ & & & & & K_{66}^{(CD)} \end{bmatrix} \begin{bmatrix} 0 & 0 & 0 & 1 & 0 & 0 \\ 0 & 0 & 0 & 0 & 1 & 0 \\ 0 & 0 & 0 & 0 & 0 & 1 \\ 0 & 0 & 0 & 0 & 0 & 0 \\ 0 & 0 & 0 & 0 & 0 & 0 \\ 0 & 0 & 0 & 0 & 0 & 0 \end{bmatrix} \tag{9.12f}$$

Performing the multiplications leads to

$$\mathbf{k}^{(CD)} = \begin{bmatrix} 0 & 0 & 0 & 0 & 0 & 0 \\ 0 & 0 & 0 & 0 & 0 & 0 \\ 0 & 0 & 0 & 0 & 0 & 0 \\ 0 & 0 & 0 & K_{11}^{(CD)} & K_{12}^{(CD)} & K_{13}^{(CD)} \\ 0 & 0 & 0 & K_{21}^{(CD)} & K_{22}^{(CD)} & K_{23}^{(CD)} \\ 0 & 0 & 0 & K_{31}^{(CD)} & K_{32}^{(CD)} & K_{33}^{(CD)} \end{bmatrix} \tag{9.12g}$$

As expected, we observe that the multiplication has moved some components of $\mathbf{K}_g^{(CD)}$ to specific locations in $\mathbf{k}^{(CD)}$ without changing any of the values. For instance, $K_{23}^{(CD)}$ appears in the location of $k_{56}^{(CD)}$, etc.

A key feature of this operation is the fact that the location of individual terms is *completely defined* by the relationships between element end displacements and system displacements. To demonstrate, recall these relationships for element CD from Example 4; in particular, we have

Element displacement	System displacement
1	4
2	5
3	6
4	—
5	—
6	—

In other words, end displacement 1 is the same as system displacement 4, and so on; end displacements 4, 5, and 6 are all zero. Note that any ordering of two end displacements, such as 2–3, corresponds to a particular ordering of system displacements, e.g., 5–6 in this case. This is exactly the same type of correspondence we observed in connection with the location of $K_{23}^{(CD)}$, i.e., $K_{23}^{(CD)}$ was moved to $k_{56}^{(CD)}$. A similar correspondence exists for all possible orderings of pairs of end displacements that have associated nonzero system displacements.

In the particular case of element CD, we must consider all possible orderings of pairs of end displacements 1, 2, and 3, i.e., 1–1, 1–2, 1–3, 2–1, 2–2, 2–3, 3–1, 3–2, 3–3;

these correspond to system displacement orderings of 4–4, 4–5, 4–6, 5–4, 5–5, 5–6, 6–4, 6–5, 6–6, respectively. The end displacement orderings define the components of $\mathbf{K}_g^{(CD)}$, that will appear in $\mathbf{k}^{(CD)}$, and the corresponding system displacement orderings define the locations of these components in $\mathbf{k}^{(CD)}$. Thus, for this element,

$$K_{11}^{(CD)} \rightarrow k_{44}^{(CD)}$$

$$K_{12}^{(CD)} \rightarrow k_{45}^{(CD)}$$

$$K_{13}^{(CD)} \rightarrow k_{46}^{(CD)}$$

$$K_{21}^{(CD)} \rightarrow k_{54}^{(CD)}$$

$$K_{22}^{(CD)} \rightarrow k_{55}^{(CD)}$$

$$K_{23}^{(CD)} \rightarrow k_{56}^{(CD)}$$

$$K_{31}^{(CD)} \rightarrow k_{64}^{(CD)}$$

$$K_{32}^{(CD)} \rightarrow k_{65}^{(CD)}$$

$$K_{33}^{(CD)} \rightarrow k_{66}^{(CD)}$$

This is exactly the same result as in Equation 9.12g since it is understood that all other components of $\mathbf{k}^{(CD)}$ are zero.

This procedure is, in essence, an algorithm for obtaining each of the individual stiffness matrices, $\mathbf{k}^{(e)}$, for any structure. As has been demonstrated, the algorithm requires only $\mathbf{K}_g^{(e)}$ and the correspondence between end displacements and system displacements for each element. Explicit compatibility matrices, $\boldsymbol{\beta}^{(e)}$, are not required, nor are any matrix multiplications; in effect, the system stiffness matrix $\mathbf{k}$ is assembled by inserting the components of the element stiffness matrices into their proper locations in $\mathbf{k}$ and adding the insertions to any other stiffness data already in place.

A similar algorithm may be developed in connection with matrix products of the type on the right side of Equation 9.12b. Consider, for instance, the matrix product $\boldsymbol{\beta}^{(CD)^T}\widehat{\mathbf{P}}_g^{(CD)}$, which we will denote as $\widehat{\mathbf{F}}^{(CD)}$. Matrix $\widehat{\mathbf{P}}_g^{(CD)}$ is defined in Equation 9.10g, so the general form of the product is

$$\boldsymbol{\beta}^{(CD)^T}\widehat{\mathbf{P}}_g^{(CD)} = \begin{bmatrix} 0 & 0 & 0 & 0 & 0 & 0 \\ 0 & 0 & 0 & 0 & 0 & 0 \\ 0 & 0 & 0 & 0 & 0 & 0 \\ 1 & 0 & 0 & 0 & 0 & 0 \\ 0 & 1 & 0 & 0 & 0 & 0 \\ 0 & 0 & 1 & 0 & 0 & 0 \end{bmatrix} \begin{Bmatrix} \widehat{Q}_1^{(CD)} \\ \widehat{Q}_2^{(CD)} \\ \widehat{Q}_3^{(CD)} \\ \widehat{Q}_4^{(CD)} \\ \widehat{Q}_5^{(CD)} \\ \widehat{Q}_6^{(CD)} \end{Bmatrix} = \begin{Bmatrix} 0 \\ 0 \\ 0 \\ \widehat{Q}_1^{(CD)} \\ \widehat{Q}_2^{(CD)} \\ \widehat{Q}_3^{(CD)} \end{Bmatrix} = \widehat{\mathbf{F}}^{(CD)} \quad \textbf{(9.12}\boldsymbol{h}\textbf{)}$$

In this case the algorithm may be apparent: the end displacements directly define the components of $\widehat{\mathbf{P}}_g^{(CD)}$, that appear in the final product, and the corresponding system displacements define the locations. Thus, the first component of $\widehat{\mathbf{P}}_g^{(CD)}$ (i.e., $\widehat{Q}_1^{(CD)}$)

appears in the fourth row of $\widehat{\mathbf{F}}^{(CD)}$, the second component of $\widehat{\mathbf{P}}_g^{(CD)}$ appears in the fifth row of $\widehat{\mathbf{F}}^{(CD)}$, and the third component of $\widehat{\mathbf{P}}_g^{(CD)}$ appears in the sixth row of $\widehat{\mathbf{F}}^{(CD)}$. Once again, the result may be obtained without performing the multiplication.

EXAMPLE 5

We will demonstrate the development of Equation 9.11*c* (Equation 9.12*b*) for the structure in Example 4 using the direct stiffness method.

The contributions of element CD to Equation 9.11*c* have been examined in the preceding discussion; therefore, we need consider only the additional contributions of elements AB and BC, and then assemble the complete solution.

For element AB, the correspondence between end displacements and system displacements is given in Table 9.5.1.

Table 9.5.1. Element AB

Element displacement	System displacement
1	—
2	—
3	—
4	1
5	2
6	3

Therefore, the components of $\mathbf{K}_g^{(AB)}$ are moved to specific locations in $\mathbf{k}^{(AB)}$ (or $\mathbf{k}$) in accordance with Table 9.5.2.

Table 9.5.2. Element AB

$$K_{44}^{(AB)} \rightarrow k_{11}^{(AB)}$$

$$K_{45}^{(AB)} \rightarrow k_{12}^{(AB)}$$

$$K_{46}^{(AB)} \rightarrow k_{13}^{(AB)}$$

$$K_{54}^{(AB)} \rightarrow k_{21}^{(AB)}$$

$$K_{55}^{(AB)} \rightarrow k_{22}^{(AB)}$$

$$K_{56}^{(AB)} \rightarrow k_{23}^{(AB)}$$

$$K_{64}^{(AB)} \rightarrow k_{31}^{(AB)}$$

$$K_{65}^{(AB)} \rightarrow k_{32}^{(AB)}$$

$$K_{66}^{(AB)} \rightarrow k_{33}^{(AB)}$$

The matrix product $\boldsymbol{\beta}^{(AB)T}\widehat{\mathbf{P}}_g^{(AB)} = \widehat{\mathbf{F}}^{(AB)}$ is also obtained from Table 9.5.1; specifically

Table 9.5.3. Element AB

$\widehat{Q}_4^{(AB)} \rightarrow \widehat{F}_1^{(AB)}$
$\widehat{Q}_5^{(AB)} \rightarrow \widehat{F}_2^{(AB)}$
$\widehat{Q}_6^{(AB)} \rightarrow \widehat{F}_3^{(AB)}$

For element BC, the correspondence between end displacements and system displacements is given in Table 9.5.4.

Table 9.5.4. Element BC

Element displacement	System displacement
1	1
2	2
3	3
4	4
5	5
6	6

This direct correspondence between end displacements and system displacements (which is entirely coincidental) means that $\mathbf{k}^{(BC)} = \mathbf{K}_g^{(BC)}$ and $\widehat{\mathbf{F}}^{(BC)} = \widehat{\mathbf{P}}_g^{(BC)}$.

We now have sufficient information to assemble $\mathbf{k}$ in Equation 9.12*c* since $\mathbf{k}^{(AB)}$ is defined in Table 9.5.2, $\mathbf{k}^{(BC)}$ was determined to be identical to $\mathbf{K}_g^{(BC)}$, and $\mathbf{k}^{(CD)}$ is given in Equation 9.12*g*. From this information we arrive at the symbolic form of $\mathbf{k}$,

$$\mathbf{k} = \begin{bmatrix} (K_{44}^{(AB)} + K_{11}^{(BC)}) & (K_{45}^{(AB)} + K_{12}^{(BC)}) & (K_{46}^{(AB)} + K_{13}^{(BC)}) & (K_{14}^{(BC)}) & (K_{15}^{(BC)}) & (K_{16}^{(BC)}) \\ & (K_{55}^{(AB)} + K_{22}^{(BC)}) & (K_{56}^{(AB)} + K_{23}^{(BC)}) & (K_{24}^{(BC)}) & (K_{25}^{(BC)}) & (K_{26}^{(BC)}) \\ & & (K_{66}^{(AB)} + K_{33}^{(BC)}) & (K_{34}^{(BC)}) & (K_{35}^{(BC)}) & (K_{36}^{(BC)}) \\ & & & (K_{44}^{(BC)} + K_{11}^{(CD)}) & (K_{45}^{(BC)} + K_{12}^{(CD)}) & (K_{46}^{(BC)} + K_{13}^{(CD)}) \\ & \textit{Sym.} & & & (K_{55}^{(BC)} + K_{22}^{(CD)}) & (K_{56}^{(BC)} + K_{23}^{(CD)}) \\ & & & & & (K_{66}^{(BC)} + K_{33}^{(CD)}) \end{bmatrix} \tag{9.13a}$$

The physical values of the components of **k** are obtained by substituting the appropriate terms from the matrices in Equations 9.10a, 9.10b, and 9.10c. For example,

$$
\begin{aligned}
k_{11} &= K_{44}^{(AB)} + K_{11}^{(BC)} = \frac{EI}{l}\left(\frac{12}{l^2}\right) + \frac{EI}{l}\left(\frac{0.2147}{l^2} + \frac{0.3578A}{I}\right) = \frac{EI}{l}\left(\frac{12.2147}{l^2} + \frac{0.3578A}{I}\right) \\
&\vdots \\
k_{13} &= K_{46}^{(AB)} + K_{13}^{(BC)} = \frac{EI}{l}\left(-\frac{6}{l}\right) + \frac{EI}{l}\left(\frac{0.5367}{l}\right) = \frac{EI}{l}\left(-\frac{5.4633}{l}\right) \\
&\vdots \\
k_{45} &= K_{45}^{(BC)} + K_{12}^{(CD)} = \frac{EI}{l}\left(-\frac{0.4293}{l^2} + \frac{0.1789A}{I}\right) + \frac{EI}{l}\left(\frac{0.4293}{l^2} - \frac{0.1789A}{I}\right) = 0
\end{aligned}
\tag{9.13b}
$$

These results agree with the corresponding expressions in Equation 9.7g! The remaining components of Equation 9.13a may be determined in a similar manner, and we will find that this stiffness matrix is, in fact, exactly the same as in Equation 9.7g.

The right-hand side of Equation 9.11c (Equation 9.12b) is assembled in the same way. Substituting **F** from Equation 9.11a, $\widehat{\mathbf{F}}^{(AB)}$ from Table 9.5.3, $\widehat{\mathbf{F}}^{(BC)}$ from Equation 9.10f, and $\widehat{\mathbf{F}}^{(CD)}$ from Equation 9.12h leads to

$$
\mathbf{F}^* = \mathbf{F} - \widehat{\mathbf{F}}^{(AB)} - \widehat{\mathbf{F}}^{(BC)} - \widehat{\mathbf{F}}^{(CD)} = \begin{Bmatrix} 0 \\ 0 \\ 0 \\ 0 \\ -F_1 \\ 0 \end{Bmatrix} - \begin{Bmatrix} \widehat{Q}_4^{(AB)} \\ \widehat{Q}_5^{(AB)} \\ \widehat{Q}_6^{(AB)} \\ 0 \\ 0 \\ 0 \end{Bmatrix} - \begin{Bmatrix} \widehat{Q}_1^{(BC)} \\ \widehat{Q}_2^{(BC)} \\ \widehat{Q}_3^{(BC)} \\ \widehat{Q}_4^{(BC)} \\ \widehat{Q}_5^{(BC)} \\ \widehat{Q}_6^{(BC)} \end{Bmatrix} - \begin{Bmatrix} 0 \\ 0 \\ 0 \\ \widehat{Q}_1^{(CD)} \\ \widehat{Q}_2^{(CD)} \\ \widehat{Q}_3^{(CD)} \end{Bmatrix}
$$

$$
= -\begin{Bmatrix} \widehat{Q}_4^{(AB)} + \widehat{Q}_1^{(BC)} \\ \widehat{Q}_5^{(AB)} + \widehat{Q}_2^{(BC)} \\ \widehat{Q}_6^{(AB)} + \widehat{Q}_3^{(BC)} \\ \widehat{Q}_4^{(BC)} + \widehat{Q}_1^{(CD)} \\ F_1 + \widehat{Q}_5^{(BC)} + \widehat{Q}_2^{(CD)} \\ \widehat{Q}_6^{(BC)} + \widehat{Q}_3^{(CD)} \end{Bmatrix}
\tag{9.13c}
$$

Substituting the appropriate expressions from Equations 9.10e, 9.10f, and 9.10g gives the final result

$$
\mathbf{F}^* = \begin{Bmatrix} F_1^* \\ F_2^* \\ F_3^* \\ F_4^* \\ F_5^* \\ F_6^* \end{Bmatrix} = -\begin{Bmatrix} \widehat{Q}_4^{(AB)} + \widehat{Q}_1^{(BC)} \\ \widehat{Q}_5^{(AB)} + \widehat{Q}_2^{(BC)} \\ \widehat{Q}_6^{(AB)} + \widehat{Q}_3^{(BC)} \\ \widehat{Q}_4^{(BC)} + \widehat{Q}_1^{(CD)} \\ F_1 + \widehat{Q}_5^{(BC)} + \widehat{Q}_2^{(CD)} \\ \widehat{Q}_6^{(BC)} + \widehat{Q}_3^{(CD)} \end{Bmatrix} = -\begin{Bmatrix} (-q_o l/2) + (-q_o l/2) \\ 0 + 0 \\ (q_o l^2/12) + (-q_o l^2/12 + EI\alpha \Delta T_o/h) \\ (-q_o l/2) + 0 \\ (F_1) + 0 + (F_2/2) \\ (q_o l^2/12 - EI\alpha \Delta T_o/h) + (-F_2 l/8) \end{Bmatrix}
\tag{9.13d}
$$

$$= \begin{Bmatrix} q_o l \\ 0 \\ -EI\alpha\Delta T_o/h \\ q_o l/2 \\ -F_1 - F_2/2 \\ F_2 l/8 - q_o l^2/12 + EI\alpha\Delta T_o/h \end{Bmatrix}$$

which is identical to Equation 9.7*h*!

The systematic nature of the direct stiffness method makes it ideal for computer implementation and it is, in fact, the basis for most modern programs for structural analysis. As illustrated by Example 5, such a program will require as input only the most basic data, i.e., information on each element (ϕ, l, E, A, and I) and ''tables'' relating element and system displacements. The element information is used to generate $\mathbf{K}_g^{(e)}$, Equation 9.8*m*, the components of which are inserted/added directly into $\mathbf{k}$ in accordance with the tabulated relationships between coordinates. Fixed-end forces may either be input directly for each element or may be computed from specified loads;[13] the components of $\widehat{\mathbf{P}}_g^{(e)}$ are similarly inserted into the appropriate locations of vector $\mathbf{F}^*$. The resulting matrix equilibrium equation is solved in the usual way, and element forces are again obtained by back-substitution.

9.6.1 General Support Motions

The direct stiffness method is based on the premise that all system displacements (i.e., all components of vector $\mathbf{u}$) are parallel to the global coordinate directions. Some structures may have supports with allowable motions that are not parallel to the global coordinates, for example, the support shown in Figure 9.12, which is capable of displacing only in a direction that makes an angle γ with the horizontal axis. In such cases, some minor modification of the basic formulation is necessary.

Let us consider a structure having m independent system displacements and assume that one of the supports for this structure is as shown in Figure 9.12. One of the m system displacements is the motion of this support in the direction parallel to the roller, which we will denote as u_s.[14]

To implement the direct stiffness method, we require the horizontal and vertical components of u_s, which we designate by u_1 and u_2, respectively,[15]

---

[13] It is important to remember that the signs of fixed-end forces (and angle ϕ) depend on the designation of the ''I-end'' of the element; thus, a computer program must distinguish the I-end and the J-end of each element to establish the orientation.

[14] Note that there is also a possible rotation at the joint of the support in Figure 9.12, and this must be included among the other independent system displacements; however, it is only the translational motion that requires special consideration.

[15] In general, these components of displacement could be assigned any integers i and $i + 1$ ($1 \leq i \leq m$); we use 1 and 2 for the sake of convenience.

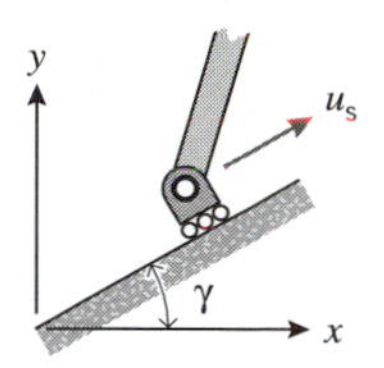

Figure 9.12 Support displacement that is not parallel to global coordinates.

$$\begin{aligned} u_1 &= u_s \cos\gamma \\ u_2 &= u_s \sin\gamma \end{aligned} \qquad \textbf{(9.14}a\textbf{)}$$

It should be recognized that u_1 and u_2 are not independent since both are functions of system displacement u_s.

The remaining independent system displacements will be labeled as $u_3, u_4, \ldots, u_{m+1}$, and will be assumed to be parallel to the global coordinates; thus, vector $\mathbf{u}$ comprises an "extended" set of $m + 1$ displacements, $u_1, u_2, \ldots, u_{m+1}$. This extended set of displacements is a function of the m independent system displacements, $u_s, u_3, \ldots, u_{m+1}$,

$$\mathbf{u} = \begin{Bmatrix} u_1 \\ u_2 \\ u_3 \\ \\ \vdots \\ u_{m+1} \end{Bmatrix} = \begin{bmatrix} \overline{C} & 0 & 0 & \vdots & 0 \\ \overline{S} & 0 & 0 & \vdots & 0 \\ 0 & 1 & 0 & \vdots & 0 \\ 0 & 0 & 1 & \vdots & 0 \\ \vdots & \vdots & \vdots & \ddots & \vdots \\ 0 & 0 & 0 & \vdots & 1 \end{bmatrix} \begin{Bmatrix} u_s \\ u_3 \\ \vdots \\ u_{m+1} \end{Bmatrix} = \mathbf{L}\tilde{\mathbf{u}} \qquad \textbf{(9.14}b\textbf{)}$$

where $\overline{C} = \cos\gamma$ and $\overline{S} = \sin\gamma$, and where $\mathbf{L}$ is an $(m + 1) \times m$ matrix and $\tilde{\mathbf{u}}$ is the vector of independent system displacements.

The procedure described in the previous section may be used to generate the matrices $\mathbf{k}$ and $\mathbf{F}^*$ in the equation $\mathbf{ku} = \mathbf{F}^*$. To modify our equations so that $\tilde{\mathbf{u}}$ is the displacement vector, we substitute Equation 9.14b into $\mathbf{ku} = \mathbf{F}^*$ and premultiply both sides of the equation by $\mathbf{L}^{\mathrm{T}}$,

$$\begin{aligned} \mathbf{ku} &= \mathbf{F}^* \\ \mathbf{L}^{\mathrm{T}}\mathbf{k}(\mathbf{L}\tilde{\mathbf{u}}) &= \mathbf{L}^{\mathrm{T}}\mathbf{F}^* \end{aligned} \qquad \textbf{(9.14}c\textbf{)}$$

The modified equation has the form

$$\tilde{\mathbf{k}}\tilde{\mathbf{u}} = \tilde{\mathbf{F}}^* \qquad \textbf{(9.14}d\textbf{)}$$

where

$$\tilde{\mathbf{k}} = \mathbf{L}^{\mathrm{T}}\mathbf{k}\mathbf{L} \qquad \textbf{(9.14}e\textbf{)}$$

and

$$\tilde{\mathbf{F}}^* = \mathbf{L}^{\mathrm{T}}\mathbf{F}^* \qquad \textbf{(9.14}f\textbf{)}$$

Substituting **L** from Equation 9.14*b* into Equation 9.14*e* leads to a general $m \times m$ expression for $\tilde{\mathbf{k}}$,

$$\tilde{\mathbf{k}} = \begin{bmatrix} (k_{11}\overline{C}^2 + 2k_{12}\overline{SC} + k_{22}\overline{S}^2) & (k_{13}\overline{C} + k_{23}\overline{S}) & (k_{14}\overline{C} + k_{24}\overline{S}) & \dots & (k_{1,m+1}\overline{C} + k_{2,m+1}\overline{S}) \\ & k_{33} & k_{34} & \dots & k_{3,m+1} \\ & & k_{44} & \dots & k_{4,m+1} \\ & Sym. & & & \vdots \\ & & & \ddots & \vdots \\ & & & & k_{m+1,m+1} \end{bmatrix} \quad \textbf{(9.14g)}$$

Similarly, from Equation 9.14*f* we obtain

$$\tilde{\mathbf{F}}^* = \begin{Bmatrix} F_1^*\overline{C} + F_2^*\,\overline{S} \\ F_3^* \\ F_4^* \\ \vdots \\ F_{m+1}^* \end{Bmatrix} \quad \textbf{(9.14h)}$$

which has dimensions $m \times 1$.

9.6.2 Direct Stiffness Method for Trusses

The stiffness matrix for bars in global coordinates is a special case of the matrix in Equation 9.8*m*. We note that the element in Figure 9.11*a*, upon which Equation 9.8*m* is based, reduces to a bar element if the transverse force/transverse displacement and moment/rotation quantities are deleted. The remaining axial forces and displacements still have the horizontal and vertical components shown in Figure 9.11*b*, but renumbered as shown in Figure 9.13. Consequently, the stiffness that relates the end forces and displacements in Figure 9.13 must be a 4×4 matrix.

The derivation is similar to the one in Section 9.5 except that we delete v_{IJ}, v_{JI}, θ_{IJ}, θ_{JI}, δ_3, and δ_6 from Equations 9.8*d* and V_{IJ}, V_{JI}, M_{IJ}, M_{JI}, Q_3, and Q_6 from Equations 9.8*e*. In this case the result, Equation 9.8*c*, is

$$\begin{Bmatrix} Q_1 \\ Q_2 \\ Q_3 \\ Q_4 \end{Bmatrix} = \frac{EA}{l} \begin{bmatrix} C^2 & SC & -C^2 & -SC \\ & S^2 & -SC & -S^2 \\ & & C^2 & SC \\ & Sym. & & S^2 \end{bmatrix} \begin{Bmatrix} \delta_1 \\ \delta_2 \\ \delta_3 \\ \delta_4 \end{Bmatrix} + \begin{Bmatrix} C\widehat{P}_{IJ} \\ S\widehat{P}_{IJ} \\ C\widehat{P}_{JI} \\ S\widehat{P}_{JI} \end{Bmatrix} \quad \textbf{(9.15a)}$$

In other words, the stiffness matrix for the bar (Equation 9.8*m*) is[16]

[16] The same stiffness matrix may be obtained directly from Equation 9.8*m* by deleting the third and sixth rows and columns and setting moment of inertia, *I*, equal to zero.

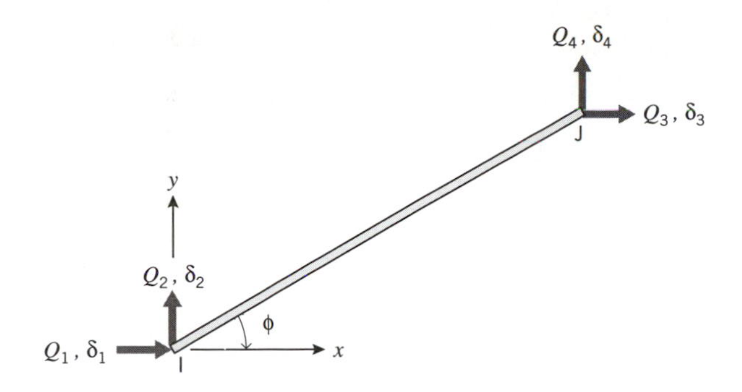

Figure 9.13 End forces/displacements, bar element.

$$\mathbf{K}_g^{(e)} = \frac{EA}{l}\begin{bmatrix} C^2 & SC & -C^2 & -SC \\ & S^2 & -SC & -S^2 \\ & & C^2 & SC \\ & Sym. & & S^2 \end{bmatrix} \tag{9.15b}$$

and the vector of fixed-end forces (Equation 9.8n) is

$$\widehat{\mathbf{P}}_g^{(e)} = \begin{Bmatrix} C\widehat{P}_{\mathrm{IJ}} \\ S\widehat{P}_{\mathrm{IJ}} \\ C\widehat{P}_{\mathrm{JI}} \\ S\widehat{P}_{\mathrm{JI}} \end{Bmatrix} \tag{9.15c}$$

Note that the 4 × 4 matrix in Equation 9.15b is of rank 1 as was the basic 1 × 1 stiffness matrix originally utilized in Chapter 8.

The analysis of trusses using the element force-displacement relation in Equation 9.15a and the direct stiffness method is quite straightforward, and is analogous to the procedure illustrated for frames in Section 9.6.

EXAMPLE 6

Reanalyze the truss in Figure 8.10a (Figure 5.11) using the direct stiffness method; note that the support at C is of the type discussed in Section 9.6.1. The truss is subjected to a vertical load F at joint D and uniform temperature changes ΔT in bars AB and CD. We use the numbering scheme for the independent system displacements shown in Figure 9.14a. This differs from the numbering in Figure 8.2a but is consistent with the formulation in Section 9.6.1. As in Section 9.6.1, we require the components of u_s

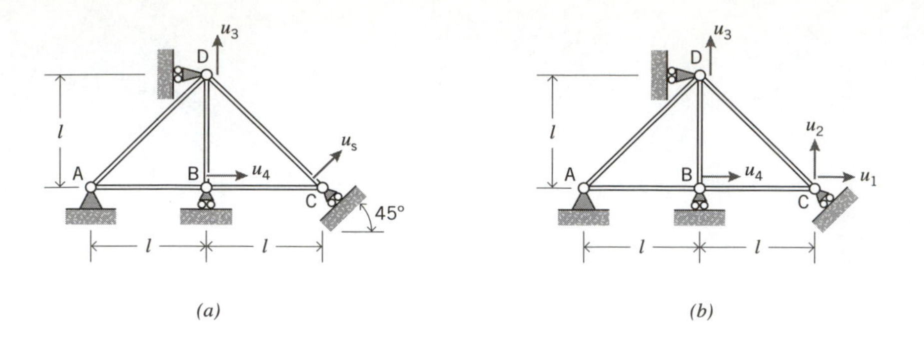

Figure 9.14 Example 6: (*a*) independent system displacements; (*b*) ‘‘extended’’ displacements.

defined in Figure 9.14*b*; for the particular geometry of the support at C, angle γ (Figure 9.12) is 45°, so these components are given by

$$u_1 = u_s \cos 45° = u_s/\sqrt{2} \tag{9.16a}$$

$$u_2 = u_s \sin 45° = u_s/\sqrt{2}$$

Thus, the relation between the ‘‘extended’’ set of displacements (Figure 9.14*b)* and the independent system displacements (Figure 9.14*a*) is

$$\mathbf{u} = \begin{Bmatrix} u_1 \\ u_2 \\ u_3 \\ u_4 \end{Bmatrix} = \begin{bmatrix} 1/\sqrt{2} & 0 & 0 \\ 1/\sqrt{2} & 0 & 0 \\ 0 & 1 & 0 \\ 0 & 0 & 1 \end{bmatrix} \begin{Bmatrix} u_s \\ u_3 \\ u_4 \end{Bmatrix} = \mathbf{L}\tilde{\mathbf{u}} \tag{9.16b}$$

We require the various element stiffness matrices. For bars AB and BC, angle $\phi = 0$ and length is l; therefore, from Equation 9.15*b* we obtain

$$\mathbf{K}_g^{(\mathrm{AB})} = \mathbf{K}_g^{(\mathrm{BC})} = \frac{EA}{l} \begin{bmatrix} 1 & 0 & -1 & 0 \\ & 0 & 0 & 0 \\ & & 1 & 0 \\ & Sym. & & 0 \end{bmatrix} \tag{9.16c}$$

For bar BD, $\phi = 90°$ and length is l,

$$\mathbf{K}_g^{(\mathrm{BD})} = \frac{EA}{l} \begin{bmatrix} 0 & 0 & 0 & 0 \\ & 1 & 0 & -1 \\ & & 0 & 0 \\ & Sym. & & 1 \end{bmatrix} \tag{9.16d}$$

For bar AD, $\phi = 45°$ and length is $\sqrt{2}l$,

$$\mathbf{K}_g^{(AD)} = \frac{EA}{\sqrt{2}l}\begin{bmatrix} 1/2 & 1/2 & -1/2 & -1/2 \\ & 1/2 & -1/2 & -1/2 \\ & & 1/2 & 1/2 \\ & Sym. & & 1/2 \end{bmatrix} \tag{9.16e}$$

and for bar CD, $\phi = 135°$ and length is $\sqrt{2}l$,

$$\mathbf{K}_g^{(CD)} = \frac{EA}{\sqrt{2}l}\begin{bmatrix} 1/2 & -1/2 & -1/2 & 1/2 \\ & 1/2 & 1/2 & -1/2 \\ & & 1/2 & -1/2 \\ & Sym. & & 1/2 \end{bmatrix} \tag{9.16f}$$

Also, fixed-end forces exist in bars AB and CD due to thermal loads. These are determined from Equation 9.4*e* and are converted to components in the global coordinate directions via Equation 9.15*c*,

$$\hat{\mathbf{P}}_l^{(AB)} = \begin{Bmatrix} EA\alpha\Delta T \\ -EA\alpha\Delta T \end{Bmatrix}, \quad \hat{\mathbf{P}}_g^{(AB)} = \begin{Bmatrix} EA\alpha\Delta T \\ 0 \\ -EA\alpha\Delta T \\ 0 \end{Bmatrix} \tag{9.16g}$$

and

$$\hat{\mathbf{P}}_l^{(CD)} = \begin{Bmatrix} EA\alpha\Delta T \\ -EA\alpha\Delta T \end{Bmatrix}, \quad \hat{\mathbf{P}}_g^{(CD)} = \begin{Bmatrix} -EA\alpha\Delta T/\sqrt{2} \\ EA\alpha\Delta T/\sqrt{2} \\ EA\alpha\Delta T/\sqrt{2} \\ -EA\alpha\Delta T/\sqrt{2} \end{Bmatrix} \tag{9.16h}$$

The relationship between the end displacements for the various bars and the extended set of displacements in Figure 9.14*b* is given in Table 9.6.1.

The stiffness matrix, **k**, is obtained directly from the relationships in the table; in general notation,

$$\mathbf{k} = \begin{bmatrix} (K_{33}^{(BC)} + K_{11}^{(CD)}) & (K_{34}^{(BC)} + K_{12}^{(CD)}) & (K_{14}^{(CD)}) & (K_{13}^{(BC)}) \\ & (K_{44}^{(BC)}) + K_{22}^{(CD)}) & (K_{24}^{(CD)}) & (K_{14}^{(BC)}) \\ & & (K_{44}^{(BD)} + K_{44}^{(AD)} + K_{44}^{(CD)}) & (K_{14}^{(BD)}) \\ & Sym. & & (K_{33}^{(AB)} + K_{11}^{(BC)} + K_{11}^{(BD)}) \end{bmatrix} \tag{9.16i}$$

Table 9.6.1

Bar	Element displacement	System displacement
AB	1	—
	2	—
	3	4
	4	—
BC	1	4
	2	—
	3	1
	4	2
BD	1	4
	2	—
	3	—
	4	3
AD	1	—
	2	—
	3	—
	4	3
CD	1	1
	2	2
	3	—
	4	3

Substituting the values in Equation 9.16*b* through Equation 9.16*f* leads to

$$\mathbf{k} = \frac{EA}{l}\begin{bmatrix} 1 + \dfrac{1}{2\sqrt{2}} & 0 + \left(-\dfrac{1}{2\sqrt{2}}\right) & \dfrac{1}{2\sqrt{2}} & -1 \\ & 0 + \dfrac{1}{2\sqrt{2}} & -\dfrac{1}{2\sqrt{2}} & 0 \\ & & 1 + \dfrac{1}{2\sqrt{2}} + \dfrac{1}{2\sqrt{2}} & 0 \\ & Sym. & & 1 + 1 + 0 \end{bmatrix} \tag{9.16j}$$

$$
= \frac{EA}{l}\begin{bmatrix} \dfrac{4+\sqrt{2}}{4} & -\dfrac{\sqrt{2}}{4} & \dfrac{\sqrt{2}}{4} & -1 \\ & \dfrac{\sqrt{2}}{4} & -\dfrac{\sqrt{2}}{4} & 0 \\ & & \dfrac{2+\sqrt{2}}{2} & 0 \\ & Sym. & & 2 \end{bmatrix}
$$

It should be noted that terms in Equation 9.16*j* from element stiffness matrices $\mathbf{K}_g^{(AD)}$ and $\mathbf{K}_g^{(CD)}$ include the multiplier $EA/\sqrt{2}l$ rather than EA/l.

The final system stiffness matrix $\tilde{\mathbf{k}}$ is obtained by substituting the components of $\mathbf{k}$ from Equation 9.16*j* into Equation 9.14*g*,

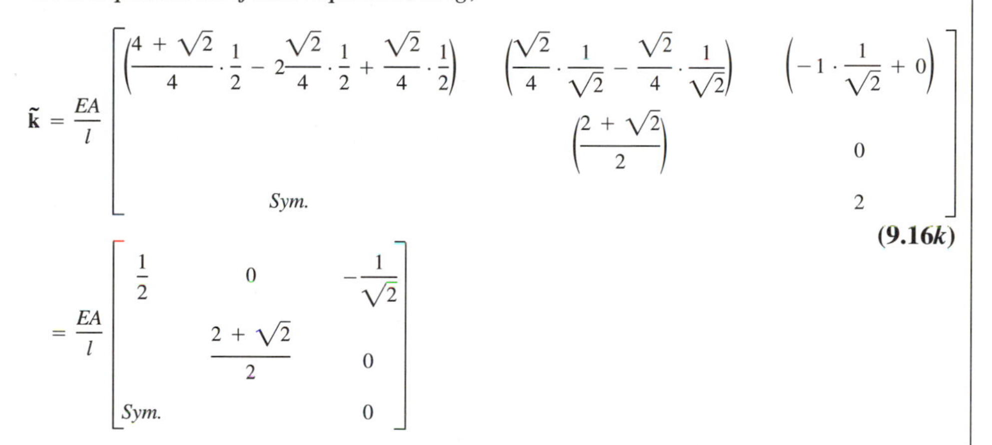

$$
\tilde{\mathbf{k}} = \frac{EA}{l}\begin{bmatrix} \left(\dfrac{4+\sqrt{2}}{4}\cdot\dfrac{1}{2} - 2\dfrac{\sqrt{2}}{4}\cdot\dfrac{1}{2} + \dfrac{\sqrt{2}}{4}\cdot\dfrac{1}{2}\right) & \left(\dfrac{\sqrt{2}}{4}\cdot\dfrac{1}{\sqrt{2}} - \dfrac{\sqrt{2}}{4}\cdot\dfrac{1}{\sqrt{2}}\right) & \left(-1\cdot\dfrac{1}{\sqrt{2}} + 0\right) \\ & \left(\dfrac{2+\sqrt{2}}{2}\right) & 0 \\ & Sym. & 2 \end{bmatrix} \tag{9.16k}
$$

$$
= \frac{EA}{l}\begin{bmatrix} \dfrac{1}{2} & 0 & -\dfrac{1}{\sqrt{2}} \\ & \dfrac{2+\sqrt{2}}{2} & 0 \\ Sym. & & 0 \end{bmatrix}
$$

This result agrees with Equation 8.8*a*. (The locations of individual terms in the two matrices differ because the numbering of the independent system displacements differs.)

Vector $\mathbf{F}^*$ is also obtained directly from the correspondences in Table 9.6.1; the general expression is

$$
\mathbf{F}^* = \begin{Bmatrix} 0 \\ 0 \\ -F \\ 0 \end{Bmatrix} - \begin{Bmatrix} \hat{Q}_3^{(BC)} + \hat{Q}_1^{(CD)} \\ \hat{Q}_4^{(BC)} + \hat{Q}_2^{(CD)} \\ \hat{Q}_4^{(BD)} + \hat{Q}_4^{(AD)} + \hat{Q}_4^{(CD)} \\ \hat{Q}_3^{(AB)} + \hat{Q}_1^{(BC)} + \hat{Q}_1^{(BD)} \end{Bmatrix} \tag{9.16l}
$$

The term $-F$ in the third row of the $\mathbf{F}$ vector is a result of the applied force at joint D, which is opposite to system displacement u_3.

Only bars AB and CD have nonzero components of fixed-end forces, Equations 9.16*g* and 9.16*h*; substituting these values into Equation 9.16*l* leads to

$$\mathbf{F}^* = \begin{Bmatrix} 0 - 0 - (-EA\alpha\Delta T/\sqrt{2}) \\ 0 - 0 - EA\alpha\Delta T/\sqrt{2} \\ -F - 0 - 0 - (-EA\alpha\Delta T/\sqrt{2}) \\ 0 - (-EA\alpha\Delta T/\sqrt{2}) - 0 - 0 \end{Bmatrix} = \begin{Bmatrix} EA\alpha\Delta T/\sqrt{2} \\ -EA\alpha\Delta T/\sqrt{2} \\ -F + EA\alpha\Delta T/\sqrt{2} \\ EA\alpha\Delta T/\sqrt{2} \end{Bmatrix} \quad \mathbf{(9.16m)}$$

Using Equation 9.14*h*,

$$\tilde{\mathbf{F}}^* = \begin{Bmatrix} \left(\frac{EA\alpha\Delta T}{\sqrt{2}}\right)\frac{1}{\sqrt{2}} + \left(-\frac{EA\alpha\Delta T}{\sqrt{2}}\right)\frac{1}{\sqrt{2}} \\ -F + \frac{EA\alpha\Delta T}{\sqrt{2}} \\ \frac{EA\alpha\Delta T}{\sqrt{2}} \end{Bmatrix} = \begin{Bmatrix} 0 \\ -F + EA\alpha\Delta T/\sqrt{2} \\ EA\alpha;\Delta T/\sqrt{2} \end{Bmatrix} \quad \mathbf{(9.16n)}$$

Equation 9.16*n* agrees with Equation 8.8*b* (again with different locations for terms because of the different numbering schemes).

9.7 MISCELLANEOUS CONSIDERATIONS

The presentation of the fundamental concepts of the displacement method is essentially complete. The direct stiffness method, in particular, is the procedure of choice for computer analysis of general structures. The method can be employed for all of the applications dealt with previously and can also be extended to model a variety of additional structural configurations. Some of these typical applications will be illustrated in the following sections.

9.7.1 Support Settlement

Support settlements are easily incorporated in the direct stiffness method by introducing additional system displacements in the same directions as the components of these motions. For example, consider the structure in Figure 9.15*a*, and assume that there are three support settlements denoted by δ_{o_1}, δ_{o_2} and δ_{o_3}, as shown. Also assume that there are no loads applied to the structure. We define a set of system displacements u_1–u_7 as in Figure 9.15*b*, where u_1–u_4 are the usual system displacements in the absence of support settlements,[17] and u_5–u_7 are specified displacements (in positive global coordinate directions) associated with the support settlements. Thus, $u_5 = -\delta_{o_1}$, $u_6 = \delta_{o_2}$, and $u_7 = -\delta_{o_3}$.

[17] We are using the same numbering sequence as in Figure 8.9.

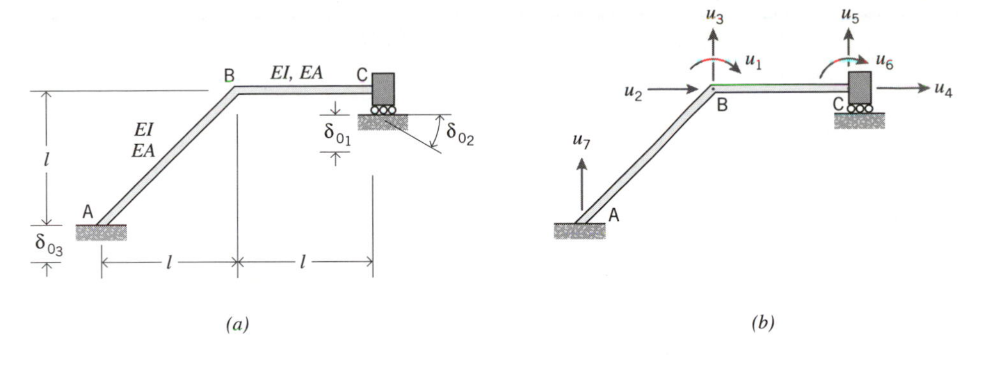

Figure 9.15 Example 7.

We denote the extended set of displacements u_1–u_7 by vector $\mathbf{u}$ and use the direct stiffness method to assemble the system stiffness matrix, $\mathbf{k}$. The resulting system equilibrium equation is

$$\mathbf{k}\mathbf{u} = \mathbf{F}^* \tag{9.17a}$$

but, it follows from the definition of $\mathbf{u}$ that Equation 9.17*a* can be written in terms of partitioned matrices,

$$\begin{bmatrix} \tilde{\mathbf{k}} & | & \mathbf{k}_o \\ - & - & - \\ \mathbf{k}_o^T & | & \mathbf{k}_s \end{bmatrix} \begin{Bmatrix} \tilde{\mathbf{u}} \\ - \\ \mathbf{u}_s \end{Bmatrix} = \begin{Bmatrix} \mathbf{0} \\ - \\ \mathbf{R} \end{Bmatrix} \tag{9.17b}$$

where $\tilde{\mathbf{k}}$ is the familiar 4×4 stiffness matrix associated with system displacements u_1–u_4 (vector $\tilde{\mathbf{u}}$), $\mathbf{k}_s$ is a 3×3 stiffness matrix associated with support displacements u_5–u_7 (vector $\mathbf{u}_s$), $\mathbf{k}_o$ is a 4×3 matrix that couples the effects of $\tilde{\mathbf{u}}$ and $\mathbf{u}_s$, and $\mathbf{R}$ is a 3×1 vector of unknown forces (in this case, the reactions in the directions of u_5–u_7); the 4×1 $\mathbf{0}$ vector in $\mathbf{F}^*$ is due to our assumption of no applied loads.

Expanding Equation 9.17*b* gives two matrix equations,

$$\begin{aligned} \tilde{\mathbf{k}}\tilde{\mathbf{u}} + \mathbf{k}_o\mathbf{u}_s &= \mathbf{0} \\ \mathbf{k}_o^T\tilde{\mathbf{u}} + \mathbf{k}_s\mathbf{u}_s &= \mathbf{R} \end{aligned} \tag{9.17c}$$

The first of these equations may be solved for the unknown system displacements,

$$\tilde{\mathbf{u}} = -\tilde{\mathbf{k}}^{-1}\mathbf{k}_o\mathbf{u}_s \tag{9.17d}$$

Once $\tilde{\mathbf{u}}$ is obtained then the reactions can be found from the second of Equations 9.17*c*.

EXAMPLE 7

We will analyze the behavior of the structure in Figure 9.15a due to support settlements.

We require the two element stiffness matrices. For element AB, $\phi = 45°$ and length is $\sqrt{2}l$; therefore, from Equation 9.8m we obtain

$$\mathbf{K}_g^{(AB)} = \frac{EI}{\sqrt{2}l}\begin{bmatrix} \left(\frac{3}{l^2}+\frac{A}{2I}\right) & \left(-\frac{3}{l^2}+\frac{A}{2I}\right) & \left(\frac{3}{l}\right) & \left(-\frac{3}{l^2}-\frac{A}{2I}\right) & \left(\frac{3}{l^2}-\frac{A}{2I}\right) & \left(\frac{3}{l}\right) \\ & \left(\frac{3}{l^2}+\frac{A}{2I}\right) & \left(-\frac{3}{l}\right) & \left(\frac{3}{l^2}-\frac{A}{2I}\right) & \left(-\frac{3}{l^2}-\frac{A}{2I}\right) & \left(-\frac{3}{l}\right) \\ & & (4) & \left(-\frac{3}{l}\right) & \left(\frac{3}{l}\right) & (2) \\ & & & \left(\frac{3}{l^2}+\frac{A}{2I}\right) & \left(-\frac{3}{l^2}+\frac{A}{2I}\right) & \left(-\frac{3}{l}\right) \\ & \mathit{Sym.} & & & \left(\frac{3}{l^2}+\frac{A}{2I}\right) & \left(\frac{3}{l}\right) \\ & & & & & (4) \end{bmatrix} \quad \textbf{(9.18a)}$$

For element BC, $\phi = 0$ and length is l,

$$\mathbf{K}_g^{(BC)} = \frac{EI}{l}\begin{bmatrix} A/I & 0 & 0 & -A/I & 0 & 0 \\ & 12/l^2 & -6/l & 0 & -12/l^2 & -6/l \\ & & 4 & 0 & 6/l & 2 \\ & & & A/I & 0 & 0 \\ & \mathit{Sym.} & & & 12/l^2 & 6/l \\ & & & & & 4 \end{bmatrix} \quad \textbf{(9.18b)}$$

The relationship between the element end displacements and the system displacements in Figure 9.15b is given in Table 9.7.1,

We obtain $\mathbf{k}$ from the relationships in the table; thus, Equation 9.17b becomes

$$\left[\begin{array}{cccc|ccc} K_{66}^{(AB)}+K_{33}^{(BC)} & K_{46}^{(AB)}+K_{13}^{(BC)} & K_{56}^{(AB)}+K_{23}^{(BC)} & K_{34}^{(BC)} & K_{35}^{(BC)} & K_{36}^{(BC)} & K_{26}^{(AB)} \\ & K_{44}^{(AB)}+K_{11}^{(BC)} & K_{45}^{(AB)}+K_{12}^{(BC)} & K_{14}^{(BC)} & K_{15}^{(BC)} & K_{16}^{(BC)} & K_{24}^{(AB)} \\ & & K_{55}^{(AB)}+K_{22}^{(BC)} & K_{24}^{(BC)} & K_{25}^{(BC)} & K_{26}^{(BC)} & K_{25}^{(AB)} \\ & & & K_{44}^{(BC)} & K_{45}^{(BC)} & K_{46}^{(BC)} & 0 \\ \hline & \mathit{Sym.} & & & K_{55}^{(BC)} & K_{56}^{(BC)} & 0 \\ & & & & & K_{66}^{(BC)} & 0 \\ & & & & & & K_{22}^{(AB)} \end{array}\right]\begin{Bmatrix} u_1 \\ u_2 \\ u_3 \\ u_4 \\ \hline u_5 \\ u_6 \\ u_7 \end{Bmatrix} = \begin{Bmatrix} 0 \\ 0 \\ 0 \\ 0 \\ \hline R_1 \\ R_2 \\ R_3 \end{Bmatrix} \quad \textbf{(9.18c)}$$

Table 9.7.1

Element	Element displacement	System displacement
AB	1	—
	2	7
	3	—
	4	2
	5	3
	6	1
BC	1	2
	2	3
	3	1
	4	4
	5	5
	6	6

After substituting the values from Equations 9.18*a* and 9.18*b*, we find the explicit forms of submatrices $\tilde{\mathbf{k}}$ and $\mathbf{k}_o$ (Equation 9.17*b*),

$$\tilde{\mathbf{k}} = \frac{EI}{l}\begin{bmatrix} (4+2\sqrt{2}) & \left(-\dfrac{3\sqrt{2}}{2l}\right) & \left(\dfrac{-12+3\sqrt{2}}{2l}\right) & 0 \\ & \left(\dfrac{3\sqrt{2}}{2l^2}+\dfrac{4+\sqrt{2}}{4C'l^2}\right) & \left(-\dfrac{3\sqrt{2}}{2l^2}+\dfrac{\sqrt{2}}{4C'l^2}\right) & \left(-\dfrac{1}{C'l^2}\right) \\ & \textit{Sym.} & \left(\dfrac{12}{l^2}+\dfrac{3\sqrt{2}}{2l^2}+\dfrac{\sqrt{2}}{4C'l^2}\right) & 0 \\ & & & \left(\dfrac{1}{C'l^2}\right) \end{bmatrix} \quad \mathbf{(9.18}d\mathbf{)}$$

and

$$\mathbf{k}_o = \frac{EI}{l}\begin{bmatrix} \dfrac{6}{l} & 2 & -\dfrac{3}{\sqrt{2}l} \\ 0 & 0 & \left(\dfrac{3}{\sqrt{2}l^2}-\dfrac{1}{2\sqrt{2}C'l^2}\right) \\ -\dfrac{12}{l^2} & -\dfrac{6}{l} & \left(-\dfrac{3}{\sqrt{2}l^2}-\dfrac{1}{2\sqrt{2}C'l^2}\right) \\ 0 & 0 & 0 \end{bmatrix} \quad \mathbf{(9.18}e\mathbf{)}$$

in which we have set $C' = I/l^2A$. We note that $\tilde{\mathbf{k}}$ is identical to the stiffness matrix in Equation 9.6*f*. In the computations that follow, we will approximate axially rigid elements by setting $C' = 0.0001$.

The components of $\mathbf{u}_s$ are specified,

$$\mathbf{u}_s = \begin{Bmatrix} u_5 \\ u_6 \\ u_7 \end{Bmatrix} = \begin{Bmatrix} -\delta_{o_1} \\ \delta_{o_2} \\ -\delta_{o_3} \end{Bmatrix} \tag{9.18f}$$

Substituting this expression for $\mathbf{u}_s$, and $\tilde{\mathbf{k}}$ and $\mathbf{k}_o$ from Equations 9.18*d* and 9.18*e* into Equation 9.17*d* leads to

$$\tilde{\mathbf{u}} = \begin{Bmatrix} u_1 \\ u_2 \\ u_3 \\ u_4 \end{Bmatrix} = \begin{Bmatrix} 0.7443\delta_{o_1}/l - 0.2224\delta_{o_2} - 0.7443\delta_{o_3}/l \\ 0.5218\delta_{o_1} - 0.2737\delta_{o_2}l - 0.5218\delta_{o_3} \\ -0.5218\delta_{o_1} + 0.2737\delta_{o_2}l - 0.4780\delta_{o_3} \\ 0.5218\delta_{o_1} - 0.2737\delta_{o_2}l - 0.5218\delta_{o_3} \end{Bmatrix} \tag{9.18g}$$

The values of u_1 and u_2 due to δ_{o_1} are identical to the values in Equation 8.31*f*. It is interesting to note that the horizontal and vertical motions of joint B have different characteristics for δ_{o_1} and δ_{o_3}. Both motions have the same magnitudes for δ_{o_1} ($0.522\delta_{o_1}$), but the horizontal motion due to δ_{o_3} ($-0.522\delta_{o_3}$) is different than the vertical motion ($-0.478\delta_{o_3}$).

9.7.2 Structures Composed of Several Types of Elements

The analysis of structures composed of different types of elements presents no conceptual difficulty. If the structure includes springs, the stiffness matrices for these elements must be given in global coordinates. For an axial spring, the stiffness matrix is the same as in Equation 9.15*b* except that the spring constant k_a replaces EA/l. For a torsional spring, we introduce two end rotations δ_1 and δ_2 (Figure 9.16) so the stiffness matrix becomes

$$\mathbf{K}_g^{(e)} = k_\theta \begin{bmatrix} 1 & -1 \\ -1 & 1 \end{bmatrix} \tag{9.19}$$

where k_θ is the spring constant.

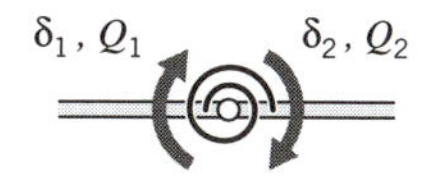

Figure 9.16 Torsional spring.

EXAMPLE 8

Reanalyze the structure in Figure 8.38 (Example 10, Chapter 8). We assume that elements AB and BC have axial stiffness *EA* in addition to bending stiffness *EI*. As in Figure 8.38, we denote the spring constants of the axial and torsional springs by k_1 and k_2, respectively.

The system displacements are shown in Figure 9.17;[18] note that u_1–u_3 are the same as in Figure 8.39. For elements AB and BC the stiffness matrices are (see Equation 9.18*b*),

$$\mathbf{K}_g^{(AB)} = \mathbf{K}_g^{(BC)} = \frac{EI}{l}\begin{bmatrix} A/I & 0 & 0 & -A/I & 0 & 0 \\ & 12/l^2 & -6/l & 0 & -12/l^2 & -6/l \\ & & 4 & 0 & 6/l & 2 \\ & & & A/I & 0 & 0 \\ & Sym. & & & 12/l^2 & 6/l \\ & & & & & 4 \end{bmatrix} \quad \textbf{(9.20a)}$$

For the axial spring (element DB) $\phi = 90°$, so the stiffness matrix (Equation 9.15*b)* is

$$\mathbf{K}_g^{(DB)} = k_1\begin{bmatrix} 0 & 0 & 0 & 0 \\ & 1 & 0 & -1 \\ & & 0 & 0 \\ & Sym. & & 1 \end{bmatrix} \quad \textbf{(9.20b)}$$

For the torsional spring (element BB) the stiffness matrix is given by Equation 9.19,

$$\mathbf{K}_g^{(BB)} = k_2\begin{bmatrix} 1 & -1 \\ -1 & 1 \end{bmatrix} \quad \textbf{(9.20c)}$$

[18] System displacements u_4 and u_5 are included for the sake of completeness. Since there are no axial loads in AB or BC, we must have $u_4 = u_5 = 0$.

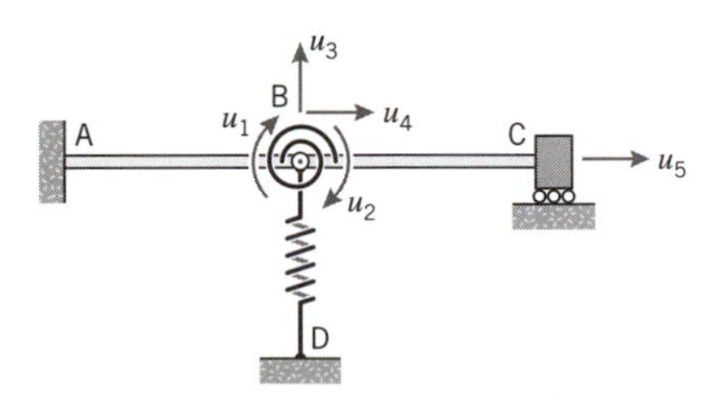

Figure 9.17 System displacements

The relationships between element displacements and system displacements are given in Table 9.8.1,

Table 9.8.1

Element	Element displacement	System displacement
AB	1	—
	2	—
	3	—
	4	4
	5	3
	6	1
BC	1	4
	2	3
	3	2
	4	5
	5	—
	6	—
DB	1	—
	2	—
	3	4
	4	3
BB	1	1
	2	2

Given the relationships in the table, we may use the direct stiffness assembly procedure to find

$$\mathbf{k} = \begin{bmatrix} K_{66}^{(AB)} + K_{11}^{(BB)} & K_{12}^{(BB)} & K_{56}^{(AB)} & K_{46}^{(AB)} & 0 \\ & K_{33}^{(BC)} + K_{22}^{(BB)} & K_{23}^{(BC)} & K_{13}^{(BC)} & K_{34}^{(BC)} \\ & & K_{55}^{(AB)} + K_{22}^{(BC)} + K_{44}^{(DB)} & K_{45}^{(AB)} + K_{12}^{(BC)} + K_{34}^{(DB)} & K_{24}^{(BC)} \\ & Sym. & & K_{44}^{(AB)} + K_{11}^{(BC)} + K_{33}^{(DB)} & K_{14}^{(BC)} \\ & & & & K_{44}^{(BC)} \end{bmatrix} \quad \textbf{(9.20}d\textbf{)}$$

After substituting the values from Eqs. 9.20*a*–9.20*c* we find

$$\mathbf{k} = \begin{bmatrix} \frac{4EI}{l} + k_2 & -k_2 & \frac{6EI}{l^2} & 0 & 0 \\ & \frac{4EI}{l} + k_2 & -\frac{6EI}{l^2} & 0 & 0 \\ & & \frac{24EI}{l^3} + k_1 & 0 & 0 \\ & Sym. & & \frac{2EA}{l} & -\frac{EA}{l} \\ & & & & \frac{EA}{l} \end{bmatrix} \quad \textbf{(9.20e)}$$

Note that the submatrix in the first three rows and columns of the matrix in Equation 9.20*e* is identical to the stiffness matrix in Equation 8.32*d*.

The fixed-end moments in element BC due to applied force F have been determined previously (see Equation 8.32*b*); fixed-end shears are obtained from equations of equilibrium. Thus,

$$\widehat{\mathbf{P}}_l^{(BC)} = \begin{Bmatrix} -4Fl/27 \\ 2Fl/27 \\ 20F/27 \\ 7F/27 \\ 0 \\ 0 \end{Bmatrix} \quad \textbf{(9.20f)}$$

We obtain the fixed-end forces in global coordinates from Equation 9.8*n*,

$$\widehat{\mathbf{P}}_g^{(BC)} = \begin{Bmatrix} 0 \\ 20F/27 \\ -4Fl/27 \\ 0 \\ 7F/27 \\ 2Fl/27 \end{Bmatrix} \quad \textbf{(9.20g)}$$

Since there are no loads applied at joints, we have $\mathbf{F} = \mathbf{0}$. Consequently, by using the direct assembly procedure we find vector $\mathbf{F}^*$,

$$\mathbf{F}^* = -\begin{Bmatrix} 0 \\ \widehat{Q}_3^{(BC)} \\ \widehat{Q}_2^{(BC)} \\ \widehat{Q}_1^{(BC)} \\ \widehat{Q}_4^{(BC)} \end{Bmatrix} = \begin{Bmatrix} 0 \\ 4Fl/27 \\ -20F/27 \\ 0 \\ 0 \end{Bmatrix} \quad \textbf{(9.20h)}$$

The first three terms in this vector are identical to Equation 8.32*e*.

The solution obtained by solving the equation $\mathbf{ku} = \mathbf{F}^*$ will provide the same values of u_1–u_3 as Equation 8.32*f* (as well as $u_4 = u_5 = 0$).

9.7.3 Effect of Joint Dimensions

In most structural analyses, and in the models we have considered thus far, it has been tacitly assumed that joints are points defined by the intersections of element axes, and that element lengths are equal to distances between joints. Since joints have physical dimensions (Figure 9.18), elements must actually be shorter than the nominal lengths. The true behavior between the intersection points depends on the stiffness of the physical element, on the stiffnesses of the connections between the element and the joints, and on the stiffnesses of the joint assemblies. These stiffnesses may be difficult to quantify, and the interactions between them may be quite complex. However, in some circumstances these effects can be important and must be examined analytically. The effects often can be accounted for in the analysis by introducing a minor modification of the element stiffness relations.

For illustrative purposes, we will consider elements that are rigidly connected to rigid joint assemblies as in Figure 9.18. A typical element is shown in Figure 9.19*a*. Specifically, I and J in Figure 9.19*a* represent joints, e.g., I or J would coincide with a point such as O in Figure 9.18. Points I′ and J′ represent the actual ends of an element of length λ,[19] and l_1 and l_2 are distances between the ends of the element and the joints; we assume that l_1 and l_2 are rigid extensions of the element.

The quantities θ_{IJ}, v_{IJ}, u_{IJ}, and θ_{JI}, v_{JI}, u_{JI} in Figure 9.19*a* are the joint displacements

[19] Here λ is the length of the physical element, and is distinct from the quantity l previously defined as the distance between I and J.

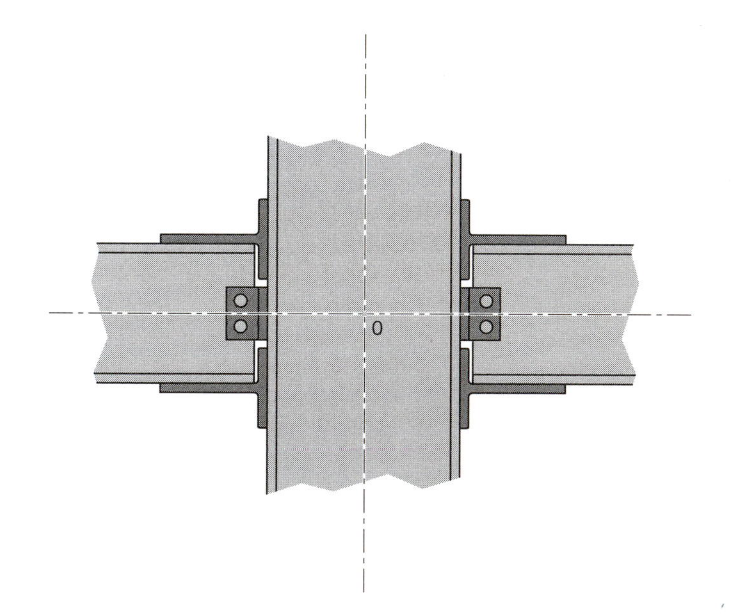

Figure 9.18 Typical rigid joint (beam/column connection, elevation view).

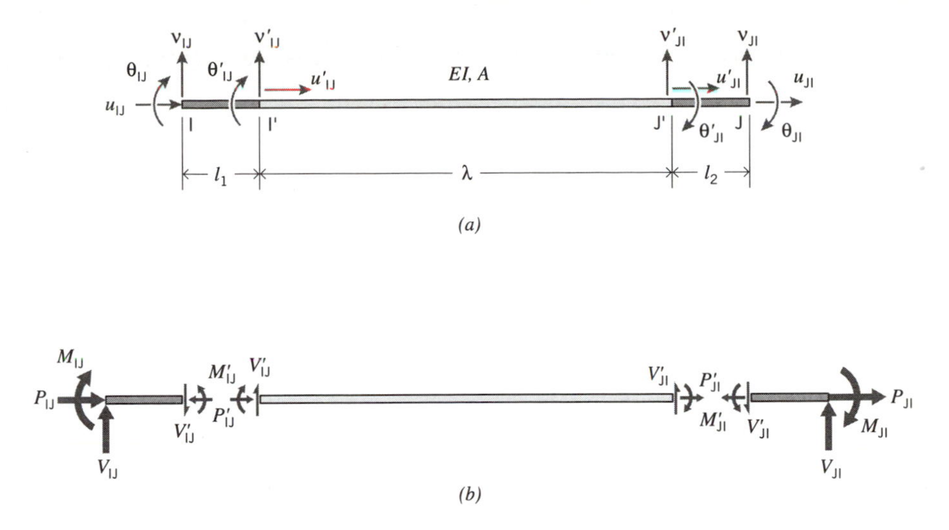

Figure 9.19 Short element.

in local coordinates. In previous formulations, these were identical to element end displacements. However, in the current situation, the relationships between the element end displacements (denoted by θ'_{IJ}, v'_{IJ}, u'_{IJ}, and θ'_{JI}, v'_{JI}, u'_{JI} in Figure 9.19*a*) and the joint displacements are slightly more complicated. These relationships are derived next.

Since l_1 and l_2 are rigid links, we must have $\theta'_{IJ} = \theta_{IJ}$, $\theta'_{JI} = \theta_{JI}$, $u'_{IJ} = u_{IJ}$ and $u'_{JI} = u_{JI}$, so the only end displacements that are effected by the offsets are v'_{IJ} and v'_{JI}. It follows that $v'_{IJ} = v_{IJ} - l_1\theta_{IJ}$ and $v'_{JI} = v_{JI} + l_2\theta_{JI}$. Thus, in matrix form,

$$\boldsymbol{\delta}_l'^{(e)} \equiv \begin{Bmatrix} \theta'_{IJ} \\ \theta'_{JI} \\ v'_{IJ} \\ v'_{JI} \\ u'_{IJ} \\ u'_{JI} \end{Bmatrix} = \begin{bmatrix} 1 & 0 & 0 & 0 & 0 & 0 \\ 0 & 1 & 0 & 0 & 0 & 0 \\ -l_1 & 0 & 1 & 0 & 0 & 0 \\ 0 & l_2 & 0 & 1 & 0 & 0 \\ 0 & 0 & 0 & 0 & 1 & 0 \\ 0 & 0 & 0 & 0 & 0 & 1 \end{bmatrix} \begin{Bmatrix} \theta_{IJ} \\ \theta_{JI} \\ v_{IJ} \\ v_{JI} \\ u_{IJ} \\ u_{JI} \end{Bmatrix} = \mathbf{T}'\boldsymbol{\delta}_l^{(e)} \qquad \textbf{(9.21a)}$$

Figure 9.19*b* shows free-body diagrams of the element and the two rigid extensions. We continue to denote the forces at the joints by M_{IJ}, V_{IJ}, P_{IJ}, M_{JI}, V_{JI}, and P_{JI}; these forces are related to those at ends of the element (M'_{IJ}, V'_{IJ}, P'_{IJ}, M'_{JI}, V'_{JI}, P'_{JI}) by equilibrium.[20] The result is

$$\mathbf{P}_l^{(e)} \equiv \begin{Bmatrix} M_{IJ} \\ M_{JI} \\ V_{IJ} \\ V_{JI} \\ P_{IJ} \\ P_{JI} \end{Bmatrix} = \begin{bmatrix} 1 & 0 & -l_1 & 0 & 0 & 0 \\ 0 & 1 & 0 & l_2 & 0 & 0 \\ 0 & 0 & 1 & 0 & 0 & 0 \\ 0 & 0 & 0 & 1 & 0 & 0 \\ 0 & 0 & 0 & 0 & 1 & 0 \\ 0 & 0 & 0 & 0 & 0 & 1 \end{bmatrix} \begin{Bmatrix} M'_{IJ} \\ M'_{JI} \\ V'_{IJ} \\ V'_{JI} \\ P'_{IJ} \\ P'_{JI} \end{Bmatrix} = \mathbf{T}'^{\mathrm{T}}\mathbf{P}_l'^{(e)} \qquad \textbf{(9.21b)}$$

[20] This relation can be directly found using ordinary statics. Alternatively, it can be found from the virtual work theorem using virtual displacements of the form in Equation 9.21*a*.

Using the current primed notation, the stiffness data for the element (Equation 9.5) are written

$$\begin{Bmatrix} M'_{IJ} \\ M'_{JI} \\ V'_{IJ} \\ V'_{JI} \\ P'_{IJ} \\ P'_{JI} \end{Bmatrix} = \frac{EI}{\lambda} \begin{bmatrix} 4 & 2 & -6/\lambda & 6/\lambda & & \\ 2 & 4 & -6/\lambda & 6/\lambda & & \\ -6/\lambda & -6/\lambda & 12/\lambda^2 & -12/\lambda^2 & & \\ 6/\lambda & 6/\lambda & -12/\lambda^2 & 12/\lambda^2 & & \\ & & & & A/I & -A/I \\ & & & & -A/I & A/I \end{bmatrix} \begin{Bmatrix} \theta'_{IJ} \\ \theta'_{JI} \\ v'_{IJ} \\ v'_{JI} \\ u'_{IJ} \\ u'_{JI} \end{Bmatrix} + \begin{Bmatrix} \widehat{M}'_{IJ} \\ \widehat{M}'_{JI} \\ \widehat{V}'_{IJ} \\ \widehat{V}'_{JI} \\ \widehat{P}'_{IJ} \\ \widehat{P}'_{JI} \end{Bmatrix} \tag{9.21c}$$

or, see Equation 9.8a,

$$\mathbf{P}_l'^{(e)} = \mathbf{K}_l'^{(e)}\boldsymbol{\delta}_l'^{(e)} + \widehat{\mathbf{P}}_l'^{(e)} \tag{9.21d}$$

We obtain the desired force-deformation relationship (in local coordinates) by following a procedure similar to the one employed in Equations 9.8j–9.8l; premultiplying Equation 9.21d by $\mathbf{T}'^{\mathrm{T}}$ leads to

$$\begin{aligned} \mathbf{T}'^{\mathrm{T}}\mathbf{P}_l'^{(e)} &= \mathbf{T}'^{\mathrm{T}}\mathbf{K}_l'^{(e)}\boldsymbol{\delta}_l'^{(e)} + \mathbf{T}'^{\mathrm{T}}\widehat{\mathbf{P}}_l'^{(e)} \\ &= \mathbf{T}'^{\mathrm{T}}\mathbf{K}_l'^{(e)}(\mathbf{T}'\boldsymbol{\delta}_l^{(e)}) + \mathbf{T}'^{\mathrm{T}}\widehat{\mathbf{P}}_l'^{(e)} \end{aligned} \tag{9.21e}$$

or

$$\mathbf{P}_l^{(e)} = \overline{\mathbf{K}}_l^{(e)}\boldsymbol{\delta}_l^{(e)} + \overline{\mathbf{P}}_l^{(e)} \tag{9.21f}$$

where

$$\overline{\mathbf{K}}_l^{(e)} = \mathbf{T}'^{\mathrm{T}}\mathbf{K}_l'^{(e)}\mathbf{T}'$$

$$= \frac{EI}{\lambda}\begin{bmatrix} 4 + \frac{12l_1}{\lambda} + \frac{12l_1^2}{\lambda^2} & 2 + \frac{6l_2}{\lambda} + \frac{12l_1l_2}{\lambda^2} & -\frac{6}{\lambda} - \frac{12l_1}{\lambda^2} & \frac{6}{\lambda} + \frac{12l_1}{\lambda^2} & 0 & 0 \\ & 4 + \frac{12l_2}{\lambda} + \frac{12l_2^2}{\lambda^2} & -\frac{6}{\lambda} - \frac{12l_2}{\lambda^2} & \frac{6}{\lambda} + \frac{12l_2}{\lambda^2} & 0 & 0 \\ & & \frac{12}{\lambda^2} & -\frac{12}{\lambda^2} & 0 & 0 \\ & & & \frac{12}{\lambda^2} & 0 & 0 \\ & \mathit{Sym.} & & & \frac{A}{I} & -\frac{A}{I} \\ & & & & & \frac{A}{I} \end{bmatrix} \tag{9.21g}$$

and

$$\overline{\mathbf{P}}_l^{(e)} = \mathbf{T}'^{\mathrm{T}}\widehat{\mathbf{P}}_l'^{(e)} = \begin{Bmatrix} \widehat{M}'_{\mathrm{IJ}} - l_1\widehat{V}'_{\mathrm{IJ}} \\ \widehat{M}'_{\mathrm{JI}} + l_2\widehat{V}'_{\mathrm{JI}} \\ \widehat{V}'_{\mathrm{IJ}} \\ \widehat{V}'_{\mathrm{JI}} \\ \widehat{P}'_{\mathrm{IJ}} \\ \widehat{P}'_{\mathrm{JI}} \end{Bmatrix} \tag{9.21h}$$

The stiffness matrix in Equation 9.21g is expressed in global coordinates by means of Equation 9.8l,

$$\overline{\mathbf{K}}_g^{(e)} = \mathbf{T}^{\mathrm{T}}\overline{\mathbf{K}}_l^{(e)}\mathbf{T}$$

$$= \frac{EI}{\lambda}\begin{bmatrix} \frac{12S^2}{\lambda^2}+\frac{A}{I}C^2 & -\frac{12SC}{\lambda^2}+\frac{A}{I}SC & \frac{6S}{\lambda}+\frac{12l_1S}{\lambda^2} & -\frac{12S^2}{\lambda^2}-\frac{A}{I}C^2 & \frac{12SC}{\lambda^2}-\frac{A}{I}SC & \frac{6S}{\lambda}+\frac{12l_2S}{\lambda^2} \\ & \frac{12C^2}{\lambda^2}+\frac{A}{I}S^2 & -\frac{6C}{\lambda}-\frac{12l_1C}{\lambda^2} & \frac{12SC}{\lambda^2}-\frac{A}{I}SC & -\frac{12C^2}{\lambda^2}-\frac{A}{I}S^2 & -\frac{6C}{\lambda}-\frac{12l_2C}{\lambda^2} \\ & & 4+\frac{12l_1}{\lambda}+\frac{12l_1^2}{\lambda^2} & -\frac{6S}{\lambda}-\frac{12l_1S}{\lambda^2} & \frac{6C}{\lambda}+\frac{12l_1C}{\lambda^2} & 2+\frac{6l_1}{\lambda}+\frac{6l_2}{\lambda}+\frac{12l_1l_2}{\lambda^2} \\ & & & \frac{12S^2}{\lambda^2}+\frac{A}{I}C^2 & -\frac{12SC}{\lambda^2}+\frac{A}{I}SC & -\frac{6S}{\lambda}-\frac{12l_2S}{\lambda^2} \\ & \textit{Sym.} & & & \frac{12C^2}{\lambda^2}+\frac{A}{I}S^2 & \frac{6C}{\lambda}+\frac{12l_2C}{\lambda^2} \\ & & & & & 4+\frac{12l_2}{\lambda}+\frac{12l_2^2}{\lambda^2} \end{bmatrix} \tag{9.21i}$$

Similarly, from Equations 9.21h and 9.8i we obtain

$$\overline{\mathbf{P}}_g^{(e)} = \mathbf{T}^{\mathrm{T}}\overline{\mathbf{P}}_l^{(e)} = \begin{Bmatrix} -S\widehat{V}'_{\mathrm{IJ}} + C\widehat{P}'_{\mathrm{IJ}} \\ C\widehat{V}'_{\mathrm{IJ}} + S\widehat{P}'_{\mathrm{IJ}} \\ \widehat{M}'_{\mathrm{IJ}} - l_1\widehat{V}'_{\mathrm{IJ}} \\ -S\widehat{V}'_{\mathrm{JI}} + C\widehat{P}'_{\mathrm{JI}} \\ C\widehat{V}'_{\mathrm{JI}} + S\widehat{P}'_{\mathrm{JI}} \\ \widehat{M}'_{\mathrm{JI}} + l_2\widehat{V}'_{\mathrm{JI}} \end{Bmatrix} \tag{9.21j}$$

EXAMPLE 9

Let's analyze the structure in Figure 9.20*a* using the rigid joints shown in Figure 9.20*b*. Assume that each joint extends a distance $0.05l$ from its center,[21] i.e., that $l_1 = l_2 = 0.05l$ (Figure 9.19*a*) for both elements. This implies that the actual lengths of elements AB and BC are $1.3142l$ and $0.9l$, respectively.

The system displacements are u_1–u_4 shown in Figure 9.15*b*. The stiffness matrices for the elements are obtained from Equation 9.21*i*; for element AB using $\phi = 45°$, $1.3142l$ for λ, and C' for I/l^2A, we find

$$\overline{\mathbf{K}}_g^{(AB)} = \frac{EI}{l}\begin{bmatrix} \frac{2.6433}{l^2}+\frac{0.3805}{C'l^2} & -\frac{2.6433}{l^2}+\frac{0.3805}{C'l^2} & \frac{2.6433}{l} & -\frac{2.6433}{l^2}-\frac{0.3805}{C'l^2} & \frac{2.6433}{l^2}-\frac{0.3805}{C'l^2} & \frac{2.6433}{l} \\ & \frac{2.6433}{l^2}+\frac{0.3805}{C'l^2} & -\frac{2.6433}{l} & \frac{2.6433}{l^2}-\frac{0.3805}{C'l^2} & -\frac{2.6433}{l^2}-\frac{0.3805}{C'l^2} & -\frac{2.6433}{l} \\ & & 3.4043 & -\frac{2.6433}{l} & \frac{2.6433}{l} & 1.8824 \\ & & & \frac{2.6433}{l^2}+\frac{0.3804}{C'l^2} & -\frac{2.6433}{l^2}+\frac{0.3805}{C'l^2} & -\frac{2.6433}{l} \\ & \textit{Sym.} & & & \frac{2.6433}{l^2}+\frac{0.3805}{C'l^2} & \frac{2.6433}{l} \\ & & & & & 3.4043 \end{bmatrix} \quad \textbf{(9.22a)}$$

[21] Note that l is a characteristic length of the structure and not the element length.

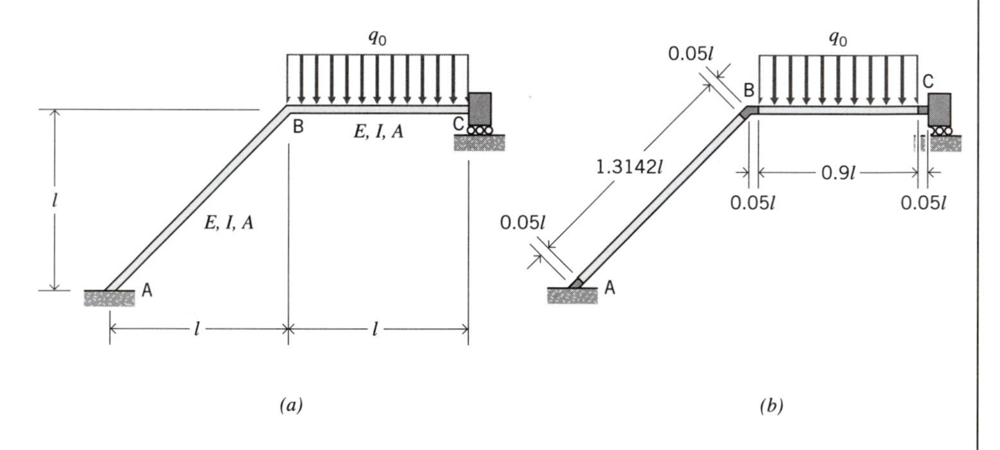

Figure 9.20 Example 9.

For element BC, $\phi = 0$ and length is $0.9l$,

$$\overline{\mathbf{K}}_g^{(BC)} = \frac{EI}{l}\begin{bmatrix} \frac{1.1111}{C'l^2} & 0 & 0 & -\frac{1.1111}{C'l^2} & 0 & 0 \\ & \frac{16.4609}{l^2} & -\frac{8.2305}{l} & 0 & -\frac{16.4609}{l^2} & -\frac{8.2305}{l} \\ & & 5.2263 & 0 & \frac{8.2305}{l} & 3.0041 \\ & & & \frac{1.1111}{C'l^2} & 0 & 0 \\ & Sym. & & & \frac{16.4609}{l^2} & \frac{8.2305}{l} \\ & & & & & 5.2263 \end{bmatrix} \quad \textbf{(9.22b)}$$

The relationships between joint displacements and system displacements are similar to those in Table 9.7.1 except that there are no system displacements u_5–u_7. The system stiffness matrix may be obtained using a direct assembly procedure similar to that employed in Example 9.7. Assembling the system stiffness matrix using the components from Equations 9.22*a* and 9.22*b* gives

$$\mathbf{k} = \frac{EI}{l}\begin{bmatrix} 8.6306 & -\frac{2.6433}{l} & -\frac{5.5871}{l} & 0 \\ & \frac{2.6433}{l^2} + \frac{1.4916}{C'l^2} & -\frac{2.6433}{l^2} + \frac{0.3805}{C'l^2} & -\frac{1.1111}{C'l^2} \\ & Sym. & \frac{19.1042}{l^2} + \frac{0.3805}{C'l^2} & 0 \\ & & & \frac{1.1111}{C'l^2} \end{bmatrix} \quad \textbf{(9.22c)}$$

In the following computations, we will set $C' = 0.0001$ to approximate axially rigid elements.

There are no concentrated forces applied to joints and no loads on element AB. The vector of fixed-end forces for element BC is

$$\widehat{\mathbf{P}}_l'^{(BC)} = \begin{Bmatrix} \widehat{M}'_{BC} \\ \widehat{M}'_{CB} \\ \widehat{V}'_{BC} \\ \widehat{V}'_{CB} \\ \widehat{P}'_{BC} \\ \widehat{P}'_{CB} \end{Bmatrix} = \begin{Bmatrix} -0.0675q_o l^2 \\ 0.0675q_o l^2 \\ 0.45q_o l \\ 0.45q_o l \\ 0 \\ 0 \end{Bmatrix} \quad \textbf{(9.22d)}$$

in which we have substituted $0.9l$ for length (see Equation 9.6*d* and Figure 9.20*b*). From Equation 9.21*j* we obtain

$$\overline{\mathbf{P}}_g^{(BC)} = \begin{Bmatrix} \widehat{Q}_1^{(BC)} \\ \widehat{Q}_2^{(BC)} \\ \widehat{Q}_3^{(BC)} \\ \widehat{Q}_4^{(BC)} \\ \widehat{Q}_5^{(BC)} \\ \widehat{Q}_6^{(BC)} \end{Bmatrix} = \begin{Bmatrix} 0 \\ 0.45q_o l \\ -0.09q_o l^2 \\ 0 \\ 0.45q_o l \\ 0.09q_o l^2 \end{Bmatrix} \tag{9.22e}$$

Using the relationships in Table 9.7.1, direct assembly gives

$$\mathbf{F}^* = -\begin{Bmatrix} \widehat{Q}_3^{(BC)} \\ \widehat{Q}_1^{(BC)} \\ \widehat{Q}_2^{(BC)} \\ \widehat{Q}_4^{(BC)} \end{Bmatrix} = \begin{Bmatrix} 0.09q_o l^2 \\ 0 \\ -0.45q_o l \\ 0 \end{Bmatrix} \tag{9.22f}$$

Solving the equilibrium equation $\mathbf{ku} = \mathbf{F}^*$ leads to

$$\mathbf{u} = \begin{Bmatrix} 0.00493q_o l^3/EI \\ 0.01609q_o l^4/EI \\ -0.01611q_o l^4/EI \\ 0.01609q_o l^4/EI \end{Bmatrix} \tag{9.22g}$$

These displacements are approximately 20–30% different from the values obtained in Example 4, Chapter 8 (Equation 8.23*c*). The previous solution (based on nominal element lengths) gives displacements $u_1 = 0.0061q_o l^3/EI$ and $u_2 = 0.0239q_o l^4/EI$. (Note that $u_2 = -u_3 = u_4$ for this particular structure when elements are axially rigid.) Differences in end forces will be similar. In other words, the analysis shows that the inclusion of realistic joint dimensions can have a significant effect on structural response. Bear in mind, however, that the formulation assumes perfectly rigid connections, an idealized upper bound to actual conditions.

9.7.4 Releases

Frame structures are often constructed with connections at which some components of element end forces (or moments) must be zero. For instance, a basic pin connection cannot transmit a moment. When an element end moment is zero, the corresponding end rotation can be determined given the remaining end displacements and the applied loads. The analysis can then be simplified because the rotation at the pin need not be treated as one of the unknown system displacements; however, the simplification can be achieved only if the element stiffness relation is modified to reflect the special end conditions. In this section, we will demonstrate the procedure for dealing with this type of release of an end force. The procedure is applicable to any element for which some end forces are known to be zero.

Consider the element in Figure 9.11*b* and assume $Q_3 = 0$, which is equivalent to assuming that the I end is pinned. We seek a relationship between the nonzero end forces and the corresponding end displacements, i.e., between Q_1, Q_2, Q_4, Q_5, Q_6 and $\delta_1, \delta_2, \delta_4, \delta_5, \delta_6$. The desired relationship is obtained by setting $Q_3 = 0$ and eliminating δ_3 from the element stiffness relation in Equation 9.8*c*. The reduction is accomplished by manipulating partitioned matrices in a manner similar to that employed in Equations 9.17*b*–9.17*c*.

We begin by moving the equation containing the released force (Q_3) to the last row of the stiffness relation (Equation 9.8*c*); this is most easily accomplished by interchanging the third and sixth equations. We also interchange the third and sixth columns of $\mathbf{K}_g^{(e)}$ to retain the symmetric form of the stiffness matrix. Note that this means that the vector of displacement components is reordered. The result, in symbolic form, is

$$\begin{Bmatrix} Q_1 \\ Q_2 \\ Q_6 \\ Q_4 \\ Q_5 \\ -- \\ Q_3 \end{Bmatrix} = \begin{bmatrix} K_{11} & K_{12} & K_{16} & K_{14} & K_{15} & | & K_{13} \\ & K_{22} & K_{26} & K_{24} & K_{25} & | & K_{23} \\ & & K_{66} & K_{64} & K_{65} & | & K_{63} \\ & & & K_{44} & K_{45} & | & K_{43} \\ & Sym. & & & K_{55} & | & K_{53} \\ -- & -- & -- & -- & -- & & -- \\ & & & & & | & K_{33} \end{bmatrix} \begin{Bmatrix} \delta_1 \\ \delta_2 \\ \delta_6 \\ \delta_4 \\ \delta_5 \\ -- \\ \delta_3 \end{Bmatrix} + \begin{Bmatrix} \hat{Q}_1 \\ \hat{Q}_2 \\ \hat{Q}_6 \\ \hat{Q}_4 \\ \hat{Q}_5 \\ -- \\ \hat{Q}_3 \end{Bmatrix} \tag{9.23a}$$

The matrices in Equation 9.23*a* are shown with the partitions required for the subsequent manipulation. For the sake of convenience, we rewrite the equation in terms of submatrices,

$$\begin{Bmatrix} \mathbf{P}_{g_c}^{(e)} \\ -- \\ \mathbf{P}_{g_r}^{(e)} \end{Bmatrix} = \begin{bmatrix} \mathbf{K}_{g11}^{(e)} & | & \mathbf{K}_{g12}^{(e)} \\ -- & - & -- \\ \mathbf{K}_{g21}^{(e)} & | & \mathbf{K}_{g22}^{(e)} \end{bmatrix} \begin{Bmatrix} \boldsymbol{\delta}_{g_c}^{(e)} \\ -- \\ \boldsymbol{\delta}_{g_r}^{(e)} \end{Bmatrix} + \begin{Bmatrix} \hat{\mathbf{P}}_{g1}^{(e)} \\ -- \\ \hat{\mathbf{P}}_{g2}^{(e)} \end{Bmatrix} \tag{9.23b}$$

where $\mathbf{P}_{g_c}^{(e)}$ and $\boldsymbol{\delta}_{g_c}^{(e)}$ are vectors containing the contracted sets of forces and displacements. The vector of released forces, $\mathbf{P}_{g_r}^{(e)}$, is $\mathbf{0}$. Also, by symmetry, $\mathbf{K}_{g21}^{(e)} = \mathbf{K}_{g12}^{(e)\mathrm{T}}$.

Expanding Equation 9.23*b* provides two matrix equations,

$$\begin{aligned} \mathbf{P}_{g_c}^{(e)} &= \mathbf{K}_{g11}^{(e)} \boldsymbol{\delta}_{g_c}^{(e)} + \mathbf{K}_{g12}^{(e)} \boldsymbol{\delta}_{g_r}^{(e)} + \hat{\mathbf{P}}_{g1}^{(e)} \\ \mathbf{0} &= \mathbf{K}_{g21}^{(e)} \boldsymbol{\delta}_{g_c}^{(e)} + \mathbf{K}_{g22}^{(e)} \boldsymbol{\delta}_{g_r}^{(e)} + \hat{\mathbf{P}}_{g2}^{(e)} \end{aligned} \tag{9.23c}$$

Solving the second of Equations 9.23*c* for $\boldsymbol{\delta}_{g_r}^{(e)}$ gives

$$\boldsymbol{\delta}_{g_r}^{(e)} = -\mathbf{K}_{g22}^{(e)^{-1}} [\mathbf{K}_{g21}^{(e)} \boldsymbol{\delta}_{g_c}^{(e)} + \hat{\mathbf{P}}_{g2}^{(e)}] \tag{9.23d}$$

Substituting this into the first of Equations 9.23c leads to the stiffness relation,

$$\begin{aligned}\mathbf{P}_{gc}^{(e)} &= \mathbf{K}_{g11}^{(e)}\boldsymbol{\delta}_{gc}^{(e)} - \mathbf{K}_{g12}^{(e)}\mathbf{K}_{g22}^{(e)^{-1}}[\mathbf{K}_{g21}^{(e)}\boldsymbol{\delta}_{gc}^{(e)} + \widehat{\mathbf{P}}_{g2}^{(e)}] + \widehat{\mathbf{P}}_{g1}^{(e)} \\ &= [\mathbf{K}_{g11}^{(e)} - \mathbf{K}_{g12}^{(e)}\mathbf{K}_{g22}^{(e)^{-1}}\mathbf{K}_{g21}^{(e)}]\boldsymbol{\delta}_{gc}^{(e)} + [\widehat{\mathbf{P}}_{g1}^{(e)} - \mathbf{K}_{g12}^{(e)}\mathbf{K}_{g22}^{(e)^{-1}}\widehat{\mathbf{P}}_{g2}^{(e)}]\end{aligned} \tag{9.23e}$$

or, in general,

$$\mathbf{P}_{gc}^{(e)} = \mathbf{K}_{gc}^{(e)}\boldsymbol{\delta}_{gc}^{(e)} + \widehat{\mathbf{P}}_{gc}^{(e)} \tag{9.23f}$$

where the modified element stiffness matrix and fixed-end force vector are defined by

$$\begin{aligned}\mathbf{K}_{gc}^{(e)} &= \mathbf{K}_{g11}^{(e)} - \mathbf{K}_{g12}^{(e)}\mathbf{K}_{g22}^{(e)^{-1}}\mathbf{K}_{g21}^{(e)} \\ \widehat{\mathbf{P}}_{gc}^{(e)} &= \mathbf{P}_{g1}^{(e)} - \mathbf{K}_{g12}^{(e)}\mathbf{K}_{g22}^{(e)^{-1}}\widehat{\mathbf{P}}_{g2}^{(e)}\end{aligned} \tag{9.23g}$$

For the element with Q_3 released, the explicit form of Equation 9.23a is[22]

$$\begin{Bmatrix} Q_1 \\ Q_2 \\ Q_6 \\ Q_4 \\ Q_5 \\ -- \\ 0 \end{Bmatrix} = \frac{EI}{l}\left[\begin{array}{cccccc|c}
\left(\frac{12S^2}{l^2}+\frac{A}{I}C^2\right) & \left(-\frac{12SC}{l^2}+\frac{A}{I}SC\right) & \left(\frac{6S}{l}\right) & \left(-\frac{12S^2}{l^2}-\frac{A}{I}C^2\right) & \left(\frac{12SC}{l^2}-\frac{A}{I}SC\right) & & \left(\frac{6S}{l}\right) \\
 & \left(\frac{12C^2}{l^2}+\frac{A}{I}S^2\right) & \left(-\frac{6C}{l}\right) & \left(\frac{12SC}{l^2}-\frac{A}{I}SC\right) & \left(-\frac{12C^2}{l^2}-\frac{A}{I}S^2\right) & & \left(-\frac{6C}{l}\right) \\
 & & (4) & \left(-\frac{6S}{l}\right) & \left(\frac{6C}{l}\right) & & (2) \\
 & Sym. & & \left(\frac{12S^2}{l^2}+\frac{A}{I}C^2\right) & \left(-\frac{12SC}{l^2}+\frac{A}{I}SC\right) & & \left(-\frac{6S}{l}\right) \\
 & & & & \left(\frac{12C^2}{l^2}+\frac{A}{I}S^2\right) & & \left(\frac{6C}{l}\right) \\
\hline
 & & & & & & (4)
\end{array}\right]\begin{Bmatrix} \delta_1 \\ \delta_2 \\ \delta_6 \\ \delta_4 \\ \delta_5 \\ -- \\ \delta_3 \end{Bmatrix} \tag{9.23h}$$

[22] Note that the interchange of the third and sixth rows and columns does not alter $\mathbf{K}_g^{(e)}$ (Equation 9.8m) in this particular case, but the interchange of rows does alter $\widehat{\mathbf{P}}_g^{(e)}$ (Equation 9.8n).

$$
+\begin{Bmatrix} -S\widehat{V}_{IJ} + C\widehat{P}_{IJ} \\ C\widehat{V}_{IJ} + S\widehat{P}_{IJ} \\ \widehat{M}_{JI} \\ -S\widehat{V}_{JI} + C\widehat{P}_{JI} \\ C\widehat{V}_{JI} + S\widehat{P}_{JI} \\ -- \\ \widehat{M}_{IJ} \end{Bmatrix}
$$

which defines the submatrices $\mathbf{K}_{g_{11}}^{(e)}$, $\mathbf{K}_{g_{12}}^{(e)}$, $\mathbf{K}_{g_{22}}^{(e)}$, $\widehat{\mathbf{P}}_{g_1}^{(e)}$, and $\widehat{\mathbf{P}}_{g_2}^{(e)}$ in Equation 9.23*b*. Substituting these submatrices into Eqs. 9.23*g* and performing the indicated operations leads to

$$
\mathbf{K}_{g_c}^{(e)} = \frac{EI}{l}\begin{bmatrix}
\left(\frac{3S^2}{l^2} + \frac{A}{I}C^2\right) & \left(-\frac{3SC}{l^2} + \frac{A}{I}SC\right) & \left(\frac{3S}{l}\right) & \left(-\frac{3S^2}{l^2} - \frac{A}{I}C^2\right) & \left(\frac{3SC}{l^2} - \frac{A}{I}SC\right) \\
 & \left(\frac{3C^2}{l^2} + \frac{A}{I}S^2\right) & \left(-\frac{3C}{l}\right) & \left(\frac{3SC}{l^2} - \frac{A}{I}SC\right) & \left(-\frac{3C^2}{l^2} - \frac{A}{I}S^2\right) \\
 & & (3) & \left(-\frac{3S}{l}\right) & \left(\frac{3C}{l}\right) \\
 & Sym. & & \left(\frac{3S^2}{l^2} + \frac{A}{I}C^2\right) & \left(-\frac{3SC}{l^2} + \frac{A}{I}SC\right) \\
 & & & & \left(\frac{3C^2}{l^2} + \frac{A}{I}S^2\right)
\end{bmatrix} \tag{9.23i}
$$

and

$$
\widehat{\mathbf{P}}_{g_c}^{(e)} = \begin{Bmatrix} -S\widehat{V}_{IJ} + C\widehat{P}_{IJ} - 3S\widehat{M}_{IJ}/(2l) \\ C\widehat{V}_{IJ} + S\widehat{P}_{IJ} + 3C\widehat{M}_{IJ}/(2l) \\ \widehat{M}_{JI} - \widehat{M}_{IJ}/2 \\ -S\widehat{V}_{JI} + C\widehat{P}_{JI} + 3S\widehat{M}_{IJ}/(2l) \\ C\widehat{V}_{JI} + S\widehat{P}_{JI} - 3C\widehat{M}_{IJ}/(2l) \end{Bmatrix} \tag{9.23j}
$$

These results have a relatively simple form because we are eliminating only one variable, $\boldsymbol{\delta}_3$, and $\mathbf{K}_{g_{22}}^{(e)}$ is thus a 1×1 matrix with a very simple inverse.

Equations 9.23*f*, 9.23*i*, and 9.23*j* define the stiffness relation between the five nonzero end forces (in the order Q_1, Q_2, Q_6, Q_4, Q_5) and the corresponding end displacements. To conveniently implement the direct stiffness algorithm, however, it is desirable to retain the 6 × 6 stiffness matrix format with the force/displacement ordering defined in Figure 9.11*b*. Therefore, we perform a final modification of the matrix in Equation 9.23*i* to recast it into the appropriate form. Specifically, we add a sixth row and sixth column of 0's, and then interchange the third and sixth rows and columns. The result is

$$\mathbf{K}_g^{(e)} = \frac{EI}{l}\begin{bmatrix} \left(\frac{3S^2}{l^2} + \frac{A}{I}C^2\right) & \left(-\frac{3SC}{l^2} + \frac{A}{I}SC\right) & 0 & \left(-\frac{3S^2}{l^2} - \frac{A}{I}C^2\right) & \left(\frac{3SC}{l^2} - \frac{A}{I}SC\right) & \left(\frac{3S}{l}\right) \\ & \left(\frac{3C^2}{l^2} + \frac{A}{I}S^2\right) & 0 & \left(\frac{3SC}{l^2} - \frac{A}{I}SC\right) & \left(-\frac{3C^2}{l^2} - \frac{A}{I}S^2\right) & \left(-\frac{3C}{l}\right) \\ & & 0 & 0 & 0 & 0 \\ & & & \left(\frac{3S^2}{l^2} + \frac{A}{I}C^2\right) & \left(-\frac{3SC}{l^2} + \frac{A}{I}SC\right) & \left(-\frac{3S}{l}\right) \\ & Sym. & & & \left(\frac{3C^2}{l^2} + \frac{A}{I}S^2\right) & \left(\frac{3C}{l}\right) \\ & & & & & (3) \end{bmatrix} \quad \textbf{(9.23k)}$$

Equation 9.23*k* is the final form of the stiffness matrix for an element that is pinned at the I end and has $Q_3 = 0$.

Equation 9.23*j* is modified in a similar manner, namely by moving the third component to a sixth row and replacing it with 0,

$$\widehat{\mathbf{P}}_g^{(e)} = \begin{Bmatrix} -S\widehat{V}_{IJ} + C\widehat{P}_{IJ} - 3S\widehat{M}_{IJ}/(2l) \\ C\widehat{V}_{IJ} + S\widehat{P}_{IJ} + 3C\widehat{M}_{IJ}/(2l) \\ 0 \\ -S\widehat{V}_{JI} + C\widehat{P}_{JI} + 3S\widehat{M}_{IJ}/(2l) \\ C\widehat{V}_{JI} + S\widehat{P}_{JI} - 3C\widehat{M}_{IJ}/(2l) \\ \widehat{M}_{JI} - \widehat{M}_{IJ}/2 \end{Bmatrix} \quad \textbf{(9.23l)}$$

This is the desired form of the fixed-end force vector for the element.

The same procedure may be used to develop special stiffness relations for elements when other end forces (or combinations of end forces) are released.

EXAMPLE 10

We will analyze the structure in Figure 9.21*a*. This is similar to the structure in Example 8 (Figures 8.38 and 9.17) except that the torsional spring at B has been deleted. Since neither element develops a moment at B, we may use the stiffness relation for an element with one end pinned.

The rotation at the pinned end has been eliminated from the element stiffness relation defined by Equations 9.23*k* and 9.23*l*. Consequently, system displacements u_1 and u_2 (Figure 9.17) are not required in the analysis. Also, axial displacements (u_4 and u_5, Figure 9.17) may be deleted because they are known to be zero, so the analysis will be conducted with only one system displacement, Figure 9.21*b*.

The stiffness relation defined by Equations 9.23*k* and 9.23*l* is based on the assumption that the I end of the element is pinned; this relation can be applied to both components of this structure provided we define them as element BA ($\phi = \pi$) and element BC ($\phi = 0$). Therefore, the modified stiffness matrices (Equation 9.23*k*) are

$$\mathbf{K}_g^{(\mathrm{BA})} = \frac{EI}{l}\begin{bmatrix} 1/C'l^2 & 0 & 0 & -1/C'l^2 & 0 & 0 \\ & 3/l^2 & 0 & 0 & -3/l^2 & 3/l \\ & & 0 & 0 & 0 & 0 \\ & & & 1/C'l^2 & 0 & 0 \\ & \mathit{Sym.} & & & 3/l^2 & -3/l \\ & & & & & 3 \end{bmatrix} \tag{9.24a}$$

and

$$\mathbf{K}_g^{(\mathrm{BC})} = \frac{EI}{l}\begin{bmatrix} 1/C'l^2 & 0 & 0 & -1/C'l^2 & 0 & 0 \\ & 3/l^2 & 0 & 0 & -3/l^2 & -3/l \\ & & 0 & 0 & 0 & 0 \\ & & & 1/C'l^2 & 0 & 0 \\ & \mathit{Sym.} & & & 3/l^2 & 3/l \\ & & & & & 3 \end{bmatrix} \tag{9.24b}$$

where $C' = I/Al^2$. The stiffness matrix for the spring is given in Equation 9.20*b*.

The basic (local coordinate) fixed-end forces for element BC are given in Equation 9.20*f*; substituting the values from Equation 9.20*f* into Equation 9.23*l* gives

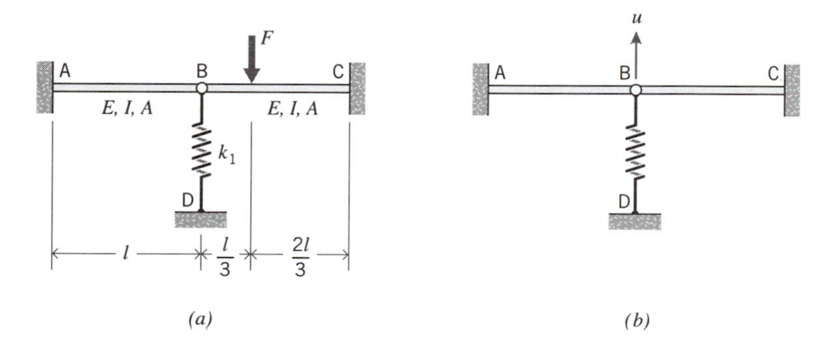

Figure 9.21 Example 10.

$$\widehat{\mathbf{P}}_g^{(BC)} = \begin{Bmatrix} 0 \\ 14F/27 \\ 0 \\ 0 \\ 13F/27 \\ 4Fl/27 \end{Bmatrix} \tag{9.24c}$$

The relationships between element displacements and system displacement are given in Table 9.10.1,

Table 9.10.1

Element	Element displacement	System displacement
BA	1	—
	2	1
	3	—
	4	—
	5	—
	6	—
BC	1	—
	2	1
	3	—
	4	—
	5	—
	6	—
DB	1	—
	2	—
	3	—
	4	1

Based on these relationships we obtain

$$\mathbf{k} = [K_{22}^{(BA)} + K_{22}^{(BC)} + K_{44}^{(DB)}] = [k_1 + 6EI/l^3] \tag{9.24d}$$

and

$$\mathbf{F}^* = -\{\widehat{Q}_2^{(BC)}\} = \{-14F/27\} \tag{9.24e}$$

Solving the equation $\mathbf{ku} = \mathbf{F}^*$ leads to

$$u = -\frac{7Fl^3}{81EI\left(1 + \dfrac{k_1 l^3}{6EI}\right)} \tag{9.24f}$$

This agrees with the limiting solution for u_3 as $k_2 \to 0$ in Equation 8.32*f*.

PROBLEMS

9.P1. (a) Reanalyze the structure shown (see Example 4, Chapter 8) using the element force-displacement relations given in Equation 9.1*d*.

(b) Explain the physical significance of the equilibrium equation associated with system displacement u_2.

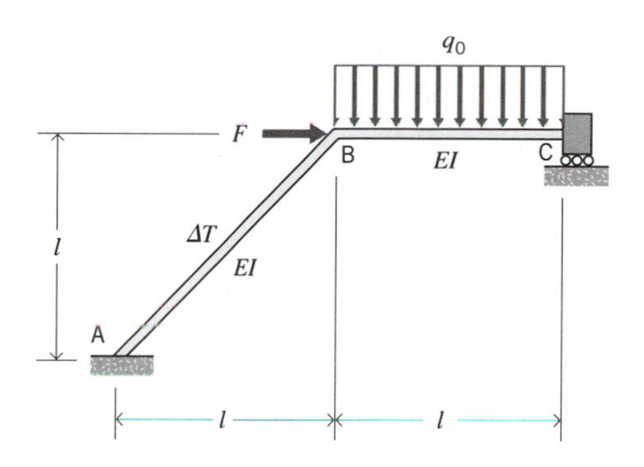

9.P2. Find the internal forces in the structure shown. Assume axially rigid elements and use the element force-displacement relations given in Equation 9.1*d*. (Compare the results to the solution obtained in Problem 8.P26, Chapter 8.)

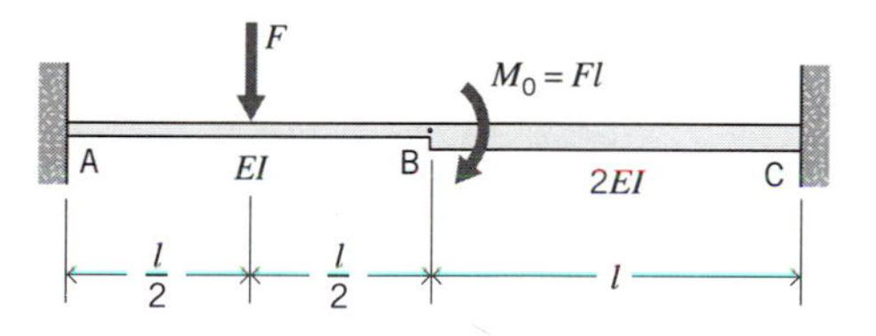

9.P3. Find the internal forces in the structure shown. Assume axially rigid elements and use the element force-displacement relations given in Equation 9.1*d*. (Compare the results to the solution obtained in Problem 8.P27, Chapter 8.)

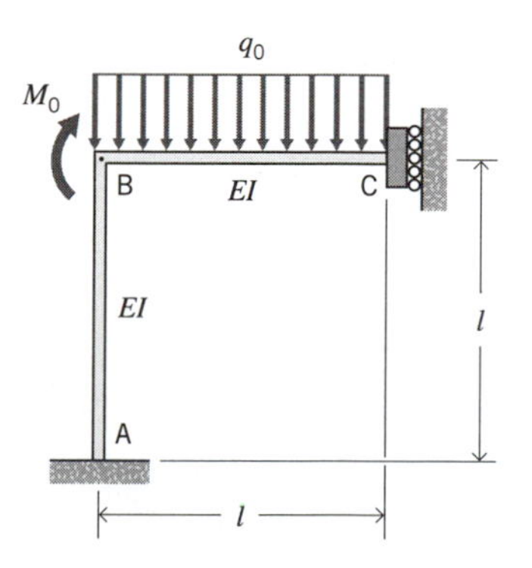

9.P4. Find the internal forces in the structure shown. Assume axially rigid elements and use the element force-displacement relations given in Equation 9.1*d*. Note: the axial component of load in element BC must be considered. (Compare the results to the solution obtained in Problem 8.P30, Chapter 8.)

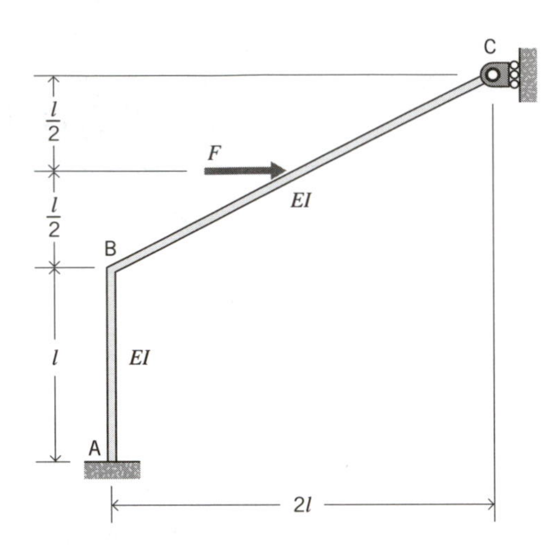

9.P5. Reanalyze the structure in Figure 8.28 (Example 5, Chapter 8) using the element force-displacement relations given in Equation 9.1*d* to characterize the axially rigid elements.

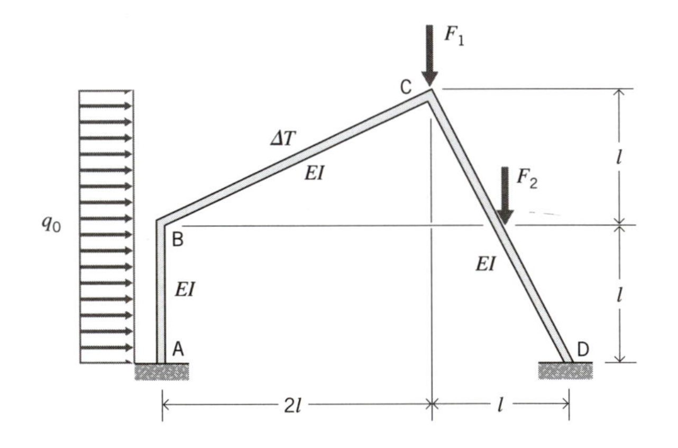

9.P6. Analyze the structure in Problem 9.P4 assuming that the elements are axially flexible, i.e., use the stiffness relation in Equation 9.5 to characterize element behavior. Find an approximate solution for $EA \rightarrow \infty$.

9.P7. Find the internal forces in the structure shown assuming that the elements are axially flexible, i.e., use the stiffness relation in Equation 9.5 to characterize element behavior. Find an approximate solution for $EA \rightarrow \infty$. (Compare the results to the solution obtained in Problem 8.P29, Chapter 8.)

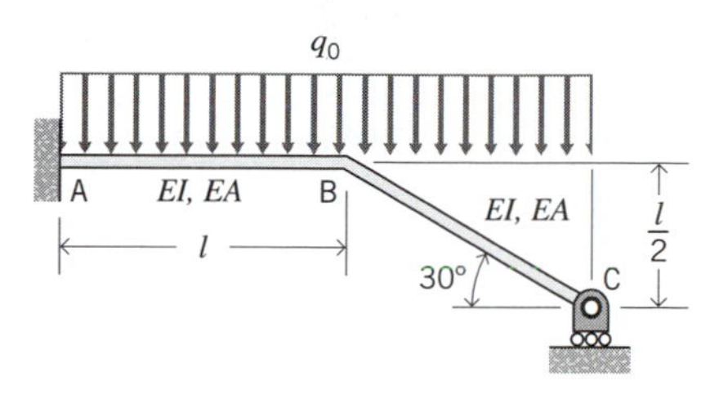

9.P8. Use the direct stiffness method to assemble the stiffness matrix and force vector for the structure in Problem 8.P6.

9.P9. Use the direct stiffness method to assemble the stiffness matrix and force vector for the structure in Problem 9.P7.

9.P10. For the structure shown, elements AB, BC, BG, DG, and GC are axially flexible beams that are rigidly connected at joints to form a frame. Elements AG and BD are truss bars with pinned ends. Assume that the axial stiffness of all elements is EA and the bending stiffness of the beams is EI. Let $EA = \alpha EI/l^2$, where α is a constant. Use the direct stiffness method to assemble the stiffness matrix and force vector. Solve for internal forces for $\alpha = 0.01$, $\alpha = 1.0$, and $\alpha = 100$. Which, if any, of the solutions do you expect to closely approximate truss-type behavior?

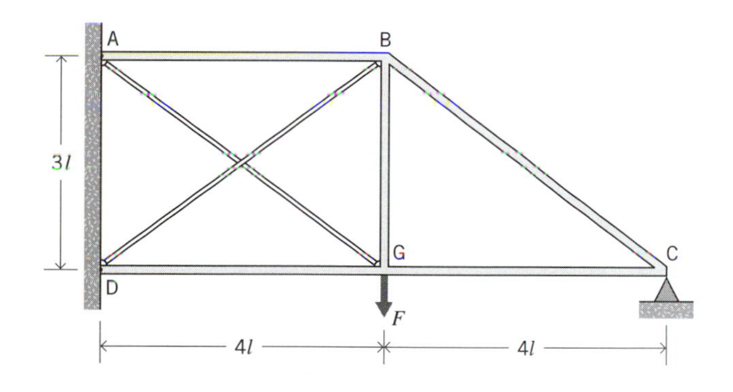

9.P11. Use the direct stiffness method to analyze the truss shown. Assume that all bars have the same *EA*. Note that bar AB has an initial length change Δ_I. (Compare the results to the solution obtained in Problem 8.P8, Chapter 8.)

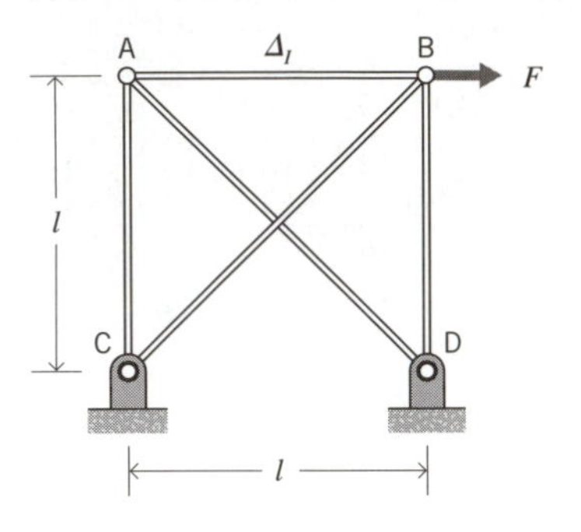

9.P12. Assume that the rigid joints in the structure in Problem 9.P10 are replaced by pins so that the structure becomes a truss. Use the direct stiffness method to analyze this truss. (Compare the results to the solution obtained in Problem 9.P10.)

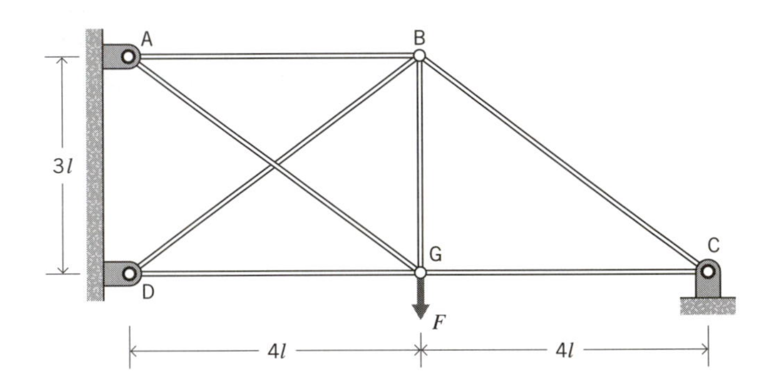

9.P13. Use the direct stiffness method to analyze the truss shown. Assume that all bars have the same *EA*, and that there is a uniform temperature change ΔT in bar DE. (Compare the results to the solution obtained in Example 1, Chapter 5.)

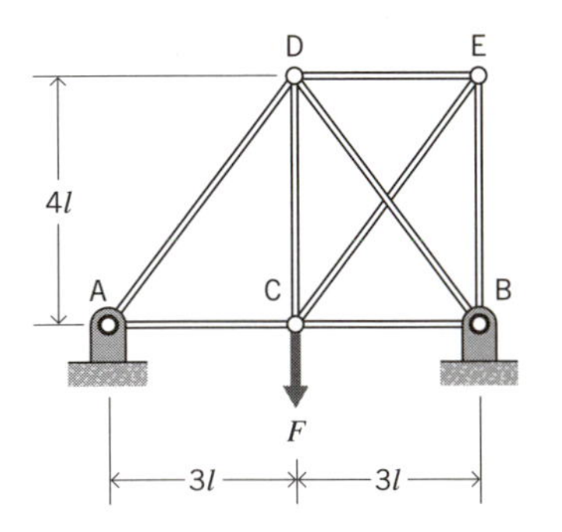

9.P14. Use the direct stiffness method to analyze the truss shown. Assume that all bars have the same EA. (Compare the results to the solution obtained in Problem 8.P10, Chapter 8.)

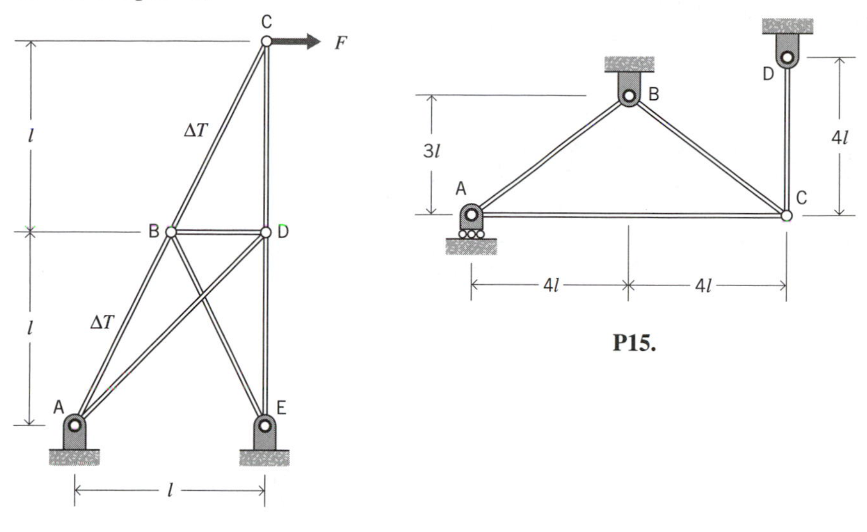

P15.

P14.

9.P15. Use the direct stiffness method to find the internal forces in the truss shown. There are no applied loads, but the support at A undergoes a small vertical settlement δ_o. (Compare the results to the solution obtained in Problem 8.P42, Chapter 8 or Problem 5.P43, Chapter 5.)

9.P16. The structure consists of a uniform beam ABC and a bar BD. There is a fixed support at C which can translate but cannot rotate. Assume that the axial stiffness of the beam is large and that for the bar $EA = 3\sqrt{2}EI/l^2$. There are no applied loads, but the support at A (only) undergoes a settlement consisting of a small counterclockwise rotation θ_o. Use the direct stiffness method to find the internal forces. (Compare the results to the solution obtained in Problem 8.P43, Chapter 8.)

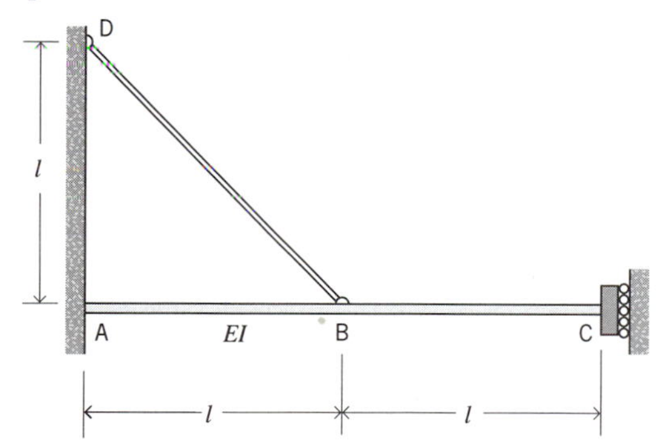

9.P17. Analyze the structure in Problem 9.P6 (see Figure 9.P4) using the stiffness relations defined by Equations 9.23k and 9.23l to characterize element BC. Assemble the equilibrium equation directly.

9.P18. Analyze the structure in Problem 9.P7 using the stiffness relations defined by Equations 9.23k and 9.23l to characterize element BC. Assemble the equilibrium equation directly.

Chapter 10

Energy Concepts and Approximations

10.0 INTRODUCTION

The first nine chapters of this text contain the basic techniques required to conduct a matrix analysis, either in stiffness (displacement) or flexibility (force) form, of structures composed of one-dimensional beam/truss elements. Emphasis has been placed on implementing a five-step definition, assembly, and solution process in which the basic physical behavior of the structure was clearly organized and delineated. In particular, equilibrium under forces, elastic stiffness at the element level, and interelement compatibility were the major constituents of the process.

The most abstract concept employed was that of virtual work, i.e., virtual force patterns were used to describe physical deformation/displacement behavior in the force method, and virtual displacement patterns were used to generate equilibrium equations in the displacement method. As powerful and convenient a tool as virtual work was seen to be, it was not essential to the formulations because it is basically only an alternative statement of equilibrium or compatibility requirements. In fact, we saw that it was always possible to directly generate either equilibrium or compatibility statements, i.e., to independently verify all the relations resulting from applications of virtual work.

As one pursues the analysis of more complex structures composed of elements that require two- or three-dimensional theory to describe behavior, it will be seen that virtual work becomes a much more important and essential tool for defining element (and therefore structural) behavior. To lay an appropriate foundation for these more advanced analyses, it is important to present some basic *energy concepts* that have not been used or discussed up to this point. These concepts, together with the prior background in virtual work, are necessary for the understanding of *finite element analysis,* which will be introduced in Chapter 11. The latter is the most widely used method of analysis in modern structural engineering.

As with virtual work, energy concepts provide nothing that is essential in the area of actual problem solving capability when dealing with simple structures composed of one-dimensional beam/truss elements. Rather, the concepts provide some additional general insights into structural behavior, which are necessary for developing effective and accurate approximate methods (such as the finite element method) for analyzing systems composed of more complex elements.

10.1 BASIC ENERGY FORMS

We shall begin our discussion by using energy concepts to provide general proofs regarding the symmetry property of stiffness and flexibility matrices, i.e., the Maxwell-Betti symmetry relations.[1] In addition, we shall expand our understanding of elastic behavior to include the idea of energy conservation. This will require the introduction of the concept of strain energy for a structure. Two additional energy forms will be introduced subsequently, namely the deformation energy for the structure and the external potential energy of a conservative external (applied) force field. Given these three energy forms, we will be able to establish several important energy theorems for elastic structures:

1. The theorem of stationary value of the total potential energy.
2. The theorem of minimum total potential energy.
3. The theorem of least work.

The first two of these theorems relate to the displacement method, the last one to the force method.

Each of the theorems provides some insight into the behavior of structures when analyses are exact; more importantly, each also provides information about the effects introduced by errors or approximations in analyses. As we shall see in Chapter 11, approximations are, in fact, unavoidable and are an essential part of finite element analysis of complex structures. There are two types of unavoidable approximations: (a) the use of a finite number of displacement patterns to represent behavior of a structure with an infinite number of displacement degrees of freedom, and (b) the use of a finite number of force patterns to represent behavior of a structure that is more highly indeterminate[2] than the model takes into account. We shall demonstrate that the first type of approximation has the effect of making the structure appear stiffer than it actually is; this implies that the true equilibrium configuration of the structure is only approximated and that the true stresses are generally underestimated. The second type of approximation has the effect of making the structure appear more flexible than it actually is; this implies that the approximate internal forces produce incompatible structural motions and that stresses are generally overestimated. Such insights are essential for an understanding of the nature of finite element analysis.

---

[1] Recall that in Chapter 6 stiffness and flexibility matrices were shown to be symmetric, but only for one-dimensional elements.

[2] Structures composed of two and three-dimensional elements may be viewed as infinitely indeterminate.

10.1.1 Strain Energy

Let's begin by considering a general structure of the form shown in Figure 10.1*a*, for which a set of system displacements (and corresponding nodal forces) is somehow defined. This structure is not necessarily assembled from one-dimensional beam/truss elements, although such a situation is encompassed as a special case, e.g., Figure 10.1*b*.

We will specify that behavior is to be taken as linearly ''elastic'' regardless of whether or not we have detailed knowledge of the elements that constitute the interior of the structure. In other words, we will assume that the equilibrium equations for the structure can be written in the form of a linear elastic force-displacement matrix relationship.[3] We will assume that the general form of such a relationship is similar to that discussed in Chapter 9, namely $\mathbf{ku} = \mathbf{F}^*$. In Chapter 9, vector $\mathbf{F}^*$ was seen to comprise two constituents, $\mathbf{F} - \boldsymbol{\beta}^T\widehat{\mathbf{P}}$, where $\mathbf{F}$ contained the effects of concentrated joint forces and $\boldsymbol{\beta}^T\widehat{\mathbf{P}}$ contained the effects of interior loads. In subsequent formulations, interior loads will be assumed to be directly replaced by ''equivalent'' joint forces; therefore, for the sake of generality, we will henceforth assume that vector $\mathbf{F}$ contains the effects of all applied forces, and that an additional component of $\mathbf{F}^*$, which we will now denote as $\mathbf{F_o}$, contains terms associated with thermal or initial loads.

Consequently, we may rewrite the force-deflection relationship as $\mathbf{ku} - \mathbf{F_o} = \mathbf{F}$. In this form, we recognize that $\mathbf{F_o}$ must be equivalent to $\mathbf{ku_o}$ where $\mathbf{u_o}$ denotes initial displacements that are present before any loads are applied, i.e., when $\mathbf{F} = \mathbf{0}$. These initial displacements are assumed known when $\mathbf{F} = \mathbf{0}$. Thus, the force-displacement relationship may be written

$$\mathbf{F} = \mathbf{k}(\mathbf{u} - \mathbf{u_o}) \tag{10.1a}$$

We shall also assume that the stiffness matrix $\mathbf{k}$ is a square matrix with constant stiffness coefficients k_{ij}, and that $\mathbf{k}^{-1} \equiv \mathbf{a}$ is a constant flexibility matrix in the inverse relation

$$\mathbf{u} = \mathbf{aF} + \mathbf{u_o} \tag{10.1b}$$

In stipulating the existence of $\mathbf{k}^{-1}$ we implicitly require that the determinant of $\mathbf{k}$ exists, so that a unique relation exists between $\mathbf{F}$ and $\mathbf{u}$, given $\mathbf{u_o}$. Stated another way, this means that there is a unique displacement $\mathbf{u}$ for any arbitrary loading $\mathbf{F}$, and vice versa. Note, however, that we have not said anything about symmetry of the stiffness or flexibility matrices.

Before proceeding with further developments, it is useful to elaborate on some implications associated with the use of force vector $\mathbf{F}$ in Equation 10.1. If the structure is loaded only by concentrated forces at the coordinates, as in Figure 10.1*a*, there is no ambiguity because the components of $\mathbf{F}$ are exactly the concentrated forces. However, if there are interior forces, as in Figure 10.1*c*, then the consequences are not so clear.

To illustrate this last statement, we note that the distributed forces q in Figure 10.1*c* can be expected to cause some nodal displacements, as also shown in Figure 10.1*c*; we denote the vector of these displacements by $\mathbf{u_q}$. There must also exist a set of nodal forces $\mathbf{F_q} \equiv \mathbf{ku_q}$, which produces identical nodal displacements, Figure 10.1*d*. It follows

[3] For example, for the general structure, such a relationship could be obtained by measurement in a laboratory test, or could simply be agreed upon as a reasonable mathematical description of structural behavior.

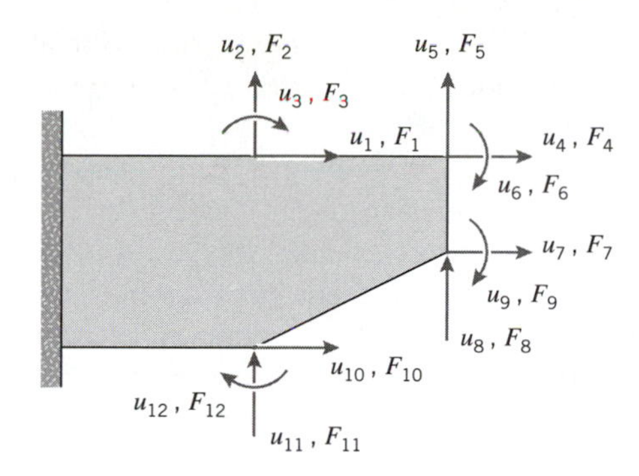

(*a*) General structure

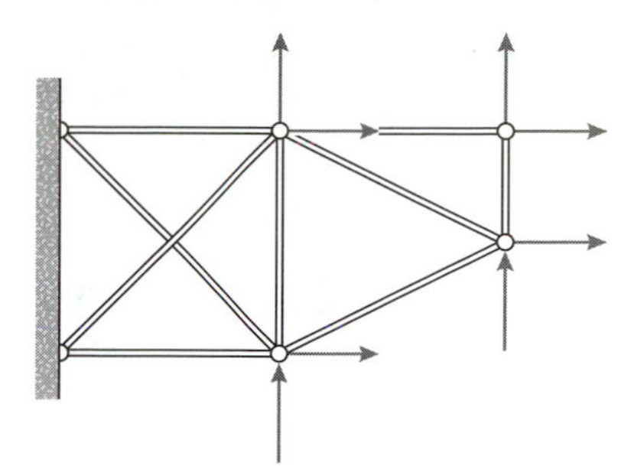

(*b*) Special case

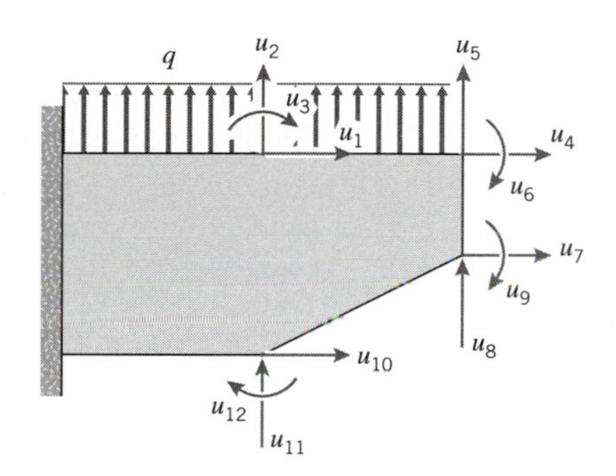

(*c*) Displacements produced by distributed load q

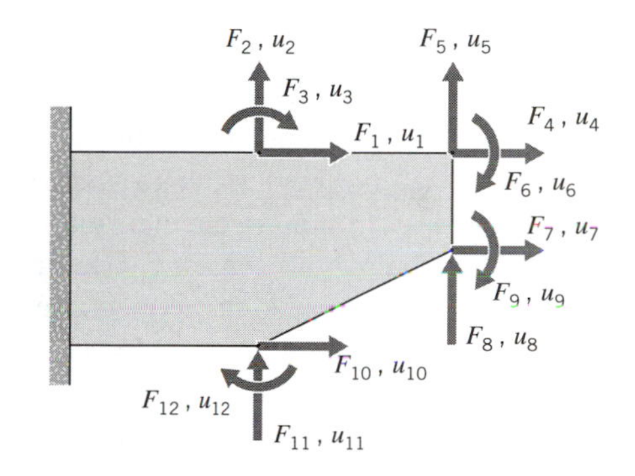

(*d*) Nodal forces that produce the same displacement as in (*c*)

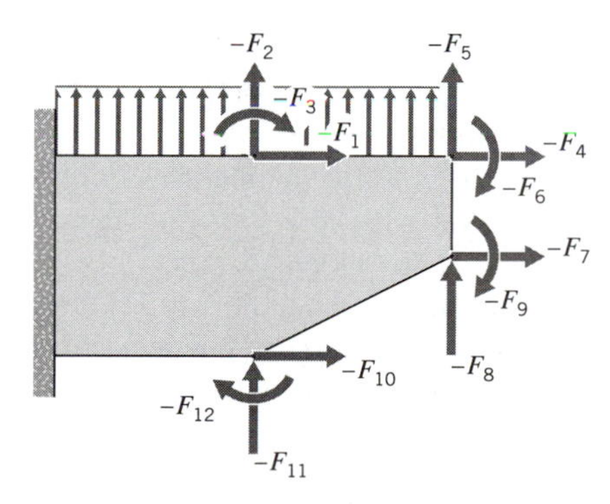

(*e*) Fixed-coordinate state ($\mathbf{u} = \mathbf{0}$)

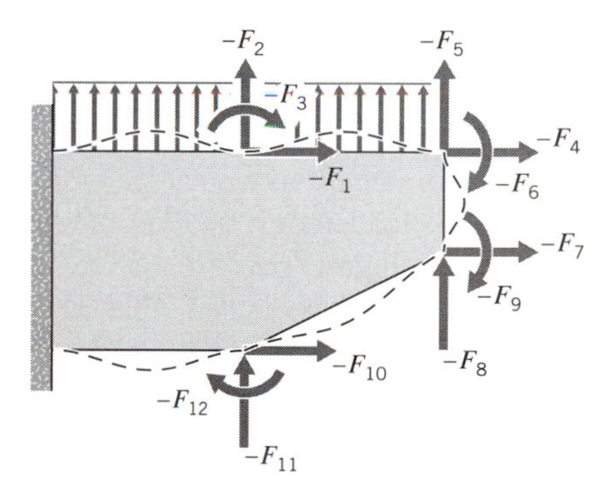

(*f*) Motion in fixed-coordinate state

Figure 10.1 System displacements and nodal forces for structure.

that if the negative of these forces, $-\mathbf{F_q}$, is added to the distributed load in Figure 10.1*c* the result will be $\mathbf{u} = \mathbf{0}$ (see Figure 10.1*e*). Figure 10.1*e* is referred to as a "fixed-coordinate" state; the nodal forces in this state are $\mathbf{F_f} = -\mathbf{F_q}$. The physical loads in Figure 10.1*e* thus constitute a self-equilibrating force system, i.e., a set of forces in equilibrium that produces no nodal displacements whatsoever.

Although the nodal displacements due to the distributed loads q and $\mathbf{F_f}$ are equal and opposite, the two sets of forces are not entirely equivalent. For example, our experience with ordinary beam elements shows us that the effect within an element due to a distributed interior load is different than that due to concentrated end forces. So, in general, there will be some displacement field in the structure of Figure 10.1*e* even though the nodal response is zero (see Figure 10.1*f*).

The nodal forces in Figure 10.1*f* may be regarded as the forces that prevent nodal displacements; in other words, these are fixed-end forces. This implies that the nodal displacements of the structure in Figure 10.1*c*, which is loaded by distributed forces, can be completely, i.e., exactly, determined from Equation 10.1*a* provided the components of **F** are taken as the negative of the fixed-end forces.[4] However, the portion of the displacement field associated with the fixed-coordinate state, namely that in Figure 10.1*f*, will be missed.

For one-dimensional elements, we can recover the complete internal stress state using simple statics; however, for complex two- and three-dimensional structures it will not be possible to find exact solutions for the self-equilibrating stresses. The only way to investigate their magnitudes will be to develop a model with refined displacement patterns that are capable of revealing the nonzero displacements, which we know are present in the fixed-coordinate state.

EXAMPLE 1

Consider a uniform axially rigid beam with one end built in and the other end simply supported. The beam is subjected to two different loads: (1) a uniform load q_o as shown in Figure 10.2*a*, and (2) a concentrated moment M_o applied at the simply supported end as shown in Figure 10.2*b*. There is only one nodal displacement in either case, namely the rotation at the simply supported end.

We wish to ascertain (1) the value of M_o in Figure 10.2*b* that causes the same nodal displacement as the distributed load in Figure 10.2*a*, and (2) transverse displacement, $v(x)$, for each of the beams when the nodal displacements are equal.

The beam in Figure 10.2*a* has been analyzed in Chapter 4 (Example 16) and in Chapter 5 (Section 5.3.1). We will denote the transverse displacement of the beam by $v_1(x)$; from the information developed previously, it may be shown that

$$v_1(x) = \frac{q_o l^3}{48EI}\left(-x + \frac{3x^3}{l^2} - \frac{2x^4}{l^3}\right) \qquad \textbf{(10.2a)}$$

[4] This is exactly the effect of the vector $-\boldsymbol{\beta}^T\widehat{\mathbf{P}}$.

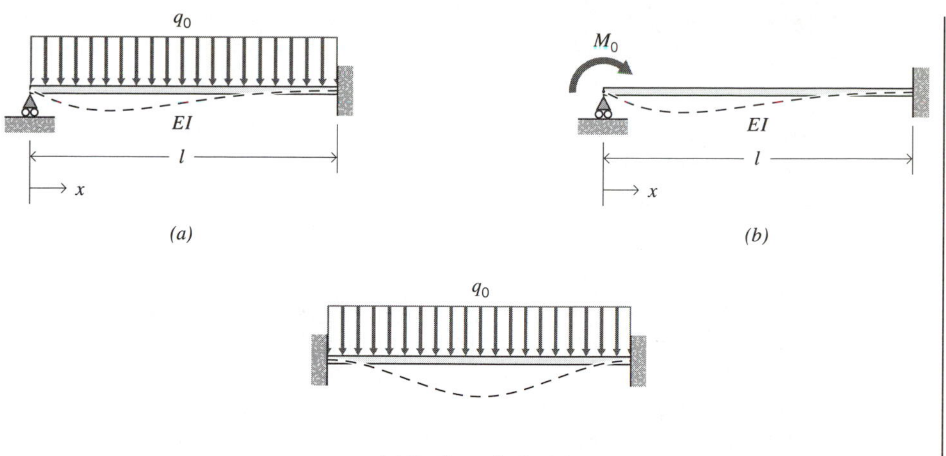

Figure 10.2 Example 1: Illustration of fixed-coordinate state.

The nodal displacement, u, is the negative of the slope at the end,

$$u = -v_1'(0) = \frac{q_o l^3}{48EI} \tag{10.2b}$$

The transverse displacement of the beam in Figure 10.2*b*, denoted by $v_2(x)$, has been given previously in Figure 8.18; in terms of the applied force M_o, we have

$$v_2(x) = \frac{M_o l}{4EI}\left(-x + \frac{2x^2}{l} - \frac{x^3}{l^2}\right) \tag{10.2c}$$

We observe that we obtain $v_2'(0) = v_1'(0)$ for the condition

$$M_o = \frac{q_o l^2}{12} \tag{10.2d}$$

The end moment in Equation 10.2*d* is the nodal force that causes the same nodal displacement as the distributed load q_o. We note that this nodal force is the negative of the fixed-end moment due to q_o (see Equation 8.22*e*), exactly as stated previously. Substituting Equation 10.2*d* into Equation 10.2*c* gives the desired expression for $v_2(x)$,

$$v_2(x) = \frac{q_o l^3}{48EI}\left(-x + \frac{2x^2}{l} - \frac{x^3}{l^2}\right) \tag{10.2e}$$

Superposing the load in Figure 10.2*a* and the negative of the load in Figure 10.2*b* produces the fixed-coordinate state, i.e., the state in which the nodal displacement is zero (Figure 10.2*c*). The transverse displacement of the beam in this state is

$$v_1(x) - v_2(x) = \frac{q_o l^3}{24EI}\left(-\frac{x^2}{l} + \frac{2x^3}{l^2} - \frac{x^4}{l^3}\right) \qquad \textbf{(10.2f)}$$

This is precisely the portion of the displacement of the beam in Figure 10.2*a* that is not contained in Equation 10.2*e*.

Returning, now, to the concept of "elastic" behavior, we can state that the existence of the simple force-displacement relationship in Equation 10.1 is not, by itself, sufficient to assure that the structure is truly elastic. This is the case despite the fact that several features of Equation 10.1 are consistent with our basic perception of what constitutes elasticity: (*a*) the force-displacement relation is unique, (*b*) the relationship is linear, (*c*) application of loads moves the structure from the initial configuration $\mathbf{u}_o$ to $\mathbf{u}$, and removal of the loads returns the structure to $\mathbf{u}_o$, and (*d*) loads can be applied and removed for an indefinite number of cycles without causing any change in the configuration when $\mathbf{F} = \mathbf{0}$. However, to fully understand elastic behavior, a general structure must be viewed from a slightly different perspective, namely that of a body that stores energy in a unique, recoverable manner.

The energy stored in a body is related to the work done by applied forces causing deformation. As the forces $\mathbf{F}$ move through an incremental change in displacements, $d\mathbf{u}$, we define an increment of physical work, dW, as

$$dW = d\mathbf{u}^T\mathbf{F} \qquad \textbf{(10.3a)}$$

Given this definition of work, we may now expand our concept of elastic behavior to include, in addition to the behavior described by Equation 10.1, the following requirements:

1. The elastic body stores all of the work done by forces $\mathbf{F}$ that produce displacements $\mathbf{u}$ in the form of "strain energy," U, where[5]

$$U \equiv \int_{\mathbf{u}_o}^{\mathbf{u}} dW = \int_{\mathbf{u}_o}^{\mathbf{u}} d\bar{\mathbf{u}}^T\mathbf{F}(\bar{\mathbf{u}}) \qquad \textbf{(10.3b)}$$

   and where $\bar{\mathbf{u}}$ is an intermediate displacement between $\mathbf{u}_o$ and $\mathbf{u}$.
2. For any $\mathbf{u} \neq \mathbf{u}_o$, i.e., for any displacement configuration other than that associated with $\mathbf{F} = \mathbf{0}$, the work done by the applied forces to deform the body must always be positive, so that $U > 0$.
3. The strain energy U is a unique function of the displacement vector $\mathbf{u}$.[6]

---

[5] For a one-degree-of-freedom system, it may be recognized that this definition has the physical interpretation of area under the force-displacement diagram.

[6] This implies that, given $\mathbf{k}$ and $\mathbf{u}_o$ in Equation 10.1, we could determine U simply by measuring final values of $\mathbf{u}$ for the loaded structure, and that we need not be concerned about the details of the motion between $\mathbf{u}_o$ and $\mathbf{u}$.

4. The strain energy is completely recovered from the elastic body when the applied loads are removed and the body is returned to its initial state $\mathbf{u}_o$.

Finally, it is important to recognize that Equation 10.3*b* considers only the energy in the structure associated with displacements $\mathbf{u}$ and therefore does not contain the energy stored in the fixed-coordinate state. This state may be viewed as a preexisting state, with an energy U_o, which is completely independent of $\mathbf{u}$.

10.1.2 Symmetry of Stiffness and Flexibility Matrices

Requirement 4 in Section 10.1.1 is very reasonable for a true elastic body since no net energy should be required to complete any load/unload cycle; otherwise net work would have to be expended after each cycle to return the body to its former "initial" state. Requirement 3 is very close to requirement 4, but states that the strain energy should be independent of the deflection path. To demonstrate the significance of this requirement, consider the simple linear structure in Figure 10.3 and suppose, for the sake of argument, two different load paths. We shall show that work done during loading is path-dependent unless the stiffness matrix is symmetric. In particular, consider a load path in which forces are applied such that only u_1 develops, and then (later) additional forces are applied such that u_2 develops while u_1 is held constant. If we assume for the sake of convenience that initial displacements are zero,[7] we see from Equation 10.1*a* that

$$\begin{aligned} f_1 &= k_{11}u_1 + k_{12}u_2 \\ f_2 &= k_{21}u_1 + k_{22}u_2 \end{aligned} \tag{10.4}$$

Thus, we may first apply forces $f_1^{(1)} = k_{11}u_1$ and $f_2^{(1)} = k_{21}u_1$ to produce u_1 while holding $u_2 = 0$. Then, later, we may add $f_1^{(2)} = k_{12}u_2$ and $f_2^{(2)} = k_{22}u_2$, which produce u_2 while u_1 is held constant. This is the loading denoted by path 1 in Figure 10.4.

Let's compute the work done by the forces during this particular loading process. During the first part of the loading, as u_1 is increased from zero to its final value (i.e., from A to B in Figure 10.4), at any intermediate displacement $0 < \bar{u}_1 < u_1$ the forces are $\bar{f}_1^{(1)} = k_{11}\bar{u}_1$ and $\bar{f}_2^{(1)} = k_{21}\bar{u}_1$. Then, the work done in deforming the structure from A to B is

[7] If the initial displacements are nonzero then we may simply introduce a new displacement $\mathbf{x} = \mathbf{u} - \mathbf{u}_o$ and carry out the following arguments.

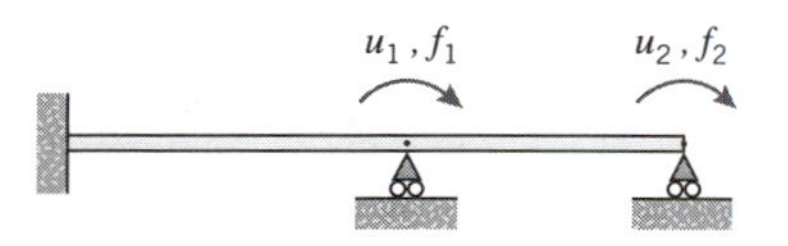

Figure 10.3 Two-degree-of-freedom linear structure ($\mathbf{u_o} = \mathbf{o}$).

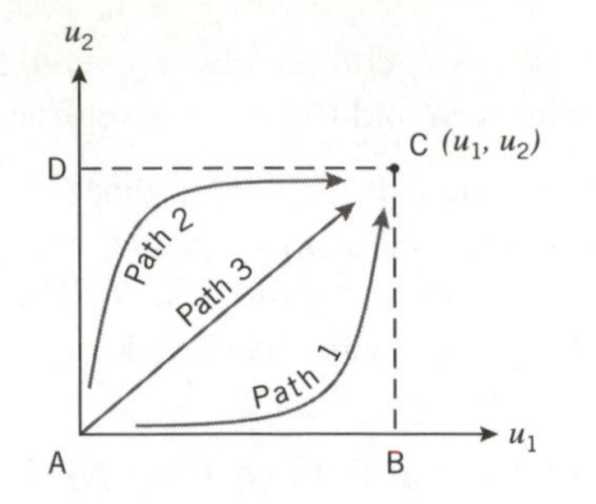

Figure 10.4 Loading paths.

$$\int_A^B dW = \int_0^{u_1} d\bar{\mathbf{u}}^{\mathrm{T}}\bar{\mathbf{F}} = \int_0^{u_1} (d\bar{u}_1\bar{f}_1 + d\bar{u}_2\bar{f}_2)$$
$$= \int_0^{u_1} (d\bar{u}_1 k_{11}\bar{u}_1 + d\bar{u}_2 k_{21}\bar{u}_1) \tag{10.5}$$

where $\bar{\mathbf{F}} \equiv \mathbf{F}(\bar{\mathbf{u}})$. Since $\bar{u}_2$ is zero it follows that $d\bar{u}_2$ is also zero, so performing the integration in Equation 10.5 gives

$$W_{AB} = \frac{1}{2}k_{11}u_1^2 \tag{10.6}$$

as the work required to produce the displacement u_1.

Now, let's apply the loads that cause the second component of **u** to go from zero to its final value u_2. The additional forces we must apply to the structure (which is already loaded by $f_1^{(1)}$ and $f_2^{(1)}$) are $f_1^{(2)} = k_{12}u_2$ and $f_2^{(2)} = k_{22}u_2$. For $0 < \bar{u}_2 < u_2$ (i.e., as displacements move from B to C in Figure 10.4), intermediate values of the forces are $\bar{f}_1^{(2)} = k_{12}\bar{u}_2$ and $\bar{f}_2^{(2)} = k_{22}\bar{u}_2$. The additional work done on the structure is

$$\int_B^C dW = \int_0^{u_2} d\bar{\mathbf{u}}^{\mathrm{T}}\bar{\mathbf{F}} = \int_0^{u_2} d\bar{u}_1(k_{11}u_1 + k_{12}\bar{u}_2) + \int_0^{u_2} d\bar{u}_2(k_{21}u_1 + k_{22}\bar{u}_2) \tag{10.7}$$

In this case, $d\bar{u}_1 = 0$ and the first integral disappears. Performing the last integration, and recognizing that $k_{21}u_1$ is constant, leads to

$$W_{BC} = k_{21}u_1u_2 + \frac{1}{2}k_{22}u_2^2 \tag{10.8}$$

Thus, the total work done in deforming the body along path 1 must be

$$W_1 = W_{AB} + W_{BC} = \frac{1}{2}k_{11}u_1^2 + k_{21}u_1u_2 + \frac{1}{2}k_{22}u_2^2 \tag{10.9}$$

The surprising result of this analysis is that the work done in deforming the structure along path 1 contains no effect of k_{12}. If we were to load the body along path 2 (Figure 10.3), producing u_2 first and then u_1, the results would be

$$W_{AD} = \frac{1}{2}k_{22}u_2^2 \tag{10.10}$$

$$W_{DC} = \frac{1}{2}k_{11}u_1^2 + k_{12}u_1u_2 \tag{10.11}$$

and

$$W_2 = W_{AD} + W_{DC} = \frac{1}{2}k_{11}u_1^2 + k_{12}u_1u_2 + \frac{1}{2}k_{22}u_2^2 \tag{10.12}$$

By comparing Equation 10.12 and Equation 10.9, we see that the work done by the applied loads in causing the displacements u_1 and u_2 will be different along path 1 and path 2 unless $k_{12} = k_{21}$. It follows that strain energy U will be a unique function of u_1 and u_2 only if the stiffness matrix is symmetric! In other words, this is a necessary and sufficient condition for the existence of a unique strain energy. This implies that $\mathbf{a} = \mathbf{k}^{-1}$ must be symmetric as well.

The symmetry requirement also insures that strain energy is completely recoverable (requirement 4 above). If the work done in following path 1 (from A to B to C, Figure 10.4) is exactly negated by unloading along path 2 (from C to D to A) then all of the work (i.e., strain energy) will be recovered.

It is worth noting what would happen if $W_1 > W_2$. If we again load along path 1 and unload along path 2, we would recover less energy than had been expended even though the structure returns to its initial configuration. However, if we load along path 2 and unload along path 1, we would recover more energy than had been expended, a neat (*but impossible*) trick! Thus, for linear structures, we really have no choice but to use stiffness matrices that are symmetric; otherwise, our analyses could produce nonsensical results such as energy generation!

10.1.3 Positive Definiteness of Strain Energy

As was noted in the preceding section, we expect to always have to do positive work and, hence, generate a positive strain energy whenever a linear elastic body is given a displacement $\mathbf{u} \neq \mathbf{u}_o$. To examine further restrictions on the force-displacement relationship that results from this requirement, we must first develop a general expression for the strain energy U.

Since we now know that any loading path must produce the same strain energy, we may simplify the subsequent computations by considering the energy associated with a path such as path 3 in Figure 10.4, i.e., a proportional loading path for which a vector of intermediate displacements is defined by

$$\bar{\mathbf{u}} = \alpha(\mathbf{u} - \mathbf{u}_o) + \mathbf{u}_o \tag{10.13}$$

where α ($0 < \alpha < 1$) is a scalar "loading factor." For any linear elastic body, the corresponding intermediate force vector, $\overline{\mathbf{F}}$, is obtained from Equation 10.1*a*,

$$\begin{aligned}\overline{\mathbf{F}} &= \mathbf{k}(\overline{\mathbf{u}} - \mathbf{u}_o) \\ &= \alpha\mathbf{k}(\mathbf{u} - \mathbf{u}_o) \\ &= \alpha\mathbf{F}\end{aligned} \tag{10.14}$$

For an incremental increase in displacements $d\overline{\mathbf{u}} = d\alpha(\mathbf{u} - \mathbf{u}_o)$, the increment of work done is given by Equation 10.3*a*,

$$dW = d\overline{\mathbf{u}}^{\mathrm{T}}\overline{\mathbf{F}} \tag{10.15a}$$

or

$$dW = \alpha\, d\alpha(\mathbf{u} - \mathbf{u}_o)^{\mathrm{T}}\mathbf{k}(\mathbf{u} - \mathbf{u}_o) \tag{10.15b}$$

Integrating between $\alpha = 0$ and $\alpha = 1$ leads to

$$W \equiv U = (\mathbf{u} - \mathbf{u}_o)^{\mathrm{T}}\mathbf{k}(\mathbf{u} - \mathbf{u}_o)\int_0^1 \alpha\, d\alpha \tag{10.15c}$$

or

$$\boxed{U = \frac{1}{2}(\mathbf{u} - \mathbf{u}_o)^{\mathrm{T}}\mathbf{k}(\mathbf{u} - \mathbf{u}_o)} \tag{10.15d}$$

Equation 10.15*d* is a general matrix expression for strain energy U for a linear elastic body. Expressions of this type are known as quadratic forms; in this case, the matrix of the quadratic form, $\mathbf{k}$, is symmetric. Furthermore, since $U > 0$ for any $\mathbf{u} \neq \mathbf{u}_o$, the quadratic form is said to be positive definite. From the mathematical theory of such forms, it is known that a necessary and sufficient condition that guarantees that $U > 0$ is that every principal minor[8] of $\mathbf{k}$ be positive. In other words, for $\mathbf{k}$ to be an appropriate stiffness matrix for a real linear elastic structure, it is necessary that the principal minors of $\mathbf{k}$ be positive.

To gain some insight into the significance of this condition, let's consider the 2×2 $\mathbf{k}$ matrix defined by Equation 10.4 for which we must have $k_{21} = k_{12}$. The strain energy for a body with this stiffness matrix is

$$U = \frac{1}{2}\begin{Bmatrix} u_1 - u_{1_o} \\ u_2 - u_{2_o}\end{Bmatrix}^{\mathrm{T}}\begin{bmatrix} k_{11} & k_{12} \\ k_{12} & k_{22}\end{bmatrix}\begin{Bmatrix} u_1 - u_{1_o} \\ u_2 - u_{2_o}\end{Bmatrix} \tag{10.16a}$$

[8] A principal minor of $\mathbf{k}$ is the determinant of a principal submatrix of $\mathbf{k}$; a principal submatrix of $\mathbf{k}$ is a square submatrix of order i, $1 \leq i \leq n$ (n being the order of $\mathbf{k}$), which is obtained by deleting any $n - i$ identically numbered rows and columns. The principal diagonal of the resulting submatrix contains only components of the principal diagonal of $\mathbf{k}$. (The principal diagonal of $\mathbf{k}$ is the set of components k_{jj}, $j = 1, \ldots, n$.) Note that the definition of principal submatrices encompasses matrices of all orders $1, \ldots, n$, i.e., from matrices that consist of single components of the principal diagonal of $\mathbf{k}$, to $\mathbf{k}$ itself.

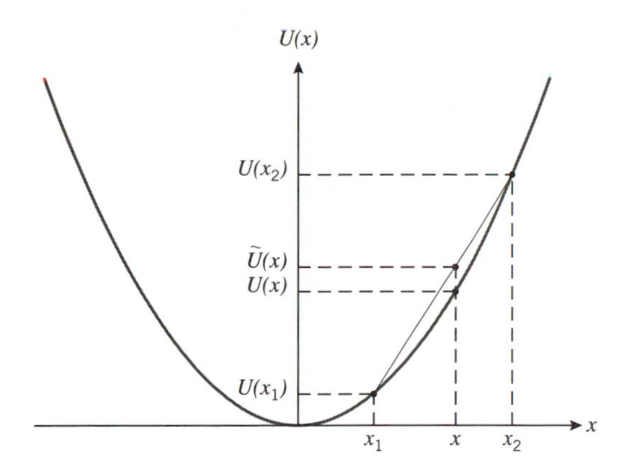

Figure 10.5 Strain energy for one degree of freedom.

Setting $x_i = u_i - u_{i_o}$, $i = 1,2$, enables us to rewrite U as

$$U = \frac{1}{2}(k_{11}x_1^2 + 2k_{12}x_1x_2 + k_{22}x_2^2) \tag{10.16b}$$

which we recognize as identical to the quadratic expressions for work in Equations 10.9 or 10.12. This equation may be rewritten in the alternate form

$$U = \frac{1}{2}k_{11}\left(x_1 + \frac{k_{12}}{k_{11}}x_2\right)^2 + \frac{1}{2k_{11}}(k_{11}k_{22} - k_{12}^2)x_2^2 \tag{10.16c}$$

Let's examine Equation 10.16*c* to see what conditions the k_{ij} must satisfy so that $U > 0$ for any $\mathbf{x} \neq \mathbf{0}$. Suppose, for instance, that $x_2 = 0$ but $x_1 \neq 0$; it follows that $U > 0$ in Equation 10.16*c* for any x_1 only if $k_{11} > 0$. Now, suppose that $x_2 \neq 0$ but $x_1 = 0$ and, also, that $k_{12} = 0$; it follows that $U > 0$ in Equation 10.16*c* for any x_2 only if $k_{22} > 0$. Finally, suppose that $k_{12} \neq 0$, and let $x_1 = -(k_{12}/k_{11})x_2 \neq 0$; in this case, the first term in Equation 10.16*c* vanishes, and it follows that $U > 0$ for any x_2 only if $k_{11}k_{22} - k_{12}^2 > 0$. We recognize $k_{11}k_{22} - k_{12}^2$ as *det* **k**, i.e., the determinant of **k**. Thus, the condition $k_{11} > 0$, $k_{22} > 0$, $k_{11}k_{22} - k_{12}^2 > 0$ is exactly the requirement that every principal minor of **k** must be positive. This concept can be extended to the general case of an $n \times n$ matrix.[9]

10.1.4 Convexity of Strain Energy

For a one-degree-of-freedom system, Equation 10.15*d* reduces to $U = kx^2/2$, i.e., a parabolic function of the type shown schematically in Figure 10.5. Similarly, for a two-

[9] The concept is also useful in a somewhat different context, namely that of monitoring the numerical behavior of a stiffness matrix during solution of equations. For example, during a solution using Gaussian elimination, the components of the principal diagonal of **k** are successively decreased in value. If any of these components becomes zero or negative, the implication is either that computational errors are excessive or that there is a flaw in the analytic model of the structure leading to a violation of the requirement that $U > 0$.

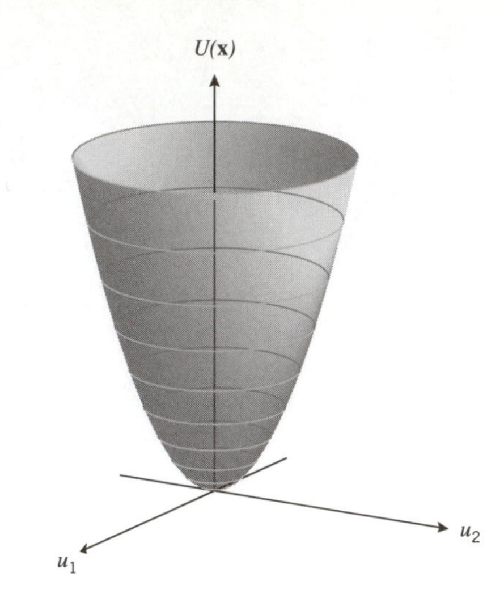

Figure 10.6 Strain energy for two degrees of freedom.

degree-of-freedom system, Equation 10.15*d* reduces to Equation 10.16*b*, which defines a parabolic surface (Figure 10.6). In fact, Equation 10.15*d* defines a ''parabolic surface'' in the general case of n degrees of freedom. Although this is impossible to represent graphically, we can prove that Equation 10.15*d* belongs to the class of convex functions; the significance is that a minimum of a convex function is unique.[10] This property is important for subsequent interpretations of the behavior of structures when approximations exist in the analytic model.

To demonstrate the nature of convexity, consider the strain energy in Figure 10.5. Let $U(x_1)$ and $U(x_2)$ denote values of strain energy for any two different values of x, and let $U(x)$ denote strain energy at any point along the line between x_1 and x_2. It is apparent from the geometry of Figure 10.5 that

$$U(x) < \tilde{U}(x) \tag{10.17}$$

where $\tilde{U}(x)$ is the linear interpolation function between $U(x_1)$ and $U(x_2)$. This is the definition of a convex function except that it must be generalized to n dimensions.

For the general case, where $\mathbf{x}$ denotes the n dimensional vector of displacement degrees of freedom, we wish to show that the strain energy at a point $\mathbf{x}$ that lies on the straight line between any two points $\mathbf{x}_1$ and $\mathbf{x}_2$ is less than that of the linear interpolation between the points. As shown in Figure 10.7, the point $\mathbf{x}$ that lies on the straight line between $\mathbf{x}_1$ and $\mathbf{x}_2$ can be defined by

$$\mathbf{x} = \alpha\mathbf{x}_1 + (1 - \alpha)\mathbf{x}_2 \tag{10.18}$$

where $0 < \alpha < 1$ is a scalar parameter. By extension of the relationship in Equation 10.17, the function U is said to be convex[11] provided

[10] This may be seen by contradiction: if there is more than one minimum (local minima), then a line joining any two local minima must lie partly outside the convex region. (See developments that follow.)

[11] More precisely, for this definition the function is said to be strictly convex.

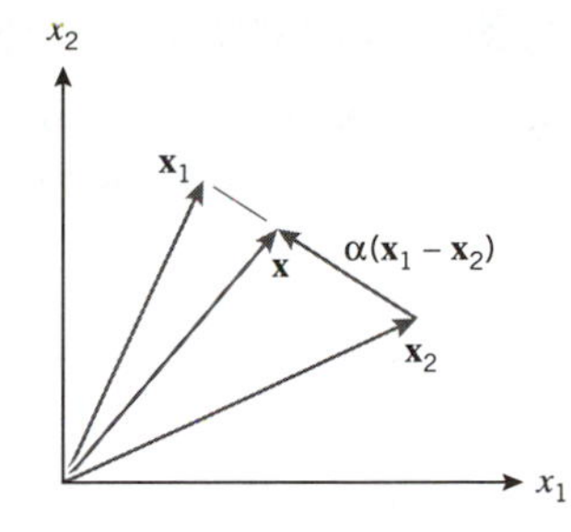

Figure 10.7 Parametric representation of point $\mathbf{x}$ (in 2-D space).

$$U[\alpha\mathbf{x}_1 + (1 - \alpha)\mathbf{x}_2] < \alpha U(\mathbf{x}_1) + (1 - \alpha)\, U(\mathbf{x}_2) \equiv \tilde{U}(\mathbf{x}) \qquad \textbf{(10.19)}$$

As noted, the right side of this inequality is the linear interpolation between $U(\mathbf{x}_1)$ and $U(\mathbf{x}_2)$, i.e., $\tilde{U}(\mathbf{x})$.

To show that inequality 10.19 holds for the strain energy function, we substitute Equation 10.15*d* into the left side of the inequality; this gives

$$\frac{1}{2}[\alpha\mathbf{x}_1 + (1 - \alpha)\mathbf{x}_2]^{\mathrm{T}}\mathbf{k}[\alpha\mathbf{x}_1 + (1 - \alpha)\mathbf{x}_2] < \alpha U(\mathbf{x}_1) + (1 - \alpha)\, U(\mathbf{x}_2) \qquad \textbf{(10.20a)}$$

Expanding leads to

$$\frac{1}{2}[\alpha^2\mathbf{x}_1^{\mathrm{T}}\mathbf{k}\mathbf{x}_1 + (1 - \alpha)^2\mathbf{x}_2^{\mathrm{T}}\mathbf{k}\mathbf{x}_2 + 2\alpha(1 - \alpha)\mathbf{x}_1^{\mathrm{T}}\mathbf{k}\mathbf{x}_2] < \alpha U(\mathbf{x}_1) + (1 - \alpha)\, U(\mathbf{x}_2) \qquad \textbf{(10.20b)}$$

or

$$\alpha^2 U(\mathbf{x}_1) + (1 - \alpha)^2 U(\mathbf{x}_2) + \alpha(1 - \alpha)\mathbf{x}_1^{\mathrm{T}}\mathbf{k}\mathbf{x}_2 < \alpha U(\mathbf{x}_1) + (1 - \alpha)\, U(\mathbf{x}_2) \qquad \textbf{(10.20c)}$$

Inequality 10.20*c* may be simplified to

$$-U(\mathbf{x}_1) + \mathbf{x}_1^{\mathrm{T}}\mathbf{k}\mathbf{x}_2 - U(\mathbf{x}_2) < 0 \qquad \textbf{(10.20d)}$$

or, combining terms

$$-\left[\frac{1}{2}(\mathbf{x}_1 - \mathbf{x}_2)^{\mathrm{T}}\mathbf{k}(\mathbf{x}_1 - \mathbf{x}_2)\right] < 0 \qquad \textbf{(10.20e)}$$

The term in brackets is equivalent to $U(\mathbf{z})$ where $\mathbf{z} = \mathbf{x}_1 - \mathbf{x}_2$, and is always positive by definition. Therefore, the original inequality, Equation 10.19, is proved.

10.1.5 Partial Derivative of Strain Energy; Castigliano's Theorem, Part I

For a linear elastic structure with n degrees of freedom, the strain energy defined by Equation 10.15d may be written in expanded form,

$$\begin{aligned} U = \frac{1}{2}(&k_{11}x_1^2 + k_{12}x_1x_2 + \ldots + k_{1j}x_1x_j + \ldots + k_{1n}x_1x_n \\ &+ k_{21}x_1x_2 + k_{22}x_2^2 + \ldots + k_{2j}x_2x_j + \ldots + k_{2n}x_2x_n \\ &\vdots \\ &+ k_{j1}x_1x_j + k_{j2}x_2x_j + \ldots + k_{jj}x_j^2 + \ldots + k_{jn}x_jx_n \\ &\vdots \\ &+ k_{n1}x_1x_n + k_{n2}x_2x_n + \ldots + k_{nj}x_jx_n + \ldots + k_{nn}x_n^2) \end{aligned} \tag{10.21a}$$

where $x_i = u_i - u_{i_o}$, $i = 1, \ldots, n$. The first partial derivative of U with respect to any displacement u_j is the same as the first partial derivative with respect to x_j,

$$\frac{\partial U}{\partial u_j} = \frac{k_{j1} + k_{1j}}{2}x_1 + \frac{k_{j2} + k_{2j}}{2}x_2 + \cdots + k_{jj}x_j + \cdots + \frac{k_{jn} + k_{nj}}{2}x_n \tag{10.21b}$$

If $k_{ji} = k_{ij}$, the right-hand side of Equation 10.21b is the stiffness portion of the jth equation of equilibrium for the structure that, by definition, is equal to the generalized force f_j associated with u_j; therefore, we must have

$$\frac{\partial U}{\partial u_j} \equiv f_j, \qquad j = 1, \ldots, n \tag{10.21c}$$

The set of all such equations has the vector form

$$\frac{\partial U}{\partial \mathbf{u}} = \mathbf{F} \tag{10.21d}$$

In other words, the partial derivatives of the strain energy function provide the equilibrium equations. This result, Equation 10.21c or Equation 10.21d, is known as Castigliano's theorem, Part I.[12] As a consequence of Equation 10.21b, the theorem is only valid if $\mathbf{k}$ is symmetric. At this point, Castigliano's theorem appears to provide little new information since we already have the equilibrium equations, Equation 10.1a.

---

[12] Also referred to as Castigliano's first theorem.

10.1.6 Deformation Energy; Castigliano's Theorem, Part II

The strain energy of a linear elastic body may also be written in terms of the force variables. If we substitute $\mathbf{u} - \mathbf{u_o} = \mathbf{aF}$ (Equation 10.1*b*) into Equation 10.15*d*, and recognize that $\mathbf{a} = \mathbf{k}^{-1}$ and $\mathbf{a}^T = \mathbf{a}$, we obtain

$$U = \frac{1}{2}\mathbf{F}^T\mathbf{aF} \qquad \textbf{(10.22}\textbf{\textit{a}}\textbf{)}$$

It is useful to define a slightly different energy form called the deformation energy, U_d, such that[13]

$$U_d = \frac{1}{2}\mathbf{F}^T\mathbf{aF} + \mathbf{F}^T\mathbf{u_o} \qquad \textbf{(10.22}\textbf{\textit{b}}\textbf{)}$$

The additional term $\mathbf{F}^T\mathbf{u_o}$ is sometimes referred to as the "free energy."[14] The deformation energy has first partial derivatives that provide the compatibility equations; since u_{jo}, $j = 1, \ldots, n$, is a specified constant (initial) displacement, it follows from Equation 10.22*b* that

$$\frac{\partial U_d}{\partial f_j} = \frac{a_{j1} + a_{1j}}{2} f_1 + \frac{a_{j2} + a_{2j}}{2} f_2 + \cdots + a_{jj} f_j + \cdots + \frac{a_{jn} + a_{nj}}{2} f_n + a_{jo} \qquad \textbf{(10.23}\textbf{\textit{a}}\textbf{)}$$

If $a_{ij} = a_{ji}$, then the right-hand side of Equation 10.23*a* is the right-hand side of the jth equation in Equation 10.1*b*, and must therefore be equal to u_j; in other words

$$\frac{\partial U_d}{\partial f_j} = u_j, \qquad j = 1, \ldots, n \qquad \textbf{(10.23}\textbf{\textit{b}}\textbf{)}$$

The set of all such equations has the vector form

$$\frac{\partial U_d}{\partial \mathbf{F}} = \mathbf{u} \qquad \textbf{(10.23}\textbf{\textit{c}}\textbf{)}$$

This result is known as Castigliano's theorem, part II.[15]

---

[13] For a one-degree-of-freedom system, it may be recognized that this definition has the physical interpretation of area above and to the left of the force-displacement diagram. In the special case when $\mathbf{u_o} = \mathbf{0}$, this form is often referred to as complementary energy.

[14] We may give this term a physical interpretation by imagining the structure as having no $\mathbf{u_o}$; the work done during deformation would be $U = (1/2)\mathbf{F}^T\mathbf{aF}$. If the initial displacements were then permitted to develop, the forces already in place would do extra work $\mathbf{F}^T\mathbf{u_o}$, a sort of "free energy."

[15] Also referred to as Castigliano's second theorem.

10.1.7 External Potential Energy

As defined in Equation 10.3*b*, the internal energy U is equal to the work done by the external forces as the body is loaded; energy U_d is also closely related to this work.

For the developments that follow, it is convenient to define a somewhat different kind of energy associated with the external forces. Rather than relating energy to actual work done by the forces, we consider the work that would be done by the forces assuming they were constant throughout the displacement process. For such constant forces, the work done during displacement would be

$$W = (\mathbf{u} - \mathbf{u}_o)^T \mathbf{F} \qquad \textbf{(10.24}\textbf{\textit{a}}\textbf{)}$$

where $\mathbf{u}_o$ denotes the initial reference point about which this work is measured, i.e., we assume $W = 0$ when $\mathbf{u} = \mathbf{u}_o$. We define an external potential energy, U_e, as exactly the negative of this work,

$$U_e = -(\mathbf{u} - \mathbf{u}_o)^T \mathbf{F} \qquad \textbf{(10.24}\textbf{\textit{b}}\textbf{)}$$

We note that

$$\frac{\partial U_e}{\partial u_j} = -f_j \qquad \textbf{(10.24}\textbf{\textit{c}}\textbf{)}$$

and

$$\frac{\partial U_e}{\partial f_j} = -(u_j - u_{j_o}) \qquad \textbf{(10.24}\textbf{\textit{d}}\textbf{)}$$

Given the energy forms defined in Equations 10.15*d* and 10.24*b*, we may proceed to develop several important classical energy theorems of structural analysis.

10.1.8 Total Potential Energy; Stationary and Minimum Values

Energy effects associated with the applied forces (i.e., the external potential energy U_e), and the strain energy (U) absorbed and stored by the elastic body, may be summed to define a total potential energy V,

$$V = U + U_e \qquad \textbf{(10.25}\textbf{\textit{a}}\textbf{)}$$

Substituting Equations 10.15*d* and 10.24*b* gives

$$V = \frac{1}{2}(\mathbf{u} - \mathbf{u}_o)^T \mathbf{k}(\mathbf{u} - \mathbf{u}_o) - (\mathbf{u} - \mathbf{u}_o)^T \mathbf{F} \qquad \textbf{(10.25}\textbf{\textit{b}}\textbf{)}$$

If we take the first partial derivative of V with respect to displacement u_j we obtain

$$\frac{\partial V}{\partial u_j} = k_{j1}(u_1 - u_{1_o}) + k_{j2}(u_2 - u_{2_o}) + \ldots$$
$$+ k_{jj}(u_j - u_{j_o}) + \ldots + k_{jn}(u_n - u_{n_o}) - f_j \qquad \textbf{(10.25}\textbf{\textit{c}}\textbf{)}$$

Assembling the set of all such equations in vector form leads to

$$\frac{\partial V}{\partial \mathbf{u}} = \mathbf{k}(\mathbf{u} - \mathbf{u}_o) - \mathbf{F} \tag{10.25d}$$

We observe that the right side of Equation 10.25*d* is identically equal to zero provided **k** is symmetric and Equation 10.1*a*, the equilibrium equation for the structure, is satisfied. It follows from Equation 10.25*d* that if the structure is in a state of equilibrium then

$$\boxed{\frac{\partial V}{\partial \mathbf{u}} = \mathbf{0}} \tag{10.25e}$$

Equation 10.25*e* is the mathematical definition of a stationary point of the function *V*, i.e., a point at which the gradient of *V* is zero. In other words, Equation 10.25*e* shows that the total potential energy (V) has a stationary value when the structure is in equilibrium. This result is referred to as the theorem of stationary value of total potential energy.

The significance of this result is illustrated in Figure 10.8 for a one-degree-of-freedom structure. Schematic plots of U, U_e, and their sum V, are shown as functions of variable u. It may be seen that, as u increases beyond u_o, the increasingly negative external potential (U_e) reduces the value of V until the reductions are offset by increases in strain energy (U); this is the stationary point at which equilibrium is satisfied.

As should be apparent from Figure 10.8, the point $\mathbf{u}^*$, which corresponds to the stationary value of V, also gives the minimum value of the total potential energy. This is a general result. It is an easy matter to show that V, Equation 10.25*a*, is a convex function just like U (see Section 10.1.4); it follows that the stationary point must, in fact, be a unique (global) minimum. Thus, we conclude that the displacements $\mathbf{u}^*$ that minimize the total potential energy are the displacements that produce equilibrium. This result is referred to as the theorem of minimum total potential energy.

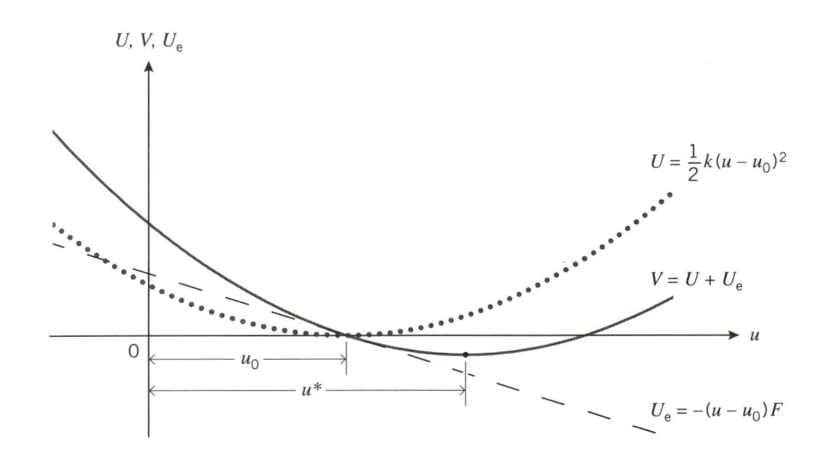

Figure 10.8 Potential surfaces for one-degree-of-freedom structure.

The actual minimum value of V associated with equilibrium is obtained by substituting Equation 10.1a into Equation 10.25b; the result is

$$V_{min} \equiv V(\mathbf{u}^*) = -\frac{1}{2}(\mathbf{u}^* - \mathbf{u}_o)^T\mathbf{k}(\mathbf{u}^* - \mathbf{u}_o) \qquad \mathbf{(10.26a)}$$

We see that this minimum value of total potential energy is the negative of the strain energy stored by the structure in its equilibrium configuration,

$$V(\mathbf{u}^*) = -U(\mathbf{u}^*) \qquad \mathbf{(10.26b)}$$

10.2 EFFECTS OF APPROXIMATIONS

Most of the conclusions reached in the preceding sections are quite reasonable when viewed from a qualitative perspective. For example, Equation 10.26b shows that the total potential energy lost during loading is stored as internal strain energy by an elastic structure. Nevertheless, apart from the introduction of various energy forms and the development of some additional ways of viewing equilibrium, so far little has been gained from energy concepts beyond the demonstration of requirements that stiffness (or flexibility) matrices must be symmetric and positive definite.

For our purposes, the most important application of energy concepts will be in assessing the effects on numerical results of approximations (or errors) in the analytic model of a structure. Initially, we will directly examine the effects of approximations in the displacement method; in a subsequent section, we will also examine the effects of approximations in the force method from the viewpoint of the displacement method.

10.2.1 Constrained Displacements

We will begin with an assessment of the effects of an incomplete description of the motion of a structure. In other words, we will assume that the behavior is described by fewer kinematic degrees of freedom than actually exist; this is equivalent to assuming that some possible motions are prevented or constrained.

Consider the two-degree-of-freedom structure in Figure 10.3, which has a total potential energy given by

$$\begin{aligned} V = {} & \frac{1}{2}k_{11}(u_1 - u_{1_o})^2 + k_{12}(u_1 - u_{1_o})(u_2 - u_{2_o}) \\ & + \frac{1}{2}k_{22}(u_2 - u_{2_o})^2 - f_1(u_1 - u_{1_o}) - f_2(u_2 - u_{2_o}) \end{aligned} \qquad \mathbf{(10.27)}$$

This potential energy function is shown schematically in Figure 10.9; it is similar to the function in Figure 10.6 except that the additional linear terms $-f_1(u_1 - u_{1_o}) - f_2(u_2 - u_{2_o})$ are included.[16] The minimum point of this function, $\mathbf{u}^*$, is the true equilibrium point.

---

[16] In fact, it is exactly the same except that the vertex of the parabolic surface is shifted to the equilibrium point $\mathbf{u}^* = \mathbf{k}^{-1}\mathbf{F} + \mathbf{u}_o$.

Now suppose that for some reason (e.g., computing limitations) we do not define the correct analytic model with two degrees of freedom given in Equation 10.27. Instead, we define a model with only one degree of freedom. This might be done by assigning any arbitrary value (including zero) to one of the system displacements. For example, we might set $u_1 = u_{1_o}$; then the total potential energy for this analytic model would reduce from the expression in Equation 10.27 to an approximate expression $\tilde{V}$, which, in this particular case, is given by

$$\tilde{V} = \frac{1}{2}k_{22}(u_2 - u_{2_o})^2 - f_2(u_2 - u_{2_o}) \tag{10.28}$$

Note that $\tilde{V}$ is a function of u_2 only.

This approximate total potential energy function has a physical interpretation in Figure 10.9; it is the curve defined by the intersection of the function V and the plane $u_1 = u_{1_o}$. We observe from the figure (and from Equation 10.28) that $\tilde{V}$ has a minimum at the value of u_2 that corresponds to the stationary point of $\tilde{V}$, i.e., the value that satisfies the equation

$$\frac{\partial \tilde{V}}{\partial u_2} = 0 = k_{22}(u_2 - u_{2_o}) - f_2 \tag{10.29}$$

We denote this value of u_2 by $\tilde{u}_2^*$, and denote the actual coordinates of the point by $\tilde{\mathbf{u}}^*$ where $\tilde{\mathbf{u}}^{*\mathrm{T}} = \{u_{1_o}\ \tilde{u}_2^*\}$.

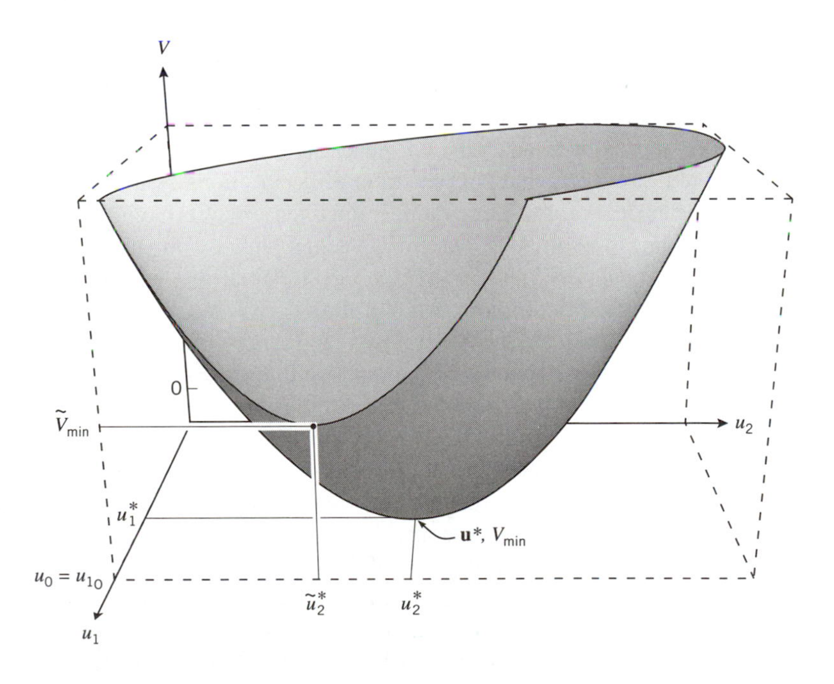

Figure 10.9 Potential energy function for two-degree-of-freedom structure; approximate one-degree-of-freedom potential energy function (for $u_1 = u_{10}$) is shown as parabola in front plane.

Unless it (coincidentally) happens that the true value of u_1 is u_{1_o} (an unlikely event), the minimum point on the curve of $\tilde{V}$ will not be the true equilibrium point for the actual structure. Thus, the point $\tilde{\mathbf{u}}^*$ represents displacements corresponding to an approximate equilibrium state.

We note that the minimum value of $\tilde{V}$ must necessarily be greater than the minimum value of V,

$$\tilde{V}_{min} \equiv V(\tilde{\mathbf{u}}^*) > V(\mathbf{u}^*) \qquad \textbf{(10.30)}$$

This is a general consequence of the theorem of minimum total potential energy. Since the minimum value of either V or $\tilde{V}$ is the negative of the strain energy in the structure in its state of true or approximate equilibrium, it also follows that

$$-U(\tilde{\mathbf{u}}^*) > -U(\mathbf{u}^*) \qquad \textbf{(10.31}\boldsymbol{a}\textbf{)}$$

The strain energy is always positive regardless of whether it is based on a true or an approximate displacement pattern, so we must have

$$U(\tilde{\mathbf{u}}^*) < U(\mathbf{u}^*) \qquad \textbf{(10.31}\boldsymbol{b}\textbf{)}$$

Thus, the effect of leaving out displacement degrees of freedom because we cannot afford to include them (or, possibly, because we forgot to include them) means that the strain energy in the constrained solution will be less than the true value.

This is a worrisome finding. The implication is that when artificial displacement constraints are imposed, the applied forces do not generate as much internal energy as in the true structure. Consequently, in an overall sense, the constrained structure will appear stiffer than the actual structure and, on average, lower internal stresses will be predicted for the constrained structure than for the actual structure. In other words, this type of approximate analysis will generally under-predict the state of stress. It should be noted, however, that stresses in specific locations in the structure may be greater than the correct values. For example, if a fixed support (having no displacement degrees of freedom) is assumed in place of an actual pin support (having a rotational degree of freedom), bending moments will be developed at the support when, in reality, none should exist (see Example 1, which follows).

Structures composed of general two- and three-dimensional elements have infinitely many displacement degrees of freedom, so in most cases the analyst has no choice but to use a smaller (finite) number to represent structural behavior. Thus, analyses will give displacement results that predict too stiff a structural response and therefore, in general, must under-predict the magnitudes of internal stresses.

To obtain some indication of how close an approximate solution may be to a true solution, analysts typically construct two or three analytic models with increasing numbers of displacement degrees of freedom. The internal strain energy is computed after each model is analyzed, and the increases in strain energy are computed as a function of model refinement. If the strain energy appears to be stable, and does not increase significantly with increasing degrees of freedom, then the analyst can have some confidence that the numerical solution is converging to a reliable and accurate result and that stresses are converging to true values.

EXAMPLE 2

Consider the structure in Figure 10.10*a*, which has the three displacement degrees of freedom, u_1, u_2, u_3, shown. The structure is loaded by moments applied at B and C.

The equilibrium equation, Equation 10.1*a*, is

$$\frac{EI}{l}\begin{bmatrix} 4 & 2 & 0 \\ 2 & 8 & 2 \\ 0 & 2 & 4 \end{bmatrix}\begin{Bmatrix} u_1 \\ u_2 \\ u_3 \end{Bmatrix} = \begin{Bmatrix} 0 \\ Fl/8 \\ -Fl/8 \end{Bmatrix} \qquad \textbf{(10.32}\boldsymbol{a}\textbf{)}$$

from which we obtain

$$\mathbf{u}^* \equiv \begin{Bmatrix} u_1 \\ u_2 \\ u_3 \end{Bmatrix} = \frac{Fl^2}{64EI}\begin{Bmatrix} -1 \\ 2 \\ -3 \end{Bmatrix} \qquad \textbf{(10.32}\boldsymbol{b}\textbf{)}$$

Therefore, the strain energy in the structure, Equation 10.15*d*, is

$$U(\mathbf{u}^*) = \left(\frac{1}{2}\right)\left(\frac{Fl^2}{64EI}\right)^2\left(\frac{EI}{l}\right)\{-1 \quad 2 \quad -3\}\begin{bmatrix} 4 & 2 & 0 \\ 2 & 8 & 2 \\ 0 & 2 & 4 \end{bmatrix}\begin{Bmatrix} -1 \\ 2 \\ -3 \end{Bmatrix} \qquad \textbf{(10.32}\boldsymbol{c}\textbf{)}$$

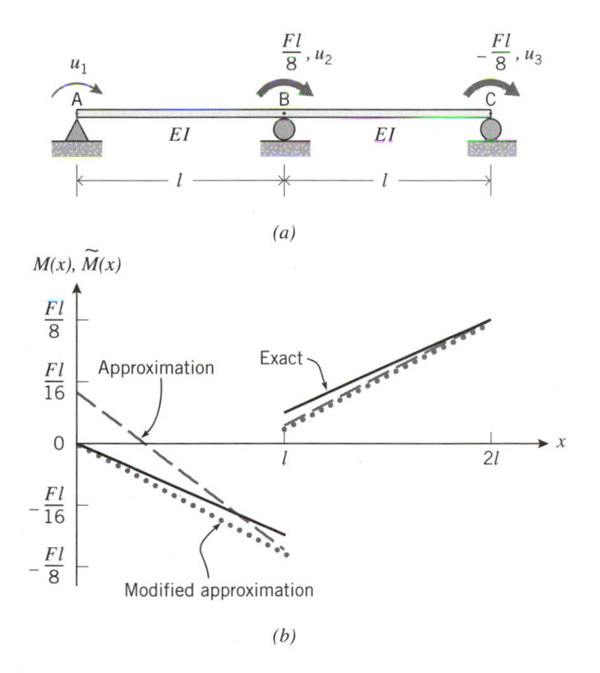

Figure 10.10 Example 2.

$$= \frac{5F^2l^3}{1024EI} = 0.004883\frac{F^2l^3}{EI}$$

Now, we will repeat the analysis, this time omitting displacement coordinate u_1, which is equivalent to setting $u_1 = 0$. This implies that there is an artificial fixed boundary condition at A. As a result, Equation 10.32*a* reduces to

$$\begin{bmatrix} 8 & 2 \\ 2 & 4 \end{bmatrix} \begin{Bmatrix} \tilde{u}_2 \\ \tilde{u}_3 \end{Bmatrix} = \frac{Fl}{8} \begin{Bmatrix} 1 \\ -1 \end{Bmatrix} \tag{10.32d}$$

We use the symbols $\tilde{u}_2$ and $\tilde{u}_3$ to denote the approximate values of the remaining nonzero system displacements, which we expect to be different from the true values of u_2 and u_3 in Equation 10.32*b*.

Solving Equation 10.32*d* leads to

$$\tilde{\mathbf{u}}^* \equiv \begin{Bmatrix} \tilde{u}_2 \\ \tilde{u}_3 \end{Bmatrix} = \frac{Fl^2}{112EI} \begin{Bmatrix} 3 \\ -5 \end{Bmatrix} \tag{10.32e}$$

from which we obtain

$$U(\tilde{\mathbf{u}}^*) = \left(\frac{1}{2}\right)\left(\frac{Fl^2}{112EI}\right)^2\left(\frac{EI}{l}\right)\{3 \quad -5\}\begin{bmatrix} 8 & 2 \\ 2 & 4 \end{bmatrix}\begin{Bmatrix} 3 \\ -5 \end{Bmatrix} = \frac{F^2l^3}{224EI} = 0.00446\frac{F^2l^3}{EI} \tag{10.32f}$$

which demonstrates that

$$U(\tilde{\mathbf{u}}^*) < U(\mathbf{u}^*) \tag{10.32g}$$

The internal bending moments may also be compared for the two solutions. From the exact analysis we obtain

$$\begin{Bmatrix} M_{AB} \\ M_{BA} \end{Bmatrix} = \frac{EI}{l}\begin{bmatrix} 4 & 2 \\ 2 & 4 \end{bmatrix}\begin{Bmatrix} u_1 \\ u_2 \end{Bmatrix} = \begin{Bmatrix} 0 \\ 3FL/32 \end{Bmatrix} \tag{10.33a}$$

and

$$\begin{Bmatrix} M_{BC} \\ M_{CB} \end{Bmatrix} = \frac{EI}{l}\begin{bmatrix} 4 & 2 \\ 2 & 4 \end{bmatrix}\begin{Bmatrix} u_2 \\ u_3 \end{Bmatrix} = \begin{Bmatrix} Fl/32 \\ -FL/8 \end{Bmatrix} \tag{10.33b}$$

The approximate solution gives

$$\begin{Bmatrix} \tilde{M}_{AB} \\ \tilde{M}_{BA} \end{Bmatrix} = \frac{EI}{l}\begin{bmatrix} 4 & 2 \\ 2 & 4 \end{bmatrix}\begin{Bmatrix} 0 \\ \tilde{u}_2 \end{Bmatrix} = \begin{Bmatrix} 3Fl/56 \\ 3FL/28 \end{Bmatrix} \tag{10.33c}$$

and

$$\begin{Bmatrix} \tilde{M}_{BC} \\ \tilde{M}_{CB} \end{Bmatrix} = \frac{EI}{l}\begin{bmatrix} 4 & 2 \\ 2 & 4 \end{bmatrix}\begin{Bmatrix} \tilde{u}_2 \\ \tilde{u}_3 \end{Bmatrix} = \begin{Bmatrix} Fl/56 \\ -FL/8 \end{Bmatrix} \tag{10.33d}$$

The bending moment diagrams are compared in Figure 10.10*b*. We see that the approximate solution does produce a somewhat lower internal moment on average than the exact solution; however, this is not the case everywhere in the structure. As expected, the approximate solution gives poor results at A, indicating an artificially high moment rather than zero. Moment equilibrium is preserved at B and C in the approximate solution, although the values of $\tilde{M}_{BA}$ and $\tilde{M}_{BC}$ are perturbed by the constraint at A.

Despite the differences, the comparison shows that approximate analyses can give good qualitative results, especially at points well removed from the locations of artificial constraints in the model. The tendency of the artificial constraint to ''attract'' stress, although fairly obvious in this case, is actually quite general. It can be used to advantage to gain insight into the accuracy of an approximate analysis. For example, if an accurate approximate analysis had somehow provided $\tilde{M}_{AB} \approx 0$, then we would have had confidence that there was relatively little error in the solution. Conversely, if $\tilde{M}_{AB} \neq 0$, then we can ask ourselves if the magnitude of the error is important compared to the other moments in the structure; the answer would determine whether or not a more accurate (and possibly more complex) analysis should be initiated. In the current example, based on the results of the two-degree-of-freedom model only, we would have to conclude that the solution might have significant errors since $|\tilde{M}_{AB}|$ is approximately 40% of the maximum moment (at C); but, after performing the exact analysis, we would see that the approximate solution is not so bad after all, except at point A where we know what the answer should be. Thus, we could have used the approximate solution and simply adjusted the bending moment in element AB based on our knowledge of the correct value at A. Such a modified approximate solution is shown in Figure 10.10*b*, and appears quite reasonable.

10.2.2 Incompatible Displacements

The converse of the situation examined above occurs when an analytic model is described by more displacement degrees of freedom than are physically present. In this case, the implication is that some existing kinematic constraints are released. This, in turn, implies that compatibility conditions will not be properly imposed in the analytic model.

For instance, imagine that the structure in Figure 10.10*a* actually has a fixed support at A, and suppose that the structure is modeled as one that has a rotational degree of freedom at A; the exact and approximate solutions in Example 2 would then be interchanged. It must follow that the minimum value of the total potential energy for a structure with extra (incompatible) displacements will be less than the true minimum value,

$$V(\tilde{\mathbf{u}}^*) < V(\mathbf{u}^*) \qquad \textbf{(10.34a)}$$

and that

$$U(\tilde{\mathbf{u}}^*) > U(\mathbf{u}^*) \qquad \textbf{(10.34b)}$$

In other words, the analytic model of the structure with the additional degrees of freedom will appear more flexible than the true model, and will have a greater strain energy for a given set of loads. Consequently, the model with too many (nonphysical or incompatible) displacement degrees of freedom will over-predict internal stresses, on average.

The fact that incompatible displacements make the structure appear too flexible, and that constrained displacements make the structure appear too stiff, invites one to ask if it might not be possible to intentionally mix both effects and get reasonable results by having one offset the other. Such a procedure is sometimes employed in conjunction with advanced techniques in finite element analysis; however, it is a difficult concept to use effectively.

10.2.3 Approximations in the Force Method

In the force method, we begin with equilibrium equations involving known applied loads and unknown redundant forces ($\mathbf{X}$) acting on a statically determinate substructure. For arbitrary values of the components of $\mathbf{X}$ (including zero), the structure exhibits incompatible motions associated with the redundant forces. An exact analysis utilizing compatibility equations leads to values of the redundant forces for which the incompatible motions vanish, i.e., are zero. It follows that any approximation of $\mathbf{X}$ that differs from the exact solution must produce incompatible motions; therefore, the strain energy obtained from the approximate solution will be greater than that for the exact solution (see Section 10.2.2) and stresses will generally be overestimated.

This type of behavior may be examined in greater detail using the displacement method approach employed in the preceding sections. Consider, for instance, the structure in Figure 10.11*a* that has two kinematic degrees of freedom. Figure 10.11*b* shows the possible reactions at B and C, and also two additional degrees of freedom associated with motions in the directions of the reactions; in the force method of analysis the reactions would be the redundants ($\mathbf{X}$) and the additional displacements would be (incompatible) motions associated with arbitrary values of $\mathbf{X}$.

Assuming any constant values of forces X_i, which we associate with redundants, a structure like the one in Figure 10.11*b* could be analyzed by the displacement method; the equilibrium equation would have the form

$$\left[\begin{array}{c|c} \mathbf{k}_{uu} & \mathbf{k}_{u\Delta} \\ \hline \mathbf{k}_{u\Delta}{}^{T} & \mathbf{k}_{\Delta\Delta} \end{array}\right] \left(\left\{\begin{array}{c} \mathbf{u} \\ \hline \boldsymbol{\Delta} \end{array}\right\} - \left\{\begin{array}{c} \mathbf{u}_o \\ \hline \boldsymbol{\Delta}_o \end{array}\right\} \right) = \left\{\begin{array}{c} \mathbf{f} \\ \hline \mathbf{X} \end{array}\right\} \qquad \mathbf{(10.35}\boldsymbol{a}\mathbf{)}$$

where the components of $\mathbf{u}$ are the physical displacements of the structure without releases (''cuts''), e.g., displacements of the structure in Figure 10.11*a*; the components of $\mathbf{u}_o$ are initial displacements; the components of $\mathbf{f}$ are the applied physical loads, e.g., f_1 and f_2 in Figure 10.11*a*; the components of $\boldsymbol{\Delta}$ are the incompatible motions at the releases, e.g., Δ_1 and Δ_2 in Figure 10.11*b*. The components associated with initial incompatibilities, $\boldsymbol{\Delta}_o$, are present if the statically determinate ''cut'' structure has initial ''lack of fit'' across the cuts. The sign of $\boldsymbol{\Delta}_o$ is the same as that of $\boldsymbol{\Delta}$. Note that there will be no internal energy when $\mathbf{f}$ and $\mathbf{X}$ are both zero and $\mathbf{u} = \mathbf{u}_o$ and $\boldsymbol{\Delta} = \boldsymbol{\Delta}_o$, consistent with Equation 10.3*b*. The stiffness matrix in Equation 10.35*a* is partitioned in a manner consistent with the dimensions of the various submatrices in the force and displacement vectors.

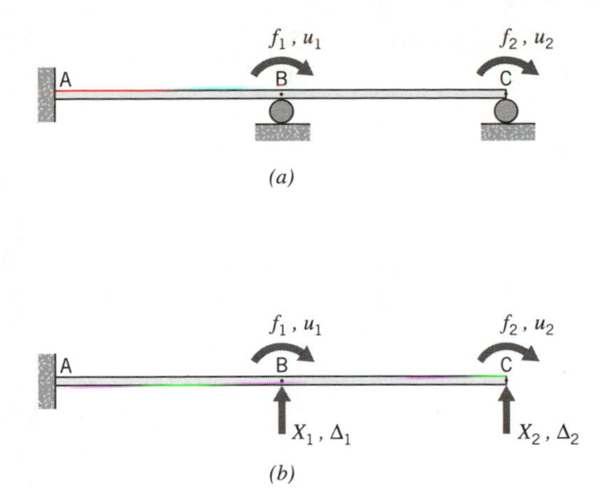

Figure 10.11 Forces and displacements for typical structure.

Equation 10.35*a* may be solved for the displacements,

$$\begin{Bmatrix} \mathbf{u} \\ -- \\ \mathbf{\Delta} \end{Bmatrix} = \begin{bmatrix} \mathbf{k_{uu}} & | & \mathbf{k_{u\Delta}} \\ -- & + & -- \\ \mathbf{k_{u\Delta}}^T & | & \mathbf{k_{\Delta\Delta}} \end{bmatrix}^{-1} \begin{Bmatrix} \mathbf{f} \\ -- \\ \mathbf{X} \end{Bmatrix} + \begin{Bmatrix} \mathbf{u_o} \\ -- \\ \mathbf{\Delta_o} \end{Bmatrix} \quad \textbf{(10.35}b\textbf{)}$$

or, in flexibility form,

$$\begin{Bmatrix} \mathbf{u} \\ -- \\ \mathbf{\Delta} \end{Bmatrix} = \begin{bmatrix} \mathbf{a_{ff}} & | & \mathbf{a_{fX}} \\ -- & + & -- \\ \mathbf{a_{fX}}^T & | & \mathbf{a_{XX}} \end{bmatrix} \begin{Bmatrix} \mathbf{f} \\ -- \\ \mathbf{X} \end{Bmatrix} + \begin{Bmatrix} \mathbf{u_o} \\ -- \\ \mathbf{\Delta_o} \end{Bmatrix} \quad \textbf{(10.35}c\textbf{)}$$

In Equation 10.35*c*, it may be shown that

$$\begin{aligned} [\mathbf{a_{ff}}] &= ([\mathbf{k_{uu}}] - [\mathbf{k_{u\Delta}}][\mathbf{k_{\Delta\Delta}}]^{-1}[\mathbf{k_{u\Delta}}]^T)^{-1} \\ [\mathbf{a_{fX}}] &= -[\mathbf{a_{ff}}][\mathbf{k_{u\Delta}}][\mathbf{k_{\Delta\Delta}}]^{-1} \\ [\mathbf{a_{fX}}]^T &= -[\mathbf{a_{XX}}][\mathbf{k_{u\Delta}}]^T[\mathbf{k_{uu}}]^{-1} \\ [\mathbf{a_{XX}}] &= ([\mathbf{k_{\Delta\Delta}}] - [\mathbf{k_{u\Delta}}]^T[\mathbf{k_{uu}}]^{-1}[\mathbf{k_{u\Delta}}])^{-1} \end{aligned} \quad \textbf{(10.36)}$$

It should be understood that the displacement vector in the above equations satisfies equilibrium exactly for any (constant) value of $\mathbf{X}$ and any value of $\mathbf{\Delta_o}$.

The minimum total potential energy corresponding to the equilibrium solution is given by Equations 10.26*a* and 10.35*a*,

$$V_{min} = -\frac{1}{2}\left(\begin{Bmatrix} \mathbf{u} \\ -- \\ \mathbf{\Delta} \end{Bmatrix} - \begin{Bmatrix} \mathbf{u_o} \\ -- \\ \mathbf{\Delta_o} \end{Bmatrix}\right)^T \begin{Bmatrix} \mathbf{f} \\ -- \\ \mathbf{X} \end{Bmatrix} \quad \textbf{(10.37}a\textbf{)}$$

or, if we substitute Equation 10.35*c*,

$$V_{min} = -U_{equil} = -\frac{1}{2}\left\{\begin{array}{c}\mathbf{f}\\ \hline \mathbf{X}\end{array}\right\}^{\mathrm{T}}\left[\begin{array}{c|c}\mathbf{a}_{\mathrm{ff}} & \mathbf{a}_{\mathrm{fX}}\\ \hline \mathbf{a}_{\mathrm{fX}}^{\mathrm{T}} & \mathbf{a}_{\mathrm{XX}}\end{array}\right]\left\{\begin{array}{c}\mathbf{f}\\ \hline \mathbf{X}\end{array}\right\} \tag{10.37b}$$

Note that V_{min} is the negative of the strain energy in the equilibrium state (U_{equil}), Equation 10.26*b*. We see that U_{equil} is a positive definite quadratic function of **X**, which we may write in expanded form as

$$U_{equil} = \frac{1}{2}\mathbf{f}^{\mathrm{T}}\mathbf{a}_{\mathrm{ff}}\mathbf{f} + \mathbf{X}^{\mathrm{T}}\mathbf{a}_{\mathrm{fX}}^{\mathrm{T}}\mathbf{f} + \frac{1}{2}\mathbf{X}^{\mathrm{T}}\mathbf{a}_{\mathrm{XX}}\mathbf{X} \tag{10.38}$$

Equation 10.38 gives the strain energy for a structure like the one in Figure 10.11*b*, i.e., a structure with arbitrary (constant) values of **X** permitted at the locations of releases. Suppose we ask the following question: Of all the values of **X** we could choose, which value will minimize the strain energy in its equilibrium state? Since U_{equil} is a positive definite quadratic function, we may find the minimum by setting the partial derivative equal to zero,

$$\frac{\partial U_{equil}}{\partial \mathbf{X}} = \mathbf{0} = \mathbf{a}_{\mathrm{fX}}^{\mathrm{T}}\mathbf{f} + \mathbf{a}_{\mathrm{XX}}\mathbf{X} \tag{10.39}$$

We recognize the expression $\mathbf{0} = \mathbf{a}_{\mathrm{fX}}^{\mathrm{T}}\mathbf{f} + \mathbf{a}_{\mathrm{XX}}\mathbf{X}$ as the second of the two vector equations in Equation 10.35*c* with $\boldsymbol{\Delta} = \boldsymbol{\Delta}_{\mathrm{o}}$. In other words, we see that the values of X_i that minimize U_{equil} are the ones for which the initial incompatibilities $\boldsymbol{\Delta} = \boldsymbol{\Delta}_{\mathrm{o}}$ are retained. This implies that **X** does not contribute to the work done in deforming the structure away from its initial ''cut'' deformation state, i.e., the unloaded structure with **f** and **X** both zero. As a consequence, the expression $\partial U_{equil}/\partial \mathbf{X} = \mathbf{0}$ in Equation 10.39 is referred to as the theorem of least work.

We can also write $U_{equil} = -V_{min}$ in another form by using Equation 10.26*a* and Equation 10.35*a*,

$$U_{equil} = \frac{1}{2}\left\{\begin{array}{c}\mathbf{u}-\mathbf{u}_{\mathrm{o}}\\ \hline \boldsymbol{\Delta}-\boldsymbol{\Delta}_{\mathrm{o}}\end{array}\right\}^{\mathrm{T}}\left[\begin{array}{c|c}\mathbf{k}_{\mathrm{uu}} & \mathbf{k}_{\mathrm{u}\Delta}\\ \hline \mathbf{k}_{\mathrm{u}\Delta}^{\mathrm{T}} & \mathbf{k}_{\Delta\Delta}\end{array}\right]\left\{\begin{array}{c}\mathbf{u}-\mathbf{u}_{\mathrm{o}}\\ \hline \boldsymbol{\Delta}-\boldsymbol{\Delta}_{\mathrm{o}}\end{array}\right\} \tag{10.40a}$$

where, again, **u** and **Δ** are no longer arbitrary but rather solutions of Equations 10.35*b*. When $\boldsymbol{\Delta} = \boldsymbol{\Delta}_{\mathrm{o}}$, the case for which the theorem of least work holds true, this reduces to

$$U_{equil} = \frac{1}{2}(\mathbf{u} - \mathbf{u}_{\mathrm{o}})^{\mathrm{T}}\mathbf{k}_{\mathrm{uu}}(\mathbf{u} - \mathbf{u}_{\mathrm{o}}) \tag{10.40b}$$

Furthermore, for the special case $\boldsymbol{\Delta} = \boldsymbol{\Delta}_{\mathrm{o}}$, the first of the vector equations in Equations 10.35*a* becomes

$$\mathbf{k}_{\mathrm{uu}}(\mathbf{u} - \mathbf{u}_{\mathrm{o}}) = \mathbf{f} \tag{10.41a}$$

from which we obtain

$$(\mathbf{u} - \mathbf{u_o}) = \mathbf{k_{uu}^{-1}} \mathbf{f} \tag{10.41b}$$

Thus, substituting Equations 10.41a and 10.41b into Equation 10.40b gives

$$U_{equil} = \frac{1}{2}\mathbf{f}^{\mathrm{T}}\mathbf{k_{uu}^{-1}}\mathbf{f} \tag{10.42}$$

Note that for any value of $\mathbf{X}$ other than the one that both minimizes U_{equil} and corresponds to $\mathbf{\Delta} = \mathbf{\Delta_o}$, the strain energy in the structure would be greater than the value in Equation 10.42.

Ideally, real structures are assembled without any initial incompatibilities or other irregularities. In reality, initial incompatibilities are not uncommon. Most incompatibilities can be eliminated during assembly by repositioning the members or, if necessary, modifying the connections slightly, for example by redrilling connection holes. If, however, the structure can only be assembled by ''forcing'' it together because of some type of uncorrectable ''mismatch,'' the analytic implication is that $\mathbf{\Delta_o} \neq \mathbf{0}$ when $\mathbf{f} = \mathbf{X} = \mathbf{0}$.

In this case, analysis requires that we find values of $\mathbf{X}$ that will force $\mathbf{\Delta} = \mathbf{0}$ and also maintain compatibility when loads $\mathbf{f}$ are applied. It may be helpful to view this as a two-step process. First, prior to applying $\mathbf{f}$, some value of $\mathbf{X_o}$ is applied to force $\mathbf{\Delta} = \mathbf{0}$; this requires work, resulting in the structure storing an initial energy U_o. When $\mathbf{f}$ is applied, an additional redundant force $\mathbf{X_f}$ is required to maintain $\mathbf{\Delta} = \mathbf{0}$; but $\mathbf{X_f}$ plays exactly the same role as $\mathbf{X}$ in the case when $\mathbf{\Delta} = \mathbf{\Delta_o}$, namely to maintain the state of the ''cut structure'' so that no changes occur in the geometry of the ''cuts.'' Then, from Equation 10.42, the additional energy in the structure is

$$U_f = \frac{1}{2}\mathbf{f}^{\mathrm{T}}\mathbf{k_{uu}^{-1}}\mathbf{f} \tag{10.43}$$

and the total energy is

$$U = U_f + U_i \tag{10.44}$$

where it may be shown that

$$U_i = \frac{1}{2}\mathbf{\Delta_o^{\mathrm{T}}}\mathbf{a_{xx}^{-1}}\mathbf{\Delta_o} \tag{10.45}$$

In effect, the $\mathbf{X_o}$ used to eliminate $\mathbf{\Delta_o}$ preloads the structure, giving rise to an initial energy U_i. If the redundant forces $\mathbf{X_o}$ are known, then the complete force state can be determined by analyzing only the effects of $\mathbf{f}$ and the corresponding $\mathbf{X_f}$.

Any approximations in the force analysis that is used to find $\mathbf{X_f}$ will lead to a greater U_f than an exact solution and, by inference, higher stresses due to $\mathbf{f}$. Bear in mind, however, that the preloaded state may already contain significant stresses, and these stresses may be of opposite sign to the stresses due to $\mathbf{f}$ only.

10.3 GENERALIZED DISPLACEMENTS AND FORCES

We observed in Section 10.1.1 that when we model a structure using a finite number of displacement degrees of freedom, a particular set of nodal displacements ($\mathbf{u}$) can be produced by more than one physical loading, *cf.* Figures 10.1*c* and 10.1*d*. We also observed that the different physical loadings produce internal stresses and strains that are not identical. The differences in stresses and strains were associated with a ''fixed-coordinate'' state (Figures 10.1*e* or 10.1*f*), i.e., a state in which the nodal displacements are zero.

Suppose we wish to use a set of nodal forces $\mathbf{F_q}$ (Figure 10.1*d*) that will produce the same nodal displacements $\mathbf{u_q}$ in the structure as that produced by the actual forces q (Figure 10.1*c*). The question that must first be answered is: How do we calculate the forces $\mathbf{F_q}$ that would produce the same $\mathbf{u_q}$ without a prior knowledge of the response $\mathbf{u_q}$, i.e., without solving for $\mathbf{F_q}$ from the equation $\mathbf{F_q} = \mathbf{k}\mathbf{u_q}$? As we shall see, we can actually find $\mathbf{F_q}$ from a knowledge of the displacement patterns associated with $\mathbf{u_q}$, rather than from the actual values of $\mathbf{u_q}$ itself. This concept, which is very important for approximate analysis, is introduced in this section.

To illustrate these concepts, consider a structure modeled with n degrees of freedom such as shown in Figure 10.12*a*. This could be a general two-dimensional structure; however, in order to relate developments to familiar concepts, we will assume a frame structure composed of uniform beam elements (Figure 10.12*b*). Both bending and axial effects will be permitted in each of the elements.

We note that the bending displacements $v_j(x_j)$ and axial displacements $u_j(x_j)$, $j = 1, \ldots, m$, where m is the number of elements, define the bending curvatures and axial strains along the axis of each element. The bending moments, transverse shears, and axial forces[17] and, finally, the internal stresses can be determined directly from these displacements. In other words, if we know the displacements in each element of the structure then we can completely determine the mechanical behavior of the structure and obtain complete information regarding the internal forces in all the elements.

We will denote the set of concentrated forces applied directly at the nodes as $\mathbf{F}$, (Figure 10.12*c*), and all other loads as part of a set of forces ''q'' (Figure 10.12*d*). We recognize that the ''q'' forces will produce motion at each of the n degrees of freedom.

Now, consider a specific set of nodal forces $\mathbf{F}^{(k)}$, which has the property of producing a nodal displacement pattern for which all the nodal displacements $u_i^{(k)}$ are zero except for $u_k^{(k)}$, which is unity, i.e., $u_i^{(k)} = 0, i=1, \ldots, k-1, k+1, \ldots, n$, but $u_k^{(k)} = 1$, for example as in Figures 10.12*e* or 10.12*f*. (This type of behavior could be achieved physically by loading the structure with a set of jacks applied at the nodes so that only one degree of freedom was nonzero, equal to unity.) Note that in each element there could be internal bending displacements $v_j^{(k)}(x_j)$ and/or axial displacements $u_j^{(k)}(x_j)$. Here, the superscript is added to emphasize that these are internal displacements associated with the kth nodal displacement set equal to unity and all others zero.

We also recognize that if the ''q'' forces are applied to a completely constrained structure in which all nodal displacements are zero, there will still be nonzero bending displacements and axial displacements in at least some of the elements; we denote these displacements by $v_j^{(q)}(x_j)$ and $u_j^{(q)}(x_j)$, respectively. This is the fixed-coordinate state for

---

[17] Recall that $M(x) = EId^2v(x)/dx^2$, $V(x) = -EId^3v(x)/dx^3$, and $P(x) = EAdu(x)/dx$.

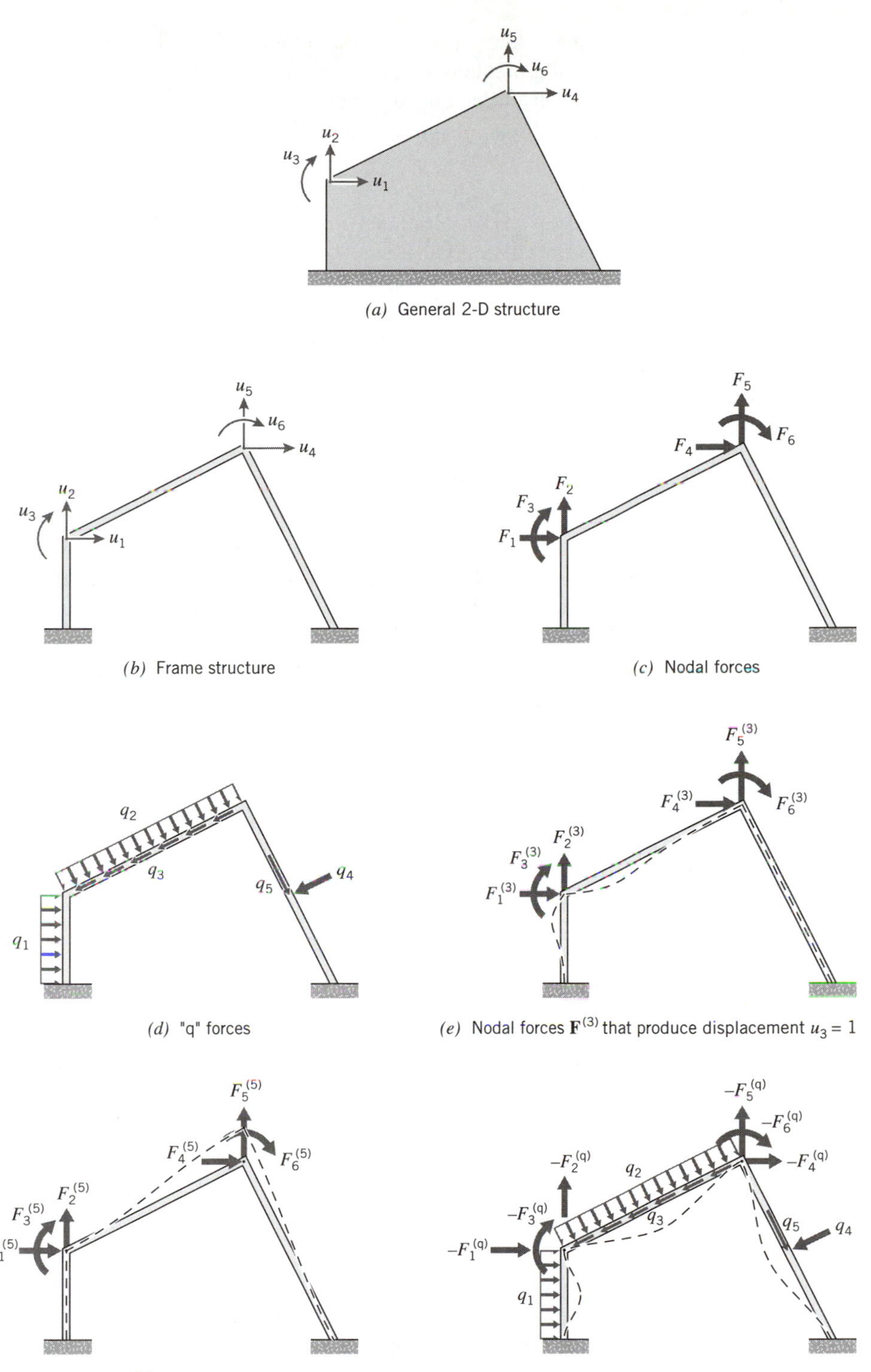

Figure 10.12 Illustration of generalized coordinates.

the "q" forces. In general, there will be a set of nodal forces $-\mathbf{F_q}$ present in this fixed-coordinate state. As noted previously, these are the negatives of the nodal forces that produce the same nodal displacements $\mathbf{u_q}$ as the "q" forces.

It follows that the exact total displacement pattern for each element in the structure may be written as

$$\begin{Bmatrix} u_j(x) \\ v_j(x) \end{Bmatrix} = u_1 \begin{Bmatrix} u_j^{(1)}(x) \\ v_j^{(1)}(x) \end{Bmatrix} + u_2 \begin{Bmatrix} u_j^{(2)}(x) \\ v_j^{(2)}(x) \end{Bmatrix} + \ldots + u_n \begin{Bmatrix} u_j^{(n)}(x) \\ v_j^{(n)}(x) \end{Bmatrix} + \begin{Bmatrix} u_j^{(q)}(x) \\ v_j^{(q)}(x) \end{Bmatrix}, j = 1, \ldots, m \tag{10.46}$$

where $u_1, \ldots, u_n$ are the actual nodal displacements produced by the applied loads. Each of the first n terms in brackets on the right side of Equation 10.46 is referred to as a generalized displacement pattern; the amplitude u_k is referred to as a generalized displacement or generalized coordinate. It is important to recognize that each generalized displacement pattern is the exact displacement response when the special set of nodal forces $\mathbf{F}^{(k)}$ is applied. The last vector is the vector of displacements in the fixed coordinate state (see Figure 10.12*g*).

Now, to find the components of $\mathbf{F_q}$, which we denote by $F_k^{(q)}$, $k = 1, \ldots, n$, we apply Betti's theorem to this structure when it is subjected to two different sets of forces. The first set of forces will be taken as $\mathbf{F}^{(k)}$, which produces the nodal displacement pattern for which $u_i^{(k)} = 0$, $i = 1, \ldots, k-1, k+1, \ldots, n$, but $u_k^{(k)} = 1$ (e.g., Figure 10.12*e* or Figure 10.12*f*); the second set of forces will be taken as that corresponding to the fixed-coordinate state for which $u_i = 0$, $i = 1, \ldots, n$ (Figure 10.12*g*). Since the nodal displacements in the fixed-coordinate state are zero, the external virtual work done by the first set of (nodal) forces going through the displacements associated with the second set of forces is zero. The external virtual work done by the forces in the fixed-coordinate state (Figure 10.12*g*) going through displacements corresponding to $u_k^{(k)} = 1$, and $u_i^{(k)} = 0$, $i = 1, \ldots, k-1, k+1, \ldots, n$, (e.g., Figures 10.12*e* or 10.12*f*) is $[-F_k^{(q)}] \cdot 1$ + work done by "q" forces going through displacements $v_j^{(k)}(x_j)$ and $u_j^{(k)}(x_j)$. Equating this external virtual work to zero leads to

$$F_k^{(q)} = \text{work done by ``}q\text{'' forces going through displacements } v_j^{(k)}(x_j) \text{ and } u_j^{(k)}(x_j), j = 1, \ldots, m \tag{10.47}$$

Equation 10.47 is the definition of a generalized (work-equivalent) force $F_k^{(q)}$ associated with a generalized displacement pattern. The definition requires knowledge of the exact displacement patterns in each element associated with nodal displacement behavior. Surprisingly, the fixed-coordinate displacement patterns are not required. Once all the components of $\mathbf{F_q}$ are determined, the desired nodal displacements $\mathbf{u_q}$ can be found by solving the equation

$$\mathbf{k}\mathbf{u_q} = \mathbf{F_q} \tag{10.48}$$

To illustrate, consider a simple beam element (axially rigid) for which the exact displacement patterns are known (see Figure 8.18 and Equation 8.16*b* in Section 8.3.3). From Equation 8.16*b* we obtain[18]

---

[18] We take the displacement in Equation 8.16*b* as real rather than virtual, and substitute the definitions of end deformations in terms of displacements, Equations 5.39*a* and 5.39*b*.

$$v(x) = \theta_{IJ}\left(-x + \frac{2x^2}{l} - \frac{x^3}{l^2}\right) + \theta_{JI}\left(\frac{x^2}{l} - \frac{x^3}{l^2}\right) + v_{IJ}\left(1 - \frac{3x^2}{l^2} + \frac{2x^3}{l^3}\right) + v_{JI}\left(\frac{3x^2}{l^2} - \frac{2x^3}{l^3}\right) \quad \textbf{(10.49)}$$

where θ_{IJ}, θ_{JI}, v_{IJ}, and v_{JI} are the nodal (end) displacements (including rotations). In this case, there is only one component of interior displacement (i.e., the transverse displacement), and we recognize that Equation 10.49 reduces to the matrix form

$$v(x) = \left[\left(-x + \frac{2x^2}{l} - \frac{x^3}{l^2}\right)\left(\frac{x^2}{l} - \frac{x^3}{l^2}\right)\left(1 - \frac{3x^2}{l^2} + \frac{2x^3}{l^3}\right)\left(\frac{3x^2}{l^2} - \frac{2x^3}{l^3}\right)\right]\begin{Bmatrix} \theta_{IJ} \\ \theta_{JI} \\ v_{IJ} \\ v_{JI} \end{Bmatrix} \quad \textbf{(10.50)}$$

Thus, the terms in parentheses in Equations 10.49 and 10.50 are the generalized displacement patterns and the end displacements are the generalized coordinates! Note that the fixed-coordinate state is not included in Equations 10.49 and 10.50.

Assume, now, that there is an arbitrary transverse load, $q(x)$, applied to the beam. The generalized forces, Equation 10.47, are

$$\begin{aligned} F_1^{(q)} &= \int_0^l q(x)\left(-x + \frac{2x^2}{l} - \frac{x^3}{l^2}\right)dx \\ F_2^{(q)} &= \int_0^l q(x)\left(\frac{x^2}{l} - \frac{x^3}{l^2}\right)dx \\ F_3^{(q)} &= \int_0^l q(x)\left(1 - \frac{3x^2}{l^2} + \frac{2x^3}{l^3}\right)dx \\ F_4^{(q)} &= \int_0^l q(x)\left(\frac{3x^2}{l^2} - \frac{2x^3}{l^3}\right)dx \end{aligned} \quad \textbf{(10.51}\textbf{\textit{a}}\textbf{)}$$

From Equations 8.17*l*, we recognize the first two of these equations as the negative of the fixed-end moments due to the physical load $q(x)$; from Equations 9.1*e* we recognize the last two of these equations as the negative of the fixed-end shears. In other words,

$$\begin{aligned} F_1^{(q)} &= -\widehat{M}_{IJ}^o \\ F_2^{(q)} &= -\widehat{M}_{JI}^o \\ F_3^{(q)} &= -\widehat{V}_{IJ}^o \\ F_4^{(q)} &= -\widehat{V}_{JI}^o \end{aligned} \quad \textbf{(10.51}\textbf{\textit{b}}\textbf{)}$$

Thus, we conclude that applying nodal forces that are work-equivalent to the actual distributed force $q(x)$ [and equal to the negative of the fixed-end forces due to $q(x)$] will have exactly the same effect on nodal displacements of the beam as the distributed force itself; this is the same result that was reached in the qualitative discussion in Section 10.1.1. In fact, this is equivalent to the analyses performed previously (Chapters

8 and 9) where, in essence, the vectors of applied nodal forces were augmented by fixed-end force terms in the form $-\boldsymbol{\beta}^{T}\widehat{\mathbf{P}}$.

For simple one-dimensional elements, it is not necessary to use Equation 10.47 for the evaluation of the equivalent nodal forces because other means are available for evaluating these forces; however, the definition in Equation 10.47 is essential for use in conjunction with more complex finite elements.

10.3.1 Approximate Displacement Patterns

If we don't know the exact displacement patterns for a structure, then we cannot define the correct generalized forces or establish the correct stiffness. In other words, we can't get the exact solution without knowing it from the outset! Nevertheless, the energy approach makes it possible for us to proceed with an approximate analysis; and, at the same time, gain some insight into the approximations we have produced.

Suppose, in particular, that we don't know the exact displacement pattern $v_j^{(k)}(x_j)$, $u_j^{(k)}(x_j)$ in each element associated with $u_k^{(k)} = 1$, but are able to define an approximate pattern $\overline{v}_j^{(k)}(x_j)$, $\overline{u}_j^{(k)}(x_j)$, which at least has the property that $u_j^{(k)} \equiv 0, j \neq k$. Then, the exact behavior for each element would be given by

$$\begin{Bmatrix} u_j^{(k)}(x_j) \\ v_j^{(k)}(x_j) \end{Bmatrix} = \overline{u}_k \begin{Bmatrix} \overline{u}_j^{(k)}(x_j) \\ \overline{v}_j^{(k)}(x_j) \end{Bmatrix} + u_{k_C}\left(\begin{Bmatrix} u_j^{(k)}(x_j) \\ v_j^{(k)}(x_j) \end{Bmatrix} - \begin{Bmatrix} \overline{u}_j^{(k)}(x_j) \\ \overline{v}_j^{(k)}(x_j) \end{Bmatrix}\right) + \begin{Bmatrix} u_j^{(q)}(x_j) \\ v_j^{(q)}(x_j) \end{Bmatrix} \qquad \textbf{(10.52}\textbf{\textit{a}}\textbf{)}$$

where u_{k_C} is a correction term that depends on the difference between the exact and approximate solutions, and $u_j^{(q)}(x_j)$ and $v_j^{(q)}(x_j)$ are the displacements produced by the actual "q" forces in the fixed-coordinate state. The same type of assumption can be used to represent every other nodal displacement pattern, leading to[19]

$$\begin{Bmatrix} u_j(x_j) \\ v_j(x_j) \end{Bmatrix} = \begin{bmatrix} \overline{u}_j^{(1)} & \cdots & \overline{u}_j^{(n)} \\ \overline{v}_j^{(1)} & \cdots & \overline{v}_j^{(n)} \end{bmatrix} \begin{Bmatrix} \overline{u}_1 \\ \vdots \\ \overline{u}_n \end{Bmatrix} + \begin{bmatrix} \Delta u_j^{(1)} & \cdots & \Delta u_j^{(n)} \\ \Delta v_j^{(1)} & \cdots & \Delta v_j^{(n)} \end{bmatrix} \begin{Bmatrix} u_{1_C} \\ \vdots \\ u_{n_C} \end{Bmatrix} + \begin{Bmatrix} u_j^{(q)}(x_j) \\ v_j^{(q)}(x_j) \end{Bmatrix} \qquad \textbf{(10.52}\textbf{\textit{b}}\textbf{)}$$

where $\Delta u_j^{(i)}$, $\Delta v_j^{(i)}$ are the (unknown) differences between the exact and approximate displacement patterns in each element.

Let us denote the vectors of generalized coordinates in Equation 10.52*b* by $\overline{\mathbf{u}}$ and $\mathbf{u_C}$, respectively. Then, the strain energy and external potential energy can be written in terms of these vectors; general forms of the expressions for these energies would be

$$U = \frac{1}{2}\begin{Bmatrix} \overline{\mathbf{u}} \\ -- \\ \mathbf{u_C} \end{Bmatrix}^{T} \begin{bmatrix} \mathbf{k}_{11} & | & \mathbf{k_{1C}} \\ --- & + & --- \\ \mathbf{k_{1C}}^{T} & | & \mathbf{k_{CC}} \end{bmatrix} \begin{Bmatrix} \overline{\mathbf{u}} \\ -- \\ \mathbf{u_C} \end{Bmatrix} + U_q \qquad \textbf{(10.53}\textbf{\textit{a}}\textbf{)}$$

and

$$U_e = -\begin{Bmatrix} \overline{\mathbf{u}} \\ -- \\ \mathbf{u_C} \end{Bmatrix}^{T} \begin{Bmatrix} \overline{\mathbf{F}}_{\mathbf{q}} \\ -- \\ \mathbf{F_{qC}} \end{Bmatrix} + U_{e_q} \qquad \textbf{(10.53}\textbf{\textit{b}}\textbf{)}$$

[19] For the sake of compactness we use $\overline{u}_j^{(i)}$ to represent $\overline{u}_j^{(i)}(x)$, etc.

where U_q and U_{e_q} are the strain energy and external potential energy, respectively, in the fixed-coordinate state.[20] Note that generalized (work-equivalent) forces appear in the expression for U_e; the components of $\overline{\mathbf{F}}_\mathbf{q}$ are the generalized forces associated with the approximate generalized displacements, and the components of $\mathbf{F}_{\mathbf{qC}}$ are the generalized forces associated with the generalized correction patterns. It should be recognized, however, that submatrix $\mathbf{k}_{11}$ in Equation 10.53*a* is the only portion of the stiffness matrix that can be explicitly determined, and depends on the specific forms of the approximate displacement patterns $\overline{u}_j^{(i)}$ and $\overline{v}_j^{(i)}$. (The computation of $\mathbf{k}_{11}$ will be demonstrated in Chapter 11.) Since we don't know $\Delta u_j^{(i)}$ and $\Delta v_j^{(i)}$, we are unable to evaluate the portions of the stiffness matrix (or, for that matter, $\mathbf{F}_{\mathbf{qC}}$) associated with $\mathbf{u}_\mathbf{C}$.

If we did have the complete stiffness matrix in Equation 10.53*a*, we could obtain the equilibrium equation either directly or by minimizing the total potential energy $(U + U_e)$,

$$\left[\begin{array}{c|c} \mathbf{k}_{11} & \mathbf{k}_{1C} \\ \hline \mathbf{k}_{1C}^{\mathrm{T}} & \mathbf{k}_{CC} \end{array}\right] \left\{\begin{array}{c} \overline{\mathbf{u}} \\ \hline \mathbf{u}_\mathbf{C} \end{array}\right\} = \left\{\begin{array}{c} \overline{\mathbf{F}}_\mathbf{q} \\ \hline \mathbf{F}_{\mathbf{qC}} \end{array}\right\} \tag{10.53c}$$

This equation could then be solved for the exact displacements $\overline{\mathbf{u}}$, $\mathbf{u}_\mathbf{C}$; but, since we only have $\mathbf{k}_{11}$, the best we can do is find an approximate equilibrium equation,

$$\mathbf{k}_{11}\overline{\mathbf{u}} = \overline{\mathbf{F}}_\mathbf{q} \tag{10.53d}$$

As usual, the solution to Equation 10.53*d* will yield a strain energy that is less than that for the exact solution, and the stresses in the structure will be underestimated, on average.

An analysis of the foregoing type, in which we assume displacement patterns and find an approximate equilibrium equation by minimizing total potential energy, is known as the Rayleigh-Ritz method. In the original form of this method, the assumed displacement patterns are usually defined throughout the entire structure; but the definition of reasonable assumed displacement patterns is difficult and cumbersome for all but the simplest structures. However, in about 1960, the modern finite element method emerged as a powerful, yet simple, modification of the Rayleigh-Ritz method. Essentially, what was done was to define generalized displacement patterns for each structural element in terms of nodal displacements, as described above. This seems very logical and reasonable for simple structures where elements and connections (nodes) are evident. It is less clear or evident for complex built-up structures such as those composed of two-dimensional elements (Figure 10.13). The most important requirement is to make sure that any two elements sharing a common edge (or joint, if the elements are one-dimensional), as in Figure 10.13*a*, are assigned displacement patterns (generalized displacements) that are identical where the elements are connected together.

By dividing the physical structure into a series of simpler finite regions, the task of developing assumed nodal displacement patterns is greatly simplified, and the technique makes it possible to formulate (approximate) element stiffness relationships having the

---

[20] Note that U_q and U_{e_q} are independent of either $\overline{\mathbf{u}}$ or $\mathbf{u}_\mathbf{C}$; thus, these components of energy have no effect on the stiffness relations which result from minimizing the total potential energy with respect to $\overline{\mathbf{u}}$ and $\mathbf{u}_\mathbf{C}$.

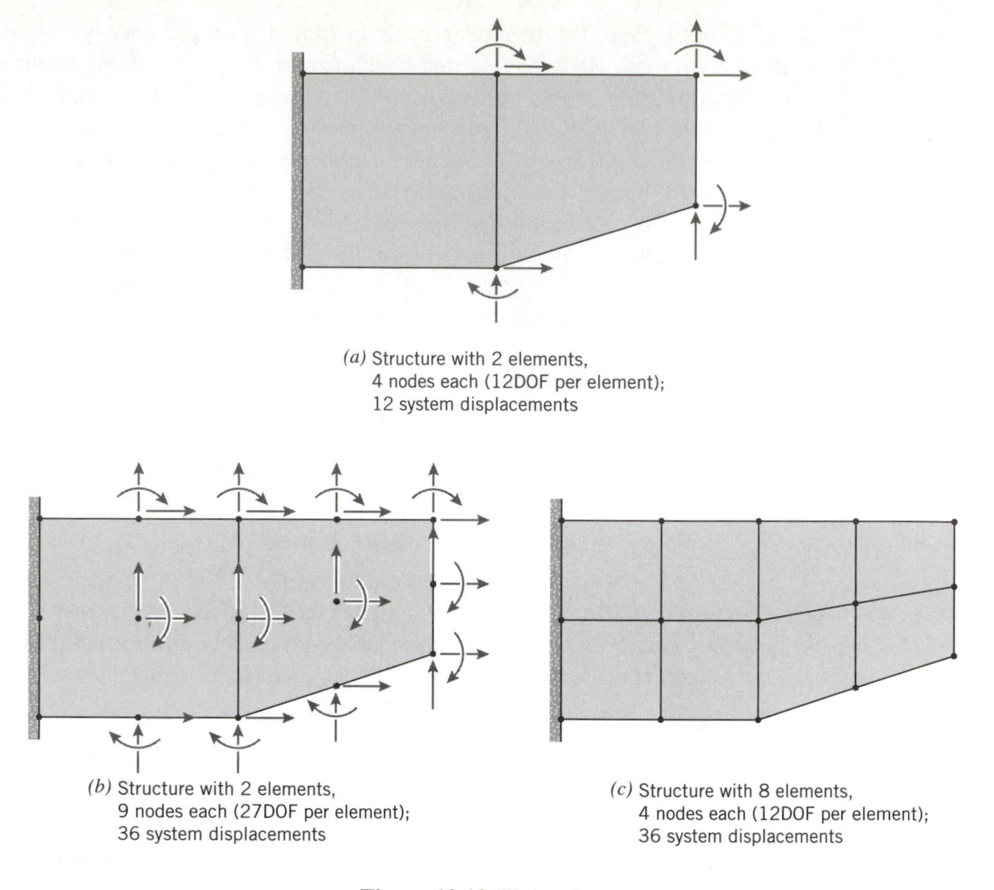

(*a*) Structure with 2 elements, 4 nodes each (12DOF per element); 12 system displacements

(*b*) Structure with 2 elements, 9 nodes each (27DOF per element); 36 system displacements

(*c*) Structure with 8 elements, 4 nodes each (12DOF per element); 36 system displacements

Figure 10.13 Finite elements.

form in Equation 10.53*d* with great ease and economy. In particular, if the element stiffness relations are written in a global *x-y* coordinate system, the direct stiffness method can be used to assemble the system equilibrium equations.

It must be reemphasized, however, that the formulation includes two distinct types of approximations: (1) the nodal displacement patterns are inherently approximate, and (2) the fixed (nodal) coordinate state contains stresses that are not known. In practice, the analyst can usually attenuate errors resulting from these approximations by preparing successively refined finite element models, and repeating computations until little change is observed in the largest stresses in the structure. This refinement is achieved by either assigning more nodes to the same element (Figure 10.13*b*) or using a greater number of simple elements to describe the structure (Figure 10.13*c*). The latter approach is more common because it does not involve any increase in mathematical complexity at the element level.

10.4 APPLICATIONS TO STRUCTURES COMPOSED OF ONE-DIMENSIONAL ELEMENTS

Prior to any detailed examination of finite element formulations in Chapter 11, it will be informative to illustrate typical applications of the Rayleigh-Ritz method to the analysis of familiar structures. This will enable us to demonstrate that energy methods will lead naturally to the "best possible" solutions. These solutions will be exact provided that (1) the assumed displacement state encompasses the exact solution, and (2) the energy stored in the structure is properly expressed as a function of the assumed displacement state; otherwise, as we have seen, solutions will be approximate.

We recognize that the energy definitions presented previously in this chapter, for example Equation 10.15*d*, are inherently approximate in the sense that they do not include energy stored in the fixed-coordinate state. Therefore, to adequately represent the behavior of one-dimensional elements in subsequent examples, it will first be necessary to derive exact analytic expressions for the energy in these elements.

10.4.1 Strain Energy of Bars and Beams

The basic behavior of linearly elastic bars and Bernoulli-Euler beams was detailed in Chapters 2 and 3. There were two essential assumptions inherent in the formulations: (1) cross sections that are originally plane and perpendicular to the undeformed reference axis remain plane and perpendicular to the reference axis after motion occurs, and (2) axial stress at any point in the cross section is proportional to axial strain in accordance with Hooke's law. The equations describing this behavior are (see the first of Equations 3.13 and 2.34*c*, respectively),

$$\epsilon_E = \frac{du(x)}{dx} - y\frac{d^2v(x)}{dx^2} = u'(x) - yv''(x) \qquad \textbf{(10.54a)}$$

and

$$\sigma_{xx} = E(\epsilon_E - \epsilon_T - \epsilon_I) \qquad \textbf{(10.54b)}$$

Considerations of statics led us to establish centroidal coordinate axes such that

$$E\int_A y\,dA = 0 \qquad \textbf{(10.54c)}$$

where the integration is performed in the *y-z* plane over the entire cross-sectional area *A*, and where the cross section has been assumed homogeneous. We then used the definition

$$P(x) = \int_A \sigma_{xx} dA \qquad \textbf{(10.54d)}$$

and substituted Equations 10.54*a* and 10.54*b*, as well as the definition $\epsilon_T \equiv \alpha\Delta T$, to obtain

$$\begin{aligned} P(x) &= EAu'(x) - P_T(x) - P_I(x) = EA\left[u'(x) - \frac{P_T(x)}{EA} - \frac{P_I(x)}{EA}\right] \\ &\equiv EA[u'(x) - u_T'(x) - u_I'(x)] \end{aligned} \qquad \textbf{(10.54e)}$$

where

$$P_T(x) = E\alpha \int_A \Delta T \, dA \equiv EA u_T'(x)$$

$$P_I(x) = E \int_A \epsilon_I \, dA \equiv EA u_I'(x) \tag{10.54f}$$

and where we have now denoted thermal and initial strains by $u_T'(x)$ and $u_I'(x)$.

The beam bending relationship was obtained from the definition

$$M(x) = -\int_A y\sigma_{xx} \, dA \tag{10.54g}$$

Substituting Equations 10.54*a*, 10.54*b*, and 10.54*c* led to

$$M(x) = EI\frac{d^2v(x)}{dx^2} - M_T(x) - M_I(x) \equiv EI[v''(x) - v_T''(x) - v_I''(x)] \tag{10.54h}$$

where

$$M_T(x) = -EA \int_A y\Delta T \, dA \equiv EIv_T''(x)$$

$$M_I(x) = -E \int_A y\epsilon_I dA \equiv EIv_I''(x) \tag{10.54i}$$

and where we have now denoted thermal and initial curvatures by $v_T''(x)$ and $v_I''(x)$.

The stress-strain form of Hooke's law in Equation 10.54*b* is actually derived from the results of an axial load test performed on a uniform bar. The test provides force-deformation data, which is then converted to the stress-strain relation in Equation 10.54*b*. In fact, we may easily reconstruct the original data for the bar test by multiplying stress in Equation 10.54*b* by bar area A and recognizing that strain is defined as Δ/l, where l is the length of the bar over which deformation Δ is measured. Thus, the force-deformation relation is

$$P = \frac{EA}{l}(\Delta - \Delta_T - \Delta_I) \tag{10.55}$$

The quantities Δ_T and Δ_I are thermal and initial deformations, respectively, and EA/l is the constant of proportionality which defines the slope of the relationship between force P and deformation Δ (Figure 10.14).

As was noted in connection with Equation 10.3*a*, the work done in deforming the bar is stored in the form of strain energy, and is physically equivalent to the shaded area in Figure 10.14. This strain energy is

$$U = \frac{1}{2}P(\Delta - \Delta_T - \Delta_I)$$

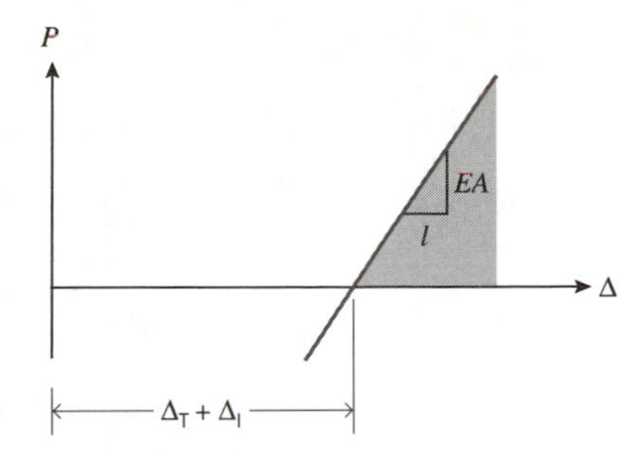

Figure 10.14 Force-deformatinon relationship for a bar; strain energy.

$$
\begin{aligned}
&= \frac{1}{2}\sigma_{xx}(\epsilon_E - \epsilon_T - \epsilon_I)\,Al \\
&= \frac{1}{2}E(\epsilon_E - \epsilon_T - \epsilon_I)^2 Al
\end{aligned}
\qquad \textbf{(10.56}\textbf{\textit{a}}\textbf{)}
$$

Since Al is the volume of the bar, we may define $U_o \equiv U/Al$ to be the strain energy per unit volume,[21]

$$
\begin{aligned}
U_o &= \frac{1}{2}\sigma_{xx}(\epsilon_E - \epsilon_T - \epsilon_I) \\
&= \frac{1}{2}E(\epsilon_E - \epsilon_T - \epsilon_I)^2
\end{aligned}
\qquad \textbf{(10.56}\textbf{\textit{b}}\textbf{)}
$$

Now, referring to a beam that is subjected to a general axial displacement $u(x)$ and bending displacement $v(x)$, we consider a differential volume, $dV = A\,dx$ (Figure 10.15). The differential volume may be viewed as an assembly of infinitesimal bars of cross-sectional area dA and length dx, each of which deforms under the action of axial stress. Each infinitesimal bar has a strain energy per unit volume given by Equation 10.56*b*, and a total strain energy that is obtained by multiplying Equation 10.56*b* by the volume of the bar which, in this case, is $dV = dA\,dx$. Then, adding all the energy contributions of all the infinitesimal bars that make up the volume, we obtain a differential strain energy,

$$
dU = \int_A U_o \, dA\, dx \qquad \textbf{(10.56}\textbf{\textit{c}}\textbf{)}
$$

The strain energy in Equation 10.56*c* is a differential quantity simply because of the length dx of the beam segment under consideration. Note that x is a constant with regard to integration over the cross-section; only y and z are variables with respect to

[21] Also referred to as strain energy density.

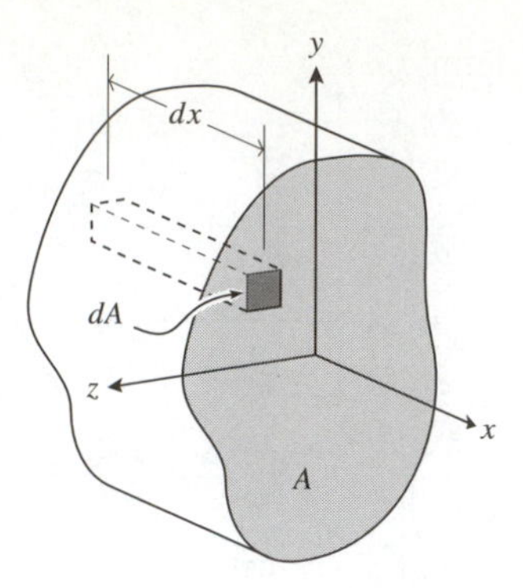

Figure 10.15 Differential volume.

this integration. Thus, dx, as well as any other quantities which are functions only of x, may be taken out of the integral.

We may now define

$$\frac{dU}{dx} \equiv U' = \int_A U_o dA \qquad \textbf{(10.56}\textbf{\textit{d}}\textbf{)}$$

as the strain energy per unit length of the Bernoulli-Euler beam. Substituting Equations 10.56*b* and 10.54*a* into Equation 10.56*d* gives

$$\begin{aligned} U' &= \frac{1}{2}E\int_A (\epsilon_E - \epsilon_T - \epsilon_I)^2 dA \\ &= \frac{1}{2}E\int_A [u'(x) - yv''(x) - \epsilon_T - \epsilon_I]^2 dA \end{aligned} \qquad \textbf{(10.56}\textbf{\textit{e}}\textbf{)}$$

Expanding the integrand in Equation 10.56*e* provides

$$\begin{aligned} U' = \frac{1}{2}E\int_A [&u'^2(x) - 2yv''(x) + y^2v''^2(x) - 2u'(x)(\epsilon_T + \epsilon_I) \\ &+ 2yv''(x)(\epsilon_T + \epsilon_I) + (\epsilon_T + \epsilon_I)^2]dA \end{aligned} \qquad \textbf{(10.56}\textbf{\textit{f}}\textbf{)}$$

We recognize that $u'(x)$ and $v''(x)$ are functions only of x and not of the cross-sectional variables y and z; therefore, integrating leads to

$$\begin{aligned} U' = \frac{1}{2}EA\,u'^2(x) &+ \frac{1}{2}EIv''^2(x) - u'(x)[P_T(x) + P_I(x)] \\ &- v''(x)[M_T(x) + M_I(x)] + \frac{1}{2}E\int_A (\epsilon_T + \epsilon_I)^2\, dA \end{aligned} \qquad \textbf{(10.56g)}$$

Substituting the definitions in Equations 10.54*f* and 10.54*i*, and adding and subtracting terms $EA[u_T'(x) + u_I'(x)]^2/2$ and $EI[v_T''(x) + v_I''(x)]^2/2$, enables us to rewrite Equation 10.56*g* in the simpler form

$$\begin{aligned} U' &= \frac{1}{2}EA[u'(x) - u_T'(x) - u_I'(x)]^2 + \frac{1}{2}EI[v''(x) - v_T''(x) - v_I''(x)]^2 + U_I' \\ &= \frac{1}{2}\frac{P^2(x)}{EA} + \frac{1}{2}\frac{M^2(x)}{EI} + U_I' \end{aligned} \tag{10.56h}$$

where

$$U_I' = \frac{1}{2}E\int_A (\alpha \Delta T + \epsilon_I)^2\, dA - \frac{1}{2}EA[u_T'(x) + u_I'(x)]^2 - \frac{1}{2}EI[v_T''(x) + v_I''(x)]^2 \tag{10.56i}$$

The quantity U_I' is an initial strain energy per unit length when $P(x) = M(x) = 0$, i.e., when no bar/beam forces exist.

If both $\epsilon_T \equiv \alpha \Delta T$ and ϵ_I are linear functions of y then U_I' is zero; otherwise an initial strain energy per unit length will exist. This initial energy can be shown to have no effect whatsoever in calculations to determine the internal forces in assemblies of these elements.

Finally, for a complete bar/beam element of length l, the strain energy is given by any of the alternate forms

$$\begin{aligned} U &= \int_0^l U'\, dx \\ &= \frac{1}{2}\int_0^l EA[u'(x) - u_T'(x) - u_I'(x)]^2\, dx + \frac{1}{2}\int_0^l EI[v''(x) - v_T''(x) - v_I''(x)]^2\, dx + U_I \\ &= \frac{1}{2}\int_0^l P(x)[u'(x) - u_T'(x) - u_I'(x)]\, dx + \frac{1}{2}\int_0^l M(x)[v''(x) - v_T''(x) - v_I''(x)]\, dx + U_I \\ &= \frac{1}{2}\int_0^l \frac{P^2(x)}{EA} dx + \frac{1}{2}\int_0^l \frac{M^2(x)}{EI} dx + U_I \end{aligned} \tag{10.57a}$$

where

$$U_I = \int_0^l U_I'\, dx \tag{10.57b}$$

10.4.2 Examples of the Rayleigh-Ritz Method

The use of assumed displacement patterns provides a straightforward means for conducting approximate analyses of structures. The implications of such an approach have been discussed previously. For instance, we have seen that an incomplete description of the displacements of a structure leads to an underestimation of the total strain energy stored in the structure, produces a response that is ‘‘stiffer’’ than the actual response, and generally under-represents the internal state of stress. If we make fortuitous choices of displacement patterns we can obtain excellent, even exact, results. We will demonstrate these effects in several examples.

EXAMPLE 3

Consider a uniform cantilever beam with a concentrated transverse force F applied at the free end, as shown in Figure 10.16. For reference purposes, we note that the exact solution for the internal bending moment $M(x)$ and transverse displacement $v(x)$ of this beam is

$$M(x) = F(l - x) \tag{10.58a}$$

and

$$v(x) = \frac{Fl}{2EI}x^2 - \frac{F}{6EI}x^3 \tag{10.58b}$$

Let's suppose we have no knowledge of this exact solution, and begin our approximate analysis by assuming that the displacement is given by a function of form

$$\overline{v}(x) = \overline{u}_1\left(\frac{x}{l}\right)^2 \tag{10.59}$$

where $\overline{u}_1$ is an unknown generalized displacement and $(x/l)^2$ is a generalized displacement pattern. Note that the assumed displacement pattern, $(x/l)^2$, explicitly satisfies all geometric boundary conditions, i.e., $\overline{v}(0) = \overline{v}'(0) = 0$. The overbars in Equation 10.59 are used to denote the approximate nature of the displacement quantities or functions.

The strain energy that would be stored in the beam in the approximate displacement state is obtained by substituting the second derivative of $\overline{v}(x)$ into the first of Equations 10.57*a*,

$$\begin{aligned} U &= \frac{1}{2}\int_0^l EI\overline{v}''^2(x)\,dx \\ &= \frac{1}{2}\int_0^l EI\left(\frac{2\overline{u}_1}{l^2}\right)^2 dx \\ &= \frac{2EI}{l^3}\overline{u}_1^2 \end{aligned} \tag{10.60a}$$

The external potential energy for this displacement state, Equation 10.24*b*, is

$$\begin{aligned} U_e &= -F\overline{v}(l) \\ &= -F\overline{u}_1 \end{aligned} \tag{10.60b}$$

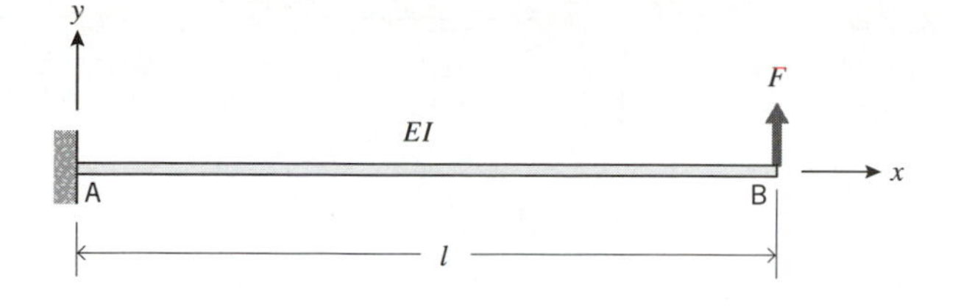

Figure 10.16 Example 3.

Therefore, for the assumed displacement state, the total potential energy, Equation 10.25*a*, is

$$
\begin{aligned}
V &= U + U_e \\
&= \frac{2EI}{l^3}\bar{u}_1^2 - F\bar{u}_1
\end{aligned}
\tag{10.60c}
$$

From the theorem of minimum total potential energy, Equation 10.25*e*, we obtain the value of the generalized displacement, which corresponds to the approximate equilibrium state,

$$\frac{\partial V}{\partial \bar{u}_1} = 0 = \frac{4EI}{l^3}\bar{u}_1 - F \tag{10.61a}$$

or

$$\bar{u}_1 = \frac{Fl^3}{4EI} \tag{10.61b}$$

Substituting this result into Equation 10.59 gives the approximate solution,

$$\bar{v}(x) = \frac{Fl^3}{4EI}\left(\frac{x}{l}\right)^2 \tag{10.62a}$$

Note that the tip deflection obtained from this solution is $\bar{v}(l) = Fl^3/4EI$, which is 75% of the exact value $Fl^3/3EI$ (Equation 10.58*b*); thus, we observe that the approximate solution does, indeed, make the beam too stiff.

The approximate internal bending moment function is obtained from Equation 10.62*a*,

$$\bar{M}(x) = EI\bar{v}''(x) = \frac{Fl}{2} \tag{10.62b}$$

In other words, the approximate solution produces a constant bending moment in the beam, whereas we recognize that the exact solution requires a linearly varying moment, Equation 10.58*a*. The exact and approximate moments are compared in Figure 10.17.

We observe that the approximate moment function, which had to be a constant based on our choice of assumed displacement pattern, underestimates the maximum moment

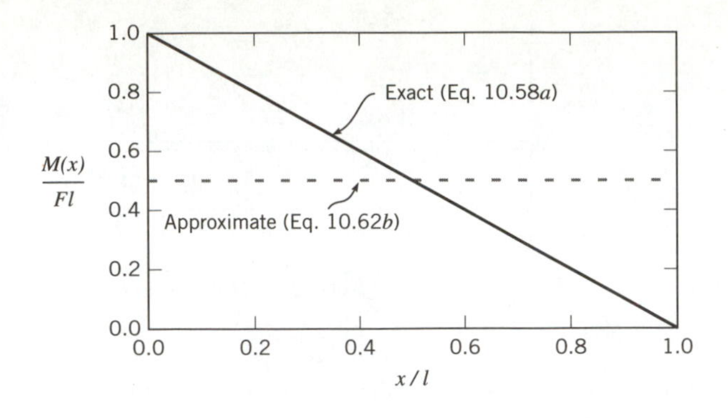

Figure 10.17 Bending moments.

by a factor of two. In fact, the approximate moment is only the average of the exact moments in the beam. More important, it is a "best fit" of the exact data in the sense that the difference between the exact and approximate moments balances the positive and negative errors. We shall see later that the method always gives a best fit approximation for which the mean square error between the true and approximate moments (or curvatures) is minimized.

EXAMPLE 4

Let's rework the preceding example assuming a more elaborate approximate displacement function,

$$\overline{v}(x) = \overline{u}_1\left(\frac{x}{l}\right)^2 + \overline{u}_2\left(\frac{x}{l}\right)^3 + \overline{u}_3\left(\frac{x}{l}\right)^4 \qquad \textbf{(10.63)}$$

for which each of the component displacement patterns $(x/l)^2$, $(x/l)^3$, $(x/l)^4$, satisfies the geometric boundary conditions $\overline{v}(0) = \overline{v}'(0) = 0$. Thus, the generalized displacements $\overline{u}_1$, $\overline{u}_2$, and $\overline{u}_3$ can be assigned any values and all of the geometric boundary conditions will still be satisfied.

The strain energy is obtained as in Equation 10.60*a*, except that we will find it convenient to express this energy in a matrix form similar to Equation 10.15*d*,

$$U = \frac{1}{2}\int_0^l \frac{EI}{l^4}\left[2\overline{u}_1 + 6\overline{u}_2\left(\frac{x}{l}\right) + 12\overline{u}_3\left(\frac{x}{l}\right)^2\right]^2 dx$$

$$= \frac{1}{2}\int_0^l \frac{EI}{l^3}\begin{Bmatrix}\overline{u}_1\\ \overline{u}_2\\ \overline{u}_3\end{Bmatrix}^T \begin{bmatrix}2\\ 6\left(\frac{x}{l}\right)\\ 12\left(\frac{x}{l}\right)^2\end{bmatrix}\begin{bmatrix}2\\ 6\left(\frac{x}{l}\right)\\ 12\left(\frac{x}{l}\right)^2\end{bmatrix}^T \begin{Bmatrix}\overline{u}_2\\ \overline{u}_2\\ \overline{u}_3\end{Bmatrix}\frac{dx}{l} \qquad \textbf{(10.64a)}$$

$$= \frac{1}{2}\begin{Bmatrix} \overline{u}_1 \\ \overline{u}_2 \\ \overline{u}_3 \end{Bmatrix}^{\mathrm{T}} \int_0^l \frac{EI}{l^3} \begin{bmatrix} 4 & 12\frac{x}{l} & 24\frac{x^2}{l^2} \\ 12\frac{x}{l} & 36\frac{x^2}{l^2} & 72\frac{x^3}{l^3} \\ 24\frac{x^2}{l^2} & 72\frac{x^3}{l^3} & 144\frac{x^4}{l^4} \end{bmatrix} \frac{dx}{l} \begin{Bmatrix} \overline{u}_1 \\ \overline{u}_2 \\ \overline{u}_3 \end{Bmatrix}$$

Performing the integration leads to

$$U = \frac{1}{2}\begin{Bmatrix} \overline{u}_1 \\ \overline{u}_2 \\ \overline{u}_3 \end{Bmatrix}^{\mathrm{T}} \frac{EI}{l^3} \begin{bmatrix} 4 & 6 & 8 \\ 6 & 12 & 18 \\ 8 & 18 & 144/5 \end{bmatrix} \begin{Bmatrix} \overline{u}_1 \\ \overline{u}_2 \\ \overline{u}_3 \end{Bmatrix} \tag{10.64b}$$

The external potential also has a matrix form,

$$\begin{aligned} U_e &= -F\overline{v}(l) \\ &= -F(\overline{u}_1 + \overline{u}_2 + \overline{u}_3) \\ &= -\begin{Bmatrix} \overline{u}_1 \\ \overline{u}_2 \\ \overline{u}_3 \end{Bmatrix}^{\mathrm{T}} \begin{Bmatrix} F \\ F \\ F \end{Bmatrix} \end{aligned} \tag{10.64c}$$

Minimizing the total potential energy now gives a set of linear equations,

$$\frac{EI}{l^3} \begin{bmatrix} 4 & 6 & 8 \\ 6 & 12 & 18 \\ 8 & 18 & 144/5 \end{bmatrix} \begin{Bmatrix} \overline{u}_1 \\ \overline{u}_2 \\ \overline{u}_3 \end{Bmatrix} = F \begin{Bmatrix} 1 \\ 1 \\ 1 \end{Bmatrix} \tag{10.65a}$$

from which we obtain

$$\begin{Bmatrix} \overline{u}_1 \\ \overline{u}_2 \\ \overline{u}_3 \end{Bmatrix} = \frac{Fl^3}{EI} \begin{Bmatrix} 1/2 \\ -1/6 \\ 0 \end{Bmatrix} \tag{10.65b}$$

Thus,

$$\overline{v}(x) = \frac{Fl^3}{EI}\left[\frac{1}{2}\left(\frac{x}{l}\right)^2 - \frac{1}{6}\left(\frac{x}{l}\right)^3\right] \tag{10.66a}$$

and

$$\overline{M}(x) = Fl\left(1 - \frac{x}{l}\right) \tag{10.66b}$$

In other words, we have recovered the exact solution, Equations 10.58*b* and 10.58*a*. Note that the method (a) changed the value of $\overline{u}_1$ from $Fl^3/4EI$ (Example 3) when $\overline{v}(x)$ was assumed to contain only the displacement pattern $(x/l)^2$, to $\overline{u}_1 = Fl^3/2EI$ when a better (exact) displacement function was chosen, and (b) rejected the $(x/l)^4$ term entirely since it was not required in the exact solution.

EXAMPLE 5

We will repeat the analyses conducted in Examples 3 and 4 for the beam in Figure 10.18. This differs from the preceding cases only in the position of the concentrated force, which is applied at $x/l = 2/3$ rather than at $x/l = 1$. For reference purposes, we note that the exact solution for this case is

$$M(x) = F\left(\frac{2}{3}l - x\right), \qquad 0 \leq x \leq \frac{2}{3}l$$
$$M(x) = 0, \qquad \frac{2}{3}l \leq x \leq l \tag{10.67a}$$

and

$$v(x) = \frac{Fl}{3EI}x^2 - \frac{F}{6EI}x^3, \qquad 0 \leq x \leq \frac{2}{3}l$$
$$v(x) = \frac{2Fl^2}{9EI}x - \frac{4Fl^3}{81EI}, \qquad \frac{2}{3}l \leq x \leq l \tag{10.67b}$$

(*a*) If we assume the same approximate displacement function as in Equation 10.59, i.e., $\overline{v}(x) = \overline{u}_1(x/l)^2$, then the strain energy will be exactly the same as in Equation 10.60*a*. Only the external potential energy changes as a result of the new location of the applied load; we now have

$$U_e = -F\overline{v}\left(\frac{2}{3}l\right) \tag{10.68}$$

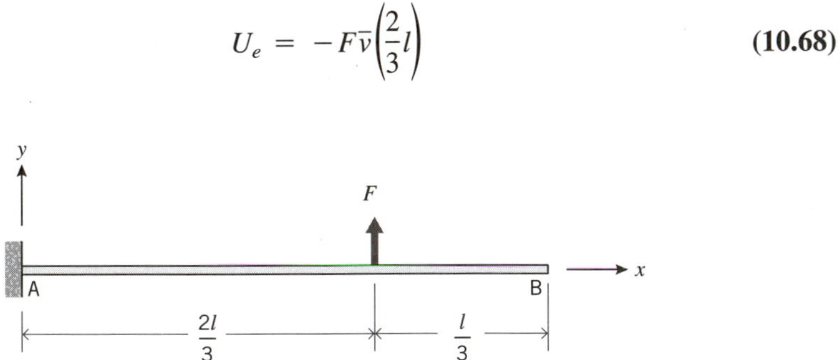

Figure 10.18 Example 5.

$$= -F\bar{u}_1\left(\frac{4}{9}\right)$$

Minimizing the total potential energy leads to

$$\bar{u}_1 = \frac{Fl^3}{9EI} \tag{10.69}$$

It follows that

$$\bar{v}(x) = \frac{Fl^3}{9EI}\left(\frac{x}{l}\right)^2 \tag{10.70a}$$

and

$$\overline{M}(x) = \frac{2Fl}{9} \tag{10.70b}$$

In this case, the assumed displacement pattern produces a solution that is less pleasing than when the load was applied at $x = l$. The assumed displacement pattern fails to recognize the fact that the beam has no curvature or internal moment for $2l/3 < x \leq l$. As a result, we end up with curvature and internal moment everywhere in the beam.

The bending moments from Equations 10.70*b* and 10.67*a* are compared in Figure 10.19. Note that the approximate constant moment is no longer equal to ½ the maximum exact moment, but is now only ⅓ this value. This is because the error in the approximation must extend over the entire beam, which means that errors for the region $2l/3 < x \leq l$ are factored into the approximate solution.

As might be expected, the transverse displacements are also not very accurate. We obtain $\bar{v}(2l/3) = 4Fl^3/81EI$ compared to the exact value $v(2l/3) = 8Fl^3/81EI$. In other words, the displacement at $x = 2l/3$ is only ½ the correct value (compared to ¾ for the beam loaded at $x = l$).

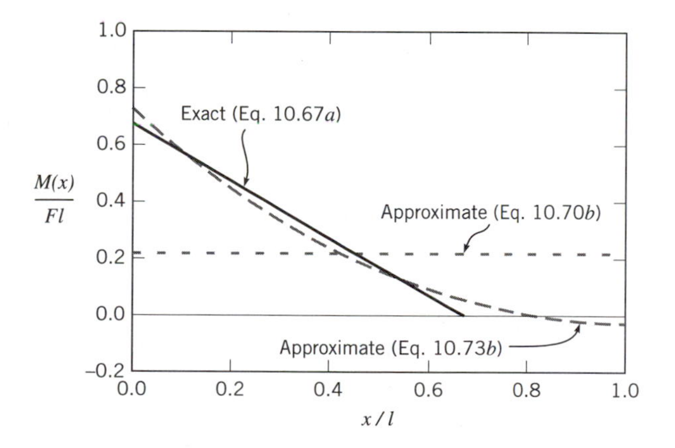

Figure 10.19 Bending moments.

(*b*) If we use the three-term approximation in Equation 10.63, results change dramatically. Strain energy is the same as in Equation 10.64*b*, but the external potential energy becomes

$$\begin{aligned} U_e &= -F\bar{v}\left(\frac{2l}{3}\right) \\ &= -F\left[\bar{u}_1\left(\frac{4}{9}\right) + \bar{u}_2\left(\frac{8}{27}\right) + \bar{u}_3\left(\frac{16}{81}\right)\right] \\ &= -\begin{Bmatrix} \bar{u}_1 \\ \bar{u}_2 \\ \bar{u}_3 \end{Bmatrix}^{\mathrm{T}} \begin{Bmatrix} 4/9 \\ 8/27 \\ 16/81 \end{Bmatrix} F \end{aligned} \quad \textbf{(10.71)}$$

Minimizing the total potential energy leads to a set of equations that resemble Equation 10.65*a* except for a different set of forces on the right side of the equation. Solving the equations gives

$$\begin{Bmatrix} \bar{u}_1 \\ \bar{u}_2 \\ \bar{u}_3 \end{Bmatrix} = \frac{Fl^3}{EI} \begin{Bmatrix} 0.3580 \\ -0.2469 \\ 0.0617 \end{Bmatrix} \quad \textbf{(10.72)}$$

which, in turn, provides

$$\bar{v}(x) = \frac{Fl^3}{EI}\left[0.3580\left(\frac{x}{l}\right)^2 - 0.2469\left(\frac{x}{l}\right)^3 + 0.0617\left(\frac{x}{l}\right)^4\right] \quad \textbf{(10.73a)}$$

and

$$\overline{M}(x) = Fl\left[0.7160 - 1.4815\left(\frac{x}{l}\right) + 0.7407\left(\frac{x}{l}\right)^2\right] \quad \textbf{(10.73b)}$$

Equation 10.73*b* is plotted in Figure 10.19 for comparison with the preceding solutions. The quadratic moment function in Equation 10.73*b* does not reproduce the exact solution, but it does constitute a much better approximate solution than Equation 10.70*b*. Although this approximate solution incorrectly implies the existence of small moments in the region $2l/3 \le x \le l$, the error in the maximum moment ($x = 0$) is now only about 7%.

In this case, the fourth-order displacement pattern does not disappear from the solution. This is because the assumed displacement function, Equation 10.63, is incorrect. As is indicated by Equation 10.67*b*, the true displacement function is expressed by different types of polynomials in the regions $0 \le x \le 2l/3$ and $2l/3 \le x \le l$. In theory, we could have chosen Equation 10.63 to represent the displacements only over the region $0 \le x \le 2l/3$, and could have recognized that only rigid-body motion occurs in the region $2l/3 \le x \le l$; the latter could be represented by a displacement function of form $\bar{v}(x) = \bar{u}_4 + \bar{u}_5(x/l)$. (This is tantamount to starting with the exact solution.) In this case, however, the constants $\bar{u}_1$–$\bar{u}_5$ are not independent because we must impose the condition that displacements and slopes are continuous at $x = 2l/3$.

Before ending our discussion of the classic Rayleigh-Ritz method, it will be useful to examine two additional examples in which displacement patterns other than polynomial forms are employed. In particular, we will demonstrate the use of Fourier series to represent the behavior of beam/structural systems.

EXAMPLE 6

Consider the simply supported, uniformly loaded beam shown in Figure 10.20. Again, for reference purposes, we note that the exact solution is (see Equation 3.19b)

$$v(x) = \frac{q_o l^4}{24EI}\left[\left(\frac{x}{l}\right) - 2\left(\frac{x}{l}\right)^3 + \left(\frac{x}{l}\right)^4\right] \tag{10.74a}$$

and

$$M(x) = -\frac{q_o l^2}{2}\left[\left(\frac{x}{l}\right) - \left(\frac{x}{l}\right)^2\right] \tag{10.74b}$$

We assume a displacement function in the form of a Fourier series of sine terms,

$$\begin{aligned}\bar{v}(x) &= \bar{u}_1 \sin\frac{\pi x}{l} + \bar{u}_2 \sin\frac{2\pi x}{l} + \bar{u}_3 \sin\frac{3\pi x}{l} + \ldots \\ &= \sum_{n=1}^{\infty} \bar{u}_n \sin\frac{n\pi x}{l}\end{aligned} \tag{10.75}$$

Note that each of the displacement patterns in Equation 10.75 satisfies the geometric boundary conditions $v(0) = v(l) = 0$.

The strain energy, Equation 10.57a, is given by

$$U = \frac{EI}{2}\int_0^l\left[-\bar{u}_1\left(\frac{\pi}{l}\right)^2 \sin\frac{\pi x}{l} - \bar{u}_2\left(\frac{2\pi}{l}\right)^2 \sin\frac{2\pi x}{l} - \bar{u}_3\left(\frac{3\pi}{l}\right)^2 \sin\frac{3\pi x}{l} - \ldots\right]^2 dx$$

$$\begin{aligned}&= \frac{EI}{2}\sum_{n=1}^{\infty}\int_0^l\left[-\bar{u}_n\left(\frac{n\pi}{l}\right)^2 \sin\frac{n\pi x}{l}\right]^2 dx \\ &= \frac{EI\pi^4}{4l^3}\sum_{n=1}^{\infty} n^4\bar{u}_n^2\end{aligned} \tag{10.76a}$$

The simple form of this result is a consequence of the orthogonality of the sine terms, i.e., of the fact that

$$\begin{aligned}&\int_0^l\left(\sin\frac{n\pi x}{l}\right)^2 dx = \frac{l}{2}, \qquad n = 1, 2, \ldots, \infty \\ &\int_0^l \sin\frac{n\pi x}{l}\sin\frac{m\pi x}{l}\, dx = 0, \qquad n, m = 1, 2, \ldots, \infty, \qquad n \neq m\end{aligned} \tag{10.76b}$$

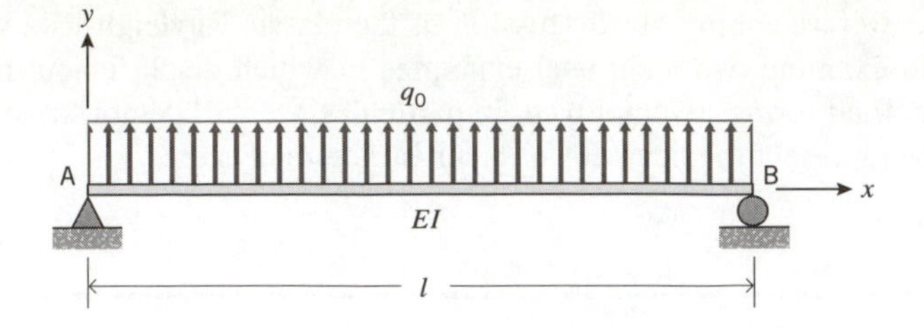

Figure 10.20 Example 6.

The external potential energy is

$$U_e = -\int_0^l q_o\left(\bar{u}_1 \sin\frac{\pi x}{l} + \bar{u}_2 \sin\frac{2\pi x}{l} + \bar{u}_3 \sin\frac{3\pi x}{l} + \dots\right)dx$$
$$= -\frac{q_o l}{\pi}\sum_{n=1}^{\infty}\frac{\bar{u}_n}{n}(1 - \cos n\pi) \qquad \textbf{(10.77)}$$

Therefore, minimizing the total potential energy leads to a set of equations of form

$$\frac{EI\pi^4}{4l^3}\cdot 2n^4\bar{u}_n - \frac{q_o l}{\pi}\cdot\frac{1}{n}(1 - \cos n\pi) = 0, \qquad n = 1, 2, \dots, \infty \qquad \textbf{(10.78}\textbf{\textit{a}}\textbf{)}$$

Solving for $\bar{u}_n$ gives

$$\bar{u}_n = \frac{2q_o l^4}{n^5\pi^5 EI}(1 - \cos n\pi), \qquad n = 1, 2, \dots, \infty \qquad \textbf{(10.78}\textbf{\textit{b}}\textbf{)}$$

We recognize that the term in parentheses is equal to zero for all even values of n, and is equal to 2 for all odd values of n; therefore, we can rewrite Equation 10.78*b* in the form

$$\bar{u}_n = \frac{4q_o l^4}{n^5\pi^5 EI}, \qquad n = 1, 3, 5, \dots, \infty \qquad \textbf{(10.78}\textbf{\textit{c}}\textbf{)}$$

Substituting Equation 10.78*c* into Equation 10.75*a* provides the solution

$$\bar{v}(x) = \frac{4q_o l^4}{\pi^5 EI}\sum_{n=1,3,\dots}^{\infty}\frac{1}{n^5}\sin\frac{n\pi x}{l} \qquad \textbf{(10.79}\textbf{\textit{a}}\textbf{)}$$

from which we obtain

$$\overline{M}(x) = -\frac{4q_o l^2}{\pi^3}\sum_{n=1,3,\dots}^{\infty}\frac{1}{n^3}\sin\frac{n\pi x}{l} \qquad \textbf{(10.79}\textbf{\textit{b}}\textbf{)}$$

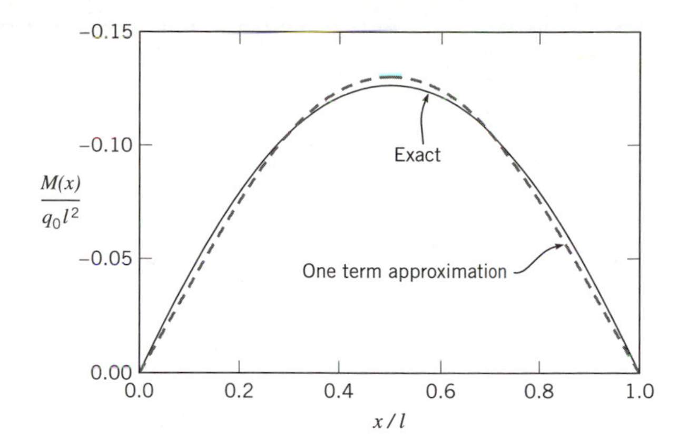

Figure 10.21 Bending moments.

It may be observed that successive terms in Equations 10.79*a* and 10.79*b* decrease very rapidly due to the factors $1/n^5$ and $1/n^3$, respectively, indicating that only a few terms of each series are required to adequately represent an approximate solution. For example, Figure 10.21 shows a comparison of the exact bending moment, Equation 10.74*b*, and an approximate solution corresponding to only the first term of Equation 10.79*b*, i.e.,

$$\overline{M}(x) = -\frac{4q_o l^2}{\pi^3} \sin\frac{\pi x}{l} \tag{10.80a}$$

The magnitude of the approximate bending moment at $x = l/2$ is $4q_o l^2/\pi^3 = 0.129 q_o l^2$, which differs from the exact value, $q_o l^2/8$, by approximately 3%.

As anticipated, an approximate solution consisting of the first two terms of Equation 10.79*b* gives even better results,

$$\overline{M}(x) = -\frac{4q_o l^2}{\pi^3}\left(\sin\frac{\pi x}{l} + \frac{1}{27}\sin\frac{3\pi x}{l}\right) \tag{10.80b}$$

In this case, Equation 10.80*b* is virtually indistinguishable from the exact solution plotted in Figure 10.21, and the approximate and exact bending moments at $x = l/2$ differ by only about 0.6%. As usual, displacements are in even better agreement. Thus, in this particular example, the accuracy of the approximate solution is excellent; in fact, the Fourier series solution converges to the exact solution and, as noted, the convergence is very rapid.

The degree of accuracy obtained in this example is certainly not universal. As will be demonstrated in the following example, complicating factors such as concentrated stiffness or loading can have a major effect on results.

EXAMPLE 7

Consider the structure shown in Figure 10.22, which is similar to the beam in Figure 10.20, but which includes a spring support at $x = l/2$. The exact solution for this one-degree indeterminate structure may be obtained by using either the force or displacement method of analysis (Chapters 5 or 8). The midspan displacement and bending moment for the beam are found to be

$$v(l/2) = \frac{\dfrac{5q_o l}{2k}}{4 + \dfrac{192EI}{kl^3}}$$

$$M(l/2) = \frac{\left(\dfrac{1}{8} - \dfrac{24EI}{kl^3}\right) q_o l^2}{4 + \dfrac{192EI}{kl^3}} \qquad \textbf{(10.81a)}$$

where k is the spring constant. For computational purposes, we will set $k = \alpha\pi^4 EI/2l^3$ where α is a constant, and let $\alpha = 20$, the case for which $k \approx 1000EI/l^3$. Equations 10.81a then reduce to

$$v(l/2) = 6.1149 \times 10^{-4} \frac{q_o l^4}{EI}$$

$$M(l/2) = 0.02391\, q_o l^2 \qquad \textbf{(10.81b)}$$

Given this value of $M(l/2)$, the exact internal bending moment in the beam may be obtained from considerations of statics,

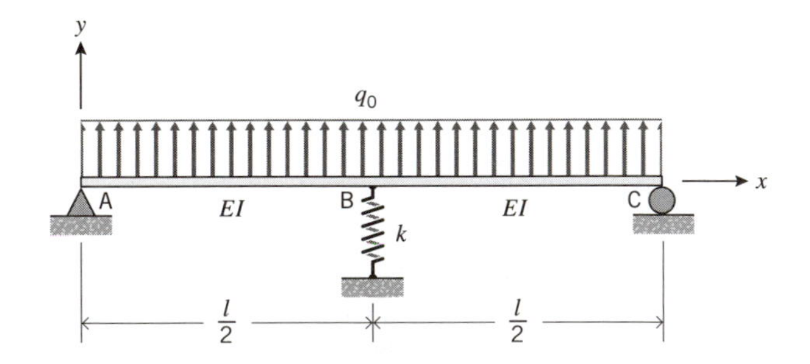

Figure 10.22 Example 7.

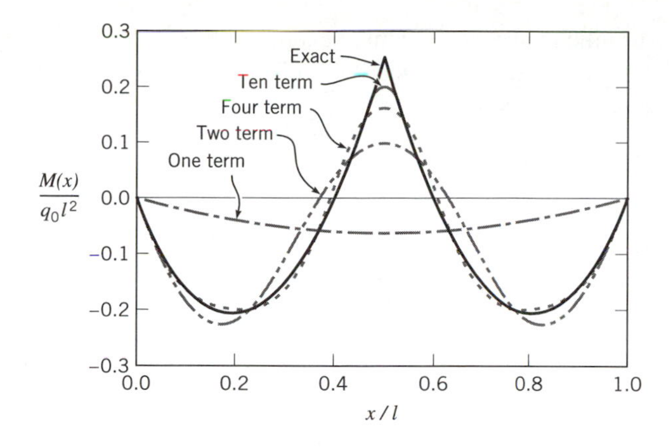

Figure 10.23 Bending moments.

$$M(x) = \frac{q_o x^2}{2} - 0.20218\, q_o l x, \qquad 0 < x < \frac{l}{2}$$

$$M(x) = \frac{q_o (l - x)^2}{2} - 0.20218\, q_o l (l - x), \qquad \frac{l}{2} < x < 1 \tag{10.81c}$$

The bending moment, Equation 10.81*c*, is plotted in Figure 10.23.

We will compare the foregoing exact result to solutions based on the approximate displacement function in Equation 10.75 (Example 6). In this case, the spring is an additional axially loaded element in which strain energy is stored during deformation. The strain energy in the spring is obtained from Equation 10.56*a*,

$$\begin{aligned} U_{spring} &= \frac{1}{2} k \Delta^2 \\ &= \frac{1}{2} k [\bar{v}(l/2)]^2 \\ &= \frac{1}{2} k \left[\sum_{n=1}^{\infty} \bar{u}_n \sin \frac{n\pi}{2} \right]^2 \\ &= \frac{1}{2} k \left[\sum_{n=1,3,\ldots}^{\infty} (-1)^{\frac{n-1}{2}} \bar{u}_n \right]^2 \end{aligned} \tag{10.82}$$

since $P = k\Delta$, where P is the axial force in the spring and Δ is the spring deformation, which is equal to $\bar{v}(l/2)$.

The strain energy in the beam is still given by Equation 10.76*a*. The external potential energy is unchanged by the addition of the spring to the structure, and is given by Equation 10.77. Thus, the total potential energy is the sum of Equations 10.76*a*, 10.77, and 10.82.

Minimizing the total potential energy leads to a set of equations, which are somewhat more complicated than Equation 10.78*a*. These equations will be easiest to solve if we represent them in matrix form,

$$\begin{bmatrix} \frac{\pi^4 EI}{2l^3} + k & -k & k & \cdots \\ -k & \frac{3^4\pi^4 EI}{2l^3} + k & -k & \cdots \\ k & -k & \frac{5^4\pi^4 EI}{2l^3} + k & \cdots \\ \cdots & \cdots & \cdots & \cdots \end{bmatrix} \begin{Bmatrix} \bar{u}_1 \\ \bar{u}_3 \\ \bar{u}_5 \\ \cdots \end{Bmatrix} = \frac{2q_o l}{\pi} \begin{Bmatrix} 1 \\ 1/3 \\ 1/5 \\ \cdots \end{Bmatrix} \tag{10.83a}$$

If, as above, we set $k = \alpha\pi^4 EI/2l^3$, then Equation 10.83*a* becomes

$$\begin{vmatrix} 1^4 + \alpha & -\alpha & \alpha & \cdots \\ -\alpha & 3^4 + \alpha & -\alpha & \cdots \\ \alpha & -\alpha & 5^4 + \alpha & \cdots \\ \cdots & \cdots & \cdots & \cdots \end{vmatrix} \begin{Bmatrix} \bar{u}_1 \\ \bar{u}_3 \\ \bar{u}_5 \\ \cdots \end{Bmatrix} = \frac{4q_o l^4}{\pi^5 EI} \begin{Bmatrix} 1 \\ 1/3 \\ 1/5 \\ \cdots \end{Bmatrix} \tag{10.83b}$$

We will again let $\alpha = 20$ and examine the solutions that are obtained for different numbers of terms in the approximate displacement function.

For a displacement function containing only one term, i.e., $n = 1$, Equation 10.83*b* gives

$$\bar{u}_1 = \frac{4q_o l^4}{21\pi^5 EI} = 6.224 \times 10^{-4}\,\frac{q_o l^4}{EI} \tag{10.84a}$$

which corresponds to

$$\begin{aligned} \bar{v}(x) &= \frac{4q_o l^4}{21\pi^5 EI}\sin\frac{\pi x}{l} \\ \overline{M}(x) &= -\frac{4q_o l^2}{21\pi^3}\sin\frac{\pi x}{l} \end{aligned} \tag{10.84b}$$

and

$$\begin{aligned} \bar{v}(l/2) &= \frac{4q_o l^4}{21\pi^5 EI} = 6.224 \times 10^{-4}\,\frac{q_o l^4}{EI} \\ \overline{M}(l/2) &= -\frac{4q_o l^2}{21\pi^3} = -6.143 \times 10^{-3}\,q_o l^2 \end{aligned} \tag{10.84c}$$

We see that the approximate midspan displacement, Equation 10.84*c*, is in good agreement with the exact value, Equation 10.81*b*; however, the approximate bending moment is very far from the correct value. The discrepancy is readily apparent from Figure

10.23 where the second of Equations 10.84*b* has been plotted for comparison with the exact solution.

For an approximate solution involving two displacement terms, Equation 10.83*b* reduces to

$$\begin{bmatrix} 21 & -20 \\ -20 & 101 \end{bmatrix} \begin{Bmatrix} \bar{u}_1 \\ \bar{u}_3 \end{Bmatrix} = \frac{4q_o l^4}{\pi^5 EI} \begin{Bmatrix} 1 \\ 1/3 \end{Bmatrix} \tag{10.85a}$$

from which we obtain

$$\begin{Bmatrix} \bar{u}_1 \\ \bar{u}_3 \end{Bmatrix} = \frac{4q_o l^4}{\pi^5 EI} \begin{Bmatrix} 6.2561 \times 10^{-2} \\ 1.5689 \times 10^{-2} \end{Bmatrix} = \frac{q_o l^4}{EI} \begin{Bmatrix} 8.1773 \times 10^{-4} \\ 2.0507 \times 10^{-4} \end{Bmatrix} \tag{10.85b}$$

This leads to

$$\begin{aligned} \bar{v}(x) &= \frac{q_o l^4}{EI}\left(8.1773 \times 10^{-4} \sin\frac{\pi x}{l} + 2.0507 \times 10^{-4} \sin\frac{3\pi x}{l}\right) \\ \overline{M}(x) &= -q_o l^2\left(0.8071 \times 10^{-2} \sin\frac{\pi x}{l} + 1.8215 \times 10^{-2} \sin\frac{3\pi x}{l}\right) \end{aligned} \tag{10.85c}$$

and

$$\begin{aligned} \bar{v}(l/2) &= 6.1267 \times 10^{-4} \frac{q_o l^4}{EI} \\ \overline{M}(l/2) &= 1.0145 \times 10^{-2}\, q_o l^2 \end{aligned} \tag{10.85d}$$

As may be seen from Figure 10.23, this two-term solution exhibits the general characteristics of the exact solution, but provides less than half the correct peak moment.

An approximate solution involving four displacement terms gives

$$\begin{Bmatrix} \bar{u}_1 \\ \bar{u}_3 \\ \bar{u}_5 \\ \bar{u}_7 \end{Bmatrix} = \frac{4q_o l^4}{\pi^5 EI} \begin{Bmatrix} 6.4092 \times 10^{-2} \\ 1.5670 \times 10^{-2} \\ -1.1775 \times 10^{-3} \\ 4.4930 \times 10^{-4} \end{Bmatrix} = \frac{q_o l^4}{EI} \begin{Bmatrix} 8.3775 \times 10^{-4} \\ 2.0482 \times 10^{-4} \\ -1.5391 \times 10^{-5} \\ 5.8728 \times 10^{-6} \end{Bmatrix} \tag{10.86a}$$

from which we obtain

$$\begin{aligned} \bar{v}(x) = \frac{q_o l^4}{EI}\bigg(&8.3775 \times 10^{-4} \sin\frac{\pi x}{l} + 2.0482 \times 10^{-4} \sin\frac{3\pi x}{l} \\ &-1.5391 \times 10^{-5} \sin\frac{5\pi x}{l} + 5.8728 \times 10^{-6} \sin\frac{7\pi x}{l}\bigg) \end{aligned} \tag{10.86b}$$

$$\overline{M}(x) = -q_o l^2\left(8.2682 \times 10^{-3} \sin\frac{\pi x}{l} + 1.8193 \times 10^{-2} \sin\frac{3\pi x}{l}\right.$$

$$\left. -3.7975 \times 10^{-3} \sin\frac{5\pi x}{l} + 2.8402 \times 10^{-3} \sin\frac{7\pi x}{l}\right)$$

and

$$\overline{v}(l/2) = 6.1167 \times 10^{-4}\frac{q_o l^4}{EI}$$

$$\overline{M}(l/2) = 1.6563 \times 10^{-2}\, q_o l^2 \qquad \textbf{(10.86c)}$$

This approximate solution illustrates the rapid convergence of the displacement values at $x = l/2$, i.e., the value $6.1167 \times 10^{-4} q_o l^4/EI$, Equation 10.86*c*, compares quite well to the exact value in Equation 10.81*b*. This implies that the force in the stiff axial spring also converges very rapidly to the correct value. However, the moment values converge much more slowly. The four-term approximation gives $M(l/2) = 1.6563 \times 10^{-2} q_o l^2$, which is only about 70% of the exact value in Equation 10.81*b* even though the overall moment diagrams appear to be in generally good agreement.

There are two explanations for the nature of the convergence characteristics exhibited by the approximate solutions: (1) in the process of minimizing the total potential energy, the flexible beam is approximated with less accuracy than the relatively stiff spring, and (2) the sine terms employed in $\overline{v}(x)$ are essentially smoothly varying functions with first derivatives that are equal to zero at $x = l/2$; such functions cannot be expected to readily converge to a function such as $M(x)$, which has a discontinuity in derivatives at $x = l/2$.

Actually, if a larger number of terms is employed in the approximate solution then the approximate moment function can be made to approach the exact function, although there will always be some imprecision in the immediate vicinity of $x = l/2$. For instance, if we take ten terms we will obtain

$$\begin{Bmatrix} \overline{u}_1 \\ \overline{u}_3 \\ \overline{u}_5 \\ \overline{u}_7 \\ \overline{u}_9 \\ \overline{u}_{11} \\ \overline{u}_{13} \\ \overline{u}_{15} \\ \overline{u}_{17} \\ \overline{u}_{19} \end{Bmatrix} = \frac{q_o l^4}{EI}\begin{Bmatrix} 8.4099 \times 10^{-4} \\ 2.0478 \times 10^{-4} \\ -1.5385 \times 10^{-5} \\ 5.8715 \times 10^{-6} \\ -1.6427 \times 10^{-6} \\ 9.1649 \times 10^{-7} \\ -3.9301 \times 10^{-7} \\ 2.5880 \times 10^{-7} \\ -1.3723 \times 10^{-7} \\ 9.9125 \times 10^{-8} \end{Bmatrix} \qquad \textbf{(10.87)}$$

Notice the relatively slow reduction in magnitude of the $\overline{u}_n$ for $n \geq 11$. Apparently, the use of additional terms provides very little overall improvement except near $x = l/2$ where this solution gives $M(l/2) = 2.0908 \times 10^{-2} q_o l^2$, which is still about 12% low (see Figure 10.23).

The preceding example illustrates that considerable algebraic effort may be required to obtain reasonable results using the Rayleigh-Ritz method even when well-known classic solution forms, such as Fourier series, are employed. The basic difficulty in this case was attributable largely to an abrupt change in internal physics of behavior near the connection between the beam and the stiff spring. In other words, when the structure consists of an assembly of elements with different characteristics, the method is more difficult to use effectively. In fact, prior to the introduction of the finite element method in 1960, the original form of the Rayleigh-Ritz procedure, in which each assumed displacement function was chosen to represent behavior over the entire body, was a practical tool only for the simplest structures. As we have seen, accurate solutions were difficult to obtain unless many displacement patterns were carried in the analysis.

By 1960, with the advent of high-speed digital computing machinery, the issue of solving large systems of algebraic equations was no longer a limiting consideration. Several hundred, or even thousand, linear equations could be readily processed. However, the basic problem of how to develop effective and accurate displacement patterns for reliable analysis still existed. The introduction of the finite element method, which we shall consider in Chapter 11, provided a great step forward in engineering analysis.

In the finite element method, the approximate (assumed) displacement patterns were developed only over individual ''elements.'' Each element was either a physical member connected between joints with other members (e.g., beam and bar elements) or an even smaller portion or subregion of the structure. The key idea was to make each element small enough so that the physical behavior of the region was quite simple (i.e., easy to guess), and with a simple enough shape to be easily connected to other elements.

The finite element method has completely revolutionized modern structural analysis. Nearly all analyses, at present, employ this method, and nearly all analyses are performed on personal computers, work stations, or main frame computers.

PROBLEMS

10.P1. As the new director of a structures testing laboratory, you are informed about the following problem. The laboratory recently conducted tests on an elastic structure to experimentally determine a 3×3 flexibility matrix $\boldsymbol{\alpha}$. Unfortunately, because of an instrumentation problem, the values of α_{12}, α_{13}, and α_{23} have been mixed up. All that is known is that two of the values are equal to 3 and one is equal to -2. Thus, the following possibilities exist:

$$\boldsymbol{\alpha} = \begin{bmatrix} 5 & 3 & -2 \\ 3 & 4 & 3 \\ -2 & 3 & 8 \end{bmatrix} \quad \text{or} \quad \boldsymbol{\alpha} = \begin{bmatrix} 5 & 3 & 3 \\ 3 & 4 & -2 \\ 3 & -2 & 8 \end{bmatrix} \quad \text{or} \quad \boldsymbol{\alpha} = \begin{bmatrix} 5 & -2 & 3 \\ -2 & 4 & 3 \\ 3 & 3 & 8 \end{bmatrix}$$

Is it possible to determine if one of these choices for the flexibility matrix is better than the others? (If so, it may not be necessary to rerun the expensive tests.) Explain how you would examine each matrix to see if it is theoretically acceptable. If possible, find the choice or choices that may be acceptable.

10.P2. The frame structure shown is subjected to two independent loads. Load 1 consists of a horizontal force F applied at joint B, load 2 consists of a moment M also applied at joint B. At joint B, load 1 causes horizontal displacement u_1 and rotation θ; load 2 causes horizontal displacement u_2 and the same rotation θ.

(a) Is $Fu_1 < M\theta$, $Fu_1 = M\theta$, or $Fu_1 > M\theta$? Explain your answer.
(b) Which is greater, u_1 or u_2? Explain your answer.

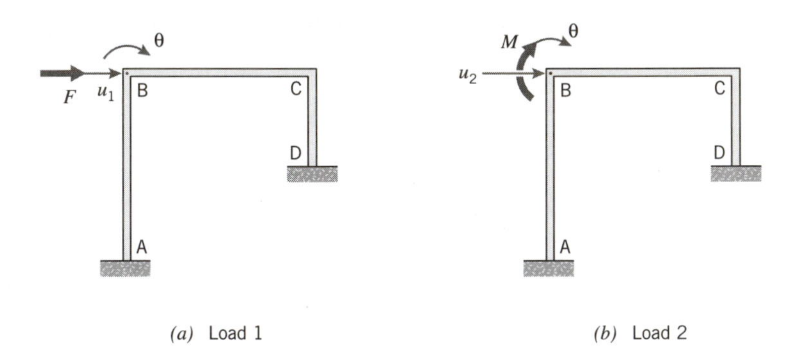

(a) Load 1   *(b)* Load 2

10.P3. Consider any equilibrium equation like Equation 10.35*a*, in which all matrices are written in a partitioned form. Assume $\mathbf{u}$ and $\boldsymbol{\Delta}$ are $m \times 1$ and $n \times 1$ subvectors of displacements, respectively; $\mathbf{u_o}$ and $\boldsymbol{\Delta_o}$ are $m \times 1$ and $n \times 1$ subvectors of initial displacements, respectively; $\mathbf{f}$ and $\mathbf{X}$ are $m \times 1$ and $n \times 1$ subvectors of applied forces, respectively; and $\mathbf{k_{uu}}$, $\mathbf{k_{u\Delta}}$, and $\mathbf{k_{\Delta\Delta}}$ are $m \times m$ and $m \times n$ and $n \times n$ submatrices, respectively, of the system stiffness. Show that the equation may be inverted and written in the form given by Equations 10.35*c* and 10.36.

10.P4. A "twisted" beam/truss structure has the configuration shown before any loads are applied. Beam element DEF is rigid; truss bars AD, BE, and CF all have stiffness EA. The dotted lines in the figure lie in the x-y plane; beam DEF lies in a plane parallel to the y-z plane; bars AD and CF lie in planes parallel to the x-z plane. The support at point E permits axial displacement u_x and rotation about the x axis θ_x; all other motions are zero at this support. Loads F_x and M_x are applied at joint E. Find the forces in bars AD, BE, and CF. Find the rotation when $M_x = 0$. Solve using Castigliano's theorem, part 1. (Compare to results obtained in Problem 8.P16, Chapter 8.)

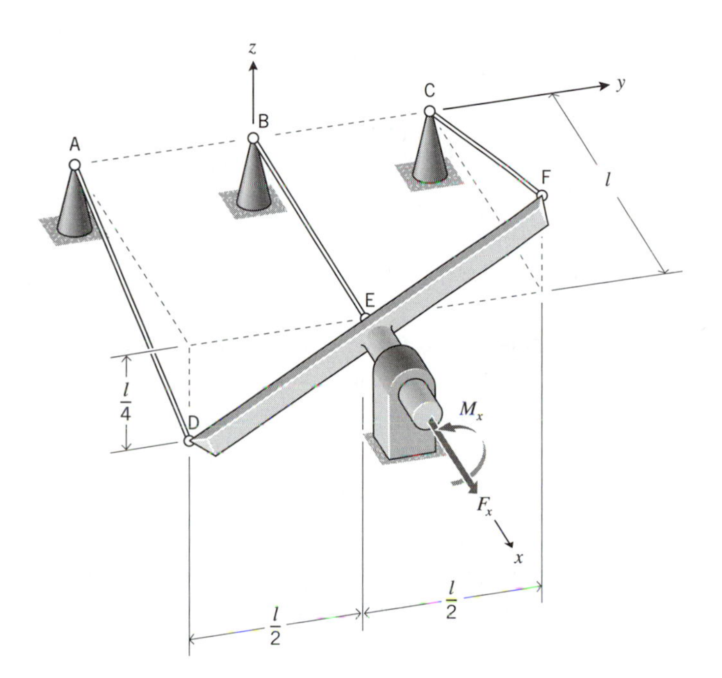

10.P5. Find the strain energy in each of the structures shown and sketch the bending moment diagrams for each.

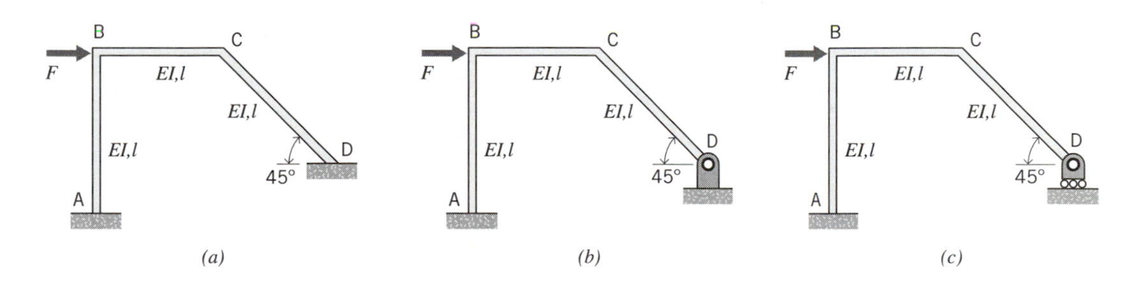

10.P6. (a) Find the exact bending moment as a function of x for the beam shown.

(b) Find the exact value of the strain energy for the beam in (a).

(c) For the purpose of conducting an approximate analysis, it is proposed to insert a hinge in the beam at an arbitrary location $x = a$, $0 < a < l/2$. This results in a statically determinate structure. Find the approximate bending moment for this case as a function of a.

(d) Find the strain energy for the beam in (c). (This strain energy is a function of a).

(e) What value of a minimizes the strain energy found in (d)? Why?

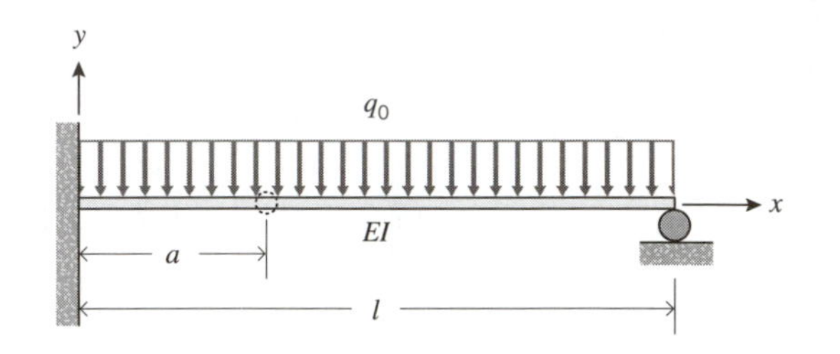

10.P7. It is proposed to analyze the beam shown in Problem 10.P6 using the following approximate displacement function:

$$\bar{v}(x) = C\frac{x}{l}\sin\frac{\pi x}{l}$$

(a) Find the strain energy in the beam for this displacement.

(b) Find the external potential, i.e., find the generalized force.

(c) Find the generalized displacement C.

(d) Sketch the bending moment diagram associated with the approximate displacement field.

(e) Compare the approximate strain energy to the exact result obtained in Problem 10.P6.

10.P8. Consider the uniform cantilever beam shown in Figure 10.18, which has a concentrated force F applied at $x = 2l/3$.

(a) Find an approximate displacement pattern using the Rayleigh-Ritz method and the following assumed displacement field, which is defined for $0 < x < l$:

$$\bar{v}(x) = C_1\left(\frac{x}{l}\right)^2 + C_2\left(\frac{x}{l}\right)^3$$

Sketch the approximate moment diagram, and compare it with the exact solution.

(b) To improve the approximate solution at $x = l$, i.e., force $M(l)$ toward a zero value, it is proposed to add a penalty function P to the total potential,

$$V = U + U_e + P$$

where

$$P = \beta\frac{M^2(l)l}{2EI} = \beta\frac{1}{2}EIl[v''(l)]^2$$

and where β is a constant. Find the approximate solution using this augmented total potential function and the strain energy U. Consider two separate cases, $\beta = 1$ and $\beta = 10$. From this study, what conclusions can you draw regarding the use of penalty functions?

Chapter 11

Introduction to Finite Element Analysis

11.0 INTRODUCTION

The concepts introduced in Chapter 10 enabled us to demonstrate that approximate displacement fields that satisfy all geometric boundary conditions lead to structural responses for which the stored strain energy is less than that of the exact equilibrium state. Also, we indicated that the approximate solution state was a type of "best fit" representation of the true structural behavior. We will elaborate further on this point as we examine the approximations that are inherent in finite element analysis.

The essence of finite element analysis is the development of assumed displacement patterns over finite portions of the physical structure. In this chapter, we will demonstrate the basic procedures for finite element analysis within the context of structures composed of beam elements.

11.1 BASIC REQUIREMENTS

Whenever a complete structure is divided into an assemblage of discrete finite elements, such as shown in Figure 10.13, it is necessary that each element satisfy certain physical conditions. We will begin by describing these conditions.

(1) Rigid Body Requirement

If free motion is permitted, the element must have the capability of being moved as a rigid body without accumulating strain energy.

For a beam bending element of length l, this means that the transverse displacement function must contain two terms corresponding to arbitrary transverse displacement and rotation,

$$\bar{v}^{(a)}(x) = C_1 + C_2\left(\frac{x}{l}\right) \tag{11.1a}$$

where C_1 and C_2 are arbitrary constants (generalized displacements).

(2) Constant Strain Requirement

If a physical object subjected to applied loads is viewed as an assembly of many small subregions or finite elements, then, when these regions or elements are made small enough, the exact strain (and hence stress) within each element will effectively be constant regardless of the complexity of the overall strain field in the object. Therefore, an element must have the capability of developing a constant strain (stress) state.

For a beam element, the strain quantity of interest is actually the bending curvature; consequently, this condition is equivalent to requiring a displacement pattern of form

$$\bar{v}^{(b)}(x) = C_3\left(\frac{x}{l}\right)^2 \tag{11.1b}$$

such that $\bar{v}''^{(b)}(x) = 2C_3/l^2 = \text{constant}$.

Thus, in order to satisfy requirements (1) and (2) above, a finite element for a beam must have an assumed displacement function of form

$$\bar{v}^{(c)}(x) = \bar{v}^{(a)}(x) + \bar{v}^{(b)}(x) = C_1 + C_2\left(\frac{x}{l}\right) + C_3\left(\frac{x}{l}\right)^2 \tag{11.1c}$$

(3) Compatibility Requirement

To use the finite element, it must be possible to connect it to other similar elements in a compatible manner.

For beam elements, each of the two ends is free to translate and rotate; consequently, we require at least four independent displacement patterns with associated generalized displacements to satisfy the arbitrary end conditions. The four patterns should include the three in Equation 11.1*c*. For example, we could choose displacement functions such as one of the three following forms,

$$\bar{v}^{(1)}(x) = C_1 + C_2\left(\frac{x}{l}\right) + C_3\left(\frac{x}{l}\right)^2 + C_4\left(\frac{x}{l}\right)^3 \tag{11.2a}$$

or

$$\bar{v}^{(2)}(x) = C_1 + C_2\left(\frac{x}{l}\right) + C_3\left(\frac{x}{l}\right)^2 + C_4\left(\frac{x}{l}\right)^4 \tag{11.2b}$$

or

$$\bar{v}^{(3)}(x) = C_1 + C_2\left(\frac{x}{l}\right) + C_3\left(\frac{x}{l}\right)^2 + C_4 \sin\frac{2\pi x}{l} \tag{11.2c}$$

Each of these displacement functions satisfies requirements (1) through (3).

The first function, Equation 11.2*a*, is our basic beam finite element model. It is also the exact solution form for any uniform beam loaded only at its ends, i.e., with no interior loads. In other words, Equation 11.2*a* is the general solution of the differential equation

$$EI\frac{d^4\bar{v}^{(1)}(x)}{dx^4} = 0 \tag{11.3}$$

The form of Equation 11.2*a* becomes more familiar when it is rewritten in terms of end displacements and rotations. If we set $\bar{v}^{(1)}(0) = v_{IJ}$, $\bar{v}^{(1)}(l) = v_{JI}$, $\bar{v}'^{(1)}(0) = -\theta_{IJ}$, and $\bar{v}'^{(1)}(l) = -\theta_{JI}$, and solve for the constants, we obtain[1]

$$\begin{aligned}\bar{v}^{(1)}(x) &= \theta_{IJ}\left(-x + \frac{2x^2}{l} - \frac{x^3}{l^2}\right) + \theta_{JI}\left(\frac{x^2}{l} - \frac{x^3}{l^2}\right) \\ &\quad + v_{IJ}\left(1 - \frac{3x^2}{l^2} + \frac{2x^3}{l^3}\right) + v_{JI}\left(\frac{3x^2}{l^2} - \frac{2x^3}{l^3}\right)\end{aligned} \tag{11.4}$$

The terms in parentheses above are the so-called "beam functions." It may be noted that Equation 11.4 is identical to Equation 10.49.

Displacement function $\bar{v}^{(2)}(x)$, Equation 11.2*b*, is the solution of

$$EI\frac{d^4\bar{v}^{(2)}(x)}{dx^4} = \frac{24}{l^4}C_4 \tag{11.5}$$

which is the governing differential equation for a beam subjected to the particular applied load $q(x) = q_o = 24C_4/l^4$. In terms of end displacements, Equation 11.2*b* becomes

$$\begin{aligned}\bar{v}^{(2)}(x) &= \theta_{IJ}\left(-x + \frac{3x^2}{2l} - \frac{x^4}{2l^3}\right) + \theta_{JI}\left(\frac{x^2}{2l} - \frac{x^4}{2l^3}\right) \\ &\quad + v_{IJ}\left(1 - \frac{2x^2}{l^2} + \frac{x^4}{l^4}\right) + v_{JI}\left(\frac{2x^2}{l^2} - \frac{x^4}{l^4}\right)\end{aligned} \tag{11.6}$$

i.e., we have $C_4 = v_{IJ} - v_{JI} - (\theta_{IJ} + \theta_{JI})l/2$; therefore, Equation 11.2*b* constitutes an exact solution only for the case of a beam element subjected to the specific load $q_o = 24[v_{IJ} - v_{JI} - (\theta_{IJ} + \theta_{JI})l/2]/l^4$, a highly unlikely circumstance! A similar conclusion can be drawn about $\bar{v}^{(3)}(x)$.

Actually, a comparison of $\bar{v}^{(1)}(x)$ and $\bar{v}^{(2)}(x)$ provides a simpler rationale for favoring the former over the latter. We observe that $\bar{v}^{(1)}(x)$ is able to model a linear variation in internal bending moment,

---

[1] As in previous chapters, we are using I and J to denote the left and right ends of the beam element, respectively.

$$\overline{M}^{(1)}(x) = EI\overline{v}''^{(1)}(x) = EI\left(\frac{2C_3}{l^2} + \frac{6C_4x}{l^3}\right) \quad \textbf{(11.7a)}$$

whereas $\overline{v}^{(2)}(x)$, because it is missing the term $(x/l)^3$, gives

$$\overline{M}^{(2)}(x) = EI\overline{v}''^{(2)}(x) = EI\left(\frac{2C_3}{l^2} + \frac{12C_4x^2}{l^4}\right) \quad \textbf{(11.7b)}$$

In other words, Equation 11.7*b* gives a quadratic moment, but without a linear component. In general, it is desirable to use assumed displacement functions that represent physical effects in a predictable manner, with an orderly increase in modeling complexity as increasingly complex displacement patterns are chosen. The fact that $\overline{v}^{(2)}(x)$ skips over the linear component is a serious shortcoming, and makes it a much less desirable displacement function than $\overline{v}^{(1)}(x)$.

Displacement function $\overline{v}^{(3)}(x)$, Equation 11.2*c*, has all integer powers of x contained in the pattern $\sin(2\pi x/l)$, but the modes are all linked to a single constant C_4. Thus, this displacement function also cannot model a linearly varying moment within the element or, for that matter, a quadratic distribution!

This line of reasoning suggests two additional conditions, which it would be desirable to satisfy in the selection of assumed displacement functions; these conditions are not required:

(4) Assumed displacement functions should model increasingly complex modes of behavior in an orderly manner, starting from constant strain, to linear strain, to quadratic strain, etc., without skipping any particular mode of behavior.

(5) The assumed displacement function should be as simple as possible, while satisfying conditions (1) through (4).

Condition (5) alone would tend to support the choice of displacement function $\overline{v}^{(1)}(x)$ over either $\overline{v}^{(2)}(x)$ or $\overline{v}^{(3)}(x)$.

11.2 STIFFNESS MATRIX AND FORCE VECTOR FOR ASSUMED DISPLACEMENTS

Each assumed displacement function leads to a particular stiffness matrix for the element and, for any applied loads, to a vector of generalized nodal forces (see Section 10.3). These quantities can be developed using either an energy approach or a direct virtual work approach.

11.2.1 Development Using Energy

The stiffness can be developed from a consideration of the strain energy, Equations 10.57*a* and 10.15*d*. We begin by writing any assumed displacement function in matrix form,

$$\overline{v}(x) = [A_1(x)\ A_2(x)\ A_3(x)\ \ldots\ A_n(x)]\begin{Bmatrix}\overline{u}_1\\ \overline{u}_2\\ \overline{u}_3\\ \vdots\\ \overline{u}_n\end{Bmatrix} = [\mathbf{A}(x)]\overline{\mathbf{u}} \quad \textbf{(11.8)}$$

where $A_i(x)$ is the displacement pattern associated with generalized displacement $\bar{u}_i$, $i = 1, \ldots, n$.[2] These generalized displacements must ultimately be related to system displacements. Thus, for a beam element, the generalized displacements are chosen as end displacements. So, for instance, if we select $\bar{v}^{(1)}(x)$, Equation 11.4, as our assumed displacement function, then $\bar{u}_1 = \theta_{IJ}$, $\bar{u}_2 = \theta_{JI}$, $\bar{u}_3 = v_{IJ}$, $\bar{u}_4 = v_{JI}$, and we may write $\bar{v}^{(1)}(x)$ in the matrix form

$$\bar{v}^{(1)}(x) = \left[\left(-x + \frac{2x^2}{l} - \frac{x^3}{l^2}\right)\left(\frac{x^2}{l} - \frac{x^3}{l^2}\right)\left(1 - \frac{3x^2}{l^2} + \frac{2x^3}{l^3}\right)\left(\frac{3x^2}{l^2} - \frac{2x^3}{l^3}\right)\right]\begin{Bmatrix} \theta_{IJ} \\ \theta_{JI} \\ v_{IJ} \\ v_{JI} \end{Bmatrix} \tag{11.9a}$$

where, for example,

$$A_1(x) = -x + \frac{2x^2}{l} - \frac{x^3}{l^2} \tag{11.9b}$$

The beam curvature may be written in a form similar to Equation 11.8,

$$\bar{v}''(x) = [\mathbf{B}(x)]\bar{\mathbf{u}} \tag{11.10a}$$

where

$$B_i(x) = \frac{d^2A_i(x)}{dx^2} \tag{11.10b}$$

Again, for the case $\bar{v}(x) = \bar{v}^{(1)}(x)$, we find

$$\bar{v}''^{(1)}(x) = \left[\left(\frac{4}{l} - \frac{6x}{l^2}\right)\left(\frac{2}{l} - \frac{6x}{l^2}\right)\left(-\frac{6}{l^2} + \frac{12x}{l^3}\right)\left(\frac{6}{l^2} - \frac{12x}{l^3}\right)\right]\begin{Bmatrix} \theta_{IJ} \\ \theta_{JI} \\ v_{IJ} \\ v_{JI} \end{Bmatrix} \tag{11.11a}$$

where, for example,

$$B_1(x) = \frac{d^2A_1(x)}{dx^2} = \frac{4}{l} - \frac{6x}{l^2} \tag{11.11b}$$

Note that the curvature, in this case, is a linear function of x.

Substituting Equation 11.10*a* into Equation 10.57*a* gives the strain energy,[3]

---

[2] We will delimit matrix or vector quantities having components that are functions of x with brackets or braces for the sake of clarity in this chapter; thus, rather than $\mathbf{A}(x)$ we write $[\mathbf{A}(x)]$.

[3] We are omitting the thermal and initial strain terms from the development. They can easily be added, and ultimately lead to the familiar thermal/initial force vector in the equilibrium equation.

$$
\begin{aligned}
U &= \frac{1}{2}\int_0^l \bar{\mathbf{u}}^{\mathrm{T}}[\mathbf{B}(x)]^{\mathrm{T}}[EI][\mathbf{B}(x)]\bar{\mathbf{u}}dx \\
&= \frac{1}{2}\bar{\mathbf{u}}^{\mathrm{T}}\int_0^l [\mathbf{B}(x)]^{\mathrm{T}}[EI][\mathbf{B}(x)]dx\bar{\mathbf{u}} \\
&= \frac{1}{2}\bar{\mathbf{u}}^{\mathrm{T}}\mathbf{k}^{(e)}\bar{\mathbf{u}}
\end{aligned}
\tag{11.12a}
$$

We recognize that the integral in the second of Equations 11.12a must be the element stiffness matrix, $\mathbf{k}^{(e)}$, because the form of the equation is identical to Equation 10.15d. The components of the element stiffness matrix are given by

$$
k_{ij}^{(e)} = \int_0^l B_i(x) \cdot EI \cdot B_j(x)dx \tag{11.12b}
$$

For example, for displacement function $\bar{v}^{(1)}(x)$,

$$
\begin{aligned}
k_{12}^{(e)} &= \int_0^l B_1(x) \cdot EI \cdot B_2(x)dx \\
&= \int_0^l EI\left(\frac{4}{l} - \frac{6x}{l^2}\right)\left(\frac{2}{l} - \frac{6x}{l^2}\right)dx \\
&= \frac{2EI}{l}
\end{aligned}
\tag{11.13a}
$$

Computing all the components of the stiffness matrix associated with $\bar{v}^{(1)}(x)$ provides

$$
\mathbf{k}^{(e)} = \frac{EI}{l}\begin{bmatrix} 4 & 2 & -6/l & 6/l \\ 2 & 4 & -6/l & 6/l \\ -6/l & -6/l & 12/l^2 & -12/l^2 \\ 6/l & 6/l & -12/l^2 & 12/l^2 \end{bmatrix} \tag{11.13b}
$$

This is the familiar element stiffness matrix presented previously in Equation 8.1d. Thus, the assumed displacement function $\bar{v}^{(1)}(x)$, which is exact except for fixed-coordinate patterns, leads to the correct stiffness matrix.

If the beam element is subjected to a general set of loads, then the external potential energy for any approximate displacement function is

$$
\begin{aligned}
U_e &= -\int_0^l q(x)\bar{v}(x)dx - F_1\bar{u}_1 - F_2\bar{u}_2 - F_3\bar{u}_3 - F_4\bar{u}_4 \\
&= -\bar{\mathbf{u}}^{\mathrm{T}}(\mathbf{F_q} + \mathbf{F})
\end{aligned}
\tag{11.14}
$$

where $q(x)$ represents applied transverse loads in the direction of $\bar{v}(x)$, and F_1–F_4 are end forces acting in the same sense as $\bar{u}_1$–$\bar{u}_4$, respectively; $\mathbf{F}$ is the vector of forces

F_1–F_4, and $\mathbf{F_q}$ is the vector of work-equivalent (generalized) forces due to the applied loads (see Equation 10.47).

For the case $\overline{v}(x) = \overline{v}^{(1)}(x)$, the conventional beam displacement model, we have $F_1 = M_{\text{IJ}}$, $F_2 = M_{\text{JI}}$, $F_3 = V_{\text{IJ}}$, $F_4 = V_{\text{JI}}$, and

$$\begin{aligned}
F_1^{(q)} &= \int_0^l A_1(x)q(x)dx = \int_0^l q(x)\left(-x + \frac{2x^2}{l} - \frac{x^3}{l^2}\right)dx \\
F_2^{(q)} &= \int_0^l A_2(x)q(x)dx = \int_0^l q(x)\left(\frac{x^2}{l} - \frac{x^3}{l^2}\right)dx \\
F_3^{(q)} &= \int_0^l A_3(x)q(x)dx = \int_0^l q(x)\left(1 - \frac{3x^2}{l^2} + \frac{2x^3}{l^3}\right)dx \\
F_4^{(q)} &= \int_0^l A_4(x)q(x)dx = \int_0^l q(x)\left(\frac{3x^2}{l^2} - \frac{2x^3}{l^3}\right)dx
\end{aligned} \tag{11.15}$$

exactly as in Equation 10.51*a*. Again, we note that these are the negatives of the fixed-end forces (see Equation 10.51*b*).

Overall, then, the total potential energy for the beam element is the sum of Equations 11.12*a* and 11.14,

$$V = \frac{1}{2}\overline{\mathbf{u}}^{\mathrm{T}}\mathbf{k}^{(e)}\overline{\mathbf{u}} - \overline{\mathbf{u}}^{\mathrm{T}}(\mathbf{F} + \mathbf{F_q}) \tag{11.16a}$$

and, in order to satisfy equilibrium, we must have

$$\frac{\partial V}{\partial \overline{\mathbf{u}}} = \mathbf{0} = \mathbf{k}^{(e)}\overline{\mathbf{u}} - (\mathbf{F} + \mathbf{F_q}) \tag{11.16b}$$

or

$$\mathbf{k}^{(e)}\overline{\mathbf{u}} = \mathbf{F} + \mathbf{F_q} \tag{11.16c}$$

Equation 11.16*c* is the same stiffness relationship that was derived previously in Equation 9.1*d* using an exact formulation; we see that the result is actually obtained in a more straightforward manner using the energy formulation, provided of course that we begin with the correct assumed displacement pattern, i.e., $\overline{v}^{(1)}(x)$ rather than $\overline{v}^{(2)}(x)$ or $\overline{v}^{(3)}(x)$.

11.2.2 Development Using Virtual Work

We can also directly derive the equilibrium equation, Equation 11.16*c*, by employing a form of the theorem of virtual work that utilizes approximate rather than exact equilibrium information. This weaker form of the theorem is referred to as the *virtual work principle*. We begin with any approximate beam displacement function, Equation 11.8, for which internal strain (i.e., curvature) is given by Equation 11.10*a*. An approximate internal stress (i.e., moment) can be determined from the displacement function,[4]

$$\overline{M}(x) = EI\overline{v}''(x) = EI[\mathbf{B}(x)]\overline{\mathbf{u}} \tag{11.17}$$

Then, we apply the virtual work principle in the form

$$\int_0^l \overline{M}(x) \cdot v_V''(x)dx = \int_0^l q(x) \cdot v_V(x)dx + F_1u_{1_V} + F_2u_{2_V} + F_3u_{3_V} + F_4u_{4_V} \tag{11.18}$$

where $v_V(x)$ is an arbitrary virtual displacement function with nodal displacements u_{1_V}–u_{4_V}. The internal virtual work in Equation 11.18 is approximate because it is based on Equation 11.17, whereas the external virtual work is based on an exact set of specified forces.

We can go one step further, and choose the virtual displacement patterns to be the same as the approximate displacement patterns,

$$v_V(x) = [\mathbf{A}(x)]\mathbf{u}_V \tag{11.19a}$$

Then

$$v_V''(x) = [\mathbf{B}(x)]\mathbf{u}_V \tag{11.19b}$$

It is important to recognize that we could choose any virtual displacement patterns; however, Equation 11.19*a* is the best choice in the sense that it will lead to approximate equilibrium equations that are exactly the same as those obtained from the energy method.

Substituting Equations 11.19*b* and 11.17 into Equation 11.18 gives

$$\int_0^l \mathbf{u}_V^T[\mathbf{B}(x)]^T[EI][\mathbf{B}(x)]\overline{\mathbf{u}}dx = \int_0^l \mathbf{u}_V^T[\mathbf{A}(x)]^T q(x)dx + \mathbf{u}_V^T\mathbf{F} \tag{11.20a}$$

or

$$\mathbf{u}_V^T\mathbf{k}^{(e)}\overline{\mathbf{u}} = \mathbf{u}_V^T(\mathbf{F_q} + \mathbf{F}) \tag{11.20b}$$

where the components of $\mathbf{k}^{(e)}$ are again given by Equation 11.12*b*, and the generalized forces $F_i^{(q)}$ are given by

$$F_i^{(q)} = \int_0^l A_i(x)q(x)dx \tag{11.20c}$$

Since the components of $\mathbf{u}_V$ are completely arbitrary, it follows that Equation 11.20*b* reduces to Equation 11.16*c*, as expected.

We note that the energy approach (a) leads to the same result as the virtual work principle, and (b) actually provides more qualitative information than virtual work

[4] We are again omitting the thermal/initial loading term for the sake of ease of presentation.

concerning the nature of the approximate equilibrium condition. One might question, then, why the virtual work principle would be considered a useful procedure in this context. The answer is that the energy approach formulated herein requires elastic behavior, but the virtual work principle is valid for any beam in equilibrium regardless of whether it is elastic or inelastic. Thus, when one is interested in developing methods for inelastic analysis (a subject that is beyond the scope of this text), the only principle available for such analysis is virtual work! Virtual work is the basis for modern finite element computer analysis of all types of structures, particularly structures that exhibit inelastic behavior.

11.2.3 Approximate Solutions as Best Fits

At this point, we have examined the derivation of equilibrium equations based on approximate displacement functions by (a) minimizing total potential energy (which also maximizes the internal strain energy), and (b) satisfying the virtual work principle. Before proceeding further, it is informative to demonstrate another perspective concerning the nature of the approximate solutions, namely, the fact that use of approximate displacements leads to approximations that are ''best fits'' of the exact behavior.

To demonstrate this point, let us assume that we are modeling beam behavior using an approximate displacement function of the form in Equation 11.8. We know that the approximate displacement can be used to find approximate curvature and approximate internal bending moment, Equations 11.10 and 11.17, respectively. Suppose we seek to determine the nodal displacements, $\overline{\mathbf{u}}$, to produce the best fit of the exact internal physics. In other words, we seek the $\overline{u}_i$ such that $\overline{M}(x)$ most closely resembles the exact moment $M(x)$.

At first glance, the task might appear unresolvable since, in general, we will not know the exact moment $M(x)$. It turns out, however, that we can generate the required information without any knowledge of $M(x)$. To demonstrate the procedure, at least for a simple beam element, let's consider a mean square error measure of how close the approximate solution is to an exact solution. In particular, let's consider an error measure $\xi(\overline{\mathbf{u}})$ of the difference in strain energy, defined by

$$\begin{aligned}\xi(\overline{\mathbf{u}}) &= \frac{1}{2}\int_0^l EI[\overline{v}''(x) - v''(x)]^2 dx \\ &= \frac{1}{2}\int_0^l EI\{[\mathbf{B}(x)]\overline{\mathbf{u}} - v''(x)\}^2 dx\end{aligned} \tag{11.21a}$$

where $v''(x)$ is the exact curvature. We note that a different error measure could have been used, for instance

$$\xi_1(\overline{\mathbf{u}}) = \frac{1}{2}\int_0^l C[\overline{v}''(x) - v''(x)]^2 dx \tag{11.21b}$$

where C is a constant. Equation 11.21b reduces to Equation 11.21a if we take $C = EI$. The choice of EI as the constant makes it possible to interpret $\xi(\bar{\mathbf{u}})$ as a kind of ''error energy'' which is the sum of all of the errors in strain energy that develop for $0 < x < l$.

We recognize that $\xi(\bar{\mathbf{u}})$ is a quadratic function of the generalized displacements $\bar{\mathbf{u}}$ since $\bar{v}''(x)$ is a linear function of the displacements, Equation 11.10a. Expanding Equation 11.21a leads to

$$\xi(\bar{\mathbf{u}}) = \frac{1}{2}\int_0^l \{\bar{\mathbf{u}}^{\mathrm{T}}[\mathbf{B}(x)]^{\mathrm{T}}EI[\mathbf{B}(x)]\bar{\mathbf{u}} - 2\bar{\mathbf{u}}^{\mathrm{T}}[\mathbf{B}(x)]EIv''(x) + EIv''^2(x)\}dx$$
$$= \frac{1}{2}\bar{\mathbf{u}}^{\mathrm{T}}\mathbf{k}^{(e)}\bar{\mathbf{u}} - \int_0^l \bar{\mathbf{u}}^{\mathrm{T}}[\mathbf{B}(x)]^{\mathrm{T}}M(x)dx + \xi_o \qquad \mathbf{(11.21}\boldsymbol{c}\mathbf{)}$$

where $M(x) = EIv''(x)$ is the exact moment, and ξ_o is a term that is independent of the $\bar{\mathbf{u}}_i$.

Finding the best fit is equivalent to minimizing the quadratic error function, Equation 11.21c; therefore, we seek the condition where

$$d\xi = \frac{\partial\xi}{\partial\bar{u}_1}d\bar{u}_1 + \frac{\partial\xi}{\partial\bar{u}_2}d\bar{u}_2 + \frac{\partial\xi}{\partial\bar{u}_3}d\bar{u}_3 + \frac{\partial\xi}{\partial\bar{u}_4}d\bar{u}_4 = 0 \qquad \mathbf{(11.22}\boldsymbol{a}\mathbf{)}$$

or

$$\begin{Bmatrix} d\bar{u}_1 \\ d\bar{u}_2 \\ d\bar{u}_3 \\ d\bar{u}_4 \end{Bmatrix}^{\mathrm{T}} \begin{Bmatrix} \partial\xi/\partial\bar{u}_1 \\ \partial\xi/\partial\bar{u}_2 \\ \partial\xi/\partial\bar{u}_3 \\ \partial\xi/\partial\bar{u}_4 \end{Bmatrix} = 0 \qquad \mathbf{(11.22}\boldsymbol{b}\mathbf{)}$$

Carrying out the differentiation gives

$$\mathbf{d}\bar{\mathbf{u}}^{\mathrm{T}}\mathbf{k}^{(e)}\bar{\mathbf{u}} - \int_0^l \mathbf{d}\bar{\mathbf{u}}^{\mathrm{T}}[\mathbf{B}(x)]^{\mathrm{T}}M(x)dx = 0 \qquad \mathbf{(11.22}\boldsymbol{c}\mathbf{)}$$

Since the components of $\mathbf{d}\bar{\mathbf{u}}$ are arbitrary differential quantities, the minimum must correspond to

$$\mathbf{k}^{(e)}\bar{\mathbf{u}} = \int_0^l [\mathbf{B}(x)]^{\mathrm{T}}M(x)dx \qquad \mathbf{(11.22}\boldsymbol{d}\mathbf{)}$$

The first term in Equation 11.22c looks very much like the left side of Equation 11.20b, except that we now have $\mathbf{d}\bar{\mathbf{u}}$ instead of $\mathbf{u}_V$. Consequently, if our argument is to lead anywhere, we must show that the remaining terms of the two equations are equal, i.e., that the right side of Equation 11.22d is equal to $\mathbf{F} + \mathbf{F}_{\mathbf{q}}$. If that is indeed the case, the conclusion will be that the best fit is identical to the approximate equilibrium solution!

Let's consider a typical term on the right side of Equation 11.22*d*. We substitute the definition of $B_i(x)$ from Equation 11.10*b* and integrate by parts,

$$\int_0^l \frac{d^2A_i(x)}{dx^2} M(x)dx = \frac{dA_i(x)}{dx} M(x)\bigg|_0^l - \int_0^l \frac{dA_i(x)}{dx}\frac{dM(x)}{dx}dx \qquad \textbf{(11.23a)}$$

We substitute $dM(x)/dx = -V(x)$ and integrate Equation 11.23*a* again by parts,

$$\int_0^l \frac{d^2A_i(x)}{dx^2} M(x)dx = \frac{dA_i(x)}{dx}\bigg|_l M(l) - \frac{dA_i(x)}{dx}\bigg|_0 M(0) + A_i(x)V(x)\bigg|_0^l - \int_0^l A_i(x)\frac{dV(x)}{dx}dx \qquad \textbf{(11.23b)}$$

Substituting $dV(x)/dx = -q(x)$ leads to

$$\int_0^l \frac{d^2A_i(x)}{dx^2} M(x)dx = \frac{dA_i(x)}{dx}\bigg|_l M(l) - \frac{dA_i(x)}{dx}\bigg|_0 M(0) + A_i(x)\bigg|_l V(l) - A_i(x)\bigg|_0 V(0) + \int_0^l A_i(x)q(x)dx \qquad \textbf{(11.23c)}$$

Since $A_i(x)$ is the nodal displacement pattern associated with $\bar{u}_i$, it has a value of ± 1 when evaluated at the node of interest for the type of motion associated with $\bar{u}_i$; it is zero when evaluated at other nodes or for other types of motion at the node of interest. For example, for the conventional beam model where $\bar{u}_i$ and $A_i(x)$ are defined in Equation 11.9*a*, we have

$$\begin{aligned}
\bar{u}_1 = \theta_{IJ}&: \quad \frac{dA_1(x)}{dx}\bigg|_0 = -1, \quad \frac{dA_1(x)}{dx}\bigg|_l = 0, \quad A_1(0) = 0, \quad A_1(l) = 0 \\
\bar{u}_2 = \theta_{JI}&: \quad \frac{dA_2(x)}{dx}\bigg|_0 = 0, \quad \frac{dA_2(x)}{dx}\bigg|_l = -1, \quad A_2(0) = 0, \quad A_2(l) = 0 \\
\bar{u}_3 = v_{IJ}&: \quad \frac{dA_3(x)}{dx}\bigg|_0 = 0, \quad \frac{dA_3(x)}{dx}\bigg|_l = 0, \quad A_3(0) = 1, \quad A_3(l) = 0 \\
\bar{u}_4 = v_{JI}&: \quad \frac{dA_4(x)}{dx}\bigg|_0 = 0, \quad \frac{dA_4(x)}{dx}\bigg|_l = 0, \quad A_4(0) = 0, \quad A_4(l) = 1
\end{aligned} \qquad \textbf{(11.23d)}$$

The integral on the right side of Equation 11.23*c* is identical to $F_i^{(q)}$. We also note that $M(0) = M_{IJ} = F_1$, $M(l) = -M_{JI} = -F_2$, $V(0) = -V_{IJ} = -F_3$, $V(l) = V_{IJ} = F_4$.

Consequently, the entire right side of Equation 11.23*c* reduces to $F_i + F_i^{(q)}$. Thus, Equation 11.22*d* is identical to Equation 11.16*c*, as anticipated. In other words, the best fit solution based on the definition of error $\xi(\overline{\mathbf{u}})$ gives precisely the same approximate equilibrium equation as either the energy method or virtual work![5]

Finally, although not derived here, another well-known numerical procedure called Galerkin's method (extended form) also provides exactly the same equilibrium equation, Equation 11.16*c*. Thus, there is ample reason to have confidence in the use of assumed displacement functions for structural analysis.

11.3 EXAMPLES OF FINITE ELEMENT ANALYSIS

In the remainder of this chapter, we will present several examples of finite element modeling applied to beam-type structures for which we have the benefit of exact solutions for comparison purposes. We will begin by using the best definition of displacements, i.e., $\overline{v}^{(1)}(x)$, Equation 11.4, to define various finite element models for some simple beam examples. In addition, we will demonstrate that a lower quality approximation, $\overline{v}^{(2)}(x)$, Equation 11.6, can also be made to provide satisfactory results, but in a much less effective manner. Finally, we will develop a finite element model for an elastically supported beam with which we have no prior experience, and for which the exact behavior is significantly different and more complex than for simple beams.

EXAMPLE 1

Consider the uniformly loaded, simply supported beam shown in Figure 11.1. This is identical to the beam analyzed in Example 6, Chapter 10, except that the direction of the load is reversed. For reference purposes, we note that the exact solution (see Equations 10.74*a* and 10.74*b*) is

$$v(x) = \frac{q_o l^4}{24EI}\left[\left(-\frac{x}{l}\right) + 2\left(\frac{x}{l}\right)^3 - \left(\frac{x}{l}\right)^4\right] \qquad \textbf{(11.24a)}$$

[5] This result may be seen from the integrations in Equations 11.23 or from the following argument. The right-hand side of Equation 11.22*d*, when premultiplied by $\mathbf{u}_V^{\mathrm{T}}$, is the internal work virtual work due to a chosen virtual displacement pattern that happens to be the exact solution for a beam with nodal end loads only; then the external virtual work must be

$$\mathbf{u}_V^{\mathrm{T}} \int_0^l [\mathbf{A}(x)]^{\mathrm{T}} q(x)dx + \mathbf{u}_V^{\mathrm{T}}\mathbf{F} = \mathbf{u}_V^{\mathrm{T}}(\mathbf{F_q} + \mathbf{F})$$

from which we see

$$\int_0^l [\mathbf{B}(x)]^{\mathrm{T}} M(x)dx = \mathbf{F_q} + \mathbf{F}$$

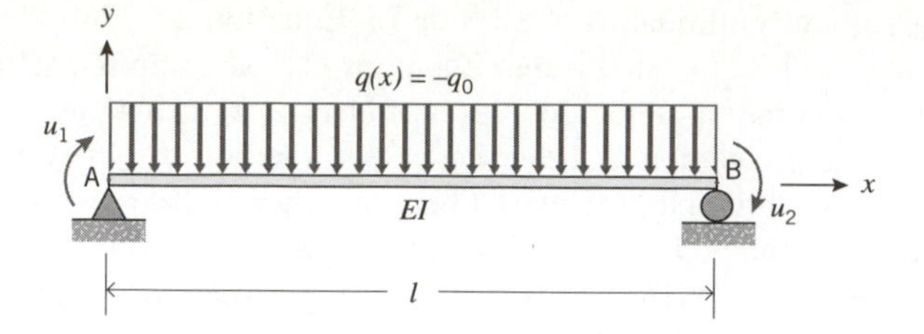

Figure 11.1 Example 1 (one element).

and

$$M(x) = \frac{q_o l^2}{2}\left[\left(\frac{x}{l}\right) - \left(\frac{x}{l}\right)^2\right] \tag{11.24b}$$

We also note that the end (nodal) displacements u_1 and u_2 (Figure 11.1) are given by

$$\begin{aligned} u_1 &= \theta_{AB} = -\left.\frac{dv(x)}{dx}\right|_0 = \frac{q_o l^3}{24EI} \\ u_2 &= \theta_{BA} = -\left.\frac{dv(x)}{dx}\right|_l = -\frac{q_o l^3}{24EI} \end{aligned} \tag{11.24c}$$

We will compare this exact solution to the results of analyses using various numbers of finite elements based on cubic polynomials (conventional beam functions), Equation 11.4.

(a) One-Element Model

We will consider the structure in Figure 11.1 to consist of a single-beam element, and begin by using the cubic polynomial displacement function (Equation 11.4) to model the behavior of the element,

$$\begin{aligned} \bar{v}^{(1)}(x) &= \theta_{AB}\left(-x + \frac{2x^2}{l} - \frac{x^3}{l^2}\right) + \theta_{BA}\left(\frac{x^2}{l} - \frac{x^3}{l^2}\right) \\ &\quad + v_{AB}\left(1 - \frac{3x^2}{l^2} + \frac{2x^3}{l^3}\right) + v_{BA}\left(\frac{3x^2}{l^2} - \frac{2x^3}{l^3}\right) \end{aligned} \tag{11.25a}$$

As explained previously, this displacement function leads to the element stiffness matrix given in Equation 11.13*b*,

$$\mathbf{k}^{(AB)} = \frac{EI}{l}\begin{bmatrix} 4 & 2 & -6/l & 6/l \\ 2 & 4 & -6/l & 6/l \\ -6/l & -6/l & 12/l^2 & -12/l^2 \\ 6/l & 6/l & -12/l^2 & 12/l^2 \end{bmatrix} \tag{11.25b}$$

Similarly, from Equations 11.15 we obtain

$$\mathbf{F}_{\mathbf{q}}^{(AB)} = \begin{Bmatrix} q_o l^2/12 \\ -q_o l^2/12 \\ -q_o l/2 \\ -q_o l/2 \end{Bmatrix} \tag{11.25c}$$

Since there are no applied end loads ($\mathbf{F} = \mathbf{0}$), the equilibrium equation for element AB is simply $\mathbf{k}^{(AB)} = \mathbf{F}_{\mathbf{q}}^{(AB)}$, Equation 11.16*c*.

The structure in Figure 11.1 has two degrees of freedom, u_1 and u_2. Therefore, we can assemble the stiffness matrix for this structure from the stiffness matrix for the element using the direct algorithm described in Section 9.6.[6] The correspondences between element end displacements and system displacements are given in Table 11.1.1.

Table 11.1.1. Element AB

Element displacement	System displacement
θ_{AB} (1)	1
θ_{BA} (2)	2
v_{AB} (3)	—
v_{BA} (4)	—

Thus, the system equilibrium equation becomes

$$\frac{EI}{l}\begin{bmatrix} 4 & 2 \\ 2 & 4 \end{bmatrix}\begin{Bmatrix} \bar{u}_1 \\ \bar{u}_2 \end{Bmatrix} = \begin{Bmatrix} q_o l^2/12 \\ -q_o l^2/12 \end{Bmatrix} \tag{11.26a}$$

from which we obtain

$$\begin{Bmatrix} \bar{u}_1 \\ \bar{u}_2 \end{Bmatrix} = \frac{q_o l^3}{24EI}\begin{Bmatrix} 1 \\ -1 \end{Bmatrix} \tag{11.26b}$$

where $\bar{u}_1 \equiv \theta_{AB}$ and $\bar{u}_2 \equiv \theta_{BA}$. We see that the nodal displacements obtained in Equation 11.26*b* are the same as the exact values in Equations 11.24*c*, as expected; however, when these displacements are substituted into Equation 11.25*a* we obtain the approximate displacement function

$$\begin{aligned} \bar{v}(x) &= \frac{q_o l^3}{24EI}\left(-x + \frac{2x^2}{l} - \frac{x^3}{l^2}\right) - \frac{q_o l^3}{24EI}\left(\frac{x^2}{l} - \frac{x^3}{l^2}\right) \\ &= \frac{q_o l^3}{24EI}\left(-x + \frac{x^2}{l}\right) \end{aligned} \tag{11.27a}$$

[6] The beam element considered here has only four degrees of freedom vs. six for the elements in Section 9.6; nevertheless, the concept of relating element end displacements to system displacements is identical.

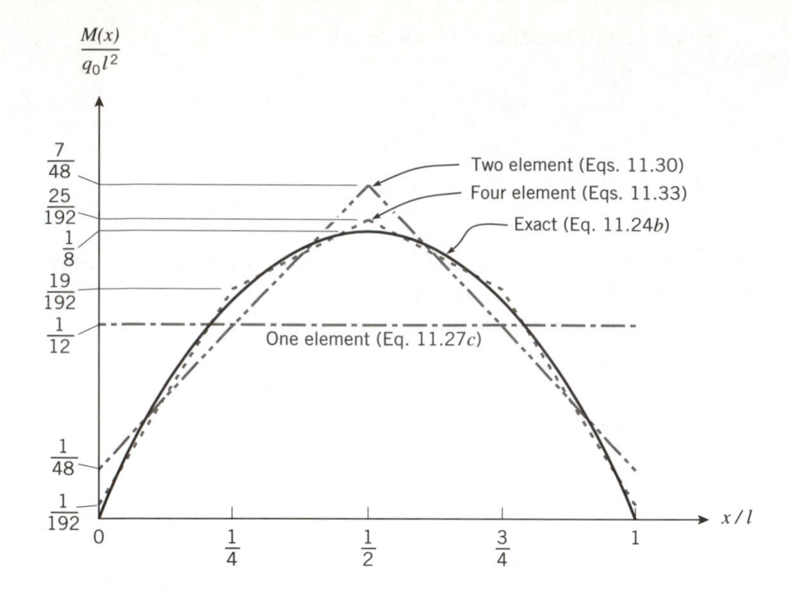

Figure 11.2 Bending moment comparison, Example 1.

which is incorrect, i.e., it is quadratic rather than the fourth-order polynomial in Equation 11.24*a*. The difference between the exact and approximate displacement functions is

$$\begin{aligned} v(x) - \bar{v}(x) &= \frac{q_o l^4}{24EI}\left[\left(-\frac{x}{l}\right) + 2\left(\frac{x}{l}\right)^3 - \left(\frac{x}{l}\right)^4\right] - \frac{q_o l^4}{24EI}\left[-\left(\frac{x}{l}\right) + \left(\frac{x}{l}\right)^2\right] \\ &= \frac{q_o l^4}{24EI}\left[-\left(\frac{x}{l}\right)^2 + 2\left(\frac{x}{l}\right)^3 - \left(\frac{x}{l}\right)^4\right] \end{aligned} \tag{11.27b}$$

This is exactly the fixed-coordinate (fixed-end) displacement previously encountered in Equation 10.2*f*. Thus, the approximate solution reveals only a portion of the true behavior.

The approximate internal moment is

$$\overline{M}(x) = EI\frac{d^2\bar{v}(x)}{dx^2} = \frac{q_o l^2}{12} \tag{11.27c}$$

which is a constant rather than the quadratic function in Equation 11.24*b*. The exact and approximate moments are compared in Figure 11.2.

(b) Two-Element Model

Clearly, the stress state obtained from the one-element model is a rather crude approximation. The results can be improved by refining the model, i.e., by taking the structure to be composed of a larger number of elements, for example the two elements shown in Figure 11.3.

Each of the two beam elements shown in Figure 11.3 will be taken to have a length

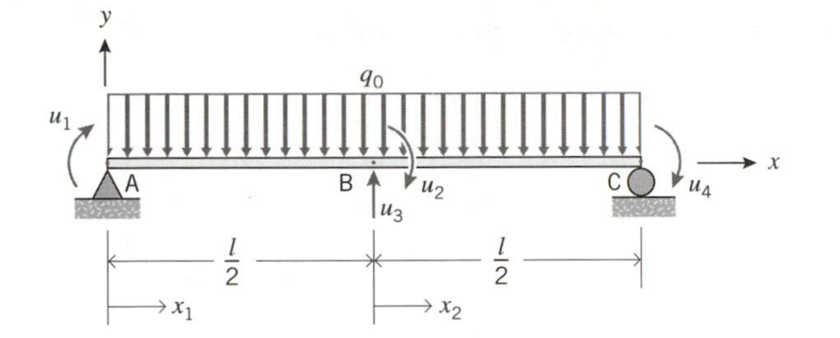

Figure 11.3 Example 1 (two elements).

$l/2$. The point at which the two elements are connected together, i.e., the midpoint of span AB, constitutes a ''joint'' or internal node in the structure. This node has two degrees of freedom, identified as system displacements u_2 and u_3 in Figure 11.3. Thus, the structure now has four degrees of freedom, including the two end rotations u_1 and u_4.

The behavior of each of the two elements is defined by the stiffness matrix in Equation 11.25*b* and the force vector in Equation 11.25*c*, except that the length quantity l is replaced by $l/2$. In other words,

$$\mathbf{k}^{(AB)} = \mathbf{k}^{(BC)} = \frac{2EI}{l}\begin{bmatrix} 4 & 2 & -12/l & 12/l \\ 2 & 4 & -12/l & 12/l \\ -12/l & -12/l & 48/l^2 & -48/l^2 \\ 12/l & 12/l & -48/l^2 & 48/l^2 \end{bmatrix} \tag{11.28a}$$

and

$$\mathbf{F}_{\mathbf{q}}^{(AB)} = \mathbf{F}_{\mathbf{q}}^{(BC)} = \begin{Bmatrix} q_o l^2/48 \\ -q_o l^2/48 \\ -q_o l/4 \\ -q_o l/4 \end{Bmatrix} \tag{11.28b}$$

Note that the elements are designated as AB and BC, as indicated in Figure 11.3.

For the structure in Figure 11.3, the correspondences between element end displacements and system displacements are given in Table 11.1.2.

Table 11.1.2

Element	Element displacement	System displacement
AB	θ_{AB} (1)	1
	θ_{BA} (2)	2
	v_{AB} (3)	—
	v_{BA} (4)	3
BC	θ_{BC} (1)	2
	θ_{CB} (2)	4
	v_{BC} (3)	3
	v_{CB} (4)	—

From the direct assembly process we obtain

$$\frac{2EI}{l}\begin{bmatrix} 4 & 2 & 12/l & 0 \\ 2 & 8 & 0 & 2 \\ 12/l & 0 & 96/l^2 & -12/l \\ 0 & 2 & -12/l & 4 \end{bmatrix}\begin{Bmatrix} \overline{u}_1 \\ \overline{u}_2 \\ \overline{u}_3 \\ \overline{u}_4 \end{Bmatrix} = \begin{Bmatrix} q_o l^2/48 \\ 0 \\ -q_o l/2 \\ -q_o l^2/48 \end{Bmatrix} \tag{11.29a}$$

Solving gives

$$\begin{Bmatrix} \overline{u}_1 \\ \overline{u}_2 \\ \overline{u}_3 \\ \overline{u}_4 \end{Bmatrix} = \frac{q_o l^3}{24EI}\begin{Bmatrix} 1 \\ 0 \\ -5l/16 \\ -1 \end{Bmatrix} \tag{11.29b}$$

As in the preceding case, these nodal displacements are exact; however, neither the element displacements nor moments are correct. The approximate moments in each element are obtained by substituting the end displacements into Equation 11.17, remembering that the length quantity l in the displacement patterns $B_i(x)$ must be replaced by $l/2$,

$$\overline{M}(x_1) = \frac{q_o l^2}{48}\left(1 + \frac{12x_1}{l}\right), \qquad 0 < x_1 < \frac{l}{2} \tag{11.30a}$$

and

$$\overline{M}(x_2) = \frac{q_o l^2}{48}\left(7 - \frac{12x_2}{l}\right), \qquad 0 < x_2 < \frac{l}{2} \tag{11.30b}$$

Equations 11.30a and 11.30b are plotted in Figure 11.2 for comparison with the exact and previous results. The fact that the approximate moment obtained from each element has the same value at joint B is due only to the symmetry of the moment about this point. Usually, the approximate moment function will not be continuous from one element to the next.

(c) Four-Element Model

As may be observed from Figure 11.2, the results obtained from the two-element model are a significant improvement over those of the one-element model, although the accuracy is still not entirely satisfactory. Further refinement of the model will provide greater accuracy.

Let's consider a four-element model, in which the length of each element is $l/4$. To simplify the computations, we will take advantage of symmetry and analyze only the half of the structure shown in Figure 11.4.

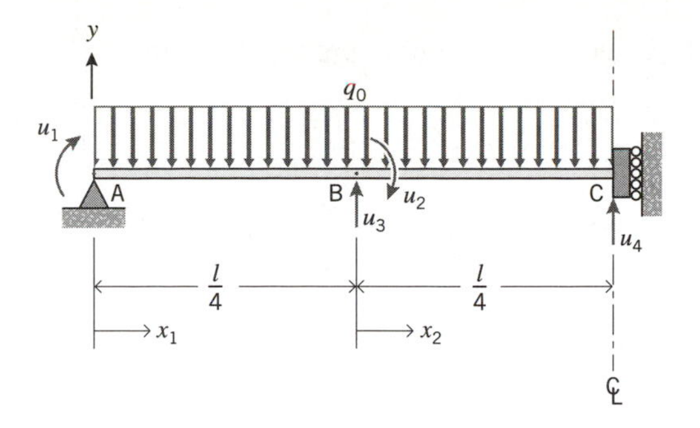

Figure 11.4 Example 1 (four elements).

The stiffness matrix for each of the two elements is

$$\mathbf{k}^{(AB)} = \mathbf{k}^{(BC)} = \frac{4EI}{l}\begin{bmatrix} 4 & 2 & -24/l & 24/l \\ 2 & 4 & -24/l & 24/l \\ -24/l & -24/l & 192/l^2 & -192/l^2 \\ 24/l & 24/l & -192/l^2 & 192/l^2 \end{bmatrix} \qquad \textbf{(11.31a)}$$

and the generalized force vector for each element is

$$\mathbf{F}_{\mathbf{q}}^{(AB)} = \mathbf{F}_{\mathbf{q}}^{(BC)} = \begin{Bmatrix} q_o l^2/192 \\ -q_o l^2/192 \\ -q_o l/8 \\ -q_o l/8 \end{Bmatrix} \qquad \textbf{(11.31b)}$$

The correspondences between element end displacements and system displacements are given in Table 11.1.3.

Table 11.1.3

Element	Element displacement	System displacement
AB	θ_{AB} (1)	1
	θ_{BA} (2)	2
	v_{AB} (3)	—
	v_{BA} (4)	3
BC	θ_{BC} (1)	2
	θ_{CB} (2)	—
	v_{BC} (3)	3
	v_{CB} (4)	4

Therefore, the direct assembly process leads to

$$\frac{4EI}{l}\begin{bmatrix} 4 & 2 & 24/l & 0 \\ 2 & 8 & 0 & 24/l \\ 24/l & 0 & 384/l^2 & -192/l^2 \\ 0 & 24/l & -192/l^2 & 192/l^2 \end{bmatrix}\begin{Bmatrix} \bar{u}_1 \\ \bar{u}_2 \\ \bar{u}_3 \\ \bar{u}_4 \end{Bmatrix} = \frac{q_o l^2}{192}\begin{Bmatrix} 1 \\ 0 \\ -48/l \\ -24/l \end{Bmatrix} \tag{11.32a}$$

from which we obtain

$$\begin{Bmatrix} \bar{u}_1 \\ \bar{u}_2 \\ \bar{u}_3 \\ \bar{u}_4 \end{Bmatrix} = \begin{Bmatrix} q_o l^3/24EI \\ 11q_o l^3/384EI \\ -19q_o l^4/2048EI \\ -5q_o l^4/384EI \end{Bmatrix} \tag{11.32b}$$

Once again, we see that the nodal displacements are exact.

The approximate moments are

$$\overline{M}(x_1) = \frac{q_o l^2}{192}\left(1 + \frac{72x_1}{l}\right), \qquad 0 < x_1 < \frac{l}{4} \tag{11.33a}$$

and

$$\overline{M}(x_2) = \frac{q_o l^2}{192}\left(19 + \frac{24x_2}{l}\right), \qquad 0 < x_2 < \frac{l}{4} \tag{11.33b}$$

Equations 11.33 (and their symmetric counterparts) are also plotted in Figure 11.2. The piecewise-linear moment obtained from the approximate solution constitutes an excellent representation of the exact moment function. The four-element model produces an error in the maximum moment (at midspan) of approximately 4%. In other words, the process of model refinement demonstrates very rapid convergence to the exact solution.

Note from Figure 11.2 the characteristic over/undershoot of all approximate solutions compared to the exact solution. Although the internal energy in the approximate solutions is slightly less than the exact strain energy, both the two- and four-element models give maximum moments that are greater than the exact value. All the approximate solutions also give nonzero moments at the ends of the beam.

The preceding introductory example demonstrates that the use of approximate displacement patterns in the finite element method constitutes a powerful modeling capability. When a computer can be used to assemble a large number of finite elements, the numerical analysis can, for all practical purposes, reproduce exact results.

Before proceeding with examples that illustrate additional ramifications of the use of assumed displacement patterns, it will be informative to apply the same finite element procedure employed in Example 1 to the structure in Figure 10.22. This structure was examined in Example 7, Chapter 10 using the Rayleigh-Ritz procedure. Recall that a displacement function consisting of ten terms of a truncated Fourier series led to an approximate solution, which was accurate except in the vicinity of $x = l/2$, at which point the bending moment was 12% low (see Figure 10.23).

EXAMPLE 2

Consider the structure in Figure 11.5 (Figure 10.22). The preceding analysis (Example 1) showed that a four-element model of a single span beam produced an excellent (piecewise linear) approximate moment diagram. Consequently, we will model each span of the beam by four elements of equal length $l/8$, as shown in Figure 11.5. We will also take advantage of symmetry and analyze only the left half of the structure. Note that the spring stiffness for the half-structure must be taken as $k/2$. For the purpose of comparison with Example 7, Chapter 10, we will set $k = \alpha\pi^4 EI/2l^3$ and take $\alpha = 20$. Also note that the distributed load acts in the positive y direction.

The half-structure we are analyzing consists of four beam elements and one spring element, and has the eight degrees of freedom identified in Figure 11.5. The stiffness matrices and force vectors for the beam elements are obtained in the same manner as in Example 1, recognizing that the length of each element is $l/8$. The correspondences between element end displacements and system displacements are given in Table 11.2.1.

Table 11.2.1

Element	Element displacement	System displacement
AB	θ_{AB} (1)	1
	θ_{BA} (2)	3
	v_{AB} (3)	—
	v_{BA} (4)	2
BC	θ_{BC} (1)	3
	θ_{CB} (2)	5
	v_{BC} (3)	2
	v_{CB} (4)	4
CD	θ_{CD} (1)	5
	θ_{DC} (2)	7
	v_{CD} (3)	4
	v_{DC} (4)	6
DE	θ_{DE} (1)	7
	θ_{ED} (2)	—
	v_{DE} (3)	6
	v_{ED} (4)	8
Sp	(1)	8

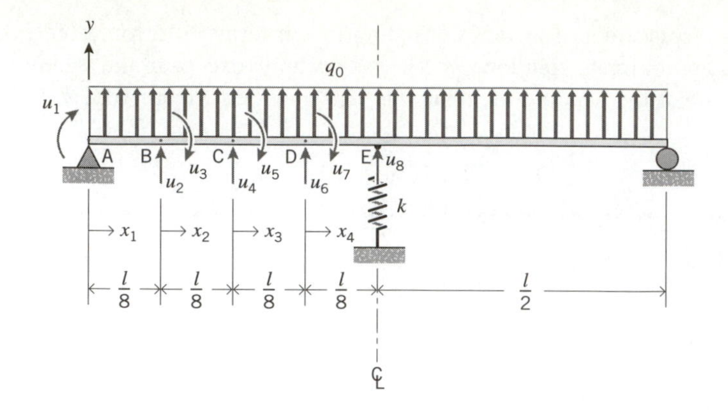

Figure 11.5 Example 2.

The basic assembly process in the direct stiffness method leads to

$$
\frac{EI}{\lambda}\begin{bmatrix}
4 & \frac{6}{\lambda} & 2 & 0 & 0 & 0 & 0 & 0 \\
 & \frac{24}{\lambda^2} & 0 & -\frac{12}{\lambda^2} & -\frac{6}{\lambda} & 0 & 0 & 0 \\
 & & 8 & \frac{6}{\lambda} & 2 & 0 & 0 & 0 \\
 & & & \frac{24}{\lambda^2} & 0 & -\frac{12}{\lambda^2} & -\frac{6}{\lambda} & 0 \\
 & & & & 8 & \frac{6}{\lambda} & 2 & 0 \\
 & & & & & \frac{24}{\lambda^2} & 0 & -\frac{12}{\lambda^2} \\
 & & & & & & 8 & \frac{6}{\lambda} \\
 & & & & & & & \frac{12}{\lambda^2} + \frac{\lambda k}{2EI}
\end{bmatrix}
\begin{Bmatrix} \bar{u}_1 \\ \bar{u}_2 \\ \bar{u}_3 \\ \bar{u}_4 \\ \bar{u}_5 \\ \bar{u}_6 \\ \bar{u}_7 \\ \bar{u}_8 \end{Bmatrix}
= q_o\lambda \begin{Bmatrix} -\frac{\lambda}{12} \\ 1 \\ 0 \\ 1 \\ 0 \\ 1 \\ 0 \\ \frac{1}{2} \end{Bmatrix}
\tag{11.34a}
$$

where $\lambda = l/8$. Substituting the values of λ and k, and solving gives

$$
\begin{Bmatrix} \bar{u}_1 \\ \bar{u}_2 \\ \bar{u}_3 \\ \bar{u}_4 \\ \bar{u}_5 \\ \bar{u}_6 \\ \bar{u}_7 \\ \bar{u}_8 \end{Bmatrix}
=
\begin{Bmatrix}
-4.4386 \times 10^{-3} q_o l^3/EI \\
4.9919 \times 10^{-4} q_o l^4/EI \\
-3.1847 \times 10^{-3} q_o l^3/EI \\
7.4592 \times 10^{-4} q_o l^4/EI \\
-7.2481 \times 10^{-4} q_o l^3/EI \\
7.1153 \times 10^{-4} q_o l^4/EI \\
9.8778 \times 10^{-4} q_o l^3/EI \\
6.1149 \times 10^{-4} q_o l^4/EI
\end{Bmatrix}
\tag{11.34b}
$$

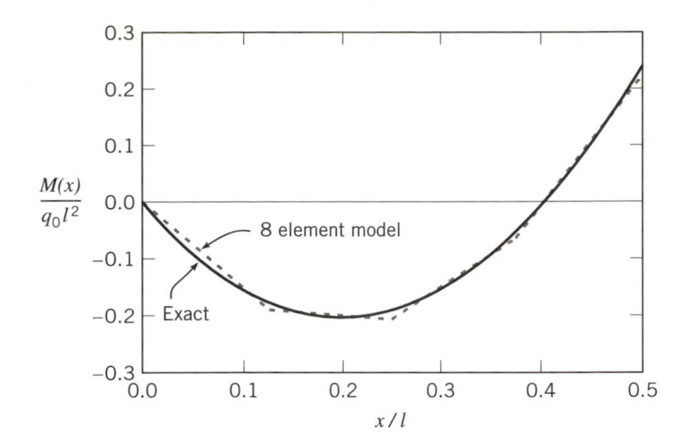

Figure 11.6 Bending moment comparison, Example 2.

from which we obtain

$$\left.\begin{aligned}\overline{M}(x_1) &= q_o l^2\left(-0.0013021 - 0.13968\frac{x_1}{l}\right)\\ \overline{M}(x_2) &= q_o l^2\left(-0.018762 - 0.014676\frac{x_2}{l}\right)\\ \overline{M}(x_3) &= q_o l^2\left(-0.020596 + 0.11032\frac{x_3}{l}\right)\\ \overline{M}(x_4) &= q_o l^2\left(-0.0068055 + 0.23532\frac{x_4}{l}\right)\end{aligned}\right\}, \quad 0 < x_i < \frac{l}{8}, \; i = 1, \ldots, 4 \qquad \textbf{(11.34c)}$$

These approximate moments for half the structure are plotted in Figure 11.6, which also shows the exact solution previously presented in Figure 10.23. The manner in which the piecewise-linear finite element solution approximates the exact solution is apparent. The maximum approximate moment (at $x_4 = l/8$, or $x = l/2$, i.e., the mid-point of the beam) is $0.02261 q_o l^2$, which is only 5.4% less than the exact value of $0.02391 q_o l^2$.

The relative accuracy of the eight-element model is comparable to that of the ten-term Rayleigh-Ritz solution generated in Chapter 10. The finite element model gives a somewhat better solution for the maximum moment, but a less accurate result at the points $x = 0$, $x = 0.125l$, and $x = 0.25l$ as a result of the piecewise-linear nature of the solution. For overall engineering analysis, either solution is satisfactory.

If a larger number of either finite elements or Fourier coefficients is used, the types of matrices that must be inverted in each approach should be considered. In particular, the Fourier approach produces a matrix that has nonzero entries in every register, while the finite element approach produces a matrix that is zero except near the diagonal. This so-called banded matrix can be exploited to minimize data storage requirements in the inversion process.

At this point, it will be informative to examine the behavior of a structure that is modeled using elements for which the assumed displacement function satisfies requirements (1) through (3), Section 11.1, but not optional conditions (4) through (5). It will be shown that solutions based on this type of "less robust" element still converge to the exact answer, but not nearly as effectively as for the element in prior examples.

EXAMPLE 3

We will repeat the analyses conducted in Example 1, but will use the approximate displacement function $\bar{v}^{(2)}(x)$ defined in Equation 11.6,

$$\bar{v}^{(2)}(x) = [\mathbf{A}(x)]\bar{\mathbf{u}}$$

$$= \left[\left(-x + \frac{3x^2}{2l} - \frac{x^4}{2l^3}\right)\left(\frac{x^2}{2l} - \frac{x^4}{2l^3}\right)\left(1 - \frac{2x^2}{l^2} + \frac{x^4}{l^4}\right)\left(\frac{2x^2}{l^2} - \frac{x^4}{l^4}\right)\right]\begin{Bmatrix} \theta_{IJ} \\ \theta_{JI} \\ v_{IJ} \\ v_{JI} \end{Bmatrix} \quad \textbf{(11.35a)}$$

Recall that this displacement function is missing the capability to develop a linearly varying moment and, instead, has an incomplete quadratic moment capability, Equation 11.7*b*. The curvature, Equation 11.10*a*, is

$$\bar{v}''^{(2)}(x) = [\mathbf{B}(x)]\bar{\mathbf{u}} = \left[\left(\frac{3}{l} - \frac{6x^2}{l^3}\right)\left(\frac{1}{l} - \frac{6x^2}{l^3}\right)\left(-\frac{4}{l^2} + \frac{12x^2}{l^4}\right)\left(\frac{4}{l^2} - \frac{12x^2}{l^4}\right)\right]\begin{Bmatrix} \theta_{IJ} \\ \theta_{JI} \\ v_{IJ} \\ v_{JI} \end{Bmatrix} \quad \textbf{(11.35b)}$$

Again, note that no linear terms exist in any of the $B_i(x)$.

Substituting Equation 11.35*b* into Equation 11.12*a* leads to the stiffness matrix for this assumed displacement function,

$$\mathbf{k}^{(e)} = \frac{EI}{5l}\begin{bmatrix} 21 & 11 & -32/l & 32/l \\ 11 & 21 & -32/l & 32/l \\ -32/l & -32/l & 64/l^2 & -64/l^2 \\ 32/l & 32/l & -64/l^2 & 64/l^2 \end{bmatrix} \quad \textbf{(11.35c)}$$

or

$$\mathbf{k}^{(e)} = \frac{EI}{l}\begin{bmatrix} 4.2 & 2.2 & -6.4/l & 6.4/l \\ 2.2 & 4.2 & -6.4/l & 6.4/l \\ -6.4/l & -6.4/l & 12.8/l^2 & -12.8/l^2 \\ 6.4/l & 6.4/l & -12.8/l^2 & 12.8/l^2 \end{bmatrix} \quad \textbf{(11.35d)}$$

We observe that this stiffness matrix differs slightly from the conventional matrix in Equation 11.13*b*.

The generalized forces are computed by substituting Equation 11.35*a* into Equations 11.15. For the particular case of a uniform load $q(x) = -q_o$ (Figure 11.1) we obtain

$$\mathbf{F}_{\mathbf{q}}^{(AB)} = \int_0^l [\mathbf{A}(x)]^T(-q_o)dx = \begin{Bmatrix} q_o l^2/10 \\ -q_o l^2/15 \\ -8q_o l/15 \\ -7q_o l/15 \end{Bmatrix} \tag{11.35e}$$

Note that the values of the generalized forces are different on the two ends of the beam, an indication that elements based on this assumed displacement function do not behave as well as the conventional element.

(a) One-Element Model

We begin by modeling the structure in Figure 11.1 using only a single beam element with stiffness defined in Equation 11.35*c* and generalized forces defined in Equation 11.35*e*. The relationship between end displacements and system displacements is the same as in Table 1.1.1, so assembling the system equilibrium equation leads to

$$\frac{EI}{5l}\begin{bmatrix} 21 & 11 \\ 11 & 21 \end{bmatrix}\begin{Bmatrix} \bar{u}_1 \\ \bar{u}_2 \end{Bmatrix} = \begin{Bmatrix} q_o l^2/10 \\ -q_o l^2/15 \end{Bmatrix} \tag{11.36a}$$

from which we obtain

$$\begin{Bmatrix} \bar{u}_1 \\ \bar{u}_2 \end{Bmatrix} = \frac{q_o l^3}{24EI}\begin{Bmatrix} 1.0625 \\ -0.9375 \end{Bmatrix} \tag{11.36b}$$

This result indicates a slight asymmetry in end rotations, with magnitudes approximately 6% different from the exact values defined in Equation 11.24*c*.

The internal moment is given by

$$\overline{M}(x) = EI[\mathbf{B}(x)]\bar{\mathbf{u}} = \frac{q_o l^2}{12}\left[\frac{9}{8} - \frac{3}{8}\left(\frac{x}{l}\right)^2\right] \tag{11.36c}$$

Equation 11.36*c* is plotted in Figure 11.7 together with the previous one-element solution, Equation 11.27*c*. The figure shows the bias in the moment (and displacement) function that is inherent in this approximation. In particular, the element always produces a moment function of form $C_1 + C_2(x/l)^2$; consequently, the moment diagram will always have zero slope at $x = 0$, and a parabolic shape. Even under a symmetric loading, as in this case, the results are not symmetric. Nevertheless, this flawed element will give results that converge to the exact solution.

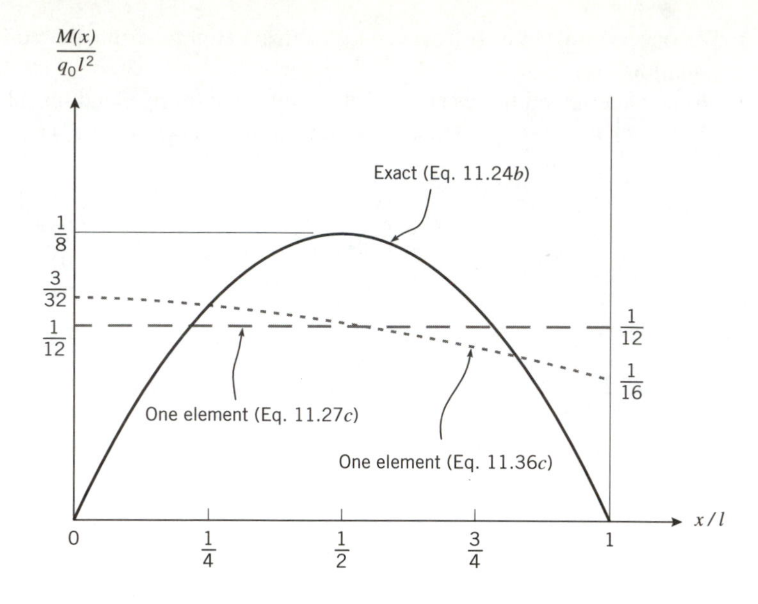

Figure 11.7 Bending moment comparison, Example 2.

(b) Two-Element Model

The analysis of the two-element model follows the same procedure employed in Example 1. The two elements and displacements are defined in Figure 11.3 and Table 11.1.2. We use the same stiffness matrix and generalized force vector as in Equations 11.35*d* and 11.35*e*, but replace l by $l/2$.

The equilibrium equation is

$$\frac{2EI}{l}\begin{bmatrix} 4.2 & 2.2 & 12.8/l & 0 \\ & 8.4 & 0 & 2.2 \\ & & 102.4/l^2 & -12.8/l \\ & & & 4.2 \end{bmatrix}\begin{Bmatrix} \bar{u}_1 \\ \bar{u}_2 \\ \bar{u}_3 \\ \bar{u}_4 \end{Bmatrix} = \begin{Bmatrix} q_o l^2/40 \\ q_o l^2/120 \\ -q_o l/2 \\ -q_o l^2/60 \end{Bmatrix} \tag{11.37a}$$

from which we obtain

$$\begin{Bmatrix} \bar{u}_1 \\ \bar{u}_2 \\ \bar{u}_3 \\ \bar{u}_4 \end{Bmatrix} = \frac{q_o l^3}{EI}\begin{Bmatrix} 0.04199 \\ 0.000326 \\ -0.01286l \\ -0.04134 \end{Bmatrix} \tag{11.37b}$$

These displacements compare very well to the exact solution, Equations 11.24*b* and 11.24*c*; the maximum difference is approximately 1%.

The approximate moment functions are given by

$$\overline{M}(x_1) = \frac{q_o l^2}{8}\left[\frac{3}{8} + \frac{7}{8}\left(\frac{x_1}{l/2}\right)^2\right], \qquad 0 < x_1 < \frac{l}{2} \tag{11.37c}$$

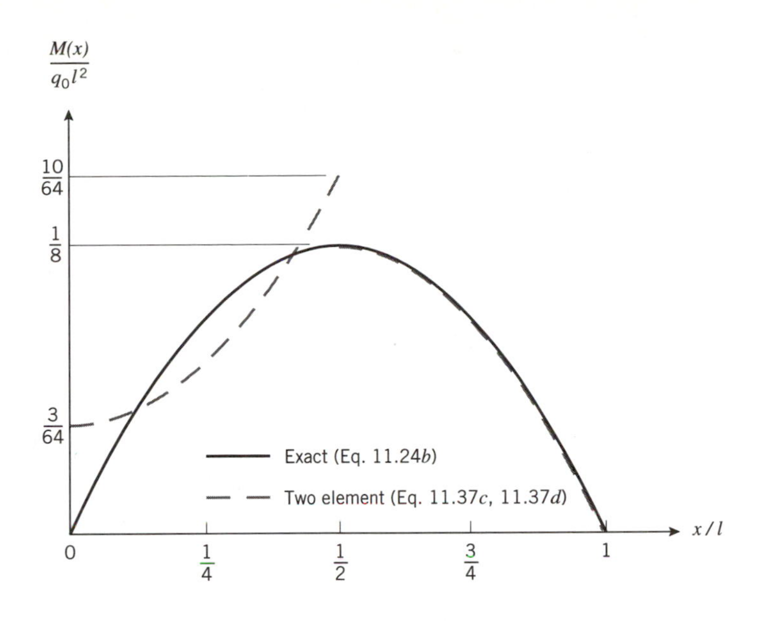

Figure 11.8 Bending moment comparison, Example 2.

and

$$\overline{M}(x_2) = \frac{q_o l^2}{8}\left[1 - \left(\frac{x_2}{l/2}\right)^2\right], \qquad 0 < x_2 < \frac{l}{2} \tag{11.37d}$$

Equations 11.37*c* and 11.37*d* are plotted in Figure 11.8. Note that Equation 11.37*d* gives exact results! This is coincidental, because this portion of the exact moment function happens to be parabolic with a zero slope at the left end. Equation 11.37*c* gives a much less accurate result but, even here, the errors at each end of the element are smaller than for the one-element model. This indicates that the procedure does appear to be converging.

Note, also, that the moment diagram, Figure 11.8, is discontinuous at $x_1 = l/2$ ($x_2 = 0$). This is a common occurrence in finite element models.

(c) Four- and Eight-Element Models

A four-element model can be developed in a straightforward manner by extension of the foregoing procedure. For the displacement function utilized here, however, the displacements will not be symmetric with respect to the midpoint of the structure. Consequently, we cannot utilize symmetry to reduce the number of elements that must be included in the computations. A four-element model will have eight nodal displacements, and an 8×8 system stiffness matrix. The details of the computations will be omitted for the sake of brevity, but the results (bending moments) are shown in Figure 11.9.

An eight-element model can be defined in an analogous manner. In this case, there will be 16 nodal displacements, and a 16×16 stiffness matrix. The results obtained from this model are illustrated in Figure 11.10.

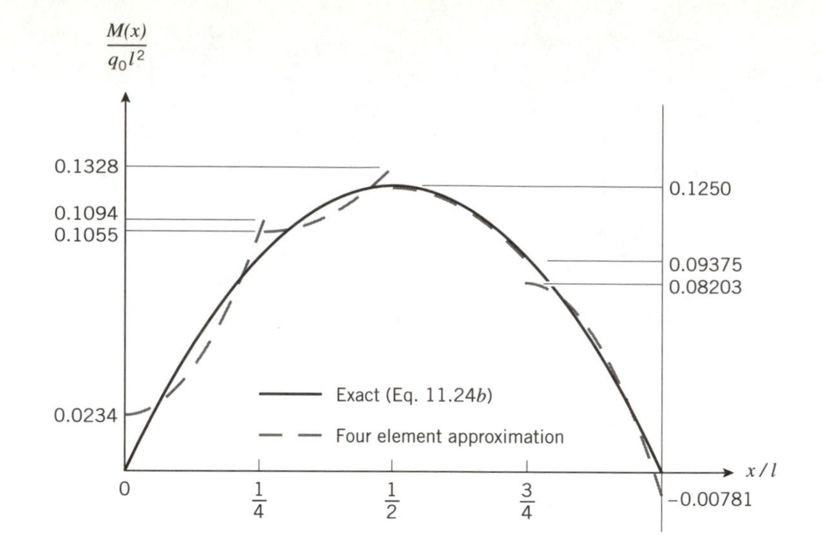

Figure 11.9 Bending moment comparison, Example 2.

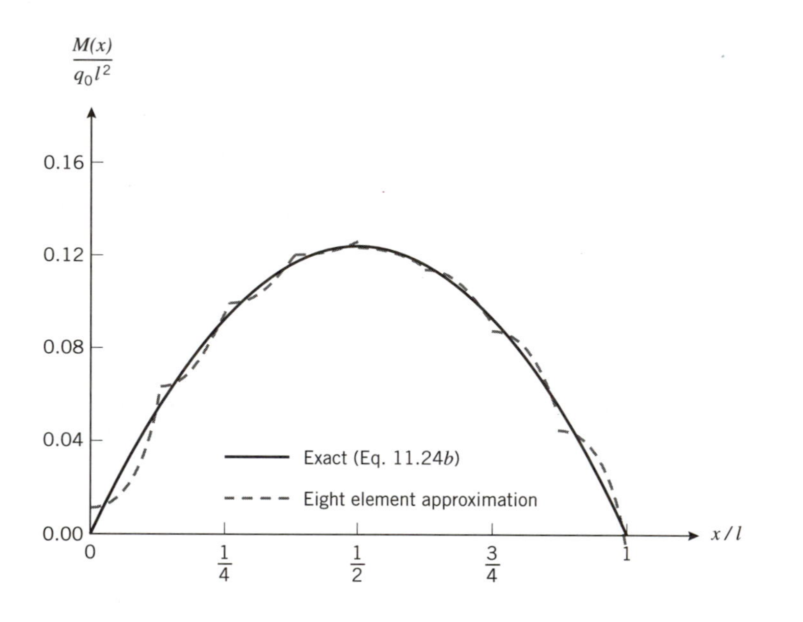

Figure 11.10 Bending moment comparison, Example 2.

The increasing accuracy that accompanies model refinement is apparent from Figures 11.7 through 11.10. Note that in each case the "right half" of the moment diagram is approximated much more closely than the "left half." Also, in each case, the element with the left end at $x = l/2$ gives an exact fit over its range.

The preceding example shows, again, that only requirements (1) through (3) (Section 11.1) are essential ingredients of element performance. Elements that have the capacity for rigid body motion, that can develop a constant strain state, and that have a sufficient number of nodal displacements to allow connection to other elements, are capable of accurately modeling the behavior of diverse structures.

As a final example, we turn our attention to a type of structure with which we have had no prior experience in this text. Specifically, we will analyze displacements and moments in a beam that is continuously supported by an elastic foundation.

EXAMPLE 4

Figure 11.11 represents a beam on an elastic foundation, with a fixed support at the left end and a roller support at the right end. We may regard the elastic foundation as a continuous support composed of an infinite number of infinitesimal springs, each occupying a length dx and providing an incremental force $dF = (kdx)v(x)$ where, in this case, k is a foundation modulus, which has the dimension of stiffness per unit length of foundation. It follows that a continuous force per unit length of magnitude $dF/dx = f(x) = kv(x)$ is developed in the foundation.

From our understanding of structural assemblies composed of beam/bar/spring elements, we recognize that the presence of infinitely many support springs means that the structure in Figure 11.11 is infinitely redundant. Thus, the structure cannot be modeled with a finite number of force degrees of freedom (because infinitely many cuts must be introduced to reduce the structure to a statically determinate configuration). The structure also cannot be modeled exactly using a finite number of conventional beam elements since, no matter how many internal joints are introduced, each span of length λ will encompass an infinite number of foundation springs.

If we knew the exact solution to the differential equations, and the boundary conditions, which describe the displacement behavior of the beam/foundation assembly, then we could model the structure using only these displacement patterns; but that is information we don't have at this stage. In general, as the complexity of the structure increases (e.g., by the introduction of a continuous support like the foundation in this example), our ability to solve the differential equations and satisfy the boundary conditions decreases to the point where closed form mathematical solutions are not prac-

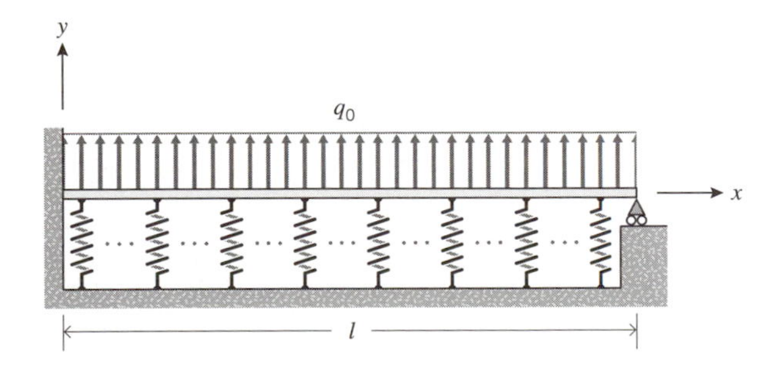

Figure 11.11 Beam on elastic foundation (EI = const.).

tically obtainable. In ordinary 2-D and 3-D elastic structures made of continuous materials, the level of difficulty of obtaining analytical solutions is so great that solutions can be found for only a few special problems of limited engineering interest. This is especially true if the physical structure has a complicated shape that differs significantly from rectangular (brick-like in 3-D) or circular (spherical in 3-D), or if the boundary conditions are at all complicated, as in most engineering problems.

Actually, the problem of a beam on elastic foundation, such as in Figure 11.11, is well known, and its solution is readily obtained by analysts familiar with ordinary differential equations. We will present the analytic solution subsequently. As we shall see, the basic character of the solution will depend on the ratio of the foundation stiffness k to the beam bending stiffness EI, specifically on a dimensionless parameter $\alpha = (kl^4/4EI)^{1/4}$.

Let us approach this problem as if we have no prior knowledge of the exact behavior of a beam on an elastic foundation (which is true at this point), and let us analyze the structure using a basic finite element formulation. This means that we must derive an expression for the total potential energy for an analytic model of a typical segment (element) of the structure, such as is shown in Figure 11.12. We will assume that the behavior of the beam and the elastic foundation can be adequately represented by an approximate displacement function that has provided good results in previous analyses, namely the cubic expression in Equations 11.4 or 11.9*a*.

For the sake of generality and computational efficiency, we will rewrite Equation 11.9*a* using λ rather than l to denote element length, and reorder the generalized displacements in accordance with the sequence defined in Figure 11.12. As in Figure 11.12, we will denote the generalized displacements by δ_1–δ_4 rather than by θ_{IJ}–v_{JI}. Thus, the assumed displacement function is

$$\bar{v}^{(1)}(x) = \left[\left(1 - \frac{3x^2}{\lambda^2} + \frac{2x^3}{\lambda^3}\right)\left(-x + \frac{2x^2}{\lambda} - \frac{x^3}{\lambda^2}\right)\left(\frac{3x^2}{\lambda^2} - \frac{2x^3}{\lambda^3}\right)\left(\frac{x^2}{\lambda} - \frac{x^3}{\lambda^2}\right)\right]\begin{Bmatrix}\delta_1\\ \delta_2\\ \delta_3\\ \delta_4\end{Bmatrix} \tag{11.38a}$$

Similarly, the curvature, Equation 11.11*a*, now has the form

$$\bar{v}''^{(1)}(x) = \left[\left(-\frac{6}{\lambda^2} + \frac{12x}{\lambda^3}\right)\left(\frac{4}{\lambda} - \frac{6x}{\lambda^2}\right)\left(\frac{6}{\lambda^2} - \frac{12x}{\lambda^3}\right)\left(\frac{2}{\lambda} - \frac{6x}{\lambda^2}\right)\right]\begin{Bmatrix}\delta_1\\ \delta_2\\ \delta_3\\ \delta_4\end{Bmatrix} \tag{11.38b}$$

To derive the stiffness of the element in Figure 11.12, we require the strain energy for both the beam and the elastic foundation. The former is computed from Equation 11.12*a*,

$$U_{beam} = \frac{1}{2}\begin{Bmatrix}\delta_1\\ \delta_2\\ \delta_3\\ \delta_4\end{Bmatrix}^{\mathrm{T}}\left\{\frac{EI}{\lambda}\begin{bmatrix} 12/\lambda^2 & -16/\lambda & -12/\lambda^2 & -6/\lambda\\ -6/\lambda & 4 & 6/\lambda & 2\\ -12\lambda^2 & 6/\lambda & 12/\lambda^2 & 6/\lambda\\ -6/\lambda & 2 & 6/\lambda & 4\end{bmatrix}\right\}\begin{Bmatrix}\delta_1\\ \delta_2\\ \delta_3\\ \delta_4\end{Bmatrix} \tag{11.38c}$$

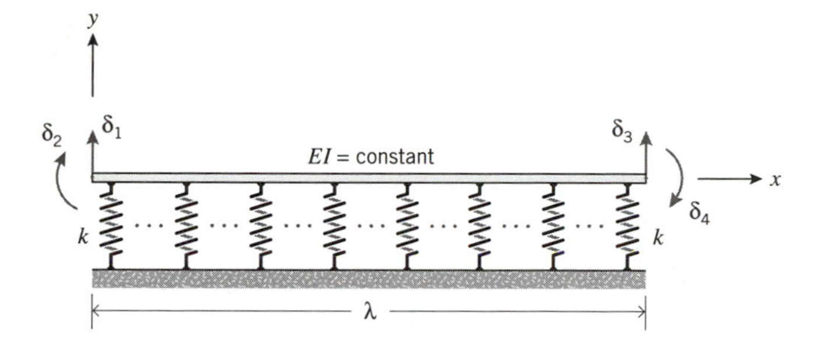

Figure 11.12 Finite element.

where the term in parenthesis is $\mathbf{k}^{(b)}$, the stiffness matrix for the beam. This is the same stiffness matrix as was obtained previously (Equation 11.13*b*), except that the data have been reordered as a result of the new numbering sequence for displacements.

The strain energy for the elastic foundation is obtained in a similar manner. For a discrete spring the strain energy is $P\Delta/2$ where $P = k\Delta$; therefore, for the foundation, which is equivalent to a distributed spring, we must have

$$\begin{aligned} U_{foundation} &= \int_0^\lambda \frac{1}{2}\bar{v}^{(1)}(x)f(x)dx \\ &= \frac{1}{2}\int_0^\lambda k\bar{v}^{(1)^2}(x)dx \\ &= \frac{1}{2}\bar{\mathbf{u}}^{\mathrm{T}}\int_0^\lambda [\mathbf{A}(x)]^{\mathrm{T}}(k)[\mathbf{A}(x)]dx\,\bar{\mathbf{u}} \\ &= \frac{1}{2}\bar{\mathbf{u}}^{\mathrm{T}}\mathbf{k}^{(f)}\bar{\mathbf{u}} \end{aligned} \qquad \textbf{(11.38}\textbf{\textit{d}}\textbf{)}$$

where $\mathbf{k}^{(f)}$ is the stiffness matrix for the foundation, which is defined by the integral in the third line of Equation 11.38*d*. Substituting Equation 11.38*a* into Equation 11.38*d* and evaluating leads to

$$U_{foundation} = \frac{1}{2}\begin{Bmatrix}\delta_1\\ \delta_2\\ \delta_3\\ \delta_4\end{Bmatrix}^{\mathrm{T}} \left\{ k\lambda^3 \begin{bmatrix} 13/35\lambda^2 & -11/210\lambda & 9/70\lambda^2 & 13/420\lambda \\ -11/210\lambda & 1/105 & -13/420\lambda & -1/140 \\ 9/70\lambda^2 & -13/420\lambda & 13/35\lambda^2 & 11/210\lambda \\ 13/420\lambda & -1/140 & 11/210\lambda & 1/105 \end{bmatrix} \right\} \begin{Bmatrix}\delta_1\\ \delta_2\\ \delta_3\\ \delta_4\end{Bmatrix} \qquad \textbf{(11.38}\textbf{\textit{e}}\textbf{)}$$

where the term in parenthesis is $\mathbf{k}^{(f)}$.

Finally, we require the external potential energy of the applied load, which in this case is a uniform load $+q_o$. The basic definition is given in Equation 11.14; substituting Equation 11.38*a* gives,

$$U_e = -\int_0^{\lambda} q_o \bar{v}^{(1)}(x)dx = -\begin{Bmatrix} \delta_1 \\ \delta_2 \\ \delta_3 \\ \delta_4 \end{Bmatrix}^{\mathrm{T}} q_o\lambda \begin{Bmatrix} 1/2 \\ -\lambda/12 \\ 1/2 \\ \lambda/12 \end{Bmatrix} = -\bar{\mathbf{u}}^{\mathrm{T}} \mathbf{F}_{\mathbf{q}} \quad \textbf{(11.38}f\textbf{)}$$

The total potential energy is the sum of the energies in Equations 11.38*c*, 11.38*e*, and 11.38*f*,

$$V = \frac{1}{2}\bar{\mathbf{u}}^{\mathrm{T}}\mathbf{k}^{(\mathrm{b})}\bar{\mathbf{u}} + \frac{1}{2}\bar{\mathbf{u}}^{\mathrm{T}}\mathbf{k}^{(\mathrm{f})}\bar{\mathbf{u}} - \bar{\mathbf{u}}^{\mathrm{T}}\mathbf{F}_{\mathbf{q}} \quad \textbf{(11.39}a\textbf{)}$$

The equilibrium equation for the element is obtained by finding the stationary value of the total potential energy,

$$\mathbf{k}^{(b)}\bar{\mathbf{u}} + \mathbf{k}^{(f)}\bar{\mathbf{u}} = \mathbf{F}_{\mathbf{q}} \quad \textbf{(11.39}b\textbf{)}$$

or

$$[\mathbf{k}^{(b)} + \mathbf{k}^{(f)}]\bar{\mathbf{u}} = \mathbf{F}_{\mathbf{q}} \quad \textbf{(11.39}c\textbf{)}$$

Thus, we see that the stiffness matrix for this element is actually the sum of the stiffness matrices of the beam and the foundation.

We are now in a position to define and analyze a model of a structure composed of finite elements of the type synthesized in Equation 11.39*c*. From the model, we can recover approximate displacements, internal bending moments, and foundation forces. For the applications that follow, we will set $\alpha = (kl^4/4EI)^{1/4} = 2\pi$, a value[7] that corresponds to a moderately stiff foundation, and which has a significant effect on the structural response.

(a) One-Element Model

If we choose to model the beam/foundation structure in Figure 11.11 using only one finite element, then the element has length l (i.e., $\lambda = l$), and there is only one system displacement, $\bar{u}_1$, the rotation at the right end (i.e., $x = l$). Thus, the relation between element and system coordinates is as defined in Table 11.4.1.

Table 11.4.1

Element	Element displacement	System displacement
1	δ_1	—
	δ_2	—
	δ_3	—
	δ_4	1

[7] The choice of this particular value of α also facilitates a subsequent comparison with the exact solution.

Assembling the equilibrium equation defined in Equation 11.39c leads to

$$\left[\frac{4EI}{l} + \frac{kl^3}{105}\right]\{\bar{u}_1\} = \left\{\frac{q_o l^2}{12}\right\} \tag{11.40a}$$

The first entry in the stiffness matrix, $4EI/l$, is the contribution from the beam, and the second, $kl^3/105$, is the contribution from the foundation. We have arbitrarily specified a relation between beam and foundation stiffnesses in terms of the parameter α; therefore, if we substitute $4EI/l = kl^3/(2\pi)^4$, we may rewrite Equation 11.40a in the form

$$\left[\frac{kl^3}{(2\pi)^4} + \frac{kl^3}{105}\right]\{\bar{u}_1\} = \left\{\frac{q_o l^2}{12}\right\} \tag{11.40b}$$

As the numbers in the matrix now indicate, the beam bending stiffness based on this particular value of α is about 1/15 of the contribution from the elastic foundation. Thus, the foundation will have a much greater effect than the beam in determining the response predicted by this model. In other words, even though our finite element model is developed using conventional beam displacement functions, the most important effect is from another source, i.e., the foundation. Thus, we may anticipate that the model will not give very accurate results. On the other hand, if the parameter α were selected to correspond to a much "softer" foundation, say $kl^4/4EI = 1$, then the stiffness contribution from the foundation would be less than 1% of that of the beam in Equation 11.40a. In that case, we would expect the overall response to be dominated by the beam, so the displacement predicted by the model should be quite accurate. The approximate moment function, however, will still be linear, and our previous experience indicates that we will need a more refined model to effectively represent the internal behavior of the beam. In other words, as we have seen from previous examples, even if there were no elastic foundation, we should have at least four elements ($\lambda = l/4$) to obtain a reasonable piecewise-linear representation of the exact quadratic moment function.

Solving Equation 11.40b gives

$$\bar{u}_1 = 8.1977\frac{q_o}{kl} \tag{11.40c}$$

Substituting this result into Equations 11.38a and 11.38b yields

$$\bar{v}(x) = 8.1977\frac{q_o}{k}\left[\left(\frac{x}{l}\right)^2 - \left(\frac{x}{l}\right)^3\right] \tag{11.41a}$$

and[8]

[8] The constant term containing α in the second line of Equation 11.41b will appear as part of a dimensionless parameter in subsequent plots of $\overline{M}(x)$, and is presented for this purpose only.

$$\overline{M}(x) = EI\overline{v}''(x) = 8.1977\frac{q_o EI}{kl^2}\left[2 - 6\left(\frac{x}{l}\right)\right]$$
$$= 0.1038\frac{q_o l^2}{2\alpha^2}\left[2 - 6\left(\frac{x}{l}\right)\right] \tag{11.41b}$$

Because we don't yet have the exact solution with which to compare the results in Equations 11.41, if we wish to compare approximate solutions we have no choice but to refine the model and reinvestigate the behavior.

(b) Two-Element Model

Figure 11.13 shows a two-element model of the structure in Figure 11.11. In this case we have the element and system coordinate relationships defined in Table 11.4.2.

Table 11.4.2

Element	Element displacement	System displacement
1	δ_1	—
	δ_2	—
	δ_3	1
	δ_4	2
2	δ_1	1
	δ_2	2
	δ_3	—
	δ_4	3

The direct assembly process leads to

$$\left(\frac{EI}{\lambda}\begin{bmatrix} \frac{24}{\lambda^2} & 0 & -\frac{6}{\lambda} \\ 0 & 8 & 2 \\ -\frac{6}{\lambda} & 2 & 4 \end{bmatrix} + k\lambda^3 \begin{bmatrix} \frac{26}{35\lambda^2} & 0 & \frac{13}{420\lambda} \\ 0 & \frac{2}{105} & -\frac{1}{140} \\ \frac{13}{420\lambda} & -\frac{1}{140} & \frac{1}{105} \end{bmatrix}\right)\begin{Bmatrix} \overline{u}_1 \\ \overline{u}_2 \\ \overline{u}_3 \end{Bmatrix} = q_o\lambda\begin{Bmatrix} \frac{1}{2} \\ 0 \\ \frac{\lambda}{12} \end{Bmatrix} \tag{11.42a}$$

where the first matrix is due to bending effects in the beam, and the second matrix is due to the foundation stiffness. It is informative to investigate the relative stiffness contributions of the beam and foundation in this two-element model compared to the preceding one-element solution. (We expect that the contributions will be different because parameter α is defined in terms of structure length l rather than element length λ.) To assess these contributions, we first manipulate the constant terms in Equation 11.42*a*. Substitute $k\lambda^4/EI = 4\alpha^4(\lambda/l)^4$, and let $\beta = l/\lambda$. Then, the equation may be written in the form

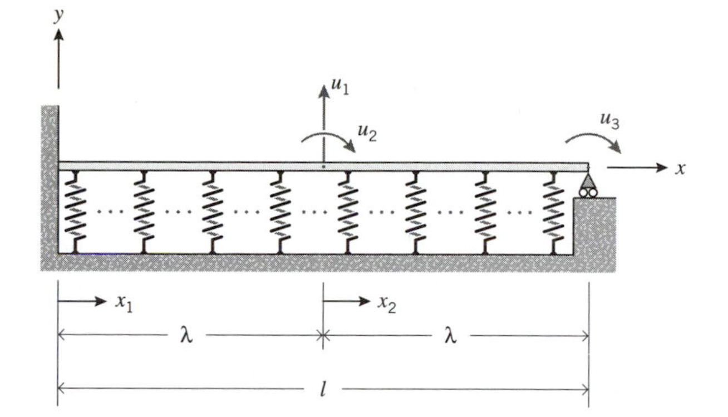

Figure 11.13 Two-element model.

$$\left\{\frac{\beta^4}{4\alpha^4}\begin{bmatrix} 24\beta^2 & 0 & -6\beta \\ 0 & 8 & 2 \\ -6\beta & 2 & 4 \end{bmatrix} + \begin{bmatrix} \frac{26}{35}\beta^2 & 0 & \frac{13}{420}\beta \\ 0 & \frac{2}{105} & -\frac{1}{140} \\ \frac{13}{420}\beta & -\frac{1}{140} & \frac{1}{105} \end{bmatrix}\right\}\begin{Bmatrix} \bar{u}_1/l \\ \bar{u}_2 \\ \bar{u}_3 \end{Bmatrix} = \frac{q_o}{kl}\begin{Bmatrix} \beta^2 \\ 0 \\ \frac{\beta}{12} \end{Bmatrix} \quad \textbf{(11.42b)}$$

Note that we have also modified the equations by replacing displacement $\bar{u}_1$ by the dimensionless quantity $\bar{u}_1/l$, and have multiplied the first equation by l.

Recall that $\alpha = 2\pi$ and that for this model $\beta = 2$; we observe that the parameter $\beta^4/4\alpha^4$ in Equation 11.42b reduces to $1/4\pi^4 \approx 1/400$. Therefore, for the last entries (3,3) in the $\mathbf{k}^{(b)}$ and $\mathbf{k}^{(f)}$ matrices, the ratio between bending and foundation stiffness effects is $105\beta^4/\alpha^4 = 1.0779 \approx 1$. In other words, the bending stiffness contribution is now relatively much greater than in the prior case where the ratio was approximately 1/15. This means that we can expect the relative importance of the two different stiffness mechanisms to be significantly readjusted.

Substituting the specified numerical values of all parameters in Equation 11.42b leads to

$$\begin{bmatrix} 3.2178 & 0 & 3.1107 \times 10^{-2} \\ 0 & 3.9580 \times 10^{-2} & -2.0099 \times 10^{-2} \\ 3.1107 \times 10^{-2} & -2.0099 \times 10^{-2} & 1.9790 \times 10^{-2} \end{bmatrix}\begin{Bmatrix} \bar{u}_1/l \\ \bar{u}_2 \\ \bar{u}_3 \end{Bmatrix} = \frac{q_o}{kl}\begin{Bmatrix} 4 \\ 0 \\ 1/6 \end{Bmatrix} \quad \textbf{(11.42c)}$$

from which we obtain

$$\begin{Bmatrix} \bar{u}_1/l \\ \bar{u}_2 \\ \bar{u}_3 \end{Bmatrix} = \frac{q_o}{kl}\begin{Bmatrix} 1.1793 \\ 0.3353 \\ 6.6023 \end{Bmatrix} \quad \textbf{(11.42d)}$$

Note that the rotation at $x = l\ (x_2 = \lambda)$ produced by this model is $6.6023\, q_o/kl$ compared to $8.1977\, q_o/kl$ for the one-element model (Equation 11.40*c*). Thus, there are significant changes occurring in the basic nodal displacements, and we can anticipate even greater changes in the internal moments.

The displacements are computed from Equation 11.38*a*,

$$\bar{v}(x_1) = \frac{q_o}{k}\left[1.1793\left(\frac{3x_1^2}{\lambda^2} - \frac{2x_1^3}{\lambda^3}\right) + 0.1676\left(\frac{x_1^2}{\lambda^2} - \frac{x_2^3}{\lambda^3}\right)\right] \tag{11.43a}$$

$$\bar{v}(x_2) = \frac{q_o}{k}\left[1.1793\left(1 - \frac{3x_2^2}{\lambda^2} + \frac{2x_2^3}{\lambda^3}\right) - 0.1676\left(\frac{x_2}{\lambda} - \frac{2x_2^2}{\lambda^2} + \frac{x_2^3}{\lambda^3}\right) + 3.3011\left(\frac{x_2^2}{\lambda^2} - \frac{x_2^3}{\lambda^3}\right)\right] \tag{11.43b}$$

and moments are obtained from Equation 11.38*b*,

$$\overline{M}(x_1) = \frac{q_o l^2}{2\alpha^2}\left[0.3754 - 0.7679\left(\frac{x_1}{\lambda}\right)\right] \tag{11.44a}$$

$$\overline{M}(x_2) = \frac{q_o l^2}{2\alpha^2}\left[9.994 \times 10^{-3} - 0.3375\left(\frac{x_2}{\lambda}\right)\right] \tag{11.44b}$$

The results for displacements obtained using the one- and two-element models are shown in Figure 11.14. Note the surprising behavior exhibited by the two-element model, in particular the increase in deflection in the left half of the structure and the decrease in the right half. In fact, the result appears to be incorrect because $\bar{u}_2$, the angular midpoint displacement, is positive (clockwise). This is not something we would expect based on our understanding of the behavior of a structure with similar end supports in the absence of the elastic foundation.

This counter-intuitive response is explained by noting the signs of entries in the $\mathbf{k}^{(b)}$ and $\mathbf{k}^{(f)}$ matrices. Referring to Equation 11.42*b*, if only the beam stiffness terms are

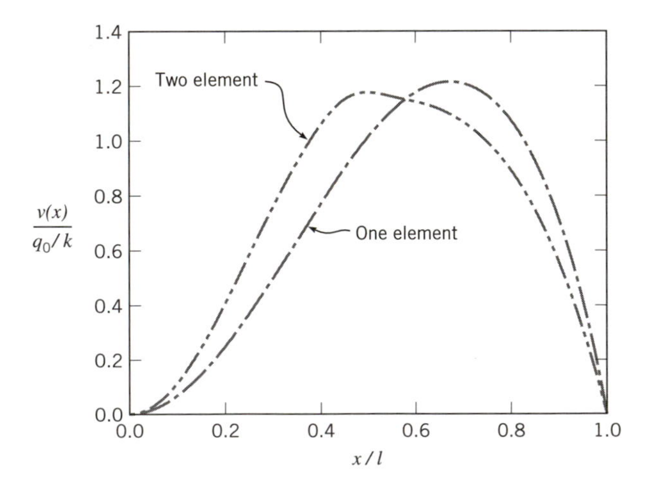

Figure 11.14 Displacements, Example 4 ($\alpha = 2\pi$).

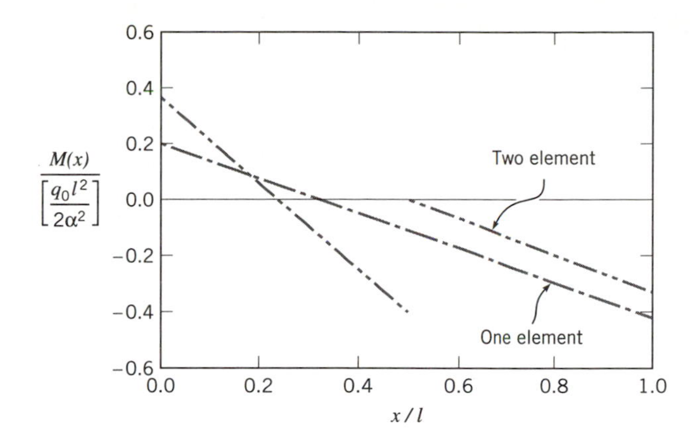

Figure 11.15 Bending moments, Example 4 ($\alpha = 2\pi$).

present, the entry in location (1, 3) is negative and the (2, 3) entry is positive; but when the foundation stiffness is added, the signs are reversed. In other words, the foundation stiffness dominates these terms, and this causes the shape of the displacement diagram produced by the analysis to be strongly influenced by the distributed elastic foundation.

The moment diagrams for our two models are shown in Figure 11.15. A number of features are worth noting. First, neither of the results can possibly be correct because neither solution gives a zero moment at the right end of the structure. Second, the two-element solution, which is piecewise linear, has a discontinuity at the midpoint of the structure, whereas the exact moment should be continuous everywhere. We note that this discontinuity is large, indicating an error of the same order as the maximum moment. Finally, we should recognize that the two-element model gives a solution for which the moment at the left end of the structure is larger than for the one-element model, while the exact opposite is the case at the right end. This, at least, appears to be reasonable!

Since we are not satisfied with the results obtained thus far, we are once again forced to consider a more refined model in the hope that (a) the displacement pattern will appear to be more realistic, and (b) the moment diagram will be more nearly continuous and will also tend to approach zero at the right end of the structure.

(c) Four-Element Model

Figure 11.16 shows a four-element model that has 7 degrees of freedom. The 7 × 7 system stiffness matrix and 7 × 1 force vector are developed exactly as in the previous case. For the sake of brevity, we will present only the basic solution,

$$\begin{Bmatrix} \bar{u}_1/l \\ \bar{u}_2 \\ \bar{u}_3/l \\ \bar{u}_4 \\ \bar{u}_5/l \\ \bar{u}_6 \\ \bar{u}_7 \end{Bmatrix} = \frac{q_o}{kl} \begin{Bmatrix} 0.7992 \\ -0.6589 \\ 1.0855 \\ -0.0648 \\ 1.0132 \\ 1.3778 \\ 6.3477 \end{Bmatrix} \tag{11.45}$$

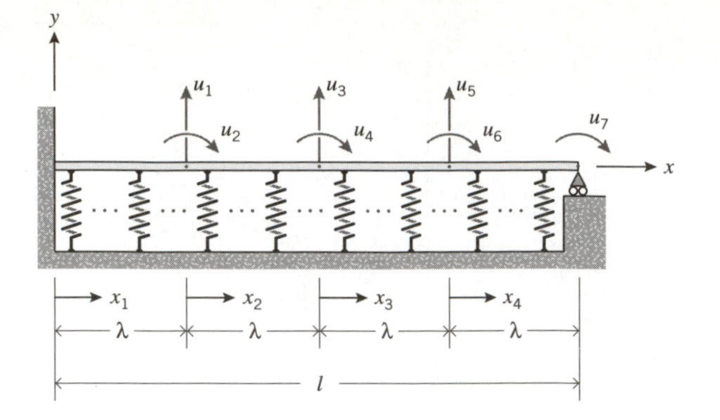

Figure 11.16 Four-element model.

The displacement and moment functions are readily obtained from these data, and will not be discussed in detail here. The results are, however, plotted in Figures 11.17 and 11.18 for comparison with the two previous models. From Figure 11.17 we see that the four-element model exhibits greater flexibility than the two-element model. The displacement for the former is markedly greater in the region $0 < x < 0.3l$, and slightly greater in the region $0.7l < x < l$. The central region of the structure, $0.3l < x < 0.7l$, appears somewhat flatter than in the two-element model, and the slope at $x = l/2$ is now positive rather than negative. In general, the overall appearance of this displacement is more reasonable than the preceding solutions.

The moment diagrams obtained from all three models are shown in Figure 11.18. For the four-element model, discontinuities still exist at $x = l/4$, $l/2$, and $3l/4$. The maximum magnitude of a discontinuity is now approximately half that of the previous case.

Finally, the moment at $x = 0$ obtained from the four-element model is nearly twice as great as that for the two-element model. At the right end of the structure, the moment is still not zero, but is significantly less than in the previous case. Also, the moment diagram now has two regions with positive slope, corresponding to elements 2 and 4.

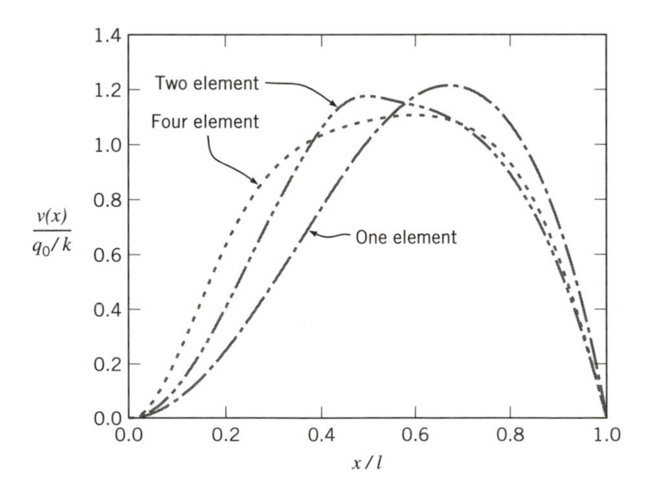

Figure 11.17 Displacements, Example 4 ($\alpha = 2\pi$).

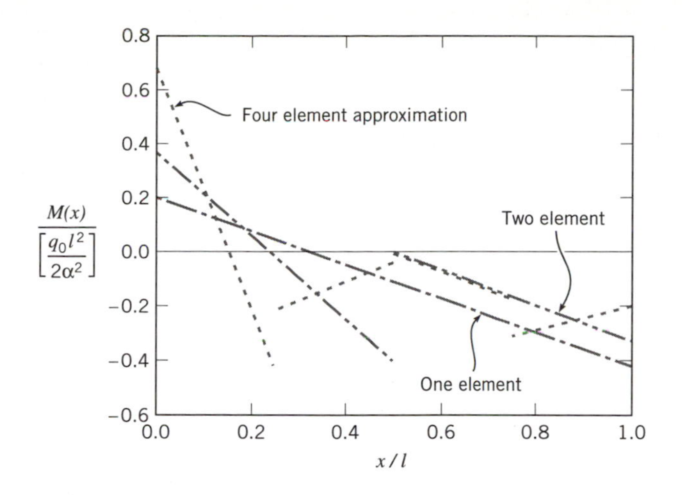

Figure 11.18 Bending moments, Example 4 ($\alpha = 2\pi$).

Thus, although the solution contains features that indicate improvement in comparison to previous results, it is still apparent that there are errors. In fact, the nature of the exact solution is still not at all evident from our analyses.

Rather than continue with additional refinements in the finite element model we will stop at this point and, for the sake of completeness, compare the results we have obtained thus far to the exact solution. The differential equation governing the behavior of the beam/foundation structure is

$$EI\frac{d^4v(x)}{dx^4} + kv(x) = q_o \tag{11.46a}$$

This equation has a solution of the form

$$\begin{aligned} v(x) = {} & A_1 \sinh\frac{\alpha x}{l}\sin\frac{\alpha x}{l} + A_2 \sinh\frac{\alpha x}{l}\cos\frac{\alpha x}{l} + \\ & A_3 \cosh\frac{\alpha x}{l}\sin\frac{\alpha x}{l} + A_4 \cosh\frac{\alpha x}{l}\cos\frac{\alpha x}{l} + \frac{q_o}{k} \end{aligned} \tag{11.46b}$$

where $\alpha = (kl^4/4EI)^{1/4}$, as defined previously. The constants $A_1, \ldots, A_4$ are obtained by imposing the four boundary conditions $v(0) = v'(0) = v(l) = 0$, and $M(l) = EIv''(l) = 0$. Solving for the constants leads to

$$\begin{Bmatrix} A_1 \\ A_2 \\ A_3 \\ A_4 \end{Bmatrix} = \frac{q_o}{k(sc - SC)}\begin{Bmatrix} (sC + cS)(Cc - 1) - (sC - cS)Ss \\ Cc(Cc - 1) + S^2s^2 \\ -Cc(Cc - 1) - S^2s^2 \\ -(sc - SC) \end{Bmatrix} \tag{11.46c}$$

where $s \equiv \sinh\alpha$, $c \equiv \cosh\alpha$, $S \equiv \sin\alpha$, and $C \equiv \cos\alpha$.

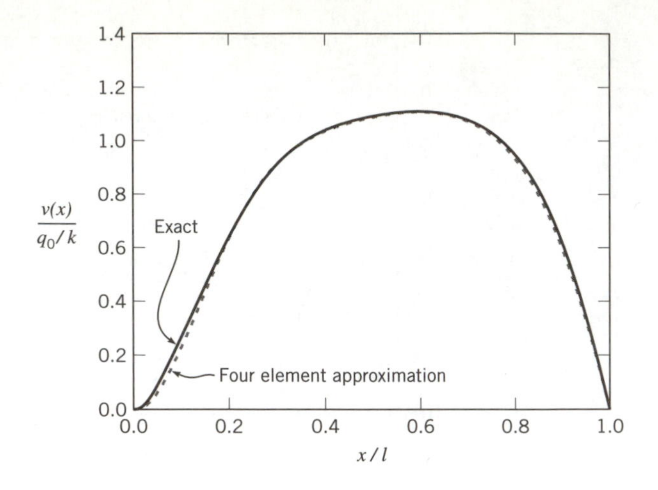

Figure 11.19 Displacements, Example 4 ($\alpha = 2\pi$).

The displacement function, Equation 11.46*b*, is far more complex than for the case of a beam without any elastic foundation. This exact solution involves transcendental functions (in this case hyperbolic and trigonometric functions) rather than the polynomial functions that we assumed would provide a reasonable basis for describing the physical response of the structure. In other words, our finite element model is based on assumed displacement patterns that are completely different from the exact solution. To make matters worse, the overall character (shape) of the exact displacement now depends on the parameter α, which appears in each of the transcendental functions. For instance, if $\alpha = 2\pi$, the value selected for our example, then $\sinh(\alpha x/l)$ ranges between zero and ~268, but if $\alpha = \pi$ the range is between zero and ~11.5, a major difference.

This example shows just how complex true behavior can be from a mathematical standpoint (when it can actually be determined). As complexity increases, we will quickly lose all hope of getting exact results, and will have no choice but to pursue a numerical analysis such as a finite element approach.

Before concluding this example, let's compare the four-element solution to the exact results stated above. The displacements are compared in Figure 11.19. The two sets of results are in remarkably close agreement! In fact, only near the left end of the structure is there any noticable difference. Thus, in spite of our concern over inaccuracies in the moment diagram, we appear to have obtained a very nice result for the displacement.

The comparison of the four-element and exact moment diagrams is presented in Figure 11.20. This finally reveals the essence of the finite element results. Note how the piecewise-linear approximation gives a best fit of the true (exact) behavior with a characteristic over/under representation in every element. The approximation is most crude in the first and fourth elements, i.e., near the left and right ends, respectively, where the true behavior is quite nonlinear.

From the vantage point of having the exact moment diagram (a luxury we will not have in more complex problems), we can see that the four-element approximate solution is very reasonable. It displays most of the important features of the true response. However, we recognize that if we seek an accurate moment diagram using the finite element approach we will have to analyze an even more refined model, say one with

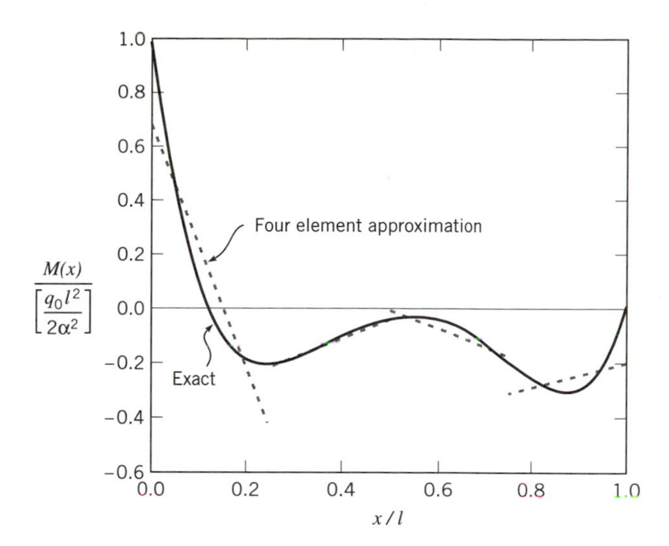

Figure 11.20 Bending moments, Example 4 ($\alpha = 2\pi$).

eight elements. In fact, if we were to do this, we would obtain excellent results for the moment, and it would be difficult to distinguish the approximate displacement from the exact. The slight discontinuities in the internal moments would persist, at a decreasing level, and the moment at $x = l$ would not be zero although it would be a lot closer than for the four-element model.

The exercise we have just gone through of building and analyzing successively more refined computational models, and comparing results, is an essential part of modern finite element analysis of complex structures. As progressively more complex structures are analyzed, we must define more complex and elaborate finite elements, especially for 2-D and 3-D continuous bodies. This type of development may be found in any of a large number of textbooks devoted to the subject. Finite element computer technology is highly advanced at this time, and commercial software programs offer a complete, powerful analysis capability with a host of modeling features, all for use on personal computers at a relatively modest cost.

Despite the availability, economy, and ease of use of programs for the numerical simulation of structural behavior, one must keep in mind two essential facts: (1) the analysis may produce errors attributable to software and/or computer hardware, and (2) the analysis itself, once infinitely redundant structures are considered, is inherently approximate. However, the concepts and perspectives presented in the last two chapters of this text regarding the basic behavior of approximate solutions can be used to gain considerable insight into the quality and, ultimately, the utility of approximate analysis.

In the final analysis, no matter how powerful the computer or the finite element software, numerical results will be meaningless unless an experienced, knowledgable structural analyst can establish the credibility of the solution, particularly in view of the fact that the solution procedure itself is approximate, and that the exact solution is unknown. This difficult task is one of the most interesting and challenging aspects of modern finite element analysis. It truly makes the subject one that extends beyond the confines of numbers ("the computer gives this result"), to one of human inference and interpretation ("based on the numerical solution, I believe . . ."). Thus, the modern field of structural analysis and engineering is, indeed, one of the most interesting and challenging of intellectual endeavors.

PROBLEMS

11.P1. It is proposed that the following displacement pattern be used to develop a finite element model for the beam shown. Does the assumed displacement pattern satisfy all basic requirements? Describe your reasoning.

$$\bar{v}(x) = C_1 + C_2\frac{x}{\lambda} + C_3\left(\frac{x}{\lambda}\right)^2 + C_4 \sin\frac{\pi x}{\lambda}$$

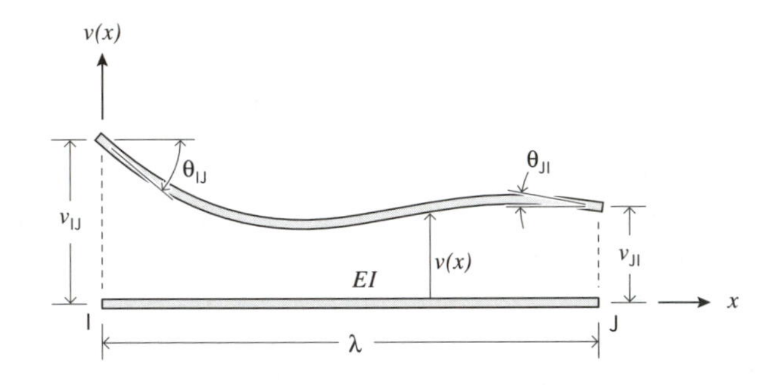

11.P2. It is proposed that the displacement pattern defined in Equation 11.9*a* be used to develop a finite element model for the beam in Problem 11.P1. Analyze the structure shown below using one element to model the entire structure. Calculate any additional data you need to complete the analysis. Sketch the approximate and exact bending moment diagrams.

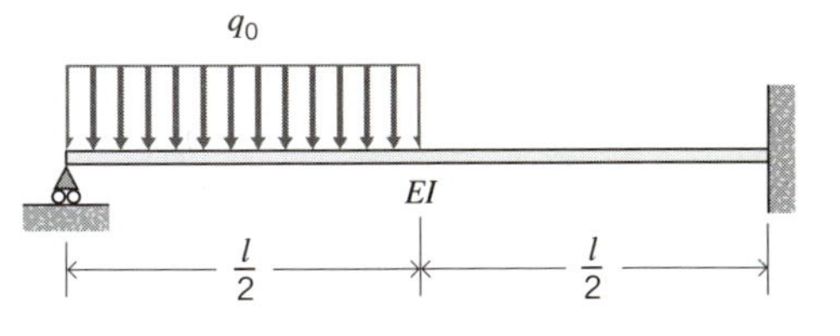

11.P3. Reanalyze the structure in Problem 11.P2 using two finite elements, each of length $l/2$. Sketch the approximate and exact bending moment diagrams.

Appendix A

Fixed-End Moments and Forces

All beams are uniform. The sign convention for positive fixed-end forces is defined by the following figures. Effects of transverse and axial loads are uncoupled. (For each case, fixed-end forces not listed are zero.)

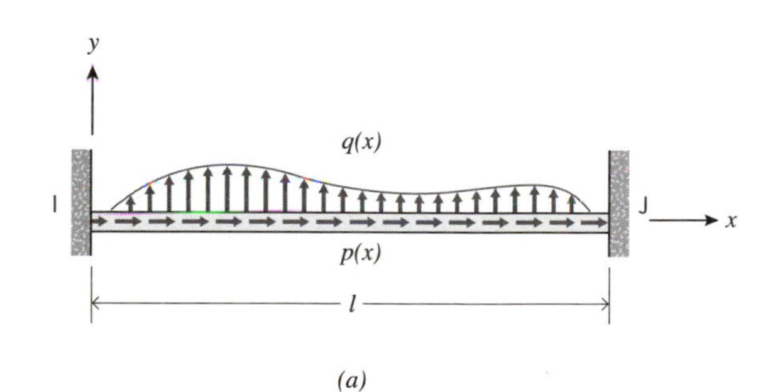

(a)

$\hat{M}_{IJ}$ $\hat{M}_{JI}$

$\hat{P}_{IJ}$ $\hat{P}_{JI}$

$\hat{V}_{IJ}$ $\hat{V}_{JI}$

(b)

1. Concentrated transverse force F

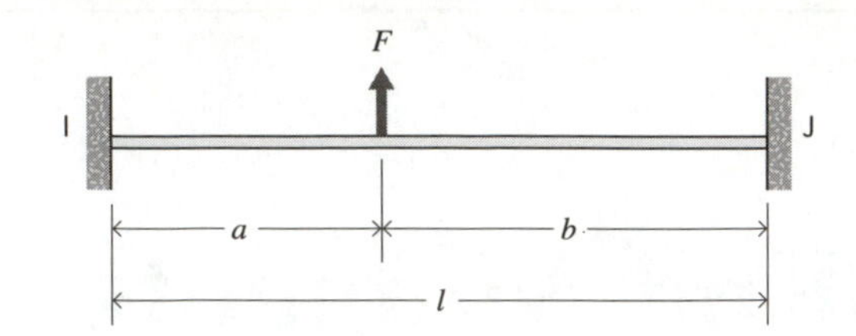

$$\widehat{M}_{IJ} = \frac{ab^2}{l^2} F \tag{A.1a}$$

$$\widehat{M}_{JI} = -\frac{a^2b}{l^2} F \tag{A.1b}$$

$$\widehat{V}_{IJ} = -\frac{(3a + b)b^2}{l^3} F \tag{A.1c}$$

$$\widehat{V}_{JI} = -\frac{(3b + a)a^2}{l^3} F \tag{A.1d}$$

2. Concentrated moment M_o

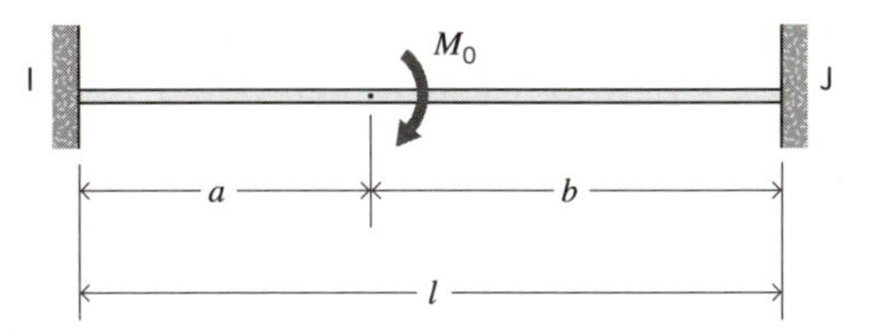

$$\widehat{M}_{IJ} = -\frac{b(l - 3a)}{l^2} M_o \tag{A.2a}$$

$$\widehat{M}_{JI} = -\frac{a(l - 3b)}{l^2} M_o \tag{A.2b}$$

$$\widehat{V}_{IJ} = -\frac{6ab}{l^3} M_o \tag{A.2c}$$

$$\widehat{V}_{JI} = \frac{6ab}{l^3} M_o \tag{A.2d}$$

3. Uniform transverse force/length q_o over entire beam

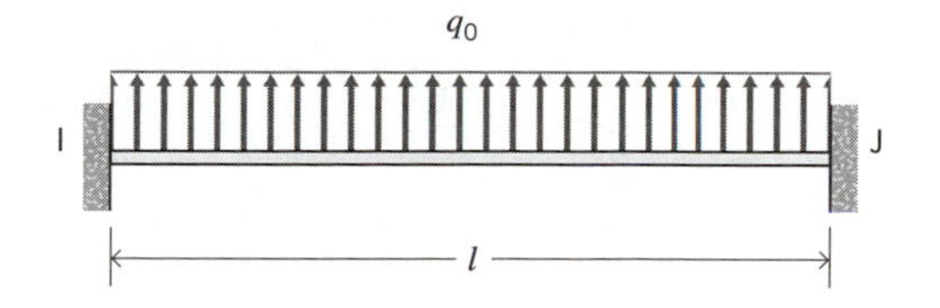

$$\widehat{M}_{IJ} = \frac{q_o l^2}{12} \qquad \textbf{(A.3}\boldsymbol{a}\textbf{)}$$

$$\widehat{M}_{JI} = -\frac{q_o l^2}{12} \qquad \textbf{(A.3}\boldsymbol{a}\textbf{)}$$

$$\widehat{V}_{IJ} = -\frac{q_o l}{2} \qquad \textbf{(A.3}\boldsymbol{c}\textbf{)}$$

$$\widehat{V}_{JI} = -\frac{q_o l}{2} \qquad \textbf{(A.3}\boldsymbol{d}\textbf{)}$$

4. Uniform transverse force/length q_o over portion of beam

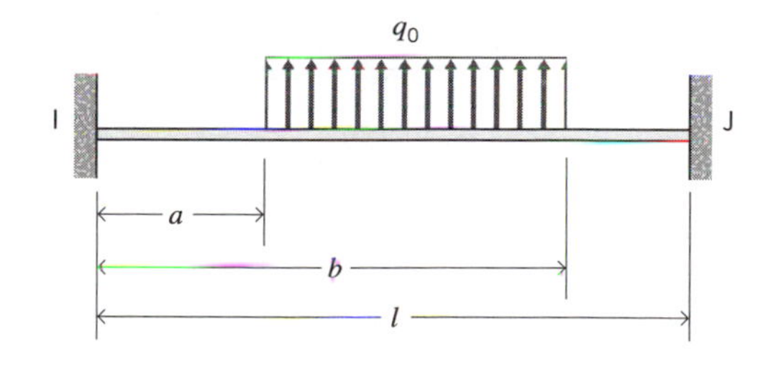

$$\widehat{M}_{IJ} = \left[\frac{(b^2 - a^2)}{2} - \frac{2(b^3 - a^3)}{3l} + \frac{(b^4 - a^4)}{4l^2}\right] q_o \qquad \textbf{(A.4}\boldsymbol{a}\textbf{)}$$

$$\widehat{M}_{JI} = \left[-\frac{(b^3 - a^3)}{3l} + \frac{(b^4 - a^4)}{4l^2}\right] q_o \qquad \textbf{(A.4}\boldsymbol{b}\textbf{)}$$

$$\widehat{V}_{IJ} = -\left[(b - a) - \frac{(b^3 - a^3)}{l^2} + \frac{(b^4 - a^4)}{2l^3}\right] q_o \qquad \textbf{(A.4}\boldsymbol{c}\textbf{)}$$

$$\widehat{V}_{JI} = \left[-\frac{(b^3 - a^3)}{l^2} + \frac{(b^4 - a^4)}{2l^3}\right] q_o \qquad \textbf{(A.4}\boldsymbol{d}\textbf{)}$$

5. Linearly varying transverse force/length over entire beam

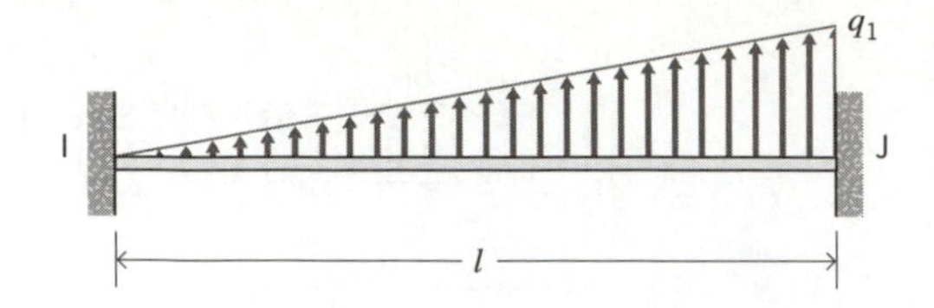

$$\widehat{M}_{IJ} = \frac{q_1 l^2}{30} \tag{A.5a}$$

$$\widehat{M}_{JI} = -\frac{q_1 l^2}{20} \tag{A.5b}$$

$$\widehat{V}_{IJ} = -\frac{3 q_1 l}{20} \tag{A.5c}$$

$$\widehat{V}_{JI} = -\frac{7 q_1 l}{20} \tag{A.5d}$$

6. Linearly varying transverse force/length over portion of beam

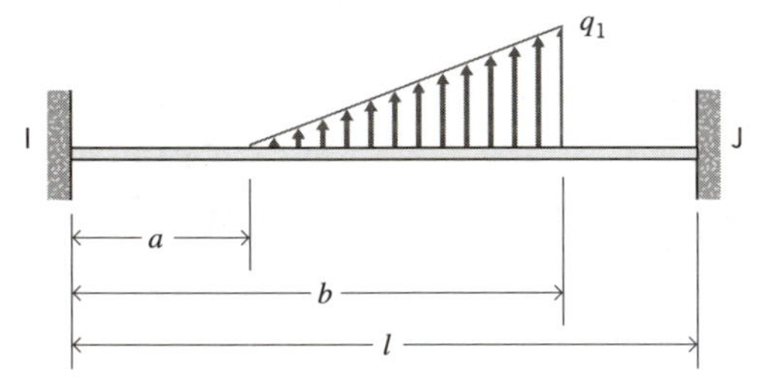

$$\widehat{M}_{IJ} = [3(a^3 + 2a^2b + 3ab^2 + 4b^3) - 10l(a^2 + 2ab + 3b^2) + 10l^2(a + 2b)]\frac{(b - a)q_1}{60l^2} \tag{A.6a}$$

$$\widehat{M}_{IJ} = [3(a^3 + 2a^2b + 3ab^2 + 4b^3) - 5l(a^2 + 2ab + 3b^2)]\frac{(b - a)q_1}{60l^2} \tag{A.6b}$$

$$\widehat{V}_{IJ} = -(2a^3 + 4a^2b + 6ab^2 + 8b^3 - 5a^2l - 10abl - 15b^2l + 10l^3)\frac{(b - a)q_1}{20l^3} \tag{A.6c}$$

$$\widehat{V}_{JI} = (2a^3 + 4a^2b + 6ab^2 + 8b^3 - 5a^2l - 10abl - 15b^2l)\frac{(b - a)q_1}{20l^3} \tag{A.6d}$$

7. Uniform temperature change ΔT_o

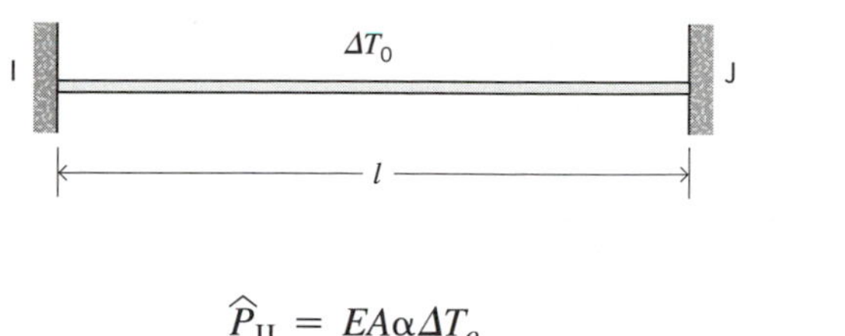

$$\widehat{P}_{IJ} = EA\alpha\Delta T_o \tag{A.7a}$$

$$\widehat{P}_{JI} = -EA\alpha\Delta T_o \tag{A.7b}$$

8. Linearly varying temperature change $\Delta T = \Delta T_o y/h$ through depth h of beam

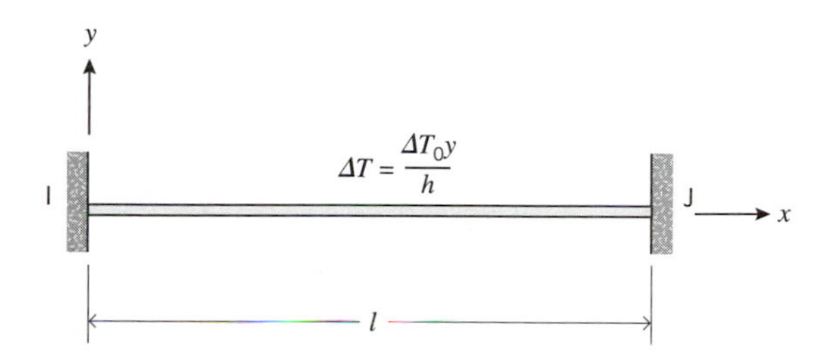

$$\widehat{M}_{IJ} = \frac{EI\alpha\Delta T_o}{h} \tag{A.8a}$$

$$\widehat{M}_{JI} = -\frac{EI\alpha\Delta T_o}{h} \tag{A.8b}$$

9. Linearly varying temperature change along length of beam (uniform through depth)

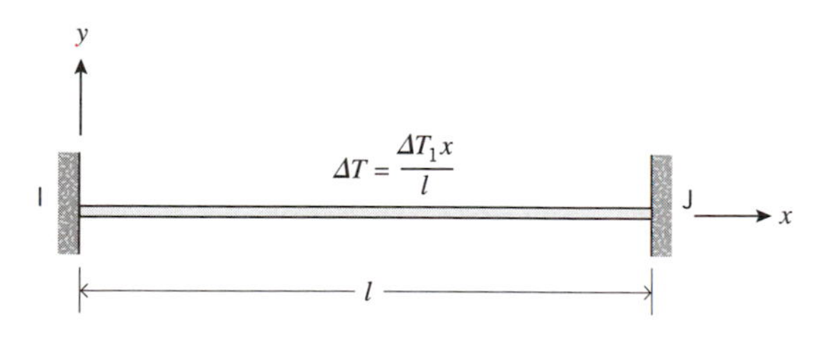

$$\widehat{P}_{IJ} = \frac{EA\alpha\Delta T_1}{2} \tag{A.9a}$$

$$\widehat{P}_{JI} = -\frac{EA\alpha\Delta T_1}{2} \tag{A.9b}$$

10. Linearly varying temperature change $\Delta T = \Delta T_o xy/lh$ through depth of h beam and along length

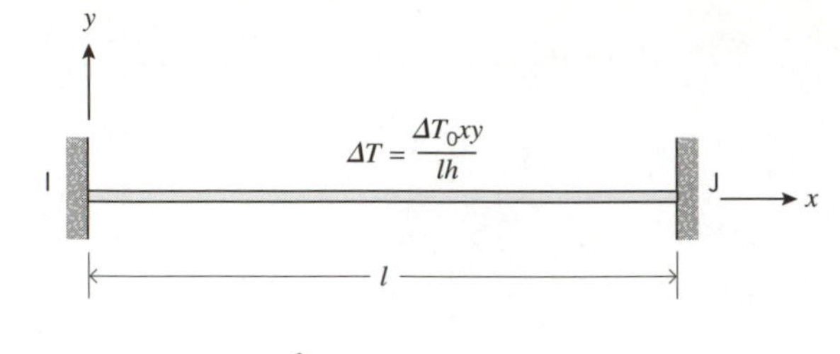

$$\widehat{M}_{IJ} = 0 \tag{A.10a}$$

$$\widehat{M}_{JI} = -\frac{EI\alpha\Delta T_o}{h} \tag{A.10b}$$

$$\widehat{V}_{IJ} = \frac{EI\alpha\Delta T_o}{lh} \tag{A.10c}$$

$$\widehat{V}_{JI} = -\frac{EI\alpha\Delta T_o}{lh} \tag{A.10d}$$

11. Concentrated axial force at $x = a$

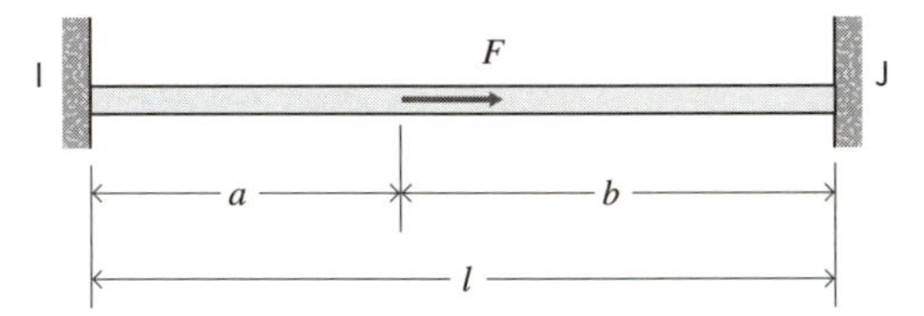

$$\widehat{P}_{IJ} = -\frac{Fb}{l} \tag{A.11a}$$

$$\widehat{P}_{JI} = -\frac{Fa}{l} \tag{A.11b}$$

12. Uniform axial force/length p_o along entire bar

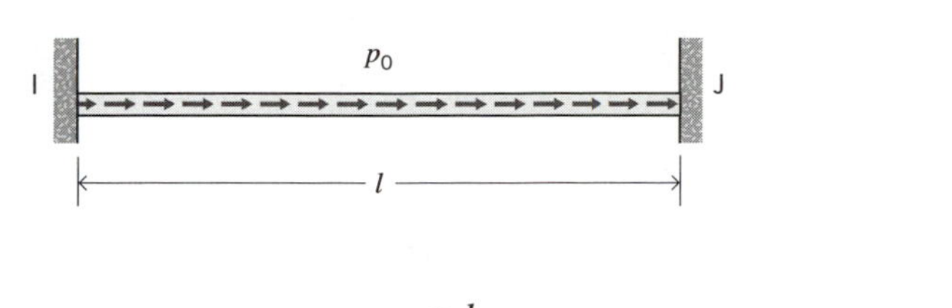

$$\widehat{P}_{IJ} = -\frac{p_o l}{2} \tag{A.12a}$$

$$\widehat{P}_{JI} = -\frac{p_o l}{2} \tag{A.12b}$$

13. Uniform axial force/length p_o along portion of bar

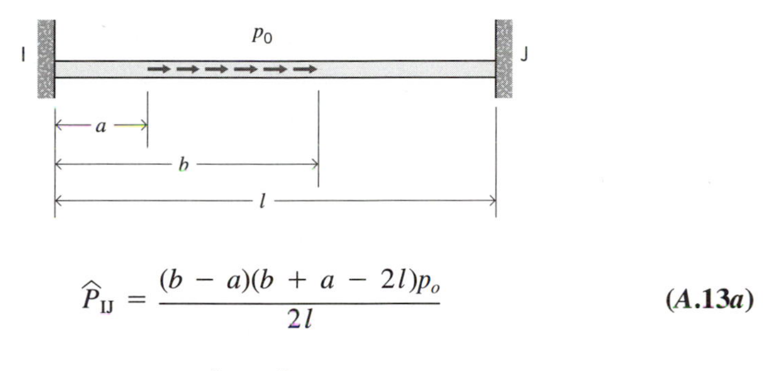

$$\widehat{P}_{IJ} = \frac{(b - a)(b + a - 2l)p_o}{2l} \tag{A.13a}$$

$$\widehat{P}_{JI} = -\frac{(b^2 - a^2)p_o}{2l} \tag{A.13b}$$

14. Linearly varying axial force/length along entire bar

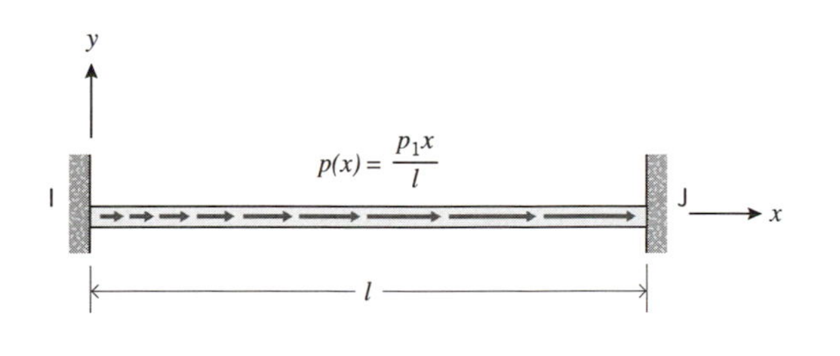

$$\widehat{P}_{IJ} = -\frac{p_1 l}{6} \tag{A.14a}$$

$$\widehat{P}_{JI} = -\frac{p_1 l}{3} \tag{A.14b}$$

15. Linearly varying axial force/length along portion of bar

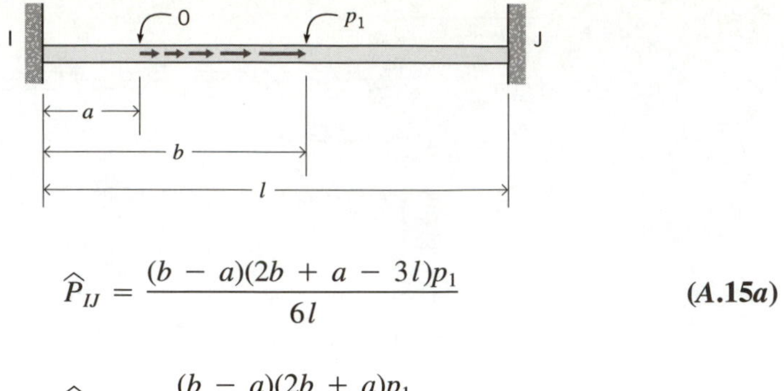

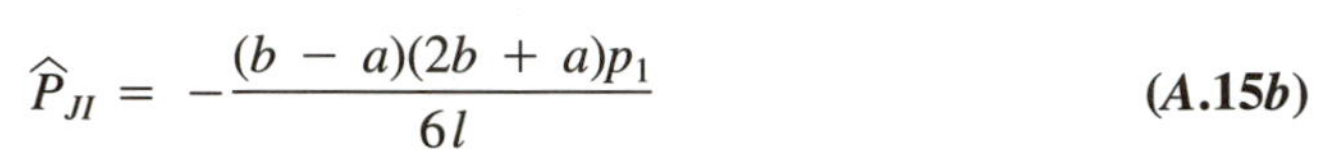

$$\hat{P}_{IJ} = \frac{(b - a)(2b + a - 3l)p_1}{6l} \tag{A.15a}$$

$$\hat{P}_{JI} = -\frac{(b - a)(2b + a)p_1}{6l} \tag{A.15b}$$

Appendix B

Moment Distribution Method for Frame Analysis

B.0 INTRODUCTION

The slope-deflection equations, developed and used in matrix form in Chapter 8, are a simple yet powerful analytic tool for conducting structural analyses of frame systems. The matrix procedure is ideally suited to the presentation of basic concepts, to organizing data, and to demonstrating the solution process. Solutions are directly and easily obtained if the analyst has access to appropriate computer software (e.g., spreadsheets or computational math programs).

Solutions can also be obtained with paper and pencil for relatively small structures if the number of displacement degrees of freedom is small enough that the required matrix operations are manageable by hand. However, for larger structures, matrix manipulation and manual solution of the resulting system of linear equations becomes a formidable task.

Fortunately, in a number of circumstances where a computer-based solution of a large system of linear equations may not be available, the analyst still has recourse to an alternative manual computational procedure based on the slope-deflection equations. This procedure is known as the moment distribution method, developed by H. Cross in 1932. For a number of different types of classical frame structures, this method is one of the most powerful, innovative and useful procedures for analysis using paper and pencil.

Before discussing details of the moment distribution method, it is helpful to understand that a numerical solution to a system of linear equations can be obtained using either (*a*) a closed-ended direct method or (*b*) an open-ended indirect procedure. The direct methods, such as Gaussian elimination or Cholesky decomposition, are the basis for nearly all computer-based algorithms for solving linear equations. Direct methods provide a numerical solution to linear equations in a finite (and countable) number of elementary mathematical operations (addition, subtraction, multiplication, and division). However, direct algorithms may require a large number of steps, all of which must be completed before a final numerical solution is attained. Significant round-off errors can occur, and these can only be dealt with by reanalyzing the equations iteratively, an expensive and time-consuming process.

As is implied by the name, open-ended methods may continue indefinitely. In essence, successive estimates of the solution are generated using very simple numerical techniques. During each stage of the iteration, solution errors are computed and used as a basis for recomputing and improving estimates for the unknown quantities. In other words, solution errors are used to drive the numerical procedure toward an accurate numerical result. Gauss-Seidel iteration and the method of relaxation are popular examples of this class of procedures.

The moment distribution method is, in fact, a relaxation method, although in such a sophisticated form that the relationship may not be evident. To demonstrate the nature of the moment distribution method, we shall begin by first considering a structure without side-sway in any component, in which case the iterative solution method and moment distribution are quite easy to follow. Subsequently, we will consider structures with side-sway.

B.1 BASIC PROCEDURE

Consider the structure in Figure B.1, which contains six beam elements and has four displacement degrees of freedom. The displacement degrees of freedom are the rotations at joints C, D, E, and G; for convenience, we will denote these system displacements by θ_C, θ_D, θ_E, and θ_G (rather than $u_1, \ldots, u_4$), respectively. We will also denote the quantity $(EI/l)_{IJ} \equiv K_{IJ} \equiv K_{JI}$ as the "stiffness factor" for beam IJ.

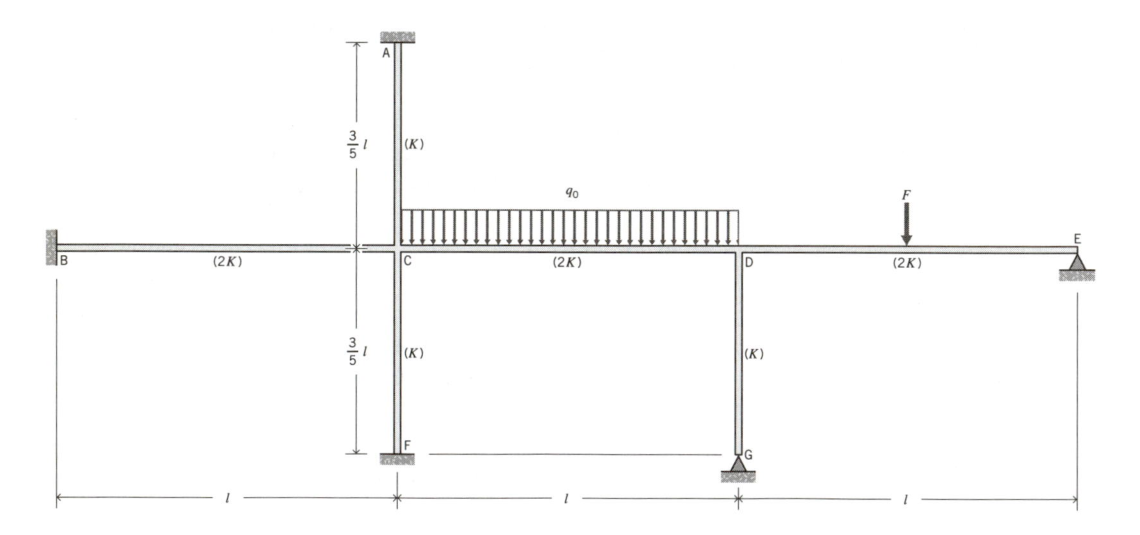

Figure B.1 Illustrative structure ($K = EI/l$).

It is apparent that there will be four equilibrium equations for this structure corresponding to moment equilibrium at joints C, D, E, and G,

$$
\begin{aligned}
M_{CA} + M_{CB} + M_{CD} + M_{CF} &= 0 \\
M_{DC} + M_{DE} + M_{DG} &= 0 \\
M_{ED} &= 0 \\
M_{GD} &= 0
\end{aligned}
\qquad (B.1a)
$$

Substituting the slope-deflection equations, Equations 8.14, with all side-sway angles $\psi = 0$, leads to

$$
\begin{aligned}
4(K_{CA} + K_{CB} + K_{CD} + K_{CF})\,\theta_C + 2K_{CD}\theta_D + \overline{M}_C &= 0 \\
2K_{DC}\theta_C + 4(K_{DC} + K_{DE} + K_{DG})\theta_D + 2K_{DE}\theta_E + 2K_{DG}\theta_G + \overline{M}_D &= 0 \\
2K_{ED}\theta_D + 4K_{ED}\theta_E + \overline{M}_E &= 0 \\
2K_{GD}\theta_D + 4K_{GD}\theta_G + \overline{M}_G &= 0
\end{aligned}
\qquad (B.1b)
$$

where

$$
\begin{aligned}
\overline{M}_C &= \widehat{M}_{CA} + \widehat{M}_{CB} + \widehat{M}_{CD} + \widehat{M}_{CF} \\
\overline{M}_D &= \widehat{M}_{DC} + \widehat{M}_{DE} + \widehat{M}_{DG} \\
\overline{M}_E &= \widehat{M}_{ED} \\
\overline{M}_G &= \widehat{M}_{GD}
\end{aligned}
\qquad (B.1c)
$$

Now, suppose some intermediate numerical values $\overline{\theta}_C$, $\overline{\theta}_D$, $\overline{\theta}_E$, and $\overline{\theta}_G$ (not the correct results) have been obtained. Equations B.1*b* may then be written in the form

$$
\begin{aligned}
4(K_{CA} + K_{CB} + K_{CD} + K_{CF})\overline{\theta}_C + 2K_{CD}\overline{\theta}_D + \overline{M}_C &= R_C \\
2K_{DC}\overline{\theta}_C + 4(K_{DC} + K_{DE} + K_{DG})\overline{\theta}_D + 2K_{DE}\overline{\theta}_E + 2K_{DG}\overline{\theta}_G + \overline{M}_D &= R_D \\
2K_{ED}\overline{\theta}_D + 4K_{ED}\overline{\theta}_E + \overline{M}_E &= R_E \\
2K_{GD}\overline{\theta}_D + 4K_{GD}\overline{\theta}_G + \overline{M}_G &= R_G
\end{aligned}
\qquad (B.2)
$$

where R_C, R_D, R_E, and R_G are errors in equilibrium at this particular step of the iterative process; physically, these are ''imbalance moments,'' which must be applied externally at the respective joints to maintain joint equilibrium.

To iterate toward the true equilibrium state, where all the R_i vanish, the equation with the largest error (in absolute value) is corrected, i.e., the imbalance moment is removed. The procedure for removing the error is quite direct and simple: the angle of rotation at the joint with the largest error is allowed to change to achieve equilibrium at that joint while all other joint rotations are maintained at their existing values.

Suppose, for example, that R_D is the current largest error in Equations B.2. We then let $\bar{\theta}_D + \Delta\bar{\theta}_D$ be a modified angle of rotation to satisfy equilibrium at joint D. In other words, when this additional increment of rotation occurs at joint D, the new value of R_D will be zero. From the second of Equations B.2 this condition corresponds to

$$2K_{DC}\bar{\theta}_C + 4(K_{DC} + K_{DE} + K_{DG})(\bar{\theta}_D + \Delta\bar{\theta}_D) + 2K_{DE}\bar{\theta}_E + 2K_{DG}\bar{\theta}_G + \bar{M}_D \equiv 0 \qquad \textbf{(B.3a)}$$

Expanding and substituting the second of Equations B.2 gives $\Delta\bar{\theta}_D$ as a function of the previous imbalance moment,

$$4(K_{DC} + K_{DE} + K_{DG})\Delta\bar{\theta}_D + R_D = 0 \qquad \textbf{(B.3b)}$$

or

$$\Delta\bar{\theta}_D = -\frac{R_D}{4(K_{DC} + K_{DE} + K_{DG})} \qquad \textbf{(B.3c)}$$

The angle change at joint D causes a change in the moment in each of the bars connected to joint D at both the ''near'' D-end and the ''far'' end, in this case end C of bar CD, end E of bar DE and end G of bar DG. The moment increments are obtained directly from Equations 8.14. At the ends connected to joint D (the ''near'' ends) we have

$$\begin{aligned}
\Delta M_{DC} &= 4K_{DC}\Delta\theta_D = -\left(\frac{K_{CD}}{K_{DC} + K_{DE} + K_{DG}}\right)R_D \\
\Delta M_{DE} &= 4K_{DE}\Delta\theta_D = -\left(\frac{K_{DE}}{K_{DC} + K_{DE} + K_{DG}}\right)R_D \qquad \textbf{(B.4a)} \\
\Delta M_{DG} &= 4K_{DG}\Delta\theta_D = -\left(\frac{K_{DG}}{K_{DC} + K_{DE} + K_{DG}}\right)R_D
\end{aligned}$$

Equations B.4*a* have an interesting interpretation. In essence, the quantity $-R_D$ is a correction in moment that is apportioned to the various beams connected to joint D to bring the joint to a condition of equilibrium. The quantities in parentheses in Equations B.4*a*, $K_{DJ}/\Sigma K_{DJ} \equiv (DF)_{DJ}$, J = C, E, G, are termed the distribution factors (*DF*) for the elements connected to joint D. They are the factors that specify exactly how the total correction in moment is apportioned among the individual beams connected to joint D. Note that

$$\sum_{J}^{C,E,G} \Delta M_{DJ} \equiv -R_D \qquad \textbf{(B.4b)}$$

and

$$\sum_{J}^{C,E,G} (DF)_{DJ} = 1 \qquad \textbf{(B.4c)}$$

At the far ends of the elements that are connected to joint D, the moment increments are

$$\Delta M_{CD} = 2K_{DC}\Delta\theta_D = \frac{1}{2}\Delta M_{DC}$$

$$\Delta M_{ED} = 2K_{DE}\Delta\theta_D = \frac{1}{2}\Delta M_{DE} \qquad \textbf{(B.5a)}$$

$$\Delta M_{GD} = 2K_{DC}\Delta\theta_D = \frac{1}{2}\Delta M_{DG}$$

These increments, termed carry-over moments, are added to current values at the far ends of the beams. This changes the total imbalance in moments at joints C, E, and G, respectively, to

$$R_C + \Delta M_{CD}$$

$$R_E + \Delta M_{ED} \qquad \textbf{(B.5b)}$$

$$R_G + \Delta M_{GD}$$

The above process is repeated, with the new greatest error used to define the joint that is to be ''unlocked'' so that its $\Delta\bar{\theta}$ can develop and bring the joint to equilibrium (with no imbalance moment). The resulting changes in moments at the near ends of bars connected to the joint are computed and recorded using relations like those in Equations B.4*a*; the carry-over moments are then applied to the far ends of the elements. The procedure is continued until the greatest residual (R_i) is below a specified level.

In the classical moment distribution method, the changes in rotation are never actually computed. Rather, only the moments acting on the ends of the beams are recorded and updated as joints are successively ''unlocked'' or ''released'' to eliminate unbalanced moments. We will demonstrate the procedure in detail for the frame in Figure B.1. For this structure, we will assume $K_{AC} = K_{CF} = K_{DG} = K$, and $K_{BC} = K_{CD} = K_{DE} = 2K$. Also, for the sake of simplicity in the following computations, we will set $q_o = 3k/ft$, $F = 20k$, and $l = 20ft$.

As the first step, we record the distribution factors (DF) at each joint by writing them in small boxes inserted directly on the drawing, as shown in Figure B.2. For example, at joint C we have

$$(DF)_{CA} = \frac{K_{CA}}{K_{CA} + K_{CB} + K_{CD} + K_{CF}} = \frac{K}{6K} = \frac{1}{6} \qquad \textbf{(B.6a)}$$

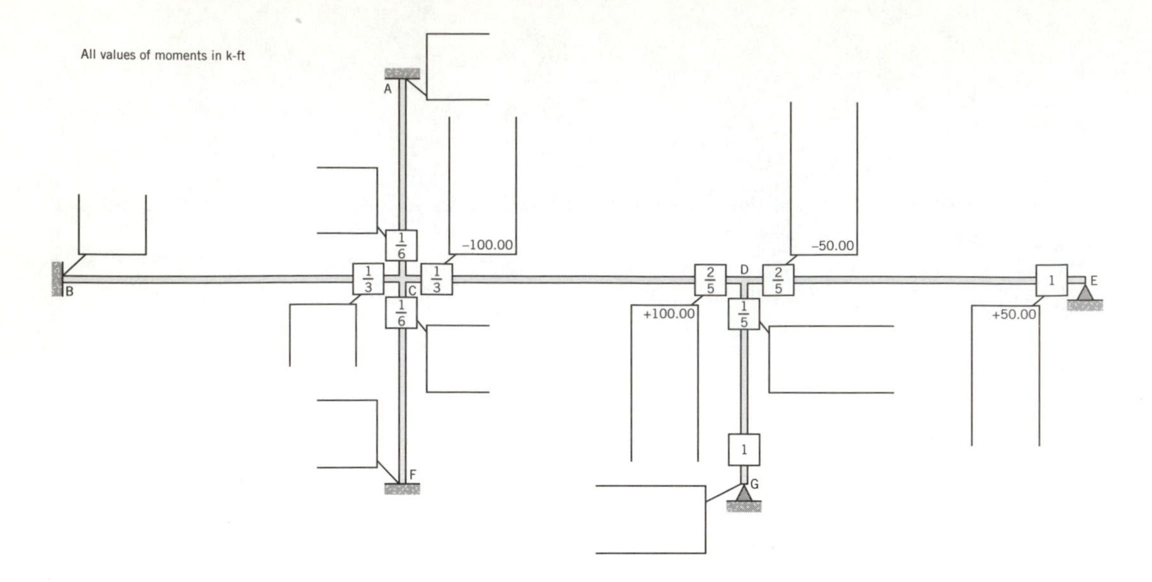

Figure B.2 Initial data for moment distribution.

Similarly, $(DF)_{\text{CB}} = (DF)_{\text{CD}} = 1/3$ and $(DF)_{\text{CF}} = 1/6$. For joint D,

$$(DF)_{\text{DC}} = \frac{K_{\text{DC}}}{K_{\text{DC}} + K_{\text{DE}} + K_{\text{DG}}} = \frac{2K}{5K} = \frac{2}{5} \qquad \textbf{(B.6b)}$$

and $(DF)_{\text{DE}} = 2/5$, $(DF)_{\text{DG}} = 1/5$. Because there is only one element at joints E and G, the distribution factors there are 1. Distribution factors at fixed joints, such as A, B, and F, are also 1, but need not be recorded because these joints are never released.

In subsequent computations, we shall also lay out on Figure B.2 the data for end moments on each beam element. These data are commonly recorded in columns perpendicular to the beam at each end, usually near the boxes containing the distribution factors. Numbers are added to these columns as joints are successively released. A typical convention locates a column containing beam end moment data in a counterclockwise position relative to the box containing the distribution factor for the beam. For instance, in Figure B.2, the column containing the data for the C-end of beam CD will be located above the box containing $(DF)_{\text{CD}} = 1/3$, and the column containing the data for the C-end of beam CA will be located to the left of the box containing $(DF)_{\text{CA}} = 1/6$, etc. The initial data in these columns consist of the fixed-end moments for the beams, i.e., the end moments when all joints are locked against rotation. For the specified loads we have $\widehat{M}_{\text{CD}} = -\widehat{M}_{\text{DC}} = -(3)(20)^2/12 = -100k\text{-}ft$, $\widehat{M}_{\text{DE}} = -M_{\text{ED}} = -(20)(20)/8 = -50k\text{-}ft$ (see Appendix A).

The next step is to determine the imbalance moment at each joint that may not be in equilibrium, in this case joint C, D, E, or G. A visual examination of Figure B.2 reveals that the imbalance moment is $-100k\text{-}ft$ at joint C and $100 - 50 = +50k\text{-}ft$ at joint E. Therefore, the greatest imbalance is at joint C. We remove this imbalance by adding $+100k\text{-}ft$ to joint C and, in accordance with Equation B.4a, distribute 1/6 of this amount ($16.67k\text{-}ft$) to both M_{CA} and M_{CF}, and 1/3 ($33.33k\text{-}ft$) to both M_{CB} and M_{CD}. These numerical values are entered in the appropriate columns at joint C on the

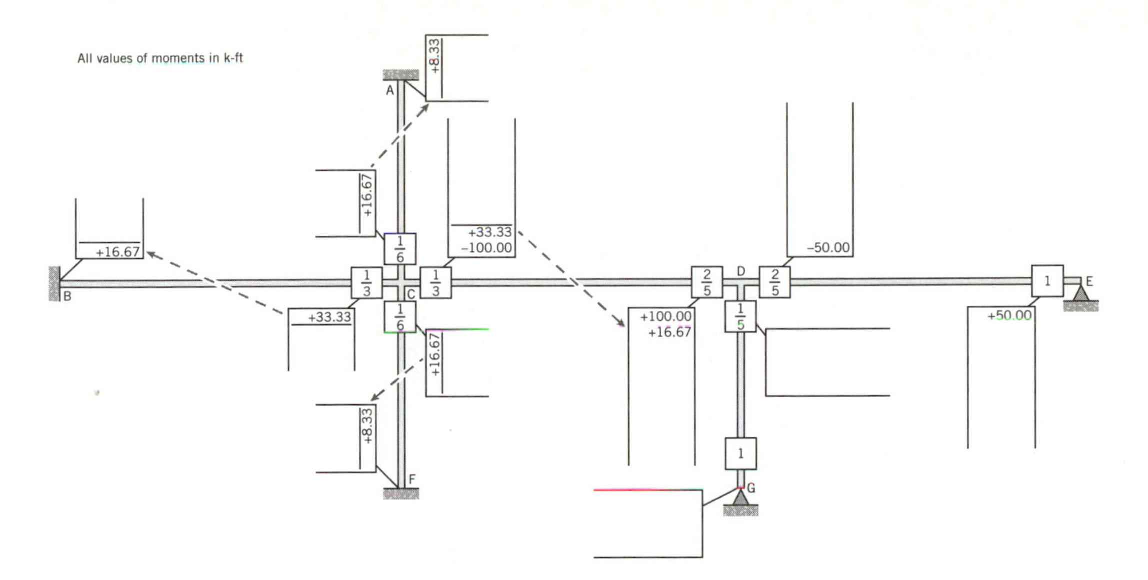

Figure B.3 Equilibrium at joint C and carry-over moments.

sketch, Figure B.3. It is customary to then draw a line across the column to indicate that all preceding data in the column correspond to a state of equilibrium at the joint.

Carry-over moments are added to the far ends of the beams in accordance with Equation B.5*a*, 8.33*k-ft* to M_{AC} and M_{FC}, 16.67*k-ft* to M_{BC} and M_{DC}, as shown by the dotted arrows in Figure B.3. This completes one iterative cycle.

The process is now repeated. Since joint C is in equilibrium, we must check only joints D, E, and G. The greatest imbalance is at joint D, 116.67 − 50 = +66.67*k-ft*. We impose equilibrium on this joint by adding −66.67*k-ft*, distributing 2/5 of this amount (−26.67*k-ft*) to M_{DC} and M_{DE}, and 1/5 (−13.33*k-ft*) to M_{DG} (Figure B.4). Carry-over moments are added at the far ends of the beams, −13.33*k-ft* to M_{CD} and M_{ED}, and −6.67*k-ft* to M_{GD}, also shown in Figure B.4. This completes the second iterative cycle.

Note, from Figure B.4, that joint C is no longer in equilibrium due to the carry-over moment M_{CD} = −13.33*k-ft* from the second iteration. Also, the imbalance moments on joints E and G are +36.67*k-ft* and −6.67*k-ft*, respectively. Therefore, for the third cycle, the moments on joint E will be balanced.

The process will continue in this fashion until all imbalance moments are determined to be negligible. Figure B.5 shows the results after 13 cycles of iteration.[1] The final end moments are indicated by double underlines.

B.1.1 Release at Simple Support

The fact that it required 13 cycles to obtain the results in Figure B.5 is indicative of relatively slow convergence. One reason for this behavior may be evident: each time

[1] The sequence of joint ''balancing'' operations in this illustration is CDEDCGDEDCGED. Also, carry-over moments are ignored after the last joint is balanced.

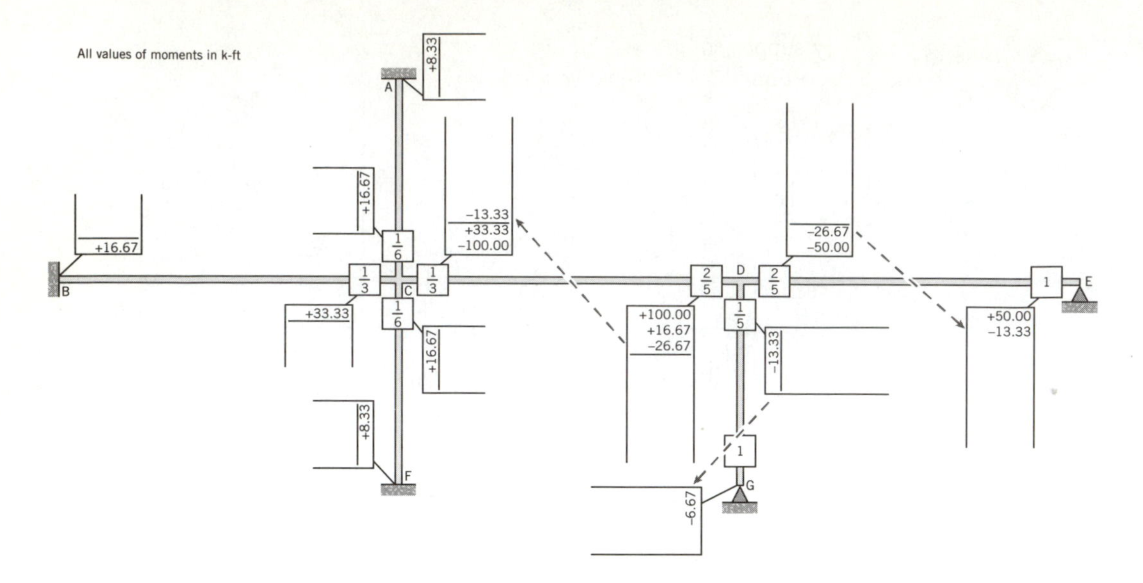

Figure B.4 Equilibrium at joint D and carry-over moments.

the moment at a simply supported joint (i.e., joint E or G) is balanced, one-half of the entire correction is carried over to the far end of the beam (in this case, joint D). Thus, moments at interior joints can persistently be unbalanced, even after a number of corrections.

Fortunately, the process can be modified and convergence improved. It is possible to alter the slope-deflection equations to reflect the fact that the moment at a simply supported end of a beam is always known. Suppose, for example, that the J-end of a

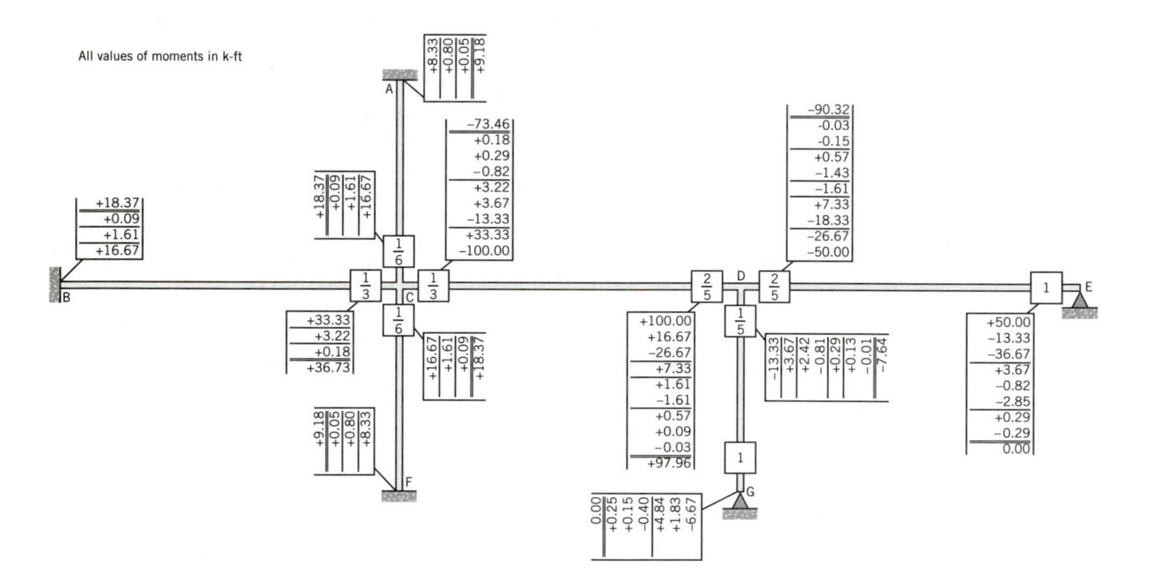

Figure B.5 Complete iteration.

beam is simply supported and unloaded (i.e., there is no applied moment); it follows that $M_{JI} = 0$, so Equations 8.14 can be written in the form

$$M_{IJ} = K_{IJ}[4(\theta_{IJ} - \psi_{IJ}) + 2(\theta_{JI} - \psi_{IJ})] + \widehat{M}_{IJ}$$
$$0 = K_{IJ}[2(\theta_{IJ} - \psi_{IJ}) + 4(\theta_{JI} - \psi_{IJ})] + \widehat{M}_{JI} \qquad \textbf{(B.7a)}$$

We solve the second equation for $(\theta_{JI} - \psi_{IJ})$,

$$(\theta_{JI} - \psi_{IJ}) = -\frac{1}{2}(\theta_{IJ} - \psi_{IJ}) - \frac{\widehat{M}_{JI}}{4K_{IJ}} \qquad \textbf{(B.7b)}$$

and substitute this result into the first of Equations B.7*a*,

$$M_{IJ} = 3K_{IJ}(\theta_{IJ} - \psi_{IJ}) + \left(\widehat{M}_{IJ} - \frac{1}{2}\widehat{M}_{JI}\right)$$
$$= \frac{3}{4}(4K_{IJ})(\theta_{IJ} - \psi_{IJ}) + \left(\widehat{M}_{IJ} - \frac{1}{2}\widehat{M}_{JI}\right) \qquad \textbf{(B.7c)}$$

Equation B.7*c* is used to define the behavior of elements such as beams DE and DG in Figure B.1. In the moment distribution method, the use of this relation implies that the beam stiffness is taken as 3/4 of its previous value in computing distribution factors at the I-end. Also, the initial fixed-end moment at the I-end is now taken as $\widehat{M}_{IJ} - \widehat{M}_{JI}/2$.

To illustrate, we will repeat the analysis of the frame in Figure B.5. The modified distribution factors at joint D are

$$(DF)_{DC} = \frac{K_{DC}}{K_{DC} + \frac{3}{4}K_{DG} + \frac{3}{4}K_{DE}} = \frac{2K}{2K + \frac{3}{4}K + \frac{3}{4}2K} = \frac{8}{17}$$

$$(DF)_{DG} = \frac{\frac{3}{4}K_{DG}}{K_{DC} + \frac{3}{4}K_{DG} + \frac{3}{4}K_{DE}} = \frac{\frac{3}{4}K}{2K + \frac{3}{4}K + \frac{3}{4}2K} = \frac{3}{17} \qquad \textbf{(B.8)}$$

$$(DF)_{DE} = \frac{\frac{3}{4}K_{DE}}{K_{DC} + \frac{3}{4}K_{DG} + \frac{3}{4}K_{DE}} = \frac{\frac{3}{4}2K}{2K + \frac{3}{4}K + \frac{3}{4}2K} = \frac{6}{17}$$

and the initial fixed-end moment at the D-end of beam DE is $-50 - (50)/2 = -75$*k-ft*. These values are shown in Figure B.6. From this point on, the moment distribution method proceeds exactly as in the previous case, except that no moments

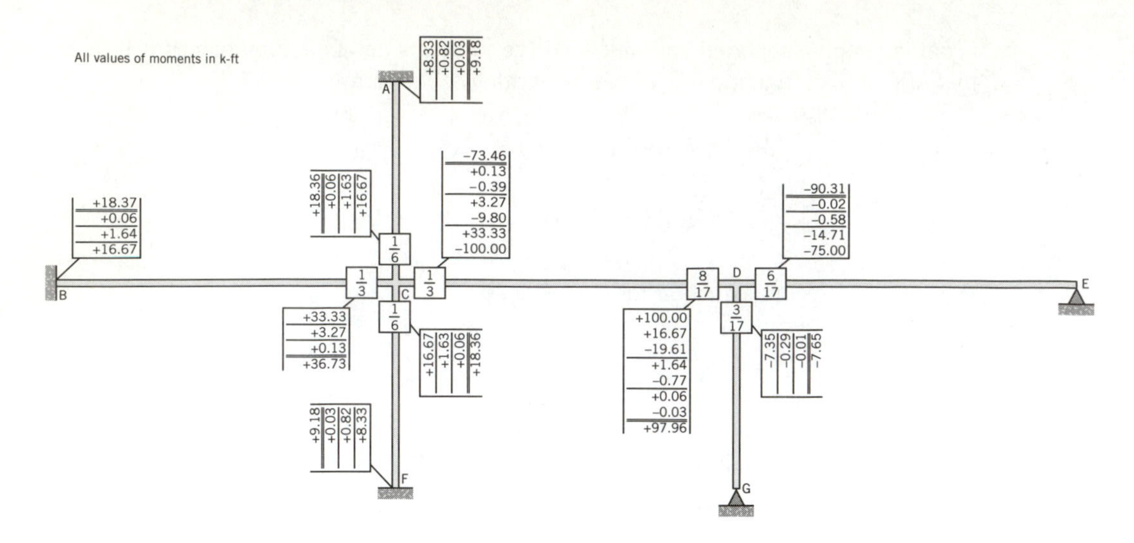

Figure B.6 Moment distribution modified for simple supports.

are carried over from joint D to either joint E or G. As may be seen from Figure B.6, convergence is obtained after only six cycles and the results are within $0.01k\text{-}ft$ of the values obtained previously, the difference being due to round-off error.

B.1.2 Free Ends

If one end of a beam in a structure is free, the beam behaves exactly like a cantilever beam. Loads on the beam are transmitted directly to the joint at the supported end, and internal forces in the cantilever are unaffected by rotation of the joint. This type of behavior is easily included in the moment distribution method by recognizing that the effective stiffness factor of the cantilever at the joint is zero.

To illustrate, consider the frame structure in Figure B.7. Even though cantilever beam CD has a physical stiffness $EI/l = 3K$, we shall set $K_{CD} = 0$. Also, for this structure we may use Equation B.7c to describe the behavior of beams BE and CF, both of which have one end simply supported. Then, noting that end E of element BE is simply supported, the distribution factors at joint B are

$$(DF)_{BA} = \frac{2K}{2K + 2K + \frac{3}{4}K} = \frac{8}{19}$$

$$(DF)_{BC} = \frac{2K}{2K + 2K + \frac{3}{4}K} = \frac{8}{19} \qquad \textbf{(B.9a)}$$

$$(DF)_{BE} = \frac{\frac{3}{4}K}{2K + 2K + \frac{3}{4}K} = \frac{3}{19}$$

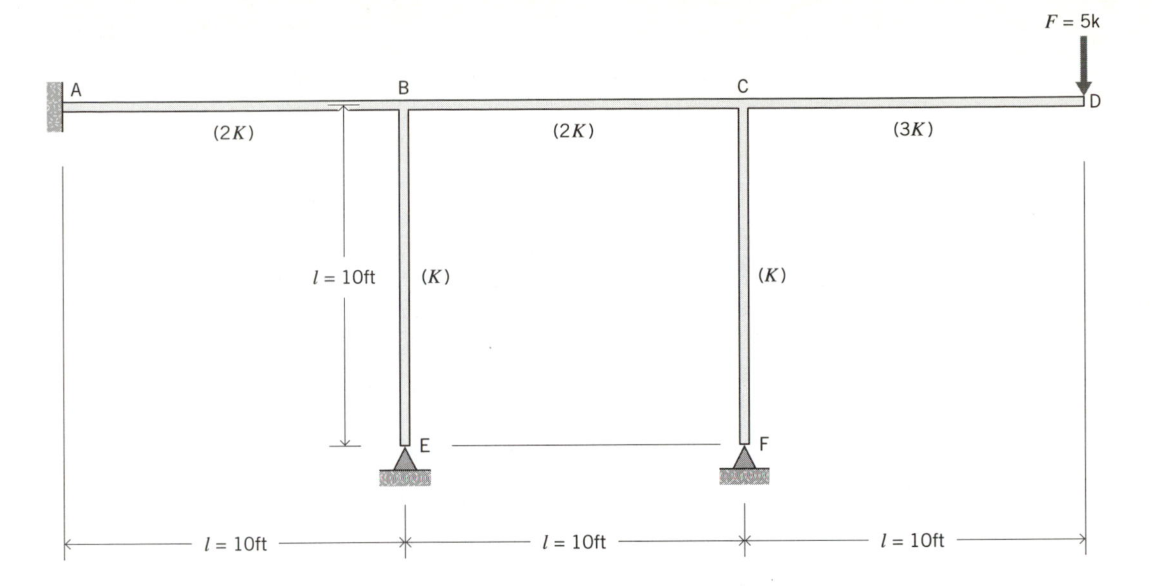

Figure B.7 Structure with cantilever beam.

and noting that end F of element CF is simply supported, the distribution factors at joint C are

$$(DF)_{CB} = \frac{2K}{2K + \frac{3}{4}K} = \frac{8}{11}$$

$$(DF)_{CF} = \frac{\frac{3}{4}K}{2K + \frac{3}{4}K} = \frac{3}{11} \tag{B.9b}$$

The only initial load on a joint is $M_{CD} = -50k\text{-}ft$ due to the $5k$ applied load on the cantilever beam. Thus, we begin by distributing a moment of $+50k\text{-}ft$ to joint C to balance the moments; 8/11 of this value $(+36.36k\text{-}ft)$ is apportioned to M_{CB} and 3/11 $(+13.64k\text{-}ft)$ is apportioned to M_{CF}. A moment of $+18.18k\text{-}ft$ is carried over to end B of beam BC, after which joint B is balanced. The process is repeated until the results shown in Figure B.8 are obtained.

The results in Figure B.8 display some noticeable effects of round-off error. In particular, note that the final value of M_{AB} is $-4.16k\text{-}ft$, which is more than one-half the final value $M_{BA} = -8.29k\text{-}ft$. The difference is attributable to the fact that we are working with a small number of decimal places, and each odd numerical value in this illustration has been rounded up when dividing by two. For example, the first value $M_{BA} = -7.65k\text{-}ft$ produces a carryover moment $M_{AB} = -3.83k\text{-}ft$; these effects are cumulative.

From these simple results it is possible to construct a worst case scenario of round-off error, namely an error $\pm$ the smallest significant figure multiplied by the number of iterations to convergence. So, for example, the results for the moments in the

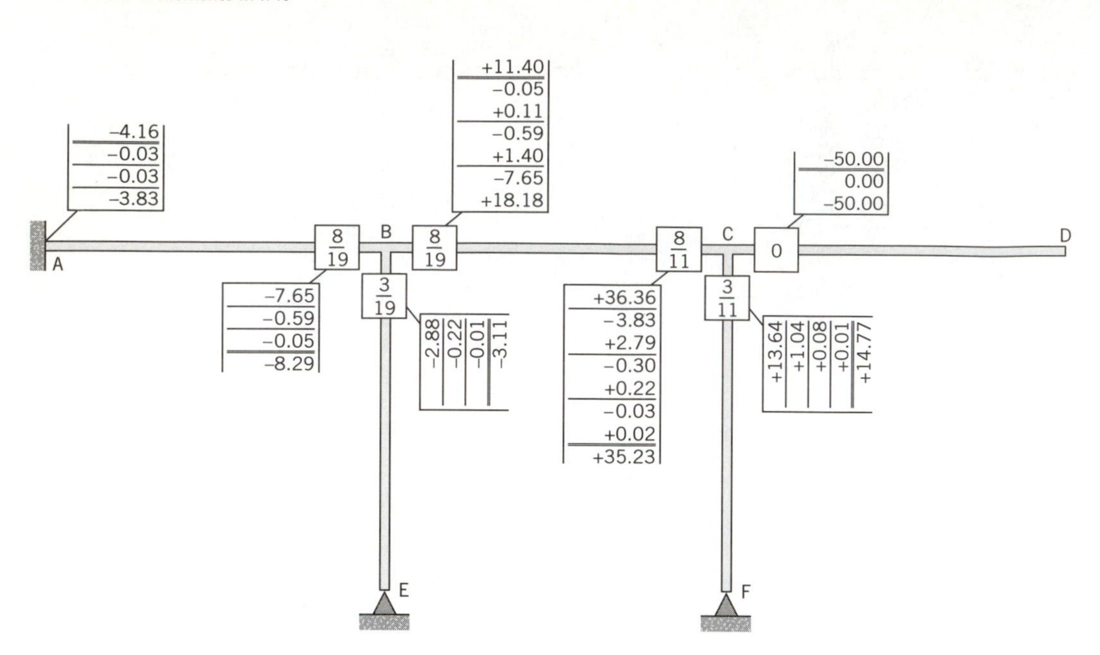

Figure B.8 Complete iteration.

preceding example are accurate to within $\pm 6 \times 0.01 = \pm 0.06 k\text{-}ft$. If this level of accuracy is acceptable, the analysis is complete; otherwise the analysis may have to be repeated using more significant figures to give a more accurate answer.

B.1.3 External Moment Loads on Joints

A moment load applied externally to a joint is easily dealt with in the moment distribution method. We define a clockwise load as positive, and distribute the moment directly to the individual beams at the joint without sign change; this produces equilibrium at the joint. One-half of each end moment is carried over to the far end of the beam, and the usual process is then initiated.

B.2 SIDE-SWAY

Up to this point, we have restricted our attention to structures having no joint translations, i.e., to structures with no side-sway in any beams. The moment distribution process in such cases was seen to be equivalent to seeking moment equilibrium by allowing joint rotations without transverse displacements, i.e., no side-sway angles. As we shall see, when side-sway motions exist, the moment distribution process becomes considerably more involved. In all but the simplest cases the complications make recourse to a modern matrix-based solution procedure preferable.

B.2.1 Prescribed Side-Sway

As a first step to describing a basic solution strategy for structures that undergo side-sway, we shall illustrate the moment distribution procedure for situations in which side-

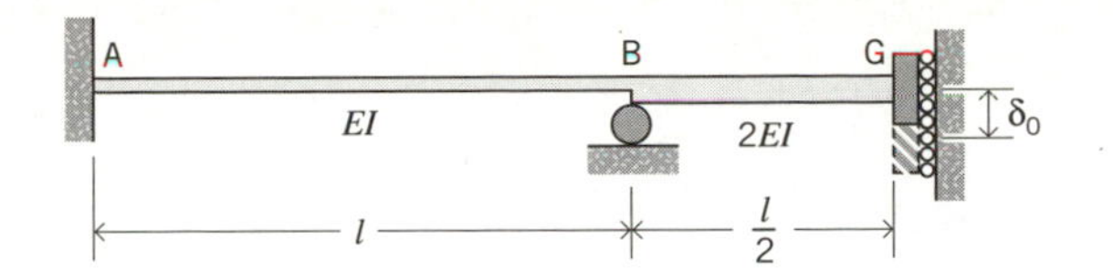

Figure B.9 Structure with prescribed side-sway.

sways are prespecified, i.e., known at the outset. We begin by recasting the slope-deflection equations, Equations 8.14, in the following form,

$$M_{IJ} = K_{IJ}(4\theta_{IJ} + 2\theta_{JI}) + \overline{M}_{IJ}$$
$$M_{JI} = K_{IJ}(2\theta_{IJ} + 4\theta_{JI}) + \overline{M}_{JI} \qquad \textbf{(B.10a)}$$

where

$$\overline{M}_{IJ} = \widehat{M}_{IJ} - 6K_{IJ}\psi_{IJ}$$
$$\overline{M}_{JI} = \widehat{M}_{JI} - 6K_{IJ}\psi_{IJ} \qquad \textbf{(B.10b)}$$

and where now ψ_{IJ} is prespecified.

Consider the structure in Figure B.9, which we have encountered previously in Chapters 8 and 9. This structure has a support at the right end (G), which can displace vertically but cannot rotate. We will analyze this structure assuming the support at G displaces a specified small amount δ_o, taken positive downward in this case. In order to facilitate eventual comparison with previous examples, we will conduct the analysis in algebraic form rather than specifying numerical values for the various parameters. We define $K_{AB} = EI/l \equiv K$ and $K_{BG} = 2EI/(l/2) \equiv 4K$.

Since the support at B does not move, we have side-sway only in element BG; the side-sway angle is

$$\psi_{BG} = \frac{v_{BG} - v_{GB}}{l_{BG}} = \frac{0 - (-\delta_o)}{l/2} = \frac{2\delta_o}{l} \qquad \textbf{(B.11)}$$

For the specified δ_o, the effective end moments, Equation B.10*b*, are $\overline{M}_{BG} = \overline{M}_{GB} = -6(4K)(2\delta_o/l) = -48K\delta_o/l$. These initial values are shown on the moment distribution sketch,[2] Figure B.10*a*, as are the distribution factors at joint B.

The moment distribution proceeds very directly; the moments at joint B are balanced, and the appropriate components are carried over to joints A and G, after which all joints are in equilibrium. Thus, for this structure, the solution is obtained in one iteration! (This fortunate occurrence was due to the fact that only one joint rotation was unknown. If more joints with unknown rotations were present, we would have to perform the usual iterative procedure.)

The displacement δ_o requires a vertical force, denoted by Q in Figure B.10*a*.[3] The

---

[2] It should be understood that all numerical values of moments in the sketch must be multiplied by the algebraic factor $K\delta_o/l$.

[3] The vertical force may be viewed as being equivalent to a reaction accompanying support settlement.

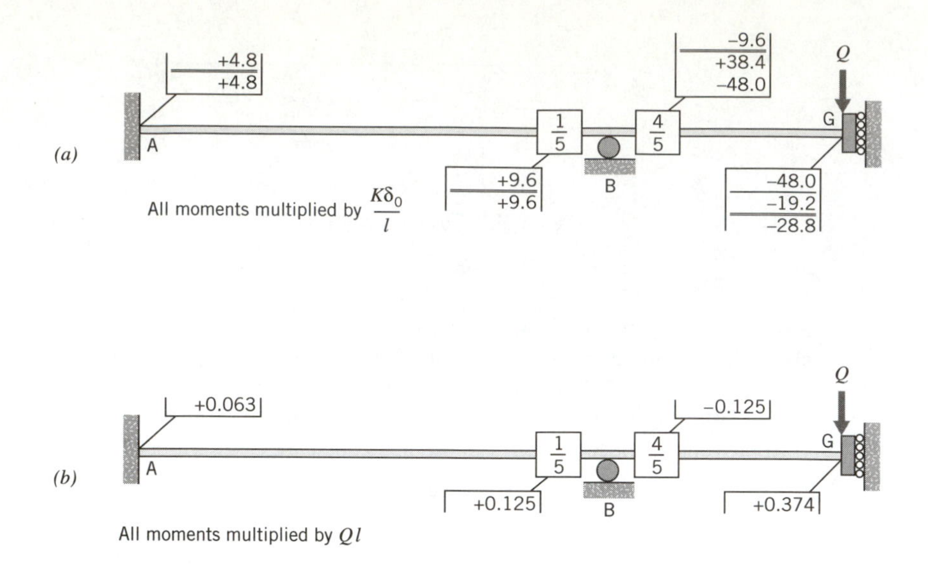

Figure B.10 Moment distribution: (*a*) prescribed side-sway; (*b*) specified Q.

magnitude of this force can be determined from equilibrium of beam BG, or by using virtual work, specifically a virtual displacement like the rigid-body portion of the side-sway motion (see Figure 8.23). For example, Figure B.11 shows a free-body diagram of beam BG; the end moments in this figure are the final values obtained in Figure B.10*a*. We see that equilibrium of moments requires $Q = 76.8K\delta_o/l^2$ (positive downward).

Suppose, now, that we consider the inverse problem, namely the structure in Figure B.10*a* subjected to a specified value of load Q. It should be apparent that we can then find the corresponding transverse displacement at G, i.e., $\delta_o = 0.0130Ql^2/K$. Similarly, we can determine the internal moments in this case by scaling the results in Figure B.10*a*. In other words, the solution enables us to compute all motions and forces due to a specified force Q. The end moments for a specified force Q are given in Figure B.10*b*.

B.2.2 Side-Sway Due to Arbitrary Loads

Let us analyze the same structure subjected to the applied force F shown in Figure B.12. We proceed by first solving for internal forces assuming the side-sway is restrained, i.e., assuming the transverse displacement at G is zero. This requires the same type of basic analysis discussed in Section B.1. The fixed-end moments in beam BG

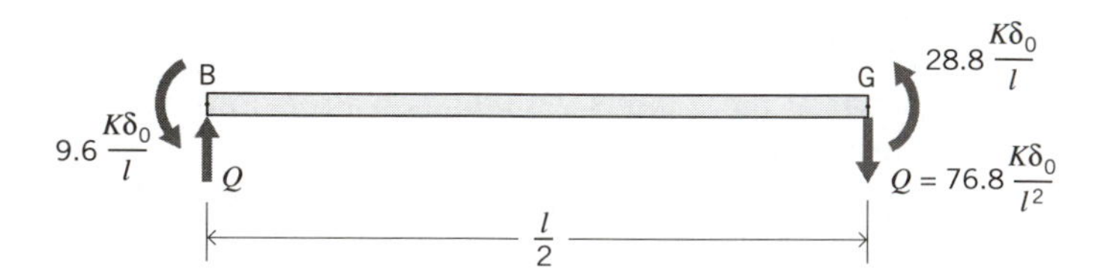

Figure B.11 Free-body diagram of beam BG.

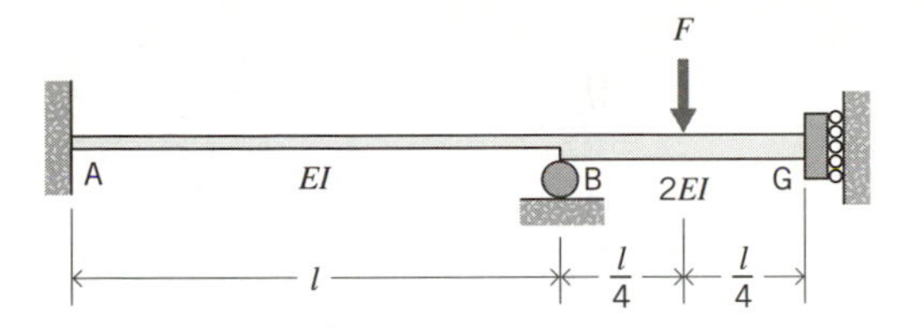

Figure B.12 Structure with applied load.

are $\widehat{M}_{\mathrm{BG}} = -\widehat{M}_{\mathrm{GB}} = -Fl/16 = -0.0625Fl$. Balancing the moments at joint B again produces a solution in one iteration, with the results shown in Figure B.13.[4]

The solution in Figure B.13 requires the existence of a transverse force Q acting at G to prevent displacement. As in the previous case (e.g., Figure B.11), we compute the magnitude of Q from either equilibrium or virtual work considerations; the result is $Q = 0.65F$ (upward).

Since Q must actually be zero in the original structure, Figure B.12, we obtain the correct solution by superposition of results from the two previous cases, as shown in Figure B.14. From the superposition we obtain

$$\delta_o = 0.0130\frac{Ql^2}{K} = 0.0130\frac{(0.65F)l^2}{K} = 0.00846\frac{Fl^2}{K} = 0.00846\frac{Fl^3}{EI} \qquad \textbf{(B.12)}$$

which is identical to the value of u_2 (due to F only) in Equation 8.21*c*.

Superposition of the moments results in

$$\begin{aligned}
M_{AB} &= 0.00625Fl + 4.8\frac{K\delta_o}{l} \\
&= 0.00625Fl + 4.8(0.00846Fl) = 0.469Fl \\
M_{BA} &= 0.0125Fl + 9.6(0.00846Fl) = 0.0937Fl \\
M_{BG} &= -0.0125Fl - 9.6(0.00846Fl) = -0.0937Fl \\
M_{GB} &= 0.0875Fl - 28.8(0.00846Fl) = -0.1561Fl
\end{aligned} \qquad \textbf{(B.13)}$$

These values agree with the results in Equation 8.21*e*.

[4] In this case, all numerical values of moments in the sketch must be multiplied by the algebraic factor Fl.

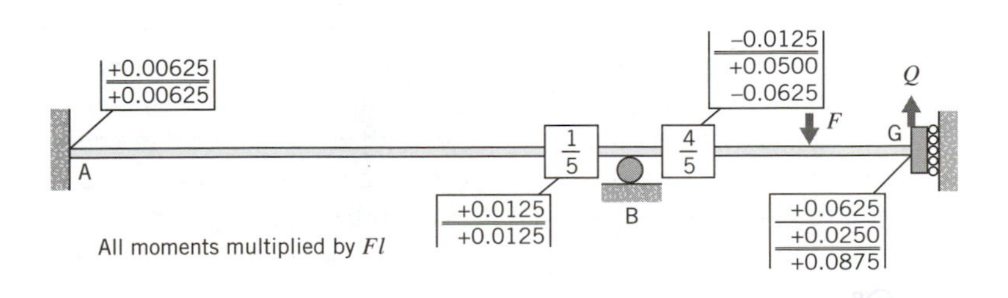

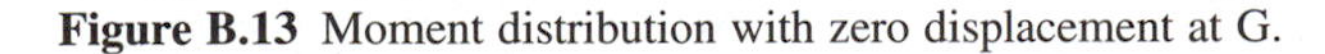

Figure B.13 Moment distribution with zero displacement at G.

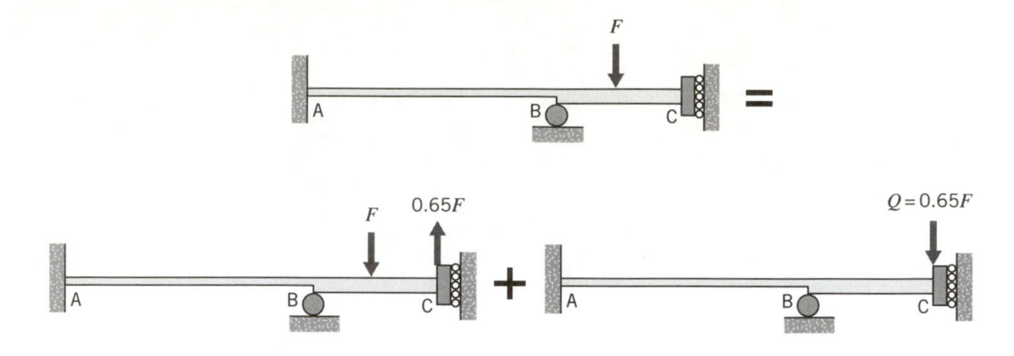

Figure B.14 Superposition of forces.

In general, the same procedure can be employed to analyze structures having N independent translational displacements. Moment distribution is first applied to obtain a solution for internal moments when all displacements are restrained to zero. Virtual work (equilibrium) is used to calculate the forces, Q_i, $i = 1, \ldots, N$, which are required to restrain the joint motions. Then, we can consider each of the joint motions, one at a time, $u_i = 1$, $u_j = 0$, $j \neq i$, and use moment distribution (and virtual work) to find internal moments and the N forces required to produce the unit value of the ith displacement, k_{ri}, $r = 1, \ldots, N$. As a last step, we would force the (artificial) reactions Q_i to disappear by superposition of N solutions, each corresponding to a particular value of u_i (other than unity). Mathematically, this requires setting up and solving N simultaneous linear equations of form[5]

$$
\begin{aligned}
k_{11}u_1 + k_{12}u_2 + \ldots + k_{1N}u_N &= -Q_1 \\
k_{21}u_1 + k_{22}u_2 + \ldots + k_{2N}u_N &= -Q_2 \\
&\ldots \\
k_{N1}u_1 + k_{N2}u_2 + \ldots + k_{NN}u_N &= -Q_N
\end{aligned}
\tag{B.14}
$$

Finally, after finding the u_i, $i = 1, \ldots, N$, we can scale the internal end moments appropriately and add the $N + 1$ sets of internal forces to obtain the solution that is consistent with the N joint translations.

This is a lengthy task if very many independent joint motions exist. Perhaps surprisingly, it involves solving a system of linear equations, a chore we were trying to avoid when we elected to use moment distribution in the first place! Therefore, there is no advantage to the moment distribution procedure unless the structure has some special characteristics that limit the translational degrees of freedom to a very small number. Otherwise, it is much simpler to just set up the matrix equations using the methods presented in Chapters 8 and 9, and let the computer do all of the numerical work.

For those problems for which it is well suited, the moment distribution method remains an interesting and powerful alternative to the formal matrix procedure. Experienced analysts can obtain very good estimates of internal moments in frame structures in a remarkably short time. The procedure also provides a good sense of exactly how the structure transmits bending loads to the support system.

---

[5] Recall that in Chapter 10 it was shown that stiffness relations of this type have symmetric coefficients, i.e., $k_{ir} = k_{ri}$.

Index

Analytic model, 3, 27, 75
Approximate analysis, 153, 432, 573, 590, 604
Approximate displacement pattern, 604
Arch, definition, 6

Bar, definition, 4
  slender, 72
Beam, definition, 6
Beam functions, 634
Beam on elastic foundation. *See* Elastic foundation
Bending moment, 119
Bending moment diagram, 127
  centroid, 158
Bendixen, 2, 436
Bernoulli-Euler hypothesis, 140
Best fit approximation, 614, 640
Betti's theorem, 348, 349, 390, 436, 438, 468, 602
Buckling, 5, 63, 94

Cable, 4, 5
Carry-over factor, 685
Castigliano's theorem
  Part I, 586
  Part II, 587
Compatibility, 56, 66, 189, 190, 214, 229, 232, 237, 250
  matrix form, 250, 251, 262, 419
  physical significance, 238, 242, 268, 272, 288, 411
Connections, 28
  flexible, 311, 321
  joint, 28
  joint, pinned (hinged), 33
  joint, rigid, 29
  kinematics, 32
Conservation of energy. *See* Energy
Constitutive relations, 10
Convex functions, 584
Convexity. *See* Convex functions
Coordinates
  global, 522
  local, 522
  transformation, 523
Counters, 94
Cross, Hardy, 2, 681
Curvature, 142
Cut, 233

Dead load. *See* Loads
Deflection diagram, 146
Deflections. *See* Displacements

Deformation energy. *See* Energy
Deformation-force relations, 232
  bar, 102, 261, 302
  beam, 261, 283
  combined bar/beam 296
  matrix form, 250, 261, 283
Deformations
  bar, 97, 102, 300
  beam, 139
  end. *See* End deformations
  hypothesis, 140
  initial, 318
  thermal, 102, 283
  unrestrained, 250, 282, 283, 302
Degree of freedom, 32, 63, 107, 409
Determinacy, 54, 95
  necessary condition, 95
Differential equation of equilibrium
  bar, 74
  beam, 121
Direct stiffness method, 523, 531, 606
  algorithm, 534
Displacement method, 66, 110, 169, 408, 499
  matrix formulation, 418, 434
  steps of, 409
Displacements
  approximate, 604
  beam, 140, 144
  consistent, 240
  constrained, 590
  coupled, 411
  end. *See* End displacements
  frame, 327
  generalized, 600
  incompatible, 595
  pattern, 602
  rigid-body, 280, 411, 448
  system, 408
  truss, 104, 182, 260, 409
Distribution factor, 684

Effective axial force, 143
Effective bending moment, 143
Elastic behavior, 15, 574, 578
Elastic foundation, 659
Element
  line, 3
  one-dimensional, 3, 4
  rigid, 295, 312, 411, 499
  slender, 72
  small, 73
  three-dimensional, 4, 9
  two-dimensional, 4, 7
End deformations, 277, 280, 435
  unrestrained, 282, 283, 302
End displacements, 280, 500, 508, 524
End forces, 278, 300, 500, 524
  fixed, 420, 434, 501, 509, 525
Energy
  conservation, 573, 578
  convexity, 583
  deformation, 573, 587
  density, 609
  external potential, 573, 588
  in bars and beams, 607
  minimum potential, theorem of, 573, 589
  per unit volume, 609
  positive definiteness, 581
  stationary potential, theorem of, 573, 589
  strain, 573, 574, 582
  total potential, 573, 588
  uniqueness, 581
Energy theorems. *See* Energy
Equilibrium
  equations, 16, 443, 505
  of bar, 72
  of beam, 119
Equivalent forces. *See* Force
Errors, 573, 590
External potential energy. *See* Energy

Fatigue, 12
Finite element, 2, 9, 10, 573, 605, 632
Fixed-coordinate state, 576
Fixed-end forces. *See* End forces
Fixed-end moments. *See* End forces
Fixed-end shears. *See* End forces
Flexibility
  coefficient, 240, 248, 269, 385
  joint, 307
  matrix, 248, 276, 277, 348
Flutter, 12
Force
  continuous, 23
  concentrated, 16, 27
  distributed, 23
  end. *See* End forces
  equivalent, 574
  generalized, 449, 600, 602
  work-equivalent, 600, 602
Force method, 66, 169, 229
  matrix formulation, 248, 277, 284
    computing displacements, 327
  steps, 232

Force-deformation relations, 409
 bar, 419
 beam, 434
Force-displacement relations, 499
 bars, 508
 beams, 500
 combined bar/beam, 509
Foundation modulus, 659
Fourier series, 619
Frame, 119
 closed, 369
 definition, 6
Free-body diagram, 22

Galerkin's method, 643
Generalized displacement. *See* Displacements
Generalized force. *See* Force

Heller, R., 2
Homogeneity, 100, 142, 607
Hooke's law, 607

Incompatible displacements, 595
Indeterminacy, 54, 95, 229
 degree, 56, 97, 263
 external, 56
 internal, 57
Inflection point, 153
Influence coefficient, 385
Influence function, 388
Influence lines, 382, 385
Initial deformations. *See* Deformations
Initial strain. *See* Strain
Instability, 59, 63, 93, 95, 230

Joint. *See* Connections
Joint dimensions
 effect, 554
Joints, method of, 77

Kinematics, 10, 32

Lack of fit, 599
Least work. *See* Work
Line element. *See* Element, line
Linear behavior, 15, 99
Live loads. *See* Loads
Load path, 696
Loads, 11
 dead (Fig. 1.12), 14
 live, 14

Maney, 2, 436
Matrix, 248
 banded, 653
 block-diagonal, 288
 compatibility, 419
 deformation-force, 250, 262, 283, 302
 diagonal, 250
 equilibrium, 249, 261, 285
 fixed-end force, 419, 420
 flexibility, 248, 277, 283
 force, 421, 422, 450
 force-deformation, 409, 419
 force-displacement, 501, 508, 522
 inverse, 249
 partitioned, 476, 547, 561
 positive definite, 582
 singular, 502
 stiffness, 422, 434
  assembly, 534
 symmetric, 249, 251, 276, 348, 422
 transformation, 525
 transpose, 250
Maxwell's reciprocal relationship, 351, 422, 573
Mechanism, 65, 95
Membrane, 7
Method of joints, 77
Method of sections, 84
Minimum total potential. *See* Energy
Moment diagram. *See* Bending moment diagram
Moment distribution, 2, 681
Moment of inertia, 142
Moment, applied, 20
Moment-area method, 127, 137, 145, 156, 203, 207, 214, 283, 286
Moment-curvature relation, 143
Moving loads. *See* Live loads
Müller-Breslau principle, 390

Navier hypothesis, 140

One-dimensional element. *See* Element, one-dimensional

Petroski, H., 30
Plate, 4, 8
Positive definiteness. *See* Quadratic form
Potential energy. *See* Energy
Power functions, 158
Principal axis, 143
Principal minor, 582

Quadratic form, 582
 positive definite, 582

Rayleigh-Ritz method, 605, 606, 611
Reaction, 35

Redundant force, 59, 229
Reference axis, 4, 102, 142, 278
Release, 59, 560
Roller. *See* Supports

Salvadori, M., 2
Sections, method of, 84
Sensitivity, 300
Settlements. *See* Supports
Shear, 119
Shear diagram, 127
Shell, 4, 8
Side-sway, 280, 435
Sign convention
  bars, 73, 300
  beams, 119, 278
  fixed-end forces, 538
  frames, 451
  torsional spring, 479
Slope-deflection
  equations, 436
  method, 436
Slope diagram, 146
Spring, 257, 312, 363, 551
  constant, 257
  torsional, 312, 550
St. Venant's principle, 29
Stability, 59, 62, 63, 95
  necessary condition, 95
  statical, 63, 95
Stationary point, 589, 591
Stationary total potential. *See* Energy
Stiffness matrix. *See* Matrix
Strain, 98, 140
  initial, 99
  thermal, 99
Strain energy. *See* Energy
Stress-strain relations, 10, 100
Structure, definition, 2
Substructure, 66, 353
Superposition, 15, 121, 138, 197, 214, 232, 296, 308, 361, 444
Support settlement. *See* Supports
Supports, 28, 34
  dependent motions, 538
  fixed with roller, 37
  kinematics, 35
  pinned, 35
  rigid (fixed), 34
  roller, 36
  settlement, 247, 307, 475, 546
Symmetry, 348
  axis, 353
  behavioral, 348, 579
  boundary conditions, 356, 373
  center, 372
  line, 353
  physical, 348, 352
  point, 372
System displacements. *See* Displacements
System stiffness matrix. *See* Matrix

Theorem of virtual work. *See* Virtual work
Thermal load
  in beams, 143, 266
  in truss bars, 101, 234
Thermal strain. *See* Strain
Three-dimensional element. *See* Element, three-dimensional
Timoshenko, S. P., 2
Total potential energy. *See* Energy
Truss, 6, 72
  definition, 6, 72
Two-dimensional element. *See* Element, two-dimensional

Unrestrained end deformation. *See* End deformations

Vector, 248
Virtual displacements, 169
Virtual forces, 169
Virtual work, 10, 110, 169
  bars, 172
  beams, 192
  external, 173
  frame joints, 192
  frames, 205
  internal, 173
  principle, 638
  theorem, 10, 171, 173, 177, 194, 205
  truss joints, 170
  trusses, 175

Williot-Mohr diagram, 110
Work, 578
  least, theorem of, 573, 598
Work-equivalent force. *See* Force

TE DUE